THE CIBA COLLECTION OF MEDICAL ILLUSTRATIONS

VOLUME 8 · PART I

Other published volumes of
THE CIBA COLLECTION OF MEDICAL ILLUSTRATIONS
Prepared by
Frank H. Netter, M.D.

Nervous System, Part I: Anatomy and Physiology
Nervous System, Part II: Neurologic and Neuromuscular Disorders
Reproductive System
Digestive System, Part I: Upper Digestive Tract
Digestive System, Part II: Lower Digestive Tract
Digestive System, Part III: Liver, Biliary Tract and Pancreas
Endocrine System and Selected Metabolic Diseases
Heart
Kidneys, Ureters, and Urinary Bladder
Respiratory System

THE CIBA COLLECTION OF MEDICAL ILLUSTRATIONS

VOLUME 8

MUSCULOSKELETAL SYSTEM

PART I

Anatomy, Physiology, and Metabolic Disorders

A compilation of paintings prepared

by FRANK H. NETTER, M.D.

Russell T. Woodburne, M.A., Ph.D., Edmund S. Crelin, Ph.D., D. Sc., and Frederick S. Kaplan, M.D., *Consulting Editors*

Regina V. Dingle, *Managing Editor*

With a foreword by Henry J. Mankin, M.D., *Edith M. Ashley Professor of Orthopaedic Surgery, Harvard Medical School*

COMMISSIONED AND PUBLISHED BY CIBA-GEIGY CORPORATION

SUMMIT, NEW JERSEY

Copies of all CIBA COLLECTION books, CLINICAL SYMPOSIA reprints, and color slides of all illustrations contained in them are available from CIBA-GEIGY Medical Education Division, 14 Henderson Drive, West Caldwell, NJ 07006

First Printing

ISBN 0-914168-14-2
Library of Congress Catalog No: 53-2151

Printed in U.S.A.

Book printed offset by R.R. Donnelley & Sons Company
Laser-scanned color separations by Pittsburgh Atlas Graphic Enterprises, Inc
Layout design by Pierre J. Lair
Text photocomposed in Mergenthaler Garamond No. 3 by Granite Graphics
Contents printed on Cameo dull text, basis 80, by S.D. Warren Company
Smyth-sewn case binding by R.R. Donnelley & Sons Company
Endpapers: White Flannel text, basis 80, by Curtis Paper Division, James River Corporation
Cover material: Buckram linen cloth by Joanna Western Mills
Front and spine cover and front matter design by Philip Grushkin
Index by Steele/Katigbak Indexers

Foreword

In 1953, as an intern at the University of Chicago, I obtained a copy of the very first of the now world-famous CIBA COLLECTION masterpieces by Frank Netter, M.D. The subject was the *Nervous System* that, after an amazing thirteen printings, is now in its second printing of a revised edition. My colleagues and I at that time and all subsequent generations of physicians over the last 35 years marveled at the artistry and extraordinary clarity of those illustrations; and how remarkably, when coupled with the short but well-written text, they provided such a clear definition of complex three-dimensional structures and confusing relationships that we had struggled sometimes in vain to comprehend. There was little doubt in our minds that we were looking at the works of a genius—not only because he saw so much and so clearly, but because he could make us see it with equal clarity. We waited, as did the world, for subsequent volumes and were not disappointed with any of the next six. The *Reproductive System*, published in 1954, the *Digestive System* in 1957, the *Endocrine System* in 1965, the *Heart* in 1969, the *Kidneys, Ureters, and Urinary Bladder* in 1973, and the *Respiratory System* in 1979 all showed the same remarkable ability to portray the anatomy and embryology, physiology, pathology, and clinical states with such extraordinary clarity and in sufficient detail as to become, for each of these disciplines, major teaching and reference texts. I wonder how many times in these past 35 years a Netter illustration has been used for a lecture or demonstration in a medical school or residency classroom, and how many copies have been made of the figures to subsequently reside in teaching collections throughout the world? Surely the number must be exceeded only by the number of physicians who hold the volumes as cherished possessions and have read them over and over in a quest for knowledge or as part of a scholarly pursuit.

Having said that, I must express a degree of disappointment on behalf of my colleagues in Orthopaedics, Rheumatology, Physiatry, and the sciences associated with connective tissue diseases, with the evident fact that with the exception of some of the plates in Volumes 1 and 4 there were few of these teaching atlases that had any relevance to our rather sizable corner of the world of medicine. It is therefore with great enthusiasm and unbridled pleasure that our specialties now greet *Volume 8: The Musculoskeletal System*. Furthermore, after consideration of the contents and study of the magnificent plates and text, I conclude that not only was the product worth waiting for but in my opinion the three parts comprising this latest work are the author's finest! Frank Netter, M.D., has not only "done it again" but he's done it better than he ever did it before!

The *Musculoskeletal System* is one of Dr. Netter's most ambitious projects. Any of the subjects covered would seem to require a separate volume, and perhaps one of the major aspects of the genius of the artist is deciding what to include. Realizing that each plate contains several main themes and multiple facts (all nicely tied together by the artistry of the author), it is not surprising that anatomy and physiology (including metabolic disorders) are included in the 214 plates that comprise Part I; and that congenital and developmental disorders can be depicted in 118 plates, neoplasms in 34, rheumatic disorders in 60 and joint replacement surgery in another 41, all in Part II. Part III, on injuries (182 plates), infections (20 plates), vascular disturbances (30 plates), and rehabilitation (20 plates), completes the set. If one totals these plates, the number exceeds 715 (what a fantastic effort even for Dr. Netter!), and with the text supplied by the numerous contributors, the three parts of Volume 8 should rapidly become classic teaching texts for our specialties.

One may wonder why Volume 8 required so many plates and so much text as compared with the other disciplines, and upon consideration, I believe the answer is self-evident. The musculoskeletal system comprises most of the body's supportive and protective elements and provides movement and prehension. The tissues included vary from the undifferentiated fibrous supporting membranes to the remarkably complex organ systems of the bones and joints, and the anatomic structures are as different as the big toe and the first cervical vertebra. While trauma is almost exclusively related to the bones and joints, metabolic bone disease involves the endocrine and renal systems; genetic disorders, other multiple organ systems; arthritis, the sciences of immunology and internal metabolism; and neoplasms, the entire field of oncology. What brings these fields together in this remarkable volume and in the scientific world is the anatomic structures and, perhaps more relevantly, the entire background framework of connective tissue chemistry, mechanical engineering, and materials science, which Dr. Netter has woven so beautifully and understandably into every section.

The students, scholars, and practitioners who deal with the musculoskeletal system have been waiting along with me since 1953 for Frank Netter's Volume 8. I don't think they will be disappointed.

HENRY J. MANKIN, M.D.
Boston, July 1987

Frank H. Netter, M.D.

Introduction

I had long looked forward to undertaking this volume on the musculoskeletal system. It deals with the most humanistic, the most soul-touching, of all the subjects I have portrayed in The Ciba Collection of Medical Illustrations. People break bones, develop painful or swollen joints, are handicapped by congenital, developmental, or acquired deformities, metabolic abnormalities, or paralytic disorders. Some are beset by tumors of bone or soft tissue, some undergo amputations, either surgical or traumatic, some occasionally have reimplantation, and many have joint replacement. The list goes on and on. These are people we see about us quite commonly and are often our friends, relatives, or acquaintances. Significantly, such ailments lend themselves to graphic representation and are stimulating subject matter for an artist.

When I undertook this project, however, I grossly underestimated its scope. This was true also in regard to the previous volumes of the Ciba Collection, but in the case of this book, it was far more marked. When we consider that this project involves every bone, joint, and muscle of the body as well as all the nerves and blood vessels that supply them and all the multitude of disorders which may affect each of them, the magnitude of the project becomes enormous. In my naivete, I originally thought I could cover the subject in a single book, but it soon became apparent that this was impossible. Even two books soon proved inadequate for such an extensive undertaking and, accordingly, three books are now planned. This book, Part I, Volume 8 of the Ciba Collection, covers basic gross anatomy, embryology, physiology, and histology of the musculoskeletal system, as well as its metabolic disorders. Part II, now in press, covers rheumatic and other arthritic disorders, as well as their conservative and surgical management (including joint replacement), congenital and developmental disorders, and both benign and malignant neoplasms of bones and soft tissues. Part III, on which I am still at work, will include fractures and dislocations, and their emergency and definitive care, amputations (both surgical and traumatic) and prostheses, sports injuries, infections, peripheral nerve and plexus injuries, burns, compartment syndromes, skin grafting, arthroscopy, and care and rehabilitation of handicapped patients.

But classification and organization of this voluminous material turned out to be no simple matter, since many disorders fit equally well into several of the above groups. For example, osteogenesis imperfecta might have been classified as metabolic, congenital, or developmental. Baker's cyst, ganglion, bursitis, and villonodular synovitis might have been considered with rheumatic, developmental, or in some instances even with traumatic disorders. Pathologic fractures might be covered with fractures in general or with the specific underlying disease which caused them. In a number of instances, therefore, empiric decisions had to be made in this connection, and some subjects were covered under several headings. I hope that the reader will be considerate of these problems. In addition, there is much overlap between the fields of orthopedics, neurology, and neurosurgery, so that the reader may find it advantageous to refer at times to my atlases on the nervous system.

I must express herewith my thanks and appreciation for the tremendous help which my very knowledgeable collaborators gave to me so graciously. In this Part I, there was first of all Dr. Russell Woodburne, a truly great anatomist and professor emeritus at the University of Michigan. It is interesting that during our long collaboration I never actually met with Dr. Woodburne and all our communications were by mail or phone. This, in itself, tells of what a fine understanding and meeting of the minds there was between us. I hope and expect that in the near future I will have the pleasure of meeting him in person.

Dr. Edmund S. Crelin, professor at Yale University, is a long-standing friend (note that I do not say "old" friend because he is so young in spirit) with whom I have collaborated a number of times on other phases of embryology. He is a profound student and original investigator of the subject, with the gift of imparting his knowledge simply and clearly, and is in fact a talented artist himself.

Dr. Frederick Kaplan (now Freddie to me), assistant professor of orthopaedics at the University of Pennsylvania, was invaluable in guiding me through the difficult subjects of musculoskeletal physiology and metabolic bone disease. I enjoyed our companionship and friendship as much as I appreciated his knowledge and insight into the subject.

I was delighted to have the cooperation of Dr. Henry Mankin, the distinguished chief of orthopaedics at Massachusetts General Hospital and professor at Harvard University, for the complex subject of rickets in its varied forms—nutritional, renal, and metabolic. He is a great but charming and unassuming man.

There were many others, too numerous to mention here individually, who gave to me of their knowledge and time. They are all credited elsewhere in this book but I thank them all very much herewith. I will write about the great people who helped me with other parts of Volume 8 when those parts are published.

Finally, I give great credit and thanks to the personnel of the CIBA-GEIGY Company and to the company itself for having done so much to ease my burden in producing this book. Specifically, I would like to mention Mr. Philip Flagler, Dr. Milton Donin, Dr. Roy Ellis, and especially Mrs. Regina Dingle, all of whom did so much more in that connection than I can tell about here.

Frank H. Netter, M.D.

Contributors and Consultants

Abass Alavi, M.D.

Professor of Radiology, University of Pennsylvania School of Medicine; Chief, Division of Nuclear Medicine, Hospital of the University of Pennsylvania, Philadelphia, Pennsylvania

Maurice F. Attie, M.D.

Assistant Professor of Medicine, University of Pennsylvania School of Medicine; Director of University of Pennsylvania Osteoporosis Center, Hospital of the University of Pennsylvania, Philadelphia, Pennsylvania

Charles S. August, M.D.

Associate Professor of Pediatrics, University of Pennsylvania School of Medicine; Clinical Director of Bone Marrow Transplantation Unit, Children's Hospital of Philadelphia, Philadelphia, Pennsylvania

Jonathan Black, Ph.D.

Professor of Research in Orthopaedic Surgery, University of Pennsylvania School of Medicine, Professor of Bioengineering, School of Engineering and Applied Science, University of Pennsylvania, Philadelphia, Pennsylvania

Carl T. Brighton, M.D., Ph.D.

Paul B. Magnuson Professor of Bone and Joint Surgery; Chairman, Department of Orthopaedic Surgery; Director of Orthopaedic Research, McKay Laboratory, University of Pennsylvania School of Medicine, Philadelphia, Pennsylvania

Joseph A. Buckwalter, M.D.

Professor of Orthopaedic Surgery, University of Iowa College of Medicine, Iowa City, Iowa

Bruce M. Carlson, M.D., Ph.D.

Professor of Anatomy and Biology, University of Michigan Medical School, Ann Arbor, Michigan

Charles C. Clark, Ph.D.

Associate Professor of Research in Orthopaedic Surgery, Biochemistry and Biophysics, University of Pennsylvania School of Medicine, Philadelphia, Pennsylvania

Edmund S. Crelin, Ph.D., D.Sc.

Chairman, Human Growth and Development Study Unit; Professor of Anatomy and Orthopaedics and Rehabilitation, Yale University School of Medicine, New Haven, Connecticut

Murray K. Dalinka, M.D.

Professor of Radiology and Orthopaedic Surgery, University of Pennsylvania School of Medicine; Chief, Division of Skeletal Radiology, Hospital of the University of Pennsylvania, Philadelphia, Pennsylvania

Michael D. Fallon, M.D.

Assistant Professor of Pathology, University of Pennsylvania School of Medicine; Attending Surgical Pathologist, Hospital of the University of Pennsylvania, Philadelphia, Pennsylvania

John G. Haddad, M.D.

Professor of Medicine and Chief, Section of Endocrinology, University of Pennsylvania School of Medicine, Philadelphia, Pennsylvania

Joseph P. Iannotti, M.D.

Assistant Professor of Orthopaedic Surgery, University of Pennsylvania School of Medicine, Philadelphia, Pennsylvania

Frederick S. Kaplan, M.D.

Assistant Professor of Orthopaedic Surgery and Medicine,
Director of Medical Education, Assistant Director of
Orthopaedic Research, McKay Laboratory,
University of Pennsylvania School of Medicine;
Chief, Division of Metabolic Bone Diseases,
Hospital of the University of Pennsylvania,
Philadelphia, Pennsylvania

Joel S. Karp, Ph.D.

Research Assistant Professor of Radiology,
Hospital of the University of Pennsylvania,
Philadelphia, Pennsylvania

Henry J. Mankin, M.D.

Edith M. Ashley Professor of Orthopaedic Surgery,
Harvard Medical School; Orthopaedist-in-Chief,
Massachusetts General Hospital,
Boston, Massachusetts

Edward A. Millar, M.D.

Associate Clinical Professor of Orthopaedic Surgery,
Northwestern University Medical School;
Chief Surgeon Emeritus,
Shriners Hospital for Crippled Children,
Chicago, Illinois

George A. G. Mitchell, Ch.M., D.Sc., F.R.C.S.

Hon. Alumnus, The University of Louvain, Belgium;
Chevalier (lst Cl.) Order of the Dannebrog;
Emeritus Professor of Anatomy and Director
of the Anatomical Laboratories,
University of Manchester,
Manchester, England

Richard G. Schmidt, M.D.

Assistant Professor of Orthopaedics, University of
Pennsylvania School of Medicine; Chief, Division of
Orthopaedic Oncology and Director of Bone Tumor Service,
Hospital of the University of Pennsylvania,
Philadelphia, Pennsylvania

H. Ralph Schumacher, Jr., M.D.

Professor of Medicine, University of Pennsylvania School
of Medicine; Director of Arthritis-Immunology Center,
Veterans Administration Medical Center,
Philadelphia, Pennsylvania

Michael E. Selzer, M.D., Ph.D.

Professor of Neurology, University of Pennsylvania School
of Medicine; Member, David Mahoney Institute
of Neurological Sciences; Attending Neurologist,
Hospital of the University of Pennsylvania,
Philadelphia, Pennsylvania

Russell T. Woodburne, M.A., Ph.D.

Chairman, International Anatomical Nomenclature Committee;
Emeritus Professor of Anatomy,
University of Michigan Medical School,
Ann Arbor, Michigan

Preface

Disorders of the musculoskeletal system comprise some of the most common and disabling afflictions experienced by man. Although not a leading cause of death, musculoskeletal disorders rank first among the disease groups in their impact on the quality of life, and second only to circulatory disorders in their economic cost.

The understanding and management of this vast array of diseases involve primarily a study of the basic disciplines underlying them—anatomy, embryology, and physiology. Hence, in this first of the three-part volume on the Musculoskeletal System, Dr. Netter has begun by portraying, in collaboration with Dr. Russell Woodburne, the pertinent gross anatomy. I know of no other source where this difficult subject is so succinctly presented. The section on embryology, done in collaboration with Dr. Edmund Crelin, is of fundamental importance since so many of these disorders are congenital or developmental in etiology.

The past 40 years have seen an explosive growth in the study of musculoskeletal physiology. Involving as it does such interrelated disciplines as systemic metabolism, molecular and cell biology, genetics, immunology, biophysics, and bioengineering, this expansion of our basic knowledge has revolutionized our concepts of health and disease and has led to the development of new diagnostic and therapeutic modalities. With the advice and guidance of Dr. Carl Brighton, Dr. Henry Mankin, Dr. Michael Selzer, a number of other physicians and scientists, and myself, Dr. Netter has presented a pictorial overview of this diverse subject. It is hoped that these books will serve as "visual guideposts" to the vast landscape of musculoskeletal medicine.

The study of musculoskeletal physiology leads directly, almost imperceptibly, into the subject of metabolic bone diseases, which is also portrayed in this book. Of all metabolic bone diseases, osteoporosis is probably the most significant because it is the most common bone disease in the world and a major risk factor for fractures and disability in the elderly. At the present time, nearly 3/4 of all fractures in menopausal women occur as a result of osteoporosis. In the United States, more than one million people are hospitalized each year for treatment of fractures; one-quarter of these are hip fractures related to osteoporosis.

Other less common metabolic disorders—even such relatively rare ones as osteogenesis imperfecta and osteopetrosis, examples of heritable disorders of bone cell metabolism—are included because they are unique windows through which we can obtain a startling perspective of normal human physiology. They are essentially the exceptions that prove the rule. As the great English physician William Harvey wrote more than three centuries ago: "Nature is nowhere accustomed more openly to display her secret mysteries than in cases where she shows traces of her workings apart from the beaten path; nor is there any better way to advance the proper practice of medicine than to give our minds to the discovery of the usual law of nature by careful investigation of rarer forms of disease."

When speaking with colleagues and friends about my participation in this project, I am most often asked, "What is it like working with Dr. Frank Netter?" The answer is simple: it is like working with a wise and thoughtful teacher, an imaginative and enthusiastic student, an astute and caring physician, a brilliant and inspired artist, and a dedicated and devoted friend. Frank Netter is all of these. He epitomizes the ideal that the best teacher must also be the best student. In fact, few persons have taught more students, and few have been granted the opportunity and privilege to extend their influence across so many generations and disciplines. Dr. Netter's unique gifts continue to inspire and enlighten, and are certain to endure.

A work of this scope and magnitude involves many people, and my sincere thanks go to all those who contributed to its evolution—Dr. and Mrs. Netter, the editorial group at CIBA-GEIGY's Medical Education Division, my students, colleagues, friends, and especially my family.

FREDERICK S. KAPLAN, M.D.
CONSULTING EDITOR

Acknowledgments

Some 40 years ago, when the original plan for THE CIBA COLLECTION OF MEDICAL ILLUSTRATIONS was being developed, Frank Netter was asked how long he thought it might take him to portray the anatomy, physiology, and pathology of all the major body systems in pictures. After much thought, he replied "ten years." An agreement was struck with the then CIBA Pharmaceutical Company for his exclusive services and the project was begun. This book, *Musculoskeletal System, Part I*, is significant because it marks the beginning of the last volume of the COLLECTION as it was originally planned way back then.

It being Part I of a three-part volume is also significant, because the three parts are indicative of how the scope of the project has changed in the ensuing 40 years and how much our knowledge of the human body and its workings has grown in that time. For newcomers to the series, this volume will probably become a logical beginning, because the musculoskeletal system is, after all, what all the other body systems "hang on." For those of us who have long known the series, however, it is a perfect culmination. In a way, it should have come first, but if it had, think of all we might have missed!

For those of us who work with Dr. Netter in bringing these volumes to fruition, Part I of the *Musculoskeletal System* is a significant milestone for another reason. Even though Parts II and III are yet to come, for the first time we can realistically begin to think about—and for the moment just think about—a project we've all wanted to attack for many years: revision of the existing books of the CIBA COLLECTION. And the reason this has been impossible to contemplate in the past is Frank Netter's amazing productivity. He has always had a brand new book ready to be produced before we could get the previous one off the press. So revisions, even though needed, have always been elusive. Now, we can see that their day may really come.

As with all the books of the CIBA COLLECTION, Dr. Netter has been at the center of this one—he is always the focal point. Without his pictures and without his careful organization, without his uncanny ability to select what is important as well as what should be left out, and without his unfailing sense for what the student and clinician alike might want or need to know, this series would not exist. We thank him once more for both his great skills and for the pleasure of working with him on another challenging project.

He is backed up, of course, by a formidable group of physician and scientist authors who not only provide consultation in their areas of specialty, but write the texts which accompany the plates. We thank them all for their many and varied contributions to this book. We particularly wish to acknowledge our consulting editors, Dr. Russell Woodburne, Dr. Edmund Crelin, and Dr. Frederick Kaplan who (in addition to preparing many of the texts) acted as medical consultants for major sections of the book and helped in the selection of many of the authors. Their good nature, patience, and understanding helped make our collaboration a personal as well as a professional pleasure.

In addition, many other persons contributed generously of their time, effort, and expertise to make this volume possible. Among them we wish to acknowledge Dr. Joseph Lane, Ms. Joan Ellis, Mr. Robert Hunt, Mr. David Smith, and Mr. Harry Schwann for advice and technical assistance to Dr. Kaplan. Dr. Eugene Black supplied valuable reference material on orthoses; Dr. Lee Peachey, the photomicrographs used in Section III, Plate 1; Dr. Crawford Campbell, the radiographs in Section IV, Plates 39 and 44; and Dr. Henry Mankin, the radiographs and pathology slide in Section IV, Plates 41 and 43. We also thank the *Journal of Bone and Joint Surgery* for permission to use material from Volume 56A, 1974 in some of the illustrations on rickets and osteomalacia.

As is customary in preparing a CIBA COLLECTION volume, wherever possible we have employed terminology commonly in use in standard American medical texts. To maintain consistency of style and presentation throughout, *Blakiston's Gould Medical Dictionary,* fourth edition, and *Nomina Anatomica,* fourth edition, were our primary resources. We are indebted to Dr. Donald Jenkins for his guidance in anglicizing certain terminology to make it consistent with English names used elsewhere in the book.

Within the CIBA-GEIGY organization, we are indebted especially to our Managing Editor, Gina Dingle, who has seen this book through all its many stages from the very beginning to the very end; to our editorial group—Sally Chichester, Nicole Friedman, and Jeffie Lemons; to our production group—Don Canter, Geoffrey Wooding, and Clark Carroll; and to our long-standing associate, Jack Cesareo. Their long hours and special care were essential in transforming Dr. Netter's pictures and the texts of our authors into the quite splendid finished book we now behold. Special thanks also to Kristine Bean for editing texts appearing in Sections III and IV and for her invaluable proofreading of all plate captions and text galleys.

Completion of a project such as this certainly gives a sense of satisfaction, a feeling of pride, and even a brief sigh of relief. It also spells the end of some long-standing professional relationships which have been pleasant and fulfilling, so there is, in addition, a certain sense of loss. Most of all, however, there is a sense of exhilaration that comes from knowing that just around the corner—thanks to Frank Netter—is another set of paintings for another new book just waiting to be transformed, as with this one, into a new masterpiece of medical understanding.

PHILIP B. FLAGLER
DIRECTOR,
MEDICAL EDUCATION DIVISION

Contents

Section I

Anatomy

Frank H. Netter, M.D.

in collaboration with

George A.G. Mitchell, Ch.M., D.Sc., F.R.C.S.
Plates 8–14, 18, 25, 35–36, 47–49, 75–77, 97–98

Russell T. Woodburne, M.A., Ph.D.
Plates 1–7, 15–17, 19–24, 26–34, 37–46, 50–74, 78–96, 99–110

Back

Muscles and Fasciae

Musculature of Back

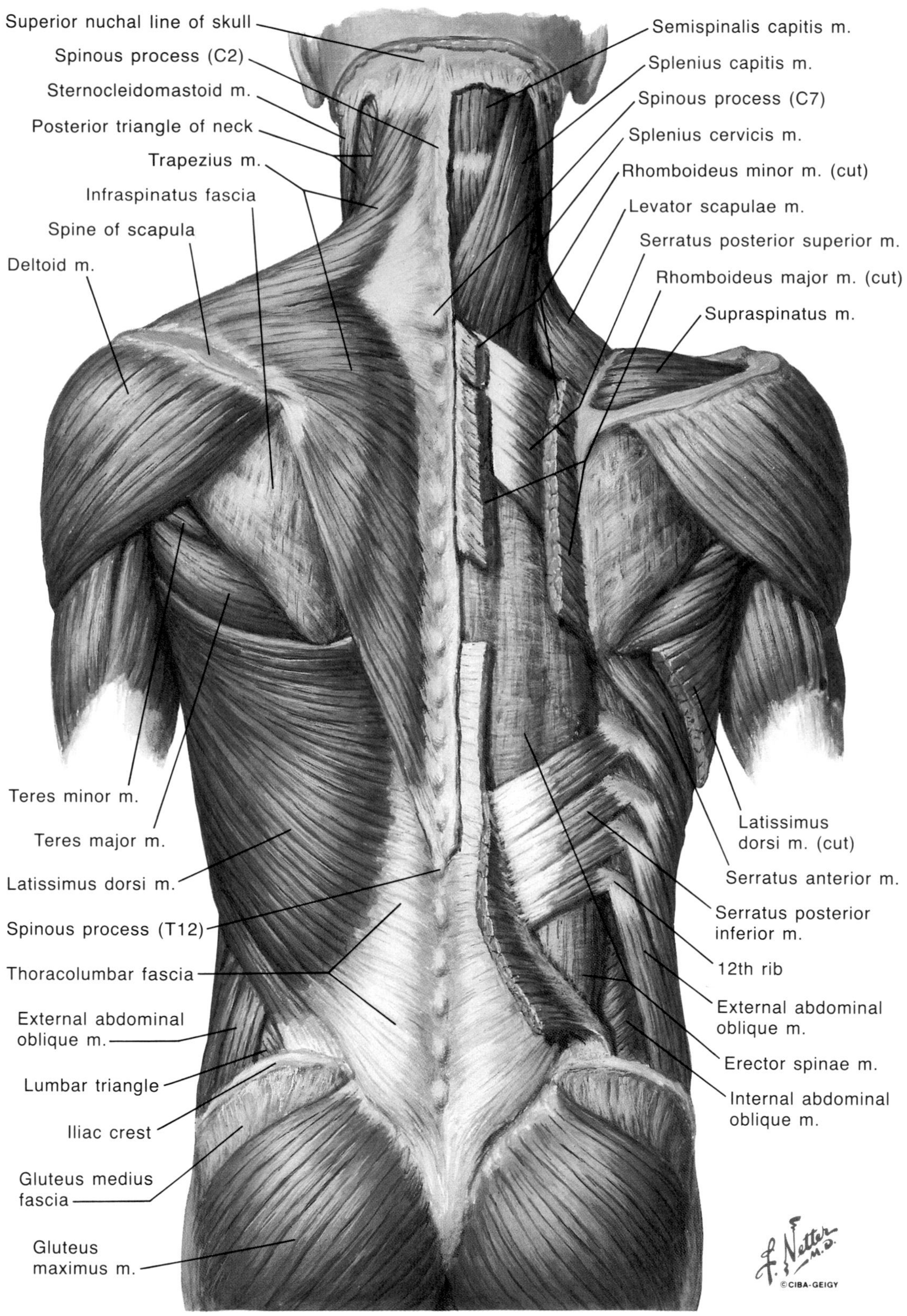
Superior nuchal line of skull
Spinous process (C2)
Sternocleidomastoid m.
Posterior triangle of neck
Trapezius m.
Infraspinatus fascia
Spine of scapula
Deltoid m.
Semispinalis capitis m.
Splenius capitis m.
Spinous process (C7)
Splenius cervicis m.
Rhomboideus minor m. (cut)
Levator scapulae m.
Serratus posterior superior m.
Rhomboideus major m. (cut)
Supraspinatus m.
Teres minor m.
Teres major m.
Latissimus dorsi m.
Spinous process (T12)
Thoracolumbar fascia
External abdominal oblique m.
Lumbar triangle
Iliac crest
Gluteus medius fascia
Gluteus maximus m.
Latissimus dorsi m. (cut)
Serratus anterior m.
Serratus posterior inferior m.
12th rib
External abdominal oblique m.
Erector spinae m.
Internal abdominal oblique m.
F. Netter M.D.
©CIBA-GEIGY

The musculature of the back is associated with the functions of the upper limb, thorax, head, and spine (Plate 1). The musculature of the back extends from the skull to the pelvis and lends support to the axial skeleton of the body. The muscles are functionally organized according to their position relative to the skull, vertebral column, rib cage, pelvis, and integument.

The superficial musculature of the back consists almost entirely of muscles associated with the shoulder and upper limb. These muscles are described individually under the upper limb. Also overlying the true back muscles are two widely separated muscles that attach to ribs, the serratus posterior superior and serratus posterior inferior muscles. These are respiratory muscles and are fully described in CIBA COLLECTION, Volume 7, page 10.

Deep Back Muscles

The deep back muscles are organized into five muscle groups based on their location relative to the vertebral column, the thoracic cage, and the pelvis. They are the splenius, erector spinae (sacrospinalis), transversospinal, interspinal, and intertransverse muscles. The deep back muscles are bilaterally symmetric to the left or right sides of the median axis of the vertebral column.

Splenius Muscle. The splenius muscle serves as a strap, covering and holding in the deeper muscles of the back of the neck (Plate 2). It takes origin from the ligamentum nuchae and the spinous processes of the seventh cervical to sixth thoracic vertebrae. The muscle may be divided into two parts—the *splenius capitis muscle*, which inserts on the mastoid process and the lateral third of the superior nuchal line of the skull, and the *splenius cervicis muscle*, which terminates in the posterior tubercles of the first two or three cervical vertebrae. The cervicis portion is the outer and lower portion of the splenius muscle, and its inserting bundles curve deeply along its lateral margin.

The splenius muscle draws the head and neck backward and rotates the face toward the side of the muscle acting. Both sides contracting together extend the head and neck. The muscle is innervated by the lateral branches of the dorsal rami of the second to fifth or sixth cervical nerves. It lies directly under the trapezius and is covered by the nuchal fascia; its mastoid insertion is deep to that of the sternocleidomastoid and it overlies the erector spinae and the semispinalis.

Erector Spinae Muscle. A complex, massive muscle, the erector spinae occupies the vertebrocostal groove of the back, lying directly under the posterior layer of thoracolumbar fascia (Plate 2). It begins below in a broad, thick tendon that is attached to the posterior surface of the sacrum, the posterior portion of the iliac crest, and the spinous processes of the lumbar vertebrae and the supraspinal ligament.

Muscular fibers, beginning on the anterior surface of the tendon, split into three columns at lumbar levels: the lateral iliocostalis, the intermediate longissimus, and the more medial spinalis muscle. The erector spinae muscle, as a whole, ascends throughout the length of the back, but its columns are composed of fascicles of shorter length. Each column contains a ropelike series of fascicles, various bundles arising as others are inserting; each fascicle spans from 6 to 10 segments between attachments.

Back

Muscles and Fasciae

(Continued)

Splenius and Erector Spinae Muscles

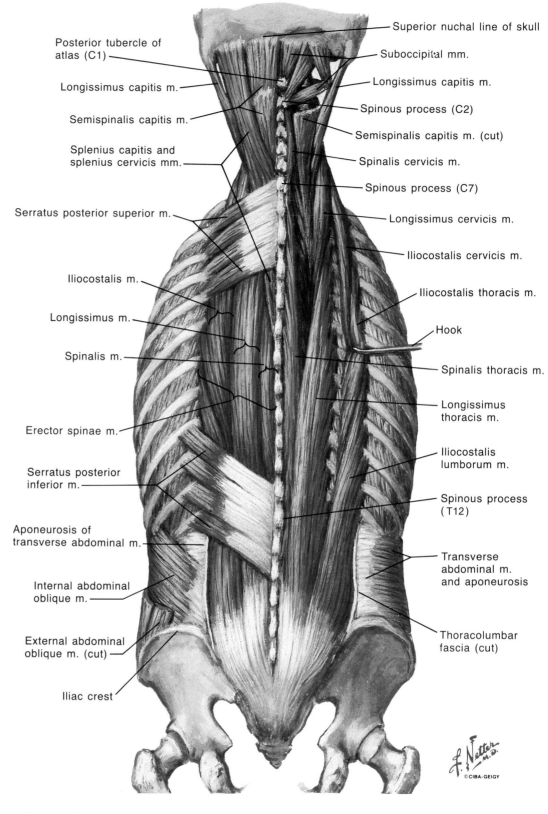

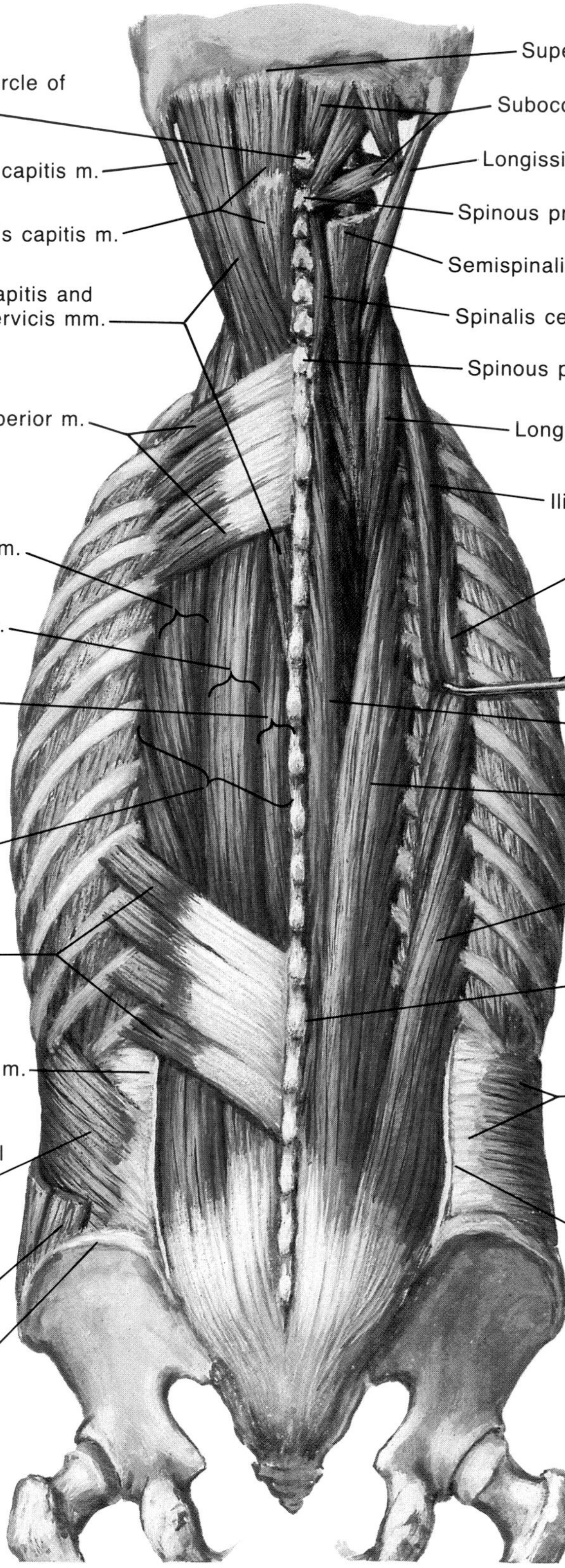

The *iliocostalis muscle*, the lateral part of the erector spinae, begins at the crest of the ilium and inserts on the angles of the lower six or seven ribs; this portion is the *iliocostalis lumborum muscle*. The succeeding *iliocostalis thoracis muscle* arises from the upper borders of the lower six ribs just medial to the insertion of the fascicles of the iliocostalis lumborum; its slips insert into the upper six ribs.

The *iliocostalis cervicis muscle* arises medial to the thoracis portion from the angles of approximately the upper six ribs and usually inserts into the transverse processes of the fourth, fifth, and sixth cervical vertebrae.

The *longissimus muscle* has thoracis, cervicis, and capitis portions. The *longissimus thoracis muscle* ascends as the intermediate part of the erector spinae muscle and inserts into the lower 9 or 10 ribs and into the transverse processes of the same levels.

The *longissimus cervicis muscle* arises medial to the upper end of the longissimus thoracis, from the transverse processes of about the upper four to six thoracic vertebrae. Its slips of insertion end in the transverse processes of the second to sixth cervical vertebrae. The *longissimus capitis muscle* connects the articular processes of the lower four cervical vertebrae with the posterior margin of the mastoid process.

The *spinalis muscle*, the most medial division of the erector spinae, is the thinnest and most poorly defined portion. It has thoracis, cervicis, and capitis portions. The *spinalis thoracis muscle* arises from the spinous processes of the last two thoracic and first two lumbar vertebrae. Its thin tendons insert into the spinous processes of between four to eight of the upper thoracic vertebrae.

The *spinalis cervicis muscle* is frequently absent or poorly developed. When completely represented, it arises from the ligamentum nuchae and from the spinous processes of the seventh cervical and, perhaps, the upper thoracic vertebrae. Its insertion may reach the spinous processes of the axis and, perhaps, of the third and fourth cervical vertebrae. The *spinalis capitis muscle* is not a separate muscle but is blended laterally with the semispinalis capitis.

The erector spinae muscle extends the vertebral column and, acting on one side, bends the column toward that side. The capitis insertion of the longissimus muscle serves to bend the head and rotate the face toward the same side. The erector spinae is also active in flexion of the trunk, controlling the degree and speed of flexion that is primarily produced by gravity and the abdominal muscles.

Transversospinal Muscles

The transversospinal muscle group consists of a series of muscles lying deep to the erector spinae (Plate 3). For the most part, their origins are in transverse processes of the vertebrae, and their insertions are in the spinous processes.

The *semispinalis muscle*, as its name implies, occupies one-half of the length of the vertebral column and contains thoracis, cervicis, and capitis portions.

Transversospinal, Interspinal, Intertransverse, and Suboccipital Muscles

Back

Muscles and Fasciae

(Continued)

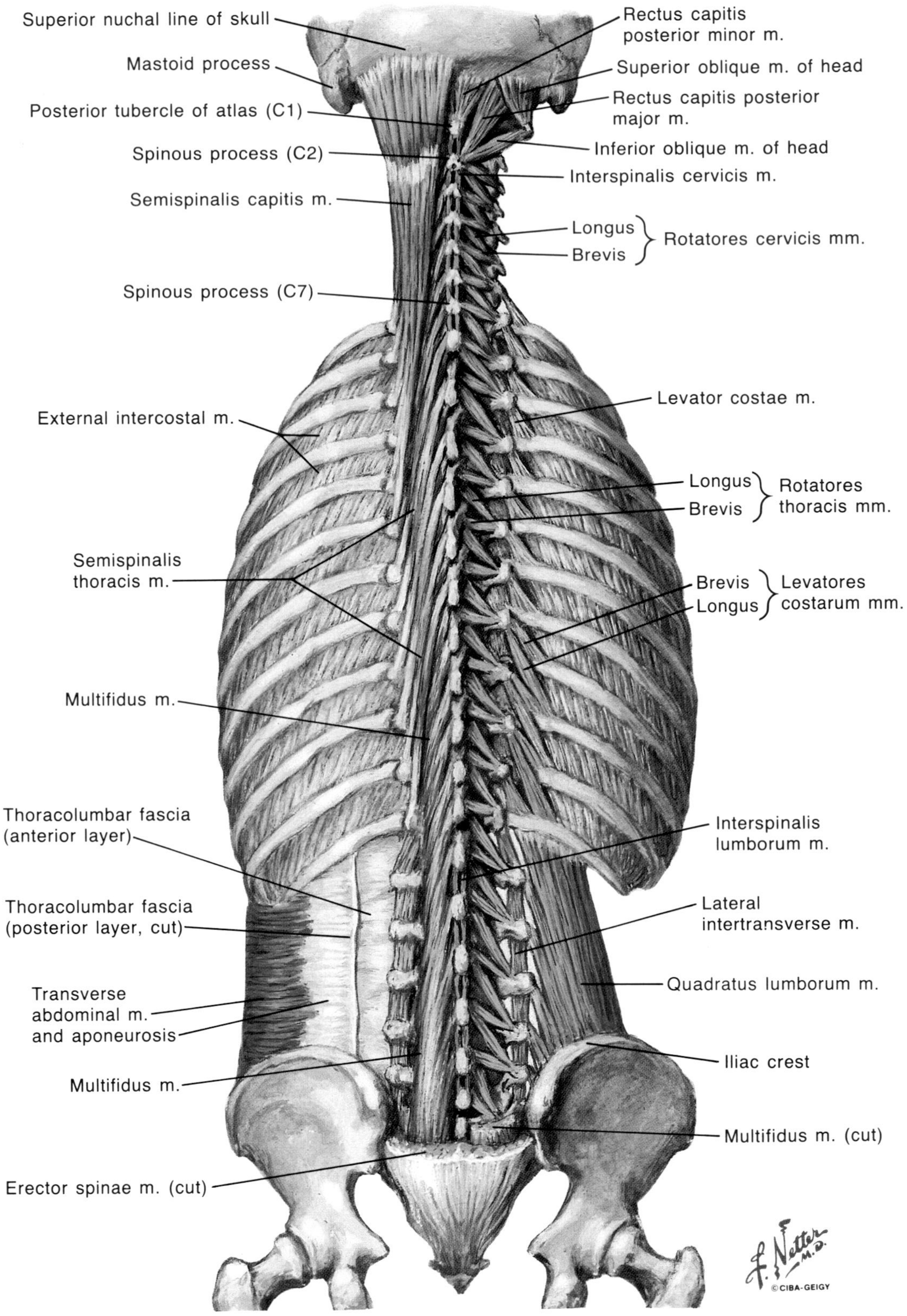
Superior nuchal line of skull
Mastoid process
Posterior tubercle of atlas (C1)
Spinous process (C2)
Semispinalis capitis m.
Spinous process (C7)
External intercostal m.
Semispinalis thoracis m.
Multifidus m.
Thoracolumbar fascia (anterior layer)
Thoracolumbar fascia (posterior layer, cut)
Transverse abdominal m. and aponeurosis
Multifidus m.
Erector spinae m. (cut)
Rectus capitis posterior minor m.
Superior oblique m. of head
Rectus capitis posterior major m.
Inferior oblique m. of head
Interspinalis cervicis m.
Longus
Brevis
Rotatores cervicis mm.
Levator costae m.
Longus
Brevis
Rotatores thoracis mm.
Brevis
Longus
Levatores costarum mm.
Interspinalis lumborum m.
Lateral intertransverse m.
Quadratus lumborum m.
Iliac crest
Multifidus m. (cut)
F. Netter M.D.
©CIBA-GEIGY

The *semispinalis thoracis* and *semispinalis cervicis muscles* form a continuous layer. They arise from the transverse processes of all thoracic vertebrae, and their fascicles insert into spinous processes four to six segments higher than their origin, culminating in a heavy muscular slip into the spine of the axis.

The *semispinalis capitis muscle* is the largest muscle mass of the back of the neck. Its fibers arise from the transverse processes of the seventh cervical and upper six or seven thoracic vertebrae, and from the articular processes of the fourth to sixth cervical vertebrae. The muscle, reinforced by the spinalis capitis medially, inserts into the underside of the occipital bone, occupying the whole area between the superior and inferior nuchal lines. It is usually traversed by an imperfect tendinous intersection at an upper cervical level. The semispinalis capitis muscle is a powerful extensor of the head. The semispinalis cervicis and semispinalis thoracis muscles extend the upper vertebral column and rotate it toward the opposite side. The muscle complex is innervated by dorsal rami of the spinal nerves of the upper half of the back.

The *multifidus muscle* extends over the whole length of the vertebral column. Its general character is also transversospinal, although it utilizes transverse processes only in the thoracic region. The corresponding attachments in the cervical region are the articular processes, and in the lumbar region, the mamillary processes. The muscle's lowest fibers originate from the posterior surface of the sacrum, and its highest origin is the fourth cervical vertebra.

Fascicles of the multifidus muscle insert two to four segments above their origin into spinous processes of all the vertebrae, from the last lumbar vertebra to the axis. The muscle is thickest in the lumbar region. It extends the vertebral column and rotates it minimally toward the opposite side. Dorsal rami of spinal nerves are its nerve supply.

The *rotatores muscles* are the shortest representatives of the transversospinal group. Similar in their attachments to the multifidus muscle, their fascicles ascend only to the adjacent or to the second vertebral spinous process above their origin (the rotatores breves and rotatores longi muscles, respectively). These muscles have the same innervation and action as the multifidus muscle. Experimental evidence suggests that the multifidus and the rotatores are more stabilizers of the vertebral column than its prime movers, adjusting the motion between individual vertebrae.

Minor Deep Back Muscles

The minor deep back muscles are the interspinal, intertransverse, and levatores costarum muscles (Plate 3).

The *interspinal muscles* are short fasciculi placed in pairs between the spinous processes of adjacent vertebrae in the cervical and lumbar regions of the vertebral column.

Transverse Section Through Lumbar Region (L2) of Back (schematic)

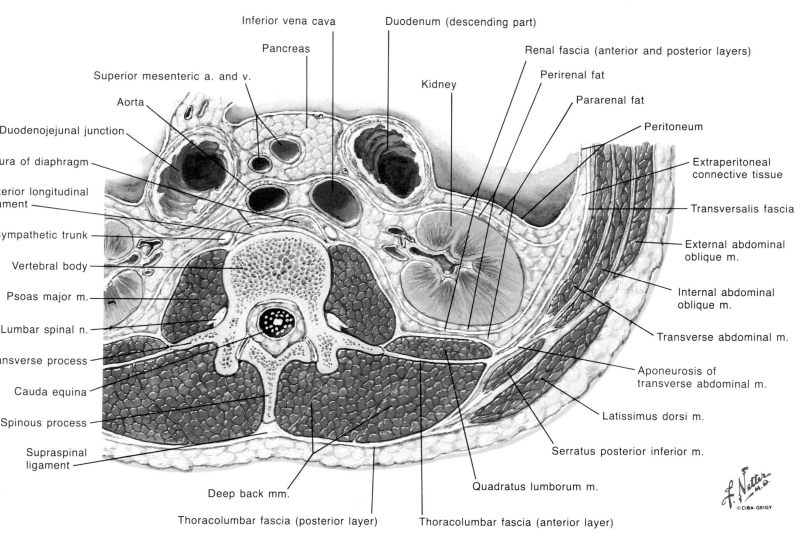

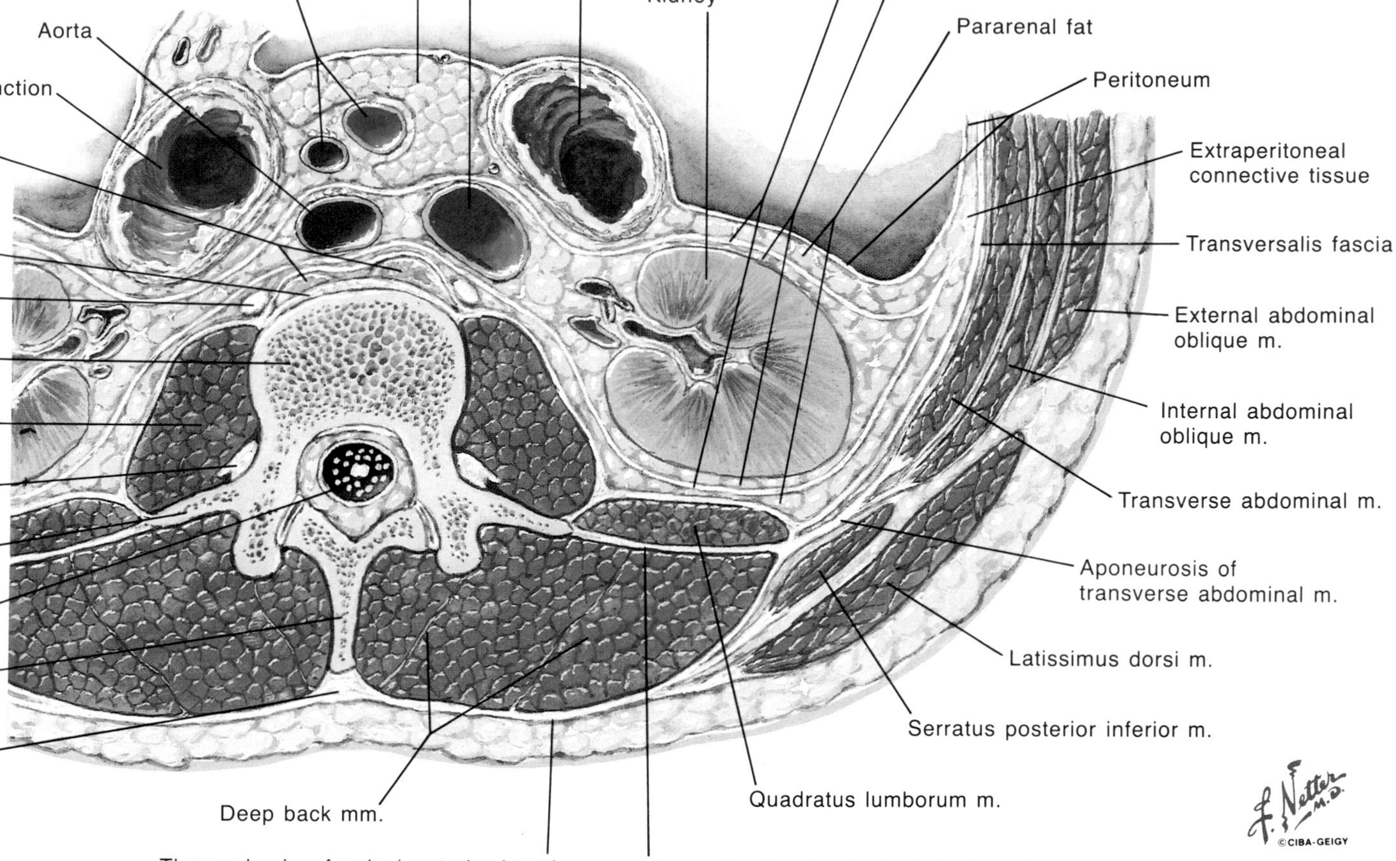

Inferior vena cava
Duodenum (descending part)
Pancreas
Renal fascia (anterior and posterior layers)
Superior mesenteric a. and v.
Perirenal fat
Kidney
Aorta
Pararenal fat
Duodenojejunal junction
Peritoneum
ura of diaphragm
Extraperitoneal connective tissue
erior longitudinal
ment
Transversalis fascia
ympathetic trunk
External abdominal oblique m.
Vertebral body
Psoas major m.
Internal abdominal oblique m.
Lumbar spinal n.
Transverse abdominal m.
nsverse process
Aponeurosis of transverse abdominal m.
Cauda equina
Latissimus dorsi m.
Spinous process
Serratus posterior inferior m.
Supraspinal ligament
Quadratus lumborum m.
Deep back mm.
Thoracolumbar fascia (posterior layer)
Thoracolumbar fascia (anterior layer)
F. Netter M.D.
©CIBA-GEIGY

ck

uscles and Fasciae

ntinued)

he *intertransverse muscles* are short pairs of mus- connecting adjacent transverse processes of ical and lumbar vertebrae. In the cervical on, the two members of a pair lie one ante- and one posterior to an emerging ventral us of a spinal nerve. In the lumbar region he vertebral column, the two members of the are medial and lateral. The medial repre- atives interconnect mamillary and accessory esses, while the lateral representatives inter- ect transverse processes.

he *levatores costarum* represent, in the thoracic on, the posterior intertransverse muscles of cervical region. They are 12 in number on each side and arise from the ends of the transverse processes of the seventh cervical and upper 11 thoracic vertebrae. They are directed obliquely downward and laterally to insert on the rib immediately below (the levatores costarum breves), between its tubercle and its angle. The four lowermost muscles divide into two fascicles, one as above, the other descending to the second rib below its origin (the levatores costarum longi). These muscles elevate the ribs, assisting in respiration, and help to produce lateral bending of the vertebral column. They are innervated, as are the interspinal and intertransverse muscles, by branches of the dorsal rami of spinal nerves.

Fasciae of Back

The muscles of the back and back of the neck are enclosed by the nuchal and thoracolumbar fasciae (Plate 4). This fascial covering has a continuous medial attachment to the ligamentum nuchae, the tips of the spinous processes, the supraspinal ligament of the seventh cervical and all the thoracic and lumbar vertebrae, and the median crest of the sacrum. This covering is deep to the superficial limb and the respiratory muscles of the back. Laterally, it attaches to the transverse processes of cervical and lumbar vertebrae, and in the thoracic region, it extends to the angles of the ribs.

The layer in the neck, the *nuchal fascia,* is the posterior portion of the prevertebral layer of the cervical fascia. It covers the splenius muscle and the deeper muscles and then attaches to the skull below the superior nuchal line.

The *thoracolumbar fascia,* as named, lies in the thoracic and lumbar regions. It is thin and transparent while covering the thoracic portions of the deep back muscles, but is thick and strong in the lumbar region. In the lumbar region, the fascia is bulged by the thick mass of back muscles located there so as to extend well beyond the line of the transverse processes. Thus, anterior and posterior layers of the thoracolumbar fascia are recognized at these levels; between them, they enclose all the deep back muscles, and from their lateral junction springs the aponeurosis of the transverse abdominal muscle. The anterior layer of the fascia provides the posterior fascia of the quadratus lumborum muscle, and from the posterior layer spring the aponeurotic sheets of origin of the latissimus dorsi and the serratus posterior inferior muscles. A thickening in the anterior layer between the twelfth rib and the transverse process of the first lumbar vertebra forms the posterior lumbocostal ligament. Below, the fascia attaches to the iliac crest and the lateral crest of the sacrum.

The *superficial fascia* of the back is well larded with fat, especially in the shoulder and low back regions. Just above the posterior part of the iliac crest, a deep indentation between the tapering lower portions of the deep back muscles and the thin origins of the abdominal muscular layers laterally is filled by a deeply running wedge of fat. Through this fat and its associated subcutaneous connective tissue, the superior cluneal nerves (L1, 2, 3) slant downward and outward to cross the posterior part of the iliac crest and enter the upper gluteal region (Plate 6).

Suboccipital Triangle

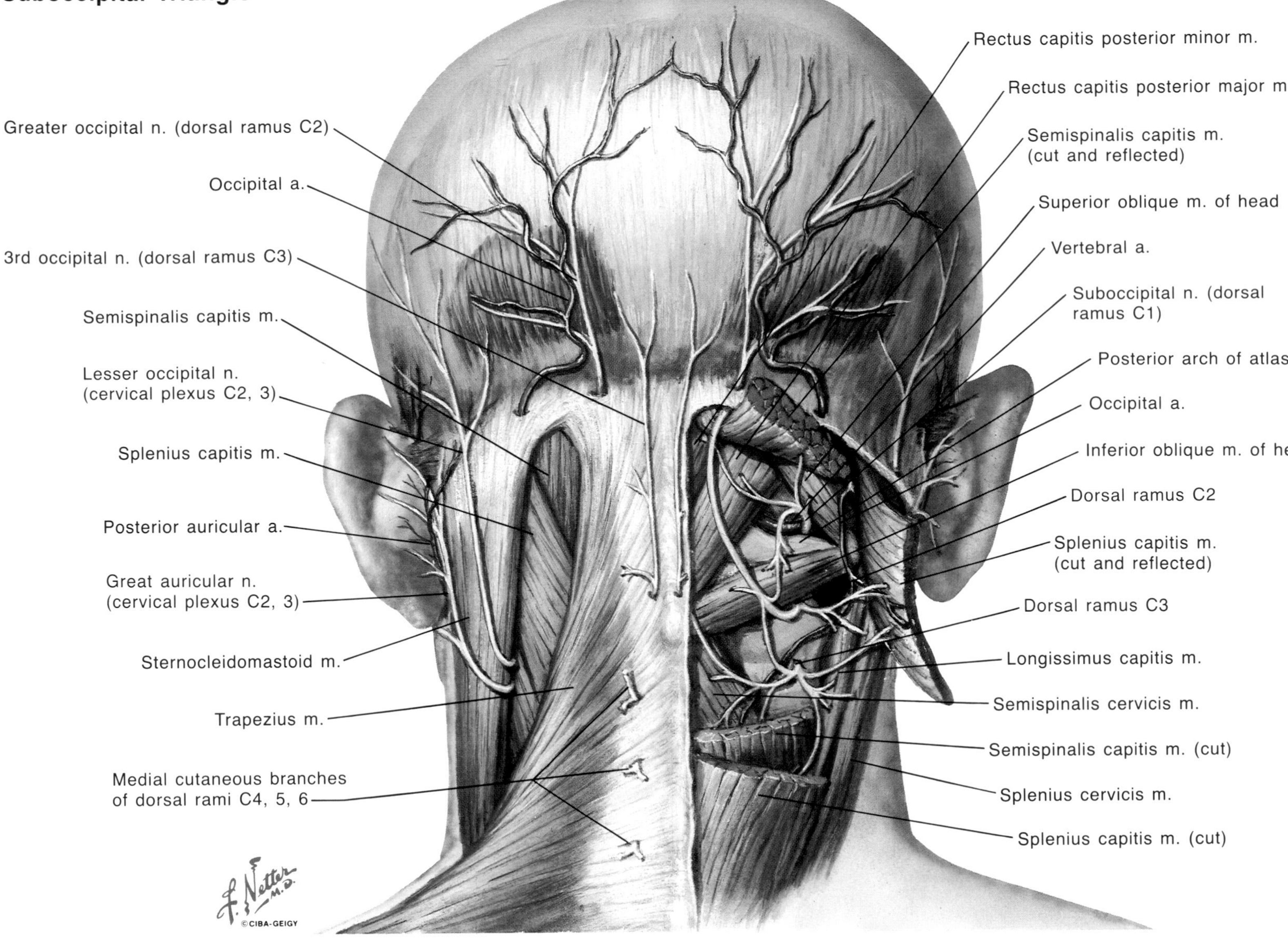

Greater occipital n. (dorsal ramus C2)
Occipital a.
3rd occipital n. (dorsal ramus C3)
Semispinalis capitis m.
Lesser occipital n. (cervical plexus C2, 3)
Splenius capitis m.
Posterior auricular a.
Great auricular n. (cervical plexus C2, 3)
Sternocleidomastoid m.
Trapezius m.
Medial cutaneous branches of dorsal rami C4, 5, 6
Rectus capitis posterior minor m.
Rectus capitis posterior major m
Semispinalis capitis m. (cut and reflected)
Superior oblique m. of head
Vertebral a.
Suboccipital n. (dorsal ramus C1)
Posterior arch of atlas
Occipital a.
Inferior oblique m. of he
Dorsal ramus C2
Splenius capitis m. (cut and reflected)
Dorsal ramus C3
Longissimus capitis m.
Semispinalis cervicis m.
Semispinalis capitis m. (cut)
Splenius cervicis m.
Splenius capitis m. (cut)
F. Netter M.D.
©CIBA-GEIGY

SECTION I PLATE 5 Slide 3570

Back

Muscles and Fasciae

(Continued)

Suboccipital Muscles

The suboccipital muscles are also members of the transversospinal muscle group. They are the rectus capitis posterior major and rectus capitis posterior minor muscles, and the inferior and superior oblique muscles of the head (Plate 5). These muscles are directly deep to the semispinalis capitis muscle, and three of them bound the suboccipital triangle.

The *rectus capitis posterior major muscle* arises from the spinous process of the axis. Broadening as it ascends, it inserts into the middle of the inferior nuchal line of the occipital bone and into the bone below this line. The *inferior oblique muscle of the head* arises from the spinous process of the axis and, passing almost horizontally lateralward, ends in the transverse process of the atlas. The *superior oblique muscle of the head* completes the triangle. It arises from the transverse process of the atlas. Passing upward and also medialward, this muscle inserts into the occipital bone above the outer part of the inferior nuchal line, where it overlaps the insertion of the rectus capitis posterior major.

The *rectus capitis posterior minor muscle* lies medial to the rectus capitis posterior major muscle and is more closely applied to the base of the skull. It takes origin from the posterior tubercle of the atlas and, widening as it ascends, inserts into the medial part of the inferior nuchal line and into the occipital bone between that line and the foramen magnum.

The suboccipital muscles extend the head and rotate it and the atlas toward the same side. They are innervated by the suboccipital nerve, the dorsal ramus of C1.

Suboccipital Triangle. The area between the two oblique muscles and the rectus capitis major muscle is defined as the suboccipital triangle (Plate 5). Its floor is the posterior atlantooccipital membrane, which is attached to the posterior margin of the posterior arch of the atlas. Deep to this tough membrane, the vertebral artery occupies the groove on the upper surface of the posterior arch of the atlas as it passes medialward toward the foramen magnum.

The suboccipital nerve emerges through the membrane between the vertebral artery and the posterior arch of the atlas. It divides in the dense fibrofatty tissue of the triangle into branche the suboccipital muscles. The greater occi nerve (dorsal ramus of the second cervical n [C2]) emerges below the inferior oblique mu of the head and turns upward to cross the occipital triangle and reach the scalp by pier the semispinalis capitis and trapezius mus The occipital artery crosses the insertion o superior oblique muscle of the head as it co medialward to join and distribute with the gr occipital nerve.

The *suboccipital nerve* (dorsal ramus of C1) no cutaneous distribution. Contrarily, the m branch of the dorsal ramus of C2 is the *gr occipital nerve*, which has a distribution as hig the vertex of the scalp and approaching th laterally. The *lesser occipital nerve* of the cer plexus (ventral ramus of C2) supplies the sk the scalp behind the ear as well as the skin o back of the ear itself.

The *third occipital nerve*, the medial bran the dorsal ramus of C3, distributes in the u neck and to the scalp, to a little beyond the s rior nuchal line. This is succeeded by a cutan distribution for the dorsal rami of C4, 5, 6 most observers have not found any distributi the neck for C7, 8 (and relatively little for C5 The lateral spread of the dorsal rami is b in the upper thoracic levels, the medial br of the dorsal ramus of T2 reaching as far a point of the shoulder. □

Back

Nerves

Nerves of Back

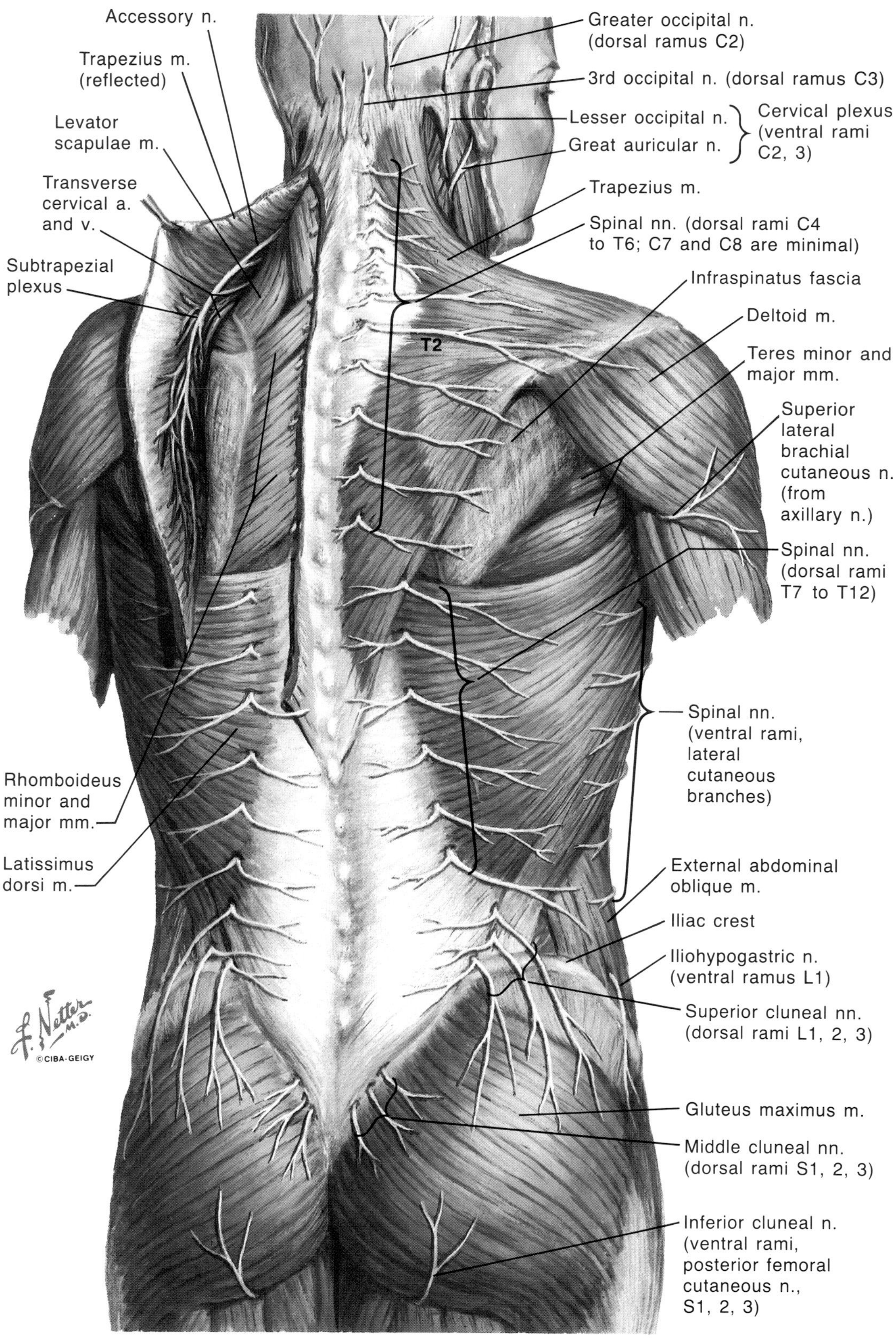
Accessory n.
Trapezius m. (reflected)
Levator scapulae m.
Transverse cervical a. and v.
Subtrapezial plexus
Rhomboideus minor and major mm.
Latissimus dorsi m.
Greater occipital n. (dorsal ramus C2)
3rd occipital n. (dorsal ramus C3)
Lesser occipital n.
Great auricular n.
Cervical plexus (ventral rami C2, 3)
Trapezius m.
Spinal nn. (dorsal rami C4 to T6; C7 and C8 are minimal)
Infraspinatus fascia
Deltoid m.
T2
Teres minor and major mm.
Superior lateral brachial cutaneous n. (from axillary n.)
Spinal nn. (dorsal rami T7 to T12)
Spinal nn. (ventral rami, lateral cutaneous branches)
External abdominal oblique m.
Iliac crest
Iliohypogastric n. (ventral ramus L1)
Superior cluneal nn. (dorsal rami L1, 2, 3)
Gluteus maximus m.
Middle cluneal nn. (dorsal rami S1, 2, 3)
Inferior cluneal n. (ventral rami, posterior femoral cutaneous n., S1, 2, 3)
F. Netter M.D.
©CIBA-GEIGY

Superficial and Cutaneous Nerves

The nerves of the back rather closely reflect the segmental nature of the construction of the body (Plate 6). There is a pair of symmetrically disposed spinal nerves for each vertebral segment, except for several reduced coccygeal segments—31 pairs in all (8 cervical, 12 thoracic, 5 lumbar, 5 sacral, and 1 coccygeal).

The spinal nerves are all mixed nerves, containing muscular, sensory, and visceral fibers. Nerves are regional entities, responsible for conveying all the various nerve components to the regions in which they distribute.

Reflecting their segmental character, the nerves begin as a series of *dorsal rootlets* and *ventral rootlets* emanating from the spinal cord. These rootlets coalesce into one spinal nerve that represents a segment of the spinal cord. The *dorsal root* is made up generally of afferent nerve fibers, both somatic and visceral. It contains, in its *spinal ganglion*, the nerve cell bodies of these afferent neurons. The *ventral root* is made up of neuronal axons whose cell bodies lie within the spinal cord, either in its ventral or in its lateral cell column. Thus, the ventral root represents motor (somatic and visceral) axons.

The dorsal and ventral roots of each segment join just lateral to the spinal ganglion in the intervertebral foramen to form the segmental *spinal nerve*. This nerve divides almost immediately into *dorsal* and *ventral rami*. The white and gray *rami communicantes*, which conduct visceral afferent and efferent neurons from and to the sympathetic trunk and ganglion, connect with the ventral ramus near its origin. Throughout the length of the body, ventral rami have the greater distribution; they form the cervical and brachial plexuses, the intercostal nerves, and the lumbosacral and coccygeal plexuses. The dorsal rami are restricted in distribution generally to the space between the spinous processes of the vertebrae and the angles of the ribs in the thoracic region. Comparable lateral boundaries exist for the other levels of the back (posterior contents of the prevertebral fascia at cervical levels and of the thoracolumbar fascia at lumbar levels).

Immediately beyond the intervertebral foramen, the dorsal ramus passes dorsally into the overlying deep musculature of the back and shortly divides into a *medial* and a *lateral branch*. Each of these branches supplies the muscles of the deep group, but as a rule, only one of them perforates to the subcutaneous tissues to become a cutaneous nerve.

In the upper half of the trunk, down to the sixth or seventh thoracic level, the medial branch is cutaneous and supplies the muscles adjacent to the corresponding spinous processes. In the lower portion of the back, the lateral branches terminate as cutaneous nerves, which enter the subcutaneous tissues in an increasingly lateral position, penetrating the musculature at about the junction of the muscular and tendinous portions of the latissimus dorsi muscle.

Back

Nerves

(Continued)

Schematic Diagram of Typical Spinal (Thoracic) Nerve

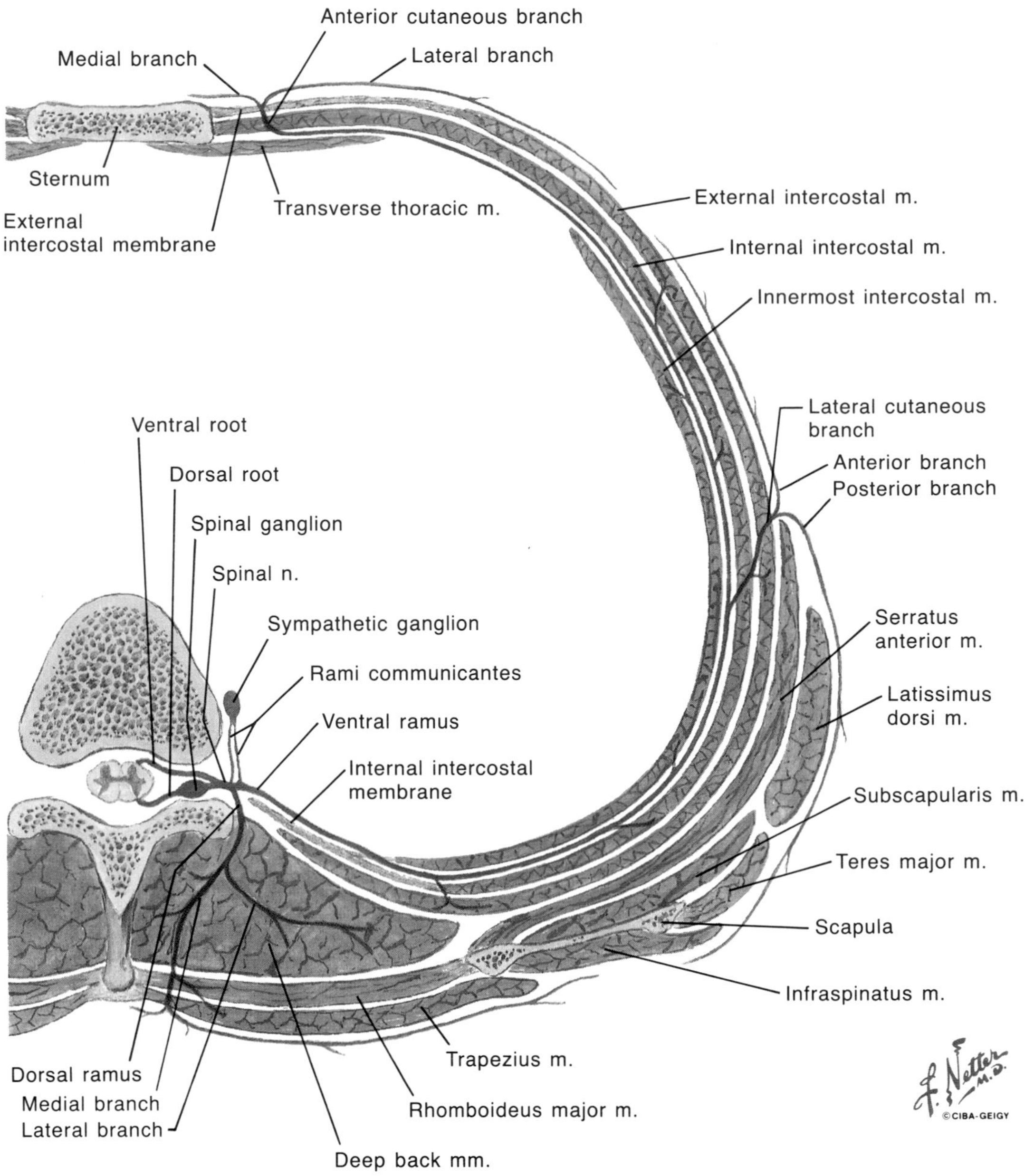
Anterior cutaneous branch
Medial branch
Lateral branch
Sternum
External intercostal membrane
Transverse thoracic m.
External intercostal m.
Internal intercostal m.
Innermost intercostal m.
Lateral cutaneous branch
Anterior branch
Posterior branch
Ventral root
Dorsal root
Spinal ganglion
Spinal n.
Sympathetic ganglion
Rami communicantes
Ventral ramus
Internal intercostal membrane
Serratus anterior m.
Latissimus dorsi m.
Subscapularis m.
Teres major m.
Scapula
Infraspinatus m.
Trapezius m.
Rhomboideus major m.
Deep back mm.
Dorsal ramus
Medial branch
Lateral branch
F. Netter M.D.
©CIBA-GEIGY

In the lumbar region, there is a cutaneous distribution of only the first three lumbar nerves (L1, 2, 3). These are known as the *superior cluneal nerves*, and they descend with a lateralward declination over the posterior part of the iliac crest to reach the upper gluteal region. The lateral branches of the dorsal rami of the first, second, and third sacral nerves (S1, 2, 3) form the *middle cluneal nerves*. They become cutaneous on a line connecting the posterior superior iliac spine and the tip of the coccyx, and they supply the skin and subcutaneous tissue over the back of the sacrum and an adjacent area of the gluteal region. The dorsal rami of the fourth and fifth sacral nerves (S4, 5) and the coccygeal nerve do not divide into medial and lateral branches. They unite to form a cutaneous nerve that distributes in the neighborhood of the coccyx.

Of nerves derived from ventral rami of spinal nerves seen in the back, the *inferior cluneal nerve* is a branch of the posterior femoral cutaneous nerve, which curves into the gluteal region around or through the lower fibers of the gluteus maximus muscle.

Thoracic Nerves

The 12 pairs of thoracic nerves resemble other typical spinal nerves in their segmental attachments to the cord by *dorsal* and *ventral nerve roots*. These roots unite to form short *spinal nerve trunks*, which emerge through the corresponding intervertebral foramina, give off recurrent meningeal filaments, establish connections through white and gray rami communicantes with the adjacent sympathetic trunk ganglia, and divide into larger *ventral* and smaller *dorsal rami*.

The *dorsal rami* of the thoracic nerves run backward near the zygapophyseal joints, which they supply, and then divide into medial and lateral branches. Both sets of branches pass through the groups of muscles constituting the erector spinae and give off branches to them. The terminations of the upper six or seven *medial branches* innervate the skin adjacent to the corresponding spinous processes, but the lower five or six often fail to reach the skin. The terminations of all the *lateral branches* usually pierce the thoracolumbar fascia over the erector spinae muscles and divide into *medial* and *lateral cutaneous branches*, which innervate much of the skin of the posterior thoracic wall and upper lumbar regions.

The *ventral rami* of most of the thoracic nerves, unlike those in other regions, do not form plexuses. They retain their segmental character, and each pair runs separately in the corresponding intercostal spaces as the *intercostal nerves*. The intercostal nerves give off *muscular, anterior cutaneous, lateral cutaneous, mammary*, and *collateral branches*. They also supply filaments to adjacent vessels, periosteum, parietal pleura, and peritoneum. The upper six pairs supply *muscular branches* to the corresponding intercostal muscles and to the subcostal, serratus posterior superior, and transverse thoracic muscles. The lower five pairs supply the lower intercostal muscles and the subcostal, serratus posterior inferior, transverse, oblique, and rectus abdominal muscles.

The *anterior cutaneous branches* supply the front of the thorax. The *lateral cutaneous branches* pierce the internal and external intercostal muscles and end by dividing into branches that extend forward and backward to innervate the skin covering the lateral sides of the thorax and abdomen. The small *lateral branch of the first intercostal nerve* supplies the skin of the axilla, and the lateral branch of the second is the *intercostobrachial nerve*, which is distributed to the skin on the medial side of the arm. The *lateral cutaneous branch of the subcostal nerve* pierces the internal and external oblique abdominal muscles and descends over the iliac crest to supply the skin of the anterior part of the gluteal region.

The mammary glands receive filaments from the lateral and anterior cutaneous branches of the fourth, fifth, and sixth intercostal nerves, which convey autonomic and sensory fibers to and from the glands. (The course of a typical thoracic nerve is shown in Plate 7.) ☐

Back

Vertebral Column and Pelvis

Vertebral Column

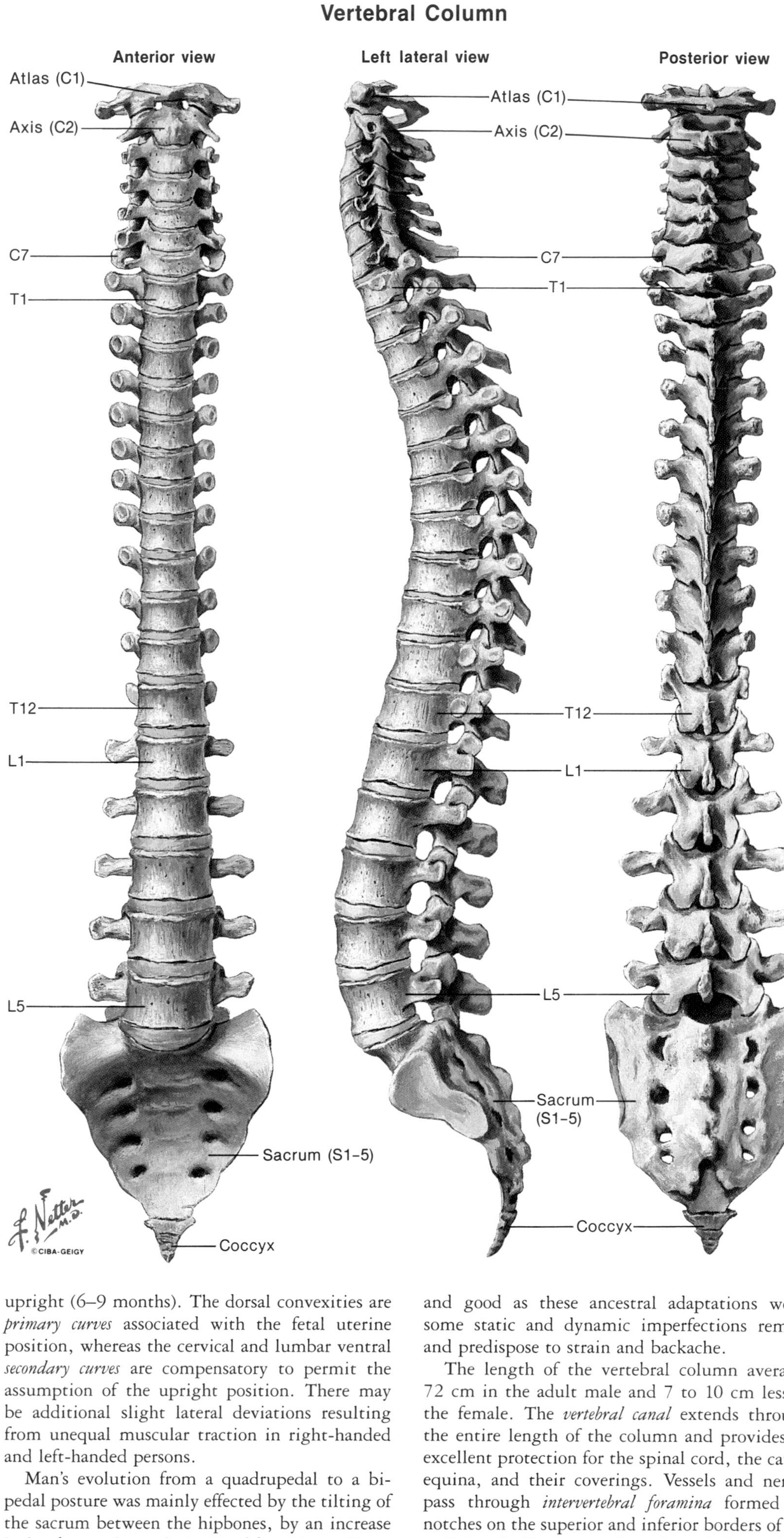
Vertebral Column
Anterior view
Left lateral view
Posterior view
Atlas (C1)
Axis (C2)
C7
T1
T12
L1
L5
Sacrum (S1-5)
Coccyx
Atlas (C1)
Axis (C2)
C7
T1
T12
L1
L5
Sacrum
(S1-5)
Coccyx
©CIBA-GEIGY

Vertebral Column

The vertebral column is built up from alternating bony vertebrae and fibrocartilaginous discs that are intimately connected by strong ligaments and supported by powerful musculotendinous masses (Plate 8). The individual bony elements and ligaments are described in Plates 9–18.

There are 33 vertebrae (7 cervical, 12 thoracic, 5 lumbar, 5 sacral, and 4 coccygeal), although the sacral and coccygeal vertebrae are usually fused to form the sacrum and coccyx. All vertebrae conform to a basic plan, but individual variations occur in the different regions. A typical vertebra is made up of an anterior, more-or-less cylindric *body* and a posterior *arch* composed of two *pedicles* and two *laminae*, the latter united posteriorly to form a *spinous process*. These processes vary in shape, size, and direction in the various regions of the spine. On each side, the arch also supports a *transverse process* and *superior* and *inferior articular processes*; the latter form synovial joints with corresponding processes on adjacent vertebrae. The spinous and transverse processes provide levers for the many muscles attached to them. The increasing size of the vertebral bodies from above downward is related to the increasing weights and stresses borne by successive segments, and the sacral vertebrae are fused to form a solid wedge-shaped base—the keystone in a bridge whose arches curve down toward the hip joints. The *intervertebral discs* act as elastic buffers to absorb the many mechanical shocks sustained by the vertebral column.

Only limited movements are possible between adjacent vertebrae, but the sum of these movements confers a considerable range of mobility on the vertebral column as a whole. Flexion, extension, lateral bending, rotation, and circumduction are all possible, and these actions are freer in the cervical and lumbar regions than in the thoracic region. Such differences exist because the discs are thicker in the cervical and lumbar areas, the splinting effect produced by the thoracic cage is lacking, the cervical and lumbar spinous processes are shorter and less closely apposed, and the articular processes are shaped and arranged differently.

At birth, the vertebral column presents a general dorsal convexity, but later, the cervical and lumbar regions become curved in the opposite directions—when the infant reaches the stages of holding up its head (3–4 months) and sitting upright (6–9 months). The dorsal convexities are *primary curves* associated with the fetal uterine position, whereas the cervical and lumbar ventral *secondary curves* are compensatory to permit the assumption of the upright position. There may be additional slight lateral deviations resulting from unequal muscular traction in right-handed and left-handed persons.

Man's evolution from a quadrupedal to a bipedal posture was mainly effected by the tilting of the sacrum between the hipbones, by an increase in lumbosacral angulation, and by minor adjustments of the anterior and posterior depths of various vertebrae and discs. An erect posture greatly increases the load borne by the lower spinal joints, and good as these ancestral adaptations were, some static and dynamic imperfections remain and predispose to strain and backache.

The length of the vertebral column averages 72 cm in the adult male and 7 to 10 cm less in the female. The *vertebral canal* extends through the entire length of the column and provides an excellent protection for the spinal cord, the cauda equina, and their coverings. Vessels and nerves pass through *intervertebral foramina* formed by notches on the superior and inferior borders of the pedicles of adjacent vertebrae, bounded anteriorly by the corresponding intervertebral discs and posteriorly, by the joints between the articular processes of adjoining vertebrae.

Back

Vertebral Column and Pelvis

(Continued)

Atlas and Axis

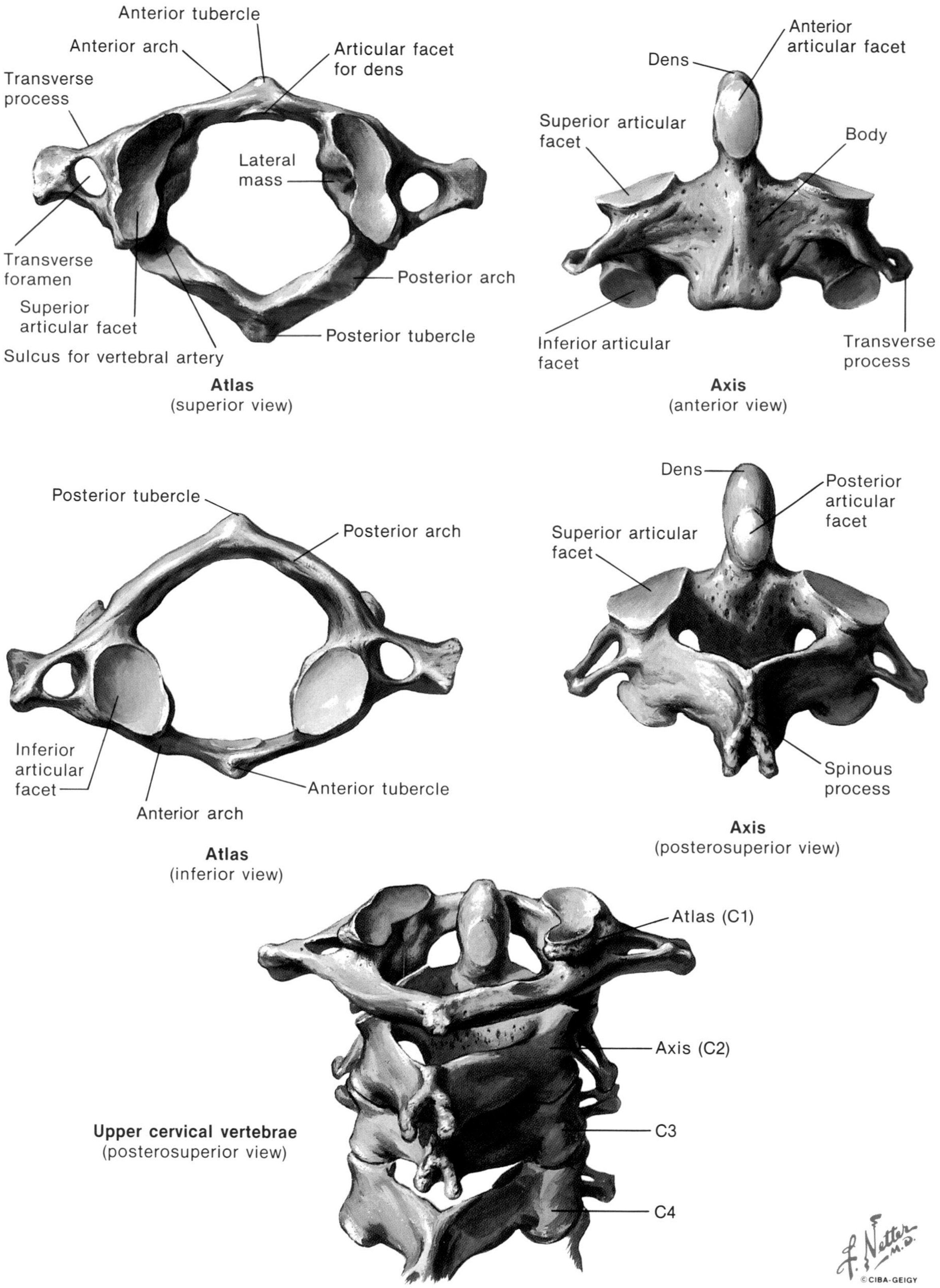

Anterior tubercle
Anterior arch
Articular facet for dens
Transverse process
Lateral mass
Transverse foramen
Posterior arch
Superior articular facet
Posterior tubercle
Sulcus for vertebral artery
Atlas
(superior view)
Dens
Anterior articular facet
Superior articular facet
Body
Inferior articular facet
Transverse process
Axis
(anterior view)
Posterior tubercle
Posterior arch
Inferior articular facet
Anterior tubercle
Anterior arch
Atlas
(inferior view)
Dens
Posterior articular facet
Superior articular facet
Spinous process
Axis
(posterosuperior view)
Atlas (C1)
Axis (C2)
C3
C4
Upper cervical vertebrae
(posterosuperior view)
F. Netter M.D.
©CIBA-GEIGY

Atlas and Axis

The atlas and axis, the first and second cervical vertebrae (C1–2), are atypical (Plate 9). They are linked together and to the skull and other cervical vertebrae by a layered pattern of craniocervical ligaments (Plates 11–12).

Atlas (C1). Named after the mythical giant who carried the earth on his shoulders, the atlas supports the globe of the skull. It lacks a body and forms a ring consisting of shorter anterior and longer posterior arches, with two lateral masses. The enclosed *vertebral foramen* is relatively large.

The *anterior arch* is slightly curved, with an anterior midline tubercle and a posterior midline facet for articulation with the dens of the axis.

The *lateral masses* bear superior and inferior articular facets and transverse processes. The *superior articular facets* are concave and ovoid (often waisted, or reniform) and are directed upward and inward as shallow cups, or foveae, for the reception of the occipital condyles. Nodding movements of the head mainly occur at these atlantooccipital joints. The *inferior articular facets* are almost circular, gently concave, and face downward and slightly medially and backward; they articulate with the superior articular facets on the axis. The *transverse processes* are each pierced by a foramen for the vertebral artery and project so far laterally that they can be easily palpated by pressing inward between the mandibular angles and the mastoid processes. They provide attachments and levers for some of the muscles involved in head rotation. On the anteromedial aspect of each lateral mass, there is a small tubercle for the attachment of the transverse ligament of the atlas.

The *posterior arch* is more curved than the anterior and has a small *posterior tubercle*, which is a rudimentary spinous process. Just behind each superior articular facet, there is a shallow *groove for the vertebral artery* and the first cervical spinal nerve, the nerve lying between the artery and the bone.

Axis (C2). A toothlike process, or dens, projects upward from the body of the axis. The *dens* is really the divorced body of the atlas that has united with the axis to form a pivot around which the atlas and the superjacent skull can rotate. Its anterior surface has an oval *anterior facet*, which articulates with the facet on the back of the anterior arch of the atlas, and a smaller *posterior facet* lower down on its posterior surface, which is separated from the transverse ligament of the atlas by a small bursa. The apex of the dens is attached to the lower end of the apical ligament, and the alar ligaments are attached to its sides.

The *body* of the axis has a lower liplike projection that overlaps the anterosuperior border of the third cervical vertebra. Its anterior surface shows a median ridge that separates slight depressions for slips of the longus colli muscles. The posteroinferior border of the body is less prominent, and attached to it are the tectorial membrane and the posterior longitudinal spinal ligament.

The *pedicles* and *laminae* are stout, and the latter end in a strong, bifid *spinous process*. The *vertebral foramen* of the axis is somewhat smaller than that of the atlas. On each side of the body are superior and inferior articular and transverse processes. The *articular processes*, unlike others in the cervical region, are offset, since the superior pair is anterior in position to the inferior pair. They articulate with the adjoining processes of the atlas and the third cervical vertebra. The *transverse processes* are smaller and shorter than those of the atlas, and their foramina are inclined superolaterally to allow the contained vertebral arteries and nerves to pass easily into the more widely spaced transverse foramina of the atlas.

Cervical Vertebrae

Back

Vertebral Column and Pelvis

(Continued)

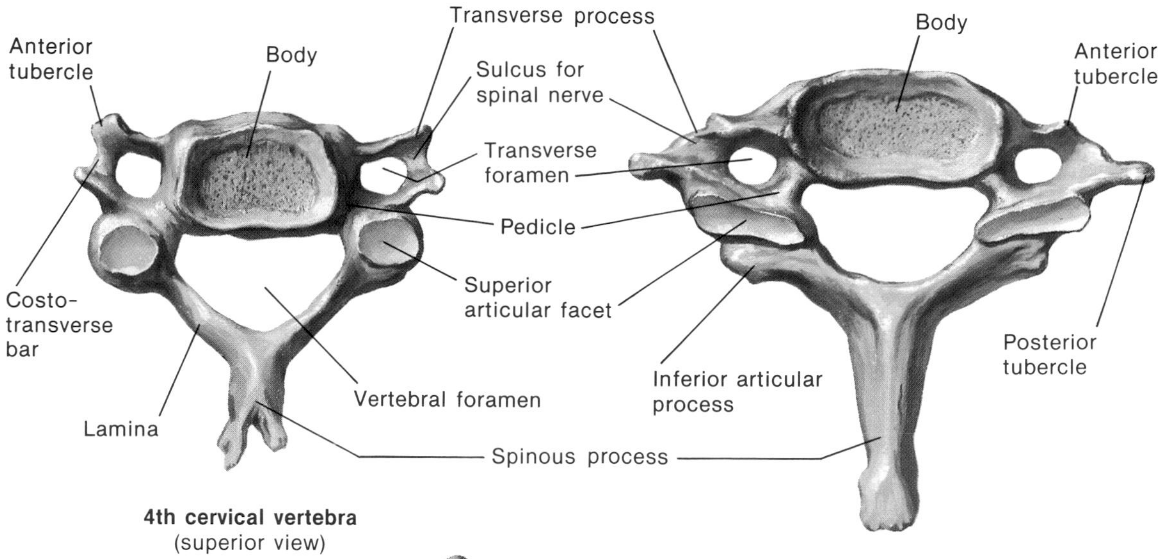

4th cervical vertebra
(superior view)

7th cervical vertebra
(superior view)

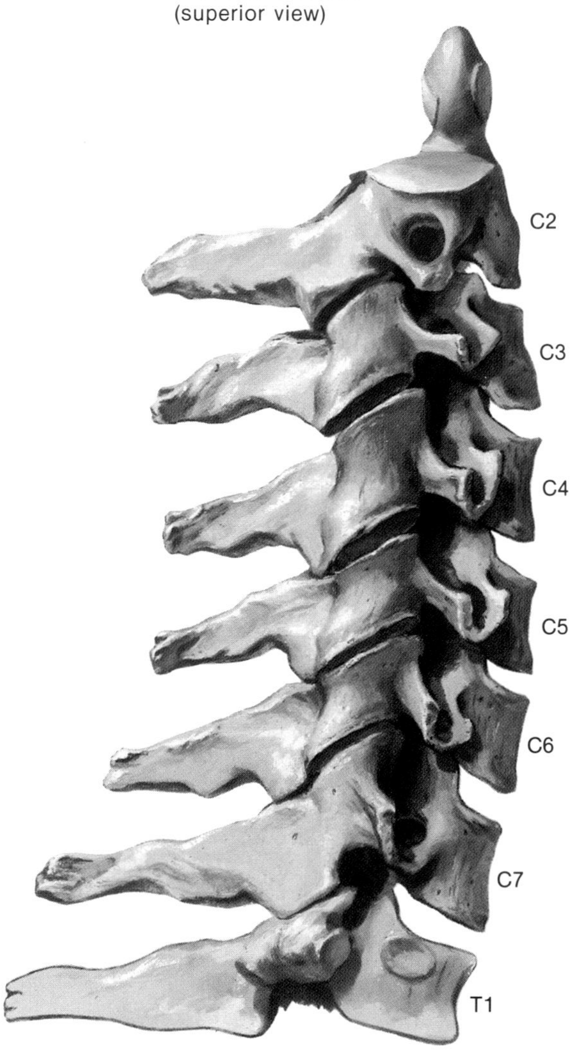

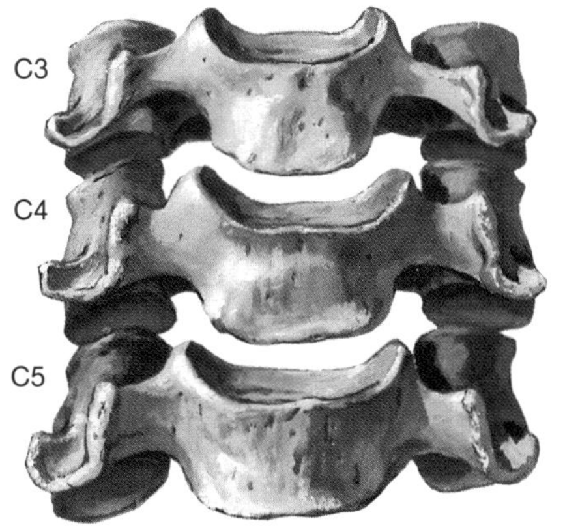

3rd, 4th, and 5th cervical vertebrae
(anterior view)

2nd cervical to 1st thoracic vertebrae
(right lateral view)

Cervical Vertebrae

The first two cervical vertebrae, the atlas and the axis (C1–2), are illustrated in Plate 9. The other five (C3–7) exhibit the general vertebral features (Plate 10). However, cervical vertebrae are easily distinguishable by the presence of foramina in their transverse processes, which (except in the case of C7) transmit the vertebral vessels and nerves.

The cervical *vertebral bodies* are smaller than those of the other movable vertebrae and increase in size from above downward; they are broader in the transverse diameter than anteroposteriorly. The superior body surfaces are concave from side to side and slightly convex from front to back, while the inferior surfaces are reciprocally curved or saddle shaped. The lateral edges of the superior body surface are raised, while those of the lower surface are beveled, and small clefts exist between them. Some claim that these clefts are miniature synovial joints, but others believe that they are merely spaces in the lateral parts of the corresponding intervertebral discs.

The *vertebral foramina* are comparatively large in order to accommodate the cervical enlargement of the spinal cord; they are bounded by the bodies, pedicles, and laminae of the vertebrae. The *pedicles* project posterolaterally from the bodies and are grooved by superior and inferior vertebral notches, almost equal in depth, which form the intervertebral foramina by connecting with similar notches on adjacent vertebrae. The medially directed *laminae* are thin and relatively long and fuse posteriorly to form short, bifid *spinous processes*. Projecting laterally from the junction of the pedicles and laminae are articular pillars supporting *superior* and *inferior articular facets*.

Each *transverse process* is pierced by a foramen, bounded by narrow bony bars ending in anterior and posterior tubercles; these are interconnected lateral to the foramen by the so-called *costotransverse bar*. Only the medial part of the posterior bar represents the true transverse process; the anterior and costotransverse bars and the lateral portion of the posterior bar constitute the costal element. Abnormally, these elements, especially in C7 and C6, or both, develop to form cervical ribs. The upper surfaces of the costotransverse bars are grooved and lodge the anterior primary rami of the spinal nerves. The anterior tubercles of C6 are large and are termed the *carotid tubercles*, because the common carotid arteries lie just anteriorly and can be compressed against them.

The seventh cervical vertebra (C7) is called the *vertebra prominens*, because its spinous process is long and ends in a tubercle that is easily palpable at the lower end of the nuchal furrow; the spinous process of the first thoracic vertebra is just as prominent. Sometimes, C7 lacks a transverse foramen on one or both sides; when present, the foramina transmit only small accessory vertebral veins.

Back

Vertebral Column and Pelvis

(Continued)

External Craniocervical Ligaments

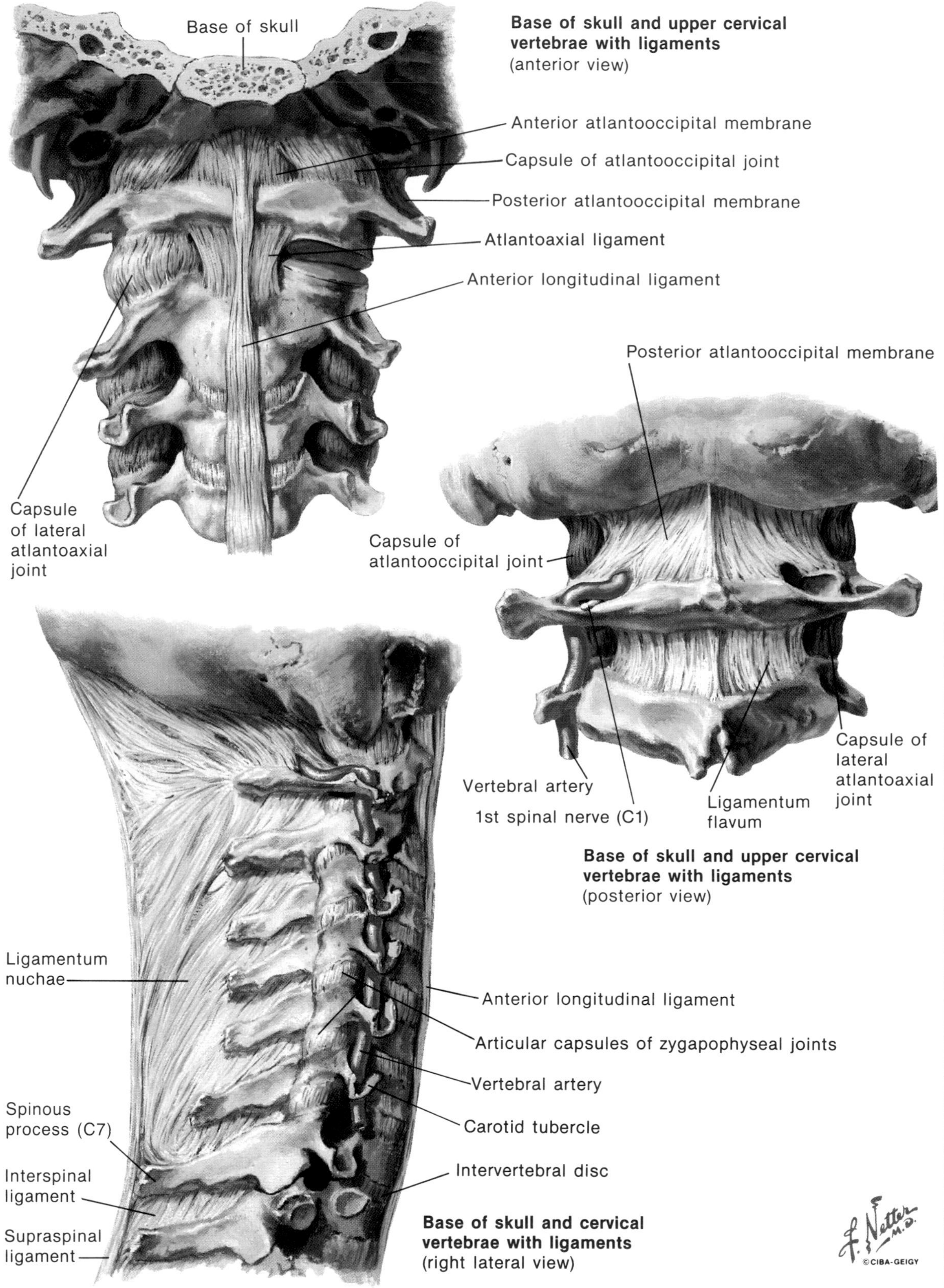
Base of skull
Anterior atlantooccipital membrane
Capsule of atlantooccipital joint
Posterior atlantooccipital membrane
Atlantoaxial ligament
Anterior longitudinal ligament
Capsule of lateral atlantoaxial joint
Posterior atlantooccipital membrane
Capsule of atlantooccipital joint
Capsule of lateral atlantoaxial joint
Vertebral artery
1st spinal nerve (C1)
Ligamentum flavum
Ligamentum nuchae
Anterior longitudinal ligament
Articular capsules of zygapophyseal joints
Vertebral artery
Carotid tubercle
Spinous process (C7)
Interspinal ligament
Intervertebral disc
Supraspinal ligament
F. Netter M.D.
©CIBA-GEIGY

Base of skull and upper cervical vertebrae with ligaments (anterior view)

Base of skull and upper cervical vertebrae with ligaments (posterior view)

Base of skull and cervical vertebrae with ligaments (right lateral view)

External Craniocervical Ligaments

The ligaments uniting the cranium, atlas, and axis allow free yet safe movement of the head, and extra security is provided by the ligamentous action of the surrounding muscles (Plate 11). Ligaments best seen from the external aspect are shown in the illustration.

The *anterior atlantooccipital membrane* is a wide, dense fibroelastic band extending between the anterior margin of the foramen magnum and the upper border of the anterior arch of the atlas. Laterally, it is continuous with the articular capsules of the atlantooccipital joints. In the midline, it is reinforced by the upward continuation of the anterior longitudinal ligament.

The *posterior atlantooccipital membrane* is broader and thinner than the anterior atlantooccipital membrane and connects the posterior margin of the foramen magnum with the upper border of the posterior arch of the atlas. On each side, the membrane arches over the groove for the vertebral artery, thus leaving an opening for the upward passage of the artery and the outward passage of the first cervical nerve.

Articular capsules surround the joints between the occipital condyles and the superior atlantal facets. They are rather loose, so as to allow nodding movements, and are thin medially; laterally, they are thickened to form *lateral atlantooccipital ligaments*, which limit lateral tilting of the head.

The *anterior longitudinal ligament* extends from the base of the skull to the sacrum. Its uppermost part reinforces the anterior atlantooccipital membrane in the midline. The part between the anterior tubercle of the atlas and the anterior median ridge on the axis may have lateral extensions, the *atlantoaxial (epistrophic) ligaments*.

The *ligamentum nuchae* is a fibroelastic membrane stretching from the external occipital protuberance and crest to the posterior tubercle of the atlas and the spinous processes of all the other cervical vertebrae. It provides areas for muscular attachments and forms a midline septum between the posterior cervical muscles. It is better developed in quadrupeds than in humans.

The *ligamenta flava* contain a high proportion of yellow elastic fibers and connect the laminae of adjacent vertebrae. They are present between the posterior arch of the atlas and the laminae of the axis, but are absent between the atlas and skull.

Intervertebral discs are lacking between the occiput and atlas and the atlas and axis.

ack

ertebral Column nd Pelvis

ontinued)

Internal Craniocervical Ligaments

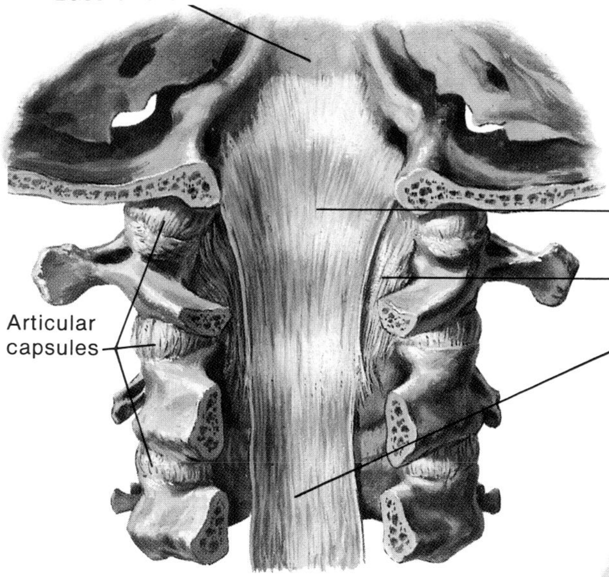
Base of skull
Tectorial membrane
Deeper portion of tectorial membrane
Articular capsules
Posterior longitudinal ligament

Upper portion of vertebral canal with spinous processes and portions of vertebral arches removed to expose ligaments on posterior aspect of vertebral bodies

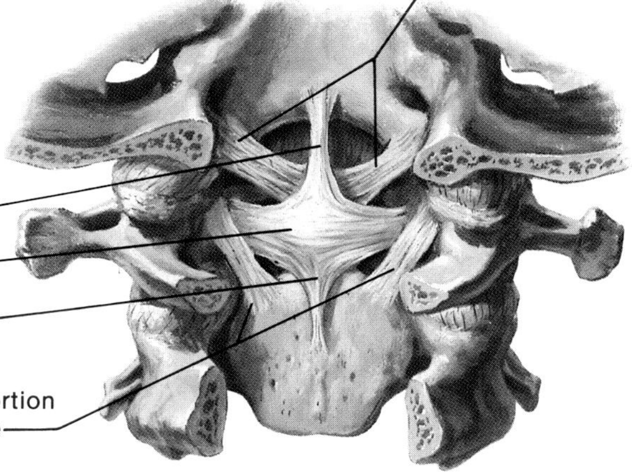
Alar ligaments
Cruciform ligament
Superior longitudinal fascicles
Transverse ligament of atlas
Inferior longitudinal fascicles
Deeper (accessory) portion of tectorial membrane

Principal portion of tectorial membrane removed to expose deeper ligaments

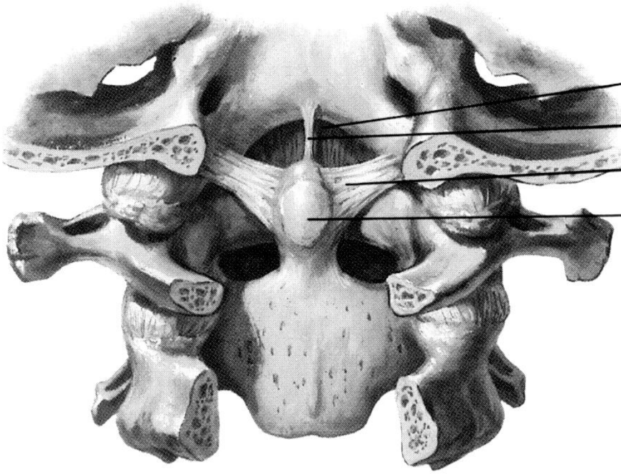
Anterior atlantooccipital membrane
Apical dental ligament
Alar ligament
Articular facet of dens for transverse ligament of atlas

Cruciform ligament removed to show deepest ligaments

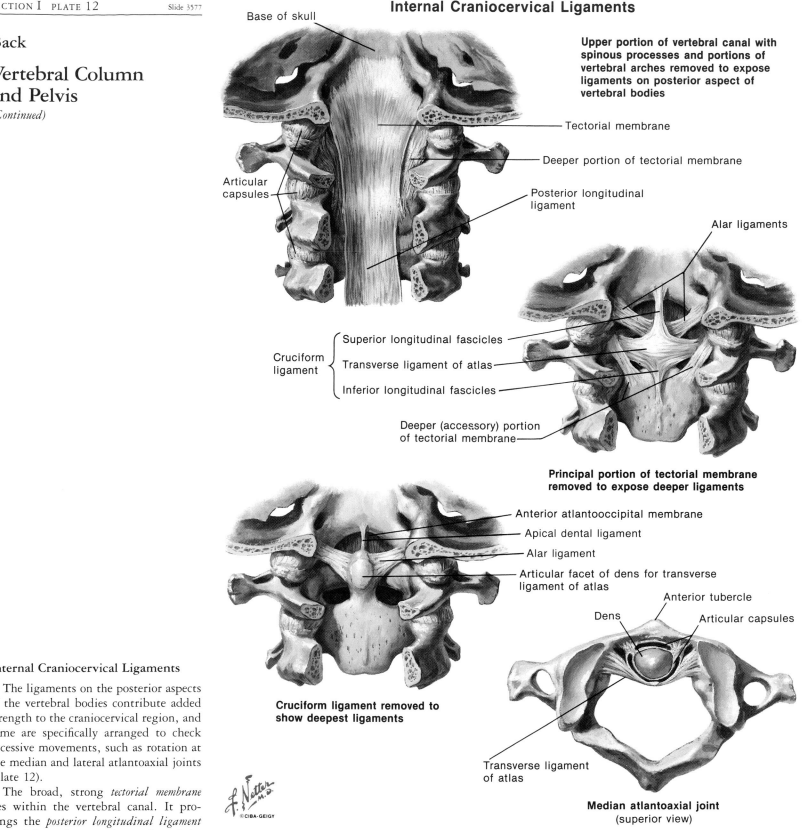

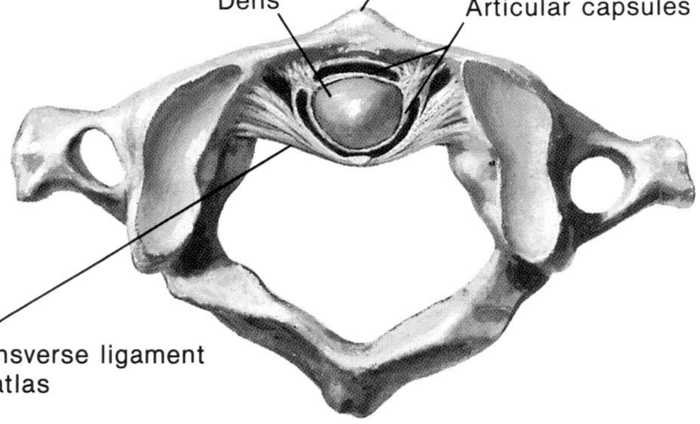
Anterior tubercle
Dens
Articular capsules
Transverse ligament of atlas

Median atlantoaxial joint
(superior view)

ternal Craniocervical Ligaments

The ligaments on the posterior aspects the vertebral bodies contribute added rength to the craniocervical region, and me are specifically arranged to check cessive movements, such as rotation at e median and lateral atlantoaxial joints late 12).

The broad, strong *tectorial membrane* s within the vertebral canal. It prongs the *posterior longitudinal ligament* oward from the posterior surface of the dy of the axis to the anterior and anteroteral margins of the foramen magnum, where blends with the dura mater. It covers the dens d its ligaments and gives added protection to e junctional area between the medulla oblonta and the spinal cord.

The *median atlantoaxial pivot joint* lies between e dens of the axis and the ring formed by the terior arch and transverse ligament of the atlas late 9). Two small synovial cavities surrounded thin articular capsules are present between the ens and the anterior arch in front and the transrse ligament of the atlas behind.

The *transverse ligament of the atlas* is a strong nd passing horizontally behind the dens and attached on each side to a tubercle on the medial side of the lateral mass of the atlas. From its midpoint, bands pass vertically upward and downward to become fixed, respectively, to the basilar part of the occipital bone between the tectorial membrane and the apical ligament of the dens and to the posterior surface of the body of the axis, the *superior* and *inferior longitudinal fascicles*. These transverse and vertical bands together form the *cruciform ligament*.

The *apical ligament* is a slender cord connecting the apex of the dens to the anterior midpoint of the foramen magnum, lying between the anterior atlantooccipital membrane and the upper limb of the cruciform ligament.

The *alar ligaments* are two fibrous bands stretching upward and outward from the superolateral aspects of the dens to the medial sides of the occipital condyles. They check excessive rotation at the median atlantooccipital joint.

Lateral atlantoaxial joints are formed between the almost-flat inferior articular facets on the lateral masses of the atlas and the superior articular facets of the axis. They are synovial joints with thin, loose articular capsules. An *accessory ligament*, deep to the tectorial membrane, extends from near the base of the dens to the lateral mass of the atlas, close to the attachment of the transverse ligament. It assists the alar ligaments in restricting atlantoaxial rotation.

Back

Vertebral Column and Pelvis

(Continued)

Thoracic Vertebrae and Ligaments

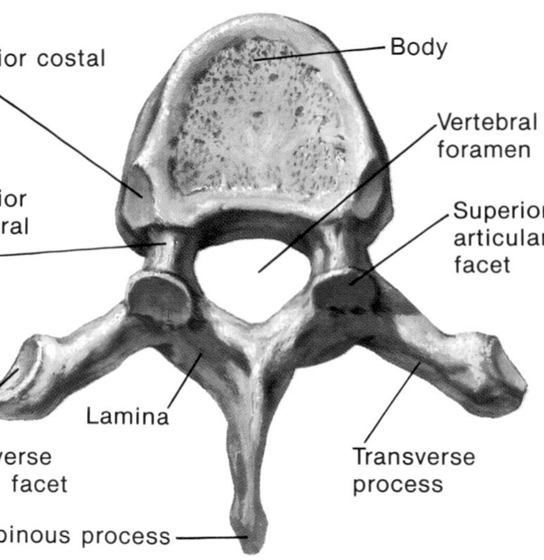

6th thoracic vertebra
(superior view)

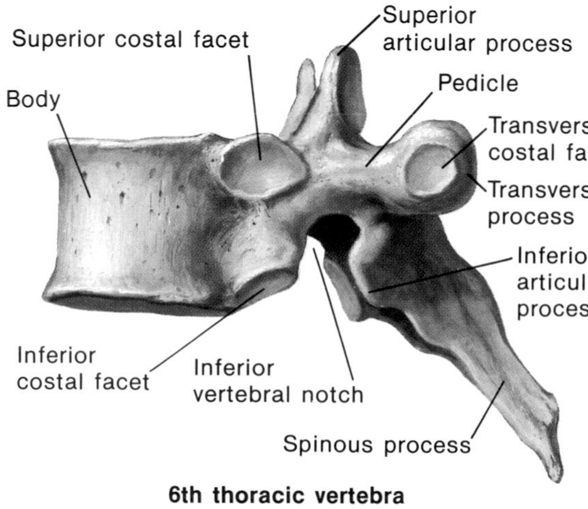

6th thoracic vertebra
(lateral view)

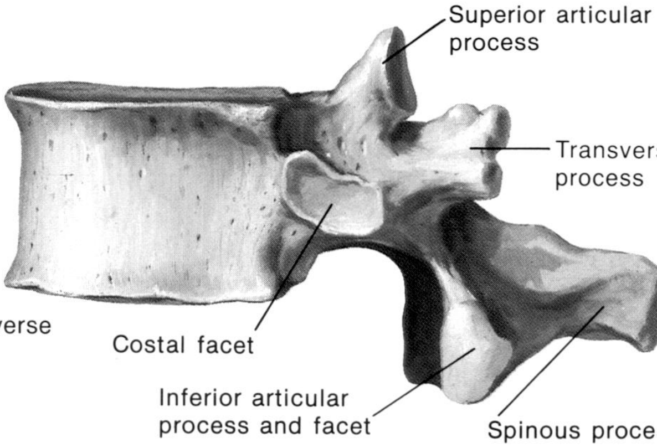

12th thoracic vertebra
(lateral view)

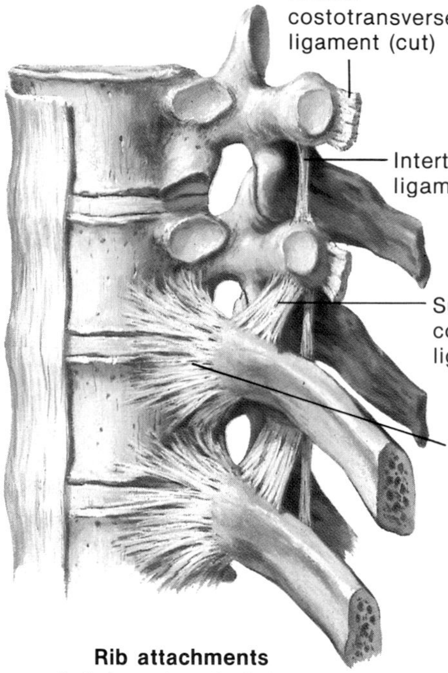

Rib attachments
(left lateral and slightly anterior view)

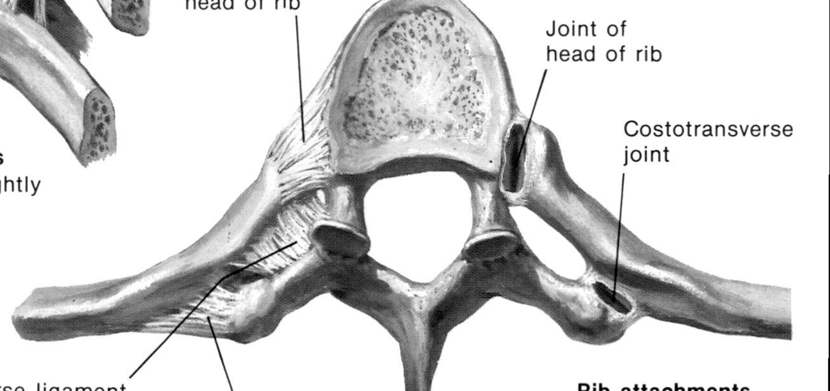

Rib attachments
(superior view with ligaments removed on right to expose joint

Thoracic Vertebrae and Ligaments

The 12 thoracic vertebrae (T1–12) are intermediate in size between the smaller cervical and the larger lumbar vertebrae. The heart-shaped *vertebral bodies* are slightly deeper posteriorly than anteriorly (Plate 13). They are easily recognized by costal facets on both sides of the bodies and on all the transverse processes (except those of T11 and T12), which articulate, respectively, with facets on the heads and tubercles of the corresponding ribs.

The *vertebral foramina* are smaller and more rounded than those in the cervical region and so conform to the reduced size and more circular shape of the spinal cord in the thoracic region. They are bounded by the posterior surfaces of the vertebral bodies and by the pedicles and laminae forming the vertebral arches. The stout *pedicles* are directed backward; they have very shallow superior and much deeper inferior vertebral notches. The *laminae* are short and relatively thick and partly overlap each other from above downward.

The typical thoracic *superior articular processes* project upward from the junctions of the pedicles and the laminae, and their facets slant backward and slightly upward and outward. The *inferior articular processes* project downward from the anterior parts of the laminae, and their facets face forward and slightly downward and inward. The processes and facets in the cervicothoracic and thoracolumbar junctional areas show gradual transitional changes.

Most of the thoracic *spinous processes* are long and inclined downward and backward. Those of the upper and lower thoracic vertebrae are more horizontal. The *transverse processes* are also relatively long and extend posterolaterally from the junctions of the pedicles and laminae. Except for those of the lowest two or, occasionally, three thoracic vertebrae, the transverse processes have small, oval facets near their tips, which articulate with similar facets on the corresponding rib tubercles.

Adjacent vertebral bodies are connected by *intervertebral discs* and by *anterior* and *posterior longitudinal ligaments*; the transverse processes, by *intertransverse ligaments*; the laminae, by *ligamenta flava*; and the spinous processes, by *supraspinal* and *interspinal ligaments*. The joints between the articular processes are surrounded by fibrous *articular capsules*.

Costovertebral Joints. The ribs are connected to the vertebral bodies and the transverse processes by various ligaments. The *costocentral joints* between the bodies and the rib heads have *articular capsules*; the second to tenth costal heads, each of which articulates with two vertebrae, are connected to the corresponding intervertebral discs by *intraarticular ligaments*. *Radiate (stellate) lig-ments* unite the anterior aspects of the rib hea with the sides of the vertebral bodies above ar below and with the intervening discs.

The *costotransverse joints* between the facets c the transverse processes and on the tubercles of t ribs are also surrounded by *articular capsules*. Th are reinforced by a (middle) *costotransverse ligame* between the rib neck and the adjoining transver process, a *superior costotransverse ligament* betwe the rib neck and the transverse process of the ve tebra above, and a *lateral costotransverse ligame* interconnecting the end of a transverse proce with the nonarticular part of the related cost tubercle.

Back

Vertebral Column and Pelvis

(Continued)

Lumbar Vertebrae and Intervertebral Discs

Vertebrae. The five lumbar vertebrae (L1–5) are the largest separate vertebrae and are distinguished by the absence of transverse foramina and costal facets (Plate 14). The *vertebral bodies* are wider from side to side than from front to back, with upper and lower surfaces that are kidney shaped and almost parallel, except in the case of the fifth vertebral body, which is slightly wedge shaped. The triangular *vertebral foramina* are larger in the thoracic vertebrae and smaller in the cervical vertebrae.

The *pedicles* are short and strong and arise from the upper and posterolateral aspects of the bodies; the superior vertebral notches are therefore less deep than the inferior notches. The *laminae* are short, broad plates that meet in the midline to form the quadrangular and almost horizontal *spinous processes*. The intervals between adjacent laminae and spinous processes are relatively wide.

The *articular processes* project vertically upward and downward from the junctional areas between the pedicles and laminae. The superior facets are gently concave and face posteromedially to embrace the inferior facets of the vertebra above, which are curved and disposed in a reciprocal fashion. This arrangement permits some flexion and extension but very little rotation. The *transverse processes* of L1 to L3 are long and slender, while those of L4, and especially of L5, are more pyramidal.

Near the roots of each transverse process are small *accessory processes*; other small, rounded *mamillary processes* protrude from the posterior margins of the superior articular processes. The former may represent the true transverse processes (or their tips), since many of the so-called transverse processes are really costal elements. In the first lumbar vertebra, these elements occasionally develop into lumbar ribs.

The *fifth lumbar vertebra* (L5) is atypical. It is the largest, its body is deeper anteriorly, its inferior articular facets face almost forward and are set more widely apart, and the roots of its stumpy transverse processes are continuous with the posterolateral parts of the body and with the entire lateral surfaces of the pedicles.

Intervertebral Discs. Interposed between the adjacent vertebral bodies from the axis to the sacrum are intervertebral discs, immensely strong fibrocartilaginous structures that provide powerful bonds and elastic buffers (Plate 14). The discs consist of outer concentric layers of fibrous tissue, the *annulus fibrosus* (the fibers in adjacent layers are arranged obliquely but in opposite directions to assist in resisting torsion), and a central springy, pulpy zone, the *nucleus pulposus*. The blood and nerve supplies to the discs are inconspicuous. If the annular fibers give way as a result of injury or disease, the enclosed turgid nucleus pulposus may prolapse and press upon related nervous and vascular structures.

In health and maturity, the intervertebral discs account for almost 25% of the length of the vertebral column; they are thinnest in the upper thoracic region and thickest in the lumbar region. In vertical section, the lumbar discs are moderately wedge shaped, with the thicker edge anteriorly. The forward lumbar spinal convexity is due more to the shape of the discs than to the disparities between the anterior and posterior depths of the lumbar vertebrae. The more defined wedge shape of the lumbosacral disc helps to minimize the effects of the marked lumbosacral angulation.

As age advances, the nucleus pulposus undergoes changes: its water content decreases, its mucoid matrix is gradually replaced by fibrocartilage, and it ultimately comes to resemble the annulus fibrosus. Although the resultant loss of depth in each disc is small, it may amount to an overall decrease of 2 to 3 cm in the length of the vertebral column.

Lumbar Vertebrae and Intervertebral Disc

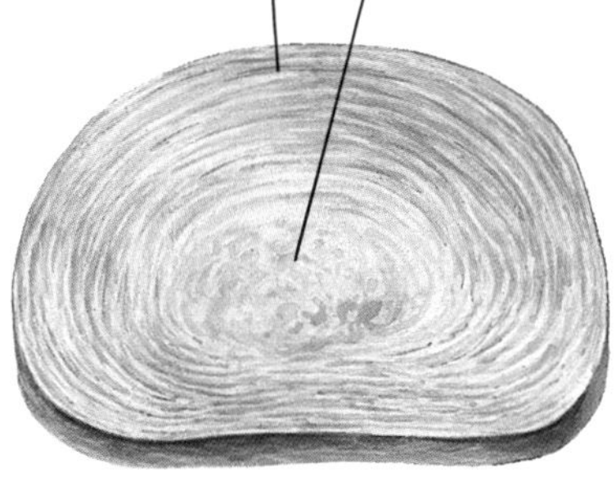
Annulus fibrosus
Nucleus pulposus

Intervertebral disc

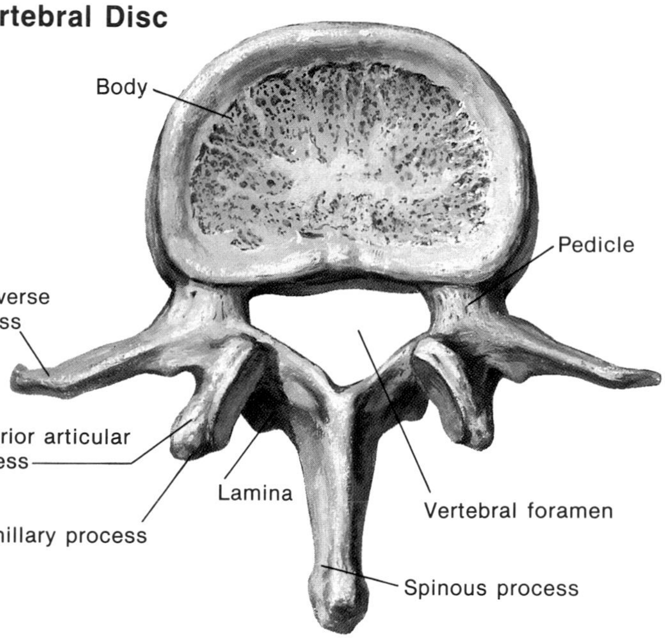
Body
Pedicle
Transverse process
Superior articular process
Lamina
Vertebral foramen
Mamillary process
Spinous process

2nd lumbar vertebra
(superior view)

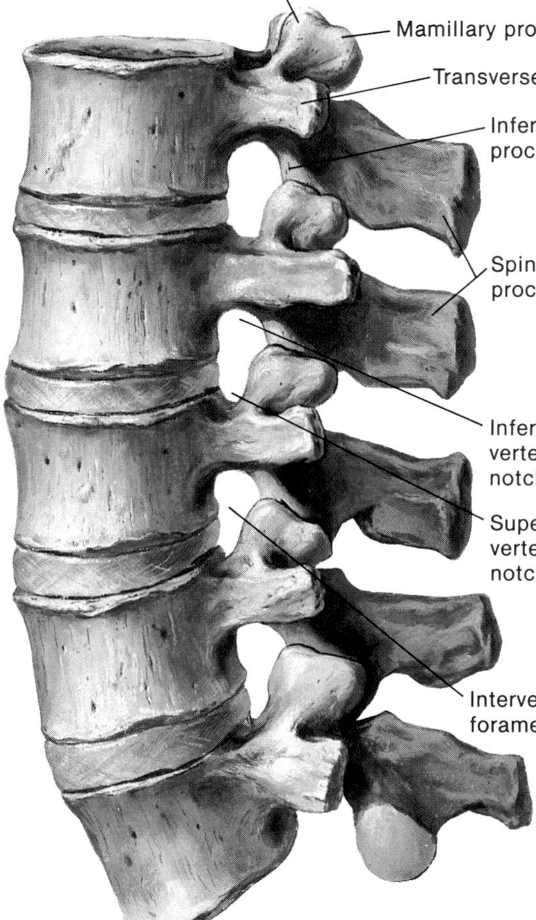
Superior articular process
Mamillary process
Transverse process
Inferior articular process
Spinous process
Inferior vertebral notch
Superior vertebral notch
Intervertebral foramen

Lumbar vertebrae
(lateral view)

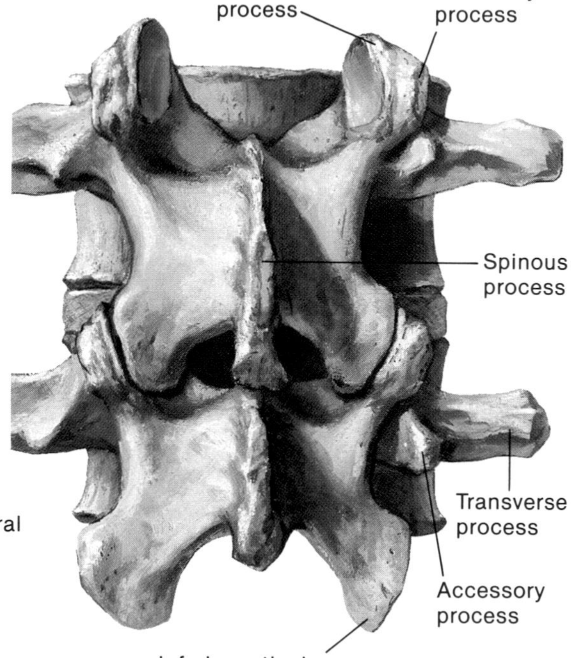

3rd and 4th lumbar vertebrae
(posterior view)

F. Netter M.D.

Back

Vertebral Column and Pelvis

(Continued)

Bones and Ligaments of Pelvis

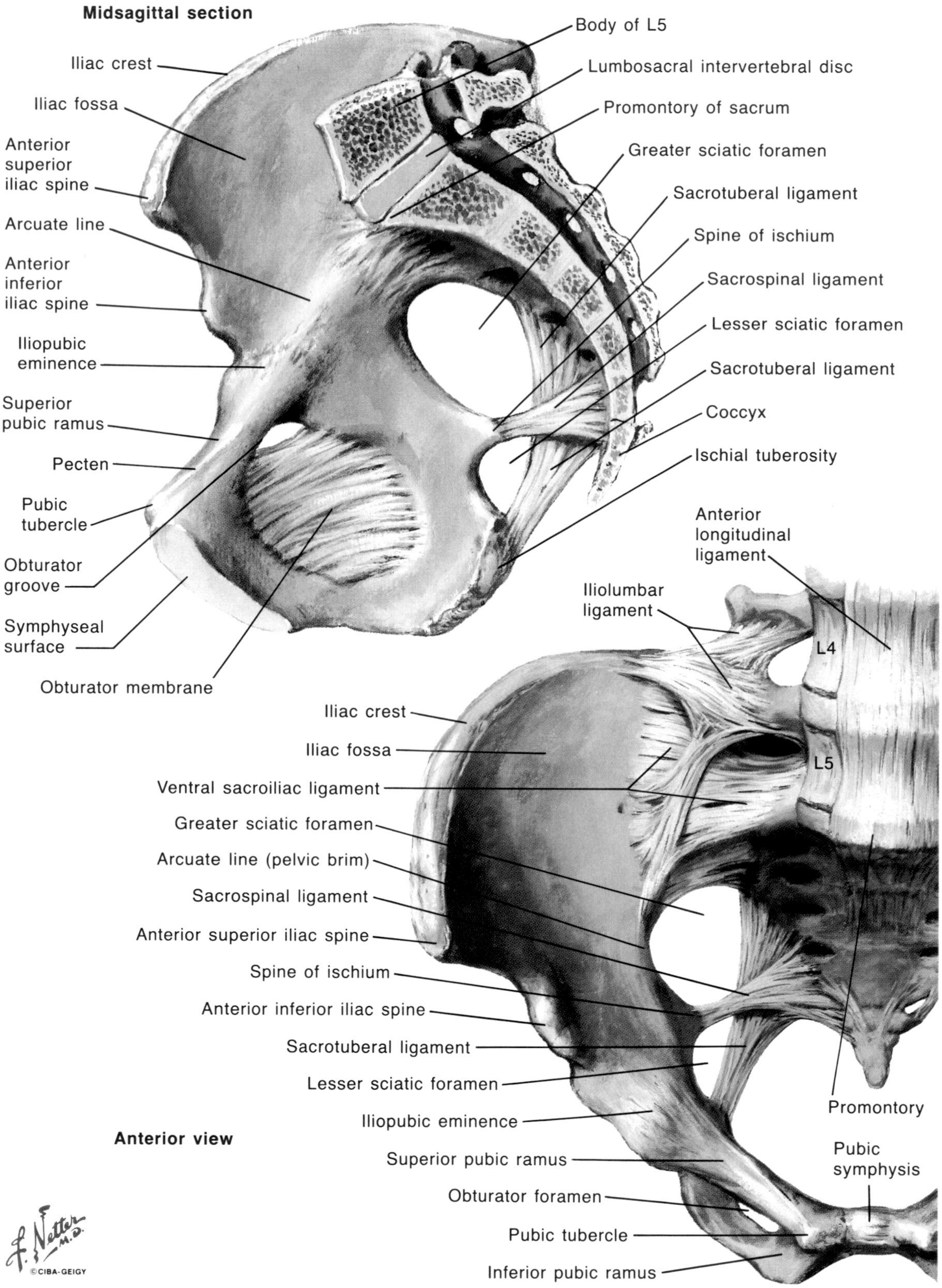
Midsagittal section
Iliac crest
Iliac fossa
Anterior superior iliac spine
Arcuate line
Anterior inferior iliac spine
Iliopubic eminence
Superior pubic ramus
Pecten
Pubic tubercle
Obturator groove
Symphyseal surface
Obturator membrane
Body of L5
Lumbosacral intervertebral disc
Promontory of sacrum
Greater sciatic foramen
Sacrotuberal ligament
Spine of ischium
Sacrospinal ligament
Lesser sciatic foramen
Sacrotuberal ligament
Coccyx
Ischial tuberosity
Anterior longitudinal ligament
Iliolumbar ligament
L4
L5
Iliac crest
Iliac fossa
Ventral sacroiliac ligament
Greater sciatic foramen
Arcuate line (pelvic brim)
Sacrospinal ligament
Anterior superior iliac spine
Spine of ischium
Anterior inferior iliac spine
Sacrotuberal ligament
Lesser sciatic foramen
Iliopubic eminence
Superior pubic ramus
Obturator foramen
Pubic tubercle
Inferior pubic ramus
Promontory
Pubic symphysis
Anterior view
F. Netter M.D.
©CIBA-GEIGY

Bones

The principal function of the pelvis is to transmit body weight to the limbs and absorb the stresses of muscular activity in the erect posture (Plate 15). The center of gravity of the body is at a point just anterior to the sacral promontory, and from here, a sacrofemoral arch of robust bone transmits this weight to the heads of the femurs. A similar sturdy sacroischial arch, ending in the ischial tuberosities, serves this weight transfer in the sitting position. The more delicate rami of the pubis and ischium act as tie-rods of these arches and resist inward collapse of the bones. Such considerations emphasize the great importance of the regions of union of the fifth lumbar vertebra and the sacrum to the coxal bone and of the pubic symphysis.

The *sacrum*, a fusion of five vertebrae, is broadened by the amalgamation of large costal elements and transverse processes into heavy lateral parts (Plate 18). It is broad above and tapered below, relatively smooth on its pelvic surface, and highly irregular dorsally for ligamentous attachments. The heavy lateral parts, one on each side, exhibit large and somewhat irregular *auricular surfaces* for articulation with the ilia. The complementary elevations and depressions of these surfaces make for antirotational and locking features at the sacroiliac joints, which, together with the extremely heavy dorsal ligaments and the contributions of the sacrotuberal and sacrospinal ligaments, provide a reasonably stable sacroiliac articulation. Architecturally, the situation of the sacrum between the coxal bones is not as a keystone; indeed, without the heavy dorsal ligaments, the sacrum would appear likely to fall into the pelvis.

The *pubic symphysis* unites the two pubic bones. The bony articular surfaces are ridged and grooved, and the sides fit closely together and are covered by a thin layer of hyaline cartilage. There is an interpubic disc of fibrocartilage, which is thicker in the female. A *superior pubic ligament* connects the pubic bones along their superior surfaces, extending as far as the pubic tubercles. The *arcuate pubic ligament* unites the bones inferiorly. Anteriorly, decussating tendinous fibers of the rectus abdominis and the external abdominal oblique muscles reinforce the capsule.

Lumbosacral Ligaments

Because the lumbosacral and sacroiliac joints transmit the entire weight of the body to the hip bones and thence to the lower limbs, their ligaments are most important (Plate 16).

The *anterior longitudinal ligament* is a straplike band increasing in width from above downward and extending from the anterior tubercle of the atlas to the sacrum. It is firmly attached to the anterior margins of the vertebral bodies and the intervertebral discs. The superficial fibers cross over several vertebrae, and the shorter, deeper fibers interconnect adjacent vertebral bodies and discs.

The *posterior longitudinal ligament* is broader above than below and lies within the vertebral canal behind the vertebral bodies. Its upper end is continuous with the tectorial membrane, and it extends from the axis to the sacrum. The edges of the ligament are serrated, especially in the lower thoracic and lumbar regions, because it spreads outward between its attachments to the borders of the vertebral bodies to blend with the annular fibers of the discs. The ligament is separated from the posterior surfaces of the vertebral bodies by the basivertebral veins that join the anterior internal vertebral venous plexus.

Back

Vertebral Column and Pelvis

(Continued)

Lumbosacral Spine and Ligaments

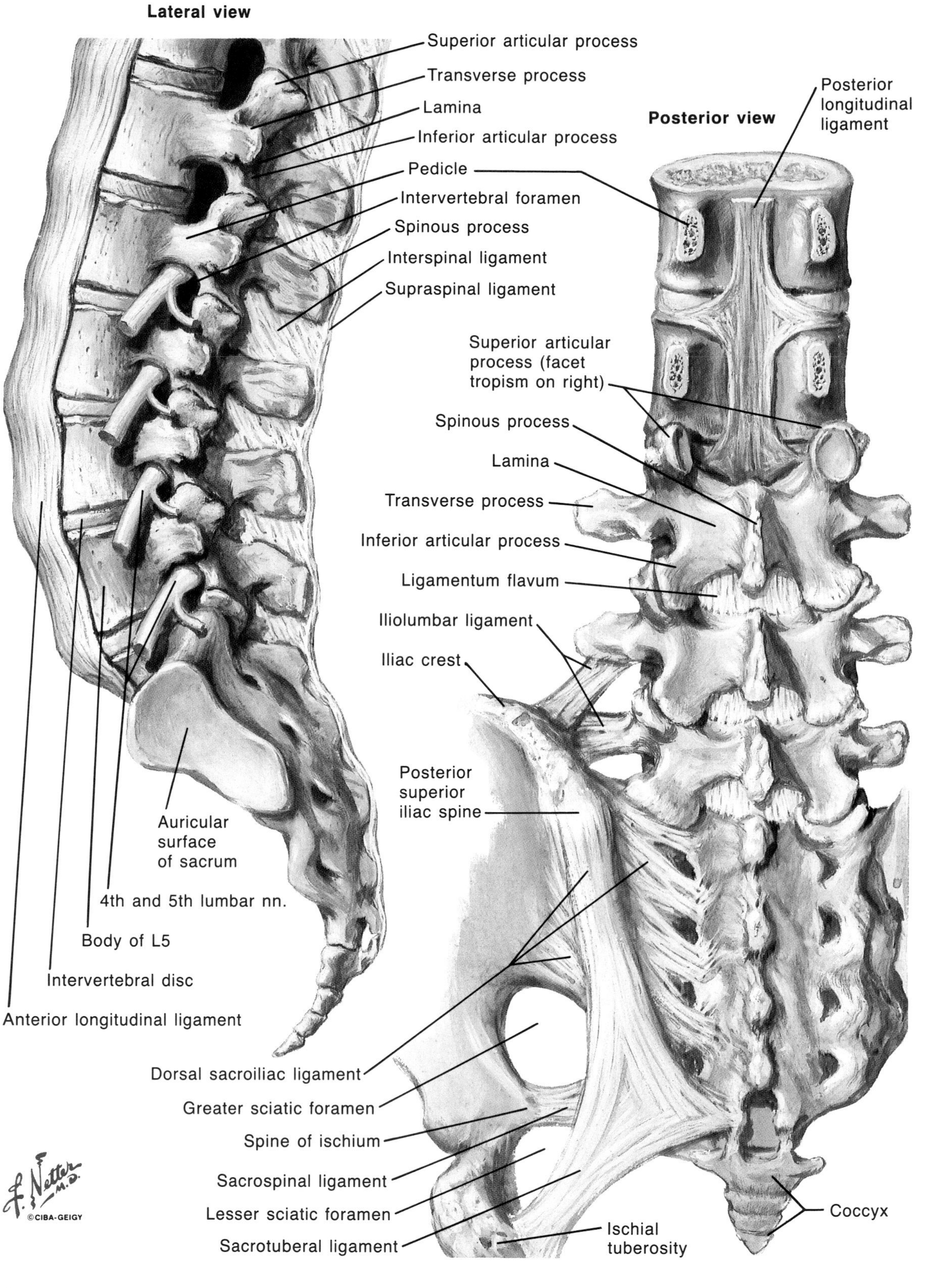
Lateral view
Superior articular process
Transverse process
Lamina
Inferior articular process
Pedicle
Intervertebral foramen
Spinous process
Interspinal ligament
Supraspinal ligament
Auricular surface of sacrum
4th and 5th lumbar nn.
Body of L5
Intervertebral disc
Anterior longitudinal ligament
Posterior view
Posterior longitudinal ligament
Superior articular process (facet tropism on right)
Spinous process
Lamina
Transverse process
Inferior articular process
Ligamentum flavum
Iliolumbar ligament
Iliac crest
Posterior superior iliac spine
Dorsal sacroiliac ligament
Greater sciatic foramen
Spine of ischium
Sacrospinal ligament
Lesser sciatic foramen
Sacrotuberal ligament
Ischial tuberosity
Coccyx
©CIBA-GEIGY

The *ligamenta flava*, largely composed of yellow elastic tissue, join adjacent laminae. They extend from the anteroinferior aspect of the lamina above to the posterosuperior surface of the lamina below, and from the midline to the articular capsules laterally. Small gaps for the passage of veins from the internal to the external vertebral venous plexuses exist between them in the midline. The ligaments increase in thickness from the cervical to the lumbar region.

The *supraspinal ligaments* interconnect the tips of the spinous processes from the seventh cervical vertebra to the sacrum. They are continuous with the ligamentum nuchae above and the interspinal ligaments in front, and they increase in thickness from above downward. The *interspinal ligaments* are thin, membranous structures extending between the roots and apexes of the spines; they are best developed in the lumbar region.

The *ventral sacroiliac ligaments* consist of numerous thin bands that close the sacroiliac joint ventrally. They especially connect the ventral margins of the auricular surfaces of the sacrum and ilia.

The *interosseous sacroiliac ligaments* are formed by short, thick bundles of fibers interconnecting the sacral and iliac tuberosities. They lie deeply in the dorsal cleft between the rough areas on the bones immediately above and behind the auricular surfaces. Succeeding them dorsally are the *short dorsal sacroiliac ligaments*, which fill the deep depression between the sacrum and the tuberosities of the ilia. They are mostly short and in the upper part of the space pass horizontally between the first and second transverse tubercles of the sacrum and the iliac tuberosities. The *long dorsal sacroiliac ligament* is oblique. Its fibers descend from the posterior superior iliac spine to the third and fourth transverse tubercles of the sacrum. Its outer fibers interdigitate with the sacrotuberal ligament. The *sacrotuberal ligament* is long, flat, and triangular. Fibers arising from the posterior superior and posterior inferior iliac spines and from the back and side of the sacrum and side of the coccyx converge below on the ischial tuberosity. Part of the origin of the gluteus maximus muscle is from the ligament's posterior surface. The *sacrospinal ligament* is attached by its apex to the spine of the ischium; its broader base arises from the side of the lower sacrum and coccyx. This ligament converts the greater sciatic notch into the *greater sciatic foramen* and, with the sacrotuberal ligament external to it, converts the lesser sciatic notch into the *lesser sciatic foramen*. These ligaments strongly resist a tendency of the sacrum to rotate under the full weight of the upper body—upper sacrum downward, lower sacrum posteriorly.

The *lumbosacral joint* unites the fifth lumbar vertebra with the first segment of the sacrum in a typical union of vertebrae, but at a strong inclination. All the usual ligaments joining vertebrae unite these vertebrae. In addition, a strong *iliolumbar ligament* passes lateralward from the transverse process of the fifth lumbar vertebra to the posterior part of the iliac crest. This ligament resists the tendency of the vertebra to descend the oblique plane of the base of the sacrum.

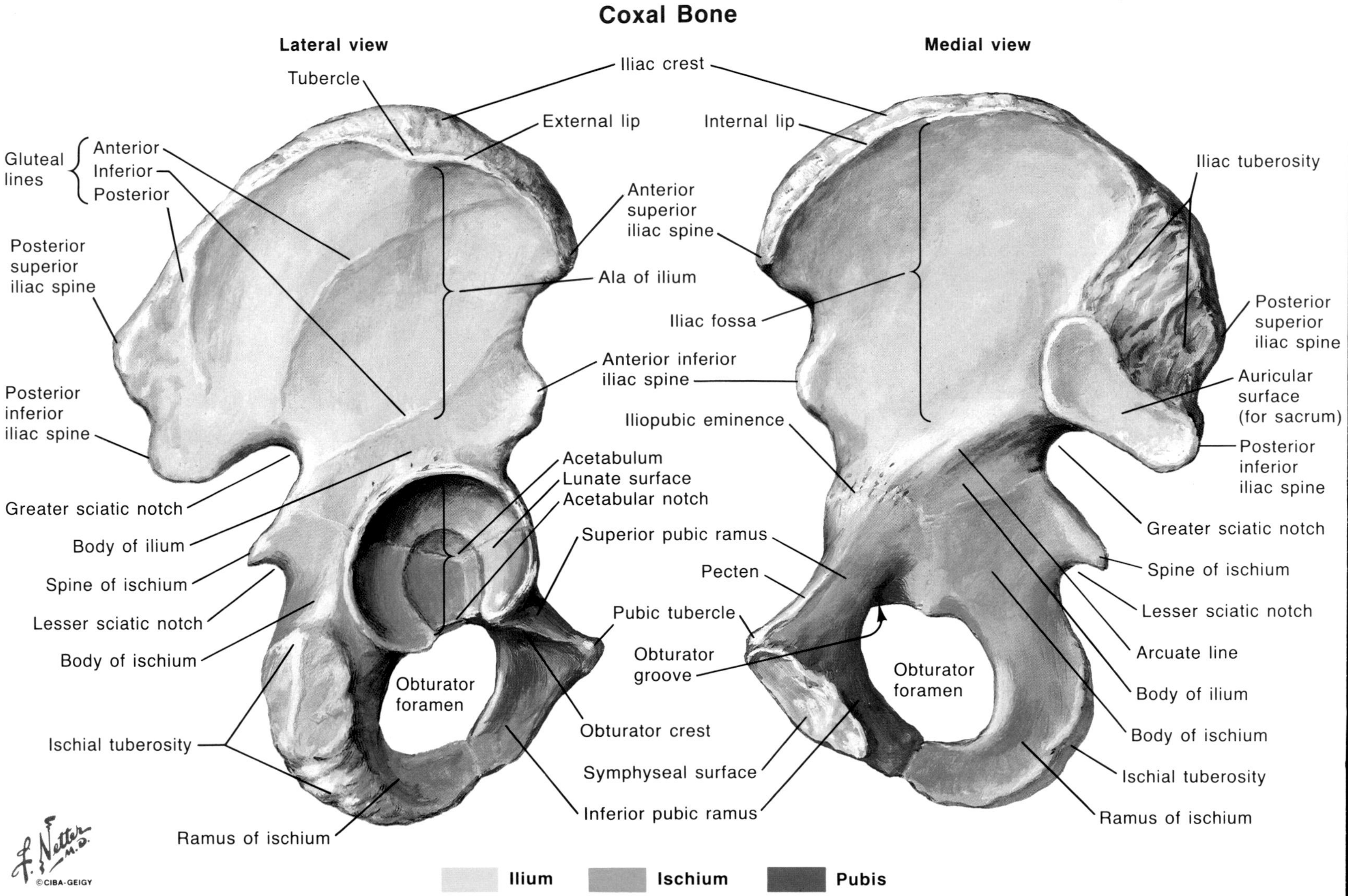
Coxal Bone
Lateral view
Medial view
Iliac crest
Tubercle
External lip
Internal lip
Gluteal lines
Anterior
Inferior
Posterior
Posterior superior iliac spine
Posterior inferior iliac spine
Greater sciatic notch
Body of ilium
Spine of ischium
Lesser sciatic notch
Body of ischium
Ischial tuberosity
Ramus of ischium
Anterior superior iliac spine
Ala of ilium
Iliac fossa
Anterior inferior iliac spine
Iliopubic eminence
Acetabulum
Lunate surface
Acetabular notch
Superior pubic ramus
Pecten
Pubic tubercle
Obturator groove
Obturator foramen
Obturator crest
Symphyseal surface
Inferior pubic ramus
Iliac tuberosity
Posterior superior iliac spine
Auricular surface (for sacrum)
Posterior inferior iliac spine
Greater sciatic notch
Spine of ischium
Lesser sciatic notch
Arcuate line
Body of ilium
Body of ischium
Ischial tuberosity
Ramus of ischium
Ilium
Ischium
Pubis
©CIBA-GEIGY

SECTION I PLATE 17 Slide 3582

Back

Vertebral Column and Pelvis

(Continued)

Coxal Bone

The ossa coxae, or hipbones, together form most of the bony pelvis (Plate 17). They are united to one another in front at the pubic symphysis; behind, they are joined to the sacrum and form a ring of bone that bounds the pelvis and unites the trunk and the lower limbs. The hipbone, large and irregular, is composed of the ilium, ischium, and pubis. The completed coxal bone has reverse curves and an apparent axis at the acetabulum, which looks outward and downward. The anterior superior spine of the ilium and the pubic tubercle lie on the same vertical plane.

The *ilium* is composed of a fan-shaped ala and a thickened and tapered body that ends as the upper two-fifths of the rim of the acetabulum. The *ala* exhibits reverse curves, and its superior curve is the *crest*, which provides attachments for abdominal muscles and fasciae. Anteriorly, the crest ends in the *anterior superior iliac spine* and posteriorly, in the *posterior superior iliac spine*. Inferior to these and separated by notches are *anterior* and *posterior inferior iliac spines*. A deep indentation just below the posterior inferior spine is the *greater sciatic notch*. The internal surface of the ala lodges the iliacus muscle. Its external surface is ridged by *anterior, inferior*, and *posterior gluteal lines*, which demarcate the bony origins of the gluteal muscles. The *iliopubic eminence* marks the area of fusion of the ilium and pubis. Posterior to the iliac fossa is the auricular area for articulation with the first two segments of the sacrum, and behind and above this is a rough area called the *tuberosity*, which provides for attachment of the *short dorsal sacroiliac ligaments* and for fibers of the erector spinae and multifidus muscles.

The *ischium* is the heavy posterior inferior portion of the coxal bone, which provides strength for both the sacrofemoral and sacroischial arches. Its blunt inferior extremity is the *ischial tuberosity*, which together with its fellow of the opposite side bears the weight of the upper body in sitting. Directly above the tuberosity is the *lesser sciatic notch*, followed by the *spine of the ischium* and the *greater sciatic notch*. The body of the ischium forms about two-fifths of the articular area of the rim of the acetabulum. The pelvic surface of the body of the ischium is smooth; it borders the obturator foramen and gives origin to some of the fibers of origin of the obturator internus muscle. The *ramus* of the ischium extends forward from the tuberosity to join the inferior ramus of the pubis and form the *ischiopubic ramus*.

The *pubis* is the smallest of the three parts of the coxal bone. It has a compressed part medially, which exhibits the *symphyseal surface* for participation in the pubic symphysis. The *pubic crest* terminates laterally in the *pubic tubercle*. The *pecten* of the pubis is a sharp crest extending from the pubic tubercle to become continuous with the arcuate line of the pelvis. Laterally, the bone is continuous with the ilium at the *iliopubic eminence*, and in the acetabulum, it provides one-fifth of the *lunate articular surface*. A short *inferior ramus* of the pubis bounds the obturator foramen below and combines with the ramus of the ischium.

The *acetabulum* is the deep, hemispheric cavity that receives the head of the femur. Its sturdy wall consists of a semilunar articular portion, the *lunate surface*, and a deep, central nonarticular portion, the *acetabular fossa*. The *acetabular notch*, the inferior opening of the cavity, is bridged by the *transverse ligament*, and its margins give attachment to the *capitis femoris ligament*.

The coxal bone is ossified from eight centers: three primary centers for the ilium, ischium, and pubis; and five secondary centers for the iliac crest, anterior inferior spine, ischial tuberosity, pubic symphysis, and a triradiate piece at the center of the acetabulum. At birth, the primary centers are still quite separate, and the secondary centers have not yet appeared. At about puberty, the major bones are complete, and the secondary centers show ossification. At age 15 or 16, the three major bones fuse through the triradiate piece of the acetabulum; the other secondary centers fuse between ages 20 and 22.

Back

Vertebral Column and Pelvis

(Continued)

Sacrum and Coccyx

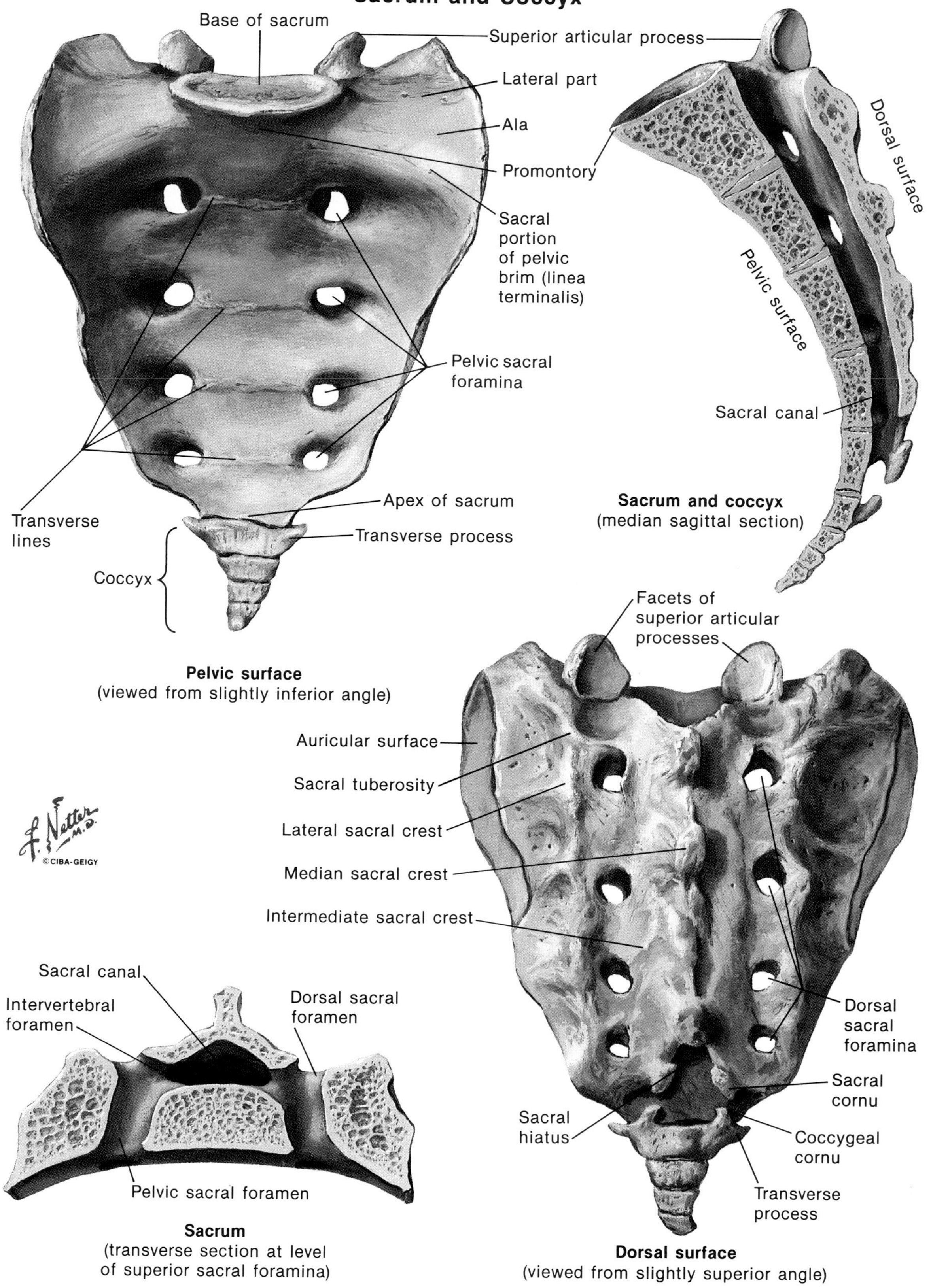
Base of sacrum
Superior articular process
Lateral part
Ala
Promontory
Sacral portion of pelvic brim (linea terminalis)
Pelvic sacral foramina
Transverse lines
Apex of sacrum
Transverse process
Coccyx
Dorsal surface
Pelvic surface
Sacral canal
Facets of superior articular processes
Auricular surface
Sacral tuberosity
Lateral sacral crest
Median sacral crest
Intermediate sacral crest
Dorsal sacral foramina
Sacral cornu
Sacral hiatus
Coccygeal cornu
Transverse process
Sacral canal
Intervertebral foramen
Dorsal sacral foramen
Pelvic sacral foramen
F. Netter M.D.
©CIBA-GEIGY

Pelvic surface (viewed from slightly inferior angle)

Sacrum and coccyx (median sagittal section)

Sacrum (transverse section at level of superior sacral foramina)

Dorsal surface (viewed from slightly superior angle)

Sacrum and Coccyx

Sacrum. The sacrum consists of five fused vertebrae (S1–5) and is wedge shaped from above down and from before back (Plate 18). It forms most of the posterior pelvic wall and is fixed between the hipbones at an angle, so that its curved pelvic surface is inclined downward and forward.

The broader *base* of the sacrum faces anterosuperiorly toward the abdomen; its elevated central third is the upper part of the first sacral vertebral body and bears a smooth oval area for the attachment of the lumbosacral intervertebral disc. Its projecting anterior border is the sacral *promontory*. On each side, the costotransverse elements of the first vertebra are fused to form a wing-shaped lateral mass (sacral *ala*), which is separated from the pelvic surface by a curved line, the sacral portion of the arcuate pelvic brim. The articular processes are fused, like most of the other components of the sacral vertebrae, but the *superior articular processes* of the first vertebra remain and project upward for articulation with the inferior articular processes of the fifth lumbar vertebra. They are flattened and face almost directly backward to assist in preventing subluxation (spondylolisthesis) of the last lumbar vertebra at the angulated lumbosacral junction.

The narrow *apex* is the lower end of the sacrum and articulates with the coccyx.

The pelvic surface is concave both vertically and horizontally and shows four *transverse ridges* indicating the lines of fusion between the bodies of the original five vertebrae. On either side of the ridges, four *pelvic sacral foramina* permit the passage of the ventral rami of the first four sacral nerves and their associated vessels.

The convex dorsal surface shows irregular *median, intermediate*, and *lateral sacral crests* representing, respectively, the fused spinous, articular, and transverse processes. The areas between the median and intermediate crests are the fused laminae, and there are four pairs of *dorsal sacral foramina* for the passage of the dorsal rami of the upper four sacral nerves. The laminae of the fifth (S5) and, occasionally, the fourth sacral vertebra (S4) fail to unite and thus leave a *hiatus*, which is exploited for the injection of epidural anesthetics. The hiatus is bounded on each side by a *cornu*, a relic of the inferior articular process, and transmits the small fifth sacral and coccygeal nerves.

The parts of the sacrum lateral to the sacral foramina are produced by the fusion of the costal, transverse, and pedicular elements of the five vertebrae. The upper, broader parts of their lateral surfaces bear uneven *auricular*, or ear-shaped, surfaces for articulation with similar surfaces on the iliac parts of the hipbones.

Transverse sections of the sacrum reveal the triangular sacral end of the *vertebral canal*. This canal surrounds and protects the terminations of the dural and arachnoid sheaths and the subarachnoid space, which end at about the level of the second sacral vertebra (S2) and enclose the sacral and coccygeal roots of the cauda equina and the lower intrathecal portion of the filum terminale. The dura mater is separated from the walls of the canal by fibrofatty tissue, fine arteries and nerves, and sacral internal vertebral venous plexuses.

Coccyx. The small, triangular coccyx is formed by the fusion of four (occasionally, three or five) rudimentary tail vertebrae (Plate 18). Its base articulates with the sacral apex, and its apex is a mere button of bone. Most of the features of a typical vertebra are lacking, but the first coccygeal vertebra has small *transverse processes* and a *cornu* on each side, which is sometimes large enough to articulate with the corresponding sacral cornu. □

Cutaneous Nerves and Superficial Veins of Shoulder and Arm

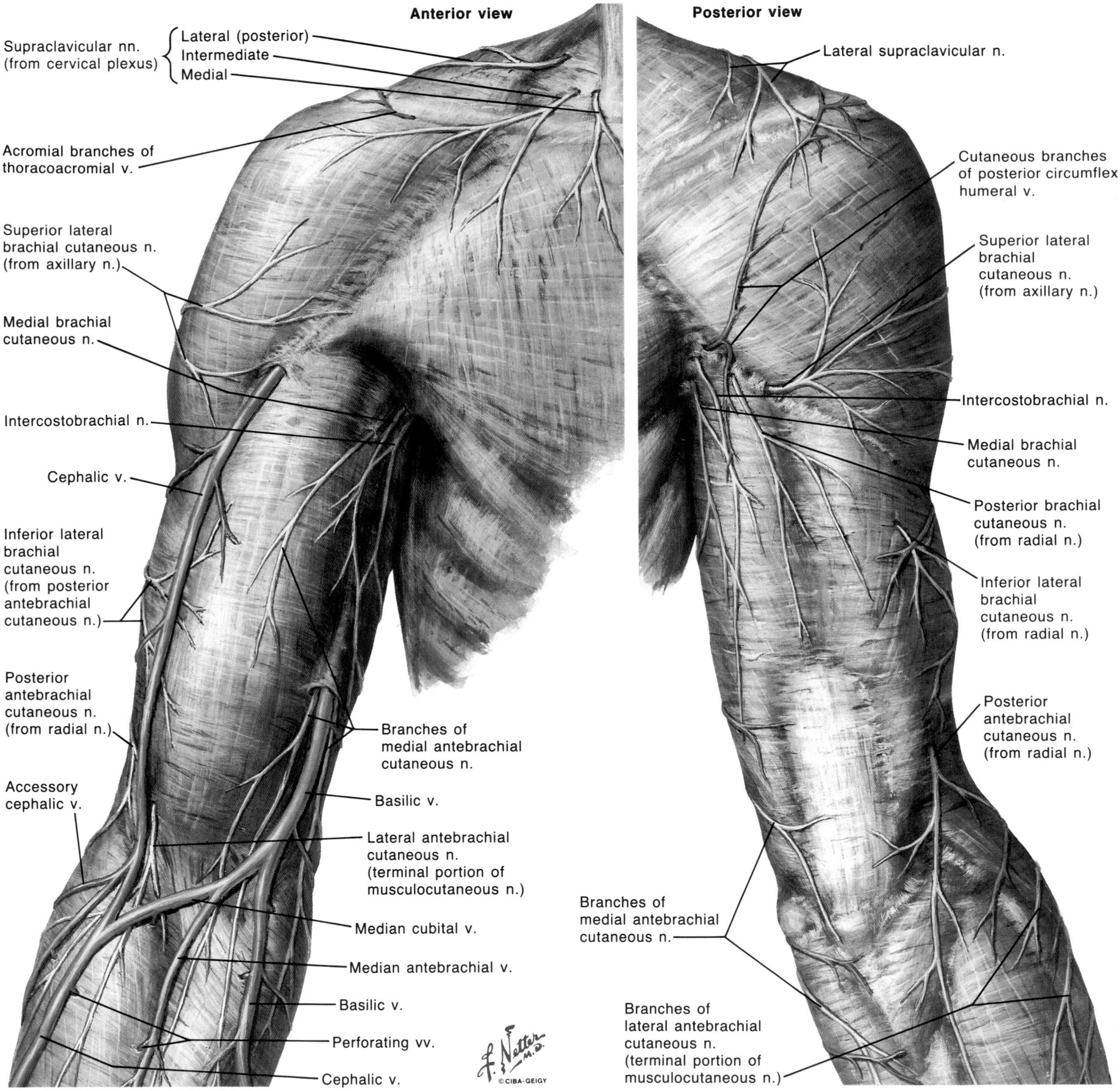
Anterior view
Posterior view
Supraclavicular nn. (from cervical plexus)
Lateral (posterior)
Intermediate
Medial
Acromial branches of thoracoacromial v.
Superior lateral brachial cutaneous n. (from axillary n.)
Medial brachial cutaneous n.
Intercostobrachial n.
Cephalic v.
Inferior lateral brachial cutaneous n. (from posterior antebrachial cutaneous n.)
Posterior antebrachial cutaneous n. (from radial n.)
Accessory cephalic v.
Branches of medial antebrachial cutaneous n.
Basilic v.
Lateral antebrachial cutaneous n. (terminal portion of musculocutaneous n.)
Median cubital v.
Median antebrachial v.
Basilic v.
Perforating vv.
Cephalic v.
Lateral supraclavicular n.
Cutaneous branches of posterior circumflex humeral v.
Superior lateral brachial cutaneous n. (from axillary n.)
Intercostobrachial n.
Medial brachial cutaneous n.
Posterior brachial cutaneous n. (from radial n.)
Inferior lateral brachial cutaneous n. (from radial n.)
Posterior antebrachial cutaneous n. (from radial n.)
Branches of medial antebrachial cutaneous n.
Branches of lateral antebrachial cutaneous n. (terminal portion of musculocutaneous n.)
F. Netter M.D.
©CIBA-GEIGY

SECTION I PLATE 19 Slide 3584

Upper Limb

Superficial Veins and Cutaneous Nerves of Shoulder and Arm

Superficial Veins

The subcutaneous veins of the limb are interconnected with the deep veins of the limb via perforating veins (Plate 19).

Certain prominent veins, unaccompanied by arteries, are found in the subcutaneous tissues of the limbs. The *cephalic* and *basilic veins*, the principal superficial veins of the upper limb, originate in venous radicals in the hand and digits (Plate 53).

Anastomosing longitudinal *palmar digital veins* empty at the webs of the fingers into longitudinally oriented dorsal digital veins. The *dorsal veins* of adjacent digits then unite to form relatively short *dorsal metacarpal veins*, which end in the *dorsal venous arch*. The radial continuation of the dorsal venous arch is the *cephalic vein*, which receives the dorsal veins of the thumb and then ascends at the radial border of the wrist (Plate 41). In the forearm, it tends to ascend at the anterior border of the brachioradialis muscle, with tributaries from the dorsum of the forearm. In the cubital space, the obliquely ascending *median cubital vein* connects the cephalic and basilic veins. Above the cubital fossa, the cephalic vein runs in the lateral bicipital groove and then in the interval between the deltoid and pectoralis major muscles, where it is accompanied by the small deltoid branch of the thoracoacromial artery. At the deltopectoral triangle, the cephalic vein perforates the costocoracoid membrane and empties into the *axillary vein*. An *accessory cephalic vein* passes from the dorsum of the forearm spirally laterally to join the cephalic vein at the elbow.

ıb

al Veins and
ıs Nerves of
and Arm

ein continues the ulnar end of the
he dorsum of the hand (Plate 41).
g the ulnar border of the forearm
ubital fossa anterior to the medial
he humerus. After receiving the
vein, the basilic vein continues
medial bicipital groove, pierces
cia a little below the middle of
nters the neurovascular compart-
dial intermuscular septum, where
l to the brachial artery. In the dis-
ns the brachial veins to form the

ntebrachial vein is a frequent col-
the middle of the anterior surface
It terminates in the cubital fossa
ubital vein or in the basilic vein.
vides into a *median basilic vein* and
c vein, which borders the biceps
and joins the cephalic vein. The
chial vein may be large or absent.

ves

ıs nerves of the upper limb are
part derived from the brachial
h the uppermost nerves to the
erived from the cervical plexus
25).

he *supraclavicular nerves* (C3, 4)
ial at the posterior border of the
toid muscle within the posterior
neck. They pierce the superficial
ical fascia and the platysma mus-
three lines: (1) over the clavicle—
vicular nerves, (2) toward the
mediate supraclavicular nerves,
e scapula—lateral, or posterior,
nerves.

erior lateral brachial cutaneous nerve
rmination of the lower branch of
e of the brachial plexus. Leaving
e, it turns superficially around the
of the lower third of the deltoid
the brachial fascia. Its cutaneous
e lower half of the deltoid muscle
ad of the triceps brachii.

teral brachial cutaneous nerve (C5, 6)
the posterior antebrachial cuta-
tly after this nerve branches from
The inferior lateral brachial cuta-
omes superficial in line with the
cular septum a little below the
deltoid muscle. It accompanies
the cephalic vein and distributes
teral and the anterior surface of

Cutaneous Innervation of Upper Limb

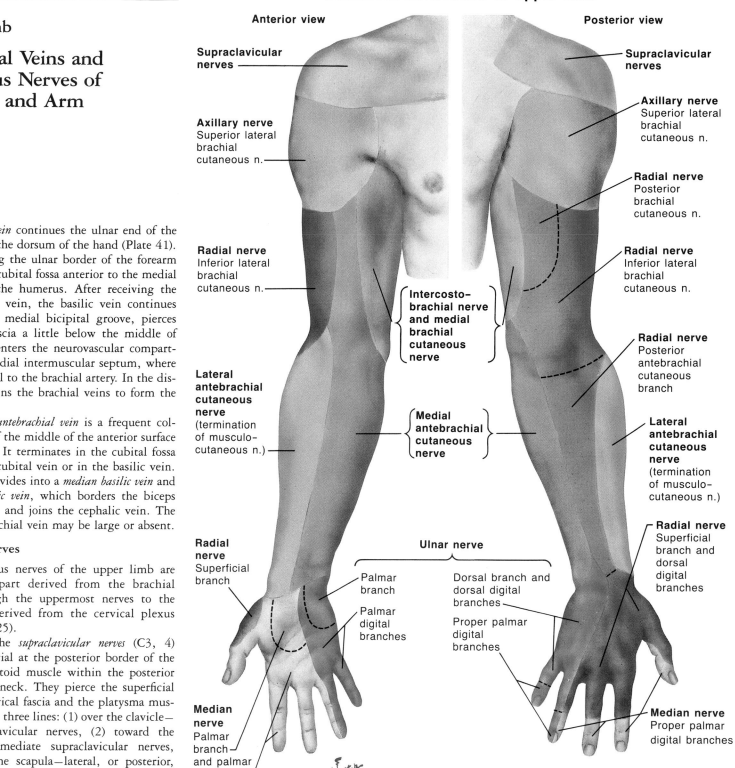

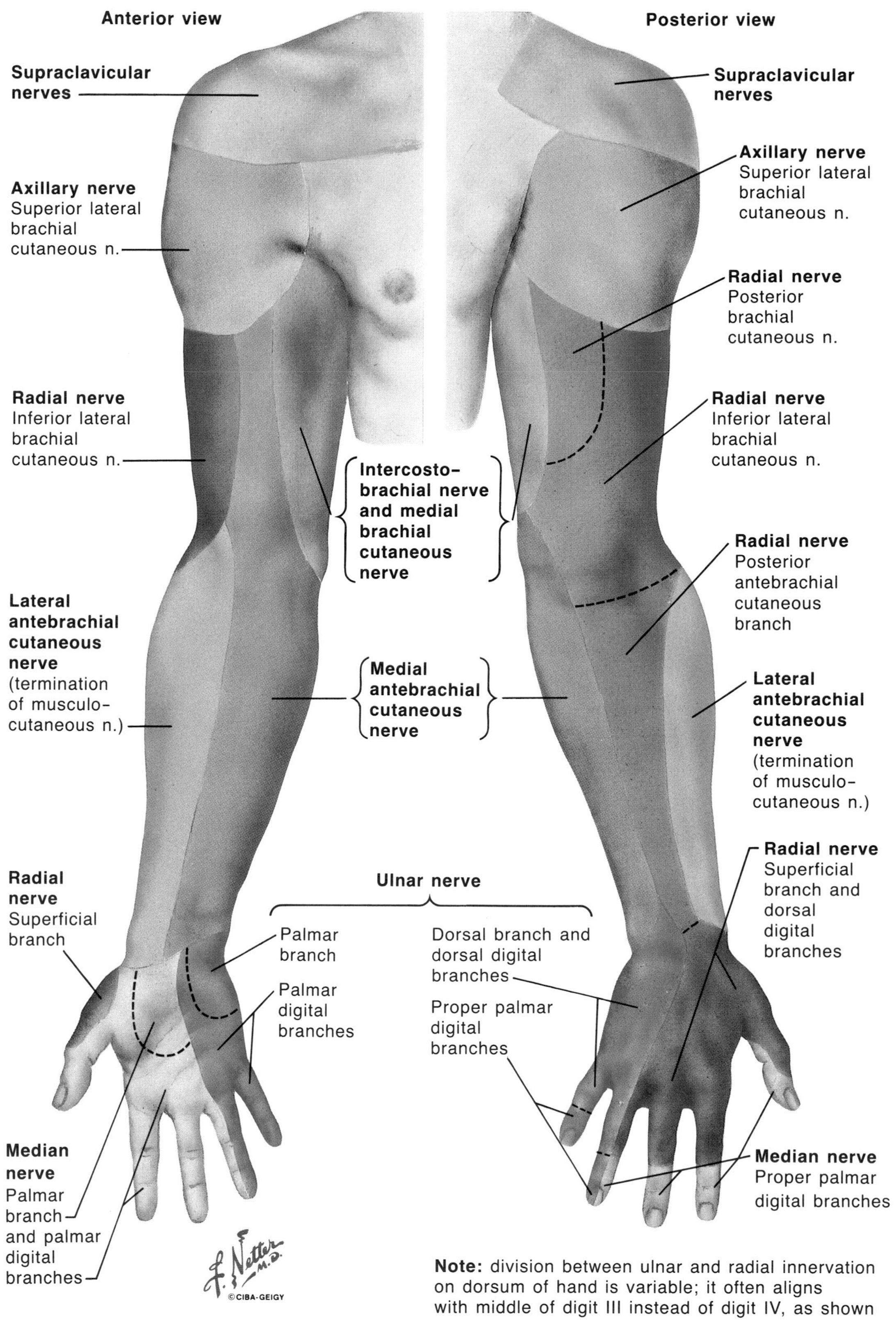
Anterior view
Posterior view
Supraclavicular nerves
Axillary nerve
Superior lateral brachial cutaneous n.
Radial nerve
Posterior brachial cutaneous n.
Radial nerve
Inferior lateral brachial cutaneous n.
Intercostobrachial nerve and medial brachial cutaneous nerve
Radial nerve
Posterior antebrachial cutaneous branch
Lateral antebrachial cutaneous nerve (termination of musculocutaneous n.)
Medial antebrachial cutaneous nerve
Radial nerve
Superficial branch
Radial nerve
Superficial branch and dorsal digital branches
Ulnar nerve
Palmar branch
Palmar digital branches
Dorsal branch and dorsal digital branches
Proper palmar digital branches
Median nerve
Palmar branch and palmar digital branches
Median nerve
Proper palmar digital branches
F. Netter M.D.
©CIBA-GEIGY

Note: division between ulnar and radial innervation on dorsum of hand is variable; it often aligns with middle of digit III instead of digit IV, as shown

The *posterior brachial cutaneous nerve* arises within the axilla as a branch of the radial nerve (C5–8). Traversing the medial side of the long head of the triceps brachii muscle, the nerve penetrates the brachial fascia to distribute in the middle third of the back of the arm above and behind the distribution of the medial brachial cutaneous nerve and the intercostobrachial nerve.

The *medial brachial cutaneous nerve* (C8; T1) arises from the medial cord of the brachial plexus in the lower axilla. It descends along the medial side of the brachial artery to the middle of the arm, where it pierces the brachial fascia and supplies the skin of the posterior surface of the lower third of the arm as far as the olecranon.

The *intercostobrachial nerve* is the larger part of the lateral cutaneous branch of the second thoracic nerve (T2). In the second intercostal space at the axillary line, it pierces the serratus anterior muscle to enter the axilla. Here, it usually anastomoses with the medial brachial cutaneous nerve and then pierces the brachial fascia just beyond the posterior axillary fold. Its cutaneous distribution is along the medial and posterior surfaces of the arm from the axilla to the elbow.

The *medial antebrachial cutaneous nerve* arises from the medial cord of the brachial plexus. A small branch pierces the axillary fascia and supplies the skin over the medial anterior area of the arm. ☐

Upper Limb

Shoulder and Axillary Regions

Muscles Connecting Upper Limb to Vertebral Column

Muscles of this group are superficially located in the back (Plate 21). They are the trapezius, latissimus dorsi, levator scapulae, rhomboideus minor, and rhomboideus major muscles. Although having their origins on the vertebral column, these muscles are innervated by ventral rami of the spinal nerves, as are all limb muscles.

Trapezius Muscle. The most superficial muscle of the back and the back of the neck, the trapezius takes its origin from the medial third of the superior nuchal line and the external occipital protuberance of the occipital bone, from the ligamentum nuchae, from the spines of the seventh cervical and all the thoracic vertebrae, and from the intervening supraspinal ligament. The muscle is relatively flat and thin, except in the low cervical and upper thoracic region, where its thickness is matched by a distinct diamond-shaped accumulation, medially, of tendinous fibers of origin. Muscular bundles converge toward the bones of the shoulder (Plate 28). Occipital and upper cervical bundles are inserted into the posterior border of the lateral third of the clavicle; lower cervical and upper thoracic bundles reach the medial border of the acromion and the upper border of the crest of the spine of the scapula; lower thoracic bundles converge to a triangular, flattened tendon inserting into the tubercle of the crest of the spine of the scapula. The trapezius is covered on both its deep and superficial surfaces by the superficial layer of cervical fascia.

The trapezius muscle assists in suspending the shoulder girdle. A low level of the activity in the upper trapezius, levator scapulae, and upper serratus anterior muscles is sufficient to suspend the shoulder girdle, but the same muscles contract vigorously with loading of the shoulder or hand. The upper thoracic part of the trapezius draws the shoulders strongly backward and contributes to pulling or extension movements of the arm. The lower and middle segments function together in pulling and in squaring the shoulders, and the middle segment is active in abduction of the arm. A force couple is effected by the joint action of the upper trapezius, levator scapulae, and upper parts of the serratus anterior muscles acting simultaneously with the lower part of the trapezius and the lower, larger part of the serratus anterior muscle. This combined action rotates the scapula on the chest wall and points the shoulder strongly upward.

The nerves reaching the trapezius muscle are the *accessory (XI) nerve* and direct branches of ventral rami of the *third* and *fourth cervical nerves* (Plate 6). The accessory nerve perforates and supplies the sternocleidomastoid muscle, then crosses the posterior triangle of the neck directly under its fascial covering, coursing diagonally downward to reach the underside of the trapezius muscle.

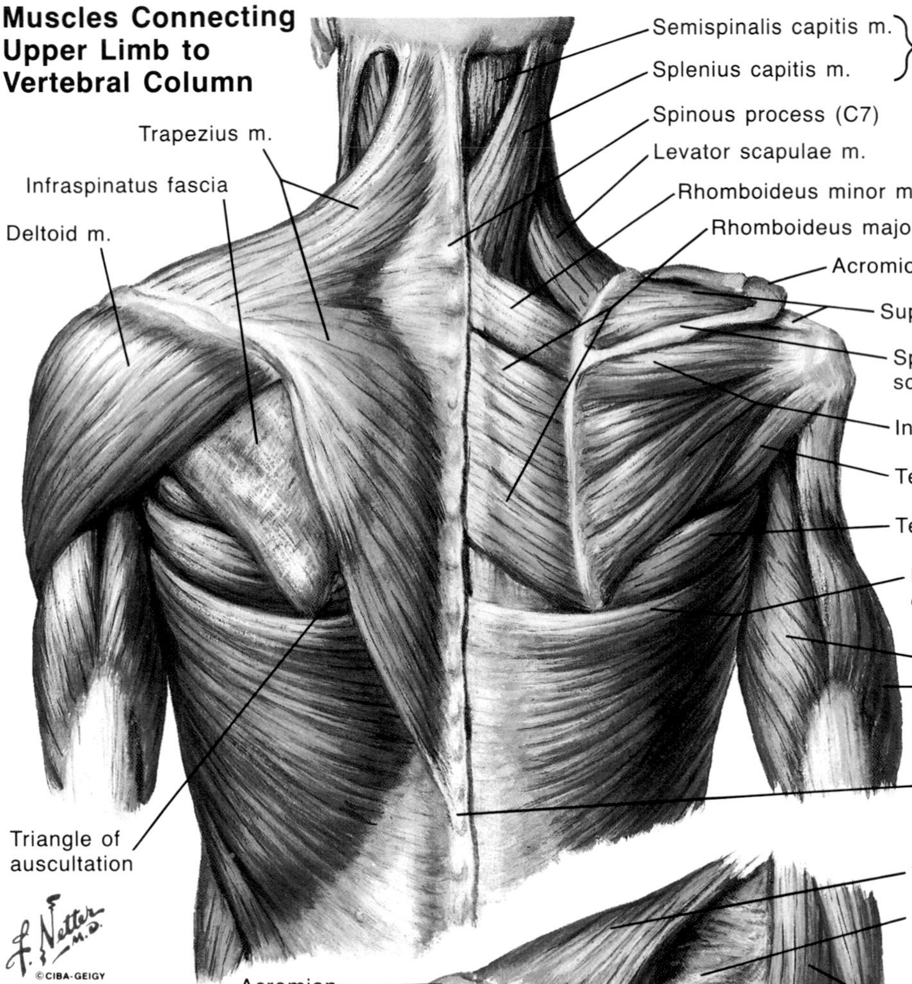
Muscles Connecting Upper Limb to Vertebral Column
Semispinalis capitis m.
Splenius capitis m.
Spinous process (C7)
Levator scapulae m.
Rhomboideus minor m
Trapezius m.
Infraspinatus fascia
Deltoid m.
Triangle of auscultation
F. Netter M.D.
©CIBA-GEIGY

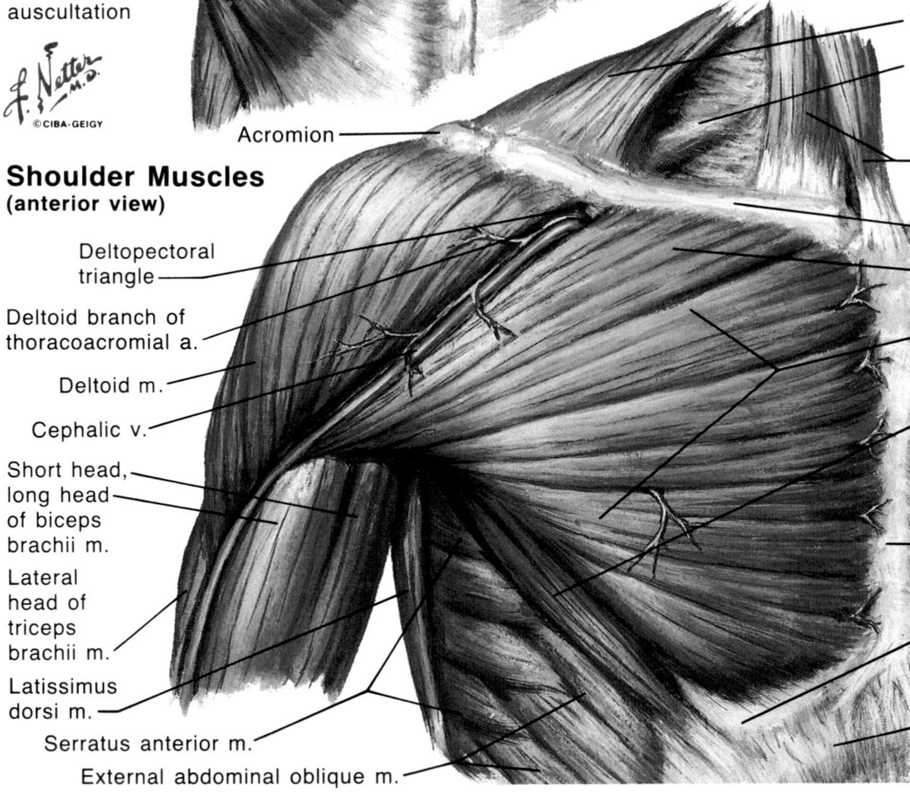
Acromion
Shoulder Muscles
(anterior view)
Deltopectoral triangle
Deltoid branch of thoracoacromial a.
Deltoid m.
Cephalic v.
Short head, long head of biceps brachii m.
Lateral head of triceps brachii m.
Latissimus dorsi m.
Serratus anterior m.
External abdominal oblique m.

With the cervical branches, a subtrapezial plexus is formed on the muscle's deep side.

The *transverse cervical artery* of the subclavian system supplies the trapezius muscle; it is supplemented in the lower third of the muscle by a muscular perforating branch of the dorsal scapular artery.

Latissimus Dorsi Muscle. This is a broad, triangular muscle of the lower part of the back. Its principal origin is aponeurotic, with attachments to the spinous processes of the lower six thoracic vertebrae and, through its fusion with the thoracolumbar fascia, to the lumbar and sacral spinous processes (Plate 1). The lower lateral portion arises additionally from the posterior third of the iliac crest, and deeply, there are of origin from the last three or f to the inferior angle of the scap narrowed muscle curves spirally major muscle in the lower edge axillary fold and ends in a b inserted in the floor of the intertu of the humerus (Plate 27).

The latissimus dorsi muscle merus, drawing the arm down ward and rotating it medialward is exemplified by the crawl strok and it is used in all pulling mover cle is innervated by the *thoracodors* posterior cord of the brachial ple

Upper Limb

Shoulder and Axillary Regions

(Continued)

from the seventh and eighth cervical nerves. The *thoracodorsal artery*, a branch of the subscapular artery, and a vein of the same name accompany the nerve.

The latissimus dorsi muscle is superficial in the lower part of the back, where it overlies abdominal musculature and the posterior inferior serratus muscle. The *lumbar triangle* is a nonmuscular interval just above the iliac crest at its lower lateral border, which is closed laterally by the margin of the external abdominal oblique muscle. The upper part of the latissimus dorsi is overlapped by the inferior portion of the trapezius. The triangle formed by the borders of the trapezius, latissimus dorsi, and rhomboideus major muscles is the *triangle of auscultation* (Plate 21).

Immediately under the trapezius lie three muscles that have a continuous insertion on the medial border of the scapula and are closely similar in action, innervation, and blood supply. These are the levator scapulae, rhomboideus minor, and rhomboideus major muscles.

Levator Scapulae Muscle. Thick and straplike, this muscle arises by four separate tendons from the transverse processes of the first three or four cervical vertebrae. It inserts into the medial border of the scapula from the superior angle to the spine. It is overlapped and partially obscured by the sternocleidomastoid and trapezius muscles.

Rhomboideus Minor Muscle. This slender muscle takes its origin from the lower part of the ligamentum nuchae, the spinous process of the last cervical and first thoracic vertebrae, and the associated segment of the supraspinal ligament. It lies parallel to the rhomboideus major muscle, directed downward and lateralward, and it is inserted on the medial border of the scapula at the root of the scapular spine.

Rhomboideus Major Muscle. Flat and sheetlike, this muscle arises from the spinous process of the second to the fifth cervical vertebrae and the corresponding segment of the supraspinal ligament. Its fibers run diagonally downward and lateralward to the medial border of the scapula below its spine.

Both rhomboideus muscles draw the scapula upward and medially and assist the serratus anterior muscle in holding it firmly to the chest wall. Their oblique traction aids in depressing the point of the shoulder. The levator scapulae muscle operates in elevation, support, and rotation of the scapula. The *dorsal scapular nerve* from the brachial plexus, chiefly from C5, is a part of the nerve supply of the levator scapulae muscle and the sole innervation of the rhomboideus muscles (Plate 36). The upper part of the levator scapulae receives an additional nerve supply from ventral rami of the third and fourth cervical nerves. The *dorsal scapular artery* crosses the neck to the lateral border of the levator scapulae muscle. Here, it joins the dorsal scapular nerve and descends toward the inferior angle of the scapula, sending branches into the muscles of this group.

Muscles and Fascia of Shoulder

The muscles especially identified with the shoulder are the deltoid, supraspinatus, infraspinatus, teres minor, teres major, subscapularis, pectoralis major, pectoralis minor, and subclavius muscles (Plates 21–22).

Deltoid Muscle. This triangular, multipennate, and coarsely fasciculated muscle caps the point of the shoulder. It has a semicircular origin, arising from the lateral third of the clavicle, the lateral border of the acromion, and the lower lip of the crest of the spine of the scapula (Plate 28). All fasciculi converge to be inserted on the deltoid tuberosity of the humerus. The clavicular and spinous portions of the muscle are composed of long, parallel fibers; the central, or acromial, part has a multipennate architecture. Four tendinous septa descend from the acromion into the muscle, and three septa ascend from the deltoid tuberosity. From the acromial bands, muscle fibers converge onto the intervening bands of insertion, resulting in a muscle of great power but short excursion.

The deltoid muscle is a principal abductor of the humerus, an action produced primarily by its powerful central portion. The supraspinatus muscle, however, also fully participates in up to 90° of abduction. The activity of the deltoid muscle increases progressively and is greatest between 90° and 180° of elevation of the limb.

Because of their position and greater fiber length, the clavicular and scapular portions of the deltoid muscle have different actions from those of the central portion of the muscle. The clavicular portion assists in flexion and medial rotation of the arm, while the scapular portion assists in extension and lateral rotation. Both portions assist in stabilizing the shoulder joint.

The *axillary nerve* (C5, 6) from the posterior cord of the brachial plexus supplies the deltoid muscle. An upper branch curves around the posterior surface of the humerus and courses from behind forward on the deep surface of the muscle, sending offshoots into the muscle. A lower branch supplies the teres minor muscle by ascending onto its lateral and superficial surface. It then becomes the *superior lateral brachial cutaneous nerve*. The *posterior circumflex humeral artery* serves the muscle.

The deltoid muscle is superficial, overlies the greater tubercle of the humerus, and forms the rounded prominence of the shoulder. It is enveloped by a muscular fascia that is an extension of the superficial layer of cervical fascia from the trapezius muscle. Anteriorly, the borders of the deltoid and pectoralis major muscles are adjacent, and their muscular fasciculi are parallel. Between these fasciculi lie the cephalic vein and the deltoid branch of the thoracoacromial artery. Deep to the central portion of the muscle is the subdeltoid, or subacromial, bursa, and in the floor of the bursa is the supraspinatus tendon.

When the deltoid muscle is reflected from its origin, the *subacromial bursa*, approximately the size of a fifty-cent piece, is exposed. Situated between the deltoid muscle and the supraspinatus tendon and the capsule of the shoulder joint, it also extends deep to the acromion and the coracoacromial ligament. The bursa facilitates the movement of the deltoid muscle over the joint capsule and tendons.

Supraspinatus Muscle. This muscle occupies the supraspinatous fossa of the scapula (Plates 28 and 30). It takes its origin from the medial two-thirds of the bony walls of this fossa, and from the dense fascia that covers the muscle. The tendon is formed within the muscle, the muscular fibers converging onto it from all sides. It blends deeply with the capsule of the shoulder joint and inserts on the highest of the three facets of the greater tubercle of the humerus. The supraspinatus muscle acts concurrently with the deltoid in the first 90° of abduction but is ineffective beyond that point. It is one of the rotator cuff muscles that hold the head of the humerus close to the glenoid fossa of the scapula and is important in resisting downward dislocation of the humerus. The *suprascapular nerve* (C5, 6) from the superior trunk of the brachial plexus enters the supraspinatous fossa through the scapular notch. Passing under the superior transverse scapular ligament, it is deep to the muscle and supplies it from its underside, together with the suprascapular artery, which passes over the ligament.

Infraspinatus Muscle. This muscle arises from the infraspinatous fossa of the scapula (except from its lateral fourth), and from the dense overlying infraspinatus fascia (Plate 28). The tendon arises within the muscle and inserts on the middle facet of the greater tubercle of the humerus. Deeply, its fibers blend with those of the capsule of the shoulder joint. The arm can be rotated about 90°, and the infraspinatus muscle is its chief lateral rotator. The muscle also assists in holding the head of the humerus in the glenoid cavity of the scapula. The *suprascapular nerve* and *artery*, having traversed the supraspinatous fossa, pass through the notch of the scapular neck and under the inferior transverse scapular ligament to enter the upper part of the infraspinatus muscle. The muscle also receives branches of the *circumflex scapular artery* (axillary system).

Teres Minor Muscle. Narrow and elongated, the teres minor muscle arises from the upper two-thirds of the lateral border of the scapula and from adjacent intermuscular septa (Plate 28). Its tendon passes upward and lateralward to insert in the lower facet of the greater tubercle and surgical neck of the humerus. It also blends deeply with the capsule of the shoulder joint. The muscle is invested by the infraspinatus fascia and is sometimes inseparable from the infraspinatus muscle. The teres minor muscle contracts with the infraspinatus in lateral rotation of the humerus and also fixes the head of the humerus to facilitate abduction and flexion of the arm. A branch of the *axillary nerve* ascends onto its lateral margin at about its midlength. The teres minor muscle is separated from the teres major by the long head of the triceps brachii and by the axillary nerve and posterior circumflex humeral vessels. It is pierced by branches of the circumflex scapular vessels along the lateral border of the scapula.

Teres Major Muscle. This thick and cylindric muscle arises from the oval area of the dorsum of the inferior angle of the scapula and from the intermuscular septa between it and adjacent muscles (Plates 27–28). It is directed along the lateral border of the scapula toward the front of the humerus. Its fibers make a half twist, like those of the latissimus dorsi muscle, and form a flat tendon that inserts into the crest of the lesser tubercle of the humerus.

Upper Limb

Shoulder and Axillary Regions

(Continued)

Scapulohumeral Region

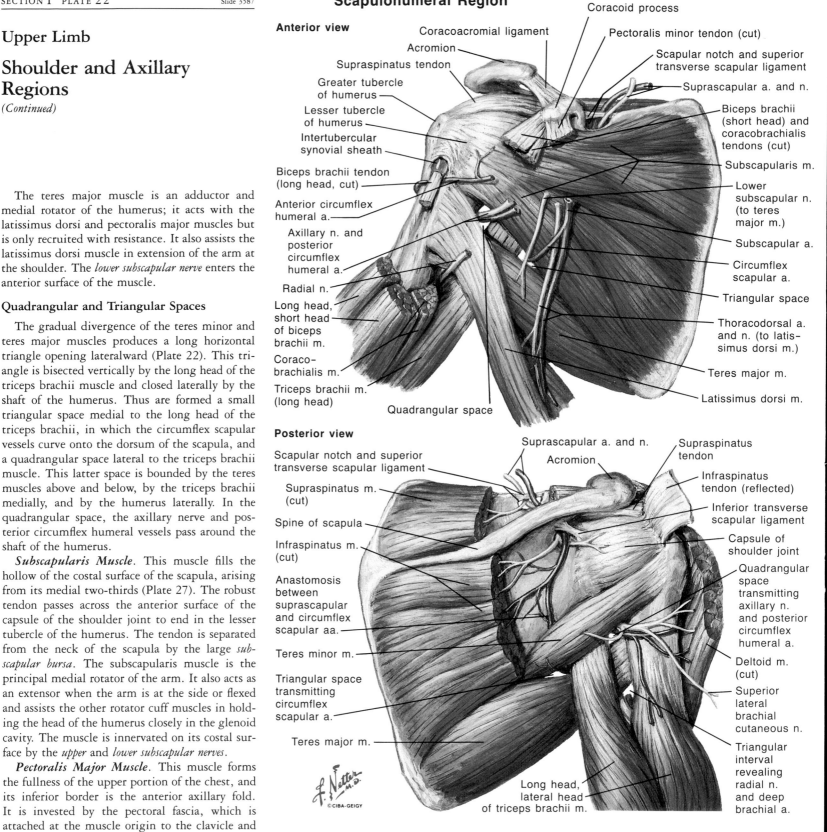

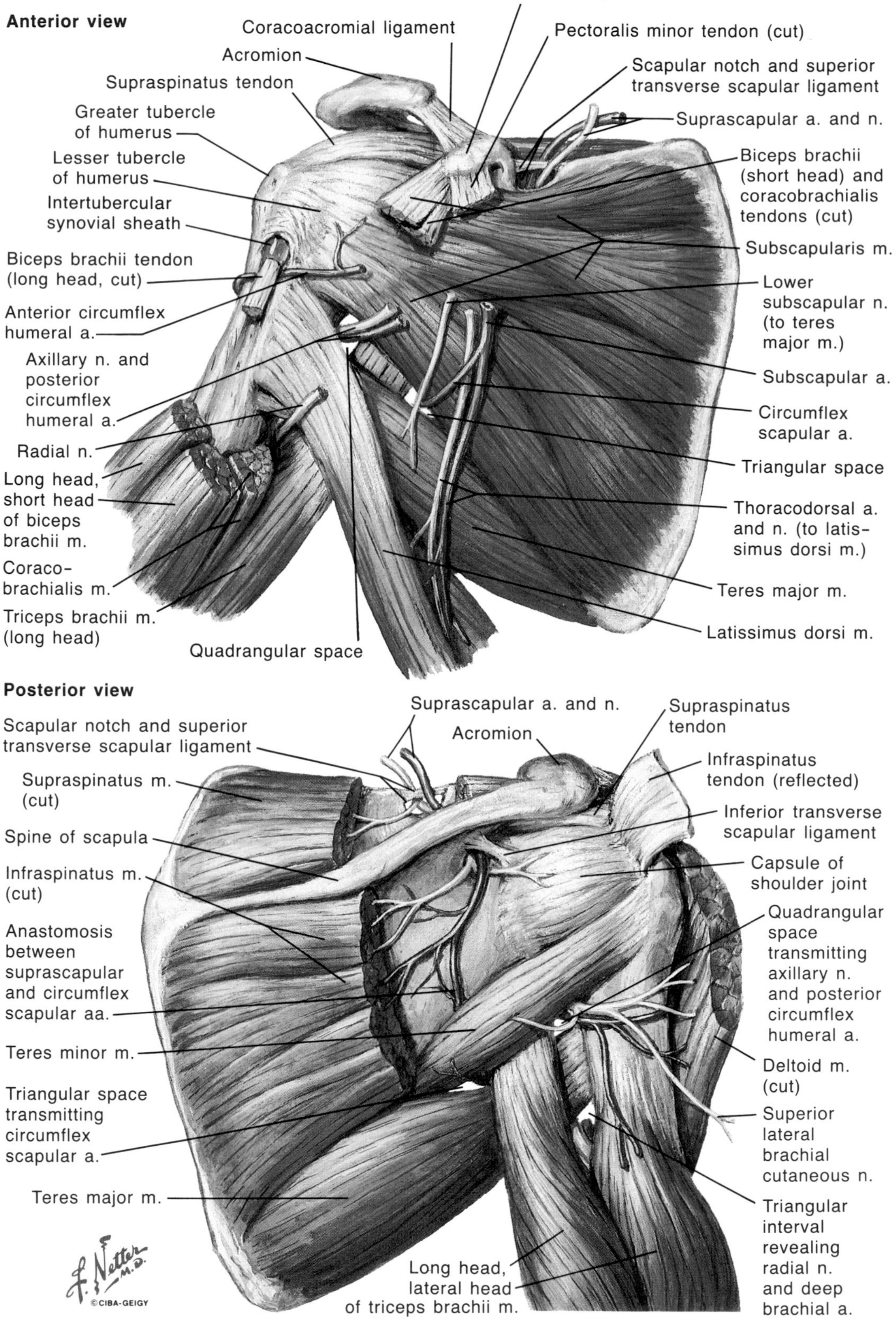
Anterior view
Coracoacromial ligament
Coracoid process
Acromion
Pectoralis minor tendon (cut)
Supraspinatus tendon
Scapular notch and superior transverse scapular ligament
Greater tubercle of humerus
Suprascapular a. and n.
Lesser tubercle of humerus
Biceps brachii (short head) and coracobrachialis tendons (cut)
Intertubercular synovial sheath
Subscapularis m.
Biceps brachii tendon (long head, cut)
Lower subscapular n. (to teres major m.)
Anterior circumflex humeral a.
Axillary n. and posterior circumflex humeral a.
Subscapular a.
Circumflex scapular a.
Radial n.
Triangular space
Long head, short head of biceps brachii m.
Thoracodorsal a. and n. (to latissimus dorsi m.)
Coraco-brachialis m.
Teres major m.
Triceps brachii m. (long head)
Latissimus dorsi m.
Quadrangular space
Posterior view
Suprascapular a. and n.
Supraspinatus tendon
Scapular notch and superior transverse scapular ligament
Acromion
Infraspinatus tendon (reflected)
Supraspinatus m. (cut)
Inferior transverse scapular ligament
Spine of scapula
Capsule of shoulder joint
Infraspinatus m. (cut)
Quadrangular space transmitting axillary n. and posterior circumflex humeral a.
Anastomosis between suprascapular and circumflex scapular aa.
Teres minor m.
Deltoid m. (cut)
Triangular space transmitting circumflex scapular a.
Superior lateral brachial cutaneous n.
Teres major m.
Triangular interval revealing radial n. and deep brachial a.
Long head, lateral head of triceps brachii m.
F. Netter M.D.
©CIBA-GEIGY

The teres major muscle is an adductor and medial rotator of the humerus; it acts with the latissimus dorsi and pectoralis major muscles but is only recruited with resistance. It also assists the latissimus dorsi muscle in extension of the arm at the shoulder. The *lower subscapular nerve* enters the anterior surface of the muscle.

Quadrangular and Triangular Spaces

The gradual divergence of the teres minor and teres major muscles produces a long horizontal triangle opening lateralward (Plate 22). This triangle is bisected vertically by the long head of the triceps brachii muscle and closed laterally by the shaft of the humerus. Thus are formed a small triangular space medial to the long head of the triceps brachii, in which the circumflex scapular vessels curve onto the dorsum of the scapula, and a quadrangular space lateral to the triceps brachii muscle. This latter space is bounded by the teres muscles above and below, by the triceps brachii medially, and by the humerus laterally. In the quadrangular space, the axillary nerve and posterior circumflex humeral vessels pass around the shaft of the humerus.

Subscapularis Muscle. This muscle fills the hollow of the costal surface of the scapula, arising from its medial two-thirds (Plate 27). The robust tendon passes across the anterior surface of the capsule of the shoulder joint to end in the lesser tubercle of the humerus. The tendon is separated from the neck of the scapula by the large *subscapular bursa*. The subscapularis muscle is the principal medial rotator of the arm. It also acts as an extensor when the arm is at the side or flexed and assists the other rotator cuff muscles in holding the head of the humerus closely in the glenoid cavity. The muscle is innervated on its costal surface by the *upper* and *lower subscapular nerves*.

Pectoralis Major Muscle. This muscle forms the fullness of the upper portion of the chest, and its inferior border is the anterior axillary fold. It is invested by the pectoral fascia, which is attached at the muscle origin to the clavicle and sternum, and is continuous below with the fascia of the anterior abdominal wall. At the lateroinferior border of the muscle, the pectoral fascia is continuous with the axillary fascia. The pectoralis major muscle has clavicular, sternocostal, and abdominal portions. The small clavicular portion arises from the medial half of the clavicle on its anterior surface. The much larger sternocostal portion arises from the anterior surface of the manubrium and body of the sternum. On the deep surface of the muscle, fascicles arise from the cartilages of the second to sixth ribs. The abdominal portion is a small slip that comes off the anterior layer of the sheath of the rectus abdominis muscle. From this long, essentially median, line on the chest, the muscular fibers converge to the upper end of the humerus. In order to concentrate the tendinous fibers of the muscle onto the crest of the greater tubercle of the humerus, the tendon folds on itself to form a bilaminar U-shaped tendon with the fold of the tendon below (Plate 23). Thus, the fibers of the clavicular part insert as the upper part of the anterior lamina; the lower sternal and abdominal fibers reach up into the superior part of the posterior limb; and the sternal fibers distribute into the anterior lamina, the fold, and the lower part of the posterior lamina.

The pectoralis major muscle flexes and adducts the humerus; it is also capable of medial rotation of the arm but usually becomes active only when this action is resisted. The clavicular portion of the pectoralis major muscle elevates the shoulder and flexes the arm, while the sternocostal portion draws the shoulder downward. The muscle is innervated by the *lateral* and *medial pectoral nerves* from both the lateral and medial cords of the brachial plexus, involving all the roots (C5–T1). The *pectoral branches* of the *thoracoacromial artery* accompany the nerves to the muscle.

The *deltopectoral triangle* is a separation just below the clavicle of the upper and adjacent fibers of the deltoid and pectoralis major muscles. Distally, the separation of these adjacent fibers is made by the cephalic vein and the deltoid branch of the thoracoacromial artery.

Upper Limb

Shoulder and Axillary Regions

(Continued)

Pectoral, Clavipectoral, and Axillary Fasciae

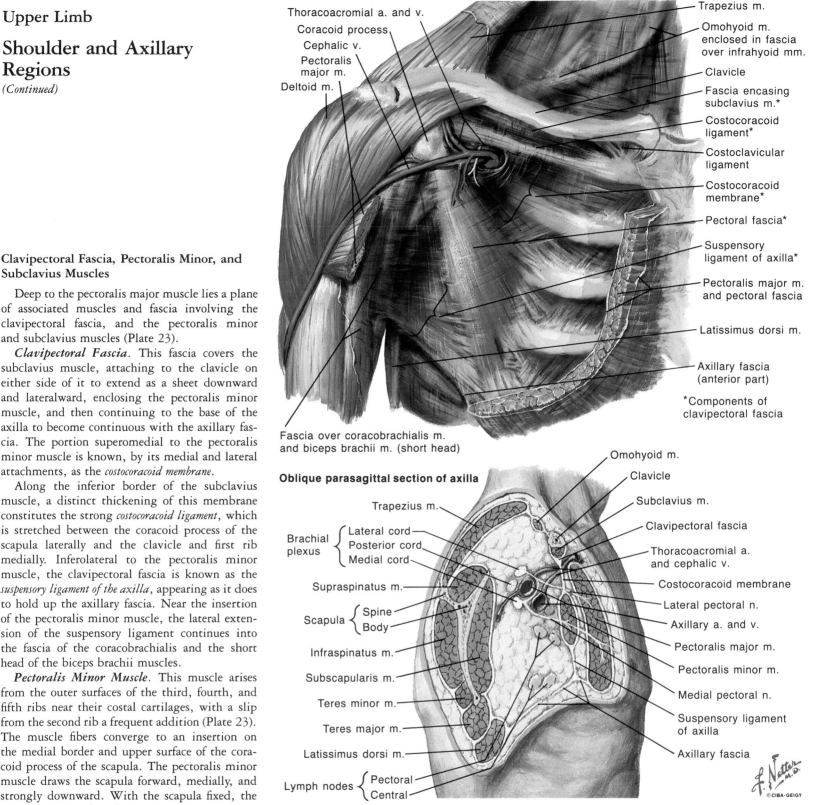

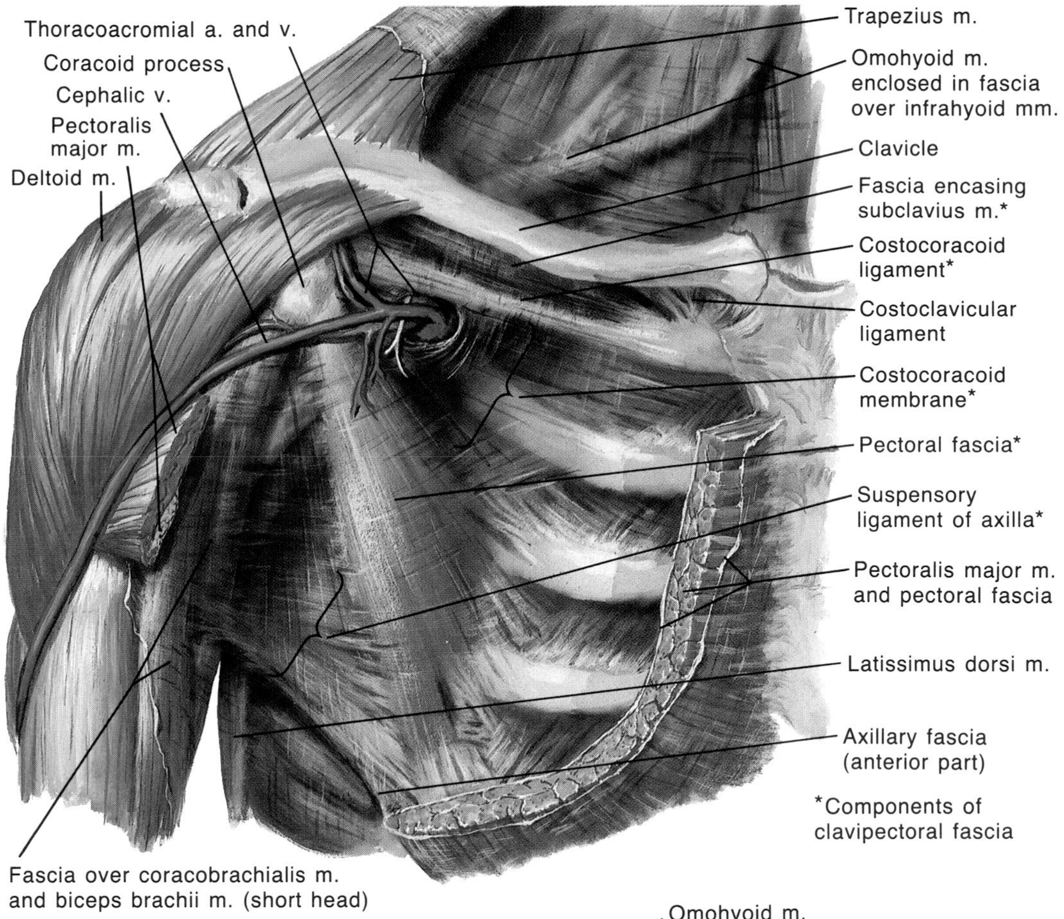
Thoracoacromial a. and v.
Coracoid process
Cephalic v.
Pectoralis major m.
Deltoid m.
Trapezius m.
Omohyoid m. enclosed in fascia over infrahyoid mm.
Clavicle
Fascia encasing subclavius m.*
Costocoracoid ligament*
Costoclavicular ligament
Costocoracoid membrane*
Pectoral fascia*
Suspensory ligament of axilla*
Pectoralis major m. and pectoral fascia
Latissimus dorsi m.
Axillary fascia (anterior part)
*Components of clavipectoral fascia
Fascia over coracobrachialis m. and biceps brachii m. (short head)

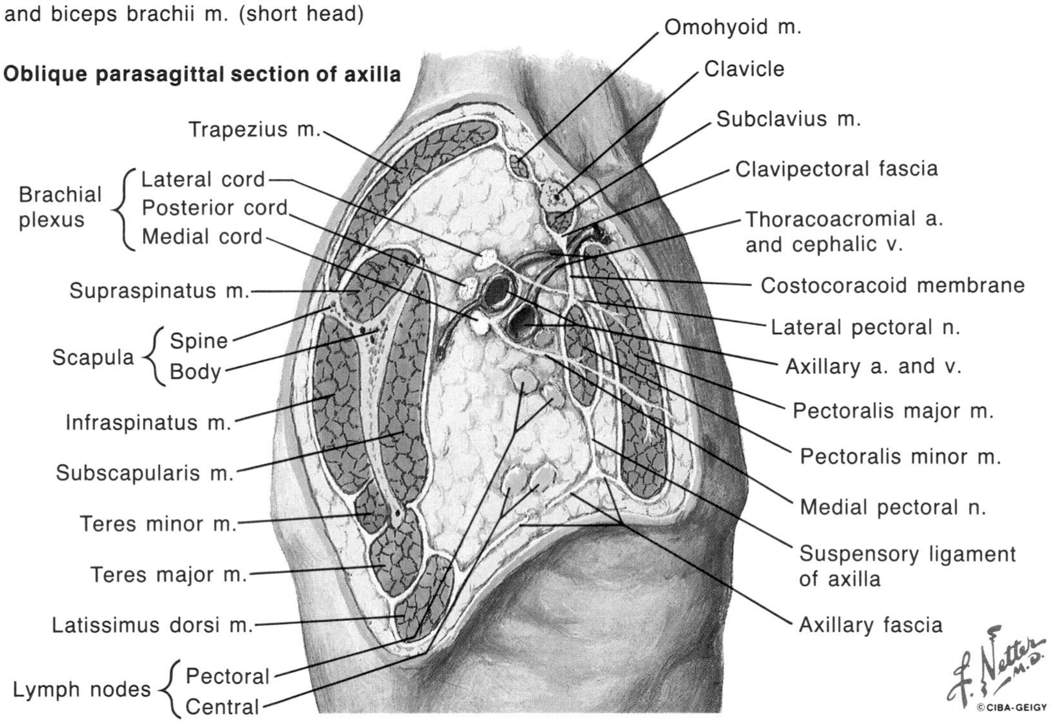
Oblique parasagittal section of axilla
Omohyoid m.
Clavicle
Subclavius m.
Clavipectoral fascia
Thoracoacromial a. and cephalic v.
Costocoracoid membrane
Lateral pectoral n.
Axillary a. and v.
Pectoralis major m.
Pectoralis minor m.
Medial pectoral n.
Suspensory ligament of axilla
Axillary fascia
Trapezius m.
Brachial plexus
Lateral cord
Posterior cord
Medial cord
Supraspinatus m.
Scapula
Spine
Body
Infraspinatus m.
Subscapularis m.
Teres minor m.
Teres major m.
Latissimus dorsi m.
Lymph nodes
Pectoral
Central
F. Netter M.D.
©CIBA-GEIGY

Clavipectoral Fascia, Pectoralis Minor, and Subclavius Muscles

Deep to the pectoralis major muscle lies a plane of associated muscles and fascia involving the clavipectoral fascia, and the pectoralis minor and subclavius muscles (Plate 23).

Clavipectoral Fascia. This fascia covers the subclavius muscle, attaching to the clavicle on either side of it to extend as a sheet downward and lateralward, enclosing the pectoralis minor muscle, and then continuing to the base of the axilla to become continuous with the axillary fascia. The portion superomedial to the pectoralis minor muscle is known, by its medial and lateral attachments, as the *costocoracoid membrane*.

Along the inferior border of the subclavius muscle, a distinct thickening of this membrane constitutes the strong *costocoracoid ligament*, which is stretched between the coracoid process of the scapula laterally and the clavicle and first rib medially. Inferolateral to the pectoralis minor muscle, the clavipectoral fascia is known as the *suspensory ligament of the axilla*, appearing as it does to hold up the axillary fascia. Near the insertion of the pectoralis minor muscle, the lateral extension of the suspensory ligament continues into the fascia of the coracobrachialis and the short head of the biceps brachii muscles.

Pectoralis Minor Muscle. This muscle arises from the outer surfaces of the third, fourth, and fifth ribs near their costal cartilages, with a slip from the second rib a frequent addition (Plate 23). The muscle fibers converge to an insertion on the medial border and upper surface of the coracoid process of the scapula. The pectoralis minor muscle draws the scapula forward, medially, and strongly downward. With the scapula fixed, the muscle assists in forced inspiration. The muscle is innervated by the *medial pectoral nerve* (C8; T1), which completely penetrates the muscle to pass across the interpectoral space into the pectoralis major muscle. *Pectoral branches of the thoracoacromial artery* are distributed with the nerve. Deep to the tendon of the pectoralis minor muscle pass the axillary artery and the cords of the brachial plexus.

Subclavius Muscle. This small, pencillike muscle arises from the junction of the first rib and its cartilage (Plate 23). It lies parallel to the underside of the clavicle and inserts in a groove on the underside of the clavicle, between the attachments of the conoid ligament laterally and the costoclavicular ligament medially. The muscle assists by its traction on the clavicle in drawing the shoulder forward and downward. The *nerve to the subclavius muscle* is a branch of the superior trunk of the brachial plexus, with fibers from the fifth cervical nerve, which reaches the upper posterior border of the muscle. There is a small, special *clavicular branch of the thoracoacromial artery* to the muscle.

Fasciae and Boundaries of Axilla

The axilla is a space at the junction of the upper limb, chest, and neck. It is shaped like a truncated pyramid and serves as the passageway for nerves, blood vessels, and lymphatics into or from the limb (Plate 23). Its walls are musculofascial. The base is the concave armpit, the actual floor being the axillary fascia. The anterior wall is composed of the two planes of pectoral muscles and the associated pectoral and clavipectoral fasciae. The lateral border of the pectoralis major muscle forms the anterior axillary fold. The posterior wall of the axilla is made up of the scapula, the scapular musculature, and the associated fasciae. The lower members of this group, together with the tendon of the latissimus dorsi muscle, form the posterior axillary fold. The chest wall, covered by the serratus anterior muscle and its fascia, forms the medial wall. The lateral wall is reduced to a mere chink by the convergence of the tendons of the anterior and posterior axillary fold muscles

Upper Limb

Shoulder and Axillary Regions

(Continued)

onto the greater tubercular crest, the intertubercular groove, and the lesser tubercular crest of the humerus. The truncated apex of the axilla is formed by the convergence of the bony members of the three major walls—the clavicle, the scapula, and the first rib. Through the triangular interval so formed pass all the neurovascular structures of the limb.

The actual boundaries of the axilla are fascial. Continuities of these fasciae into the neck are with the superficial layer of the cervical fascia covering the trapezius muscle and with the subjacent infrahyoid fascia. The axillary fascia is continuous medially with the fascia covering the serratus anterior muscle; laterally, it is continuous with the brachial fascia investing the structures of the arm. The fascia of the serratus anterior muscle is continuous with the pectoral and clavipectoral fasciae anteriorly and with the fasciae of the scapular muscles posteriorly.

The *contents* of the axilla are the brachial plexus, the axillary artery, the axillary vein, the axillary lymphatics, and fat (Plates 23–25, 37).

The vascular and nerve trunks are enclosed in the axillary sheath, a fascial extension of the prevertebral layer of cervical fascia that covers the scalene muscles. The axillary sheath adheres to the clavipectoral fascia behind the pectoralis minor muscles and continues along the vessels and nerves as far as their entrance into the neurovascular compartment of the medial intermuscular septum of the arm.

Technically within the axilla but not discussed here is the tendon of the long head of the biceps brachii muscle as it lies in the intertubercular groove of the humerus. The short head of the biceps brachii muscle and the coracobrachialis muscle are adjuncts of the anterior wall, because of their origin from the coracoid process of the scapula.

Axillary Artery, Axillary Vein, and Lymphatics

Axillary Artery. The axillary artery is the continuation of the subclavian artery, and the same vessel is the brachial artery beyond the inferior limit of the axilla (Plate 24). Thus, it is a regional name, and the artery extends from the outer border of the first rib to the lower border of the teres major muscle.

With the arm abducted, the course of the artery can be indicated on the surface by a line extending from the middle of the clavicle to the groove just behind the coracobrachialis and biceps brachii muscles. Proximally, it is deeply placed under the pectoral muscles, and the shoulder joint and neck of the humerus are lateral to it. Distally, it is superficial under the brachial fascia. The artery is large, about 1 cm in diameter and about 12 cm long. The axillary vein is anterior and inferior to the artery in normal posture, but rises and is more completely anterior to the artery when the arm is abducted. The artery is crossed by the tendon of the pectoralis minor muscle, thus allowing a description of the artery in three parts: the first part (about 2.5 cm long) has one branch and is proximal to the tendon; the second part (about 3 cm long) has two branches and is behind the tendon; and the third part (about 6.5 cm long) has three branches and is distal to the tendon (Plate 38).

1. The *superior thoracic artery* is a small vessel arising at the lower border of the subclavius muscle. It descends behind the axillary vein to the intercostal muscles of the first and second intercostal spaces, and to the upper portion of the serratus anterior muscle.

The *second part of the axillary artery* is overlain by the pectoralis minor muscle. The cords of the brachial plexus are named in relation to this portion of the artery. The lateral cord is lateral to the artery, and the medial cord is medial and inferior to it. The posterior cord of the plexus is behind the artery. Branches of the second part are the thoracoacromial and lateral thoracic arteries.

2. The *thoracoacromial artery* arises beneath the upper portion of the pectoralis minor tendon. It turns the border of the tendon, pierces the costocoracoid membrane, and immediately divides into four branches deep to the clavicular head of the pectoralis major muscle—the acromial, deltoid, pectoral, and clavicular branches.

The *acromial branch* passes lateralward across the coracoid process to the acromion. It gives branches to the deltoid muscle and participates with branches of the anterior and posterior circumflex humeral and suprascapular vessels in the formation of the acromial network of small vessels on the surface of the acromion. The *deltoid branch* (often arising not separately but as a branch of the acromial artery) occupies the interval between the deltoid and pectoralis major muscles in company with the cephalic vein. It sends branches into these muscles. The *pectoral branch* is large and descends between the pectoralis major and minor muscles. It gives branches to these muscles, anastomoses with intercostal and lateral thoracic arteries and, in the female, supplies the mammary gland in its deep aspect. The *clavicular branch* is a slender vessel ascending medialward to supply the subclavius muscle and the sternoclavicular joint.

3. The *lateral thoracic artery* is rather variable. It may arise directly from the axillary artery, from the thoracoacromial artery, or from the subscapular artery; it is frequently represented by several vessels. Typically (in 65% of cases), it arises from the axillary artery, descends along the lateral border of the pectoralis minor muscle, and sends branches to the serratus anterior and pectoral muscles and axillary lymph nodes. In the female, it provides lateral mammary rami. The lateral thoracic artery anastomoses with intercostal arteries and with radicles of the subscapular and thoracoacromial arteries. It may give an offshoot into the subscapularis muscle, which accompanies the upper subscapular nerve.

The *third part of the axillary artery* lies against the coracobrachialis muscle laterally and the axillary vein medially. The median nerve is formed proximally on the surface of this part of the artery by the junction of contributing roots from the medial and lateral cords of the brachial plexus. The ulnar nerve and the medial brachial and medial antebrachial cutaneous nerves are found medial to it, and the radial and axillary nerves lie posteriorly.

4. The *subscapular artery* is the largest branch of the axillary artery. It arises at the lower border of the subscapularis muscle and descends along its axillary border, supplying the subscapularis, teres major, and serratus anterior muscles. At 3 or 4 cm from its origin, the artery divides into the circumflex scapular and thoracodorsal branches.

The *circumflex scapular artery*, the larger branch, passes posteriorly through the triangular space, turns onto the dorsum of the scapula (which it grooves), and ramifies in the infraspinatous fossa. Here, it supplies the muscles of the dorsum of the scapula and anastomoses with the dorsal scapular artery and the terminals of the suprascapular artery. By branches given off in the triangular space, it supplies the subscapularis and the two teres muscles. The *thoracodorsal artery* continues the general course of the subscapular artery to the inferior angle of the scapula, supplying adjacent muscles and anastomosing with the circumflex scapular, dorsal scapular, lateral thoracic, and intercostal arteries. The thoracodorsal artery is the principal supply of the latissimus dorsi muscle, entering it on its deep (axillary) surface in company with the thoracodorsal nerve. It frequently has a thoracic branch that substitutes for the inferior portion of the distribution of the lateral thoracic artery.

5. The *anterior circumflex humeral artery* arises closely juxtaposed to the posterior circumflex humeral artery but is much smaller. It runs laterally across the intertubercular groove and the surgical neck of the humerus and anastomoses with the posterior circumflex humeral artery. It supplies adjacent muscles, one branch ascending in the intertubercular groove to supply the tendon of the long head of the biceps brachii muscle and the shoulder joint.

6. The *posterior circumflex humeral artery* passes backward with the axillary nerve through the quadrangular space and deep to the deltoid muscle. Numerous branches supply this muscle; others reach the shoulder joint and anastomose in the acromial network. A nutrient artery is supplied to the greater tubercle of the humerus. A descending branch follows the lateral head of the triceps brachii muscle, distributes to the long and lateral heads of this muscle, and anastomoses with an ascending branch of the deep brachial artery of the arm. Occasional enlargement of this anastomosis results in the origin of the deep brachial artery from the posterior circumflex humeral artery or of the posterior circumflex humeral from the deep brachial artery. The posterior circumflex humeral artery encircles the surgical neck of the humerus and anastomoses with the anterior circumflex humeral artery.

Axillary Vein. The veins of the limbs are usually double, lying on either side of the artery (Plate 24). They can thus be designated *venae comitantes* of the artery or by the name of the artery. There are, accordingly, two brachial veins

Shoulder and Axilla: Deep Dissection (anterior view)

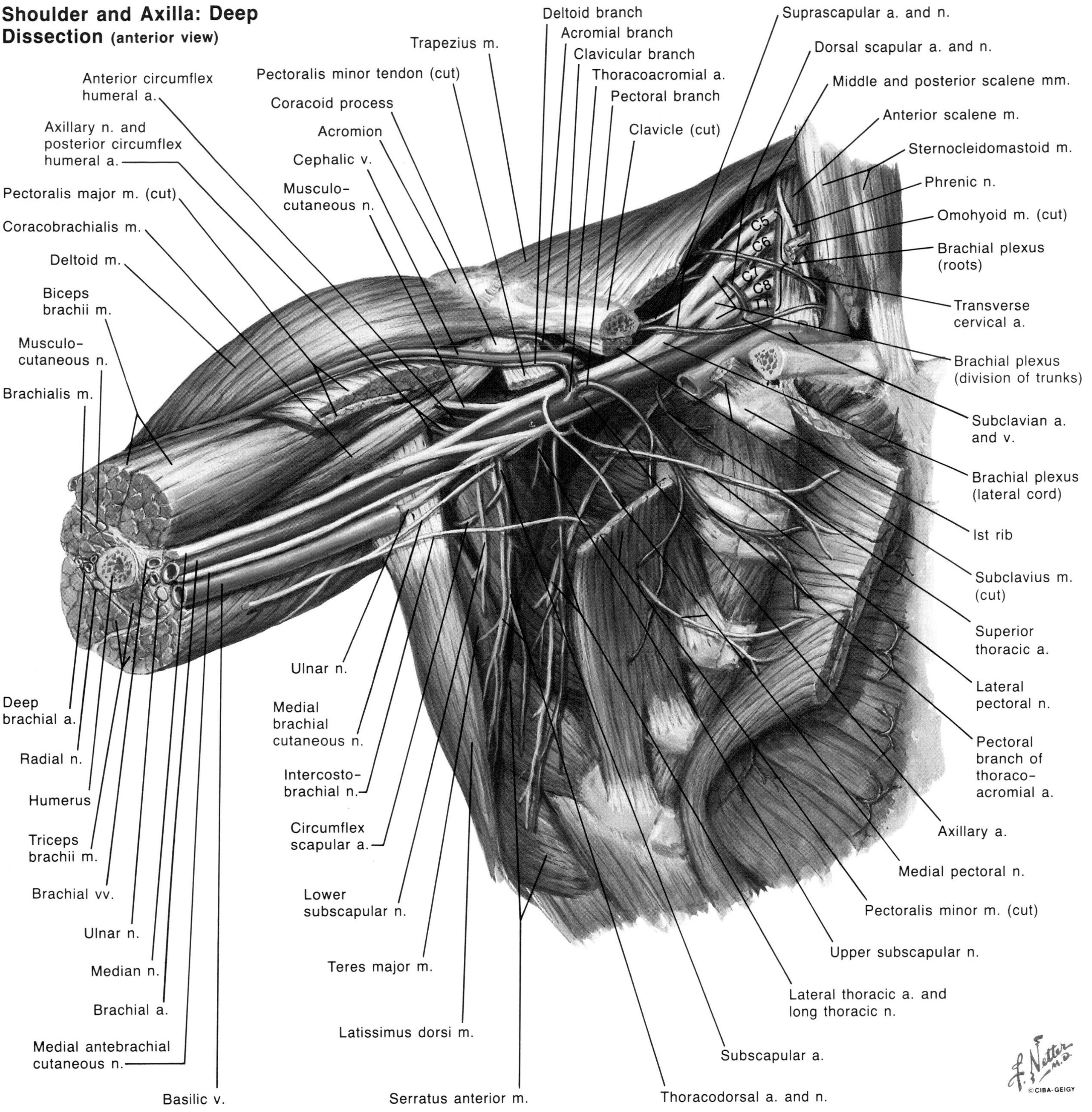

Deltoid branch
Acromial branch
Clavicular branch
Thoracoacromial a.
Pectoral branch
Clavicle (cut)
Trapezius m.
Pectoralis minor tendon (cut)
Coracoid process
Acromion
Cephalic v.
Musculo-cutaneous n.
Anterior circumflex humeral a.
Axillary n. and posterior circumflex humeral a.
Pectoralis major m. (cut)
Coracobrachialis m.
Deltoid m.
Biceps brachii m.
Musculo-cutaneous n.
Brachialis m.
Suprascapular a. and n.
Dorsal scapular a. and n.
Middle and posterior scalene mm.
Anterior scalene m.
Sternocleidomastoid m.
Phrenic n.
Omohyoid m. (cut)
Brachial plexus (roots)
Transverse cervical a.
Brachial plexus (division of trunks)
Subclavian a. and v.
Brachial plexus (lateral cord)
Ist rib
Subclavius m. (cut)
Superior thoracic a.
Lateral pectoral n.
Pectoral branch of thoraco-acromial a.
Axillary a.
Medial pectoral n.
Pectoralis minor m. (cut)
Upper subscapular n.
Lateral thoracic a. and long thoracic n.
Subscapular a.
Thoracodorsal a. and n.
C5
C6
C7
C8
T1
Ulnar n.
Medial brachial cutaneous n.
Intercosto-brachial n.
Circumflex scapular a.
Lower subscapular n.
Teres major m.
Latissimus dorsi m.
Serratus anterior m.
Deep brachial a.
Radial n.
Humerus
Triceps brachii m.
Brachial vv.
Ulnar n.
Median n.
Brachial a.
Medial antebrachial cutaneous n.
Basilic v.
F. Netter M.D.
© CIBA-GEIGY

SECTION I PLATE 24 Slide 3589

Upper Limb

Shoulder and Axillary Regions

(Continued)

associated with the brachial artery, which ascend in the medial intermuscular septum of the arm (Plate 34). At the lower border of either the teres major or the subscapularis muscle, the lateral of the brachial veins crosses the axillary artery to join the medial vein. About the same level, the basilic vein joins the medial brachial vein, and as a result of these somewhat variable junctions, the single axillary vein is formed.

The axillary vein is large, and at the outer border of the first rib it becomes the subclavian vein. It lies medial to the axillary artery. Toward the apex of the axilla, the vein is inferior and somewhat anterior to the artery, separated from it by the medial cord of the brachial plexus. It contains a pair of valves at its origin, and valves are also found at the terminations of the subscapular and cephalic veins. One or two lateral axillary lymph nodes lie closely medial to the vein. Accidental wounds of the axillary vein are dangerous because of the vessel's large size, exposed position, and nearness to the thorax.

The tributaries of the axillary vein correspond to the branches of the axillary artery, with certain exceptions. The large pectoral radicles of the thoracoacromial vein usually empty into the distal portion of the subclavian vein. The cephalic vein usually receives all but the pectoral radicles of the thoracoacromial vein, just before it perforates the costocoracoid membrane (upper border of the tendon of the pectoralis minor muscle) to terminate in the axillary vein. The lateral thoracic vein is involved in certain anastomotic connections with chest and abdominal veins via the costoaxillary and the thoracoepigastric veins, respectively. □

Right Brachial Plexus: Common Arrangement
(from ventral rami of C5, 6, 7, 8, and T1 with contributions from C4 and T2)

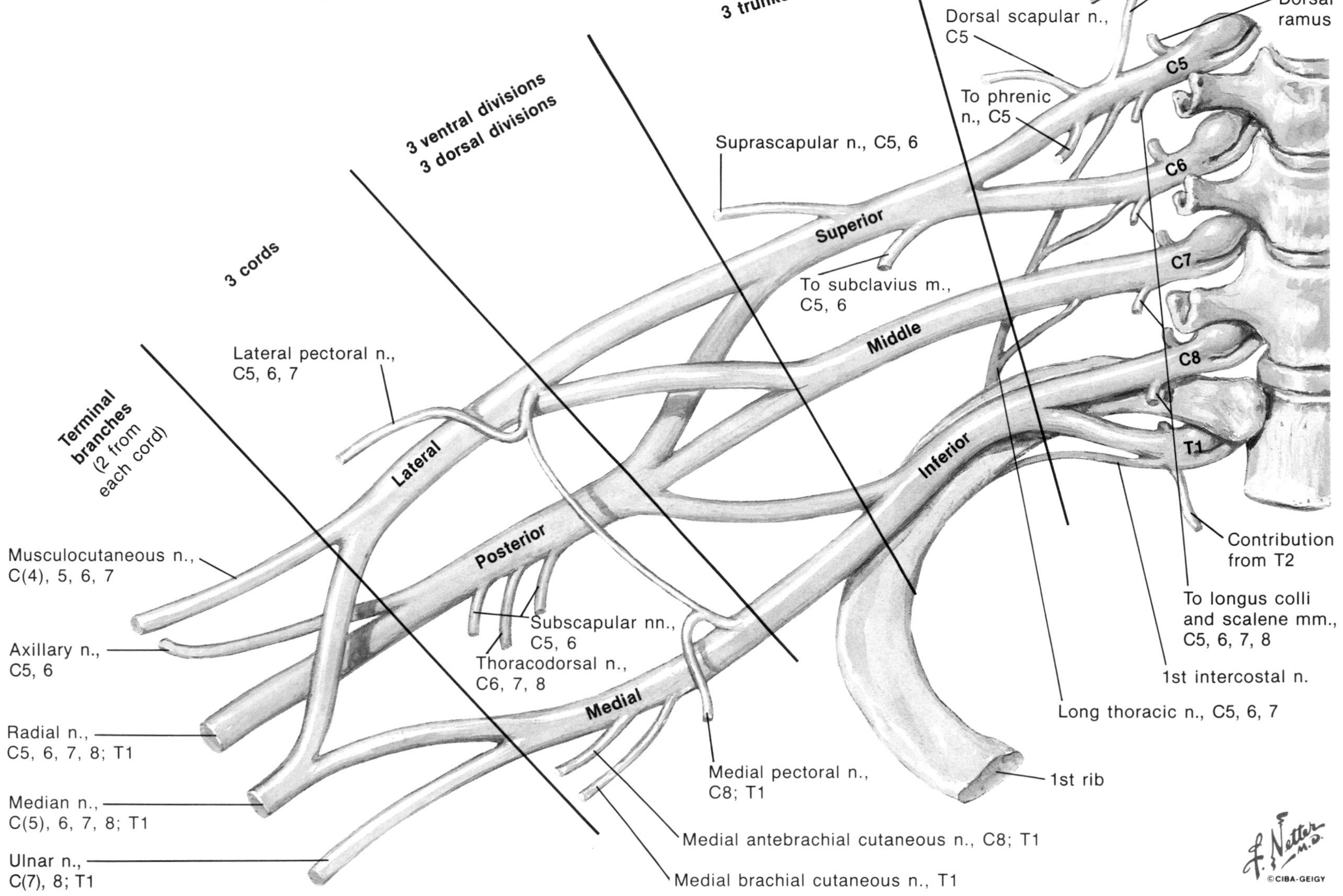

SECTION I PLATE 25 Slide 3590

Upper Limb

Brachial Plexus and Cutaneous Nerves

Brachial Plexus

This large nerve complex does not originate in the axilla, although the greater part of its branching and the formation of the definitive nerves of the limb do take place in this region (Plate 25). The terminal branches of the brachial plexus are fully described in their regions of distribution.

Roots. The brachial plexus is formed by the *ventral rami* of the fifth to the eighth cervical nerves (C5–8) and the greater part of the first thoracic nerve (T1). Small contributions may come from the fourth cervical nerve (C4) and the second thoracic nerve (T2). The ventral rami, designated as the *roots* of the plexus, emerge from between the anterior and middle scalene muscles of the neck. The upper roots descend toward the first rib, but the ventral ramus of T1 has to ascend to pass across the first rib. The sympathetic fibers conducted by each root are added as they pass between the scalene muscles. Each of the ventral rami of C5 and C6 receives a gray ramus communicans from the middle cervical ganglion. The cervicothoracic ganglion (inferior cervical plus first thoracic ganglia) contributes gray rami to the C7, C8, and T1 roots of the plexus.

Trunks. The ventral rami of C5 and C6 unite to form the *superior trunk*, the ramus of C7 continues alone as the *middle trunk*, and the rami of C8 and T1 form the *inferior trunk*.

Divisions. Each trunk separates into an anterior and a posterior division. The *anterior division* supplies the originally ventral parts of the limb, and the *posterior division* supplies the dorsal parts.

Cords. All the posterior divisions unite to form the *posterior cord* of the plexus, the anterior divisions of the superior and middle trunks form the *lateral cord*, and the *medial cord* is the continuation of the anterior division of the inferior trunk. Thus, the posterior cord contains nerve bundles from C5 to T1 destined for the back of the limb, the lateral cord is formed of nerve bundles from C5 to C7 for the anterior portion of the limb, and the medial cord carries anterior nerve components from C8 and T1. The cords are named to show their relationships to the axillary artery.

Branches. The terminal branches regroup further and form the terminal nerves of the plexus. Large portions of the lateral and medial cords form the *median nerve* (Plate 48). The remainder of the lateral cord constitutes the *musculocutaneous nerve* (Plate 35); the rest of the medial cord is the *ulnar nerve* (Plate 47). The contribution of the lateral cord to the median nerve is frequently made in several separated bundles. The posterior cord, positioned posterior to the axillary artery, gives off the *axillary nerve* at the lower border of the subscapularis muscle. The remainder is the large *radial nerve* (Plate 49).

The *nerves of distribution* of the brachial plexus also include nerves that arise from the roots and cords of the plexus. These are divided regionally as supraclavicular and infraclavicular branches.

Supraclavicular Branches. The *dorsal scapular nerve*, supplying the rhomboideus and levator scapulae muscles, arises from the posterior aspect of the ventral ramus of C5 (with a frequent contribution from C4). The nerve pierces the middle scalene muscle to enter the posterior triangle of the neck and then descends anterior to the levator scapulae and rhomboid muscles adjacent to the medial border of the scapula (Plate 36).

The *long thoracic nerve*, the nerve to the serratus anterior muscle, arises from the back of C5 to C7, those from C5 and C6 piercing the middle scalene muscle. The nerve passes behind the roots of the brachial plexus between the axillary artery and the serratus anterior muscle, on which it ramifies in company with the lateral thoracic artery.

xus and
Jerves

Dermal Segmentation of Upper Limb
(after Keegan and Garrett)

Although indicated schematically as distinctly demarcated zones, in actuality each dermatome overlaps its adjacent segment

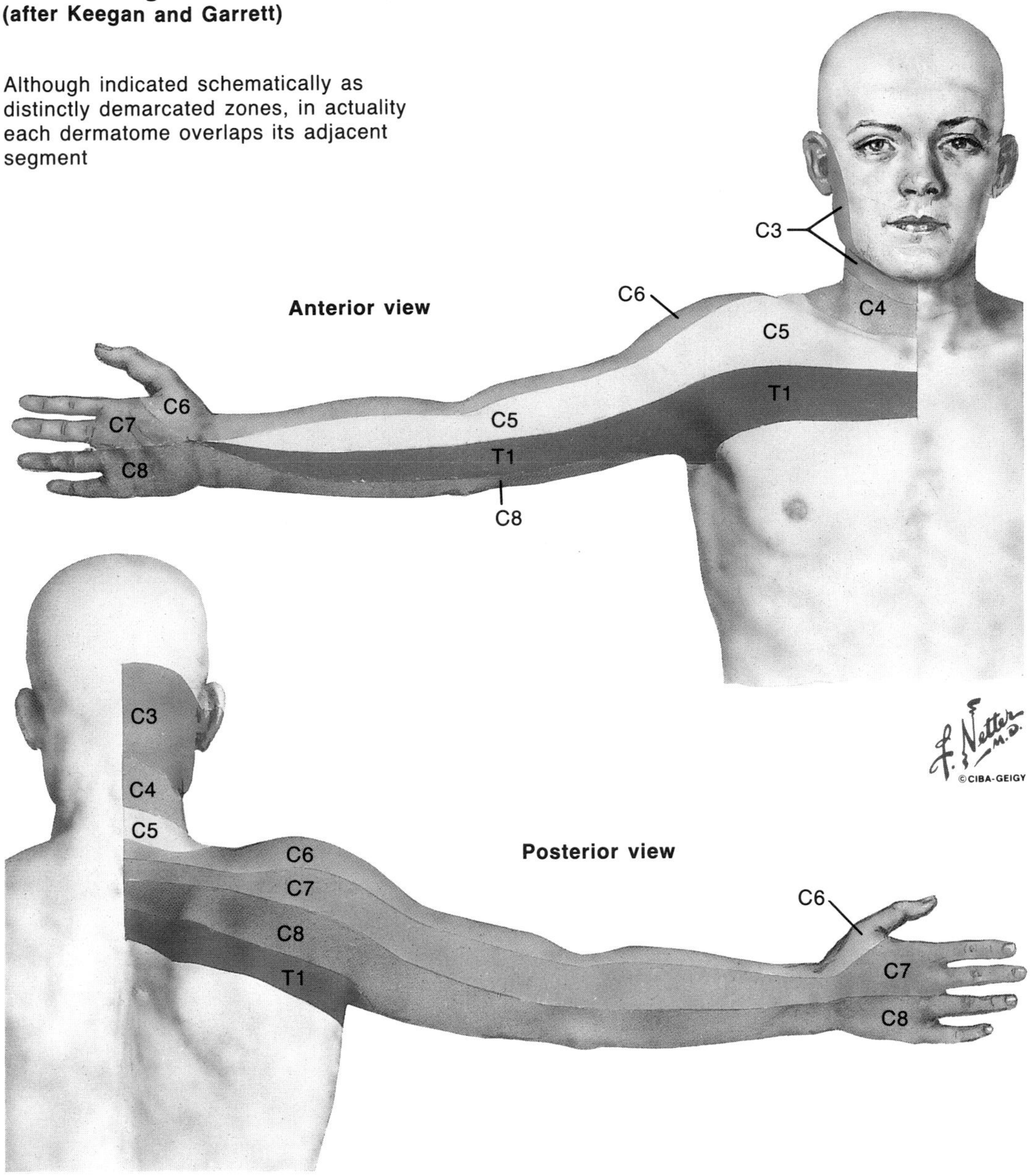

ubclavius muscle is a slender
e anterior surface of the supe-
y from C5, it may have addi-
C6. It descends across the
ne plexus and in front of the
d vein to reach the subcla-
erve frequently contains nerve
the phrenic nerve contribu-
at leave the nerve to the sub-
y phrenic nerve.)
r nerve arises from the pos-
superior trunk and contains
and C6, with frequent addi-
nerve passes lateralward across
l triangle to reach the scapu-
hrough this notch under the
capular ligament, by which
the suprascapular artery. The
ranches to the supraspinatus
ulder joint; it then descends
f the scapular neck under the
apular ligament to terminate
muscle.
Branches. The lateral pectoral
om the anterior divisions of
n the lateral cord of the bra-
sses anterior to the axillary
loop of communication with
nerve. Branches then pierce
embrane and distribute to the
scle.
l nerve arises from the medial
erve fibers from C8 and T1.
e axillary vein and artery, it
oop communicating with the
ve and enters the pectoralis
ches also pass on, through the
nto the pectoralis major mus-
lateral designations of these
reflect the cords from which

al cutaneous nerve arises from
tends through the axilla be-
in, piercing the brachial fas-

rachial cutaneous nerve arises
d between the medial brachial
the ulnar nerve. It also carries

nerve fibers representing C8 and T1. It first lies medial to the axillary artery and then medial and anterior to the brachial artery. At the aperture in the brachial fascia admitting the basilic vein, the medial antebrachial cutaneous nerve enters the subcutaneous tissue.

The *subscapular nerves* are three branches of the posterior cord. They arise in a rather close grouping and are distinguished as the upper, middle, and lower subscapular nerves. The *upper*, or *short*, *subscapular nerve*, derived from C5 and C6, enters the upper portion of the subscapularis muscle to supply it. The *middle subscapular nerve*, usually designated as the *thoracodorsal nerve*, is composed of fibers from C7 and C8. It passes behind the axillary artery along the lower border of the subscapularis muscle, reaching the axillary portion of the latissimus dorsi muscle and ramifying into the muscle to supply it. The *lower subscapular nerve* carries fibers from C5 and C6. It courses downward behind the subscapular vessels to the teres major muscle, which it supplies, giving off a few twigs to the subscapularis muscle.

The *axillary nerve* arises from the posterior cord of the brachial plexus at the lower border of the subscapularis muscle. It passes across the upper edge of the teres major muscle to enter the quadrangular space in company with the posterior circumflex humeral artery. Providing an articular twig to the inferior portion of the capsule of the shoulder joint, the nerve then divides into a superior and an inferior branch. The *superior branch* circles the surgical neck of the humerus, supplying the deltoid muscle along with radicles of the posterior circumflex humeral artery. The *inferior branch* supplies the teres minor muscle and the posterior fibers of the deltoid muscle; then, turning the posterior border of the deltoid muscle in its lower third, it ends as the *superior lateral brachial cutaneous nerve*.

Dermatomes of Upper Limb

There is an obvious serial order in the nerves of the brachial plexus as they arise from the spinal nerves of C5 to T1. This order is laid down in the primitive serial morphology of the embryo and is retained into adulthood. Thus, as seen in the earlier limb bud stage, C5 in the adult distributes to the cranial part of the limb, and T1 distributes to its caudal part (see Section II, Plate 9).

However, there is overlap of innervation, and each cutaneous nerve overlaps into the primary zone of distribution of each of its neighbors. This overlap has made precise designation of primary dermatomes difficult, but recent studies based on clinical examination of single nerve defects have resulted in the dermatome chart shown in Plate 26. □

Upper Limb

Bones and Joints of Shoulder and Arm

The bones of the shoulder are the scapula, the clavicle, and the humerus (Plates 27–29).

Scapula

The scapula is a flat bone of triangular form, which exhibits certain prominent and rather heavy processes (Plates 27–28). Most of these processes, and therefore the greatest mass and weight of the bone, lie laterally in the acromion, the spinous process, the coracoid process, the glenoid portion, and the lateral border.

The *body* of the scapula is thin and translucent. It has a large shallow concavity on its costal surface, the *subscapular fossa*, for the subscapularis muscle. The dorsum is convex and is separated by the prominent spinous process into a *supraspinatous fossa* above, lodging the supraspinatus muscle, and an *infraspinatous fossa* below, for the infraspinatus muscle. The heavy *lateral border* of the body exhibits, at and above the inferior angle, a flattened area of origin for the teres major muscle, and then in succession, areas of origin for the teres minor muscle and the long head of the triceps brachii muscle. The lateral border ends in the glenoid process above. The *medial border* is relatively straight, lying parallel and adjacent to the vertebral column. It is the longest of the borders and serves for the attachment of the levator scapulae and the two rhomboideus muscles. The *superior border* is thinnest and shortest.

At its coracoid extremity, the bone is indented sharply to form the *scapular notch*. The *spinous process* is a large triangular projection of the dorsum of the bone, extending from the medial border to just short of the glenoid process. It increases its elevation and weight as it goes lateralward and ends in a concave border, the origin of which is the neck of the scapula. The spinous process continues freely to arch above the head of the humerus as the *acromion*, which overhangs the shoulder joint. Its lateral surface provides origin for the deltoid muscle, and its smooth, concave underside is related to the subacromial bursa. The forward-directed extremity of the acromion has a medially located smooth facet, which participates in the acromioclavicular articulation (Plate 30).

The *coracoid process* is a thick, upward projection from the neck of the scapula, which then turns

Humerus and Scapula (anterior view)

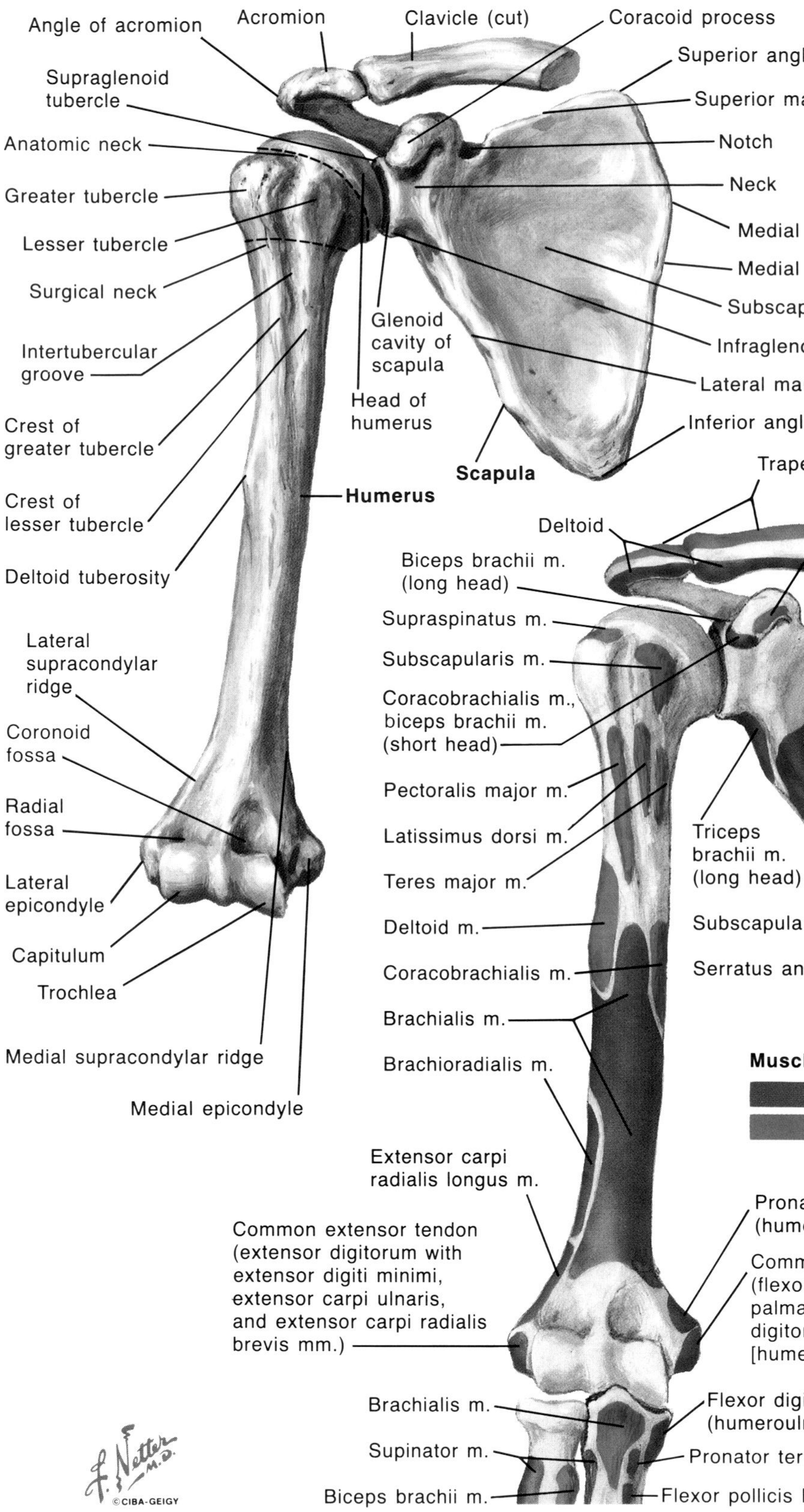

Angle of acromion
Acromion
Clavicle (cut)
Coracoid process
Supraglenoid tubercle
Anatomic neck
Notch
Neck
Greater tubercle
Lesser tubercle
Surgical neck
Intertubercular groove
Glenoid cavity of scapula
Head of humerus
Crest of greater tubercle
Scapula
Crest of lesser tubercle
Humerus
Deltoid tuberosity
Deltoid
Biceps brachii m. (long head)
Supraspinatus m.
Subscapularis m.
Lateral supracondylar ridge
Coracobrachialis m., biceps brachii m. (short head)
Coronoid fossa
Pectoralis major m.
Radial fossa
Latissimus dorsi m.
Triceps brachii m. (long head)
Lateral epicondyle
Teres major m.
Deltoid m.
Capitulum
Coracobrachialis m.
Trochlea
Brachialis m.
Medial supracondylar ridge
Brachioradialis m.
Medial epicondyle
Extensor carpi radialis longus m.
Common extensor tendon (extensor digitorum with extensor digiti minimi, extensor carpi ulnaris, and extensor carpi radialis brevis mm.)
Brachialis m.
Supinator m.
Biceps brachii m.
F. Netter M.D.
©CIBA-GEIGY

forward and lateralward. It gives attachment to the pectoralis minor muscle and to the short head of the biceps brachii muscle and the coracobrachialis muscle, and to many ligaments associated with the acromioclavicular and shoulder joints. The broadened lateral angle of the scapula has a shallow *glenoid cavity*, supported by a slightly constricted *neck*. This cavity is directed forward and laterally for reception of the head of the humerus. It is somewhat pear shaped, narrowed above and bulbous below, with a greater vertical than horizontal dimension. The margin provides attachment for the fibrocartilaginous *glenoid labrum*; above, it is elevated into the *supraglenoid tubercle* for the long head of the biceps brachii muscle.

Ossification of the scapula i
centers: one for the body, two
cess, two for the acromion,
border, and one for the inferi
in the body begins during
intrauterine life. At birth,
scapula is bony, except the
those parts with separate oss
center for the coracoid pro
after 1 year of age and fuses
14 to 15 years of age. Cent
inferior angle, and medial
puberty and fuse with the re
16 to 17 for the acromion an
the inferior angle and media

Upper Limb

Bones and Joints of Shoulder and Arm

(Continued)

Humerus and Scapula (posterior view)

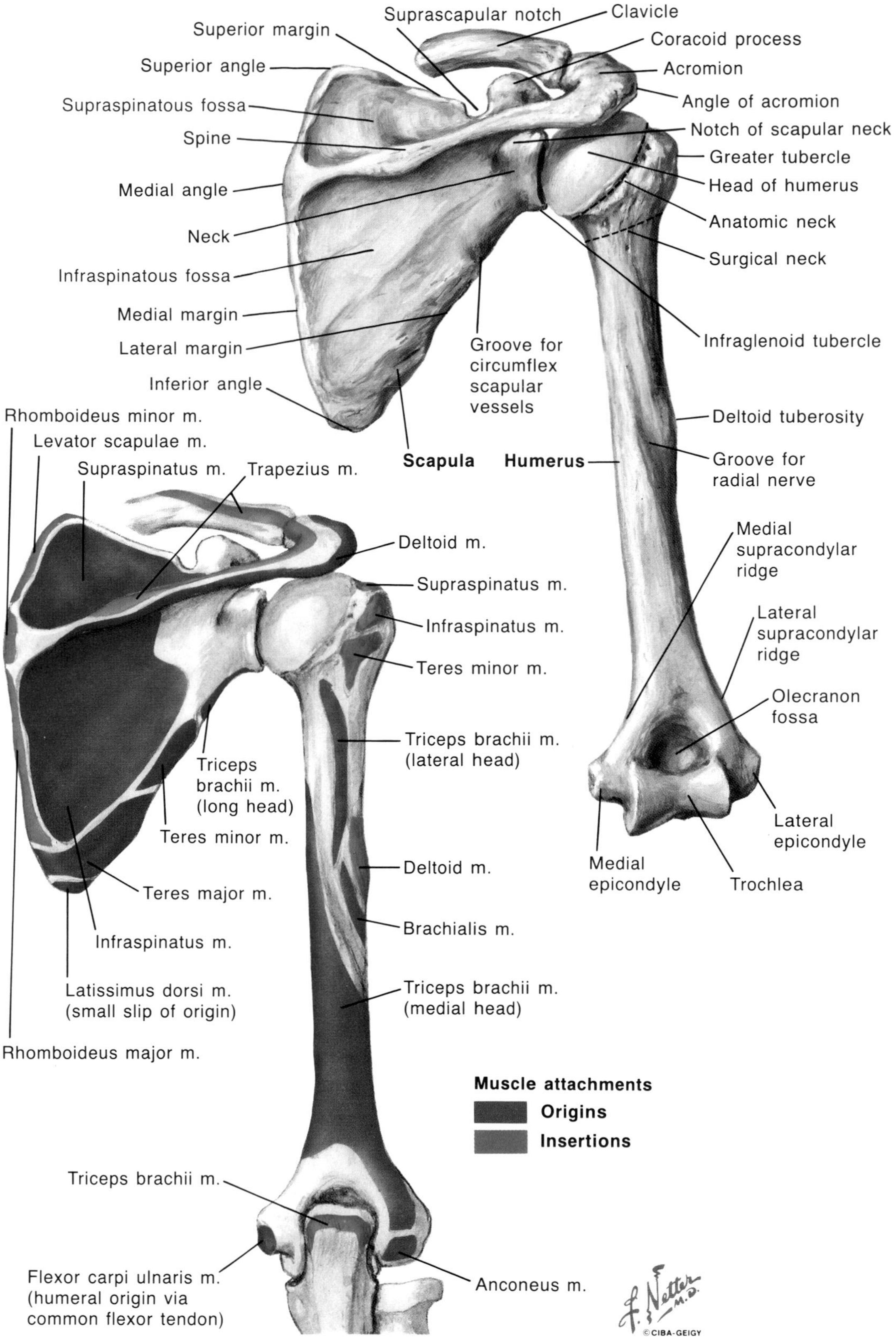
Superior margin
Suprascapular notch
Clavicle
Coracoid process
Superior angle
Acromion
Supraspinatous fossa
Angle of acromion
Notch of scapular neck
Spine
Greater tubercle
Medial angle
Head of humerus
Neck
Anatomic neck
Surgical neck
Infraspinatous fossa
Medial margin
Infraglenoid tubercle
Lateral margin
Groove for circumflex scapular vessels
Inferior angle
Deltoid tuberosity
Rhomboideus minor m.
Levator scapulae m.
Supraspinatus m.
Trapezius m.
Scapula
Humerus
Groove for radial nerve
Medial supracondylar ridge
Deltoid m.
Supraspinatus m.
Infraspinatus m.
Lateral supracondylar ridge
Teres minor m.
Olecranon fossa
Triceps brachii m. (lateral head)
Triceps brachii m. (long head)
Teres minor m.
Lateral epicondyle
Deltoid m.
Medial epicondyle
Trochlea
Teres major m.
Brachialis m.
Infraspinatus m.
Latissimus dorsi m. (small slip of origin)
Triceps brachii m. (medial head)
Rhomboideus major m.
Muscle attachments
Origins
Insertions
Triceps brachii m.
Flexor carpi ulnaris m. (humeral origin via common flexor tendon)
Anconeus m.
F. Netter M.D.
©CIBA-GEIGY

Humerus

The humerus is a long bone composed of a shaft and two articular extremities (Plates 27–28). The *head*, or upper end, is a large segment of a sphere that articulates in the glenoid cavity of the scapula; its articular surface faces upward, medialward, and backward. The *anatomic neck* of the bone is the slight indentation of the head margin for attachment of the articular capsule. The *surgical neck* is the narrowed area just distal to the tubercles, where fractures frequently occur. The *greater tubercle* represents the most laterally projecting part of the skeleton at the shoulder; it has three flattened surfaces for the attachment of the supraspinatus, infraspinatus, and teres minor tendons. The *lesser tubercle*, anteriorly placed and separated by the intertubercular groove from the greater tubercle, is the insertion of the subscapularis tendon. Each of the tubercles is prolonged downward by bony crests, the *crest of the greater tubercle* receiving the tendon of the pectoralis major muscle; the *crest of the lesser tubercle* receives that of the teres major muscle. The *intertubercular groove*, lodging the long tendon of the biceps brachii muscle, also receives the tendon of the latissimus dorsi muscle into its floor.

The *body*, or shaft, of the humerus is somewhat rounded above and prismatic in its lower portion. Above and medially, the coracobrachialis muscle is received near the middle of the shaft; about opposite laterally is the prominent *deltoid tuberosity*. This is continued upward in a V-shaped roughening for the insertion of the deltoid muscle. Just below the deltoid tuberosity, a *groove for the radial nerve* indents the bone posteriorly, spiraling lateralward as it descends. Sharp *lateral* and *medial supracondylar ridges* spring from the respective borders inferiorly and continue into the lateral and medial epicondyles of the humerus. The *inferior extremity* of the bone is flattened anteroposteriorly, and mediolaterally, it is widened by the medial and lateral epicondyles. The *lateral epicondyle* is not conspicuous, but the *medial epicondyle* forms a marked medial projection above the elbow. Projecting somewhat backward, it is grooved behind for the ulnar nerve.

The *articular surfaces* for the radius, ulna, capitulum, and trochlea are directed somewhat forward; consequently, the inferior extremity of the humerus appears to curve anteriorly. The *capitulum* is roughly globular in shape. Smaller than the trochlea, it articulates with the cupped upper surface of the radius. Above it is a shallow fossa, the *radial fossa*, for the reception of the edge of that bone during full flexion of the elbow. The *trochlea* is shaped like a spool, with a deep depression between two well-marked margins. The depression is slightly spiral and receives the central ridge of the trochlear notch of the ulna. The medial rim of the trochlea is the more prominent; the lateral rim is only a small elevation separating the trochlea from the capitulum. Above the trochlea is the *coronoid fossa*, for the reception of the coronoid process of the ulna in front, and the *olecranon fossa*, for the olecranon behind.

The humerus *ossifies* from eight centers of ossification: one for the shaft and seven for the processes—head, greater and lesser tubercles, trochlea, capitulum, lateral epicondyle, and medial epicondyle. The center for the body appears near the middle of the bone in the eighth week of fetal life and then extends toward its extremities. At birth, the humerus is ossified in nearly its whole length; only its extremities remain cartilaginous. Shortly after birth, ossification begins in the head of the bone, followed by the appearance of the centers in the greater and lesser tubercles at 3 to 5 years of age, respectively. By age 6, all these centers have merged into one large epiphysis. The conical end of the proximal extremity of the body fits into this epiphysis, and fusion takes place at ages 18 to 20 in the female, and at ages 20 to 22 in the male. In the lower extremity, secondary centers appear for the capitulum at age 2, for the trochlea at age 9 or 10, and in the lateral epicondyle at ages 13 to 14. These centers unite and fuse with the shaft at about age 13 in females and age 15 in males. The separate center for the medial epicondyle appears at ages 6 to 8 and fuses with the shaft at ages 14 to 16.

Upper Limb

Bones and Joints of Shoulder and Arm

(Continued)

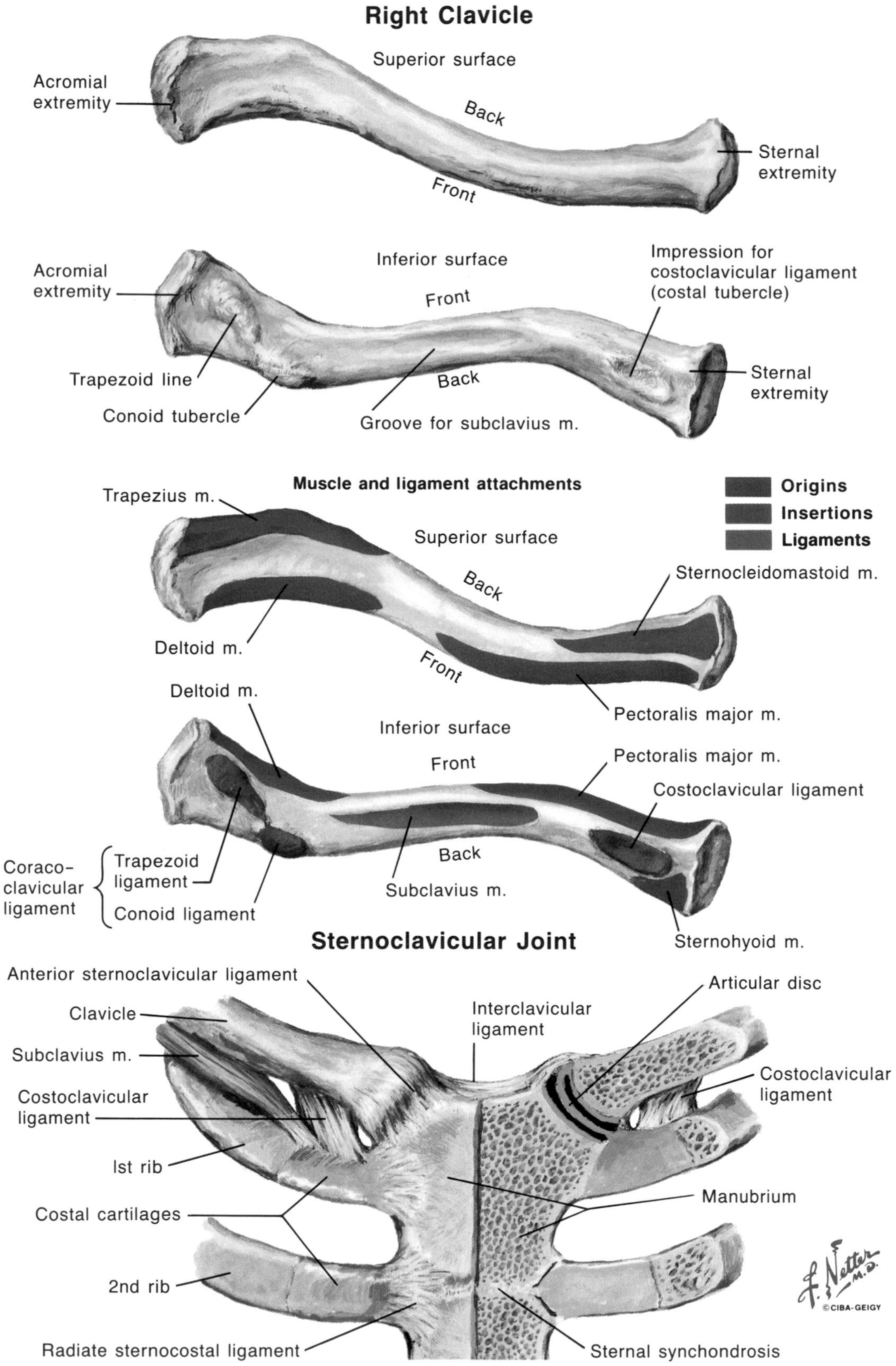

Right Clavicle
Superior surface
Acromial extremity
Back
Front
Sternal extremity
Inferior surface
Acromial extremity
Front
Impression for costoclavicular ligament (costal tubercle)
Trapezoid line
Conoid tubercle
Back
Groove for subclavius m.
Sternal extremity
Muscle and ligament attachments
Origins
Insertions
Ligaments
Trapezius m.
Superior surface
Back
Sternocleidomastoid m.
Deltoid m.
Front
Pectoralis major m.
Deltoid m.
Inferior surface
Front
Pectoralis major m.
Costoclavicular ligament
Coraco-clavicular ligament
Trapezoid ligament
Conoid ligament
Back
Subclavius m.
Sternohyoid m.
Sternoclavicular Joint
Anterior sternoclavicular ligament
Clavicle
Subclavius m.
Costoclavicular ligament
Ist rib
Costal cartilages
2nd rib
Radiate sternocostal ligament
Interclavicular ligament
Articular disc
Costoclavicular ligament
Manubrium
Sternal synchondrosis
F. Netter M.D.
©CIBA-GEIGY

Clavicle

The clavicle is the anterior member of the shoulder girdle and exhibits gentle complementary curvatures (Plate 29). The medial third is related to the first rib; the middle third, to the axillary vessels and the brachial plexus; and the lateral third, to the coracoid process of the scapula and the acromioclavicular joint. The medial two-thirds, roughly triangular in section, is convex forward, whereas the lateral third is flattened and appears quite concave when viewed from the front. To the medial third attach the *pectoralis major* and *sternocleidomastoid muscles*; the *deltoid muscle* arises in the anterior concavity of the lateral third. The *trapezius muscle* takes origin from the posterior border of the lateral half of the superior surface of the bone.

The underside of the clavicle shows prominent markings: the *costal tubercle* medially, for the *costoclavicular ligament*; and the *conoid tubercle* and *trapezoid line* laterally, for the attachment of the two parts of the *coracoclavicular ligament*. Between these prominences is a long groove, which receives the *subclavius muscle*. The *sternal extremity* of the bone is triangular and exhibits a saddle-shaped articular surface, which is received into the clavicular fossa of the manubrium of the sternum. The *acromial extremity* has an oval articular facet, directed lateralward and slightly downward, for the acromion.

The clavicle has no medullary cavity; it consists of trabecular bone within a shell of compact bone. It begins ossification before all other bones. Two primary centers, which later coalesce, appear near the center of the bone at the fifth week of fetal life. A third center for the sternal end appears at about age 17 and fuses with the shaft at about age 25 (or later). This center is the last of the epiphyses of the body to fuse.

Sternoclavicular Joint

The sternoclavicular joint represents the only bony connection between the trunk and the upper limb (Plate 29). The scapula articulates with the clavicle at the acromioclavicular joint, but it is joined to the trunk by muscles only. The sternoclavicular joint simulates in its actions a ball and socket joint, although its rather incongruent surfaces scarcely take the true form of such a joint. The sternal end of the clavicle is applied to an articular fossa formed by the superolateral angle of the manubrium of the sternum and the medial part of the cartilage of the first rib. An articular disc interposed between the surfaces greatly increases the capacity for movement. The articular cartilage of this joint is partly fibrous in nature. An *articular capsule* surrounds the joint and is attached around the clavicular and sternochondral articular surfaces. It is weak below but reinforced above, in front, and behind by capsular ligaments.

The *anterior sternoclavicular ligament* is a broad anterior band of fibers attached to the upper and anterior borders of the sternal end of the clavicle, and below, it is attached to the upper anterior surface of the manubrium of the sternum. This strong band is reinforced by the tendinous origin of the sternocleidomastoid muscle. The *posterior sternoclavicular ligament* has a similar orientation

Upper Limb

Bones and Joints of Shoulder and Arm

(Continued)

Shoulder and Acromioclavicular Joints (from above)

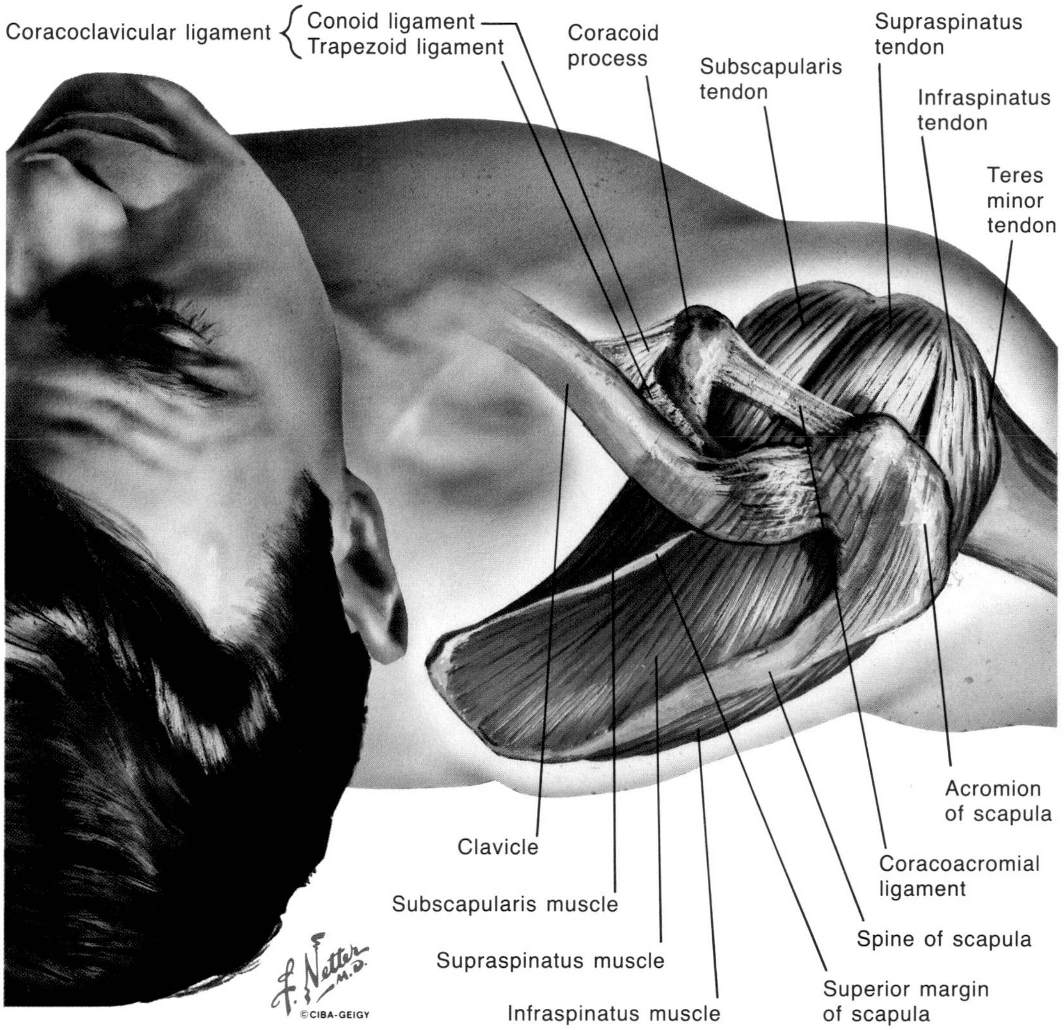
Coracoclavicular ligament
Conoid ligament
Trapezoid ligament
Coracoid process
Subscapularis tendon
Supraspinatus tendon
Infraspinatus tendon
Teres minor tendon
Acromion of scapula
Coracoacromial ligament
Spine of scapula
Superior margin of scapula
Infraspinatus muscle
Supraspinatus muscle
Subscapularis muscle
Clavicle
F. Netter M.D.
©CIBA-GEIGY

on the back of the capsule and has similar bony attachments. The *interclavicular ligament* strengthens the capsule above. It passes from clavicle to clavicle and is attached to the upper border of the sternum between. The *costoclavicular ligament* is a short, flat band of fibers running between the cartilage of the first rib and the costal tuberosity on the undersurface of the clavicle. The tendon of the subclavius muscle inserts adjacent to and in front of the ligament.

The *articular disc* is a flat circular disc of fibrocartilage that separates the joint into two synovial cavities. It is attached at its circumference to the capsule, but more importantly, it has strong attachments to the upper border of the sternal end of the clavicle, and below, it ends in the cartilage of the first rib. The disc cushions the forces transmitted into the joint from the shoulder and compensates for the incongruities of the articulating surfaces. More fibrous at its circumference, it acts like a ligament to hold the sternal end of the clavicle down into the articular fossa, thus resisting tendencies to upward and medialward dislocation of the clavicle. The clavicle is seldom dislocated; it is more commonly fractured.

The anterior supraclavicular nerve gives the sternoclavicular joint its nerve supply. Blood supply is derived from branches of the internal thoracic artery, the superior thoracic artery, and the clavicular branch of the thoracoacromial artery.

Acromioclavicular Joint

A small synovial joint, the acromioclavicular joint lies between the oval articular surface of the lateral end of the clavicle and the medial border of the acromion process of the scapula (Plate 30).

The plane joint surfaces, covered by *fibrous articular cartilage*, slope downward and medially, favoring displacement of the acromion downward and under the clavicle. The end of the clavicle rides higher than the acromion and is readily palpable. An *articular capsule* encloses the joint, attaching at the articular margins. It is stoutest above, where it is reinforced by the fibers of the trapezius muscle. A wedge-shaped articular disc dips into the joint from the superior part of the capsule. The synovial membrane lines the inner surface of the capsule.

The *coracoclavicular ligament*, although separated medially from the joint, provides principally for its stability. Its two parts, named by their shapes, are the posteromedial *conoid ligament* and the anterolaterally placed *trapezoid ligament*. The triangular conoid ligament has its apex downward, attached posteromedially to the base of the coracoid process. Its broadened base is fixed to the conoid tubercle on the underside of the clavicle. The trapezoid ligament, strong, flat, and quadrilateral, springs from the oblique trapezoid line on the underside of the clavicle and is attached below for about 2 cm to a rough ridge on the upper surface of the coracoid process.

Movements of Shoulder Girdle

The clavicle forms a strut, holding the point of the shoulder well out from the trunk and thereby facilitating its free movement. The clavicle is also the radius about which the shoulder may be moved, the sternoclavicular joint being its pivot point (Plate 29).

The acromioclavicular joint allows sufficient movement to permit the independent turning of the glenoid fossa in directions favorable for the humerus and the remainder of the limb. Many of the axes around which movement takes place fall within the coracoclavicular ligament. The freedom of movement of the clavicle is considerable; in full elevation of the limb, the clavicle can be raised to a 60° angle or more from its usual horizontal position. The totally muscular suspension of the scapula makes this bone primary in movements of the shoulder girdle, the clavicle (except as it is pulled on by the trapezius and certain other muscles) following and accommodating to the position and orientation of the scapula. The last 30° of scapular rotation is allowed by rotation of the clavicle on its long axis.

Scapular Ligaments

The *coracoacromial ligament* is stretched between the acromion and the coracoid process. Its broad base is attached to the lateral border of the coracoid process; its apex, to the tip of the acromion. The ligament is thicker at its margins, and occasionally an actual gap exists in its center. This strong triangular ligament completes, with the bony members, a bony and ligamentous arch above the head of the humerus and contributes stability to the shoulder joint.

The *superior transverse scapular ligament* converts the scapular notch into a foramen. It is a thin, flat band attached at one extremity to the base of the coracoid process and, at the other extremity, to the medial border of the scapular notch. The suprascapular nerve runs beneath the ligament; the suprascapular vessels pass over it.

The *inferior transverse scapular ligament*, when present, extends from the lateral aspect of the root of the spine of the scapula to the margin of the glenoid cavity. With the bone, it forms a foramen for passage of the suprascapular vessels and nerve into the infraspinatous fossa.

Upper Limb

Bones and Joints of Shoulder and Arm

(Continued)

Shoulder Joint

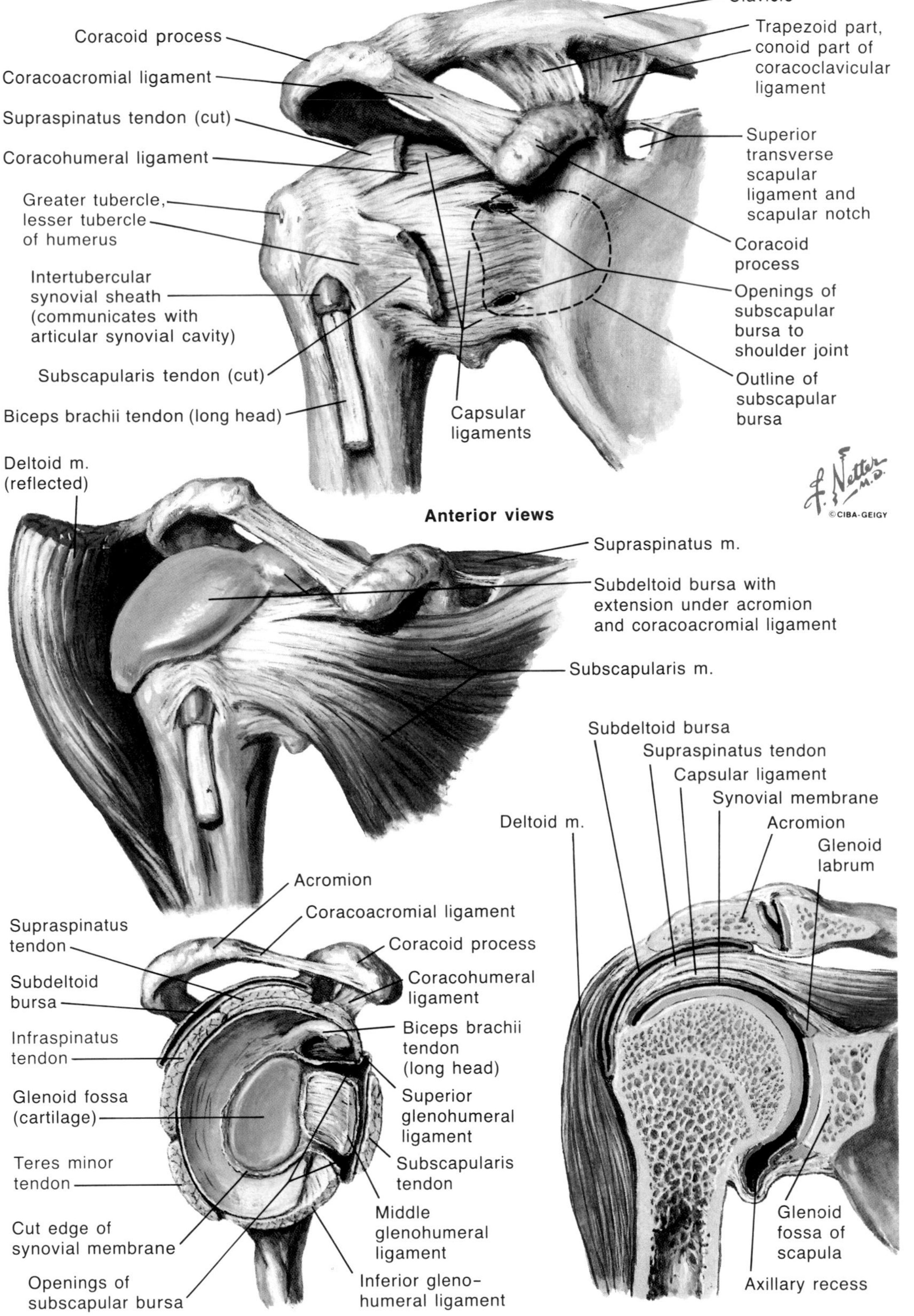
Clavicle
Coracoid process
Trapezoid part, conoid part of coracoclavicular ligament
Coracoacromial ligament
Supraspinatus tendon (cut)
Superior transverse scapular ligament and scapular notch
Coracohumeral ligament
Greater tubercle, lesser tubercle of humerus
Coracoid process
Intertubercular synovial sheath (communicates with articular synovial cavity)
Openings of subscapular bursa to shoulder joint
Subscapularis tendon (cut)
Outline of subscapular bursa
Biceps brachii tendon (long head)
Capsular ligaments
Deltoid m. (reflected)
Supraspinatus m.
Subdeltoid bursa with extension under acromion and coracoacromial ligament
Subscapularis m.
Subdeltoid bursa
Supraspinatus tendon
Capsular ligament
Synovial membrane
Deltoid m.
Acromion
Glenoid labrum
Acromion
Coracoacromial ligament
Supraspinatus tendon
Coracoid process
Subdeltoid bursa
Coracohumeral ligament
Infraspinatus tendon
Biceps brachii tendon (long head)
Glenoid fossa (cartilage)
Superior glenohumeral ligament
Teres minor tendon
Subscapularis tendon
Middle glenohumeral ligament
Glenoid fossa of scapula
Cut edge of synovial membrane
Openings of subscapular bursa
Inferior gleno-humeral ligament
Axillary recess
F. Netter M.D.
©CIBA-GEIGY

Anterior views

Shoulder joint opened (lateral view)

Coronal section through shoulder joint

Glenohumeral Joint

There is great freedom of motion in the shoulder joint (Plate 31). As it is a ball-and-socket joint, movement can take place around an infinite number of axes intersecting in the head of the humerus. The combination movement of circumduction is also free.

There is correlated activity in most movements between the glenohumeral joint and the joints of the shoulder girdle, shoulder girdle movement often contributing one-third of the total excursion.

The great freedom of movement of the glenohumeral joint is inevitably accompanied by a considerable loss of stability. The large humeral head articulates against a glenoidal surface only a little more than one-third its size, and the joint has a loose articular capsule. The head of the humerus is held into the glenoid cavity by an "articular cuff" of short scapular muscles—the supraspinatus, infraspinatus, teres minor, and subscapularis muscles. The tendons of these muscles blend with the capsule and reinforce it. The glenoid cavity of the scapula is deepened and enlarged by a *glenoid labrum*, a fibrocartilaginous rim attached around the margin of the glenoid process. This labrum is triangular in cross section, with a free edge that is thin and sharp.

The *articular capsule* of the shoulder forms a loose, cylindric sleeve that encloses the articular parts of the bones. It attaches to the scapula outside of the glenoid labrum, and partly to the labrum itself, especially above and behind. On the humerus, the capsule attaches to the anatomic neck immediately medial to the tubercles; below, it extends onto the medial surface of the shaft of the bone a little below the articular head. There are two openings in the capsule; the opening at the upper end of the intertubercular groove allows for the passage of the tendon of the long head of the biceps brachii muscle. The other opening is an anterior communication of the joint cavity with the subscapular bursa. The *synovial membrane* extends from the margin of the glenoid cavity and lines the capsule to the limits of the articular cartilage of the humerus. It extends into the subscapular bursa and also forms the intertubercular synovial sheath on the tendon of the biceps brachii muscle.

Glenohumeral Ligaments. The superior, middle, and inferior glenohumeral ligaments are thickenings in the anterior wall of the articular capsule (Plate 31). Really visible only on the inner aspect of the capsule, they radiate from the anterior glenoid margin adjacent to and extending downward from the supraglenoid tubercle of the scapula.

The *superior glenohumeral ligament* is slender, arises immediately anterior to the attachment of the tendon of the long head of the biceps brachii muscle, and parallels that tendon to end near the upper end of the lesser tubercle of the humerus. The *middle glenohumeral ligament* arises next to the superior ligament and reaches the humerus at the front of the lesser tubercle and just inferior to the insertion of the subscapularis muscle. It has an oblique course immediately inferior to the opening of the subscapular bursa. The *inferior glenohumeral ligament* arises from the scapula directly below the notch in the anterior border of the glenoidal process of the scapula and descends to the underside of the neck of the humerus. The latter two ligaments may be poorly separated.

The *coracohumeral ligament*, partly continuous with the articular capsule, is a broad band arising from the lateral border of the coracoid process. Flattening, it blends with the upper and posterior part of the capsule and ends in the anatomic neck of the humerus adjacent to the greater tubercle.

Arteries and Nerves. The arterial supply of the shoulder joint includes branches of the suprascapular, anterior and posterior circumflex humeral, and circumflex scapular arteries. Its nerves are branches of the suprascapular, axillary, and subscapular nerves. □

Upper Limb

Muscles of Arm

Arm Muscles With Portions of Arteries and Nerves (anterior view)

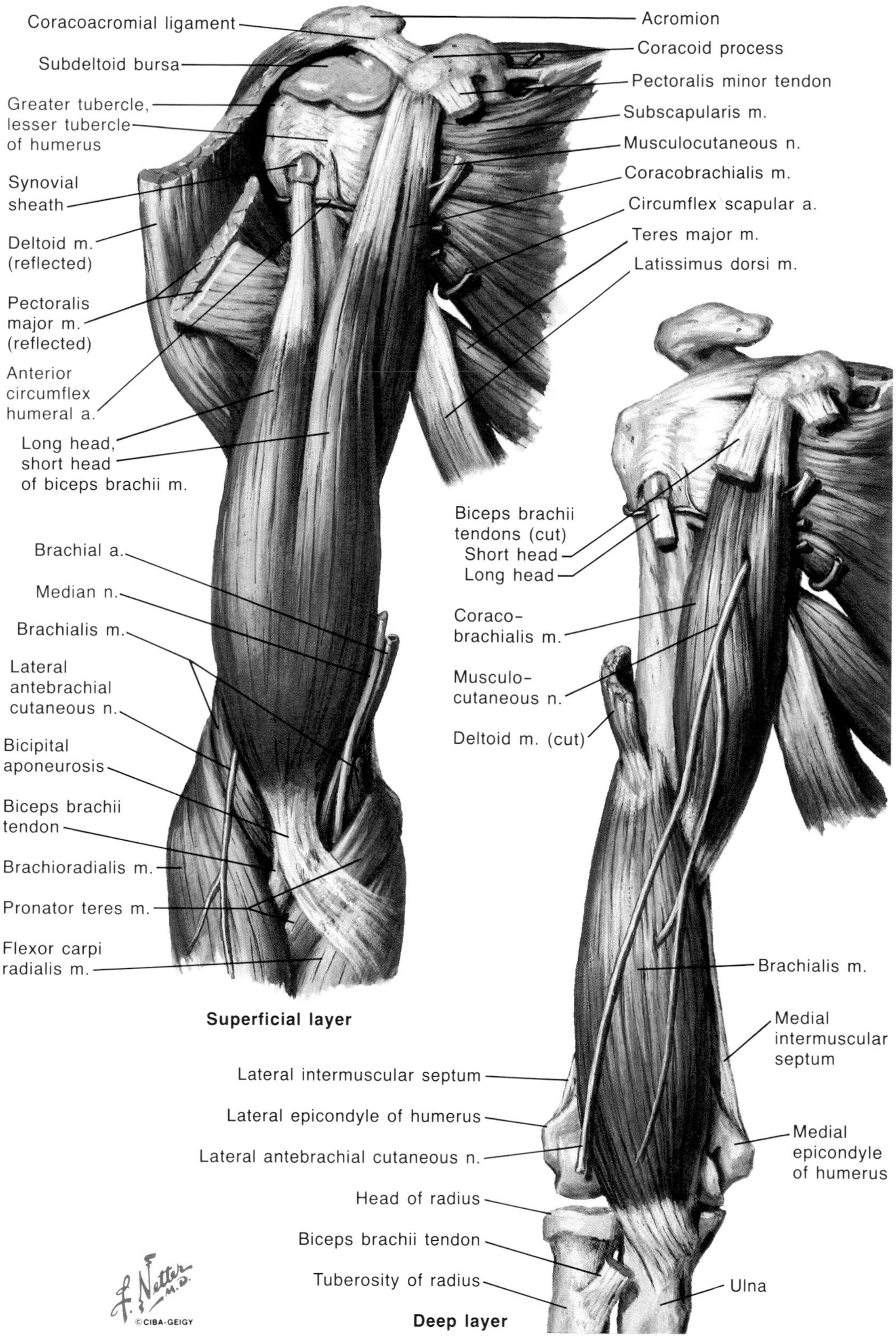
Coracoacromial ligament
Subdeltoid bursa
Greater tubercle, lesser tubercle of humerus
Synovial sheath
Deltoid m. (reflected)
Pectoralis major m. (reflected)
Anterior circumflex humeral a.
Long head, short head of biceps brachii m.
Brachial a.
Median n.
Brachialis m.
Lateral antebrachial cutaneous n.
Bicipital aponeurosis
Biceps brachii tendon
Brachioradialis m.
Pronator teres m.
Flexor carpi radialis m.
Acromion
Coracoid process
Pectoralis minor tendon
Subscapularis m.
Musculocutaneous n.
Coracobrachialis m.
Circumflex scapular a.
Teres major m.
Latissimus dorsi m.
Superficial layer
Biceps brachii tendons (cut)
Short head
Long head
Coraco-brachialis m.
Musculo-cutaneous n.
Deltoid m. (cut)
Brachialis m.
Medial intermuscular septum
Lateral intermuscular septum
Lateral epicondyle of humerus
Lateral antebrachial cutaneous n.
Medial epicondyle of humerus
Head of radius
Biceps brachii tendon
Tuberosity of radius
Ulna
Deep layer
F. Netter M.D.
©CIBA-GEIGY

The arm, or brachium, is the region between the shoulder joint and the elbow. The arm muscles are few, and they are served by certain of the terminal branches of the brachial plexus and portions of the great vascular channels of the limb (Plates 32–34).

Brachial Fascia

A strong tubular investment of the deeper parts of the arm, the brachial fascia is continuous above with the pectoral and axillary fasciae and with the fascial covering of the deltoid and latissimus dorsi muscles. Below, the brachial fascia is attached to the epicondyles of the humerus and to the olecranon, and then is continuous with the antebrachial fascia. It is perforated for the passage of the basilic vein, the medial antebrachial cutaneous nerve, and for many lesser nerves and vessels.

Two *intermuscular septa* are prolonged upward from the epicondylar attachments of the brachial fascia. These blend with the periosteum of the humerus along its supracondylar ridges and borders and fuse peripherally with the brachial fascia to form the *anterior* and *posterior compartments* of the arm. Above, the *lateral intermuscular septum* ends at the insertion of the deltoid muscle; the medial intermuscular septum ends in continuity with the fascia of the coracobrachialis muscle. The *medial intermuscular septum* has an additional, weaker anterior lamina, and the anterior and posterior laminae together with the brachial fascia form the *neurovascular compartment* of the arm (Plate 34).

Muscles

The muscles of the arm are separated both positionally and functionally by the humerus and the intermuscular septa into an anterior and a posterior group (Plates 32–34). The anterior group comprises the coracobrachialis, biceps brachii, and brachialis muscles. The posterior group muscles are the triceps brachii and anconeus muscles.

Coracobrachialis Muscle. A short, bandlike muscle of the upper arm, the coracobrachialis arises from the tip of the coracoid process in common with the short head of the biceps brachii muscle. It inserts by a flat tendon into the medial surface of the humerus just proximal to its midlength (Plate 27). The musculocutaneous nerve supplies the coracobrachialis muscle and passes diagonally through the muscle at its midlength.

Biceps Brachii Muscle. The biceps brachii is a long, fusiform muscle of the anterior aspect of the arm. Its *long head* arises as a rounded tendon from

Upper Limb

Muscles of Arm

(Continued)

Arm Muscles With Portions of Arteries and Nerves (posterior view)

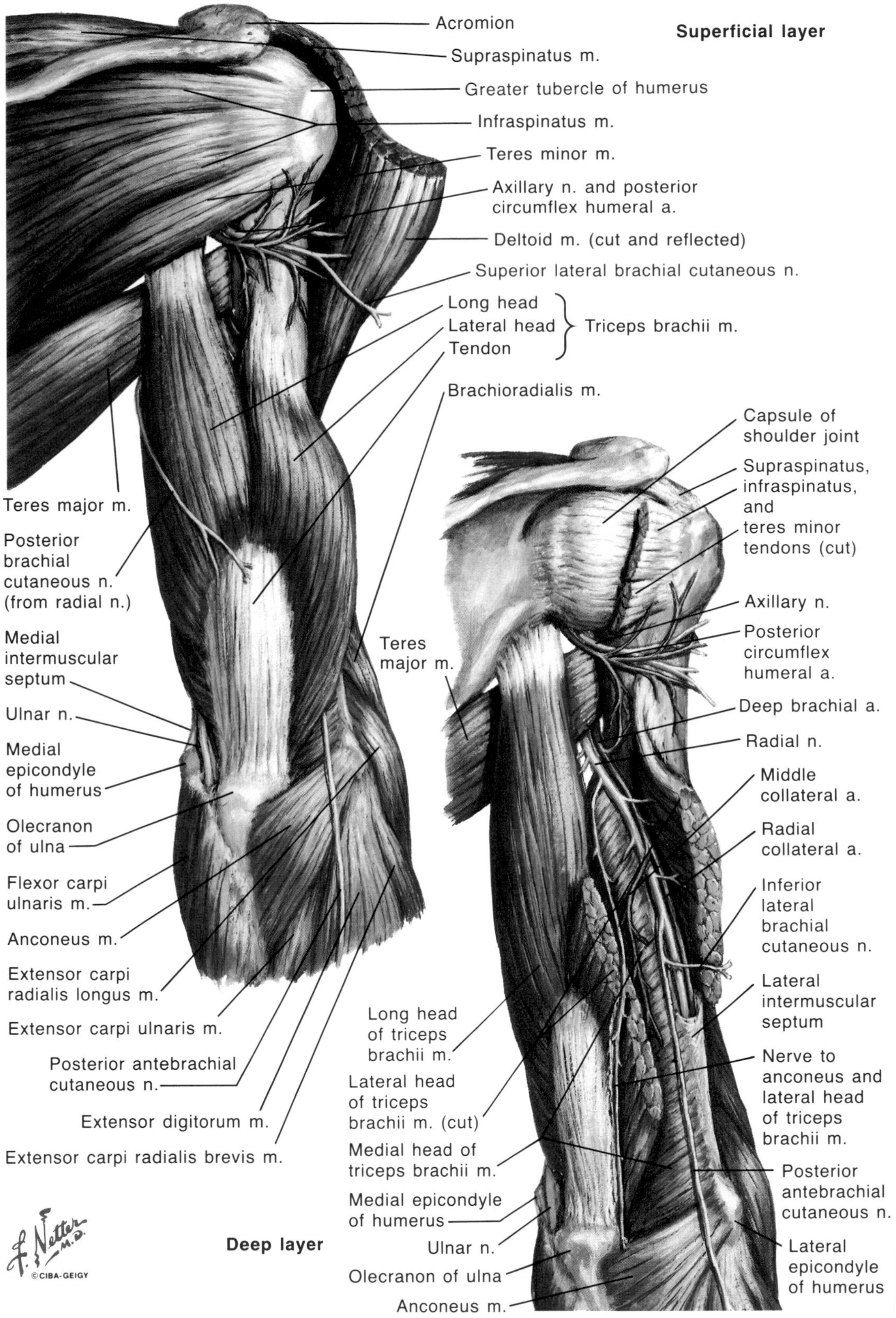
Superficial layer
Acromion
Supraspinatus m.
Greater tubercle of humerus
Infraspinatus m.
Teres minor m.
Axillary n. and posterior circumflex humeral a.
Deltoid m. (cut and reflected)
Superior lateral brachial cutaneous n.
Long head
Lateral head
Tendon
Triceps brachii m.
Brachioradialis m.
Teres major m.
Posterior brachial cutaneous n. (from radial n.)
Medial intermuscular septum
Ulnar n.
Medial epicondyle of humerus
Olecranon of ulna
Flexor carpi ulnaris m.
Anconeus m.
Extensor carpi radialis longus m.
Extensor carpi ulnaris m.
Posterior antebrachial cutaneous n.
Extensor digitorum m.
Extensor carpi radialis brevis m.
Deep layer
Capsule of shoulder joint
Supraspinatus, infraspinatus, and teres minor tendons (cut)
Axillary n.
Posterior circumflex humeral a.
Teres major m.
Deep brachial a.
Radial n.
Middle collateral a.
Radial collateral a.
Inferior lateral brachial cutaneous n.
Lateral intermuscular septum
Long head of triceps brachii m.
Lateral head of triceps brachii m. (cut)
Nerve to anconeus and lateral head of triceps brachii m.
Medial head of triceps brachii m.
Posterior antebrachial cutaneous n.
Medial epicondyle of humerus
Ulnar n.
Lateral epicondyle of humerus
Olecranon of ulna
Anconeus m.
F. Netter M.D.
©CIBA-GEIGY

the supraglenoid tubercle of the scapula, crosses the head of the humerus within the capsule of the shoulder joint, and emerges from that capsule covered by the intertubercular synovial sheath. The *short head* of the biceps brachii muscle arises by a thick, flattened tendon from the tip of the coracoid process, in common with the coracobrachialis muscle.

The two bellies of the biceps brachii muscle unite at about the middle of the arm to form the most prominent muscle of the anterior compartment. The tendon of insertion is a strong, vertical cord palpable down the center of the cubital fossa. Here, its deeper part turns its anterior surface lateralward to end on the tuberosity of the radius, separated from the anterior part of the tuberosity by the small *bicipitoradial bursa*. The variable *interosseous cubital bursa* may separate the tendon from the ulna and its covering muscles.

The *bicipital aponeurosis*, formed of the more anterior and medial tendon fibers of the muscle, arises at the bend of the elbow and passes obliquely over the brachial artery and median nerve to blend with the antebrachial fascia over the flexor group of the forearm (Plate 32). The pull of the bicipital aponeurosis is largely exerted on the ulna.

Brachialis Muscle. This muscle arises from the lower half of the anterior surface of the humerus and the two intermuscular septa. Its upper extremity has two pointed processes placed on either side of the insertion of the deltoid muscle (Plate 27). The muscular fibers converge to a thick tendon, which adheres to the capsule of the elbow joint and inserts on the tuberosity of the ulna and on the anterior surface of its coronoid process. This muscle bulges beyond the biceps brachii muscle on either side, and anterior to its medial border lie the brachial vessels and the median nerve.

Triceps Brachii Muscle. This large muscle of three heads occupies the entire dorsum of the arm. The *long head* arises by a strong tendon from the infraglenoid tubercle of the humerus. Its belly descends between the teres major and teres minor muscles (separating the triangular and quadrangular spaces), and joins the lateral and medial heads in a common insertion. The *lateral head* takes origin from the posterior surface and lateral

Upper Limb

Muscles of Arm

(Continued)

Cross-Sectional Anatomy of Right Arm

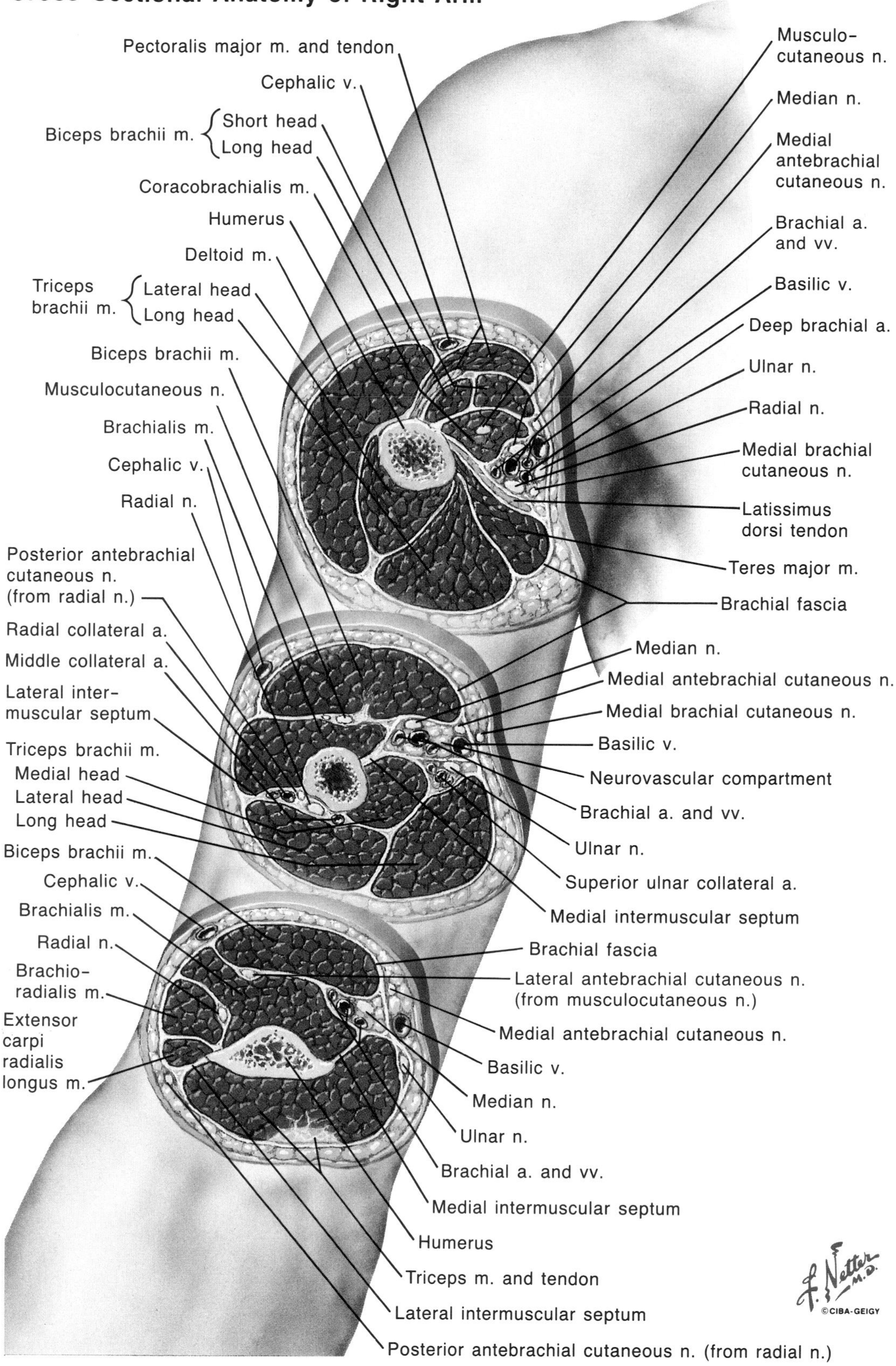

Pectoralis major m. and tendon
Cephalic v.
Biceps brachii m.
Short head
Long head
Coracobrachialis m.
Humerus
Deltoid m.
Triceps brachii m.
Lateral head
Long head
Biceps brachii m.
Musculocutaneous n.
Brachialis m.
Cephalic v.
Radial n.
Posterior antebrachial cutaneous n. (from radial n.)
Radial collateral a.
Middle collateral a.
Lateral inter-muscular septum
Triceps brachii m.
Medial head
Lateral head
Long head
Biceps brachii m.
Cephalic v.
Brachialis m.
Radial n.
Brachio-radialis m.
Extensor carpi radialis longus m.
Musculo-cutaneous n.
Median n.
Medial antebrachial cutaneous n.
Brachial a. and vv.
Basilic v.
Deep brachial a.
Ulnar n.
Radial n.
Medial brachial cutaneous n.
Latissimus dorsi tendon
Teres major m.
Brachial fascia
Median n.
Medial antebrachial cutaneous n.
Medial brachial cutaneous n.
Basilic v.
Neurovascular compartment
Brachial a. and vv.
Ulnar n.
Superior ulnar collateral a.
Medial intermuscular septum
Brachial fascia
Lateral antebrachial cutaneous n. (from musculocutaneous n.)
Medial antebrachial cutaneous n.
Basilic v.
Median n.
Ulnar n.
Brachial a. and vv.
Medial intermuscular septum
Humerus
Triceps m. and tendon
Lateral intermuscular septum
Posterior antebrachial cutaneous n. (from radial n.)
F. Netter M.D.
©CIBA-GEIGY

border of the humerus above and lateral to the radial groove, and from the lateral intermuscular septum. Crossing the groove and concealing the radial nerve and deep brachial vessels, its fibers join in the common tendon of insertion.

The *medial head* arises from the humerus entirely medial and below the radial groove from as high as the insertion of the teres major muscle to as low as the olecranon fossa of the humerus (Plate 28). It also takes origin from the entire length of the medial intermuscular septum and from the lateral septum below the radial nerve groove. The medial head is deep to the other heads and is hidden by them. The tendon of the muscle appears as a flat band covering its distal two-fifths. It inserts on the posterior part of the olecranon and into the deep fascia of the forearm on either side of it.

Anconeus Muscle. This is a small, triangular muscle, arising from the lateral epicondyle of the humerus. Its fibers diverge from this origin and insert into the side of the olecranon and the adjacent one-fourth of the posterior surface of the ulna. The muscle is deep to the dorsal antebrachial fascia and extends across the elbow and the superior radioulnar joints.

Muscle Actions

The principal movements produced by the muscles of the arm are flexion and extension of the forearm at the elbow. The brachialis and biceps brachii muscles are the principal flexors. In this action, the brachialis muscle is always active; the biceps brachii muscle becomes active against resistance and is most effective when flexion of the forearm is combined with supination. It is a powerful supinator of the forearm. Extension of the forearm is produced by the triceps brachii muscle, assisted by the anconeus muscle. The medial head of the triceps brachii muscle is usually active, and the lateral and long heads are recruited for extra power.

Certain heads of these muscles are active at the shoulder joint. The long head of the biceps brachii muscle flexes the arm at the shoulder, and its tendon aids in stabilization of the joint. The long head of the triceps brachii muscle assists in extension and adduction of the arm. □

Upper Limb

Nerves of Arm

Musculocutaneous Nerve (C5, 6, 7)

(only muscles innervated by musculocutaneous nerve are depicted)

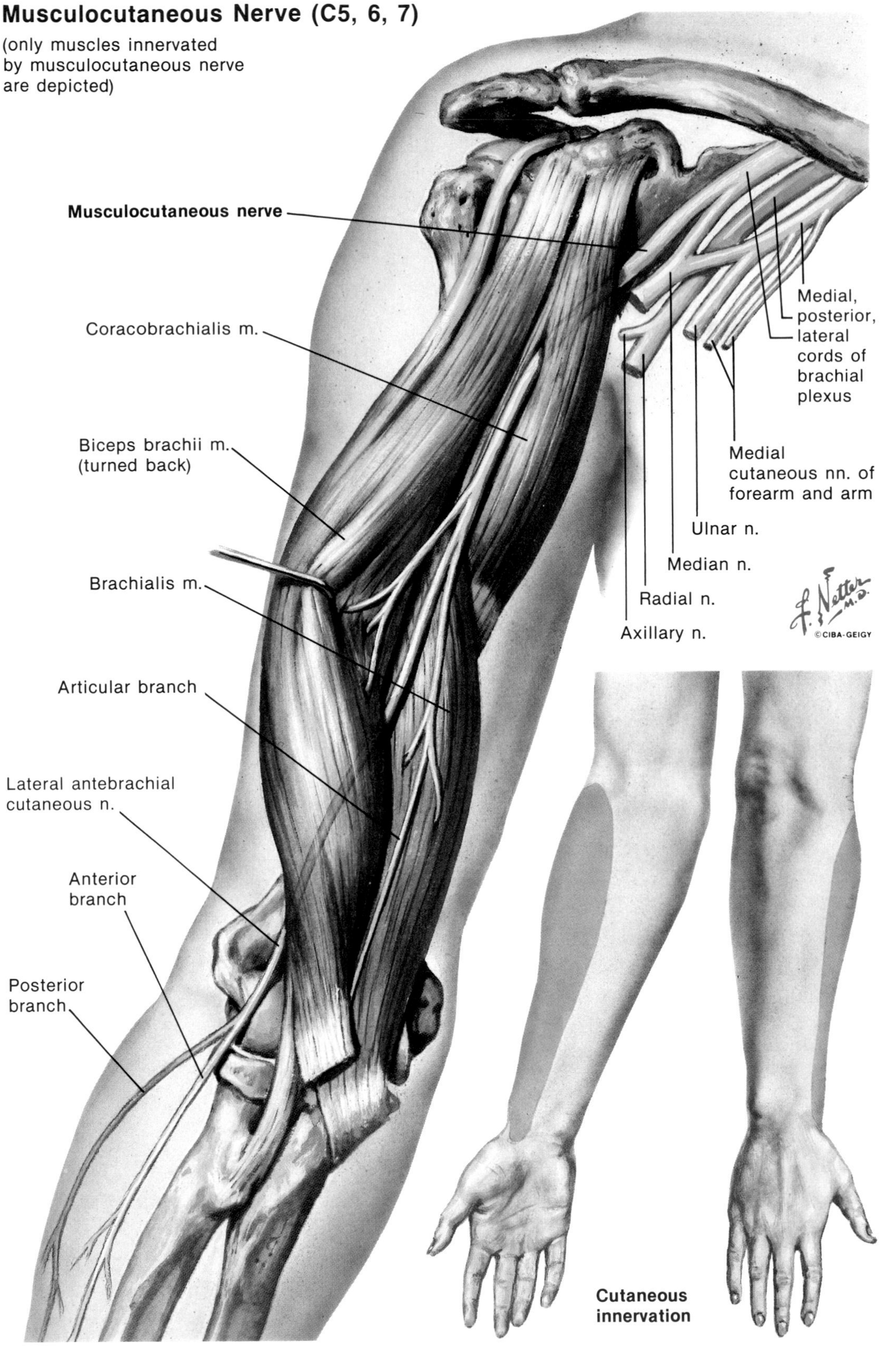

Musculocutaneous nerve
Coracobrachialis m.
Biceps brachii m.
(turned back)
Brachialis m.
Articular branch
Lateral antebrachial cutaneous n.
Anterior branch
Posterior branch
Medial, posterior, lateral cords of brachial plexus
Medial cutaneous nn. of forearm and arm
Ulnar n.
Median n.
Radial n.
Axillary n.
F. Netter M.D.
©CIBA-GEIGY
Cutaneous innervation

The terminal branches of the brachial plexus (the musculocutaneous, median, ulnar, and radial nerves) provide the entire nerve supply to the limb below the shoulder (Plate 25). Of these, only the musculocutaneous and radial nerves distribute to the muscles of the arm.

Musculocutaneous Nerve

The *musculocutaneous nerve* (C[4], 5, 6, 7), a branch of the *lateral cord* of the brachial plexus, arises opposite the lower border of the pectoralis minor muscle (Plate 35). This nerve is the principal motor nerve of the anterior (flexor) compartment of the arm. It continues into the forearm as a cutaneous nerve, the lateral antebrachial cutaneous nerve. The nerve lies between the axillary artery and the coracobrachialis muscle, which it perforates and supplies. Continuing downward, it runs between the biceps brachii and brachialis muscles, supplying branches to both heads of the biceps brachii muscle and most of the brachialis muscle, and often communicating with the median nerve. In this part of its course, the musculocutaneous nerve inclines gradually toward the lateral side of the arm; at about the level of the elbow joint, it passes between the biceps brachii and the brachioradialis muscles to pierce the deep fascia and become the *lateral antebrachial cutaneous nerve*.

The branch supplying the coracobrachialis muscle derives its fibers from C7 and usually arises from the main nerve before that nerve penetrates the muscle. Occasionally, the branch comes directly from the lateral cord of the brachial plexus. The branches to both heads of the biceps brachii muscle and to the brachialis muscle arise from the nerve after it has emerged from the coracobrachialis muscle.

The branch supplying the brachialis muscle subdivides and descends to help in the innervation of the elbow joint; other filaments supply the brachial artery and the deep brachial artery and its nutrient humeral branch. Fibers innervating the periosteum on the distal anterior aspect of the humerus are reputedly distributed with these vascular filaments.

The lateral antebrachial cutaneous nerve passes deep to the cephalic vein and soon divides into anterior and posterior branches. The *anterior branch* descends along the anterior aspect of the radial side of the forearm to the wrist and ends at the base of the thenar eminence. At the wrist, it lies in front of the radial artery and gives off branches that penetrate the deep fascia to supply this part of the artery. The terminal branches of the anterior branch of the lateral antebrachial cutaneous nerve communicate with corresponding branches from the palmar cutaneous branch of the median nerve. The *posterior branch* is smaller. It curves around the radial border of the forearm and breaks up into branches that supply a variable area of skin and fascia over the back of the forearm. These branches also communicate with branches of the posterior antebrachial cutaneous nerve and with the superficial terminal branch of the radial nerve.

The areas of skin supplied by the lateral antebrachial cutaneous nerve include sensory receptors, hairs, arrectores pilorum muscles, glands, and vessels. However, these terminal cutaneous branches show considerable individual variation in the territories they supply.

Upper Limb

Nerves of Arm

(Continued)

Scapular Nerves

Dorsal Scapular Nerve (C5). Arising from the uppermost root of the brachial plexus, this nerve pierces the middle scalene muscle, runs deep to and helps to supply the levator scapulae, and ends by supplying the minor and major rhomboideus muscles (Plate 36).

Suprascapular Nerve (C5, 6). This nerve arises from the superior trunk of the brachial plexus. It runs outward, deep to the trapezius muscle, enters the supraspinatous fossa through the suprascapular notch, and winds around the lateral border of the spine of the scapula to reach the infraspinatous fossa. It supplies the supraspinatus and infraspinatus muscles and also sends branches to the shoulder and acromioclavicular joints and suprascapular vessels.

Axillary Nerve (C5, 6). The axillary nerve arises from the posterior cord of the brachial plexus. Descending behind the axillary vessels, it curves posteriorly alongside the posterior circumflex humeral vessels and below the subscapularis muscle and the capsule of the shoulder joint, giving a twig to the joint. It then divides into anterior and posterior branches. The *anterior branch* passes between the surgical neck of the humerus and the deltoid muscle, which it supplies, and gives off small branches to the skin covering its lower half. The *posterior branch* also gives branches to the deltoid muscle, plus a branch to the teres minor muscle, and terminates as the *superior lateral brachial cutaneous nerve*.

Radial Nerve (C5, 6, 7, 8; T1). The largest branch of the brachial plexus, this nerve is the main continuation of its *posterior cord*. In the axilla, it lies behind the outer end of the axillary artery on the subscapularis, latissimus dorsi, and teres major muscles. Leaving the axilla, it enters the arm between the brachial artery and the long head of the triceps brachii muscle.

Continuing downward and accompanied by the deep brachial artery, the nerve pursues a spiral course behind the humerus, lying close to the bone in the shallow radial nerve sulcus. It passes between the long and medial and medial and lateral heads of the triceps brachii muscle and then lies deep to the lateral head. On reaching the distal third of the arm at the lateral margin of the humerus, it pierces the lateral intermuscular septum to enter the anterior compartment of the arm. Then it descends anterior to the lateral epicondyle of the humerus and the articular capsule of the elbow joint, lying deep in the furrow between the brachialis muscle medially and the brachioradialis and extensor carpi radialis longus muscles laterally. At this point, it divides into its *deep* and *superficial branches*.

In the axilla, the radial nerve gives off the small *posterior brachial cutaneous nerve* and a muscular branch to the long head of the triceps brachii muscle.

Dorsal Scapular Nerve (C5), Suprascapular Nerve (C5, 6), Axillary Nerve (C5, 6), and Radial Nerve (C5, 6, 7, 8; T1) Above Elbow (viewed from behind)

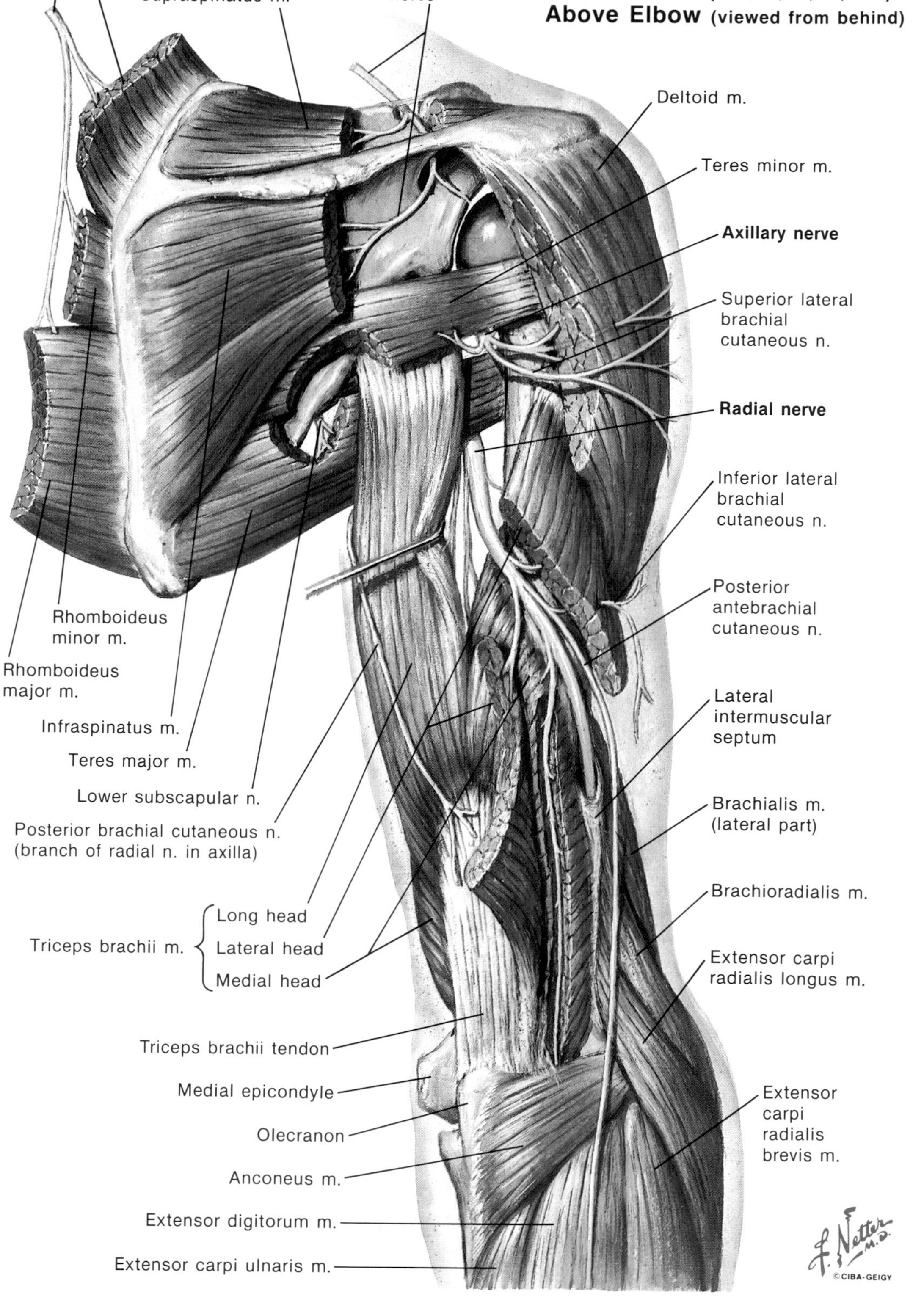

Dorsal scapular nerve
Levator scapulae m. (supplied also by branches from C5 and C6)
Supraspinatus m.
Suprascapular nerve
Deltoid m.
Teres minor m.
Axillary nerve
Superior lateral brachial cutaneous n.
Radial nerve
Inferior lateral brachial cutaneous n.
Posterior antebrachial cutaneous n.
Lateral intermuscular septum
Brachialis m. (lateral part)
Brachioradialis m.
Extensor carpi radialis longus m.
Extensor carpi radialis brevis m.
Rhomboideus minor m.
Rhomboideus major m.
Infraspinatus m.
Teres major m.
Lower subscapular n.
Posterior brachial cutaneous n. (branch of radial n. in axilla)
Triceps brachii m.
Long head
Lateral head
Medial head
Triceps brachii tendon
Medial epicondyle
Olecranon
Anconeus m.
Extensor digitorum m.
Extensor carpi ulnaris m.
F. Netter M.D.
©CIBA-GEIGY

In the arm, the radial nerve supplies muscular, cutaneous, vascular, articular, and osseous branches. The first muscular branch is long and slender, arising as the nerve enters the radial nerve sulcus; it accompanies the ulnar nerve to the lower arm to supply the distal part of the medial head of the triceps brachii muscle and to furnish twigs to the elbow joint. A second, larger branch arises from the nerve as it lies in the radial nerve sulcus; it soon subdivides into smaller branches that enter the medial head of the triceps brachii muscle, with some twigs to the humeral periosteum and bone. A stouter subdivision supplies the lateral head of the triceps brachii muscle. It descends through the muscle accompanied by the medial branch of the deep brachial artery. It then penetrates and supplies the anconeus muscle and sends branches to the humerus and the elbow joint.

Anterior to the lateral intermuscular septum, the radial nerve gives muscular branches to the lateral part of the brachialis, brachioradialis, and extensor carpi radialis longus muscles, and, occasionally, to the extensor carpi radialis brevis. Vascular branches and twigs are furnished to the elbow joint.

Three cutaneous branches arise from the radial nerve above the elbow—the *posterior brachial cutaneous*, *inferior lateral brachial cutaneous*, and *posterior antebrachial cutaneous nerves*. □

Upper Limb

Blood Supply of Arm

Brachial Artery With Related Nerves and Muscles

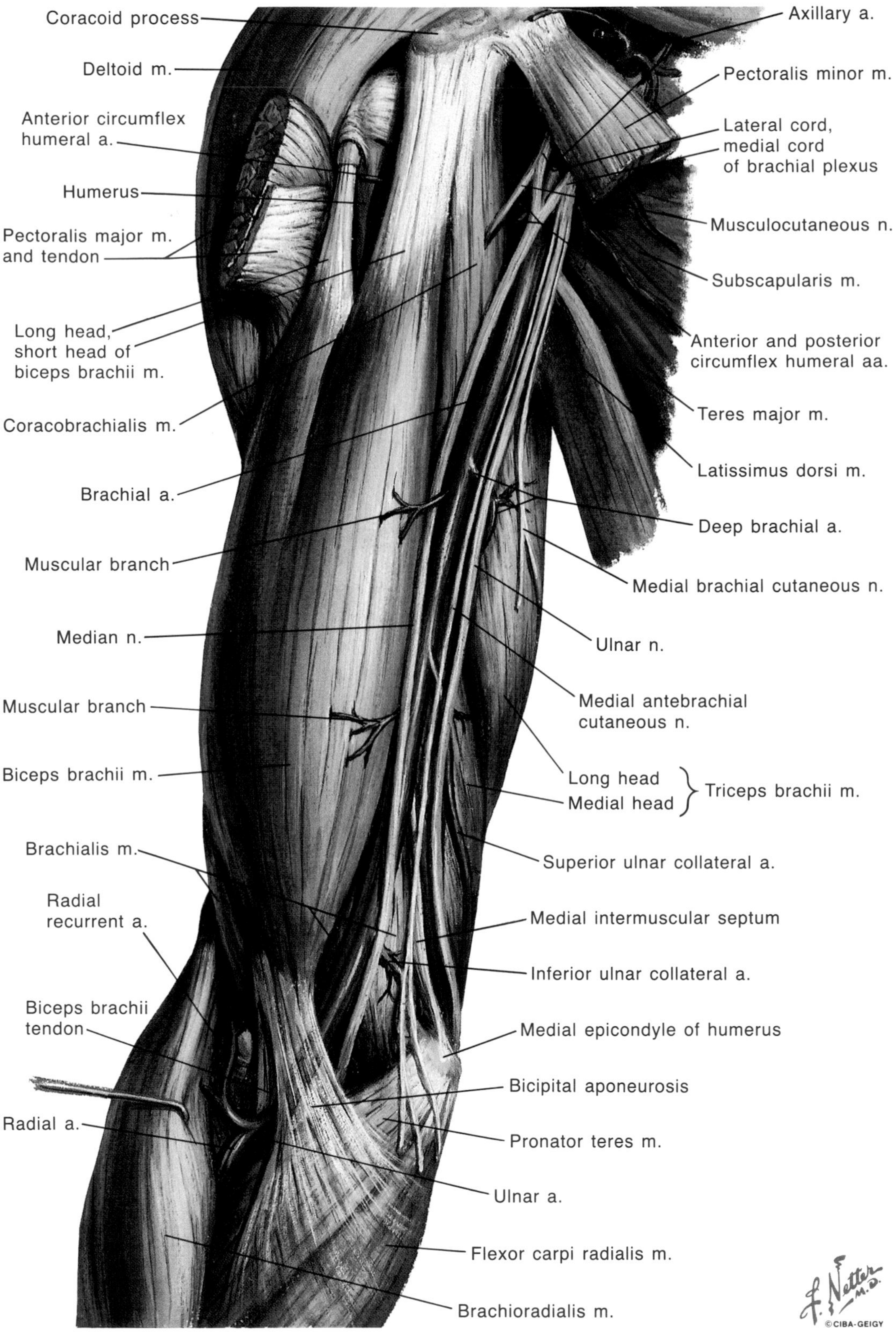
Coracoid process
Axillary a.
Deltoid m.
Pectoralis minor m.
Anterior circumflex humeral a.
Lateral cord, medial cord of brachial plexus
Humerus
Pectoralis major m. and tendon
Musculocutaneous n.
Subscapularis m.
Long head, short head of biceps brachii m.
Anterior and posterior circumflex humeral aa.
Coracobrachialis m.
Teres major m.
Latissimus dorsi m.
Brachial a.
Deep brachial a.
Muscular branch
Medial brachial cutaneous n.
Median n.
Ulnar n.
Muscular branch
Medial antebrachial cutaneous n.
Biceps brachii m.
Long head
Medial head
Triceps brachii m.
Brachialis m.
Superior ulnar collateral a.
Radial recurrent a.
Medial intermuscular septum
Inferior ulnar collateral a.
Biceps brachii tendon
Medial epicondyle of humerus
Bicipital aponeurosis
Radial a.
Pronator teres m.
Ulnar a.
Flexor carpi radialis m.
Brachioradialis m.
F. Netter M.D.
©CIBA-GEIGY

Brachial Artery

The brachial artery, the continuation of the axillary artery, extends from the lower border of the teres major muscle to its bifurcation opposite the neck of the radius in the lower part of the cubital fossa (Plates 37–38). The course of the vessel may be marked out with the limb in right-angled abduction, when the vessel lies on a line connecting the middle of the clavicle with the midpoint between the epicondyles of the humerus. The brachial artery lies deep in the neurovascular compartment of the arm, flanked by the brachial veins on either side and by the median nerve anterior to it. The median nerve gradually crosses the artery to lie medial to it in the cubital fossa. These structures are crossed by the bicipital aponeurosis at the elbow (Plate 38).

The brachial artery is a single vessel in 80% of cases. In the other 20% of cases, a superficial brachial artery arises at the level of the upper arm and descends through the arm anterior to the median nerve. Based on its forearm distribution, this artery is a high radial artery in 10% of cases, a high ulnar artery in 3%, and forms both radial and ulnar arteries in 7%. In the last case, the brachial artery is likely to become the common interosseous artery of the forearm.

The brachial artery provides numerous muscular branches in the arm, principally from its lateral side. An especially large branch supplies the biceps brachii muscle. The branches are named as follows:

1. The *deep brachial artery* arises from the medial and posterior aspect of the brachial artery, below the tendon of the teres major muscle. It is the largest branch of the brachial artery and accompanies the radial nerve in its diagonal course around the humerus. At the back of the humerus, the artery provides an *ascending (deltoid) branch*, which reaches up to anastomose with the descending branch of the posterior circumflex humeral artery. The deep brachial artery then divides into the middle collateral artery and the radial collateral arteries. The *middle collateral artery* plunges into the medial head of the triceps brachii muscle and descends to the anastomosis of vessels at the level of the elbow. The *radial collateral artery* continues with the radial nerve, both perforating the lateral intermuscular septum to enter the anterior compartment. The artery ends in the elbow joint anastomosis, connecting in particular with the radial recurrent artery from the radial artery. All these branches nourish the muscles of the arm to which they are adjacent.

2. The *nutrient humeral artery* arises about the middle of the arm and enters the nutrient canal on the anteromedial surface of the humerus.

3. The *superior ulnar collateral artery* arises from the brachial artery at or a little below the middle of the arm. It pierces the medial intermuscular septum, descending behind it with the ulnar nerve. With the nerve, it passes behind the medial epicondyle of the humerus to anastomose with the inferior ulnar collateral artery and the posterior ulnar recurrent branch of the ulnar artery.

Upper Limb

Blood Supply of Arm

(Continued)

Axillary and Brachial Arteries and Anastomoses Around Elbow Joint

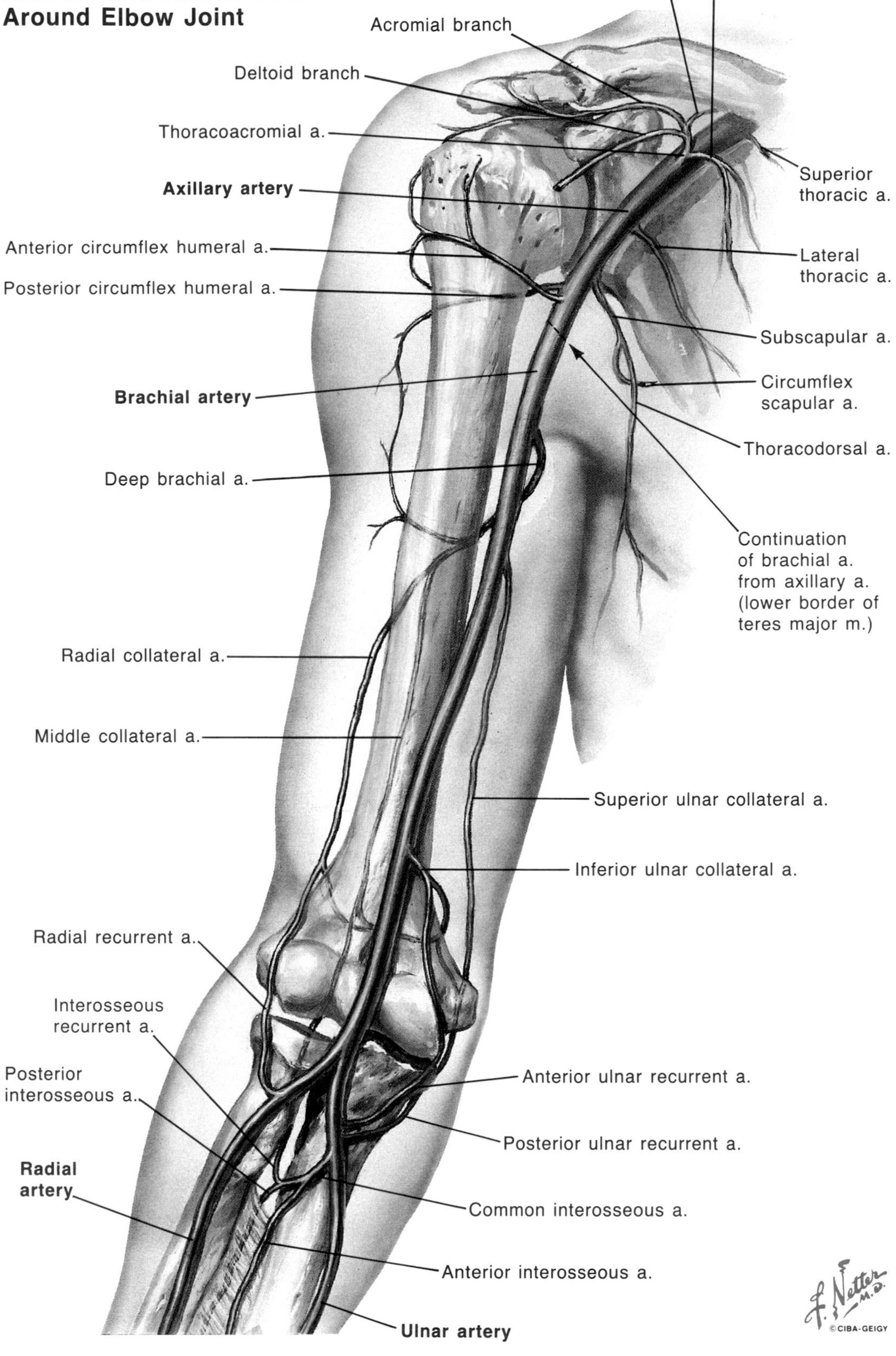

4. The *inferior ulnar collateral artery* arises from the brachial artery about 3 cm above the medial epicondyle. It divides on the brachialis muscle into anterior and posterior branches. Both these branches reach the anastomosis around the elbow joint, anterior and posterior to it, respectively.

Brachial Veins

Brachial veins accompany the artery, one on either side of it (Plates 37–38). They are formed from the venae comitantes of the radial and ulnar arteries and have tributaries that accompany the branches of the brachial artery, draining the areas supplied by the arteries. The brachial veins contain valves and frequently anastomose with one another. At the lower border of the teres major muscle, the lateral of the two veins crosses the artery to join the more medial one; then, joined by the basilic vein, they form the axillary vein.

Cubital Fossa

Like the axilla, the cubital fossa is a space at the bend of the elbow where it is helpful to note the important relationships of structures that overlie the elbow joint (Plate 37). It is described as a triangular space, apex downward, and is bounded above by a line connecting the epicondyles of the humerus. The converging side borders are muscular, the pronator teres muscle medially and the brachioradialis muscle laterally. The floor of the space is also muscular, consisting of the brachialis muscle of the arm and the supinator muscle of the forearm; deep to these muscles is the elbow joint.

The readily palpable *tendon of the biceps brachii muscle* descends centrally through the space, and its *bicipital aponeurosis* spans medialward across the brachial artery and median nerve to blend with the forearm fascia over the flexor muscle mass. Directly medial to the biceps brachii tendon, the *brachial artery* divides into the *radial* and *ulnar arteries* in the inferior part of the cubital fossa opposite the neck of the radius.

Although submerged between the brachioradialis and brachialis muscles, the *radial nerve* can be exposed by drawing the brachioradialis muscle lateralward and can be followed to its bifurcation into deep and superficial branches. Superficially, the *medial cubital vein* crosses obliquely, overlying the bicipital aponeurosis; and a *medial cephalic vein* may, on occasion, lie subcutaneously toward the lateral side of the fossa. The *medial antebrachial cutaneous nerve* crosses the median cubital vein and the *lateral antebrachial cutaneous nerve* passes deep to the median cephalic vein, if it is present. □

Upper Limb

Elbow Joint

Bones of Right Elbow Joint

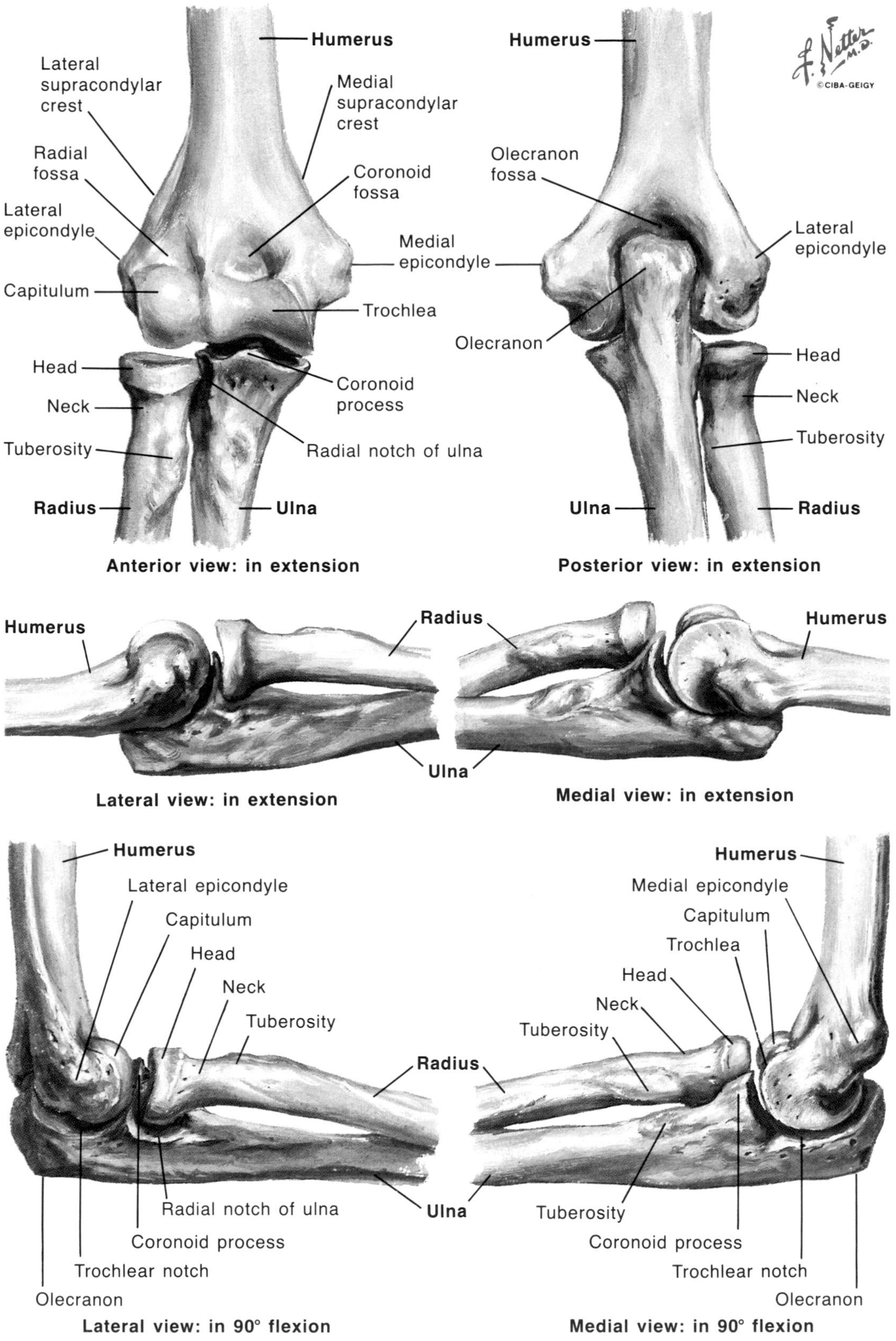

The elbow joint, comprising the humeroradial, humeroulnar, and proximal radioulnar joints within a common capsule, necessarily involves the proximal portions of the radius and ulna, as well as the distal part of the humerus (Plates 39–40).

Bones

In the region of the elbow joint, the special parts of the *radius* are the head, neck, and tuberosity (Plate 39). The head is round; it is a thick disc, articular both on its circumference and over its cupped upper free surface. The latter surface articulates with the capitulum of the humerus. The *articular circumference of the head* is broader medially for contact with the radial notch of the ulna, and narrower where it is held by the annular ligament. The *neck* of the radius is the constriction below the head, and the *tuberosity* is an oval prominence just distal to the neck. Its posterior portion is roughened for the reception of the biceps brachii tendon; its anterior part is smooth and is in contact with the *bicipitoradial bursa*.

The proximal end of the ulna has a more complex architecture. Its heavy proximal extremity exhibits the opened jaws of the trochlear notch, the olecranon, the coronoid process, and the radial notch.

The *trochlear notch* is a concavity that describes about one-third of a circle and is divided by a longitudinal ridge into medial and lateral parts. The waist of the notch is constricted, and a roughness across the waist separates the part deriving from the olecranon from the part formed by the coronoid process. The trochlear notch receives the trochlea of the humerus. In extreme flexion, the coronoid process of the ulna enters the coronoid fossa of the humerus; in extreme extension, the olecranon enters the olecranon fossa.

The *olecranon* contributes part of the trochlear notch and forms the posterior projection of the elbow. Its blunt end receives the tendon of the triceps brachii muscle and is attached to the capsule of the elbow joint along the bounding margin of the trochlear notch. Between these attachments, the bone is smooth for the *subtendinous bursa* of the triceps brachii muscle.

The *coronoid process* is a strong, triangular projection of the anterior surface of the ulna; it also forms the anterior part of the trochlear notch. The anterior surface of the coronoid process is rough for the insertion of the tendon of the brachialis muscle. The junction of this surface with the shaft is the location of the *tuberosity of the ulna*, which receives the oblique cord of the radius.

The *radial notch of the ulna*, a shallow concavity on the lateral aspect of the coronoid process, receives the circumferential articular surface of the head of the radius. Its prominent edges give attachment to the ends of the annular ligament of the radius.

Ossification. The radius is ossified by means of a center of ossification that appears in the shaft (which is bony at birth) at the eighth week of intrauterine life. Ossification for the head begins at 4 or 5 years of age, and this portion fuses with the shaft at 14 to 16 years of age.

Ossification of the ulna begins near the middle of the shaft at the eighth week of intrauterine life, and the shaft is bony at birth.

Ligaments and Capsule

The elbow joint is essentially a hinge joint (humeroradial and humeroulnar), but included within its joint capsule is the proximal radioulnar articulation (Plate 40). Like other hinge joints, there are reciprocally convex and concave articular surfaces; a capsule, loose on the sides toward which movement takes place; strong collateral ligaments; and a grouping of muscle masses at the borders where they are not in the direction of movement.

The *articular surfaces* are the spool-shaped *trochlea* and the rounded *capitulum* of the humerus proximally, and the *trochlear notch* of the ulna and the cupped upper surface of the *head* of the radius distally. The capitulum of the humerus is directed forward and downward, with the articular surfaces most completely in contact when the elbow is flexed to a 90° angle. Contact is weak between the humerus and the radius, and both the stability of the joint and its limitation of motion to flexion and extension are due to the ridged and grooved relationship of the humerus and the ulna.

Upper Limb

Elbow Joint

(Continued)

Ligaments of Right Elbow Joint

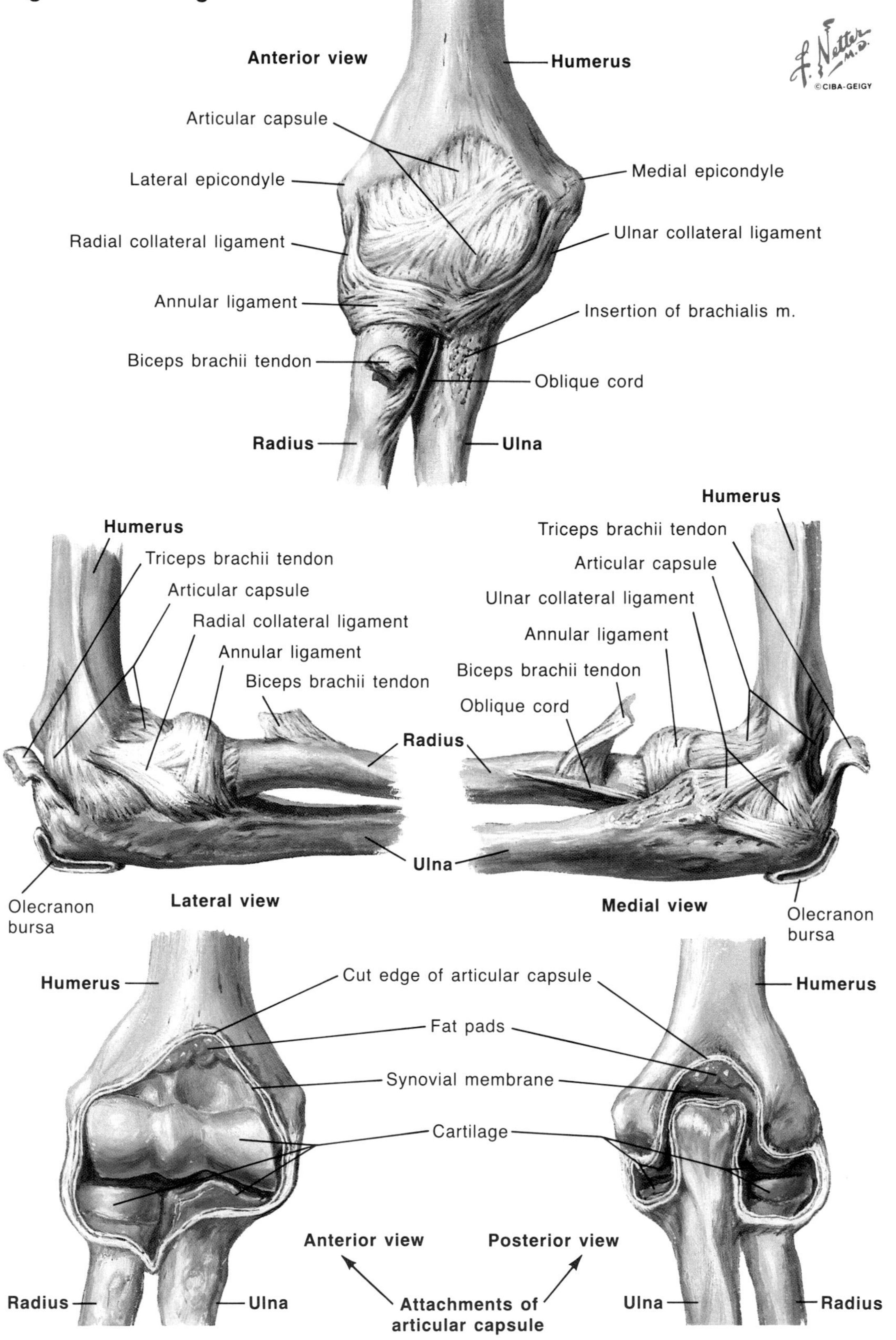
Anterior view
Humerus
Articular capsule
Lateral epicondyle
Medial epicondyle
Radial collateral ligament
Ulnar collateral ligament
Annular ligament
Insertion of brachialis m.
Biceps brachii tendon
Oblique cord
Radius
Ulna
Humerus
Triceps brachii tendon
Articular capsule
Radial collateral ligament
Annular ligament
Biceps brachii tendon
Olecranon bursa
Lateral view
Radius
Ulna
Humerus
Triceps brachii tendon
Articular capsule
Ulnar collateral ligament
Annular ligament
Biceps brachii tendon
Oblique cord
Medial view
Olecranon bursa
Humerus
Cut edge of articular capsule
Fat pads
Synovial membrane
Cartilage
Humerus
Anterior view
Posterior view
Radius
Ulna
Attachments of articular capsule
Ulna
Radius

The *articular capsule* is weak in front and behind but strengthened at the sides by the ulnar and radial collateral ligaments. In front, it is attached on the humerus from the medial to the lateral epicondyles along the superior borders of the coronoid and radial fossae. Distally, it is attached to the anterior border of the coronoid process of the ulna and to the annular ligament of the radius; it is continuous on either side with the collateral ligaments. The posterior portion of the capsule is membranous. Its attachments are the margins of the olecranon and the edges of the olecranon fossa, the lateral epicondyle, the annular ligament, and the posterior border of the radial notch of the ulna.

The *collateral ligaments* are strong, triangular thickenings of the articular capsule, attached by their apices to the medial and lateral epicondyles of the humerus. Their broader distal attachments are to the forearm bones and the annular ligament of the radius. These ligaments place strict limitations on side-to-side displacements of the joint.

The *ulnar collateral ligament* has thickened borders, the *anterior band* reaching the medial edge of the coronoid process, the *posterior band* attaching to the corresponding edge of the olecranon. The thinner *intermediate portion* ends below, in transverse fibers stretched between the coronoid process and the olecranon.

The *radial collateral ligament*, a narrower, less distinct thickening, is stretched between the underside of the lateral epicondyle above and the annular ligament and the margins of the radial notch of the ulna below.

The *synovial membrane* of the elbow joint lines the capsule and is reflected onto the borders of the radial and coronoid fossae of the humerus in front and the olecranon fossa behind. Below, it continues into the proximal radioulnar articulation.

The *blood supply* of the elbow joint comes from the anastomosis of the collateral branches of the brachial artery and the recurrent branches of the radial and ulnar arteries (Plate 38). *Nerves* reach the joint anteriorly from the musculocutaneous, median, and radial nerves, and posteriorly, from the ulnar nerve and the radial nerve branch to the anconeus muscle.

Movements

The hinge action at the elbow joint is not exactly in the line of the long axis of the humerus. In extension, the forearm deviates from a straight line with the arm, a circumstance described as the "carrying angle" of the forearm, which is obliterated when the hand is pronated. The forearm bones lie at about a 170° angle from the humerus in extension and supination. Because of a slight spiral orientation of the ridge of the trochlear notch and of the groove of the trochlea, flexion does not bring the forearm bones medial to the humerus. The habitual ease with which the hand is carried to the mouth in elbow flexion is due to the slight medial rotation of the humerus and the semipronated position of the hand.

Proximal Radioulnar Articulation

The head of the radius rotates in a ring formed by the radial notch of the ulna and the annular ligament of the radius (Plate 40). The *annular ligament of the radius* is a strong, curved band attaching to the anterior and posterior margins of the radial notch of the ulna. It serves as a restraining ligament, which prevents withdrawal of the head of the radius from its socket. The annular ligament receives the radial collateral ligament and blends with the capsule of the elbow joint. Below, a lax band, called the *quadrate ligament*, passes from the lower border of the radial notch of the ulna to the adjacent medial surface of the neck of the radius.

The *synovial membrane* of this joint is continuous with that of the elbow joint. A reflection of the membrane below the annular ligament forms a loose sac around the neck of the radius, which accommodates to the rotation of the head of the radius. □

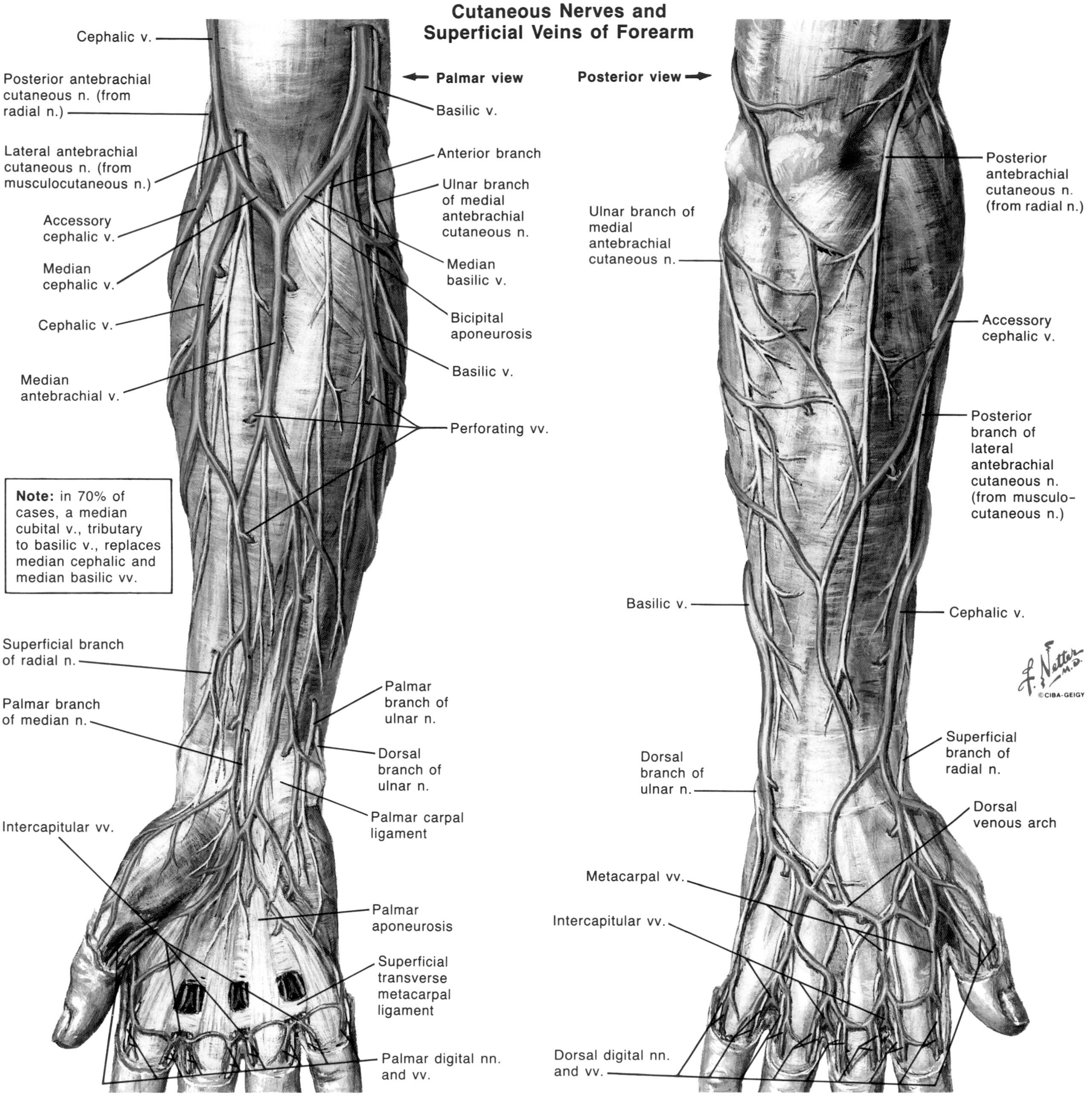

Cutaneous Nerves and Superficial Veins of Forearm
Palmar view
Posterior view
Cephalic v.
Posterior antebrachial cutaneous n. (from radial n.)
Lateral antebrachial cutaneous n. (from musculocutaneous n.)
Accessory cephalic v.
Median cephalic v.
Cephalic v.
Median antebrachial v.
Note: in 70% of cases, a median cubital v., tributary to basilic v., replaces median cephalic and median basilic vv.
Superficial branch of radial n.
Palmar branch of median n.
Intercapitular vv.
Basilic v.
Anterior branch
Ulnar branch of medial antebrachial cutaneous n.
Median basilic v.
Bicipital aponeurosis
Basilic v.
Perforating vv.
Palmar branch of ulnar n.
Dorsal branch of ulnar n.
Palmar carpal ligament
Palmar aponeurosis
Superficial transverse metacarpal ligament
Palmar digital nn. and vv.
Ulnar branch of medial antebrachial cutaneous n.
Posterior antebrachial cutaneous n. (from radial n.)
Accessory cephalic v.
Posterior branch of lateral antebrachial cutaneous n. (from musculo-cutaneous n.)
Basilic v.
Cephalic v.
F. Netter M.D.
©CIBA-GEIGY
Dorsal branch of ulnar n.
Superficial branch of radial n.
Dorsal venous arch
Metacarpal vv.
Intercapitular vv.
Dorsal digital nn. and vv.

SECTION I PLATE 41 Slide 3606

Upper Limb

Cutaneous Nerves and Superficial Veins of Forearm

The *medial antebrachial cutaneous nerve* (C8; T1) descends in the arm anterior and medial to the brachial artery. At the junction of the middle and lower thirds of the arm, the nerve emerges from the cover of the brachial fascia through the aperture by which the basilic vein passes deeply. By several anterior branches, it distributes to the skin of the anterior and medial surfaces of the forearm as far as the wrist. A smaller ulnar branch passes in front of the medial epicondyle of the humerus and supplies skin on the posteromedial surface of the forearm.

The *lateral antebrachial cutaneous nerve* (C5, 6) is the forearm continuation of the musculocutaneous nerve, from which it arises between the biceps brachii and brachialis muscles. It pierces the brachial fascia lateral to the biceps brachii tendon just above the elbow, passes behind the cephalic vein, and then divides into anterior and posterior branches. The larger anterior branch accompanies the cephalic vein into the forearm and supplies the skin of the radial half of the anterior surface as far as the thenar eminence. The smaller posterior branch serves the skin of the radial border and posterior aspect of the forearm as far as the wrist.

The *posterior antebrachial cutaneous nerve* arises from the radial nerve as it lies in its groove on the humerus. A small upper branch constitutes the inferior lateral brachial cutaneous nerve. The lower branch, the definitive posterior antebrachial cutaneous nerve (C5, 6, 7, 8), is large. It emerges through the brachial fascia about 7 cm above the elbow, descends behind the lateral epicondyle of the humerus, and distributes to the skin of the middle of the dorsum of the forearm as far as the wrist. □

cles of Forearm With Arteries and Nerves (anterior view)

Superficial layer

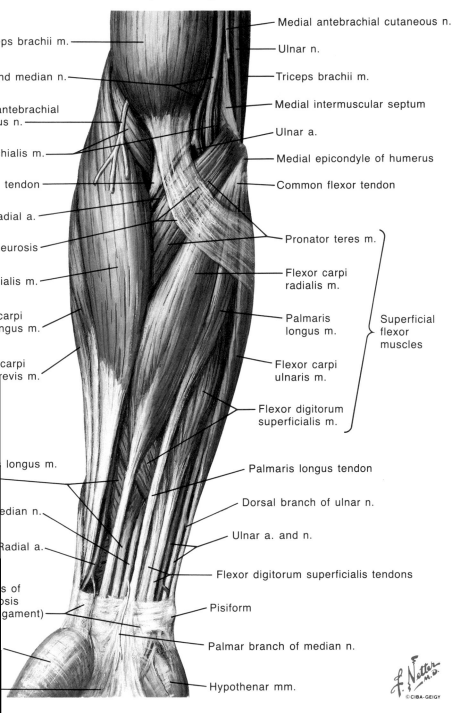

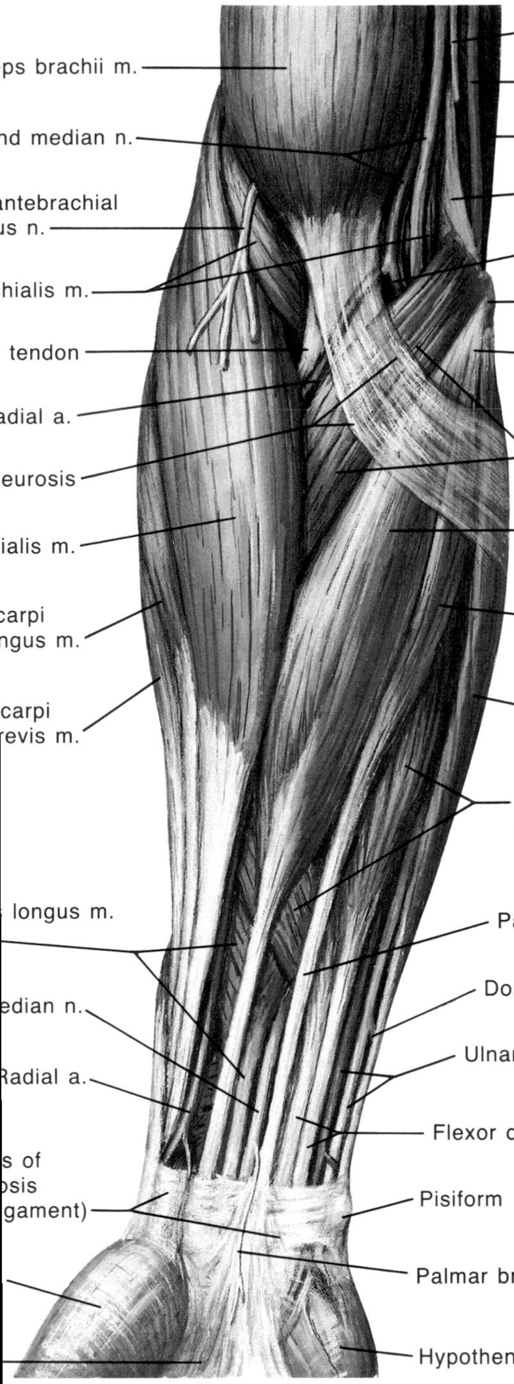

ps brachii m.
nd median n.
ntebrachial
ıs n.
hialis m.
tendon
adial a.
eurosis
ialis m.
carpi
ngus m.
carpi
evis m.
longus m.
edian n.
Radial a.
s of
sis
gament)
Medial antebrachial cutaneous n.
Ulnar n.
Triceps brachii m.
Medial intermuscular septum
Ulnar a.
Medial epicondyle of humerus
Common flexor tendon
Pronator teres m.
Flexor carpi radialis m.
Palmaris longus m.
Flexor carpi ulnaris m.
Flexor digitorum superficialis m.
Superficial flexor muscles
Palmaris longus tendon
Dorsal branch of ulnar n.
Ulnar a. and n.
Flexor digitorum superficialis tendons
Pisiform
Palmar branch of median n.
Hypothenar mm.
F. Netter M.D.
©CIBA-GEIGY

and at the wrist:
arpi radialis
arpi ulnaris
s longus

gits:
igitorum superficialis
igitorum profundus
ollicis longus

hand at the wrist:
r carpi radialis longus
r carpi radialis brevis
r carpi ulnaris

digits, except the thumb:
r digitorum
r indicis
r digiti minimi

thumb:
or pollicis longus
r pollicis brevis
r pollicis longus

The muscles of the first three groups lie in the anterior compartment of the forearm; those of the last three groups are located in the posterior compartment.

Further, there is a morphologic as well as a functional balance between these muscle groups. One group flexes the hand at the wrist, the other group extends it. Their structural balance is met by insertions on the opposite side of the base of the same bones. Thus, the flexor carpi ulnaris muscle inserts on the palmar side of the fifth metacarpal, and the extensor carpi ulnaris muscle inserts on the dorsal side of this bone. The extensor carpi radialis longus and brevis muscles insert, respectively, on the dorsum of the bases of the second and third metacarpals. Balance of these muscles is effected by the insertion of the flexor carpi radialis muscle on the palmar aspect of the bases of the same two metacarpals, plus the continuity of the tendon of the palmaris longus muscle with the palmar aponeurosis in the center of

Muscles of Forearm With Arteries and

Middle layer

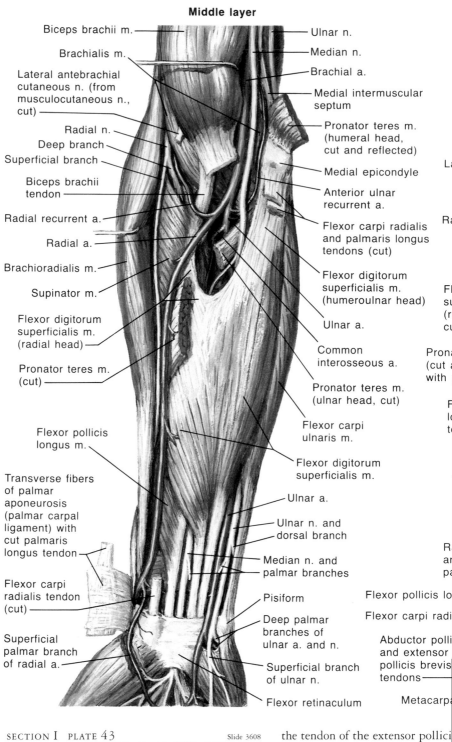

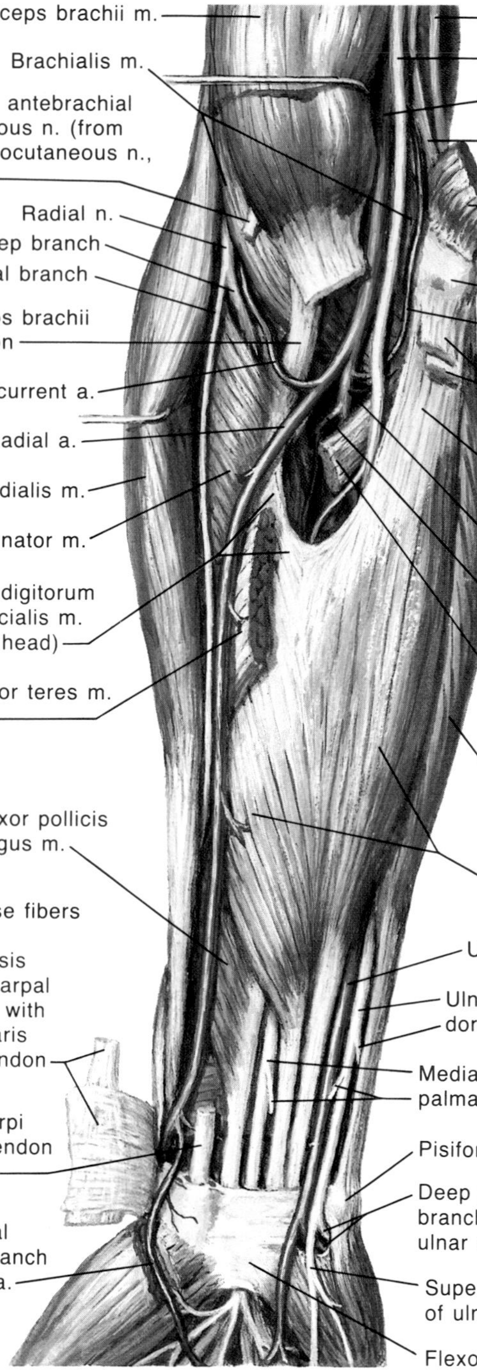

Biceps brachii m.
Brachialis m.
Lateral antebrachial cutaneous n. (from musculocutaneous n., cut)
Radial n.
Deep branch
Superficial branch
Biceps brachii tendon
Radial recurrent a.
Radial a.
Brachioradialis m.
Supinator m.
Flexor digitorum superficialis m. (radial head)
Pronator teres m. (cut)
Flexor pollicis longus m.
Transverse fibers of palmar aponeurosis (palmar carpal ligament) with cut palmaris longus tendon
Flexor carpi radialis tendon (cut)
Superficial palmar branch of radial a.
Ulnar n.
Median n.
Brachial a.
Medial intermuscular septum
Pronator teres m. (humeral head, cut and reflected)
Medial epicondyle
Anterior ulnar recurrent a.
Flexor carpi radialis and palmaris longus tendons (cut)
Flexor digitorum superficialis m. (humeroulnar head)
Ulnar a.
Common interosseous a.
Pronator teres m. (ulnar head, cut)
Flexor carpi ulnaris m.
Flexor digitorum superficialis m.
Ulnar a.
Ulnar n. and dorsal branch
Median n. and palmar branches
Pisiform
Deep palmar branches of ulnar a. and n.
Superficial branch of ulnar n.
Flexor retinaculum
Flexor pollicis lo
Flexor carpi radi
Abductor polli
and extensor
pollicis brevis
tendons
Metacarp

SECTION I PLATE 43 Slide 3608

Upper Limb

Muscles of Forearm

(Continued)

the palm. There is structural balance between the digital flexors and extensors, maintained by their insertions into the palmar and dorsal aspects, respectively, of the middle and distal phalanges of the same digits.

Muscles in the sixth group have a nice logical sequence: the tendon of the abductor pollicis longus muscle ends on the dorsum of the metacarpal of the thumb, that of the extensor pollicis brevis ends on the dorsum of the proximal phalanx, and the tendon of the extensor pollici the dorsum of the distal phalanx

Flexor Muscles

The muscles of the second a and the two pronator muscles o comprise the anterior antebrachi of these belong to a superficial la a deep layer (Plates 42–43, 45–

Superficial Layer. The muscle order in which they lie from the r side of the forearm; the flexor di cialis, however, is deep to the ot

Pronator teres
Flexor carpi radialis
Palmaris longus
Flexor carpi ulnaris
Flexor digitorum superficial

There is a *common tendon of orig* cles, which is attached to the m of the humerus. The intermuscu

Muscles of Forearm With Arteries and Nerves (posterior view)

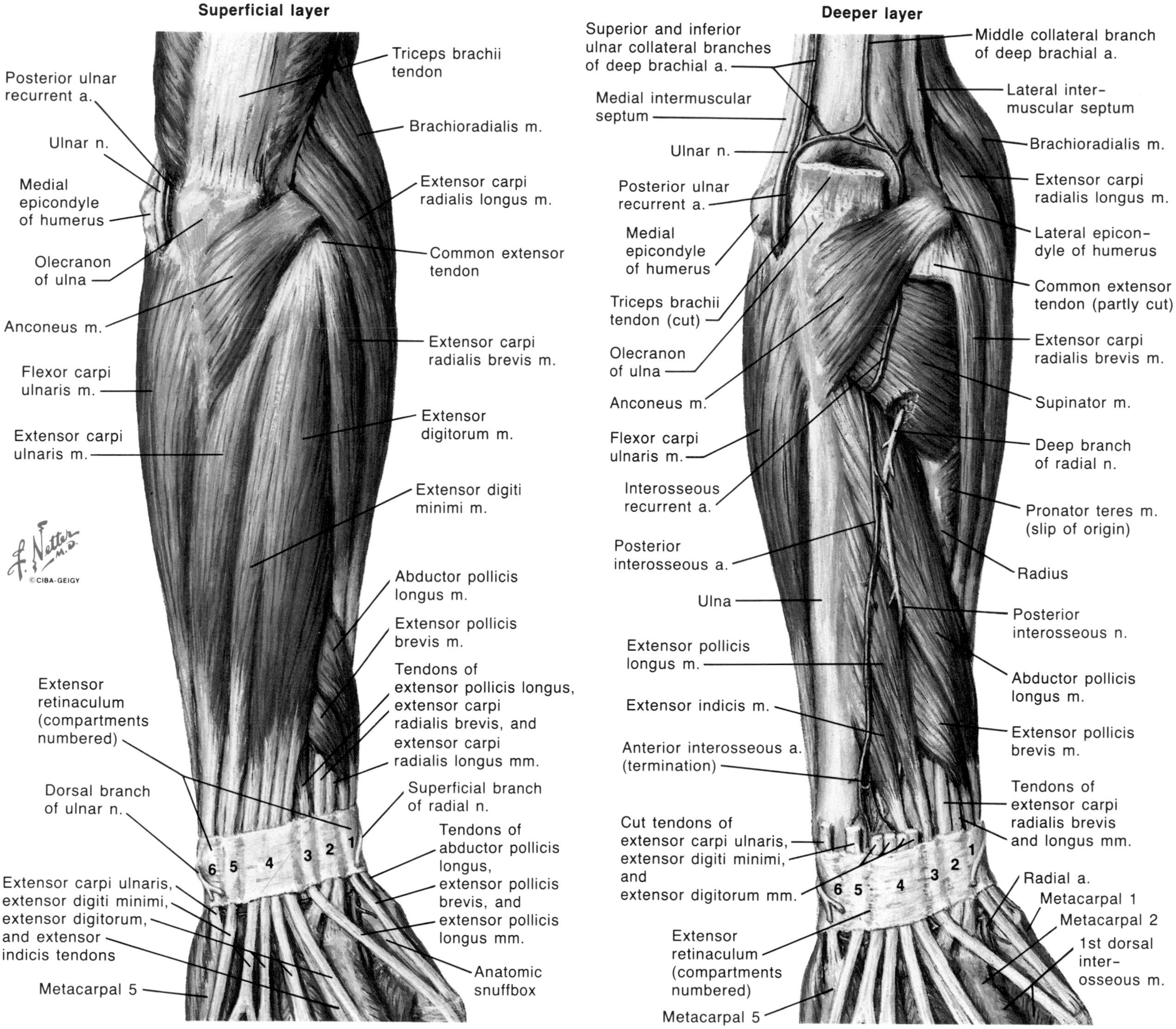
Superficial layer
Posterior ulnar recurrent a.
Ulnar n.
Medial epicondyle of humerus
Olecranon of ulna
Anconeus m.
Flexor carpi ulnaris m.
Extensor carpi ulnaris m.
Extensor retinaculum (compartments numbered)
Dorsal branch of ulnar n.
Extensor carpi ulnaris, extensor digiti minimi, extensor digitorum, and extensor indicis tendons
Metacarpal 5
Triceps brachii tendon
Brachioradialis m.
Extensor carpi radialis longus m.
Common extensor tendon
Extensor carpi radialis brevis m.
Extensor digitorum m.
Extensor digiti minimi m.
Abductor pollicis longus m.
Extensor pollicis brevis m.
Tendons of extensor pollicis longus, extensor carpi radialis brevis, and extensor carpi radialis longus mm.
Superficial branch of radial n.
Tendons of abductor pollicis longus, extensor pollicis brevis, and extensor pollicis longus mm.
Anatomic snuffbox
6 5 4 3 2 1
©CIBA-GEIGY
Deeper layer
Superior and inferior ulnar collateral branches of deep brachial a.
Medial intermuscular septum
Ulnar n.
Posterior ulnar recurrent a.
Medial epicondyle of humerus
Triceps brachii tendon (cut)
Olecranon of ulna
Anconeus m.
Flexor carpi ulnaris m.
Interosseous recurrent a.
Posterior interosseous a.
Ulna
Extensor pollicis longus m.
Extensor indicis m.
Anterior interosseous a. (termination)
Cut tendons of extensor carpi ulnaris, extensor digiti minimi, and extensor digitorum mm.
Extensor retinaculum (compartments numbered)
Metacarpal 5
Middle collateral branch of deep brachial a.
Lateral intermuscular septum
Brachioradialis m.
Extensor carpi radialis longus m.
Lateral epicondyle of humerus
Common extensor tendon (partly cut)
Extensor carpi radialis brevis m.
Supinator m.
Deep branch of radial n.
Pronator teres m. (slip of origin)
Radius
Posterior interosseous n.
Abductor pollicis longus m.
Extensor pollicis brevis m.
Tendons of extensor carpi radialis brevis and longus mm.
Radial a.
Metacarpal 1
Metacarpal 2
1st dorsal interosseous m.
6 5 4 3 2 1

Upper Limb

Muscles of Forearm

(Continued)

The *palmaris longus muscle* also uses the common tendon of origin, when present (it is absent in 13% of cases). It terminates in a slender, flattened tendon, crossing the wrist superficial to the flexor retinaculum. It constitutes, by its spreading tendinous fibers, the chief part of the *palmar aponeurosis*.

The *flexor carpi ulnaris muscle* has a humeral and an ulnar head, the humeral head coming from the common flexor tendon. The ulnar head springs from the medial border of the olecranon and the upper two-thirds of the posterior border of the ulna. The tendon of the muscle inserts on the pisiform of the wrist and, through it by two ligaments, into the hamulus of the hamate and the base of the fifth metacarpal.

The *flexor digitorum superficialis muscle* arises by a humeroulnar and a radial head of origin; these are connected by a fibrous band that crosses the median nerve and the ulnar blood vessels (Plate 43). The larger humeroulnar head arises from the common tendon, the intermuscular septa, the ulnar collateral ligament, and the medial border of the coronoid process. The radial head is a thin layer arising from the upper two-thirds of the anterior border of the radius. The muscle forms two planes; the tendons of its superficial plane pass to the middle and ring fingers, and the deep lamina ends in tendons for digits I and V (see schema in Plate 57). These tendons terminate in the palmar aspect of the shafts of the middle phalanges of digits II to V (their relationships are described in the discussion of the wrist and hand).

Deep Layer. The deep layer contains the following muscles:

- Flexor digitorum profundus
- Flexor pollicis longus
- Pronator quadratus

The *flexor digitorum profundus muscle* arises from the posterior border of the ulna (with the flexor carpi ulnaris), the proximal two-thirds of the medial surface of the ulna, and adjacent areas of the interosseous membrane. The muscle produces, near the wrist, four discrete tendons that pass side by side under the flexor retinaculum and dorsal to the tendons of flexor digitorum superficialis muscle. The tendons terminate on the bases of the distal phalanges of digits II to V. In the palm, the tendons give origin to the small lumbrical muscles.

The *flexor pollicis longus muscle* arises principally from the anterior surface of the radius (just below

Bony Attachments of Muscles of Forearm

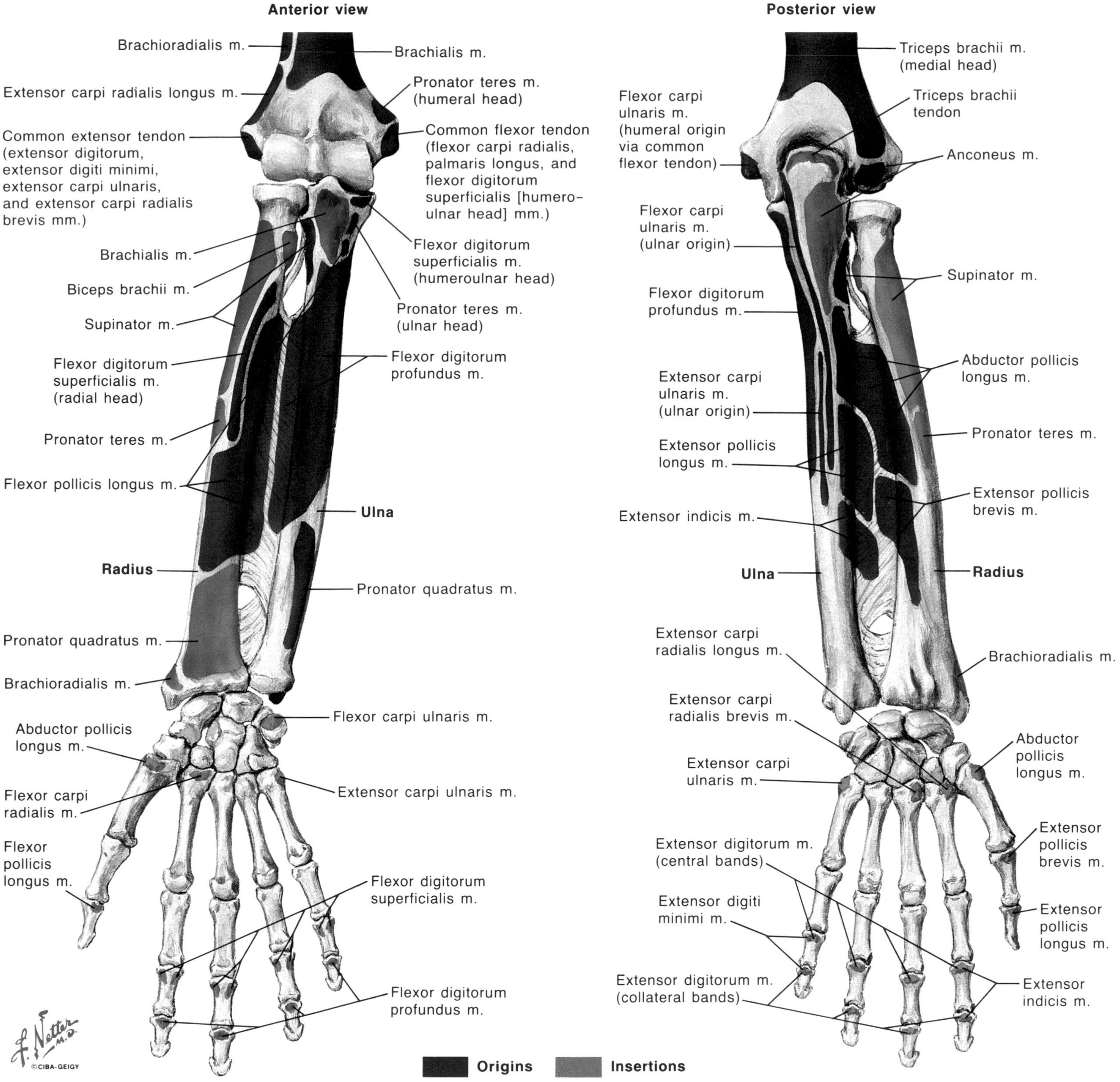

Anterior view
Brachioradialis m.
Brachialis m.
Extensor carpi radialis longus m.
Pronator teres m. (humeral head)
Common extensor tendon (extensor digitorum, extensor digiti minimi, extensor carpi ulnaris, and extensor carpi radialis brevis mm.)
Common flexor tendon (flexor carpi radialis, palmaris longus, and flexor digitorum superficialis [humero-ulnar head] mm.)
Brachialis m.
Flexor digitorum superficialis m. (humeroulnar head)
Biceps brachii m.
Pronator teres m. (ulnar head)
Supinator m.
Flexor digitorum superficialis m. (radial head)
Flexor digitorum profundus m.
Pronator teres m.
Flexor pollicis longus m.
Ulna
Radius
Pronator quadratus m.
Pronator quadratus m.
Brachioradialis m.
Flexor carpi ulnaris m.
Abductor pollicis longus m.
Flexor carpi radialis m.
Extensor carpi ulnaris m.
Flexor pollicis longus m.
Flexor digitorum superficialis m.
Flexor digitorum profundus m.
Posterior view
Triceps brachii m. (medial head)
Flexor carpi ulnaris m. (humeral origin via common flexor tendon)
Triceps brachii tendon
Anconeus m.
Flexor carpi ulnaris m. (ulnar origin)
Supinator m.
Flexor digitorum profundus m.
Abductor pollicis longus m.
Extensor carpi ulnaris m. (ulnar origin)
Pronator teres m.
Extensor pollicis longus m.
Extensor pollicis brevis m.
Extensor indicis m.
Ulna
Radius
Extensor carpi radialis longus m.
Brachioradialis m.
Extensor carpi radialis brevis m.
Abductor pollicis longus m.
Extensor carpi ulnaris m.
Extensor pollicis brevis m.
Extensor digitorum m. (central bands)
Extensor digiti minimi m.
Extensor pollicis longus m.
Extensor digitorum m. (collateral bands)
Extensor indicis m.
Origins
Insertions
F. Netter M.D.
©CIBA-GEIGY

SECTION I PLATE 45 Slide 3610

Upper Limb

Muscles of Forearm

(Continued)

its tuberosity nearly to the upper border of the pronator quadratus) and from the adjacent interosseous membrane. Its tendon, passing between the two sesamoids of the metacarpophalangeal joint of the thumb, inserts on the base of the distal phalanx of the thumb.

The *pronator quadratus muscle* is a quadrilateral muscle located just above the wrist and deep to the flexor digitorum profundus and flexor pollicis longus tendons. It arises from the anterior surface of the distal one-fourth of the ulna, its fibers running transversely across the wrist and inserting into the anterior surface of the radius.

Extensor Muscles

The muscles of the fourth, fifth, and sixth groups, the supinator muscle of the first group, and the brachioradialis muscle make up the posterior antebrachial muscles. Six of these comprise the superficial layer, and five lie in the deep layer (Plates 44–46).

Superficial Layer. The six muscles of the superficial layer are listed in the order in which they lie across the back of the forearm, from the radial to the ulnar side:

Brachioradialis
Extensor carpi radialis longus
Extensor carpi radialis brevis
Extensor digitorum
Extensor digiti minimi
Extensor carpi ulnaris

As for the flexor muscles, there is a common tendon of origin from the lateral epicondyle for all muscles arising below the lateral epicondyle.

The *brachioradialis muscle* arises from the upper two-thirds of the supracondylar ridge of the

Cross-Sectional Anatomy of Right Forearm

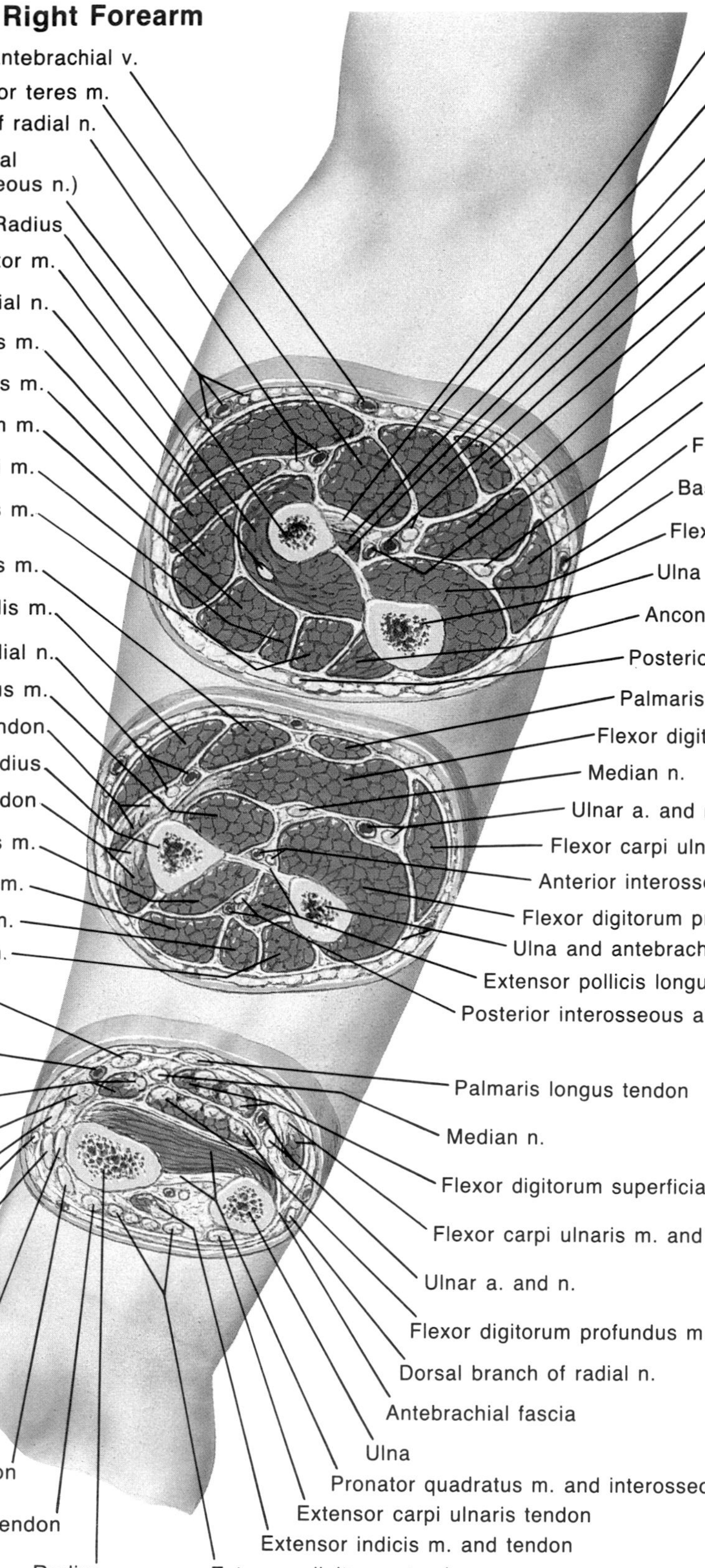

Median antebrachial v.
Pronator teres m.
Radial a. and superficial branch of radial n.
Cephalic v. and lateral antebrachial cutaneous n. (from musculocutaneous n.)
Radius
Supinator m.
Deep branch of radial n.
Extensor carpi radialis longus m.
Extensor carpi radialis brevis m.
Extensor digitorum m.
Extensor digiti minimi m.
Extensor carpi ulnaris m.
Flexor carpi radialis m.
Brachioradialis m.
Radial a. and superficial branch of radial n.
Flexor pollicis longus m.
Extensor carpi radialis longus m. and tendon
Radius
Extensor carpi radialis brevis m. and tendon
Abductor pollicis longus m.
Extensor digitorum m.
Extensor digiti minimi m.
Extensor carpi ulnaris m.
Flexor carpi radialis tendon
Radial a.
Flexor pollicis longus m. and tendon
Brachioradialis tendon
Abductor pollicis longus tendon
Superficial branch of radial n.
Extensor pollicis brevis tendon
Extensor carpi radialis longus tendon
Extensor carpi radialis brevis tendon
Extensor pollicis longus tendon
Radius
Flexor digitorum superficialis m. (radial head)
Anterior branch of medial antebrachial cutaneous n.
Flexor pollicis longus m.
Interosseous membrane
Flexor carpi radialis m.
Ulnar a. and median n.
Palmaris longus m.
Flexor digitorum superficialis m. (humeroulnar head)
Common interosseous a.
Ulnar n.
Flexor carpi ulnaris m.
Basilic v.
Flexor digitorum profundus m.
Ulna and antebrachial fascia
Anconeus m.
Posterior antebrachial cutaneous n. (from radial n.)
Palmaris longus m.
Flexor digitorum superficialis m.
Median n.
Ulnar a. and n.
Flexor carpi ulnaris m.
Anterior interosseous a. and n.
Flexor digitorum profundus m.
Ulna and antebrachial fascia
Extensor pollicis longus m. and interosseous membrane
Posterior interosseous a. and n.
Palmaris longus tendon
Median n.
Flexor digitorum superficialis m. and tendons
Flexor carpi ulnaris m. and tendon
Ulnar a. and n.
Flexor digitorum profundus m. and tendons
Dorsal branch of radial n.
Antebrachial fascia
Ulna
Pronator quadratus m. and interosseous membrane
Extensor carpi ulnaris tendon
Extensor indicis m. and tendon
Extensor digitorum tendons
F. Netter M.D.
©CIBA-GEIGY

SECTION I PLATE 46 Slide 3611

Upper Limb

Muscles of Forearm

(Continued)

humerus. Its tendon appears at about the middle of the forearm and descends to insert into the lateral side of the base of the styloid process of the radius.

The *extensor carpi radialis longus muscle* arises from the lower third of the supracondylar ridge of the humerus. It has a flat tendon that reaches into the hand to insert on the dorsum of the second metacarpal.

The *extensor carpi radialis brevis muscle* uses the common tendon of origin for the extensors. Its tendon appears in the lower third of the forearm, closely applied to the overlying tendon of the extensor carpi radialis longus, and inserts on the dorsum of the base of the third metacarpal.

The *extensor digitorum muscle* also uses the common tendon of origin for the extensors. Above the wrist, it provides four tendons that spread out on the dorsum of the hand, joined side to side in a variable manner by intertendinous connections. Participating in the rather complex "extensor expansion" described in the section on the wrist and hand, these tendons terminate on the bases of the middle and distal phalanges of digits II to V.

The *extensor digiti minimi muscle* is a slender muscle that is sometimes only incompletely separated from the extensor digitorum muscle. Its tendon joins the ulnar side of the tendon of the extensor digitorum muscle to the fifth digit. It provides independent extensor action for the fifth digit.

The *extensor carpi ulnaris muscle* arises by the common tendon from the lateral epicondyle but also from the middle two-fourths of the posterior border of the ulna. It inserts on the ulnar side of the base of the fifth metacarpal.

Upper Limb

Muscles of Forearm

(Continued)

Deep Layer. The muscles of the deep layer are generally submerged under those of the superficial group, although certain of their tendons and parts of their fleshy bellies outcrop just above the wrist:

Supinator
Abductor pollicis longus
Extensor pollicis brevis
Extensor pollicis longus

The *supinator muscle* has a complex origin from the lateral epicondyle of the humerus, the radial collateral ligament, the annular ligament of the radius, and the supinator crest and fossa of the ulna. Its fibers form a flat sheet, directed downward and lateralward, which wraps itself almost completely around the radius and inserts on the lateral surface of the upper third of this bone. As the muscle courses into the posterior compartment of the forearm, it is separated into superficial and deep laminae by the deep branch of the radial nerve.

The *abductor pollicis longus muscle* lies immediately distal to the supinator. It arises from the middle third of the posterior surface of the radius and the lateral part of the posterior surface of the ulna below the anconeus muscle. The fibers of the muscle converge onto its tendon, which, with the tendon of extensor pollicis brevis closely applied to its medial side, crosses the tendons of the extensor carpi radialis longus and brevis muscles and inserts on the radial side of the base of the metacarpal of the thumb (Plate 45).

The *extensor pollicis brevis muscle*, with origins from the radius and the interosseous membrane distal to that of the abductor pollicis longus muscle, inserts on the base of the proximal phalanx of the thumb. It is a specialization of the distal part of the abductor pollicis longus muscle.

The *extensor pollicis longus muscle* arises from the ulna and the interosseous membrane distal to the abductor pollicis longus muscle. Its tendon passes to the ulnar side of the dorsal tubercle of the radius, then obliquely across the tendons of both radial carpal extensors, and terminates on the base of the distal phalanx of the thumb.

The *extensor indicis muscle* arises just below the extensor pollicis longus from the ulna and from the interosseous membrane. In the hand, the tendon joins the ulnar side of the tendon of the digital extensor muscle for the index finger and participates with it in forming the extensor expansion over that digit.

Muscle Actions

The *rotary movements of the forearm* result in pronation and supination of the hand. Of the two pronators, the *pronator quadratus muscle* is the principal mover; the *pronator teres muscle* is added for speed and against resistance. The *supinator muscle* is the principal muscle in that action, but it is supplemented by the more powerful *biceps brachii* muscle against resistance and for speed of movement.

The *brachioradialis muscle* is an elbow flexor, recruited especially for speed and against resistance; its rotary potential is seen in its action in returning the forearm to its usual midpronation-midsupination position. It aids in maintaining the close-packed position of the bones participating in the elbow joint.

In wrist movements, flexion is produced by the "carpal flexors" (including the palmaris longus muscle), which are assisted by the "digital flexors" after these muscles have accomplished digital flexion. The carpal flexors are also synergistic for the digital extensor muscles. Extension of the wrist is produced by the "carpal extensors," assisted by the "digital extensors" after they have completed their primary pull on the digits. The carpal extensors, especially the extensor carpi radialis longus muscle, also stabilize the wrist for digital flexion. The carpal muscles on the ulnar and radial borders produce, respectively, adduction and abduction, with the flexor and extensor muscles of the same side cooperating in the action. Abduction is assisted by the long abductor and extensor muscles of the thumb.

The action of the forearm muscles on the digits can best be discussed after the intrinsic muscles of the hand are described in order to incorporate the supplemental functions of hand muscles into the account. □

Upper Limb

Nerves of Forearm

Ulnar Nerve

The ulnar nerve (C[7], 8; T1) is the main continuation of the medial cord of the brachial plexus (Plate 47).

Course in Arm. Initially, the ulnar nerve lies between the axillary artery and vein; as it enters the arm, it runs on the medial side of the brachial artery. At about the middle of the arm, it pierces the medial intermuscular septum and descends anterior to the medial head of the triceps brachii muscle, alongside the superior ulnar collateral artery. In the lower third of the arm, it inclines posteriorly to reach the interval between the medial humeral epicondyle and the olecranon. As the nerve enters the forearm, it lies in the groove behind the medial epicondyle, between the humeral and ulnar heads of the flexor carpi ulnaris muscle. Above the elbow, the ulnar nerve supplies no constant branches.

Course in Forearm and Hand. The ulnar nerve runs downward on the medial side of the forearm, lying first on the ulnar collateral ligament of the elbow joint and then on the flexor digitorum profundus muscle, deep to the flexor carpi ulnaris muscle. At the elbow, the ulnar nerve and artery are separated by a considerable gap, but they are closely apposed in the lower two-thirds of the forearm, with the artery on the lateral side. At the flexor carpi ulnaris tendon, the nerve and artery emerge from under its lateral edge and are covered only by skin and fascia. They reach the hand by crossing the anterior surface of the flexor retinaculum lateral to the pisiform, and the nerve splits, under cover of the palmaris brevis muscle, into its superficial and deep terminal branches.

Branches. In the forearm and hand, the ulnar nerve gives off articular, muscular, palmar, dorsal, superficial and deep terminal, and vascular branches.

Fine articular branches for the elbow joint arise from the main nerve as it runs posterior to the medial epicondyle; before splitting into its terminal branches, it supplies filaments to the wrist joint.

In the upper forearm, branches are given off to the flexor carpi ulnaris and the medial half of the flexor digitorum profundus muscles. The *palmar branch* arises 5 to 7 cm above the wrist, descends near the ulnar artery, pierces the deep fascia, and supplies the skin over the hypothenar eminence; it communicates with the medial antebrachial cutaneous nerve and the palmar branch of the median nerve. The *dorsal ulnar branch* arises 5 to 10 cm above the wrist, passes posteriorly, and, deep to the flexor carpi ulnaris tendon, pierces the deep fascia and continues along the dorsomedial side of the wrist. Here, it divides into branches for the areas of skin on the medial side of the back of the hand and fingers. There are usually two or three *dorsal digital nerves*, one supplying the medial side of the little finger, the second splitting into *proper dorsal digital nerves* to supply adjacent sides of the little and ring fingers, and the third (when present) supplying contiguous sides of the ring and middle fingers.

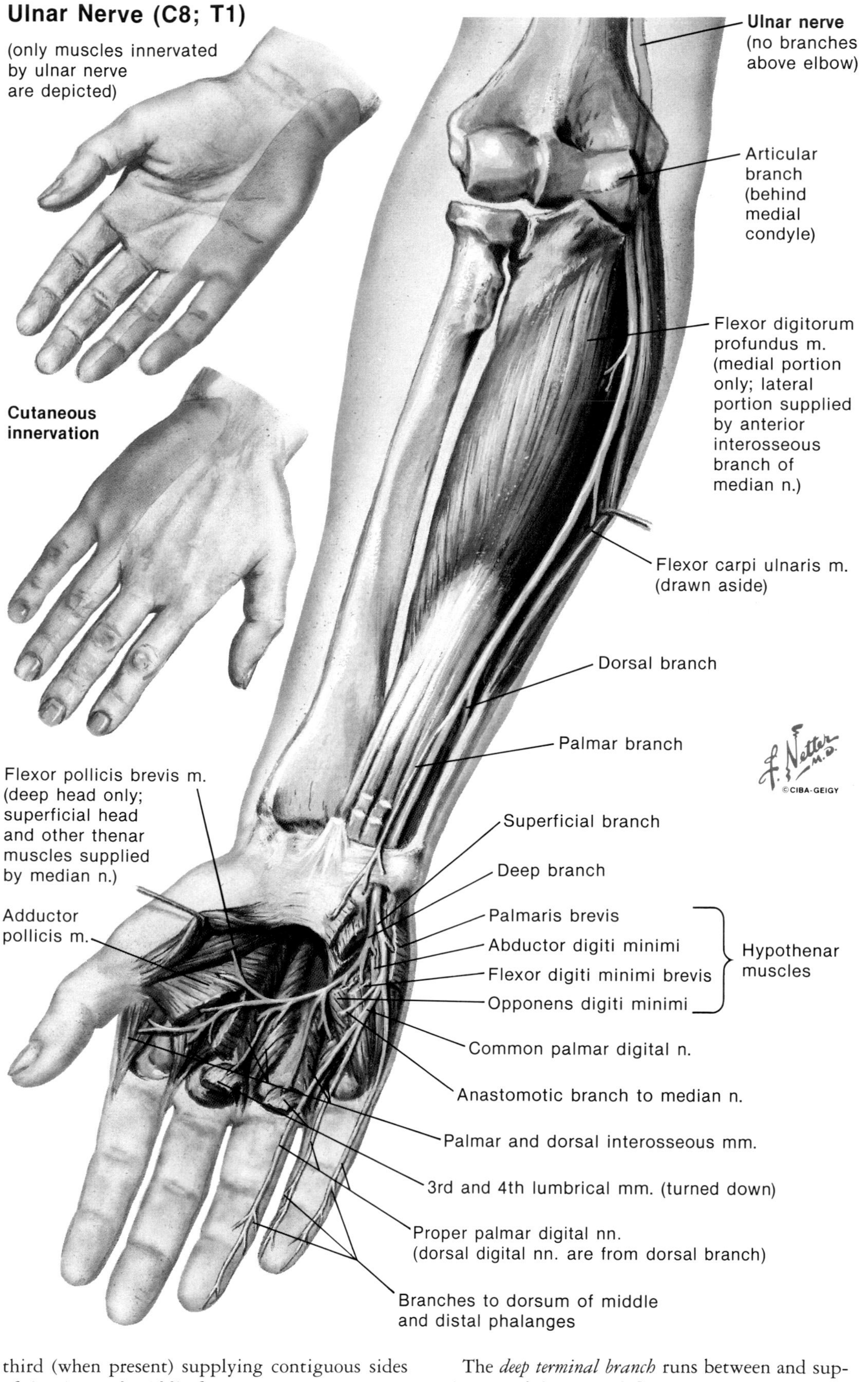

Ulnar Nerve (C8; T1)
(only muscles innervated by ulnar nerve are depicted)
Cutaneous innervation
Ulnar nerve (no branches above elbow)
Articular branch (behind medial condyle)
Flexor digitorum profundus m. (medial portion only; lateral portion supplied by anterior interosseous branch of median n.)
Flexor carpi ulnaris m. (drawn aside)
Dorsal branch
Palmar branch
Flexor pollicis brevis m. (deep head only; superficial head and other thenar muscles supplied by median n.)
Superficial branch
Deep branch
Adductor pollicis m.
Palmaris brevis
Abductor digiti minimi
Flexor digiti minimi brevis
Opponens digiti minimi
Hypothenar muscles
Common palmar digital n.
Anastomotic branch to median n.
Palmar and dorsal interosseous mm.
3rd and 4th lumbrical mm. (turned down)
Proper palmar digital nn. (dorsal digital nn. are from dorsal branch)
Branches to dorsum of middle and distal phalanges
F. Netter M.D.
©CIBA-GEIGY

The *superficial terminal branch* supplies the palmaris brevis muscle, innervates the skin on the medial side of the palm, and gives off two palmar digital nerves. The first is the *proper palmar digital nerve* for the medial side of the little finger; the second, the *common palmar digital nerve*, communicates with the adjoining common palmar digital branch of the median nerve before dividing into the two *proper palmar digital nerves* for the adjacent sides of the little and ring fingers. Rarely, the ulnar nerve supplies 2½ rather than 1½ digits, and the areas supplied by the median and radial nerves are reciprocally reduced.

The *deep terminal branch* runs between and supplies the abductor and flexor muscles of the little finger, perforates and supplies the opponens digiti minimi, and then accompanies the deep palmar arterial arch behind the flexor digitorum tendons. In the palm, it gives muscular branches to the third and fourth lumbrical and the interosseous muscles and ends by supplying the adductor pollicis muscle and, sometimes, the deep head of the flexor pollicis brevis muscle.

Variations in the nerve supplies of the palmar muscles are as common as the variations in the cutaneous distribution; they are due to the variety of interconnections between the ulnar and median nerves.

Upper Limb

Nerves of Forearm

(Continued)

Median Nerve (C[5], 6, 7, 8; T1)

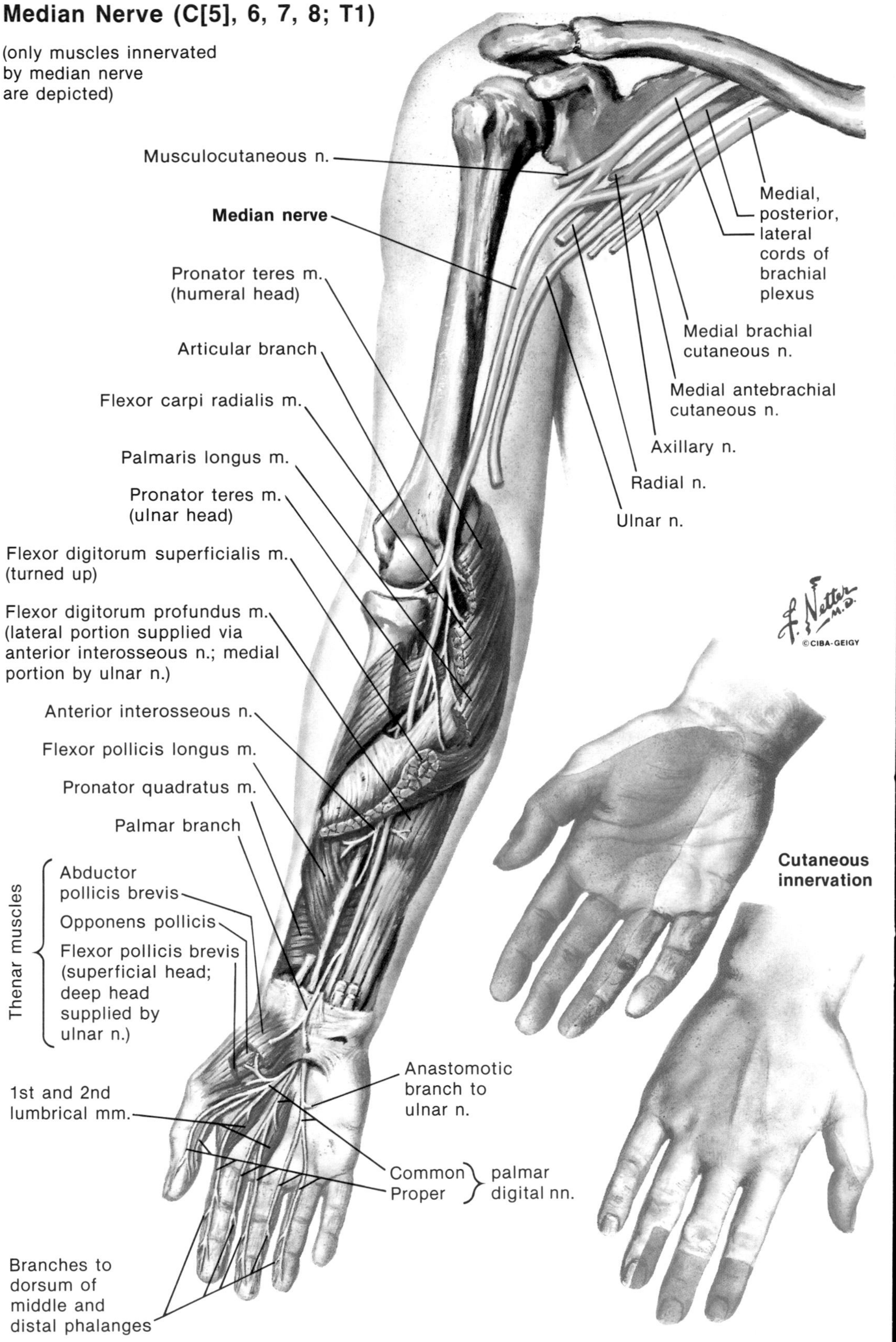

Median Nerve

The median nerve (C[5], 6, 7, 8; T1) is formed by the union of *medial* and *lateral roots* arising from the corresponding cords of the brachial plexus (Plate 48).

Course in Arm. The median nerve runs from the axilla into the arm, lateral to the brachial artery. At about the level of the insertion of the coracobrachialis muscle, the nerve inclines medially over the brachial artery and then descends along its medial side to the cubital fossa. Here, it lies behind the bicipital aponeurosis and the median cubital vein and in front of the insertion of the brachialis muscle and the elbow joint. (The close proximity of the vein, artery, and nerve should be remembered when performing venipuncture in this area.) The only branches given off in the arm are filaments to the brachial vessels and an inconstant twig to the pronator teres muscle.

Course in Forearm. The nerve passes into the forearm between the humeral and ulnar heads of the pronator teres muscle, the latter separating it from the ulnar artery. It then runs deep to the aponeurotic arch between the humeroulnar and radial heads of the flexor digitorum superficialis muscle and continues downward between this muscle and the flexor digitorum profundus muscle. In the forearm, the nerve supplies branches to the pronator teres, flexor digitorum superficialis, flexor carpi radialis, and palmaris longus muscles and articular twigs to the elbow and proximal radioulnar joints.

The longest branch is the *anterior interosseous nerve*, which, accompanied by the corresponding artery, runs downward on the interosseous membrane between the flexor pollicis longus and the flexor digitorum profundus muscles; it supplies the former muscle and the lateral part of the latter and ends under the pronator quadratus, supplying this muscle and the distal radioulnar, radiocarpal, and carpal joints. Vascular filaments help to innervate the ulnar and anterior interosseous vessels and the nutrient vessels of the radius and ulna. A *palmar branch* arises about 3 to 4 cm above the flexor retinaculum and descends over it to supply the skin of the median part of the palm and the thenar eminence. In the forearm, the median and ulnar nerves are occasionally interconnected by strands, which may explain certain anomalies in the nerve supply of the hand.

In the lower forearm, the median nerve becomes more superficial between the tendons of the palmaris longus and the flexor carpi radialis muscles. Together with the tendons of the digital flexor muscles, it enters the palm through the carpal tunnel that is bounded anteriorly by the tough flexor retinaculum and posteriorly by the carpal bones. Emerging from the tunnel, the nerve splays out into its terminal muscular and palmar digital branches. The muscular branch arises close to, or is initially united with, the common palmar digital nerve to the thumb; it curves outward over or through the flexor pollicis brevis muscle to supply it before dividing to supply the abductor pollicis brevis and opponens pollicis muscles. The muscular branch may also supply all or part of the first dorsal interosseous muscle. In rare instances, it arises in the carpal tunnel and pierces the flexor retinaculum—an arrangement of potential clinical concern.

The *common* and *proper palmar digital nerves* vary in their origins and distributions, but the usual arrangement is that shown in Plates 47 and 48. The proper palmar digital nerves give off dorsal twigs, which innervate the skin (including the nail beds) over the distal and dorsal aspects of the lateral 3½ digits. Occasionally, they supply only 2½ digits. The proper palmar digital branches to the radial side of the index finger and to the contiguous sides of the index and middle fingers also carry motor fibers to supply the first and second lumbrical muscles, respectively. Therefore, the digital nerves are not concerned solely with cutaneous sensibility. They contain an admixture of efferent and afferent somatic and autonomic fibers, which transmit impulses to and from sensory endings, vessels, sweat glands, and arrectores pilorum muscles and between fascial, tendinous, osseous, and articular structures in their areas of distribution.

Upper Limb

Nerves of Forearm

(Continued)

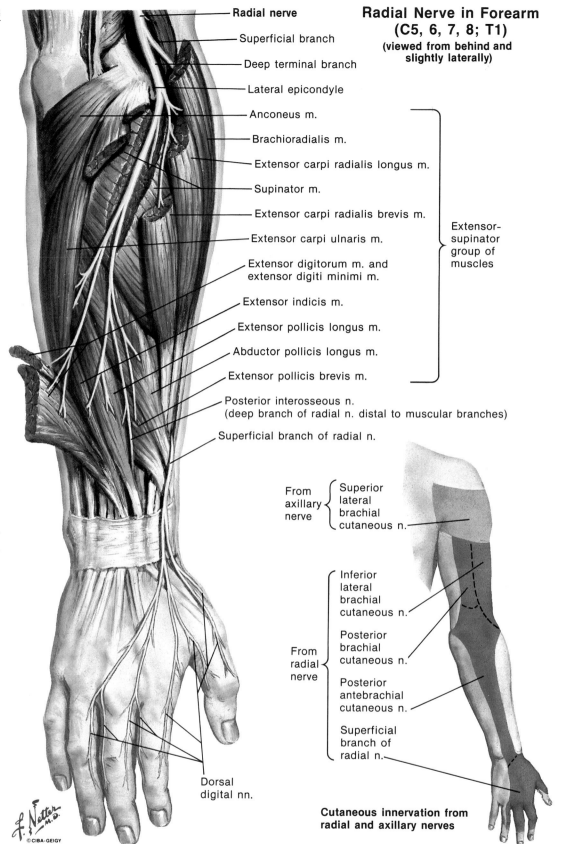

Radial Nerve in Forearm (C5, 6, 7, 8; T1)
(viewed from behind and slightly laterally)

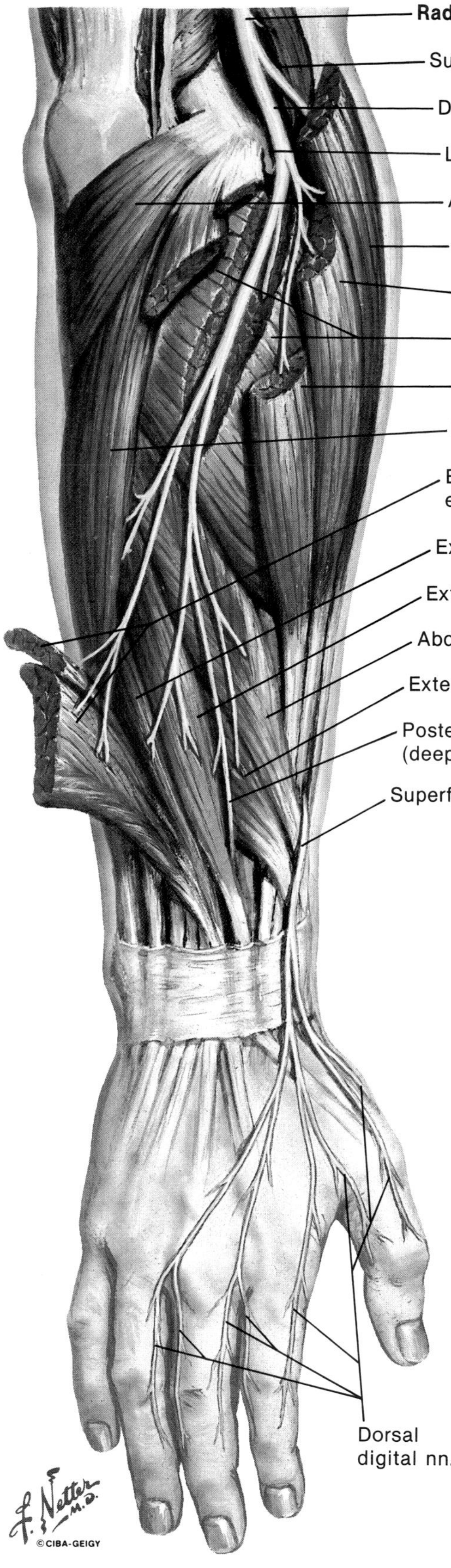
Radial nerve
Superficial branch
Deep terminal branch
Lateral epicondyle
Anconeus m.
Brachioradialis m.
Extensor carpi radialis longus m.
Supinator m.
Extensor carpi radialis brevis m.
Extensor carpi ulnaris m.
Extensor digitorum m. and extensor digiti minimi m.
Extensor indicis m.
Extensor pollicis longus m.
Abductor pollicis longus m.
Extensor pollicis brevis m.
Extensor-supinator group of muscles
Posterior interosseous n. (deep branch of radial n. distal to muscular branches)
Superficial branch of radial n.
Dorsal digital nn.
F. Netter M.D.
©CIBA-GEIGY

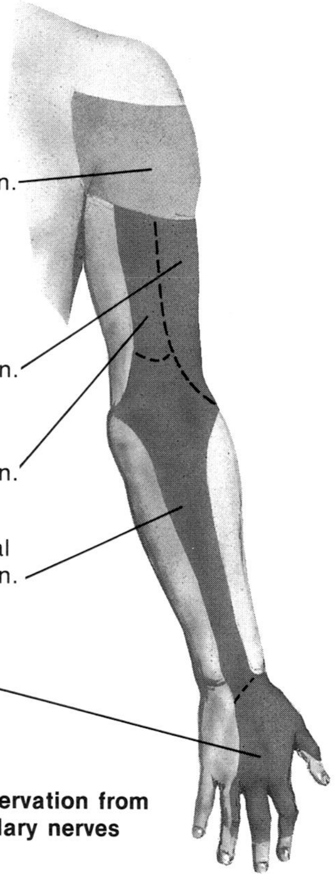
From axillary nerve
Superior lateral brachial cutaneous n.
From radial nerve
Inferior lateral brachial cutaneous n.
Posterior brachial cutaneous n.
Posterior antebrachial cutaneous n.
Superficial branch of radial n.

Cutaneous innervation from radial and axillary nerves

Radial Nerve

In its course to the forearm, the radial nerve (C5, 6, 7, 8; T1) divides anterior to the lateral humeral epicondyle into superficial and deep terminal branches (Plates 36 and 49).

The superficial branch descends along the anterolateral side of the forearm deep to the brachioradialis muscle, lying successively on the supinator, pronator teres, flexor digitorum superficialis, and flexor digitorum longus muscles. In the upper third of the forearm, the superficial branch and the radial artery converge; in the middle third, they are close together, with the nerve lying laterally; and in the lower third, they diverge as the nerve inclines posterolaterally, deep to the tendon of the brachioradialis muscle.

The superficial branch now pierces the deep fascia and commonly subdivides into two branches, which usually split into four or five *dorsal digital nerves*. The cutaneous area of supply is shown in the lower part of Plate 49. The dorsal digital nerves also supply filaments to the adjacent vessels, joints, and bones. (Note that the dorsal digital nerves extend only to the levels of the distal interphalangeal joints and that the first dorsal digital nerve gives off a twig that curves around the radial side of the thumb to supply the skin over the lateral part of the thenar eminence.) The cutaneous areas on the hand supplied by the radial, median, and ulnar nerves show wide individual variations; communications exist between their branches, and considerable marginal overlaps are found in their zones of distribution.

The deep branch courses posteroinferiorly around the lateral side of the radius and may supply additional twigs to the brachioradialis and extensor carpi radialis longus muscles; such twigs supplement twigs given off the main stem of the radial nerve in the lower third of the arm. The deep branch supplies branches to the extensor carpi radialis brevis and supinator muscles before passing between the humeral and radial heads of the supinator, or between this muscle and the upper end of the radial shaft, to reach the back of the forearm. On emerging from or beneath the supinator muscle, the nerve becomes closely related to the posterior interosseous artery, and the resulting neurovascular bundle lies at first between the superficial and deep extensor muscles in the forearm, where the nerve gives off short branches to the nearby muscular bellies of the extensor digitorum, extensor digiti minimi, and extensor carpi ulnaris muscles. Longer branches run distally to supply the extensor pollicis longus, extensor indicis, abductor pollicis longus, and extensor pollicis brevis muscles.

Distal to the muscular branches, the nerve is designated as the posterior interosseous nerve. It passes deep to the extensor pollicis longus muscle and then descends on the interosseous membrane to the back of the wrist. Here, it ends in a small nodule, a pseudoganglion, from which filaments are distributed to the distal radioulnar, radiocarpal, and carpal bones, joints, and ligaments, and to the posterior interosseous and carpal vessels. □

Upper Limb

Blood Supply of Forearm

Arteries and Nerves of Upper Limb
(anterior view)

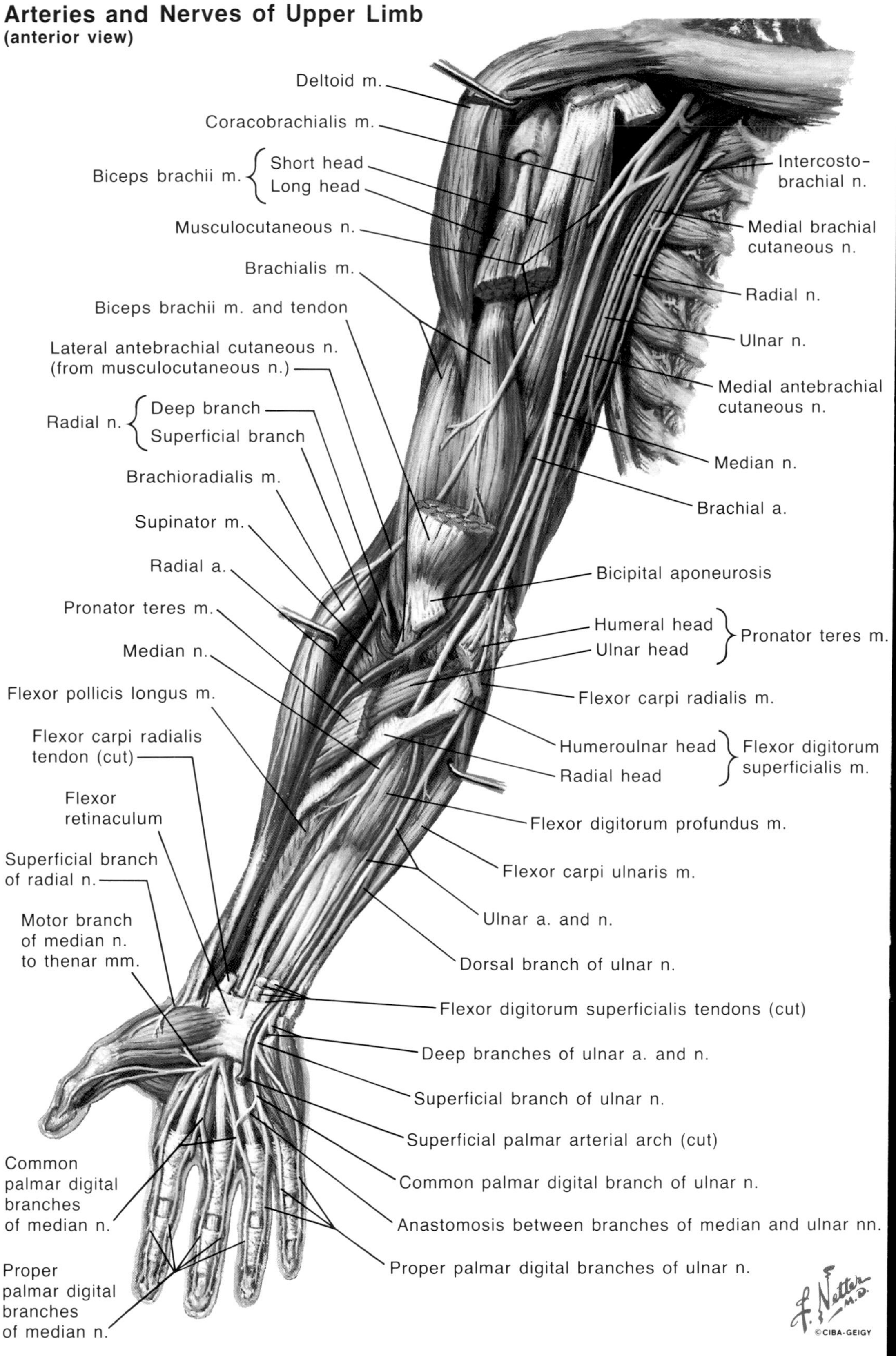

The brachial artery divides into radial and ulnar arteries. Their named branches lie near the elbow and wrist, and only unnamed muscular branches arise in the middle forearm.

Radial Artery. This smaller branch continues the direct line of the brachial artery, ending at the wrist at the "pulse position." It first crosses the tendon of the pronator teres muscle and descends adjacent to the superficial branch of the radial nerve. It branches into the radial recurrent artery and muscular, palmar carpal, and superficial palmar branches.

The *radial recurrent artery* arises shortly below the origin of the radial artery. Ascending on the supinator muscle, the artery supplies this muscle and the brachioradialis and brachialis muscles and anastomoses with the radial collateral artery (Plate 38).

Muscular branches supply the muscles of the radial side of the forearm. The *palmar carpal branch* arises near the distal border of the pronator quadratus muscle. It ends by anastomosing with the palmar carpal branch of the ulnar artery, the termination of the anterior interosseous artery, and recurrent branches of the deep palmar arterial arch.

The *superficial palmar branch* leaves the radial artery just before it turns from the radial border of the wrist onto the back of the hand. It descends over or through the muscles of the thumb and joins the superficial branch of the ulnar artery to form the superficial palmar arterial arch.

Ulnar Artery. This larger branch describes a gentle curve to the ulnar side of the forearm. It passes deep to both heads of the pronator teres and all the other superficial flexor muscles of the forearm and is crossed by the median nerve (Plate 38). It lies deep to the flexor carpi ulnaris muscle and enters the hand in company with the ulnar nerve.

The *anterior ulnar recurrent artery* turns upward from just below the elbow joint between the brachialis and pronator teres muscles. It supplies these muscles and anastomoses with the anterior branches of both ulnar collateral arteries.

The *posterior ulnar recurrent artery*, larger than the anterior, arises near or in common with it. It ascends between the flexor digitorum superficialis and flexor digitorum profundus muscles to supply the elbow joint. It ends in anastomoses with posterior branches of both ulnar collateral arteries and the interosseous recurrent artery and in the network of the olecranon.

The *common interosseous artery* arises from the radial side of the ulnar artery and divides into anterior and posterior interosseous arteries. The *anterior interosseous artery* descends on the anterior surface of the interosseous membrane as far as the upper border of the pronator quadratus muscle in company with veins and the anterior interosseous branch of the median nerve. It gives off nutrient arteries to the radius and ulna and a long slender median artery to the palm. At the upper border of the pronator quadratus muscle, a small *palmar carpal branch* is given off. The artery terminates in the dorsal carpal network.

The *posterior interosseous artery* passes to the back of the upper forearm, emerging between the supinator and abductor pollicis longus muscles with the deep branch of the radial nerve. Sending twigs to the extensor muscles of the forearm, it descends to anastomose with the dorsal terminal branch of the anterior interosseous artery. An interosseous recurrent branch ascends deep to the supinator and anconeus muscles to the interval between the lateral epicondyle of the humerus and the olecranon; there, it communicates with the middle collateral, inferior ulnar collateral, and posterior ulnar recurrent arteries.

Muscular branches of the ulnar artery reach the muscles of the ulnar side of the forearm. The *palmar carpal branch* arises at the upper border of the flexor retinaculum, passes across the wrist deep to the flexor tendons, and unites with the palmar carpal branch of the radial artery. The *dorsal carpal branch* arises just above the pisiform. It winds around the border of the wrist, deep to the tendons, to help form the dorsal carpal arterial arch.

Antebrachial Veins. These venae comitantes of the arteries of the forearm arise from the venous arcades of the hand. They ascend with the radial and ulnar arteries, communicating with superficial veins, and merge into the venae comitantes of the brachial artery at the elbow. □

Upper Limb

Wrist and Hand

Wrist and Hand: Superficial Dissection (palmar view)

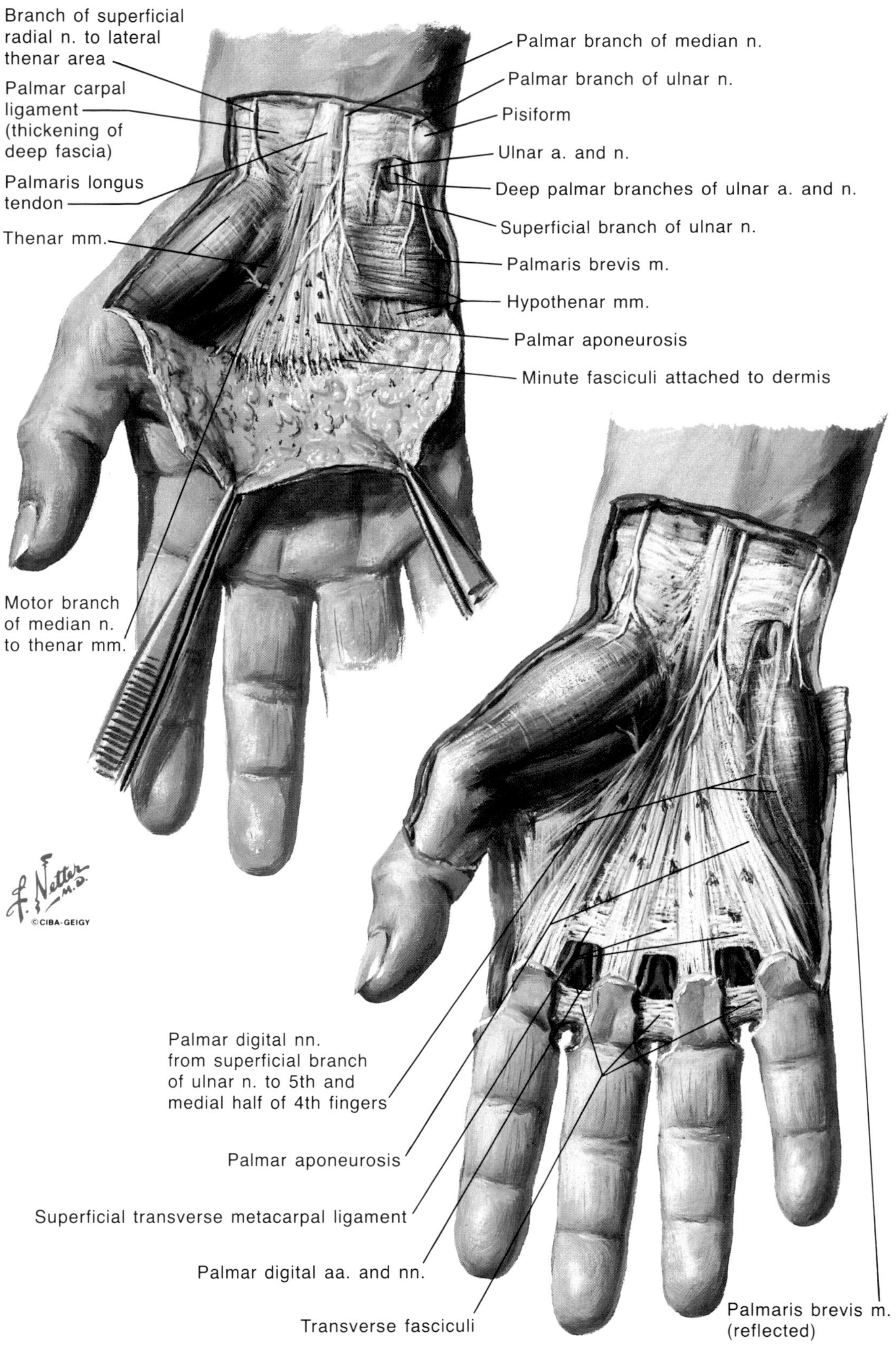
Branch of superficial radial n. to lateral thenar area
Palmar carpal ligament (thickening of deep fascia)
Palmaris longus tendon
Thenar mm.
Palmar branch of median n.
Palmar branch of ulnar n.
Pisiform
Ulnar a. and n.
Deep palmar branches of ulnar a. and n.
Superficial branch of ulnar n.
Palmaris brevis m.
Hypothenar mm.
Palmar aponeurosis
Minute fasciculi attached to dermis
Motor branch of median n. to thenar mm.
©CIBA-GEIGY
Palmar digital nn. from superficial branch of ulnar n. to 5th and medial half of 4th fingers
Palmar aponeurosis
Superficial transverse metacarpal ligament
Palmar digital aa. and nn.
Transverse fasciculi
Palmaris brevis m. (reflected)

Deep Fasciae

The antebrachial fascia is thickened at the wrist to form the extensor retinaculum and the palmar carpal ligament (Plate 51). The dorsal antebrachial fascia is strengthened to form the *extensor retinaculum* by the addition of oblique fibers that extend from the lateral border of the radius to the styloid process of the ulna and to the pisiform and the triquetrum (Plate 56). The extensor retinaculum has deep attachments to the ridges on the dorsum of the radius. The *palmar carpal ligament* is attached to the styloid processes of the radius and ulna.

The *fascia of the palm of the hand* is continuous with the antebrachial fascia of the flexor aspect of the forearm and with the palmar carpal ligament. At the borders of the hand, it is continuous with the fascia of the dorsum at attachments to the first and fifth metacarpals. The *hypothenar fascia* invests the muscles of the little finger and bounds the *hypothenar compartment* of the hand by means of a palmar attachment to the radial side of the fifth metacarpal. In a similar manner, the fascia over the thumb muscles dips deeply to attach to the palmar aspect of the first metacarpal and bounds, with the metacarpal, a *thenar compartment* in the hand. The *central compartment of the palm* is covered by the intervening part of the fascia of the palm, but this portion is reinforced superficially by the *palmar aponeurosis*, an expansion of the tendon of the palmaris longus muscle.

Recognizable in the palmar aponeurosis are a superficial stratum of longitudinally running fibers (which is continuous with the tendon of the palmaris longus muscle) and a deeper layer of transverse fibers. The transverse fibers are continuous with the thenar and hypothenar fasciae; proximally, they are continuous with the flexor retinaculum and the transverse carpal ligament. The palmar aponeurosis broadens distally in the palm and divides into four digital slips, some of its fibers meanwhile attaching to the overlying skin at the skin creases of the palm. The central parts of these slips pass into the digits, attaching superficially to the skin of the crease at the base of each digit; deeply, they attach to the fibrous sheath of the digit. The marginal fibers sink deeply between the heads of the metacarpals and attach to the metacarpophalangeal joint capsules, the deep transverse metacarpal ligaments, and the proximal phalanges of the digits. There is usually no digital slip for the thumb, but longitudinal fibers of the aponeurosis usually curve over onto the thenar fascia.

The deep attachments of the margins of the digital slips of the palmar aponeurosis define the entrance to the fibrous sheath of each digit, but they are also continued proximally into the palm for varying distances. They attach to the palmar interosseous fascia and to the shafts of the metacarpals, thus providing communicating subcompartments for each pair of flexor tendons and the associated lumbrical muscles (Plate 60). The septum reaching the third metacarpal is stronger and more constant; it separates a surgical *thenar space* under the aponeurosis to its radial side and a *midpalmar space* to its ulnar side.

Upper Limb

Wrist and Hand

(Continued)

Wrist and Hand: Superficial Dissection (radial view)

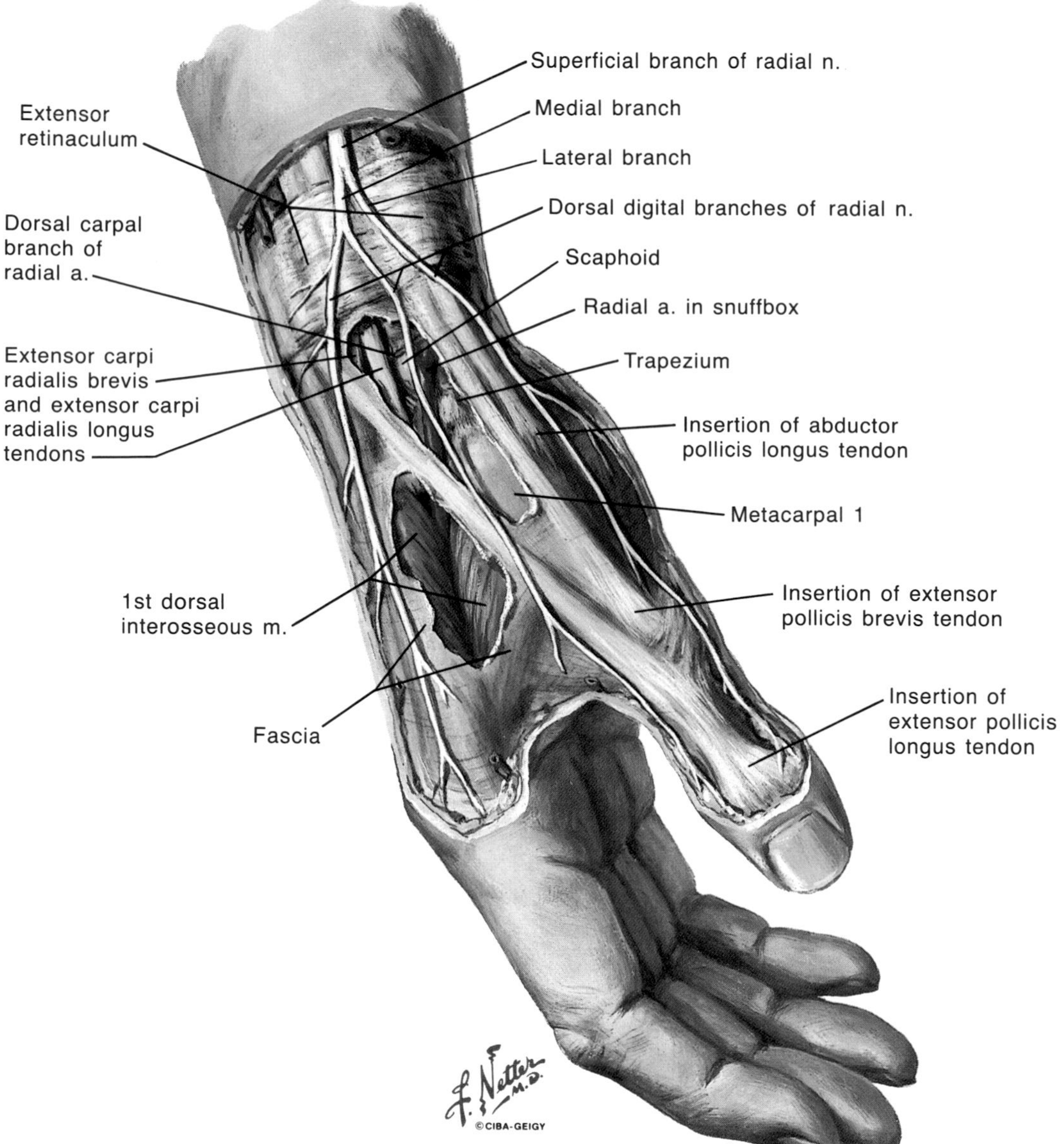
Superficial branch of radial n.
Medial branch
Lateral branch
Dorsal digital branches of radial n.
Scaphoid
Radial a. in snuffbox
Trapezium
Insertion of abductor pollicis longus tendon
Metacarpal 1
Insertion of extensor pollicis brevis tendon
Insertion of extensor pollicis longus tendon
Extensor retinaculum
Dorsal carpal branch of radial a.
Extensor carpi radialis brevis and extensor carpi radialis longus tendons
1st dorsal interosseous m.
Fascia
F. Netter M.D.
©CIBA-GEIGY

Accumulations of the deeper transverse fibers of the aponeurosis appear between the diverging digital slips. Located at the level of the heads of the metacarpals, these fibers are designated as the *superficial transverse metacarpal ligament*. Distally, the webs of the fingers are reinforced by another accumulation of transverse fibers designated as *transverse fasciculi*.

The *fascia of the dorsum of the hand* is continuous with the antebrachial fascia of the extensor surface of the forearm and with the extensor retinaculum. It encloses the tendons of the extensor muscles as they pass to the digits and continues into the extensor expansions on the dorsum of the digits; deep to it is a *subaponeurotic space*. This interfascial cleft separates the fascia of the dorsum from the deeper *dorsal interosseous fascia* covering the dorsal interosseous muscles and the descending branches of the dorsal carpal arterial arch (Plate 59).

Cutaneous Nerves

The *superficial branch of the radial nerve* arises in the cubital fossa by the division there of the radial nerve into deep and superficial branches (Plate 52). The superficial branch, which is entirely cutaneous, courses through the forearm under cover of the brachioradialis muscle and is accompanied by the radial artery. At the distal third of the forearm, the superficial branch of the radial nerve perforates the antebrachial fascia along the lateral border of the forearm and divides into two branches.

The smaller *lateral branch* supplies the skin of the radial side and eminence of the thumb and communicates with the lateral antebrachial cutaneous nerve. The larger *medial branch* divides into four dorsal digital nerves. The first dorsal digital nerve supplies the ulnar side of the thumb; the second supplies the radial side of the index finger; the third distributes to the adjoining sides of the index and middle fingers; and the fourth supplies the adjacent sides of the middle and ring fingers.

There is usually an anastomosis on the back of the hand between the superficial branch of the radial nerve and the dorsal branch of the ulnar nerve, and there is some variability in the apparent source of the last (more median) branch of either nerve. In some such cases, the adjacent sides of the middle and ring fingers are in the territory of the ulnar nerve. Dorsal digital nerves fail to reach the extremities of the digits. They reach to the base of the nail of the thumb, to the distal interphalangeal joint of the second digit, and not quite as far as the proximal interphalangeal joints of the third and fourth digits. The distal areas of the dorsum of the digits not supplied by the radial nerve receive branches from the stout palmar digital branches of the median nerve.

The *dorsal branch of the ulnar nerve* completes the cutaneous supply of the dorsum of the hand and digits (Plates 53–54). It arises about 5 cm above the wrist, passes dorsalward from beneath the flexor carpi ulnaris tendon, and then pierces the forearm fascia. At the ulnar border of the wrist, the nerve divides into three dorsal digital branches.

The first branch courses along the ulnar side of the dorsum of the hand and supplies the ulnar side of the little finger as far as the root of the nail. The second branch divides at the cleft between the fourth and fifth digits and supplies their adjacent sides. The third branch may divide similarly; it may supply the adjacent sides of the third and fourth digits, or it may simply anastomose with the fourth dorsal digital branch of the superficial branch of the radial nerve. The dorsal branches to the fourth digit usually extend only as far as the base of the second phalanx, with the more distal parts of the fourth and fifth digits supplied by palmar digital branches of the ulnar nerve.

The *palmar branch of the ulnar nerve* arises about the middle of the forearm, descending under the antebrachial fascia in front of the ulnar artery (Plate 41). It perforates the fascia just above the wrist and supplies the skin of the hypothenar eminence and the medial part of the palm.

The palmar branch of the median nerve arises just above the wrist (Plate 41). It perforates the palmar carpal ligament between the tendons of the palmaris longus and flexor carpi radialis muscles and distributes to the skin of the central depressed area of the palm and the medial part of the thenar eminence.

Wrist and Hand: Superficial Dissection (dorsal view)

Upper Limb

Wrist and Hand

(Continued)

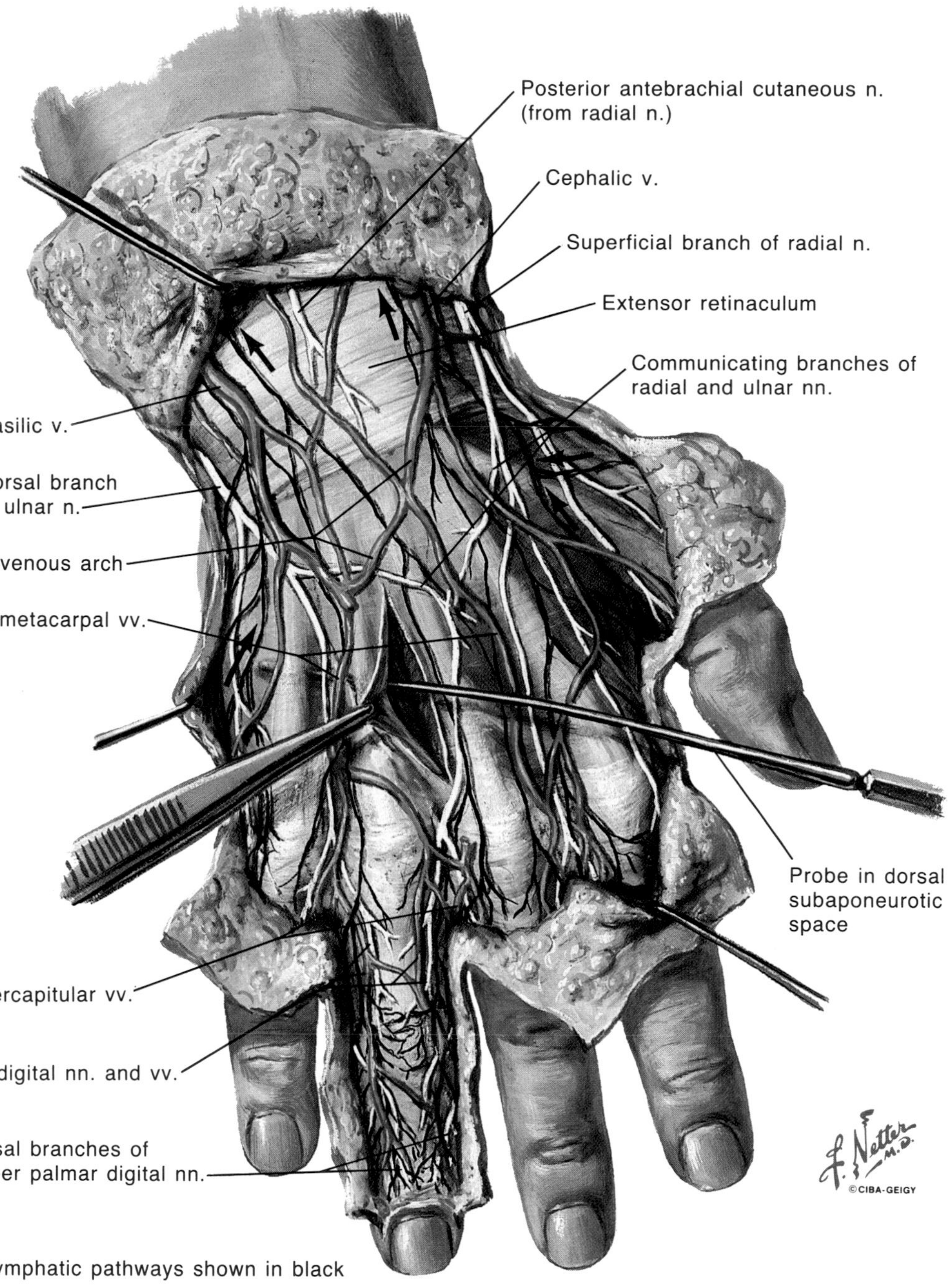
Posterior antebrachial cutaneous n. (from radial n.)
Cephalic v.
Superficial branch of radial n.
Extensor retinaculum
Communicating branches of radial and ulnar nn.
Basilic v.
Dorsal branch of ulnar n.
Dorsal venous arch
Dorsal metacarpal vv.
Probe in dorsal subaponeurotic space
Intercapitular vv.
Dorsal digital nn. and vv.
Dorsal branches of proper palmar digital nn.
©CIBA-GEIGY

Note: lymphatic pathways shown in black

The *digital branches of the median nerve*, the proper palmar digital nerves, lie subcutaneously along the margins of each of the digits distal to the webs of the fingers (Plate 55). They arise from common palmar digital nerves, which lie under the dense palmar aponeurosis of the central palm. The first common palmar digital nerve gives rise to the muscular branch to the short muscles of the thumb and then divides into three *proper palmar digital nerves*. The first two of these run to the radial and ulnar sides of the thumb, respectively, giving numerous branches to the pad and small, dorsally running branches to the nail bed of the thumb. The third proper digital branch supplies the radial side of the second digit. The second common palmar digital branch provides two proper palmar digital nerves, which reach the adjacent sides of the second and third digits. The third common palmar digital nerve communicates with a digital branch of the ulnar nerve in the palm and divides into two proper palmar digital nerves supplying adjacent sides of the third and fourth digits.

Proper palmar digital nerves are large because of the density of nerve endings in the fingers. They lie superficial to the corresponding proper palmar digital arteries and veins. As each nerve passes toward its termination in the pad of the finger, it gives off branches for the innervation of the skin of the dorsum of the digits and the matrices of the fingernails. These dorsal branches innervate the dorsal skin of the distal segment of the index finger, the two terminal segments of the third finger, and the radial side of the fourth finger (Plate 69).

The *digital branches of the ulnar nerve* are, likewise, common palmar and proper palmar, and are terminal branches of the superficial branch of the ulnar nerve (Plate 55). The latter nerve gives rise to a proper palmar digital branch to the ulnar side of the fifth digit and to the nail bed of its dorsal side and also to one common palmar digital nerve. This common palmar digital nerve divides into two proper palmar digital nerves, which supply the contiguous surfaces of the fourth and fifth digits and send branches to the dorsal surfaces of the second and third segments of these digits. A communication with the third common palmar digital branch of the median nerve is made in the palm.

The cutaneous innervation of the wrist and hand is schematically depicted in Plate 54.

Arteries and Nerves of Hand

Arteries of Dorsum of Hand. At the wrist, the *radial artery* shifts from the expanded palmar surface of the radius, through the floor of the anatomic snuffbox, to reach the dorsum of the hand at the proximal end of the first dorsal interosseous space (Plate 54). As it passes under the tendon of the abductor pollicis longus muscle, it gives origin to its *dorsal carpal branch*; continuing distally over the first dorsal interosseous space, it gives origin to the *first dorsal metacarpal artery*. (The radial artery then turns deeply into the palm of the hand and participates in forming the deep palmar arterial arch.) The dorsal carpal branch of the radial artery passes ulnarward across the distal row of carpal bones and under the extensor tendons and joins the dorsal carpal branch of the ulnar artery. Thus is formed the *dorsal carpal arterial arch*.

Three *dorsal metacarpal arteries* descend from this arch on the dorsal interosseous muscles of the second, third, and fourth intermetacarpal intervals, respectively. Opposite the heads of the metacarpals, these vessels divide into *proper dorsal digital arteries*, which proceed distally along the dorsal borders of contiguous digits. These vessels are small and fail to reach the distal phalanges of the digits. Anastomoses are formed between the dorsal metacarpal arteries and the palmar arterial system in two locations: by perforating branches at the bases of the metacarpals and at the division into proper dorsal digital arteries.

Nerves of Palm of Hand. The nerves supplying the palm of the hand are terminal branches of the median and ulnar nerves (Plates 54–55). The *median nerve* enters the palm of the hand under the flexor retinaculum, radial to the tendon of the palmaris longus muscle. Just distal to the flexor retinaculum, its *motor,* or *recurrent, branch* curves sharply into the thenar eminence and supplies the abductor pollicis brevis, flexor pollicis brevis (sometimes only its superficial head), and opponens pollicis muscles. This branch frequently arises from the median nerve together with its first common digital branch. Three common digital branches arise from the median nerve and descend toward the digits.

The *ulnar nerve* enters the hand to the radial side of the pisiform between the palmar carpal ligament and the flexor retinaculum. Just distal to the pisiform, the ulnar nerve divides into superficial and deep branches. The deep branch of the ulnar nerve, with the deep branch of the ulnar

Upper Limb

Wrist and Hand

(Continued)

artery, sinks between the origins of the abductor digiti minimi and the flexor digiti minimi brevis muscles and perforates the origin of the opponens digiti minimi muscle. It supplies these muscles and then curves around the hamulus of the hamate into the central part of the palm of the hand in conjunction with the deep palmar arterial arch. As it crosses the hand deep to the flexor tendons to the digits, the nerve gives twigs to the ulnar two lumbrical muscles and to all the interosseous muscles, both dorsal and palmar. It then supplies the adductor pollicis muscle and gives articular twigs to the wrist joint, and it may send a terminal branch into the deep head of the flexor pollicis brevis muscle.

Arteries of Palm of Hand. Reflection of the palmar aponeurosis brings immediately into view the *superficial palmar arterial arch*, its sources and its branches (Plate 55). The *ulnar artery*, with its accompanying nerve, enters the hand superficial to the flexor retinaculum and to the radial side of the pisiform. It descends, curving radially, to about the midpalm and there anastomoses with the *superficial palmar branch of the radial artery*. This branch passes across or through the muscles of the thenar eminence, supplies this group of muscles, and emerges medial to the eminence to help form the superficial palmar arterial arch. The arch is convex distalward and crosses the palm at the level of the line of the completely abducted thumb.

The branches of the superficial arch supply the medial 3½ digits; the radial 1½ digits are supplied from the deep palmar arterial arch. The superficial arch gives origin to three *common palmar digital arteries*, which proceed distalward on the flexor tendons and lumbrical muscles and superficial to the digital nerves of the palm. They unite at the webs of the fingers with the palmar metacarpal arteries and with distal perforating branches of the dorsal metacarpal arteries. From the short trunks thus formed spring *proper palmar digital arteries*.

Two proper palmar digital arteries run distalward along the adjacent margins of the second to fifth digits. A proper digital branch to the ulnar side of the fifth digit arises from the ulnar artery in the hypothenar compartment. At the webs of the fingers, the digital nerves cross the

Wrist and Hand: Deeper Dissection (dorsal view)

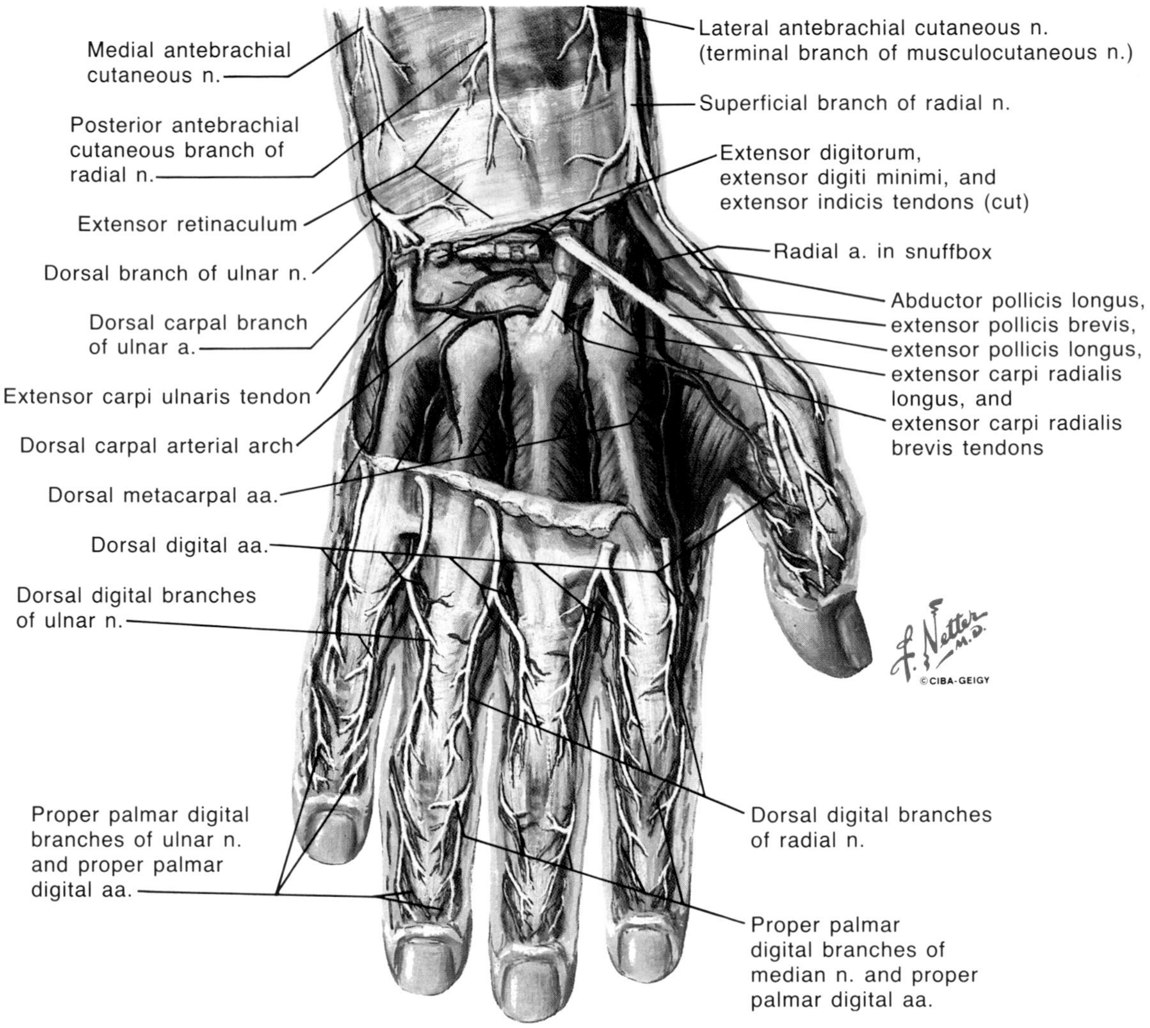
Medial antebrachial cutaneous n.
Posterior antebrachial cutaneous branch of radial n.
Extensor retinaculum
Dorsal branch of ulnar n.
Dorsal carpal branch of ulnar a.
Extensor carpi ulnaris tendon
Dorsal carpal arterial arch
Dorsal metacarpal aa.
Dorsal digital aa.
Dorsal digital branches of ulnar n.
Proper palmar digital branches of ulnar n. and proper palmar digital aa.
Lateral antebrachial cutaneous n. (terminal branch of musculocutaneous n.)
Superficial branch of radial n.
Extensor digitorum, extensor digiti minimi, and extensor indicis tendons (cut)
Radial a. in snuffbox
Abductor pollicis longus, extensor pollicis brevis, extensor pollicis longus, extensor carpi radialis longus, and extensor carpi radialis brevis tendons
Dorsal digital branches of radial n.
Proper palmar digital branches of median n. and proper palmar digital aa.
F. Netter M.D.
©CIBA-GEIGY

Cutaneous Innervation of Wrist and Hand

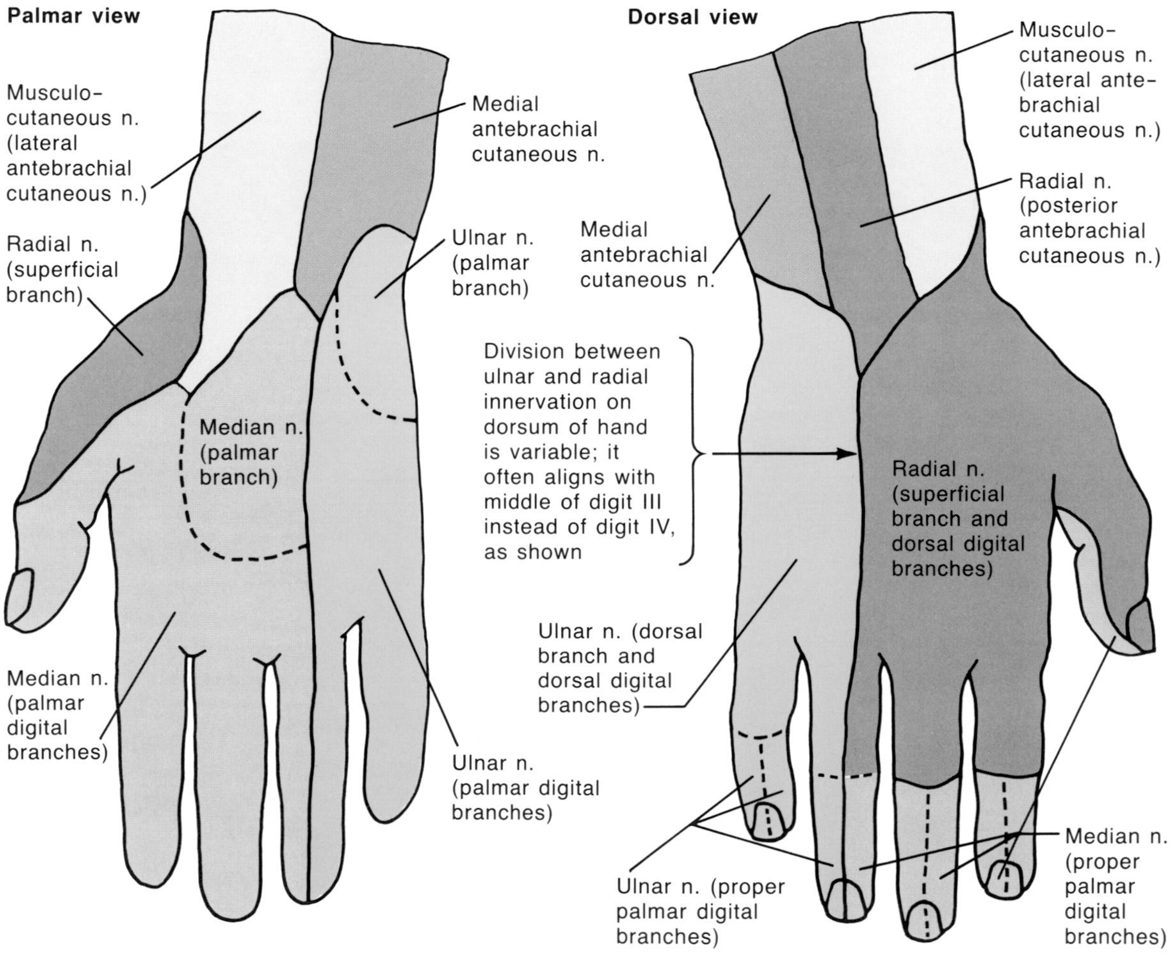
Palmar view
Musculo-cutaneous n. (lateral antebrachial cutaneous n.)
Radial n. (superficial branch)
Median n. (palmar branch)
Median n. (palmar digital branches)
Medial antebrachial cutaneous n.
Ulnar n. (palmar branch)
Ulnar n. (palmar digital branches)
Dorsal view
Medial antebrachial cutaneous n.
Division between ulnar and radial innervation on dorsum of hand is variable; it often aligns with middle of digit III instead of digit IV, as shown
Ulnar n. (dorsal branch and dorsal digital branches)
Ulnar n. (proper palmar digital branches)
Musculo-cutaneous n. (lateral antebrachial cutaneous n.)
Radial n. (posterior antebrachial cutaneous n.)
Radial n. (superficial branch and dorsal digital branches)
Median n. (proper palmar digital branches)

Arteries and Nerves of Hand (palmar view)

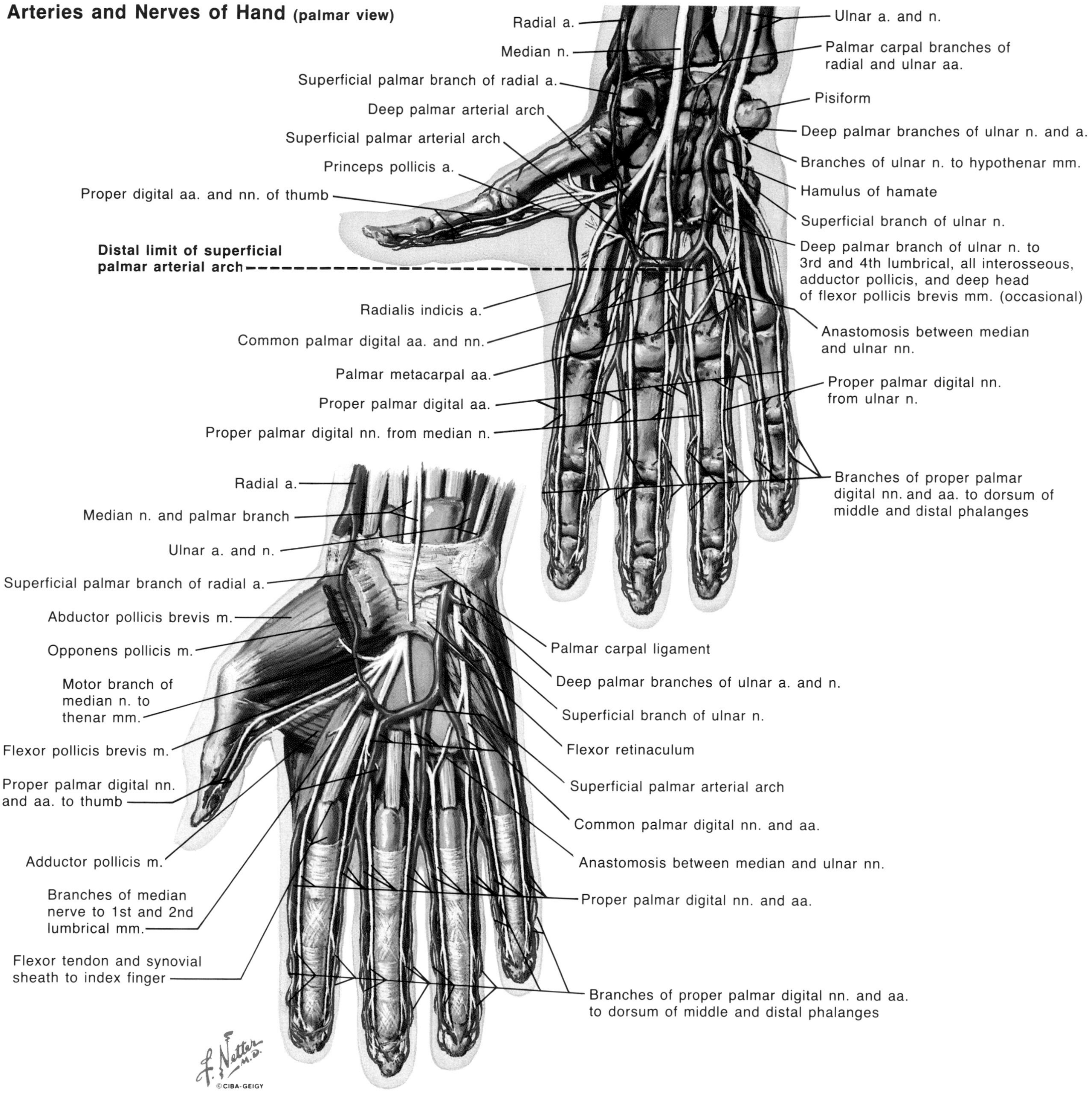
Radial a.
Median n.
Superficial palmar branch of radial a.
Deep palmar arterial arch
Superficial palmar arterial arch
Princeps pollicis a.
Proper digital aa. and nn. of thumb
Distal limit of superficial palmar arterial arch
Radialis indicis a.
Common palmar digital aa. and nn.
Palmar metacarpal aa.
Proper palmar digital aa.
Proper palmar digital nn. from median n.
Ulnar a. and n.
Palmar carpal branches of radial and ulnar aa.
Pisiform
Deep palmar branches of ulnar n. and a.
Branches of ulnar n. to hypothenar mm.
Hamulus of hamate
Superficial branch of ulnar n.
Deep palmar branch of ulnar n. to 3rd and 4th lumbrical, all interosseous, adductor pollicis, and deep head of flexor pollicis brevis mm. (occasional)
Anastomosis between median and ulnar nn.
Proper palmar digital nn. from ulnar n.
Branches of proper palmar digital nn. and aa. to dorsum of middle and distal phalanges
Radial a.
Median n. and palmar branch
Ulnar a. and n.
Superficial palmar branch of radial a.
Abductor pollicis brevis m.
Opponens pollicis m.
Motor branch of median n. to thenar mm.
Flexor pollicis brevis m.
Proper palmar digital nn. and aa. to thumb
Adductor pollicis m.
Branches of median nerve to 1st and 2nd lumbrical mm.
Flexor tendon and synovial sheath to index finger
Palmar carpal ligament
Deep palmar branches of ulnar a. and n.
Superficial branch of ulnar n.
Flexor retinaculum
Superficial palmar arterial arch
Common palmar digital nn. and aa.
Anastomosis between median and ulnar nn.
Proper palmar digital nn. and aa.
Branches of proper palmar digital nn. and aa. to dorsum of middle and distal phalanges
F. Netter M.D.
©CIBA-GEIGY

Upper Limb

Wrist and Hand

(Continued)

arteries to become superficial to them along the margins of the digits. Thus, in each digit, the palmar and dorsal digital arteries lie within the span of the corresponding cutaneous nerves (Plate 69). The proper palmar digital arteries anastomose to form terminal plexuses in the fingers. They also give off branches that supply the last two dorsal segments of the digits.

The *deep palmar arterial arch* is formed by the junction of the terminal portion of the radial artery and the deep branch of the ulnar artery. The radial artery enters the palm at the base of the first intermetacarpal space by penetrating between the two heads of origin of the first dorsal interosseous muscle. Passing then between the transverse and oblique heads of the adductor pollicis muscle, it joins the deep branch of the ulnar artery.

The *princeps pollicis artery* arises from the radial artery as it emerges from the first dorsal interosseous muscle. At the head of the first metacarpal, it provides two proper palmar digital branches for the thumb. The *radialis indicis artery* arises with the princeps pollicis to run along the radial side of the index finger. It is a proper digital artery to the radial side of the index finger.

From the convexity of the arch itself arise three *palmar metacarpal arteries*. These descend under the palmar interosseous fascia of the second to fourth intermetacarpal intervals. At the webs of the fingers, they join the common digital arteries from the superficial arch. *Recurrent carpal branches* are small. They ascend to and help form the palmar carpal network. *Perforating branches* anastomose with the dorsal metacarpal arteries on the dorsum of the hand.

Upper Limb

Wrist and Hand

(Continued)

Extensor Tendons at Wrist

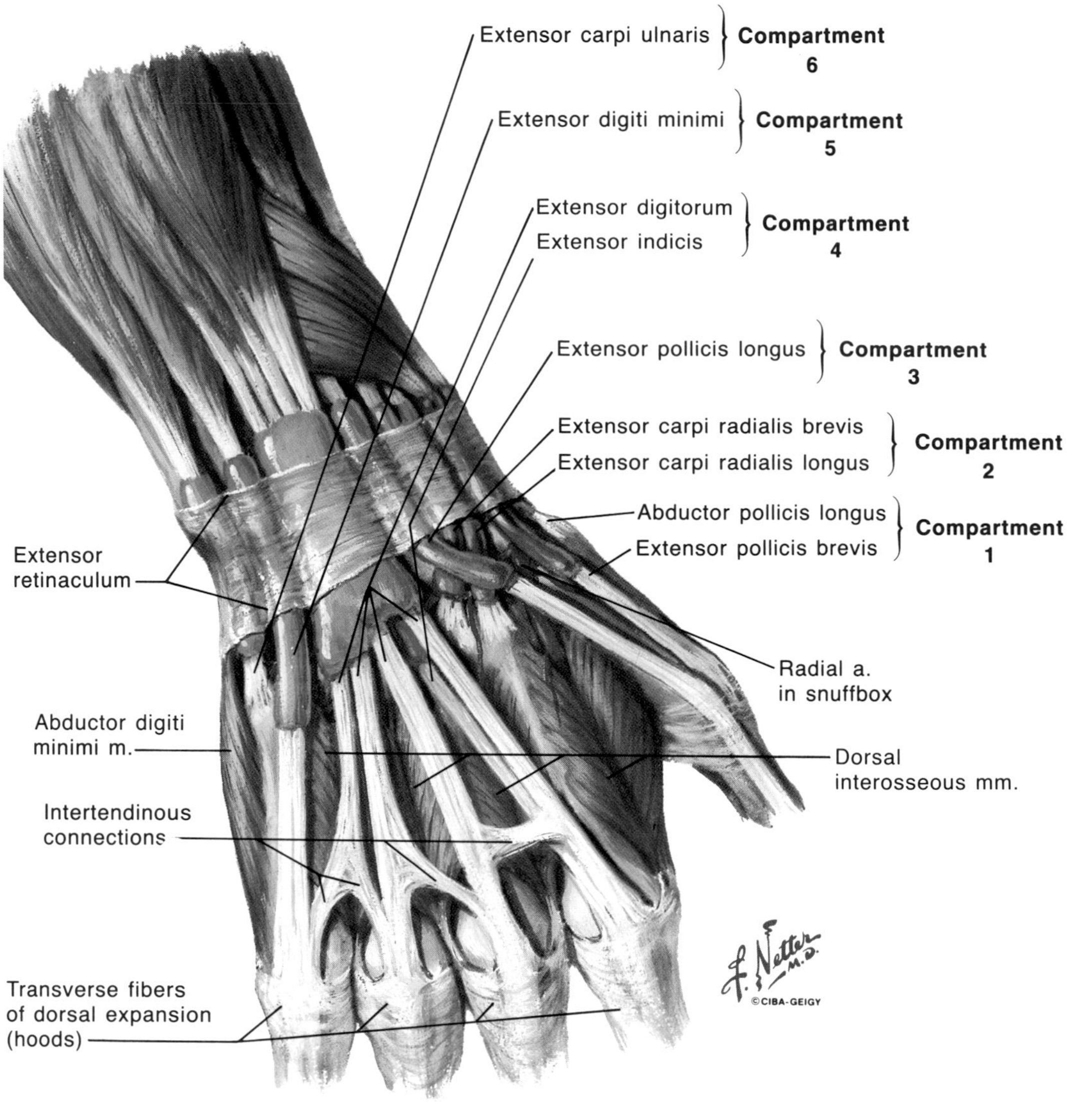
Extensor carpi ulnaris
Compartment 6
Extensor digiti minimi
Compartment 5
Extensor digitorum
Extensor indicis
Compartment 4
Extensor pollicis longus
Compartment 3
Extensor carpi radialis brevis
Extensor carpi radialis longus
Compartment 2
Abductor pollicis longus
Extensor pollicis brevis
Compartment 1
Extensor retinaculum
Radial a. in snuffbox
Abductor digiti minimi m.
Dorsal interosseous mm.
Intertendinous connections
Transverse fibers of dorsal expansion (hoods)
F. Netter M.D.
©CIBA-GEIGY

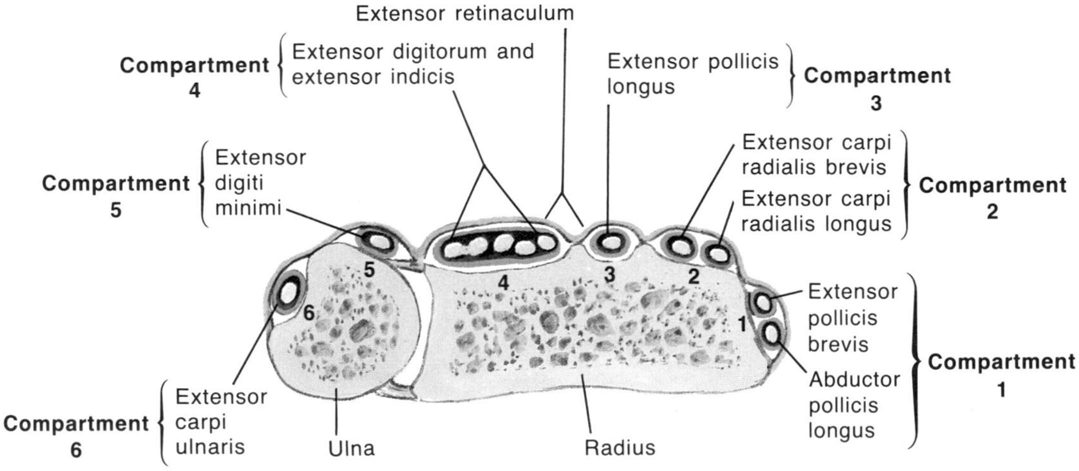
Extensor retinaculum
Compartment 4
Extensor digitorum and extensor indicis
Extensor pollicis longus
Compartment 3
Compartment 5
Extensor digiti minimi
Extensor carpi radialis brevis
Extensor carpi radialis longus
Compartment 2
Extensor pollicis brevis
Abductor pollicis longus
Compartment 1
Compartment 6
Extensor carpi ulnaris
Ulna
Radius
1
2
3
4
5
6

Extensor Tendons at Wrist

The extensor retinaculum defines, by its deep attachments (radius, ulna, and capsular tissues of the joints), six compartments. These accommodate the tendons of the nine extensor muscles that cross the wrist (Plate 56).

The tendons of the abductor pollicis longus and extensor pollicis brevis muscles occupy the first (radialmost) compartment, which is located on the styloid process of the radius. The second compartment lies over the rather smooth area of the radius, radial to its dorsal tubercle; it accommodates the tendons of the extensor carpi radialis longus and extensor carpi radialis brevis muscles. The third compartment, on the ulnar side of the tubercle, is occupied by the tendon of the extensor pollicis longus muscle. This tendon, passing obliquely to its insertion in the distal phalanx of the thumb, forms a prominent border limit for the anatomic snuffbox.

The fourth compartment is large, and the smooth ulnar third of the dorsum of the radius is its floor. In it lie the four tendons of the extensor digitorum muscle, and deep to them, the tendon of the extensor indicis (a fellow traveler). The small fifth compartment is located directly over the distal radioulnar joint and transmits only the tendon of the extensor digiti minimi muscle. The sixth compartment overlies the head of the ulna, limited by the attachment of the extensor retinaculum to the styloid process of the ulna. The tendon of the extensor carpi ulnaris muscle lodges in this compartment.

Synovial Sheaths. Friction between tendons and compartments or bony surfaces is reduced by synovial sheaths. A sheath is formed like a double-walled tube: the delicate inner wall is closely applied to the tendon, and the outer wall is the lining of the compartment in which the tendon lies. The layers are continuous with each other at the ends of the tube (as elsewhere); their facing surfaces are smooth and separated by a small amount of synovial fluid. Each of the compartments on the dorsum of the wrist contains a synovial sheath for the tendon or tendons within it. The upper limits of these sheaths are just above the extensor retinaculum, and the lower limits extend for variable distances—shorter on the "carpi" muscle tendons, where the insertion is not far beyond the retinaculum, and longer on the "digital" muscle tendons, where the excursion of the tendon is greater.

As the extensor digitorum tendons diverge over the dorsum of the hand, they are interconnected by *intertendinous connections.* These prominently interconnect the tendons for the third, fourth, and fifth digits, and severely limit the independent action of these digits, especially the fourth. Independent extensor action is retained for the index finger. The convergence of the tendon of the extensor pollicis longus muscle toward the tendons of the abductor pollicis longus and extensor pollicis brevis muscles defines a hollow known as the *anatomic snuffbox* (Plate 52). In the floor of this hollow, the radial artery passes toward the dorsum of the hand and gives off its dorsal carpal branch.

Hand

Flexor Tendons, Arteries, and Nerves at Wrist

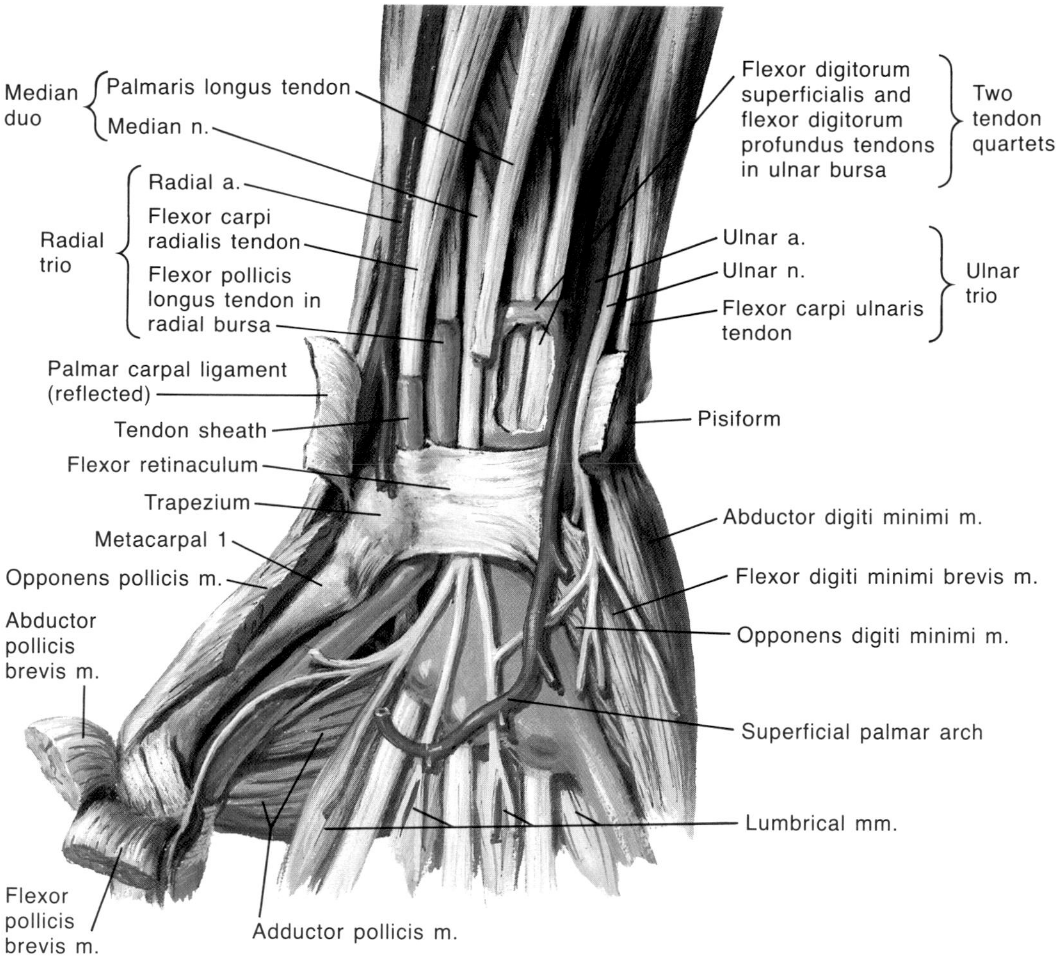
Median duo
Palmaris longus tendon
Median n.
Radial trio
Radial a.
Flexor carpi radialis tendon
Flexor pollicis longus tendon in radial bursa
Palmar carpal ligament (reflected)
Tendon sheath
Flexor retinaculum
Trapezium
Metacarpal 1
Opponens pollicis m.
Abductor pollicis brevis m.
Flexor pollicis brevis m.
Adductor pollicis m.
Flexor digitorum superficialis and flexor digitorum profundus tendons in ulnar bursa
Two tendon quartets
Ulnar a.
Ulnar n.
Flexor carpi ulnaris tendon
Ulnar trio
Pisiform
Abductor digiti minimi m.
Flexor digiti minimi brevis m.
Opponens digiti minimi m.
Superficial palmar arch
Lumbrical mm.

Cross section just above flexor retinaculum

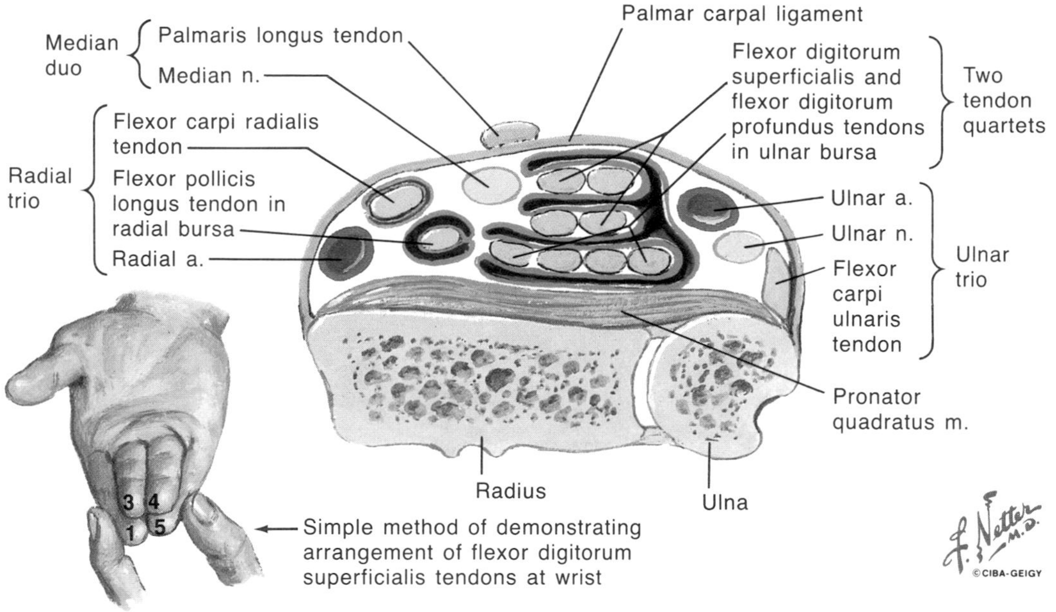
Palmar carpal ligament
Median duo
Palmaris longus tendon
Median n.
Radial trio
Flexor carpi radialis tendon
Flexor pollicis longus tendon in radial bursa
Radial a.
Flexor digitorum superficialis and flexor digitorum profundus tendons in ulnar bursa
Two tendon quartets
Ulnar a.
Ulnar n.
Flexor carpi ulnaris tendon
Ulnar trio
Pronator quadratus m.
Radius
Ulna
3
4
1
5
Simple method of demonstrating arrangement of flexor digitorum superficialis tendons at wrist

Wrist

ulum is deep and partially dis- carpal ligament. It stretches of the concavity of the carpal their arch into a fibroosseous nnel, through which pass into rable number of tendons and ve (Plate 57).

aculum, 2 to 3 cm long, is is long. It is attached radially the scaphoid and to both lips trapezium. Toward the ulna, aches to the hamulus (hook) of he pisiform. The double radial its a split of the flexor retinac- bs, between which the groove commodates the tendon of the muscle.

ger space of the carpal tunnel of the flexor pollicis longus, uperficialis, and flexor digi- muscles. The tendons of the perficialis muscle destined for digits lie superficially in the endons for the second and fifth . Still deeper in the compart- ndons of the flexor digitorum lined up side by side. The pollicis longus muscle passes l tunnel to the radial side unnel. The median nerve also under the flexor retinaculum; perficial row of flexor tendons. de of the wrist, the tendons of and flexor carpi ulnaris mus- d with synovial sheaths. The ever, are protected in the wrist

and hand by rather complex synovial coverings. A long synovial sheath for the tendon of the flexor pollicis longus muscle extends along this tendon from several centimeters above the flexor retinaculum to just proximal to its insertion on the distal phalanx of the thumb. This sheath is commonly, but not officially, designated the radial bursa.

The ulnar bursa, the complex covering of the digital flexor tendons, exhibits the more primitive, wrapped-around (invaginated) character of synovial sheaths, for the tendons of the flexor digitorum superficialis muscle are folded into it superficially from the radial side, and those of the flexor digitorum profundus muscle are folded into the sac more deeply. The general sheath for these eight tendons continues to about the middle of the palm and ends there, except for the sheath on the tendons for the fifth digit, which continues until just short of the insertion of the flexor digitorum profundus muscle. The radial and ulnar bursae occasionally communicate with one another.

Separate synovial sheaths cover the digital parts of the flexor tendons for the second, third, and fourth digits. The tendon of the flexor carpi radialis muscle also has a synovial sheath protecting it in the Y formed by the radial split in the flexor retinaculum. This sheath extends as far as the insertion of the tendon.

Upper Limb

Wrist and Hand

(Continued)

Intrinsic Muscles of Hand

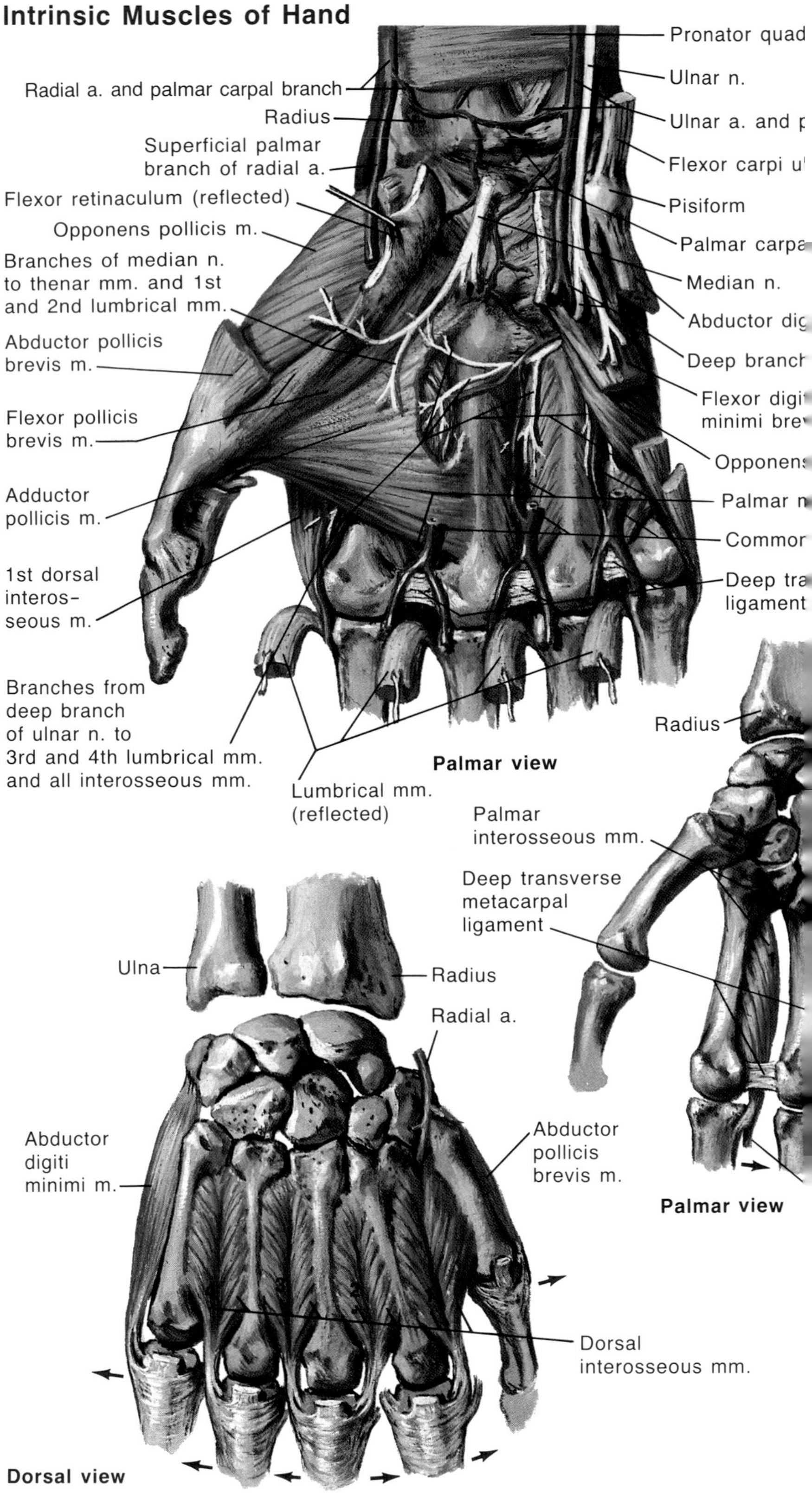
Pronator quad
Radial a. and palmar carpal branch
Ulnar n.
Radius
Ulnar a. and p
Superficial palmar branch of radial a.
Flexor carpi u
Flexor retinaculum (reflected)
Pisiform
Opponens pollicis m.
Palmar carpa
Branches of median n. to thenar mm. and 1st and 2nd lumbrical mm.
Median n.
Abductor dig
Abductor pollicis brevis m.
Deep branch
Flexor digi minimi bre
Flexor pollicis brevis m.
Opponens
Adductor pollicis m.
Palmar n
Common
1st dorsal interos-seous m.
Deep tra ligament
Branches from deep branch of ulnar n. to 3rd and 4th lumbrical mm. and all interosseous mm.
Lumbrical mm. (reflected)
Palmar view
Radius
Palmar interosseous mm.
Deep transverse metacarpal ligament
Ulna
Radius
Radial a.
Abductor digiti minimi m.
Abductor pollicis brevis m.
Palmar view
Dorsal interosseous mm.
Dorsal view

Note: arrows indicate a

Intrinsic Muscles of Hand

The intrinsic muscles of the hand are palmar in location, and are therefore innervated by either the median or the ulnar nerve (Plates 58–60). Specific sets of muscles of the thumb and little finger, respectively, occupy the thenar and hypothenar compartments.

Each compartment contains an *abductor*, an *opponens*, and a *flexor* muscle for its specific digit (abductor pollicis brevis, flexor pollicis brevis, opponens pollicis, abductor digiti minimi, flexor digiti minimi brevis, and opponens digiti minimi muscles). In each compartment, the positions and attachments of these muscles are similar. The flexor retinaculum and the bones to which it attaches (the scaphoid and trapezium radially and the hamate and pisiform on the ulnar side) provide the sites of origin for these muscles. The insertions of comparable muscles on the two sides are also the same: the base of the proximal phalanx for the abductor and flexor muscles and the shaft of the metacarpal for the opponens muscles.

The central compartment contains four slender *lumbrical muscles* associated with the flexor digitorum profundus tendon. The *interosseous muscles* located in the intervals between the metacarpals occupy, with the *adductor pollicis muscle*, a deeply placed interosseous-adductor compartment that is bounded by the dorsal and palmar interosseous fasciae. To complete these generalizations, the rule of nervous innervation may also be stated: the *median nerve* supplies the abductor pollicis brevis, opponens pollicis, flexor pollicis brevis, and the radialmost two lumbrical muscles; the *ulnar nerve* supplies all the other intrinsic muscles of the hand.

The *adductor pollicis muscle* has two heads of origin, separated by a gap through which the radial artery enters the palm. The oblique head arises from the capitate and from the bases of the second and third metacarpals. The transverse head arises from the palmar ridge (shaft) of the third metacarpal. The two heads insert together by a tendon that ends in the ulnar side of the base of the proximal phalanx of the thumb.

This tendon usually contains a sesamoid that together with the sesamoid in the tendon of the flexor pollicis brevis muscle forms a pair of small sesamoids on either side of the tendon of the flexor pollicis longus muscle. The adductor pollicis muscle overlies the interosseous muscles on the radial side of the third metacarpal. This muscle is supplied by the deep branch of the ulnar nerve.

The *interosseous muscles* occupy the intermetacarpal intervals and are of two types, dorsal and palmar. Each intermetacarpal space contains one palmar and one dorsal interosseous muscle. The four dorsal interosseous muscles are abductors of the digits and are bipennate in form; the three palmar interosseous muscles are adductors and are unipennate. The plane of reference for abduction and adduction of the fingers is the midplane of the third digit. This is evident on simultaneously spreading and then approximating the extended digits. The placement of these muscles follows

from the above consideration
erence plane for abduction an

A *dorsal interosseous muscle*
of the third metacarpal, sinc
the third digit away from it
is abduction. The other two
muscles occupy the space be
second metacarpals for the
and between the fourth and
the fourth dorsal muscle. Th
cles abduct digits II and IV.
interosseous muscles arise by
adjacent sides of the metaca
they lie. The first dorsal
is considerably larger than t

Upper Limb

Wrist and Hand

(Continued)

Spaces, Bursae, and Tendon and Lumbrical Sheaths of Hand

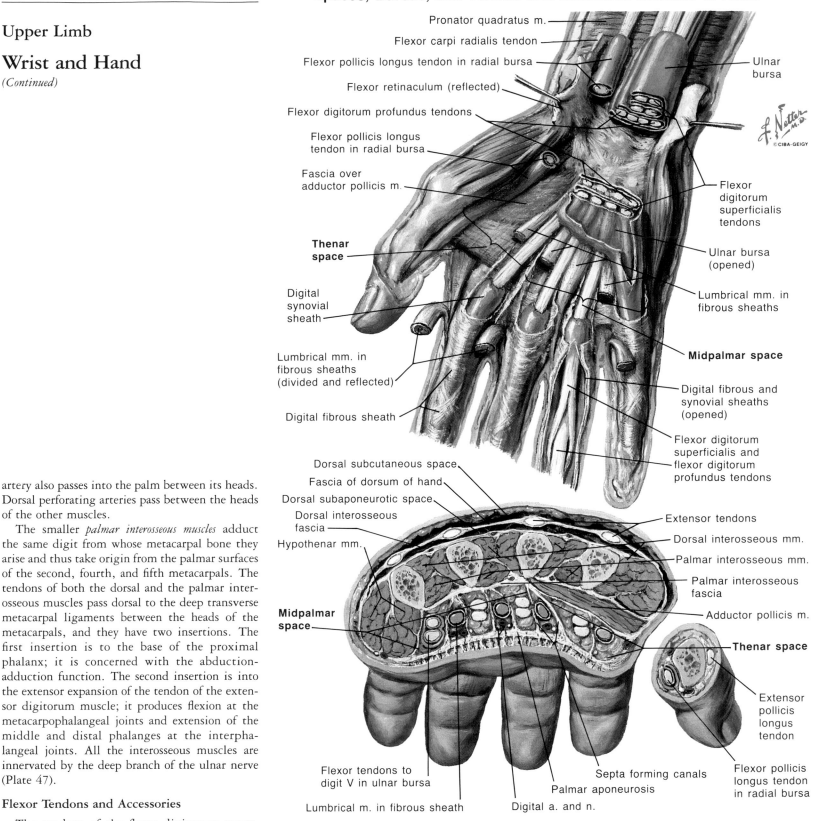

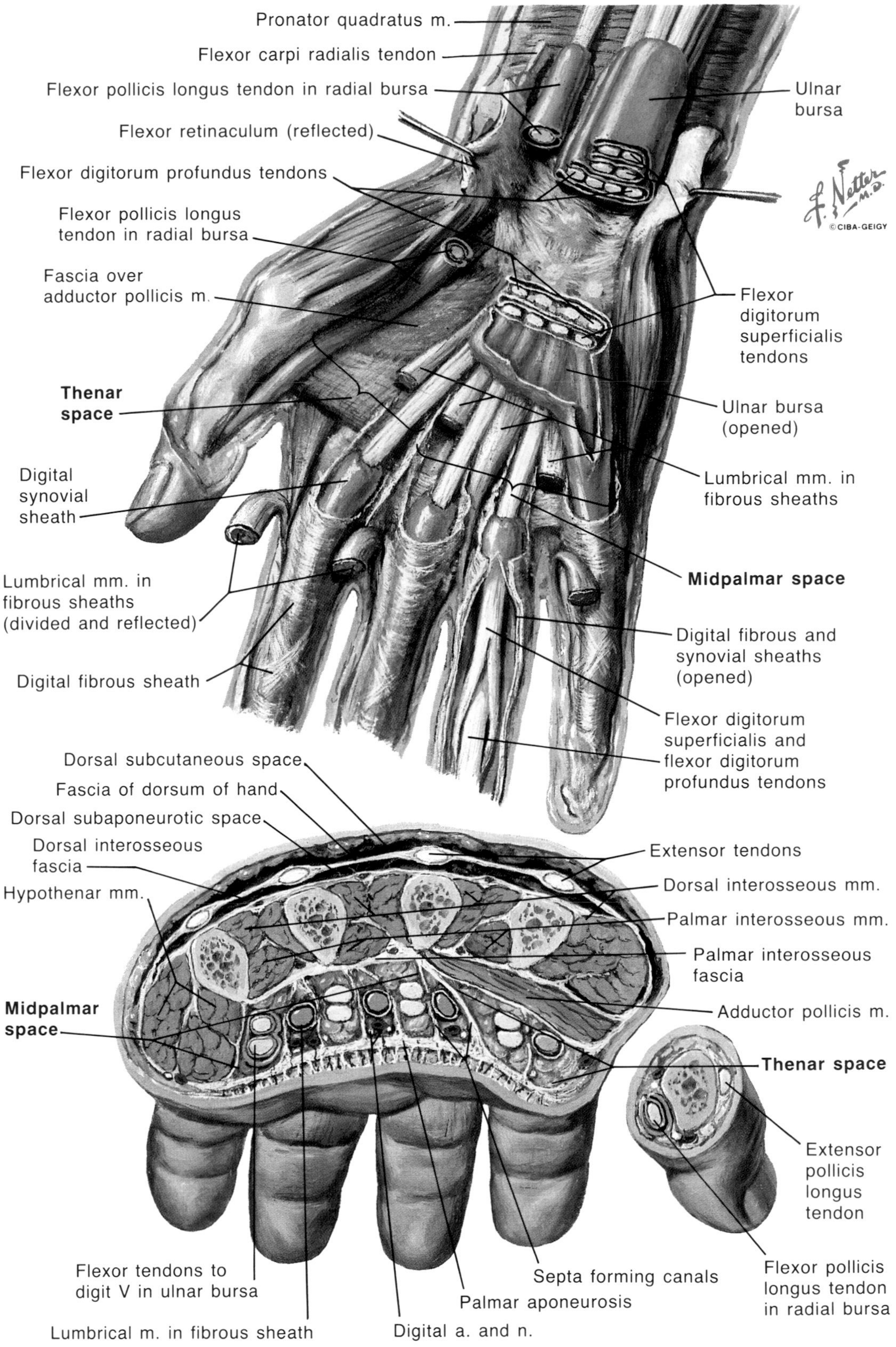

artery also passes into the palm between its heads. Dorsal perforating arteries pass between the heads of the other muscles.

The smaller *palmar interosseous muscles* adduct the same digit from whose metacarpal bone they arise and thus take origin from the palmar surfaces of the second, fourth, and fifth metacarpals. The tendons of both the dorsal and the palmar interosseous muscles pass dorsal to the deep transverse metacarpal ligaments between the heads of the metacarpals, and they have two insertions. The first insertion is to the base of the proximal phalanx; it is concerned with the abduction-adduction function. The second insertion is into the extensor expansion of the tendon of the extensor digitorum muscle; it produces flexion at the metacarpophalangeal joints and extension of the middle and distal phalanges at the interphalangeal joints. All the interosseous muscles are innervated by the deep branch of the ulnar nerve (Plate 47).

Flexor Tendons and Accessories

The tendons of the flexor digitorum superficialis and flexor digitorum profundus muscles emerge from the wrist at the distal border of the flexor retinaculum and enter the central compartment of the palm (Plate 60). Here, they fan out toward their respective digits, arranged in pairs, superficial and deep. They are invested by the ulnar bursa through the upper part of the palm, except that the extension of the bursa along the tendons for the fifth digit continues to the base of its distal phalanx. Digital synovial sheaths, after a gap in the midpalm, pick up over the heads of the metacarpals and continue over the pairs of tendons to the base of the distal phalanges of digits II, III, and IV. Except for about 5 mm of their proximal ends, these synovial sheaths (and the tendons) are contained within the fibrous sheaths of the digits of the hand.

The *fibrous sheaths of the digits* are strong coverings of the flexor tendons, which extend from the heads of the metacarpals to the base of the distal phalanges and serve to prevent "bowstringing" of the tendon away from the bones during flexion. They attach along the borders of the proximal and middle phalanges, the capsules of the interphalangeal joints, and the palmar surface of the distal phalanx. They form strong semicylindric sheaths that, with the bones, produce fibroosseous tunnels through which the flexor tendons pass to their insertions. Over the shafts of the proximal and middle phalanges, the sheaths exhibit thick accumulations of transversely running fibers (sometimes called annular ligaments, or pulleys), whereas opposite the joints, an obliquely crisscrossing arrangement is characteristic (cruciate ligaments). These latter portions of the fibrous sheaths are thin and do not interfere with flexion at the joints. Proximally, the digital slips of the palmar aponeurosis attach to the fibrous digital sheaths.

The tendons of the flexor digitorum profundus muscle insert on the bases of the distal phalanges of digits II to V, while the tendons of the flexor digitorum superficialis muscle end on the shafts of the middle phalanges of these digits. It is thus necessary for the tendons of the flexor digitorum

Wrist and Hand: Deeper Dissection (palmar view)

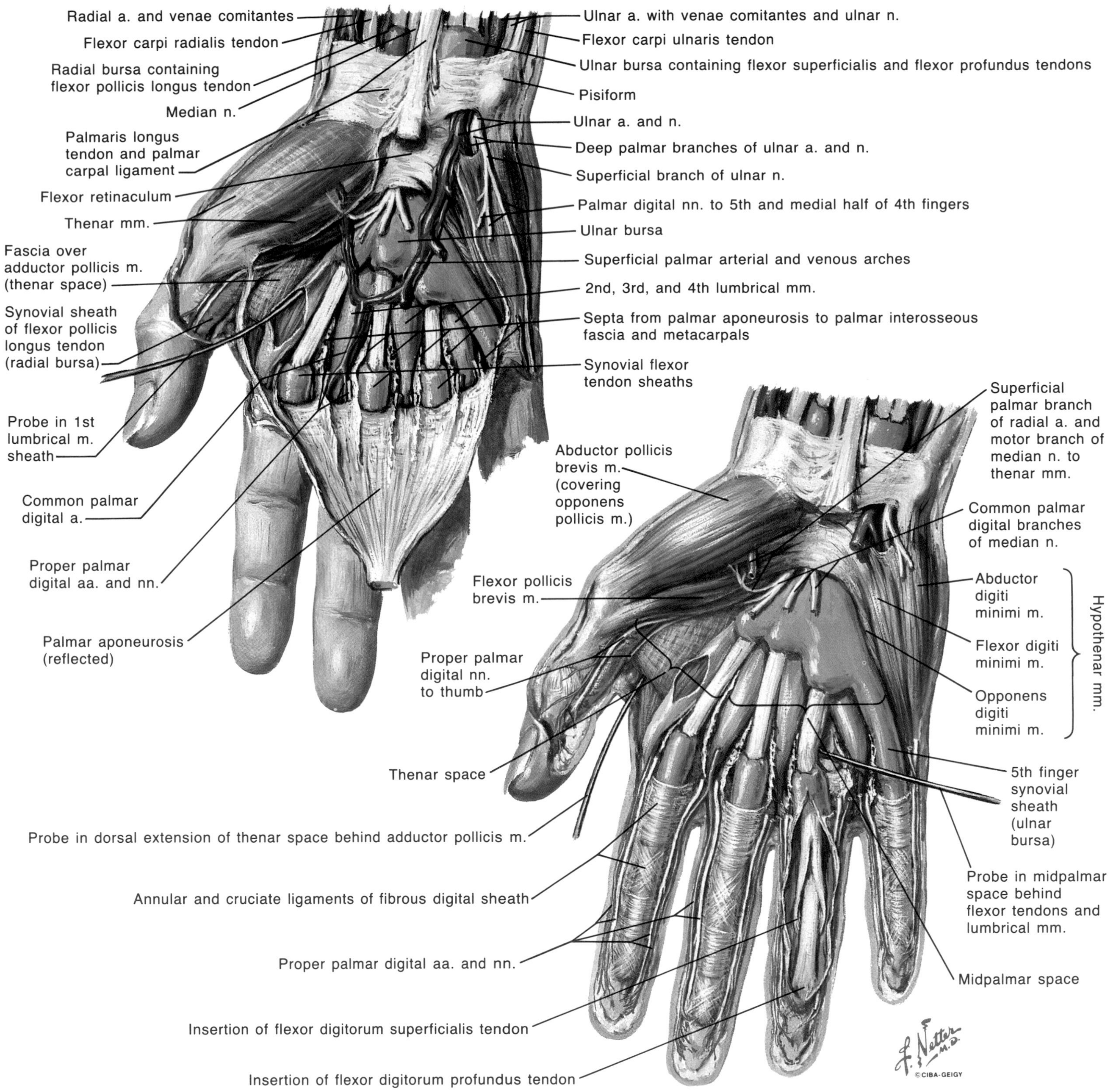

Radial a. and venae comitantes
Flexor carpi radialis tendon
Radial bursa containing flexor pollicis longus tendon
Median n.
Palmaris longus tendon and palmar carpal ligament
Flexor retinaculum
Thenar mm.
Fascia over adductor pollicis m. (thenar space)
Synovial sheath of flexor pollicis longus tendon (radial bursa)
Probe in 1st lumbrical m. sheath
Common palmar digital a.
Proper palmar digital aa. and nn.
Palmar aponeurosis (reflected)
Ulnar a. with venae comitantes and ulnar n.
Flexor carpi ulnaris tendon
Ulnar bursa containing flexor superficialis and flexor profundus tendons
Pisiform
Ulnar a. and n.
Deep palmar branches of ulnar a. and n.
Superficial branch of ulnar n.
Palmar digital nn. to 5th and medial half of 4th fingers
Ulnar bursa
Superficial palmar arterial and venous arches
2nd, 3rd, and 4th lumbrical mm.
Septa from palmar aponeurosis to palmar interosseous fascia and metacarpals
Synovial flexor tendon sheaths
Superficial palmar branch of radial a. and motor branch of median n. to thenar mm.
Abductor pollicis brevis m. (covering opponens pollicis m.)
Common palmar digital branches of median n.
Flexor pollicis brevis m.
Abductor digiti minimi m.
Flexor digiti minimi m.
Opponens digiti minimi m.
Hypothenar mm.
Proper palmar digital nn. to thumb
Thenar space
5th finger synovial sheath (ulnar bursa)
Probe in dorsal extension of thenar space behind adductor pollicis m.
Probe in midpalmar space behind flexor tendons and lumbrical mm.
Annular and cruciate ligaments of fibrous digital sheath
Proper palmar digital aa. and nn.
Midpalmar space
Insertion of flexor digitorum superficialis tendon
Insertion of flexor digitorum profundus tendon
F. Netter M.D.
©CIBA-GEIGY

SECTION I PLATE 60 Slide 3625

Upper Limb

Wrist and Hand

(Continued)

profundus muscle to pass those of the flexor digitorum superficialis muscle, and this is accomplished by a splitting of the tendon of the superficialis to allow that of the profundus to pass distalward. The division of the flexor digitorum superficialis tendon takes place over the proximal phalanx, and the two halves separate and roll in under the flexor digitorum profundus tendon to reach the bone of the middle phalanx, their fibers crisscrossing as they attach to that phalanx.

The *vincula tendinum* spring from the internal surface of the digital sheaths of these muscles (Plate 61). They are folds of synovial membrane strengthened by some fibrous tissue, which conduct blood vessels to the tendons. The smaller *vinculum breve* is at the distal end of the sheath; the *vincula longa* are narrow strands that reach the tendons more proximally.

The *lumbrical muscles* are four small, cylindric muscles associated with the tendons of the flexor digitorum profundus muscle. The two lateral muscles arise distal to the flexor retinaculum from the radial sides and palmar surfaces of the flexor digitorum profundus muscle destined for the second and third digits. These are supplied by the median nerve. The two medial muscles arise from the contiguous sides of the tendons for the third and fourth and fourth and fifth digits. These are innervated by the deep branch of the ulnar nerve. Each lumbrical tendon passes distalward on the palmar side of the deep transverse metacarpal ligament and then shifts toward the dorsum. It inserts, at the level of the proximal phalanx, into the radial border of the expansion of the extensor digitorum muscle.

Upper Limb

Wrist and Hand

(Continued)

Flexor and Extensor Tendons in Fingers

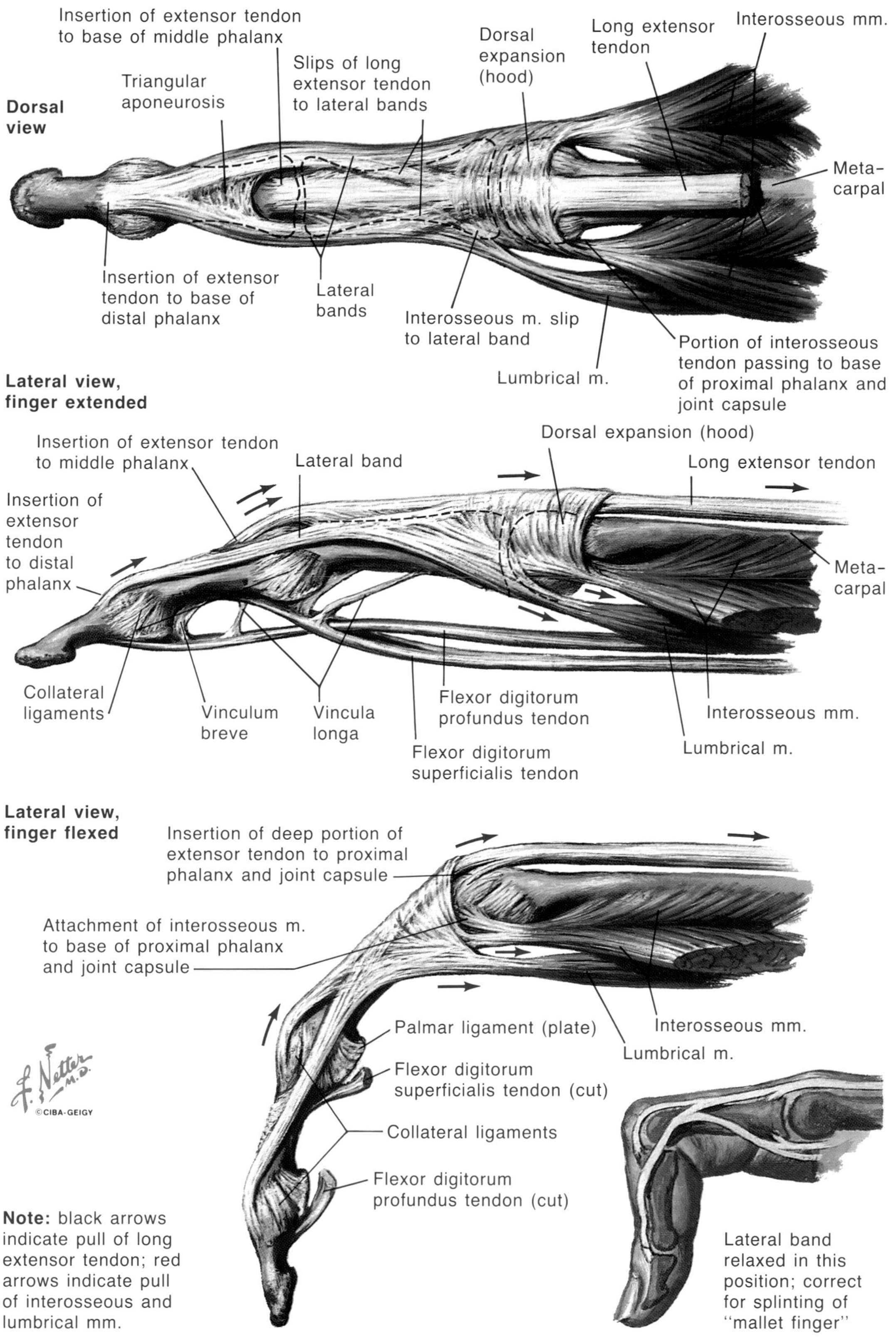

Dorsal view
Insertion of extensor tendon to base of middle phalanx
Triangular aponeurosis
Slips of long extensor tendon to lateral bands
Dorsal expansion (hood)
Long extensor tendon
Interosseous mm.
Meta- carpal
Insertion of extensor tendon to base of distal phalanx
Lateral bands
Interosseous m. slip to lateral band
Lumbrical m.
Portion of interosseous tendon passing to base of proximal phalanx and joint capsule
Lateral view, finger extended
Insertion of extensor tendon to middle phalanx
Lateral band
Dorsal expansion (hood)
Long extensor tendon
Insertion of extensor tendon to distal phalanx
Meta- carpal
Collateral ligaments
Vinculum breve
Vincula longa
Flexor digitorum profundus tendon
Flexor digitorum superficialis tendon
Interosseous mm.
Lumbrical m.
Lateral view, finger flexed
Insertion of deep portion of extensor tendon to proximal phalanx and joint capsule
Attachment of interosseous m. to base of proximal phalanx and joint capsule
Palmar ligament (plate)
Interosseous mm.
Lumbrical m.
Flexor digitorum superficialis tendon (cut)
Collateral ligaments
Flexor digitorum profundus tendon (cut)
Note: black arrows indicate pull of long extensor tendon; red arrows indicate pull of interosseous and lumbrical mm.
Lateral band relaxed in this position; correct for splinting of "mallet finger"
F. Netter M.D.
©CIBA-GEIGY

Extensor Mechanism of Fingers

The four tendons of the extensor digitorum muscle of the forearm pass across the metacarpophalangeal joints, become flattened and closely attached to the joint capsules, and substitute as dorsal ligaments for these capsules (Plate 61). At the metacarpophalangeal joint and over the proximal two phalanges, an extensor expansion is formed for each tendon by the participation of the tendons of the lumbrical and interosseous muscles of the hand. Opposite the metacarpophalangeal joints, a band of fibers passes from each side of the digital extensor tendon anteriorly on either side of the joint and attaches to the palmar ligament of the joint. This proximal spreading of the extensor expansion appears like a hood of fibers over the metacarpophalangeal joint.

Over the dorsum of the proximal phalanx, the digital extensor tendon divides into three slips. Of these, the central, broader slip passes directly forward and inserts on the dorsum of the middle phalanx. The diverging bundles on either side, the lateral bands, receive and combine with the broadening tendon of a lumbrical muscle on the radial side of the digit, and with interosseous tendons on both sides of the digit. These tendons unite into a common band that proceeds distalward, the bands of the two sides forming a triangular aponeurosis over the distal end of the middle phalanx. The apex of this aponeurosis attaches to the base of the distal phalanx.

Muscle Actions in Digital Movement. Certain forearm muscles participate in movements of the digits. The *flexor digitorum superficialis muscle* is a flexor of the proximal interphalangeal and metacarpophalangeal joints of the medial four fingers and is the principal flexor of the wrist. The *flexor digitorum profundus muscle* primarily flexes the terminal phalanx but, continuing to act, also flexes the middle and proximal phalanges. This muscle flexes the digits in slow action, the flexor digitorum superficialis muscle being recruited for speed and against resistance. The *extensor digitorum muscle*, assisted by the extensors of the index and fifth fingers, is the extensor of the fingers. Interconnecting tendinous bands between the tendons of digits III, IV, and V prevent completely independent extension of these digits, but the index finger can be moved quite separately.

The interosseous and lumbrical muscles of the hand are essential for full extension of the digits. The *interosseous muscles* act most effectively when there is combined metacarpophalangeal flexion and interphalangeal extension, principally producing interphalangeal extension. The *lumbrical muscles* are silent during total flexion but are very active in extension of the proximal or distal interphalangeal joints and also when these joints are being maintained in extension during metacarpophalangeal flexion.

Free movement of the thumb is most important in the more precise activities of the hand. The *flexor pollicis longus muscle* flexes the thumb, and the *extensor pollicis longus* and *extensor pollicis brevis muscles* extend it. The *abductor pollicis longus muscle* is an accessory flexor of the wrist; it abducts and extends its metacarpal. The short muscles of the thumb provide flexion, abduction, adduction, and opposition. Abduction of the thumb carries it anteriorly out of the plane of the palm because of the rotated position of the first metacarpal, which directs its palmar surface medially. The abductor pollicis brevis muscle also assists in flexion. The *opponens pollicis muscle* acts solely on the metacarpal of the thumb, drawing the digit across the palm and rotating it medially.

The components of opposition are abduction, flexion, and medial rotation, the tip of the thumb reaching contact with the pads of the other slightly flexed digits. In firm grasp, the *flexor pollicis brevis muscle* is especially active. The motor, or recurrent, branch of the median nerve innervates the three muscles involved. The *adductor pollicis muscle* adducts the thumb. The *abductor digiti minimi* and the *flexor digiti minimi brevis muscles* produce their characteristic movements. The *opponens digiti minimi muscle* rotates the fifth metacarpal medially and deepens the hollow of the hand. □

Upper Limb

Bones and Joints of Forearm and Wrist

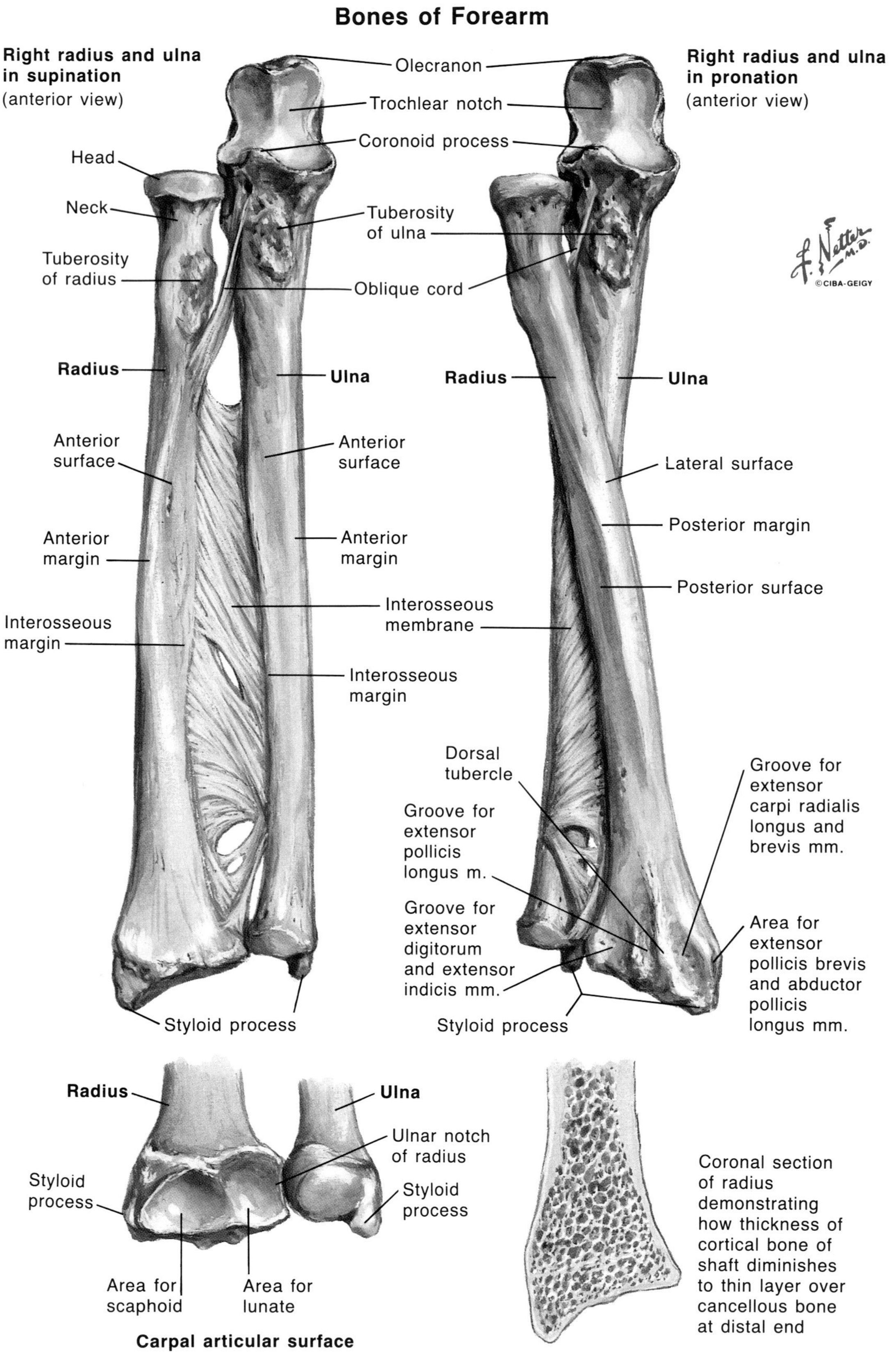
Bones of Forearm
Right radius and ulna in supination (anterior view)
Right radius and ulna in pronation (anterior view)
Olecranon
Trochlear notch
Coronoid process
Head
Neck
Tuberosity of ulna
Tuberosity of radius
Oblique cord
Radius
Ulna
Radius
Ulna
Anterior surface
Anterior surface
Lateral surface
Posterior margin
Anterior margin
Anterior margin
Posterior surface
Interosseous margin
Interosseous membrane
Interosseous margin
Dorsal tubercle
Groove for extensor carpi radialis longus and brevis mm.
Groove for extensor pollicis longus m.
Groove for extensor digitorum and extensor indicis mm.
Area for extensor pollicis brevis and abductor pollicis longus mm.
Styloid process
Styloid process
Radius
Ulna
Ulnar notch of radius
Styloid process
Styloid process
Area for scaphoid
Area for lunate
Carpal articular surface
Coronal section of radius demonstrating how thickness of cortical bone of shaft diminishes to thin layer over cancellous bone at distal end
©CIBA-GEIGY

Distal Parts of Radius and Ulna

The distal end of the radius is broadened, for its *carpal articular surface* is the bony contact of the forearm with the wrist and hand (Plate 62). This surface is concave transversely and anteroposteriorly; it is divided by a surface constriction and a slight ridging into a larger triangular portion laterally and a smaller quadrangular part medially, which are for the reception of the scaphoid and the lunate of the wrist, respectively.

The medial surface of the distal extremity of the radius is also concave and articular; as the *ulnar notch of the radius*, it receives the rounded head of the ulna. Dorsally, the distal part of the radius exhibits its *tubercle* and is otherwise somewhat ridged and grooved for the passage of the tendons of the forearm extensor muscles. Laterally, the bone ends in a downwardly projecting *styloid process*.

Ossification begins in the distal extremity of the radius at the end of the first year, and fusion takes place at about ages 19 to 20.

The *ulna* has a small distal extremity. There is a small, rounded *styloid process* in line with the posterior border of the bone, and a larger, rounded *head*. The distal surface of the head is smooth for contact with the articular disc of the inferior radioulnar joint; it is continuous with the articular surface of the circumference of the head, which is received into the ulnar notch of the head of the radius.

An *ossification center* for the distal end of the ulna appears at age 5 or 6 and fuses with the shaft at about ages 18 to 20.

Carpal Bones

The skeleton of the wrist consists of eight small bones arranged in two rows, proximal and distal (Plate 63). The bones of the proximal row, from the radial to the ulnar side, are the scaphoid, the lunate, the triquetrum, and the pisiform. Those of the distal row, in the same order, are the trapezium, the trapezoid, the capitate, and the hamate. Fundamentally, these bones may be thought of as cubes, each of which has six surfaces. Their dorsal and palmar surfaces are nonarticular and provide for the attachment of the dorsal and palmar ligaments that hold them closely together.

The other surfaces are articular, except for the subcutaneous surfaces of the bones that form the borders of the wrist. These surfaces also mostly lodge ligaments. The proximal articular surfaces are generally convex; the distal surfaces are usually concave. Foramina for the entrance of blood vessels are found on nonarticular areas of each bone.

The boat-shaped *scaphoid* is the largest bone of the proximal row. Its smooth radial articular surface is convex and triangular in form. The smooth distal surface is triangular but concave and receives both the trapezium and the trapezoid. The medial surface presents two articular facets—one for the lunate and a larger inferior concavity for part of the head of the capitate.

The *lunate* is crescentic; its proximal convexity is for the more medial of the articular surfaces of the distal end of the radius. The distal surface is deeply concave for the capitate and for a small contact with the hamate. On its radial surface, this bone contacts the scaphoid; medially, it has a surface for the base of the triquetrum.

The *triquetrum* is pyramidal in form, with the base of the pyramid toward the lunate and the apex downward and ulnarward on the ulnar border of the wrist. The inferior surface is sinuously curved for articulation with the hamate, and the palmar surface has an oval facet for the pisiform.

Upper Limb

Bones and Joints of Forearm and Wrist

(Continued)

Carpal Bones

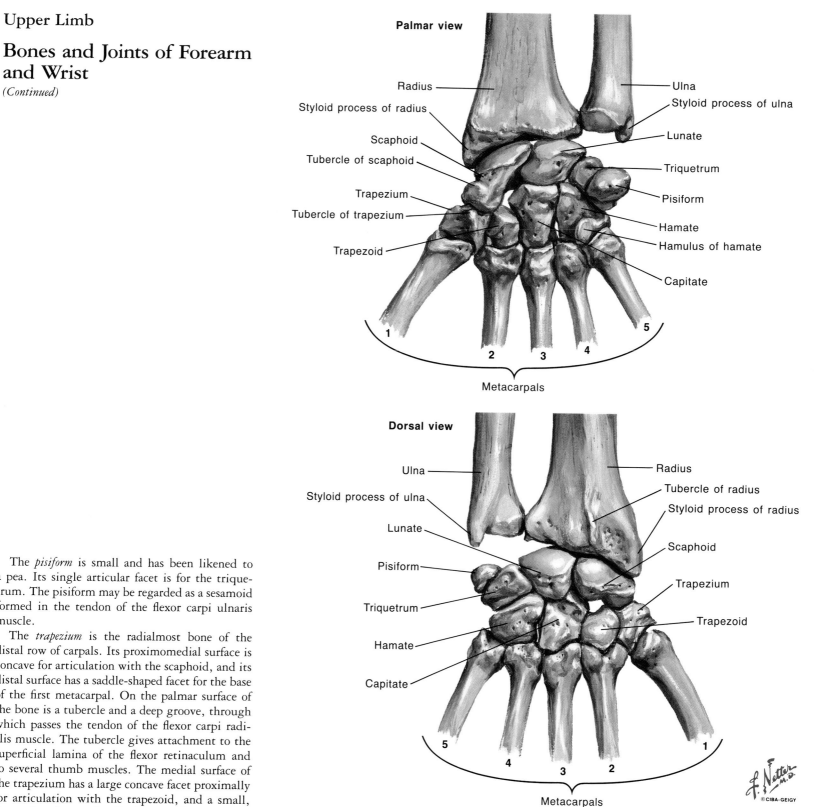

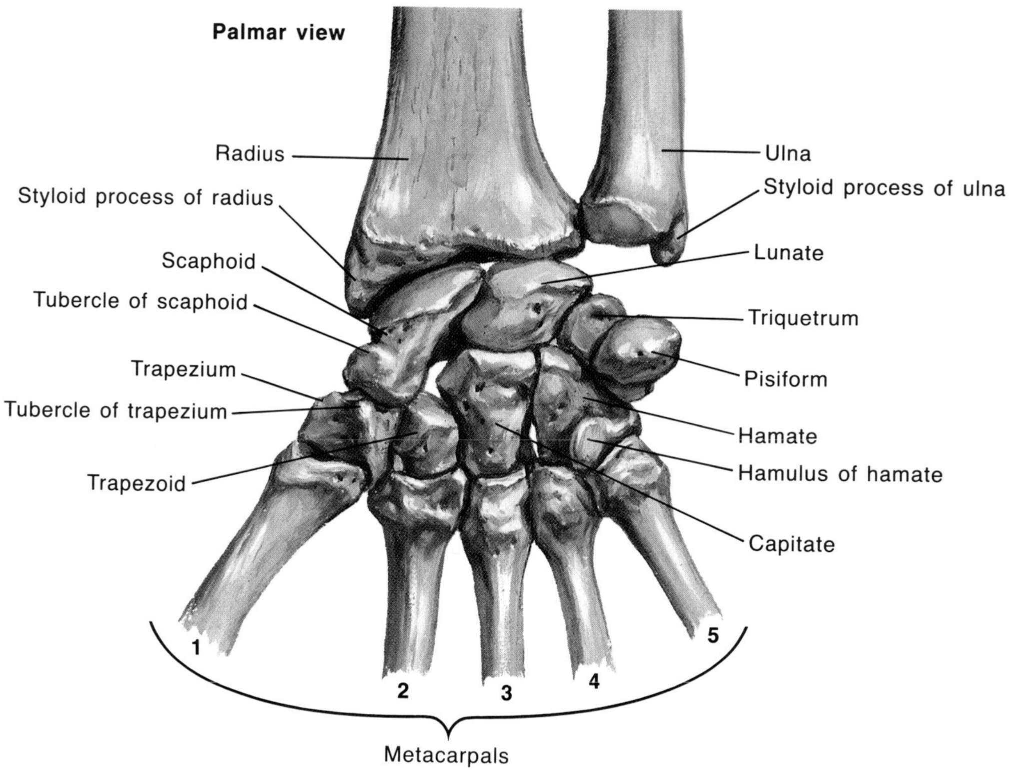
Palmar view
Radius
Styloid process of radius
Scaphoid
Tubercle of scaphoid
Trapezium
Tubercle of trapezium
Trapezoid
Ulna
Styloid process of ulna
Lunate
Triquetrum
Pisiform
Hamate
Hamulus of hamate
Capitate
1
2
3
4
5
Metacarpals

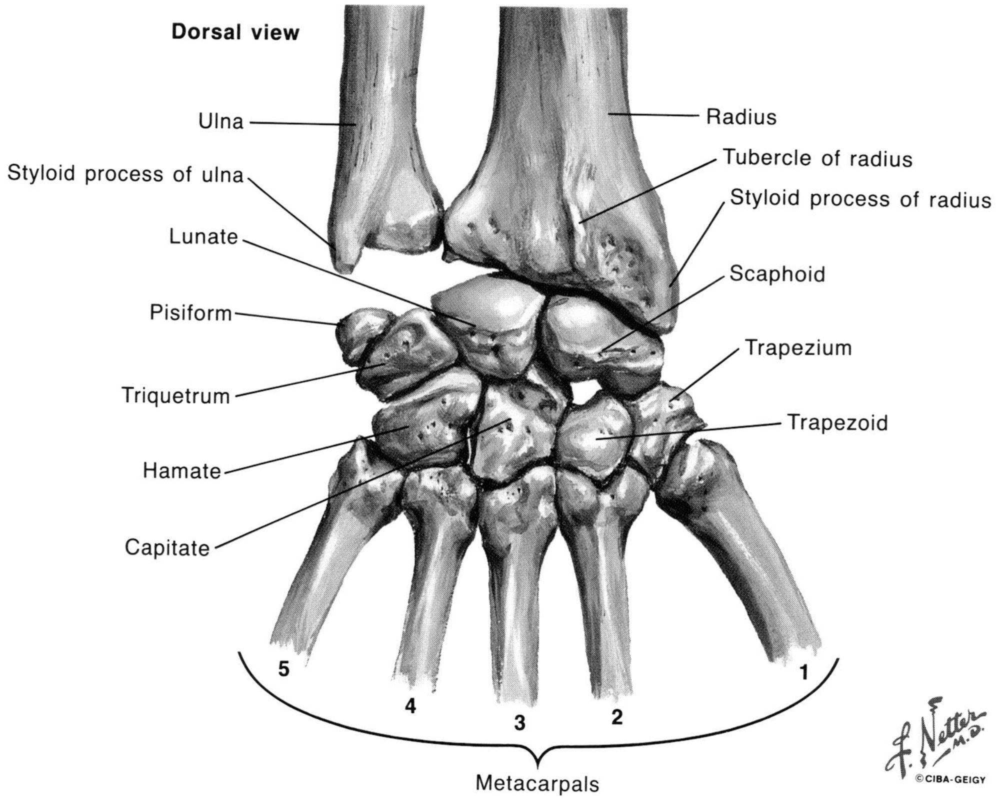
Dorsal view
Ulna
Styloid process of ulna
Lunate
Pisiform
Triquetrum
Hamate
Capitate
Radius
Tubercle of radius
Styloid process of radius
Scaphoid
Trapezium
Trapezoid
5
4
3
2
1
Metacarpals
F. Netter M.D.
©CIBA-GEIGY

The *pisiform* is small and has been likened to a pea. Its single articular facet is for the triquetrum. The pisiform may be regarded as a sesamoid formed in the tendon of the flexor carpi ulnaris muscle.

The *trapezium* is the radialmost bone of the distal row of carpals. Its proximomedial surface is concave for articulation with the scaphoid, and its distal surface has a saddle-shaped facet for the base of the first metacarpal. On the palmar surface of the bone is a tubercle and a deep groove, through which passes the tendon of the flexor carpi radialis muscle. The tubercle gives attachment to the superficial lamina of the flexor retinaculum and to several thumb muscles. The medial surface of the trapezium has a large concave facet proximally for articulation with the trapezoid, and a small, flat oval surface at the distal angle of the bone for the second metacarpal.

The *trapezoid* is somewhat wedge shaped, with the broader base of the wedge dorsally. The quadrilateral proximal surface articulates with the scaphoid, while distally, there is a large saddle-shaped articular surface for the base of the second metacarpal. The lateral surface is convex for the trapezium, while the medial surface has a smooth, flat facet for the capitate.

The *capitate* is the largest of the carpal bones and occupies the center of the wrist. Its rounded head is received into the concavity of the scaphoid and the lunate. The distal, somewhat cuboidal extremity articulates chiefly with the base of the third metacarpal, but by means of small lateral and medial facets, it also makes contact with the bases of the second and third metacarpals. The lateral surface has, distally, a small, smooth facet for the distal extremity of the trapezoid, and the medial surface has an oblong articular surface for the hamate.

The *hamate* is wedge shaped and has a characteristic hooklike process, the hamulus, or hook. The apical proximal part of the wedge articulates with the lunate; the broad distal surface has two concave facets for the bases of the fourth and fifth metacarpals. Articular surfaces laterally and medially are for the capitate and triquetrum, respectively. The hamulus gives attachment to the flexor retinaculum and the tendon of the flexor carpi ulnaris muscle and provides origin for several small finger muscles.

Ossification takes place from a single center in each bone. Ossification begins first in the capitate and then in the hamate early in the first year; in the triquetrum, during the third year; in the lunate, in the fourth year; in the trapezium, trapezoid, and scaphoid, in rather close sequence, in the fourth to sixth years; and in the pisiform, in the eleventh or twelfth year. Ossification starts earlier in the female and is completed between ages 14 and 16. The hamulus of the hamate may have a separate center. An os centrale, normally part of the scaphoid, may be present between the scaphoid, capitate, and trapezoid.

Upper Limb

Bones and Joints of Forearm and Wrist

(Continued)

Ligaments of Wrist

Palmar view with structures passing through and over carpal tunnel

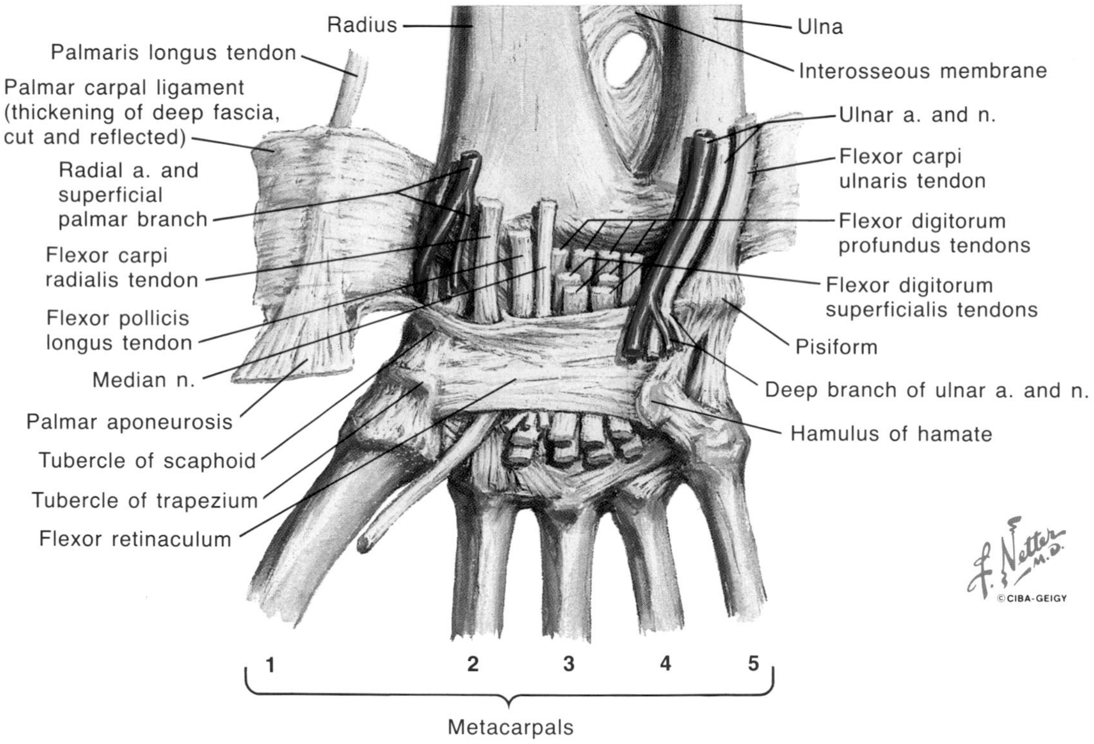
Radius
Ulna
Palmaris longus tendon
Interosseous membrane
Palmar carpal ligament (thickening of deep fascia, cut and reflected)
Ulnar a. and n.
Flexor carpi ulnaris tendon
Radial a. and superficial palmar branch
Flexor digitorum profundus tendons
Flexor carpi radialis tendon
Flexor digitorum superficialis tendons
Flexor pollicis longus tendon
Pisiform
Median n.
Deep branch of ulnar a. and n.
Palmar aponeurosis
Hamulus of hamate
Tubercle of scaphoid
Tubercle of trapezium
Flexor retinaculum
F. Netter M.D.
©CIBA-GEIGY
1
2
3
4
5
Metacarpals

Palmar view with flexor retinaculum and structures removed

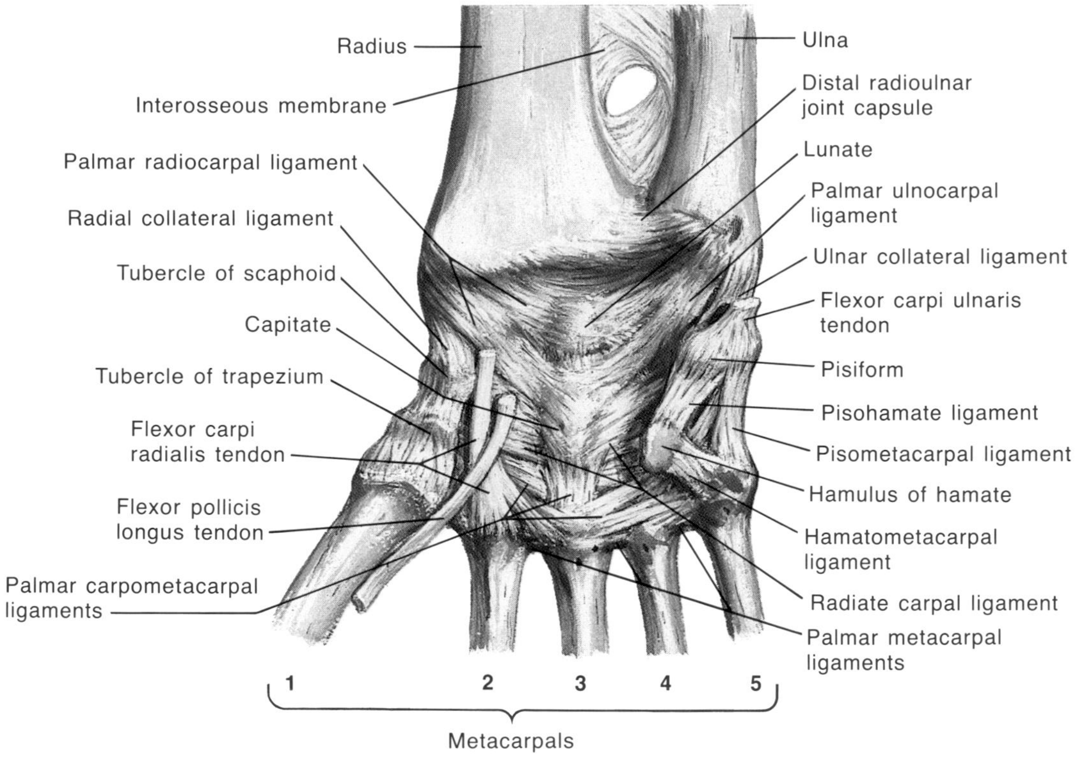
Radius
Ulna
Interosseous membrane
Distal radioulnar joint capsule
Lunate
Palmar radiocarpal ligament
Palmar ulnocarpal ligament
Radial collateral ligament
Ulnar collateral ligament
Tubercle of scaphoid
Flexor carpi ulnaris tendon
Capitate
Pisiform
Tubercle of trapezium
Pisohamate ligament
Flexor carpi radialis tendon
Pisometacarpal ligament
Hamulus of hamate
Flexor pollicis longus tendon
Hamatometacarpal ligament
Palmar carpometacarpal ligaments
Radiate carpal ligament
Palmar metacarpal ligaments
1
2
3
4
5
Metacarpals

Distal Radioulnar Joint

This articulation is a pivot joint between the head of the ulna and the ulnar notch of the radius, with the joint cavity also extending between the distal surface of the ulna and the articular disc (Plates 64–66). The *articular disc* is the chief uniting structure of the joint, attaching by its base to the sharp medial margin of the distal end of the radius and by its apex to the inner surface of the root of the ulnar styloid process. An *articular capsule* is represented by transverse bands of no great strength, which span from the anterior and posterior edges of the ulnar notch of the radius to the corresponding surfaces of the head of the ulna.

The *joint cavity* is L shaped in vertical section, and the *synovial membrane* reaches proximally beyond the articulating surfaces to form the *sacciform recess*. The *blood supply* to the joint is provided from the anterior and posterior interosseous arteries and from the dorsal and palmar networks of the wrist. *Nerves* come from the posterior interosseous branch of the radial nerve and the anterior interosseous branch of the median nerve.

A strong *interosseous membrane* also connects the interosseous borders of the radius and the ulna (Plate 62). Proximally, it extends to within 2 to 3 cm of the tuberosity of the radius and is supplemented there by the *oblique cord* stretched between the lateral border of the tuberosity of the ulna and the radius distal to its tuberosity. Distally, the interosseous membrane blends with the fascia of the posterior surface of the pronator quadratus muscle. Principally, the fibers of the interosseous membrane run downward and medialward from the radius to the ulna, thus aiding in the transmission of forces from the hand to the ulna, the bone of firm connection with the humerus.

Movements. The radioulnar joints contribute the movements of *pronation* and *supination* to the forearm. The longitudinal axis of this movement passes proximally through the center of the head of the radius and distally through the apical attachment of the articular disc to the head of the ulna. Prolonged, this axis is represented by the fourth digit, around which the hand appears to move as it follows the forearm. The ulna remains relatively stationary because of its fixation on the humerus; and the radius rotates, its head revolving within the circle of the annular ligament and the radial notch of the ulna, proximally. Distally, the radius travels around the relatively fixed head of the ulna. About 135° of rotation is possible from full pronation to full supination.

Radiocarpal Joint

This articulation is formed by the distal extremity of the radius and the articular disc of the inferior radioulnar joint above and the proximal row of carpals and their interosseous ligaments below (Plate 65). The curvatures of the proximal carpals are combined into a convex, egg-shaped, or ellipsoidal, surface that fits into the transversely elongated concavity of the radius and articular disc.

Upper Limb

Bones and Joints of Forearm and Wrist

(Continued)

Ligaments of Wrist (continued)

Dorsal view

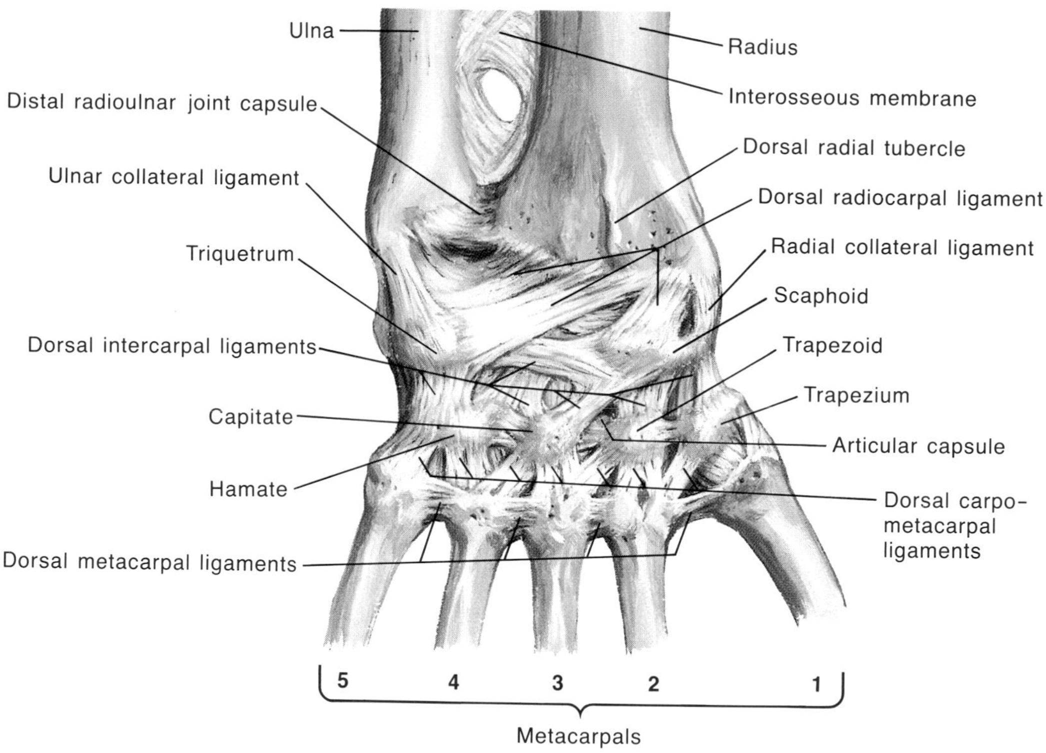
Ulna
Radius
Distal radioulnar joint capsule
Interosseous membrane
Dorsal radial tubercle
Ulnar collateral ligament
Dorsal radiocarpal ligament
Radial collateral ligament
Triquetrum
Scaphoid
Dorsal intercarpal ligaments
Trapezoid
Trapezium
Capitate
Articular capsule
Hamate
Dorsal carpo-metacarpal ligaments
Dorsal metacarpal ligaments
5 4 3 2 1
Metacarpals

Carpal articulations, coronal section (dorsal view)

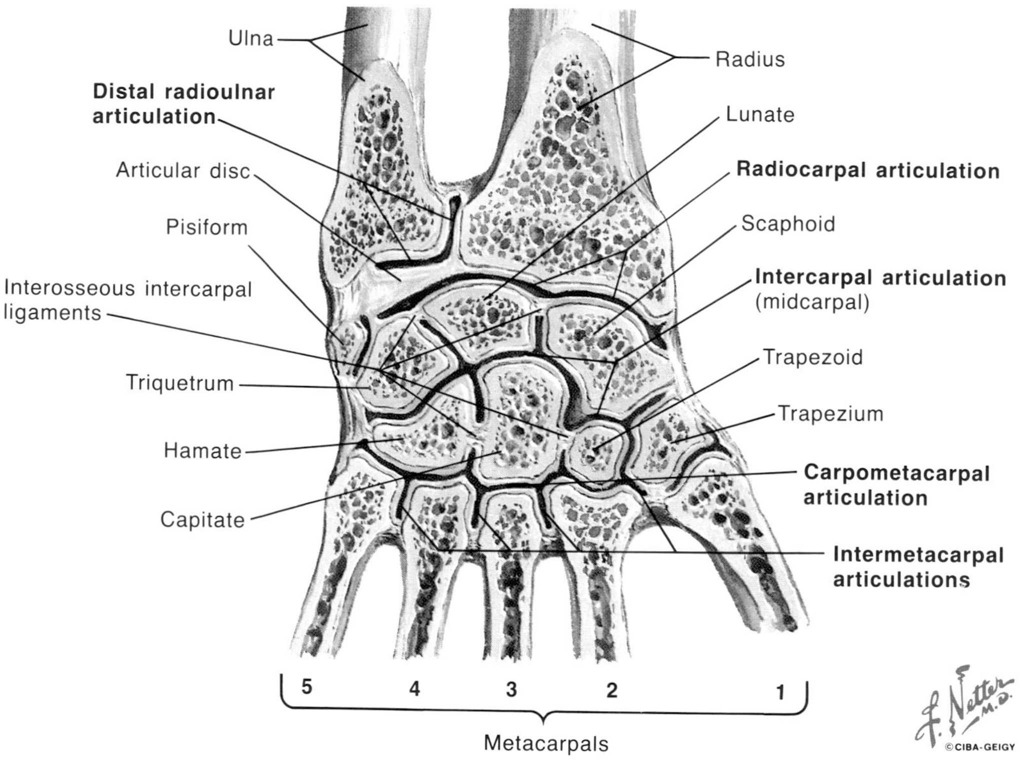
Ulna
Radius
Distal radioulnar articulation
Lunate
Articular disc
Radiocarpal articulation
Pisiform
Scaphoid
Interosseous intercarpal ligaments
Intercarpal articulation (midcarpal)
Trapezoid
Triquetrum
Trapezium
Hamate
Carpometacarpal articulation
Capitate
Intermetacarpal articulations
5 4 3 2 1
Metacarpals
©CIBA-GEIGY

With the hand straight from the forearm, the scaphoid lies below the lateral part of the radius, and the lunate is opposite the medial part and then the articular disc. The triquetrum lies in relation to the articular disc and the medial part of the joint capsule. Abduction and adduction, shown in Plate 66, change these relationships. The *articular capsule* encloses the joint and is strengthened by dorsal and palmar radiocarpal and radial and ulnar collateral ligaments.

The *dorsal* and *palmar radiocarpal ligaments* are broad bands of fibers that extend from the corresponding borders of the lower end of the radius obliquely downward and ulnarward to the scaphoid, lunate, and triquetrum. This course and attachments ensure that the hand primarily follows the radius in its movements. Some fibers of the *palmar carpal ligament* reach the capitate, and there is also an *ulnocarpal ligament*, composed of bundles, which stretches from the head of the ulna and the base of its styloid process to the carpals below.

The *radial* and *ulnar collateral ligaments* arise principally from the styloid processes of the respective bones and end below in the marginal carpals of either side (Plate 64). The *synovial cavity* of the wrist joint is restricted to the radiocarpal space. The lax *synovial membrane* lines the deep surface of the capsule and has numerous folds, especially dorsally. The *arteries* of the joint are derived from the dorsal and palmar carpal networks; its *nerves* are derived from the anterior and posterior interosseous nerves and the dorsal and deep branches of the ulnar nerve.

Movements. Flexion and extension, abduction and adduction, and circumduction are movements of the radiocarpal joint (Plate 66). No rotary movement occurs here. The limits of abduction (15°) and adduction (40°) of the hand are set by the styloid processes of the radius and ulna, the radial styloid extending farther toward the hand. Flexion is freer than extension at the wrist, but this is due to a considerable contribution from the intercarpal joints, the radiocarpal joint itself being freer in extension.

Intercarpal Articulations

Arthrodial (gliding) articulations exist between the carpals of the proximal row, those of the distal row, and those of the proximal row with those of the distal row (Plates 65–66).

Ligaments reinforcing the intercarpal joint capsule are *dorsal* and *palmar intercarpal* and, within the cavity, *interosseous ligaments*. The surface ligaments of the proximal row generally run transversely—from scaphoid to lunate and lunate to triquetrum. Similarly, for the distal row, intercarpal ligaments, dorsal and palmar, unite trapezium to trapezoid, trapezoid to capitate, and capitate to hamate. The *intercarpal interosseous ligaments* of the

Upper Limb

Bones and Joints of Forearm and Wrist

(Continued)

Movements at Wrist Joint

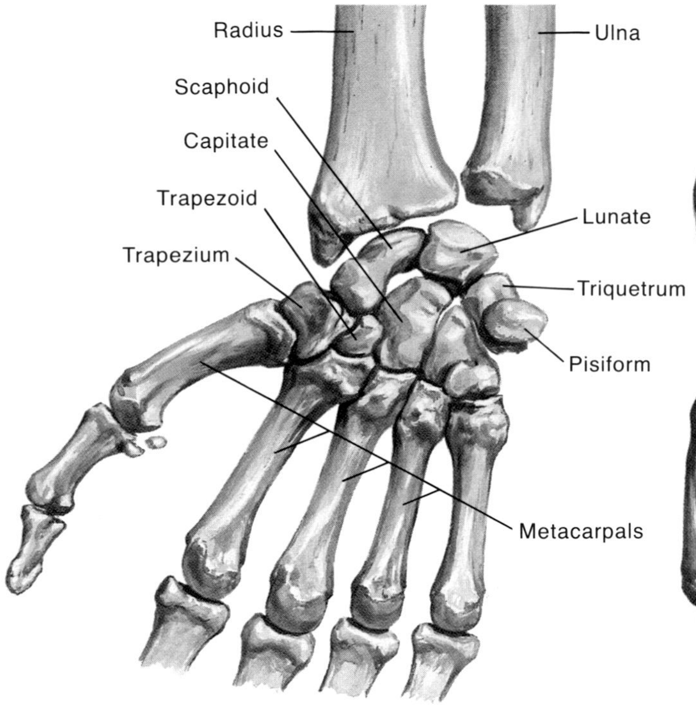

Position of carpals with hand in abduction (palmar view)

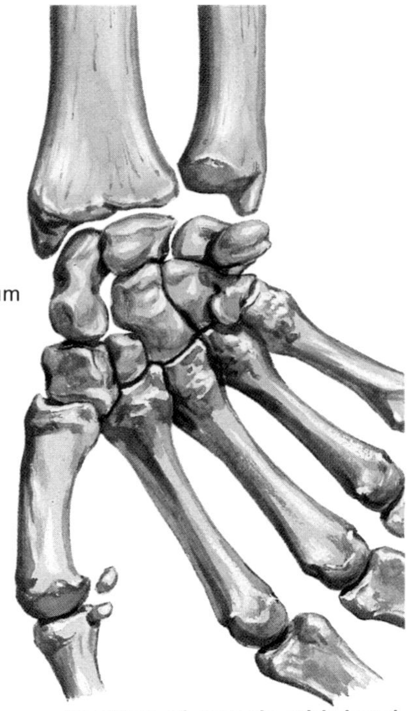

Position of carpals with hand in adduction (palmar view)

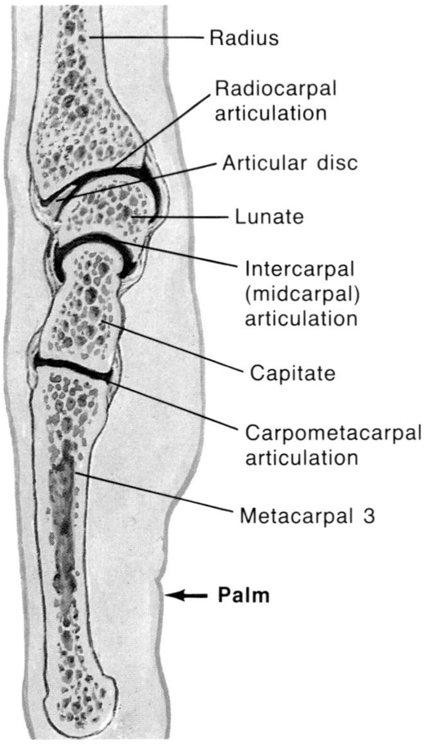

Sagittal section through wrist and middle finger with hand straight

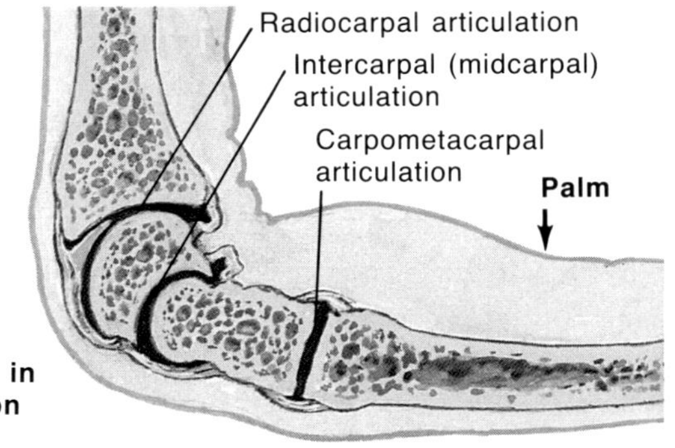

Hand in flexion

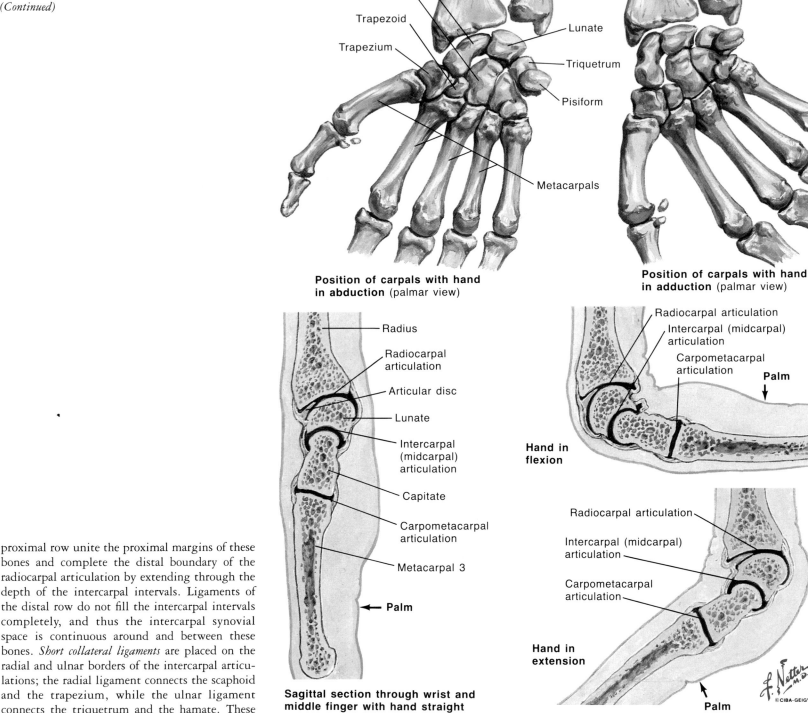

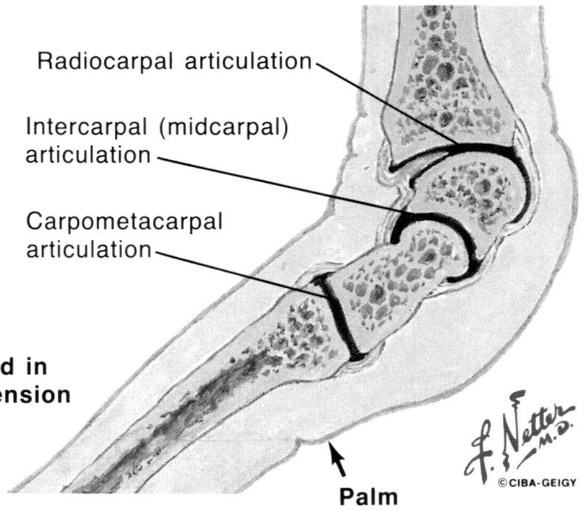

Hand in extension

proximal row unite the proximal margins of these bones and complete the distal boundary of the radiocarpal articulation by extending through the depth of the intercarpal intervals. Ligaments of the distal row do not fill the intercarpal intervals completely, and thus the intercarpal synovial space is continuous around and between these bones. *Short collateral ligaments* are placed on the radial and ulnar borders of the intercarpal articulations; the radial ligament connects the scaphoid and the trapezium, while the ulnar ligament connects the triquetrum and the hamate. These ligaments are continuous with the collateral ligaments of the wrist joint.

The *intercarpal synovial cavity* is large and complex. It fills the midcarpal interval and is continuous between the adjacent carpals of each row. It also extends onto the distal surfaces of the carpals of the second row and thus includes the carpometacarpal joint spaces (except for the thumb). The space is further prolonged to include the intermetacarpal articulations between the bases of the second, third, fourth, and fifth metacarpals. However, the intercarpal synovial cavity does not include the joint between the pisiform and triquetrum or that between the trapezium and the base of the first metacarpal, nor does it communicate with the wrist joint.

The *arterial supply* of the intercarpal articulations comes from the palmar and dorsal carpal networks. The *nerves* are twigs from the anterior and posterior interosseous nerves and from the dorsal and deep branches of the ulnar nerve.

Movements of the intercarpal articulations occur simultaneously with and augment the movements at the radiocarpal articulation (Plates 65–66). The principal region of intercarpal movement is that of the sinuous *midcarpal joint*. Here, the head of the capitate and the apex of the hamate are received into the cup-shaped cavity of the scaphoid and the lunate, a virtual ball-and-socket arrangement. The movements here, as well as the gliding movements of the bones on either side, contribute a considerable degree of flexion and some abduction to the hand. The midcarpal joint and the other intercarpal joints improve the grasp of the hand with slight rotary and gliding movements between the bones. Extension is freer in the radiocarpal joint, and flexion is freer in the midcarpal joint.

A separate intercarpal joint exists in the articulation between the triquetrum and the pisiform, with a thin but strong articular capsule uniting them. The pisiform is anchored to the hamulus of the hamate and the base of the fifth metacarpal by the pisohamate and pisometacarpal ligaments, which resist the pull of the flexor carpi ulnaris muscle and form part of its insertion. □

Upper Limb

Bones and Joints of Hand

Bones

Metacarpal Bones. Five metacarpals form the skeleton of the hand. They are miniature "long" bones, comprising a shaft, a head, and a base. They are palpable on the dorsum of the hand and terminate distally in the knuckles, which are their heads (Plate 67).

The *shafts* are curved longitudinally so as to be convex dorsally and concave on their palmar aspects. The *head*, the distal extremity, has a rounded smooth surface for articulation with the base of the proximal phalanx. The sides of the head exhibit pits, or tubercles, for the attachment of ligaments. The articular surface of the head is also convex transversely, although less so than dorsopalmarward so that the head fails to be a sphere; however, flexion and extension and abduction and adduction are permitted. The *base* is cuboidal and broader dorsally than palmarward. Its ends and sides are articular, and the dorsal and palmar surfaces are rough for ligamentous attachments.

The *first metacarpal* is shorter and stouter than the others, and its palmar surface faces toward the center of the palm. It has a proximal saddle-shaped articular surface for contact with the trapezium. Apart from its regular head configuration, it has two palmar articular eminences for the sesamoids of the thumb.

The *second metacarpal* is the longest, and its base is the largest of the metacarpals. There is a deep dorsopalmar groove in the base, which accepts the trapezoid, and the ridges bounding the groove make contact with the trapezium and the capitate. On the ulnar side of the base, there is an incompletely divided facet for the base of the third metacarpal.

The *third metacarpal* is distinguished by its styloid process, a dorsally and radially placed proximal eminence. The carpal surface of the bone is concave for the capitate. Subdivided facets exist on the sides of the base for articulation with the bases of the second and fourth metacarpals.

The *fourth metacarpal* has a square base that, proximally, has a large facet for the hamate and, laterally, a small facet for the capitate. Two facets on the lateral side of the base make contact with the base of the third metacarpal, and a single, oval facet on the other side faces the fifth metacarpal.

The *fifth metacarpal* has a single, concavo-convex facet on the proximal surface of its base for articulation with the hamate. A slightly convex facet on the radial side is received into the matching oval facet of the fourth metacarpal. On the ulnar side of the base, there is a prominent tubercle for the attachment of the tendon of the extensor carpi ulnaris muscle.

Ossification of the metacarpals proceeds from two centers—one for the body of the bone, and one for the distal extremity in each of the four fingers, except for the proximal extremity in the thumb. Ossification begins in the shafts in the eighth or ninth week of fetal life. The centers for the extremity epiphyses appear during the second year, and fusion takes place between ages 16 and 18.

Bones of Wrist and Hand

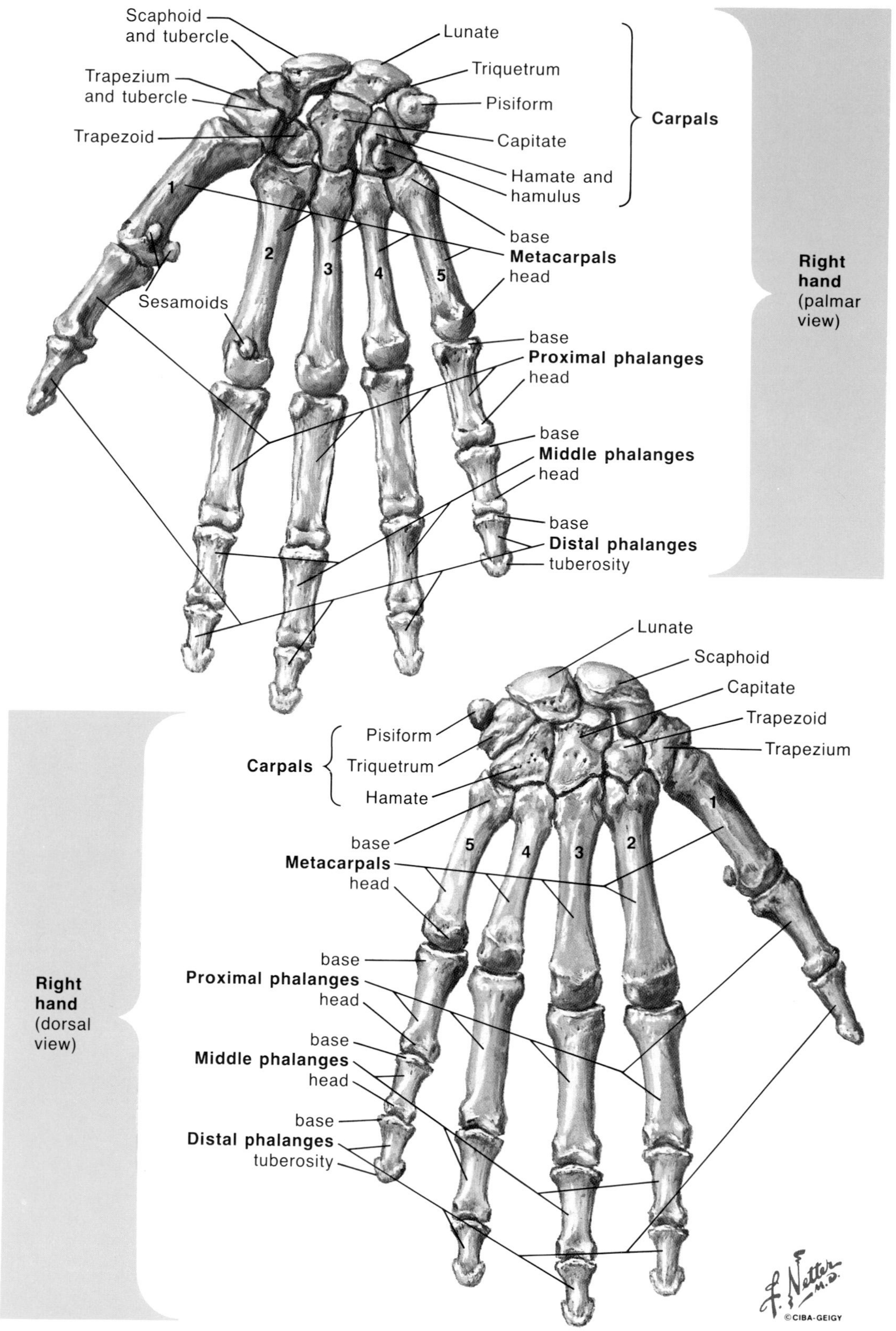
Scaphoid and tubercle
Lunate
Triquetrum
Trapezium and tubercle
Pisiform
Carpals
Trapezoid
Capitate
Hamate and hamulus
1
base
Metacarpals
head
2
3
4
5
Sesamoids
base
Proximal phalanges
head
base
Middle phalanges
head
base
Distal phalanges
tuberosity
Right hand (palmar view)
Lunate
Scaphoid
Capitate
Trapezoid
Trapezium
Pisiform
Carpals
Triquetrum
Hamate
base
Metacarpals
head
5
4
3
2
1
base
Proximal phalanges
head
base
Middle phalanges
head
base
Distal phalanges
tuberosity
Right hand (dorsal view)
F. Netter M.D.
©CIBA-GEIGY

Phalanges. The phalanges are fourteen in number, being one short in the thumb (Plate 67). These, too, are miniature long bones, with a shaft and two extremities. The dorsum of the shaft is markedly convex from side to side; its palmar surface is nearly flat. The margins of the palmar surfaces are ridged for attachment of the fibrous flexor sheaths of the digits. The proximal extremity of the first phalanx of each digit is concave and oval and broader from side to side for articulation with the head of the metacarpal. Distally, the proximal extremities of the middle and distal phalanges have two shallow concavities separated by an intervening ridge, which articulate with the pulleylike surfaces on the distal ends of the middle and distal phalanges. The distal phalanges exhibit terminal elevated roughened surfaces, which support the pulp of the fingers. The sides of the bases of the phalanges show tubercles for ligaments; the sides of the heads (except on the distal phalanx) exhibit shallow pits for ligamentous attachments.

Ossification of the phalanges proceeds from two centers—one for the body and one for the proximal extremity. Ossification in the shaft begins about the eighth week of fetal life and in the epiphysis during the second and third years, and fusion takes place between 14 and 18 years of age.

Upper Limb

Bones and Joints of Hand

(Continued)

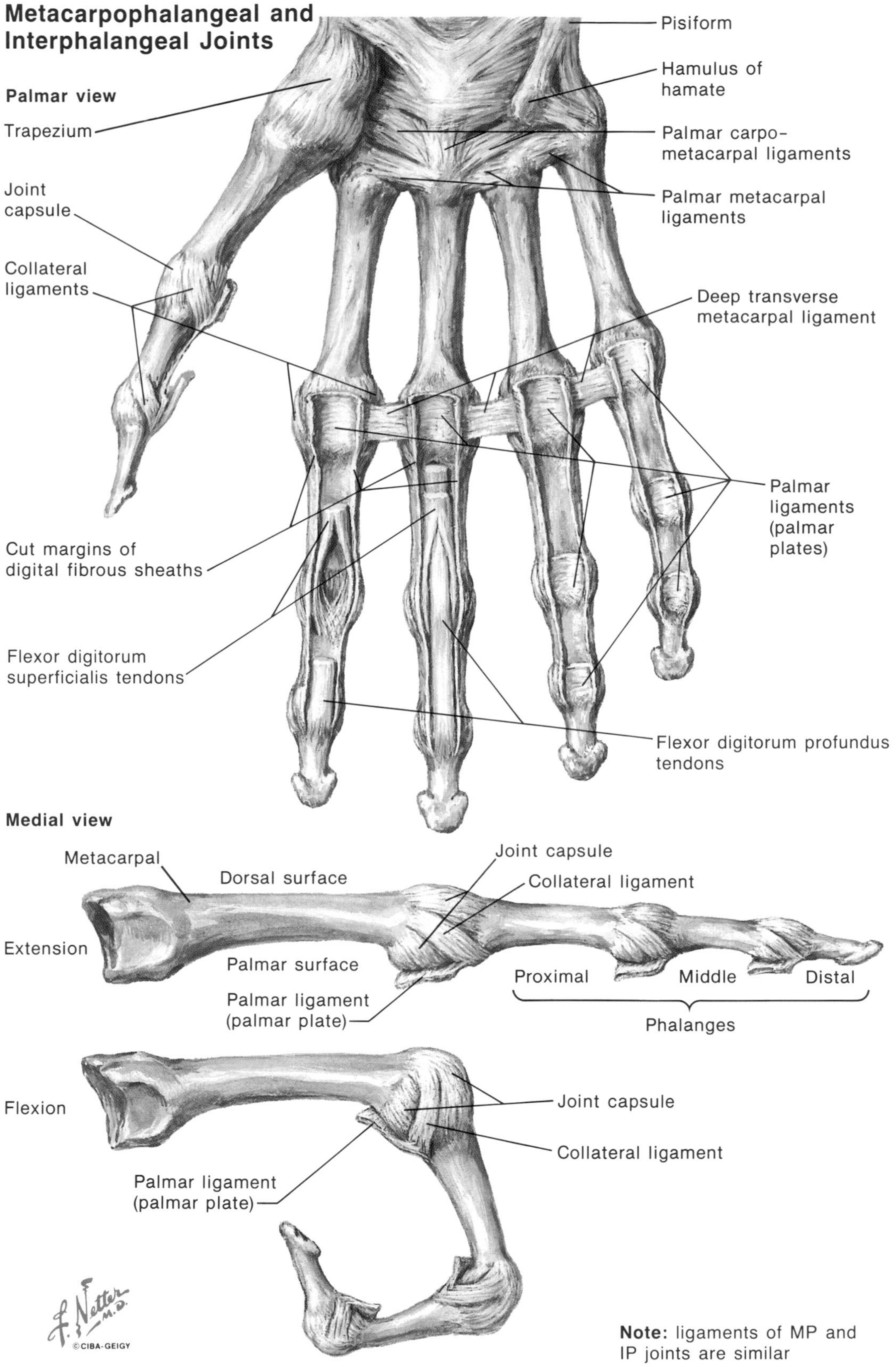

Metacarpophalangeal and Interphalangeal Joints
Palmar view
Pisiform
Hamulus of hamate
Trapezium
Palmar carpometacarpal ligaments
Joint capsule
Palmar metacarpal ligaments
Collateral ligaments
Deep transverse metacarpal ligament
Palmar ligaments (palmar plates)
Cut margins of digital fibrous sheaths
Flexor digitorum superficialis tendons
Flexor digitorum profundus tendons
Medial view
Metacarpal
Dorsal surface
Joint capsule
Collateral ligament
Extension
Palmar surface
Proximal
Middle
Distal
Palmar ligament (palmar plate)
Phalanges
Flexion
Joint capsule
Collateral ligament
Palmar ligament (palmar plate)
©CIBA-GEIGY
Note: ligaments of MP and IP joints are similar

Joints

Carpometacarpal Joint. The *carpometacarpal joint of the thumb* is the independent joint between the trapezium and the base of the first metacarpal (Plate 68). The articular surfaces are reciprocally concavo-convex, and a loose but strong articular capsule joins the bones. The biaxial nature of this joint provides for flexion and extension and abduction and adduction, and the looseness of its capsule allows opposition of the thumb that involves a small amount of rotary movement.

The *carpometacarpal joints of the four fingers* participate with the intercarpal and intermetacarpal joints in a common synovial cavity. Dorsal and palmar carpometacarpal ligaments run from the carpals of the second row to the various metacarpals. Short interosseous ligaments are usually present between contiguous angles of the capitate and the hamate and the third and fourth metacarpals.

Intermetacarpal Joints. These joints occur between the adjacent sides of the bases of the four metacarpals of the fingers. Here, also, there are dorsal and palmar ligaments, and interosseous ligaments close off the common synovial cavity by connecting the bones just distal to their articular facets. Only slight gliding movements occur between the metacarpals and between them and the carpals to which they are related. However, the articulation between the hamate and the fifth metacarpal allows that bone to flex appreciably during a tight grasp and also to rotate slightly under the traction of the opponens digiti minimi muscle.

The *deep transverse metacarpal ligaments* are short and connect the palmar surfaces of the heads of the second, third, fourth, and fifth metacarpals. They are continuous with the palmar interosseous fascia and blend with the palmar ligaments of the metacarpophalangeal joints and the fibrous sheaths of the digits. They limit the spread of the metacarpals, and the tendons of the interosseous and lumbrical muscles pass on either side of them.

Metacarpophalangeal Joints. These joints are condyloid in character, and both the rounded head of the metacarpal and the oval concavity of the proximal end of the phalanx have unequal curvatures along their transverse and vertical axes. An articular capsule and collateral and palmar ligaments unite the bones. The articular capsule is rather loose. Dorsally, it is reinforced by the expansion of the digital extensor tendon.

The *palmar ligament* is a dense, fibrocartilaginous plate, which, by means of its firm attachment to the proximal palmar edge of the phalanx, extends and deepens the phalangeal articular surface. It is loosely attached to the neck of the metacarpal; in flexion, it passes under the head of the metacarpal and serves as part of the articular contact of the bones. At its sides, the palmar ligament is continuous with the deep transverse metacarpal ligaments and the collateral ligaments. The *collateral ligaments* are strong, cordlike bands attached proximally to the tubercle and adjacent pit of the head of the metacarpals and distally to the palmar surface of the side of the phalanx. Their fibers spread fanlike to attach to the palmar ligaments. Movements of flexion and extension, abduction and adduction, and circumduction are permitted at these joints. With extension is associated abduction, as in fanning the fingers; with flexion is associated adduction, as in making a fist. The metacarpophalangeal joint of the thumb is limited in abduction and adduction; its special freedom of motion derives from its carpometacarpal joint.

Interphalangeal Joints. Structurally similar to the metacarpophalangeal series, the interphalangeal joints have the same loose capsule, palmar and collateral ligaments, and dorsal reinforcement from the extensor expansion. However, due to the pulleylike form of their articular surfaces, action here is limited to flexion and extension. Flexion is freer than extension and may reach 115° at the proximal interphalangeal joint. Arteries and nerves serving these joints are twigs of adjacent proper digital branches.

Upper Limb

Bones and Joints of Hand

(Continued)

Fingers (2, 3, 4, and 5)

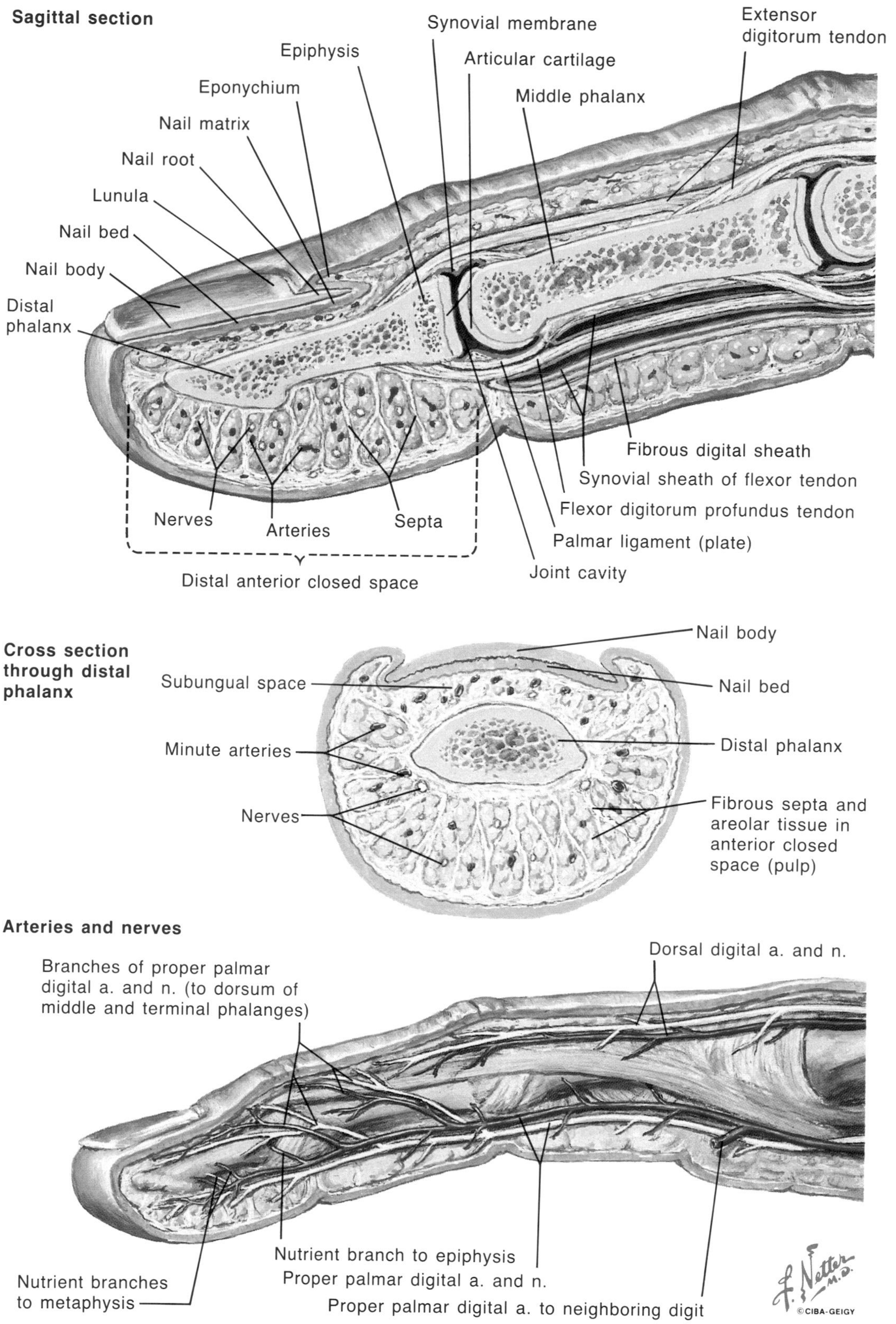
Sagittal section
Synovial membrane
Extensor digitorum tendon
Epiphysis
Articular cartilage
Eponychium
Middle phalanx
Nail matrix
Nail root
Lunula
Nail bed
Nail body
Distal phalanx
Fibrous digital sheath
Synovial sheath of flexor tendon
Flexor digitorum profundus tendon
Nerves
Arteries
Septa
Palmar ligament (plate)
Joint cavity
Distal anterior closed space
Cross section through distal phalanx
Nail body
Subungual space
Nail bed
Minute arteries
Distal phalanx
Nerves
Fibrous septa and areolar tissue in anterior closed space (pulp)
Arteries and nerves
Dorsal digital a. and n.
Branches of proper palmar digital a. and n. (to dorsum of middle and terminal phalanges)
Nutrient branch to epiphysis
Proper palmar digital a. and n.
Nutrient branches to metaphysis
Proper palmar digital a. to neighboring digit
F. Netter M.D.
©CIBA-GEIGY

Fingers

The specializations of the fingers frequently have clinical importance. The bones, joints, and tendon attachments of the fingers have already been described. It remains to add other specific items of interest or importance (Plate 69).

Nails. The fingernail is an approximately rectangular horny plate, the *nail plate*, composed of closely welded, horny scales, or cornified epithelial cells. Its semitransparency allows the pink of the highly vascular *nail bed* to show through. The nail is partially surrounded by a fold of skin, the *nail wall*, and adheres to the subjacent nail bed where strong fibers pass to the periosteum of the distal phalanx, providing the firm attachment necessary for the prying and scratching functions of the nail. The nail is formed from the proximal part of the nail bed, where the epithelium is particularly thick and extends as far distally as the whitened lunula. Developing from this *nail matrix*, the nail moves out over the longitudinal dermal ridges of the nail bed at a growth rate of approximately 1 mm/wk. Sensory nerve endings and blood vessels are abundant in the nail bed.

Anterior Closed Space. To the palmar aspect of the distal phalanx lies the anterior closed space. Areolar tissue of mixed forms lies in this region. Fiber bundles surround fatty collections and support the finer arterial and nerve branchings. More discrete septa of connective tissue fibers pass from the periosteum of the distal phalanx to blend with the underside of the dermis. An especially abundant collection of fibers attaches to form the distal skin crease of the finger and thus serves to bound the anterior closed space of the finger pad.

Small Arteries of Digits. The general origin and distribution of the dorsal and palmar digital arteries have been fully discussed (see pages 57–59), and it has been emphasized that the palmar digital arteries are the major arteries, since they send dorsal terminal branches over the distal and middle phalanges to supply the dorsum of the fingers and thumb. The dorsal digital arteries are poorly developed, except in the thumb. The proper palmar arteries are not necessarily of equal size on the two sides of the digit, although they are essentially so for the middle and ring fingers. However, in the thumb and the index and fifth digits, the larger artery is on the median side of the digit; the more diminutive artery is on the opposite side.

These proper palmar digital arteries have cross anastomoses or transverse interconnections. There is a pair of proximal transverse digital arteries that anastomoses at the level of the neck of the proximal phalanx; a pair of distal transverse digital arteries also anastomoses at the level of the neck of the middle phalanx. These arteries run close to the bone and deep to the flexor tendons. There is a rich terminal anastomosis of the palmar digital arteries, which forms a profuse tuft of small vessels in each finger pad. The proximal edge of this tuft of vessels lies on the palmar surface of the distal phalanx at about its epiphyseal line.

Digital Nerves. The cutaneous nerves parallel the arteries in course and distribution. In their course along the fingers, the proper digital nerves are outside the arteries; that is, as the digit is viewed from the side, the arteries are within the span of the dorsal and palmar nerves. Cutaneous nerves are of two types. Included are afferent somatic fibers mediating general sensation (pain, touch, pressure, and temperature), and efferent autonomic fibers supplying the smooth muscles, sweat glands, and sebaceous glands.

Both free and encapsulated nerve endings are involved in various sensations. Of the encapsulated endings, the Meissner tactile corpuscles are richly represented in the dermal papillae, and Pacinian corpuscles lie in the subcutaneous connective tissue, especially along the sides of the digits. The relatively large size of the proper palmar digital nerves suggests the high density of nerve endings in the fingers, especially in the finger pads. The tactile corpuscles are most numerous in the fingertips, less so on the palm, and rare on the dorsum of the fingers or hand. □

Lymphatic Drainage of Upper Limb

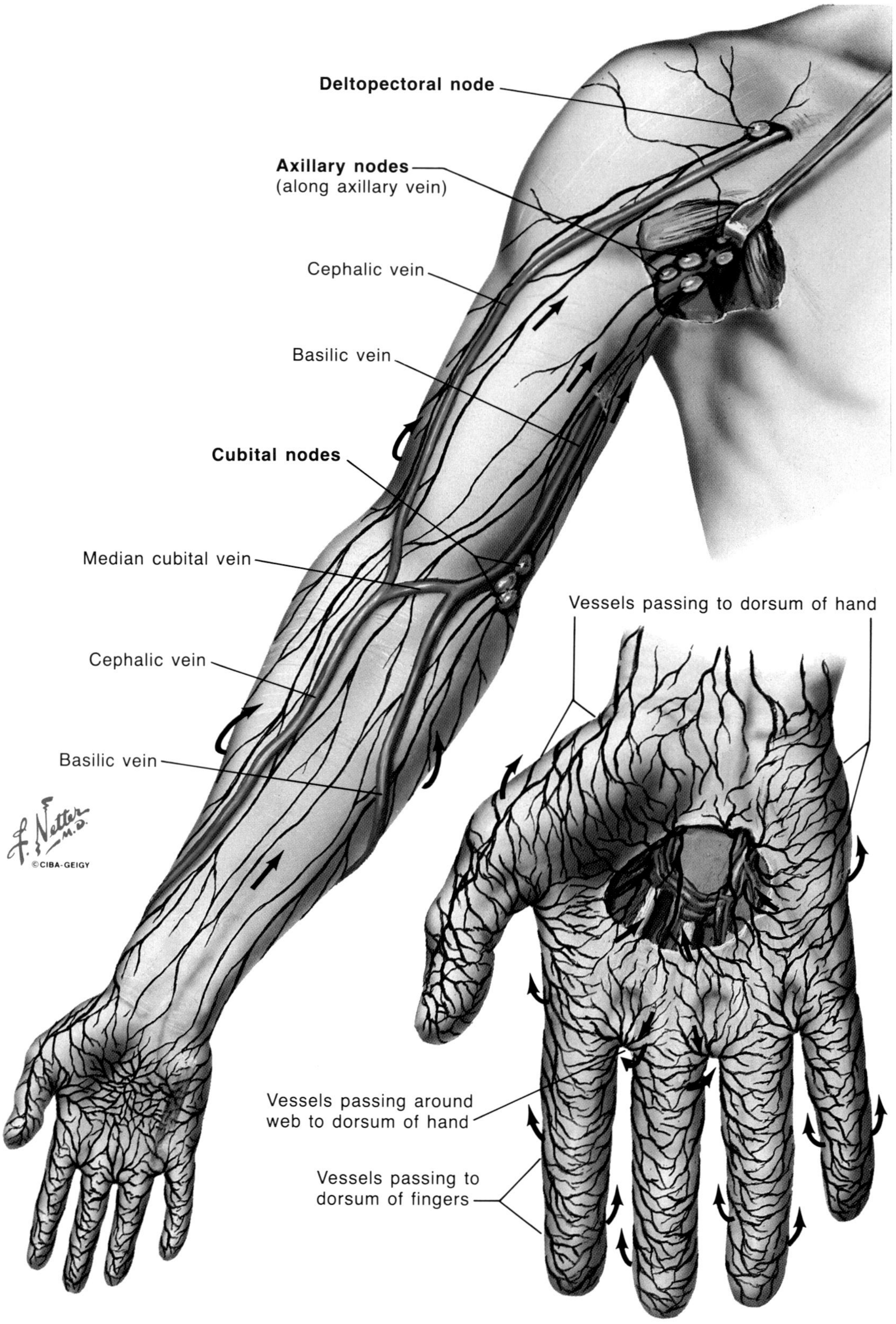
Deltopectoral node
Axillary nodes
(along axillary vein)
Cephalic vein
Basilic vein
Cubital nodes
Median cubital vein
Cephalic vein
Basilic vein
©CIBA-GEIGY
Vessels passing to dorsum of hand
Vessels passing around web to dorsum of hand
Vessels passing to dorsum of fingers

Upper Limb

Lymphatic Drainage

The superficial lymphatic vessels of the upper limb begin in the hand and pervade the skin and subcutaneous tissues (Plate 70). The dense digital lymphatic plexuses are drained by channels accompanying the digital arteries. At the interdigital clefts (and also more distally), collecting vessels of the palmar surfaces of the fingers pass to join dorsal collecting vessels and empty into the *plexus of the dorsum of the hand* (Plate 53).

Drainage of the thumb, index finger, and radial portion of the third finger is by collecting vessels that ascend along the radial side of the forearm; channels draining the ulnar fingers ascend along the ulnar side. Vessels from the *lymphatic plexus of the palm* radiate to the sides of the hand and also upward through the wrist, coalescing into two or three collecting vessels that ascend in the middle of the anterior surface of the forearm. The radial and ulnar channels turn onto the anterior surface of the forearm, lying parallel to the middle group, and all continue subcutaneously through the forearm and arm to reach the axillary nodes.

Some of the ulnar lymphatic channels are efferent to the *cubital lymph nodes*. This superficial group of one or two nodes is located 3 to 4 cm above the medial epicondyle of the humerus and below the aperture in the brachial fascia for the basilic vein. The afferent vessels of these nodes include channels originating in the ulnar three fingers and the ulnar portion of the forearm. The efferent vessels accompany the basilic vein under the brachial fascia and reach the lateral and central groups of axillary lymph nodes.

Several lymphatic channels collecting from the dorsal surface of the arm follow the upper course of the cephalic vein to the deltopectoral triangle, perforate the costocoracoid membrane with the vein, and terminate in an apical node of the axillary group. In about 10% of cases, this channel is interrupted in the deltopectoral triangle by one or two small *deltopectoral nodes*.

Axillary Lymph Nodes. The axillary lymph nodes, usually large and numerous, are arranged in five subgroups, some related to the axillary walls and others to vessels (Plate 23).

A *lateral group* of three to five nodes lies medial and posterior to the distal segment of the axillary vein. These nodes are in the direct line of lymph drainage from the upper limb, except for the drainage lymphatics along the cephalic vein. Efferent vessels from these nodes drain to the central and apical nodes.

A *pectoral group* is located along the lateral thoracic artery adjacent to the axillary border of the pectoralis minor muscle. These three to five nodes receive the lymphatic drainage of the anterolateral part of the thoracic wall, including most of the lateral drainage from the mammary gland, and of the skin and muscles of the supraumbilical part of the abdominal wall. Efferent lymphatic vessels reach the central and apical groups.

A *subscapular group* of five or six nodes is stretched along the subscapular blood vessels, from their origin in the axillary vessels to their contact with the chest wall. These nodes drain the skin and muscles of the posterior thoracic wall and shoulder region and also the lower part of the back of the neck. Their efferent lymph channels pass to the central axillary nodes.

A *central group* of four or five nodes lies under the axillary fascia, embedded in its fat. Among the largest of the axillary nodes, these nodes receive some lymphatic vessels directly from the arm and mammary regions; but primarily, they receive lymph from the lateral, pectoral, and subscapular groups. Their efferent channels pass to the apical nodes.

The *apical group*, consisting of 6 to 12 nodes, lies along the axillary vein at the apex of the axilla and adjacent to the superior border of the pectoralis minor muscle. The apical nodes receive efferent vessels of all other axillary groups, lymphatic vessels that accompany the cephalic vein, and lymphatic vessels from the mammary gland. From lymph vessels interconnecting the apical nodes arises a larger common channel, the *subclavian lymphatic trunk*.

Deep Lymphatics. These vessels serve the upper limb, draining joint capsules, periosteum, tendons, nerves, and, to a lesser extent, muscles. Collecting vessels accompany the major arteries, along whose paths lie small intercalated lymph nodes. The deep lymphatics are afferent to the central and lateral axillary nodes. □

Superficial Veins and Cutaneous Nerves of Lower Limb

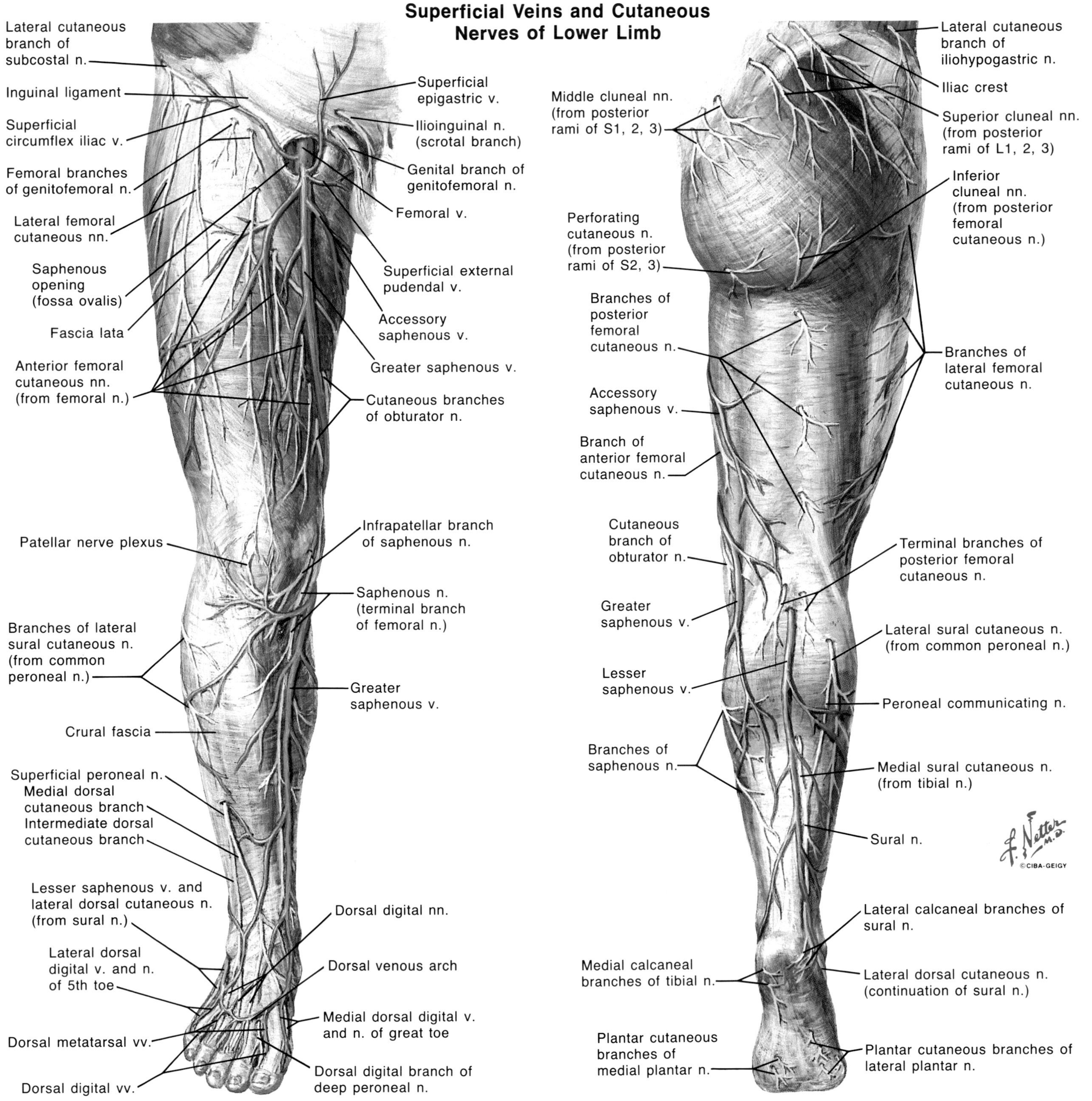
Lateral cutaneous branch of subcostal n.
Inguinal ligament
Superficial circumflex iliac v.
Femoral branches of genitofemoral n.
Lateral femoral cutaneous nn.
Saphenous opening (fossa ovalis)
Fascia lata
Anterior femoral cutaneous nn. (from femoral n.)
Superficial epigastric v.
Ilioinguinal n. (scrotal branch)
Genital branch of genitofemoral n.
Femoral v.
Superficial external pudendal v.
Accessory saphenous v.
Greater saphenous v.
Cutaneous branches of obturator n.
Patellar nerve plexus
Infrapatellar branch of saphenous n.
Saphenous n. (terminal branch of femoral n.)
Branches of lateral sural cutaneous n. (from common peroneal n.)
Greater saphenous v.
Crural fascia
Superficial peroneal n.
Medial dorsal cutaneous branch
Intermediate dorsal cutaneous branch
Lesser saphenous v. and lateral dorsal cutaneous n. (from sural n.)
Lateral dorsal digital v. and n. of 5th toe
Dorsal metatarsal vv.
Dorsal digital vv.
Dorsal digital nn.
Dorsal venous arch
Medial dorsal digital v. and n. of great toe
Dorsal digital branch of deep peroneal n.
Middle cluneal nn. (from posterior rami of S1, 2, 3)
Perforating cutaneous n. (from posterior rami of S2, 3)
Branches of posterior femoral cutaneous n.
Accessory saphenous v.
Branch of anterior femoral cutaneous n.
Cutaneous branch of obturator n.
Greater saphenous v.
Lesser saphenous v.
Branches of saphenous n.
Medial calcaneal branches of tibial n.
Plantar cutaneous branches of medial plantar n.
Lateral cutaneous branch of iliohypogastric n.
Iliac crest
Superior cluneal nn. (from posterior rami of L1, 2, 3)
Inferior cluneal nn. (from posterior femoral cutaneous n.)
Branches of lateral femoral cutaneous n.
Terminal branches of posterior femoral cutaneous n.
Lateral sural cutaneous n. (from common peroneal n.)
Peroneal communicating n.
Medial sural cutaneous n. (from tibial n.)
Sural n.
F. Netter M.D.
©CIBA-GEIGY
Lateral calcaneal branches of sural n.
Lateral dorsal cutaneous n. (continuation of sural n.)
Plantar cutaneous branches of lateral plantar n.

SECTION I PLATE 71 Slide 3636

Lower Limb

Superficial Veins and Cutaneous Nerves

Superficial Veins

Certain prominent veins, unaccompanied by arteries, are found in the subcutaneous tissue of the lower limb (Plate 71). The principal ones are the greater and lesser saphenous veins, which arise in the venous radicles in the feet and toes. Dorsal digital veins lie along the dorsal margins of each digit, uniting at the webs of the toes into short dorsal metatarsal veins that empty into the dorsal venous arch. There are also plantar digital veins, which drain into the dorsal metatarsal veins.

The *greater saphenous vein* continues the medial end of the dorsal venous arch and is the longest named vein of the body. It turns upward anterior to the medial malleolus at the ankle and, ascending immediately posterior to the medial margin of the tibia, passes the knee against the posterior border of the medial femoral condyle. In the thigh, the vein inclines anteriorly and lateralward; in the femoral triangle, it turns deeply through the saphenous opening to empty into the femoral vein. In the leg, the greater saphenous vein has tributaries from the heel of the foot, the front of the leg, and the calf. It also communicates with radicles of the lesser saphenous vein.

In the thigh, the greater saphenous vein also receives a large *accessory saphenous vein*, which collects the superficial radicles of the medial and posterior parts. Just before it turns through the saphenous opening, it receives the superficial external pudendal, superficial epigastric, and superficial circumflex iliac veins. On occasion, these veins pierce the cribriform fascia of the saphenous opening independently and empty directly into the femoral vein.

Lower Limb

Superficial Veins and Cutaneous Nerves

(Continued)

Valves in the greater saphenous vein vary from 10 to 20 in number and are more numerous in the leg than in the thigh. Perforating communications to deep veins pass through the deep fascia at all levels of the limb. Blood passes from superficial to deep through these communications, valves in the communications determining the direction of drainage. In the leg, the greater saphenous vein is accompanied by a branch of the saphenous nerve; in the thigh, an anterior femoral cutaneous nerve lies next to it.

The *lesser saphenous vein*, continuing the lateral extension of the dorsal venous arch, receives lateral marginal veins in the foot and passes backward along the lateral border of the foot in company with the sural nerve. It then turns upward to ascend through the middle of the calf. The vein pierces the crural fascia, mostly in the middle third of the leg but frequently in the upper third, and ascends deep to or in a split of the deep fascia. It usually (75% of cases) terminates in the popliteal vein. The lesser saphenous vein has from 6 to 12 valves. It communicates with radicles of the greater saphenous vein and also with deep veins of the leg. It frequently gives rise to a radicle, which communicates with the accessory saphenous vein. The lesser saphenous vein is accompanied by the sural nerve in the lower half of the leg; in its terminal course in the popliteal fossa, it has a close anatomic relationship to the tibial nerve.

Cutaneous Nerves

The deep nerves of the lower limb are discussed separately (Plates 75–77 and 97–98).

Almost all the cutaneous nerves of the lower limb originate from the lumbosacral plexus, which is formed from the ventral rami of the first lumbar to third sacral nerves (L1–S3) (Plates 71–73). However, the superior and middle cluneal nerves are lateral cutaneous branches of dorsal rami of certain of these nerves. The *superior cluneal nerves* are such branches of L1, 2, 3. They distribute to the skin of the gluteal region as far as the greater trochanter of the femur. The *middle cluneal nerves* are lateral branches of dorsal rami of S1, 2, 3 and provide cutaneous innervation to the skin over the back of the sacrum and adjacent gluteal region.

Iliac Crest Region. Certain lateral cutaneous branches of primarily abdominal nerves cross the crest of the ilium and distribute in the upper thigh. Thus, the *lateral cutaneous branch of the subcostal nerve* (T12) distributes to the skin and subcutaneous tissue of the thigh to as low as the greater trochanter of the femur. The *lateral cutaneous branch of the iliohypogastric nerve* (L1) supplies the skin of the gluteal region posterior to the area of supply of the lateral cutaneous branch of the subcostal nerve. The ilioinguinal nerve of the lumbar plexus has a small femoral distribution through its *anterior scrotal* (or *anterior labial*) *branch* (L1). Twigs of this nerve reach the skin of the thigh adjacent to the scrotum (or labium majus). The *femoral branch of the genitofemoral nerve* (L1, 2) arises from the lumbar plexus to supply the skin over the femoral triangle.

Hip and Thigh. *Anterior femoral cutaneous nerves* (L2, 3) are multiple. They usually arise from the femoral nerve in the femoral triangle on the lateral surface of the femoral artery. Medially distributing representatives, the *medial femoral cutaneous nerves*, supply the skin and subcutaneous tissues in the distal two-thirds of the medial portion of the thigh. Other branches arising in the femoral triangle, the *intermediate femoral cutaneous nerves*, supply the skin of the distal three-fourths of the front of the thigh and extend to the front of the patella, where they assist in forming the patellar plexus.

The *lateral femoral cutaneous nerve* (L2, 3) is a direct branch of the lumbar plexus. It becomes subcutaneous about 10 cm below the anterior iliac spine and distributes to anterior and lateral aspects of the thigh. Its larger anterior distribution may reach the patellar plexus.

The *posterior femoral cutaneous nerve* (S1, 2, 3) from the sacral plexus descends in the posterior midline of the thigh deep to the fascia lata, giving off branches that pierce the fascia and distribute both medially and laterally. The nerve finally reaches the levels of the popliteal fossa and the upper calf. The posterior femoral cutaneous nerve, lying subgluteally alongside the sciatic nerve, gives rise to the *inferior cluneal nerve*. This nerve turns around the lower border of the gluteus maximus muscle and supplies the skin over the lower and lateral parts of the muscle.

A *perforating cutaneous nerve* (S2, 3) arises from the sacral plexus. It gets its name because it perforates the sacrotuberal ligament and the lower fibers of the gluteus maximus muscle and distributes to the skin over the medial part of the fold of the buttock.

The *obturator nerve* (L2, 3, 4) from the lumbar plexus is largely muscular in the thigh, but its anterior branch usually ends as a *cutaneous branch*. This is distributed to the skin of the distal third of the thigh on its medial surface.

Leg, Ankle, and Foot. The *saphenous nerve* (L3, 4) is the terminal branch of the femoral nerve, even though it arises in the femoral canal. It traverses the whole length of the adductor canal, at its lower end piercing the vastoadductor membrane to become superficial along with the saphenous branch of the descending genicular artery. It descends in the leg in company with the greater saphenous vein. An infrapatellar branch curves downward below the patella, forming the patellar plexus with terminals of the medial and lateral femoral cutaneous nerves. The saphenous nerve continues distally along the medial surface of the leg, distributing therefrom and finally reaching the posterior half of the dorsum and the medial side of the foot.

The *lateral sural cutaneous nerve* (L5; S1, 2) arises in the popliteal space from the common peroneal nerve. It distributes to the skin and subcutaneous connective tissue on the lateral part of the leg in its proximal two-thirds. The *peroneal communicating branch* is a small nerve that usually arises from the lateral sural cutaneous nerve in the popliteal space or over the lateral calf. It runs downward and medially to join the medial sural cutaneous nerve in the middle third of the leg to form the sural nerve.

The *medial sural cutaneous nerve* (S1, 2) arises from the tibial nerve in the popliteal fossa. It descends as far as the middle of the leg. Here, it is joined by the peroneal communicating nerve to form the sural nerve. The level of junction of these nerves is quite variable, and in about 20% of cases, they fail to unite; the peroneal communicating branch then distributes in the leg, and the medial sural cutaneous nerve supplies the heel and foot areas.

The *sural nerve* (S1, 2), formed as described above, usually becomes superficial at the middle of the length of the leg. It descends in company with the lesser saphenous vein, turning with it under the lateral malleolus onto the side of the foot. As the *lateral dorsal cutaneous nerve* of the foot, it is cutaneous to the lateral side of the foot, having previously given off *lateral calcaneal branches* to the ankle and heel. It extends to the little toe, has articular branches to the ankle and tarsal joints, and communicates with the intermediate dorsal cutaneous branch of the superficial peroneal nerve.

The *superficial peroneal nerve* (L4, 5; S1), a branch of the common peroneal nerve, descends to the distal third of the leg, where it almost immediately divides into two terminal branches. The *medial dorsal cutaneous nerve* supplies cutaneous twigs in the distal third of the leg and then, crossing the extensor retinaculum, divides over the dorsum of the foot into two or three branches for the supply of the dorsum and sides of the medial 2½ toes. The *deep peroneal nerve* (L4, 5) has cutaneous terminals to the web and adjacent sides of the great and second toes, and in this interval, the terminals of the medial dorsal cutaneous nerve merely communicate with the branches of the deep peroneal nerve.

The *intermediate dorsal cutaneous nerve* passes more laterally over the dorsum of the foot and divides into cutaneous branches to the lateral side of the ankle and the foot. It terminates in *dorsal digital branches* for the adjacent sides of the third and fourth and fourth and fifth toes. The more lateral branch communicates with the lateral dorsal cutaneous nerve. As in the fingers, the dorsal digital branches to the toes are smaller than the corresponding plantar digital nerves and distribute distalward only over the middle phalanges. The terminal segment of the toes is supplied by dorsal terminals of the plantar nerves.

The *tibial nerve* supplies the musculature of the back of the leg and continues into the foot behind the medial malleolus. Its *medial calcaneal branches* (S1, 2) distribute to the heel and the posterior part of the sole of the foot. Derived from the tibial nerve deep to the abductor hallucis muscle are the medial and lateral plantar nerves. The *medial plantar nerve* (L4, 5) provides a *proper digital nerve* to the medial side of the great toe and three *common digital branches*. Each of the latter splits into two proper digital nerves, which supply the skin of the adjacent sides of digits I and II, II and III, and III and IV, respectively. The *lateral plantar nerve* (S1, 2) provides a common digital nerve, which also divides into proper digital nerves; two of them reach the adjacent sides of the fourth and fifth toes, and one reaches the lateral side of the little toe. The plantar digital nerves reach the whole plantar surface of the digits and also furnish small dorsal twigs for the nail bed and tip of each toe. □

Lower Limb

Lumbosacral Plexus

Nerve Supply

In describing the lower limb, the lumbosacral plexus may be said to be formed from the ventral rami of the first lumbar to third sacral nerves (L1–S3), with a common small contribution from the twelfth thoracic nerve (T12) (Plates 72–74). The lumbar portion of the plexus arises from the four upper lumbar nerves and gives rise to the iliohypogastric, ilioinguinal, genitofemoral, lateral femoral cutaneous, obturator, accessory obturator, and femoral nerves.

Lumbar Plexus

As with the brachial plexus, the spinal nerves contributing to the lumbar plexus divide into anterior and posterior branches, but the plexus lacks some of the complexity of the brachial plexus, since the definitive nerves usually arise from combinations of looping contributions from adjacent spinal nerves. The lumbar plexus is formed deep to the psoas major muscle and lies anterior to the transverse processes of the lumbar vertebrae (Plate 72). Only the first two lumbar nerves contribute preganglionic sympathetic fibers to the sympathetic chain through white rami communicantes; all lumbar nerves receive postganglionic fibers through gray rami communicantes.

The *iliohypogastric nerve* arises from L1 together with a frequent contribution from T12. It emerges from the psoas major muscle at its lateral border and crosses the quadratus lumborum muscle to penetrate the transverse abdominal muscle near the iliac crest. This nerve, primarily motor to abdominal musculature, ends in an anterior cutaneous branch to the skin of the suprapubic region and a lateral cutaneous branch that crosses the iliac crest to distribute in the hip region. A lateral cutaneous branch of the subcostal nerve (T12) also supplies the upper thigh. (The cutaneous distribution of the iliohypogastric nerve is described on page 76.)

The *ilioinguinal nerve* from L1 has a similar course to the iliohypogastric nerve in the abdominal wall but enters the lateral end of the inguinal canal and accompanies the spermatic cord through that canal. Emerging at the superficial inguinal ring, it ends as the anterior scrotal (or anterior labial) nerve as a cutaneous nerve to the scrotum and adjacent area of the thigh. In about 35% of cases, the ilioinguinal nerve combines with the genitofemoral nerve in the abdomen, running with the latter on the surface of the psoas major muscle, but distributes finally in its typical cutaneous distribution.

Lumbar Plexus

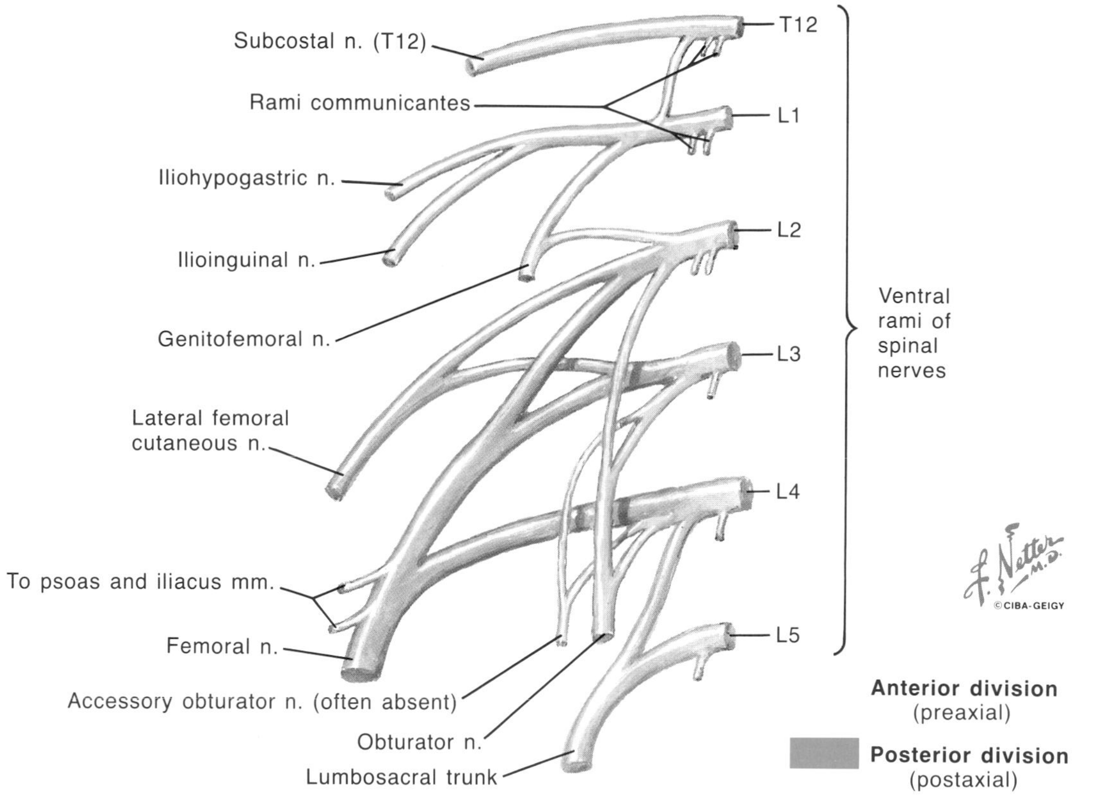

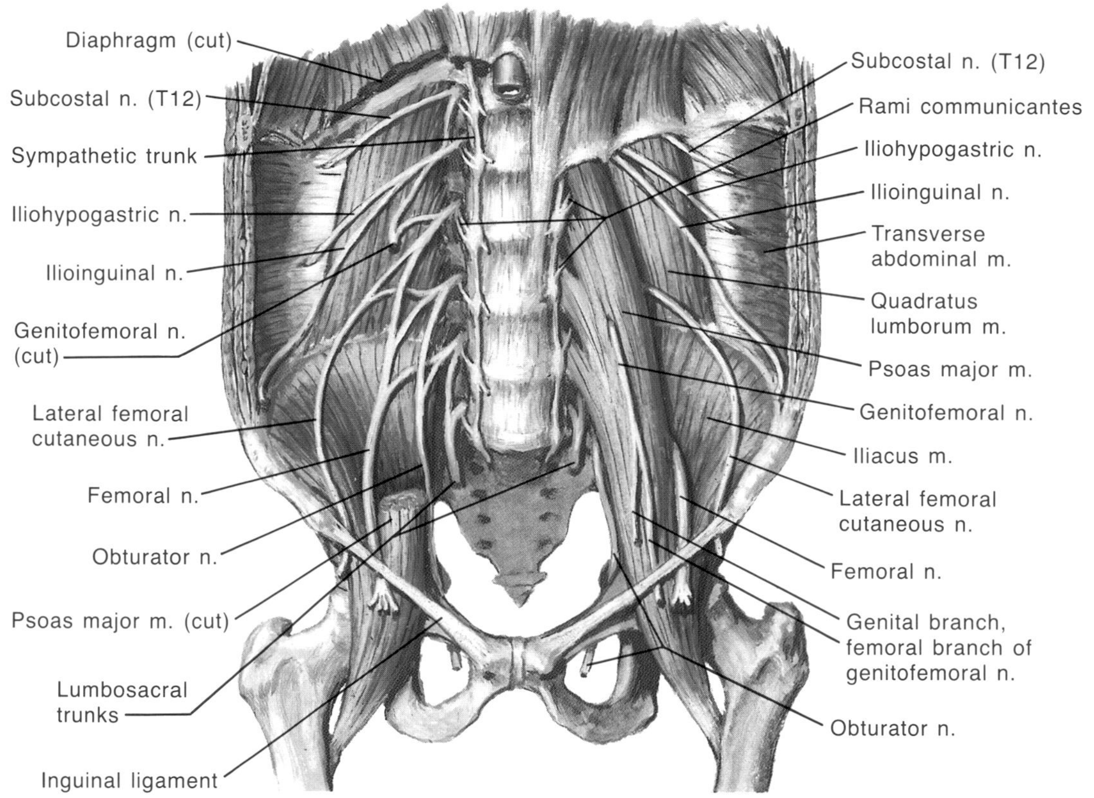

The *genitofemoral nerve* arises by a union of branches from the anterior portions of L1 and L2. In the abdomen, it descends on the ventral surface of the psoas major muscle and then divides into genital and femoral branches. The genital branch innervates the cremaster muscle and gives twigs to the scrotum and adjacent thigh; the more medial femoral branch descends under the inguinal ligament on the surface of the external iliac artery to supply the skin of the femoral triangle.

The *lateral femoral cutaneous nerve* arises from the posterior branches of L2 and L3. (This nerve is fully described in Plate 75 and on page 76.)

The *obturator nerve* is the largest nerve formed from the anterior divisions of the lumbar plexus, specifically from those of L2, 3, 4. The *accessory obturator nerve* is small and is present in only 9% of cases. (The obturator nerve is fully described in Plate 76 and on page 76.)

The *femoral nerve*, the largest branch of the lumbar plexus, is formed from the posterior branches of L2, 3, 4. Passing under the inguinal ligament, it shortly breaks up in the femoral triangle into its numerous branches. (This nerve is fully described in Plate 75.)

Muscular branches of the lumbar plexus distribute to the quadratus lumborum muscle (T12; L1, 2, 3 [4]), the psoas major muscle ([L1], 2, 3, 4), the psoas minor muscle (L1, 2), and the iliacus muscle (L2, 3, 4).

Lower Limb

Lumbosacral Plexus

(Continued)

Sacral Plexus

The sacral portion of the lumbosacral plexus, commonly designated the sacral plexus, is formed from the ventral rami of a part of the fourth lumbar nerve (L4), and the fifth lumbar and first, second, and third sacral nerves (L5; S1, 2, 3) (Plate 73). The descending portion of L4 joins L5 over the ala of the sacrum to form the lumbosacral trunk, which then descends across the sacroiliac articulation to join the ventral ramus of S1. The lumbosacral trunk contains anterior and posterior branches of the ventral rami of L4 and L5. The ventral rami of S1, 2, 3 pass lateralward from the pelvic sacral foramina and divide into anterior and posterior branches in front of the piriformis muscle. The largest nerves of the sacral plexus are L5 and S1, and the superior gluteal artery usually leaves the pelvis by passing between them.

Converging toward the lower portion of the greater sciatic foramen, the plexus forms a broad triangular band, the apex of which passes through the foramen into the gluteal region. The pelvic splanchnic nerves, arising from the ventral rami of S2, 3, 4, represent the important sacral part of the craniosacral (parasympathetic) portion of the autonomic nervous system. They join the inferior hypogastric plexus and have a largely pelvic and perineal distribution. All the nerves of the plexus receive gray rami communicantes from the sympathetic chain ganglia or trunk.

The principal nerve of the sacral plexus is the sciatic (Plates 73 and 77). It is composed of an anteriorly derived nerve, the tibial segment, and a nerve formed from posterior branches, the common peroneal nerve. These two nerves are usually combined in a single sheath, but in 10% of cases, the two parts are separated in the greater sciatic foramen by all or part of the piriformis muscle. They are occasionally separate throughout the thigh. The nerves of the plexus and their sources can be usefully tabulated as follows:

Sacral and Coccygeal Plexuses

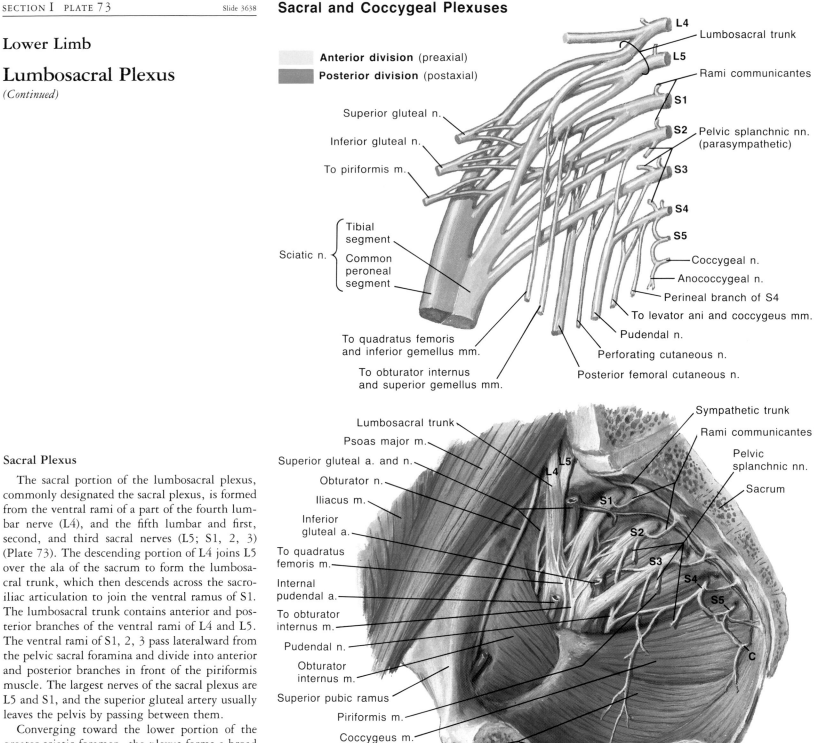

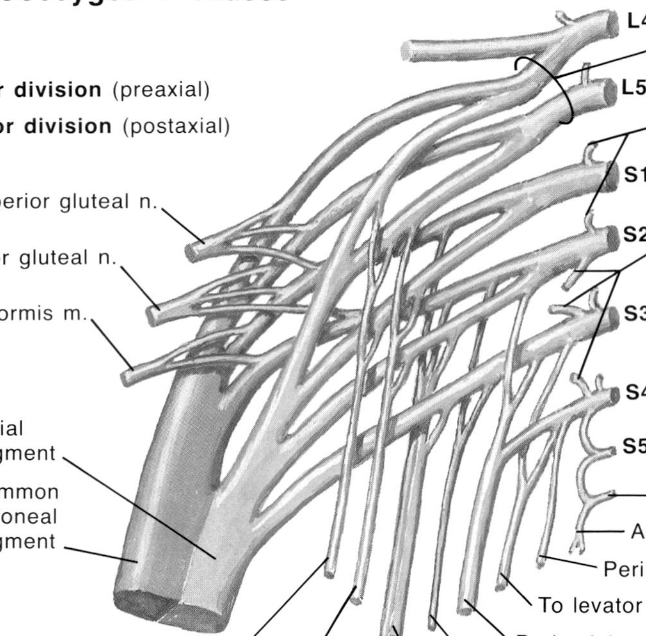

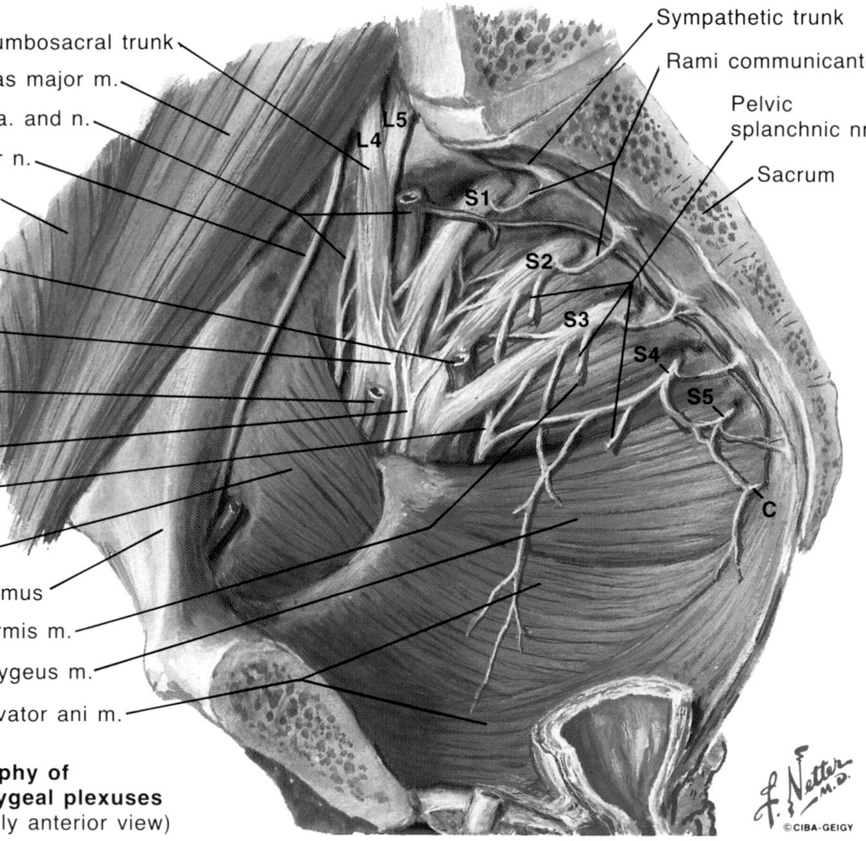

General topography of sacral and coccygeal plexuses
(lateral and slightly anterior view)

Nerves	Anterior Branches	Posterior Branches
Sciatic	Tibial—L4, 5; S1, 2, 3	Common Peroneal—L4, 5; S1, 2
Muscular branches to piriformis, levator ani, coccygeus	S3, 4	S1, 2
Superior gluteal		L4, 5; S1
Inferior gluteal		L5; S1, 2
To quadratus femoris, inferior gemellus	L4, 5; S1	
To obturator internus, superior gemellus	L5; S1, 2	
Posterior femoral cutaneous	S2, 3	S1, 2
Perforating cutaneous		S2, 3

Nerves of Buttock

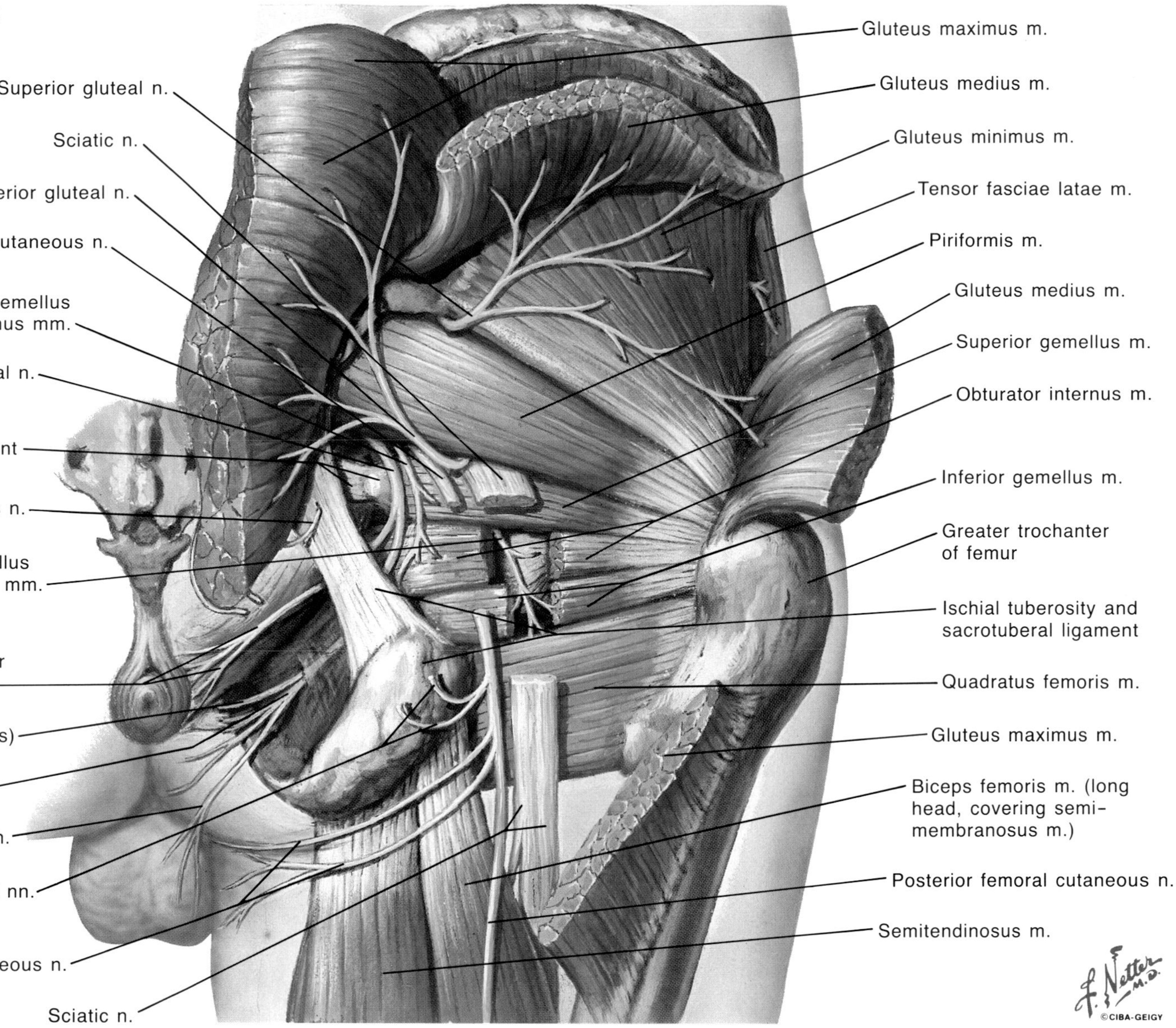

Superior gluteal n.
Sciatic n.
Inferior gluteal n.
Posterior femoral cutaneous n.
Nerve to superior gemellus and obturator internus mm.
Pudendal n.
Ischial spine and sacrospinal ligament
Perforating cutaneous n.
Nerve to inferior gemellus and quadratus femoris mm. (twig to hip joint)
Inferior rectal (inferior hemorrhoidal) nn.
Dorsal n. of penis (clitoris)
Perineal n.
Posterior scrotal (labial) n.
Inferior cluneal nn.
Perineal branches of posterior femoral cutaneous n.
Sciatic n.
Gluteus maximus m.
Gluteus medius m.
Gluteus minimus m.
Tensor fasciae latae m.
Piriformis m.
Gluteus medius m.
Superior gemellus m.
Obturator internus m.
Inferior gemellus m.
Greater trochanter of femur
Ischial tuberosity and sacrotuberal ligament
Quadratus femoris m.
Gluteus maximus m.
Biceps femoris m. (long head, covering semimembranosus m.)
Posterior femoral cutaneous n.
Semitendinosus m.
F. Netter M.D.
©CIBA-GEIGY

Lower Limb

Lumbosacral Plexus

(Continued)

Nerves of Gluteal Region

The *sciatic nerve* usually emerges from the pelvis at the lower border of the piriformis muscle and enters the thigh in the hollow between the ischial tuberosity and the greater trochanter of the femur (Plates 73–74). Its distribution as the tibial and common peroneal nerves is detailed on pages 104–105.

The *nerve to the piriformis muscle* may be represented by separate contributions from S1 and S2. Twigs arise from the dorsal aspect of these nerves and immediately enter the pelvic surface of the muscle. Muscular nerves to the levator ani and coccygeus muscles arise from the loop between the rami of S3 and S4 and descend to enter the pelvic surface of these muscles.

The *superior gluteal nerve*, from the posterior branches of L4, 5 and S1, passes from the pelvis above the piriformis muscle. Deep to the gluteus maximus and gluteus medius muscles, the nerve accompanies the superior gluteal vessels anteriorly over the surface of the gluteus minimus muscle. It supplies the gluteus medius and minimus muscles and, continuing beyond them, the tensor fasciae latae muscle.

The *inferior gluteal nerve*, formed from the posterior branches of L5 and S1, 2, passes from the pelvis below the piriformis muscle. It enters the deep surface of the gluteus maximus muscle, to which it is the sole supply.

The *nerve to the quadratus femoris* and *inferior gemellus muscles* is formed from the anterior branches of L4, 5 and S1. In the gluteal region, it is deep to the sciatic nerve and descends over the back of the ischium anterior to the gemellus muscles and the tendon of the internal obturator muscle. It provides articular branches to the hip joint and a branch to the inferior gemellus muscle and ends in the anterior surface of the quadratus femoris muscle.

The *nerve to the obturator internus* and *superior gemellus muscles* arises from anterior branches of L5 and S1, 2. In the gluteal region, it is inferomedial to the sciatic nerve and on the lateral side of the internal pudendal vessels. It crosses the superior gemellus muscle and supplies a small nerve to it. The remaining nerve to the obturator internus muscle crosses the ischial spine and enters the ischiorectal fossa through the lesser sciatic foramen. It ends in the perineal surface of the muscle.

The *posterior femoral cutaneous nerve* is a mixed nerve, formed by posterior branches from S1 and S2 and anterior branches from S2 and S3; its cutaneous distributions are described on page 76. In the gluteal region, it lies alongside the sciatic nerve and descends in the midline of the thigh. It also provides perineal branches that are cutaneous in the perineum and the back of the scrotum.

The *perforating cutaneous nerve* arises from posterior branches of S2 and S3 and is associated at its origin with the lower roots of the posterior femoral cutaneous nerve. Its cutaneous distribution is described on page 76.

Dermatomes of Lower Limb

As in the upper limb, the serial order of distribution of lower limb nerves as seen in the lumbosacral plexus is retained in the cutaneous zones of appropriate nerves in the limb. The lumbar nerves have cutaneous terminals that distribute from above down and lateromedially. Sacral segments are restricted to the posterior aspect of the limb and the lateral side of the foot. The spiraling of nerve distribution is a consequence of the medial rotation of the lower limbs of almost 90° that takes place in development, so that the future knees point ventrolaterally (see Section II, Plate 9). There is always an overlap of adjacent segments; therefore the lines of separation are indistinct (see CIBA COLLECTION, Volume 1/I, page 55). □

Lower Limb

Nerves of Thigh

Femoral Nerve (L2, 3, 4) and Lateral Femoral Cutaneous Nerve (L2, 3)

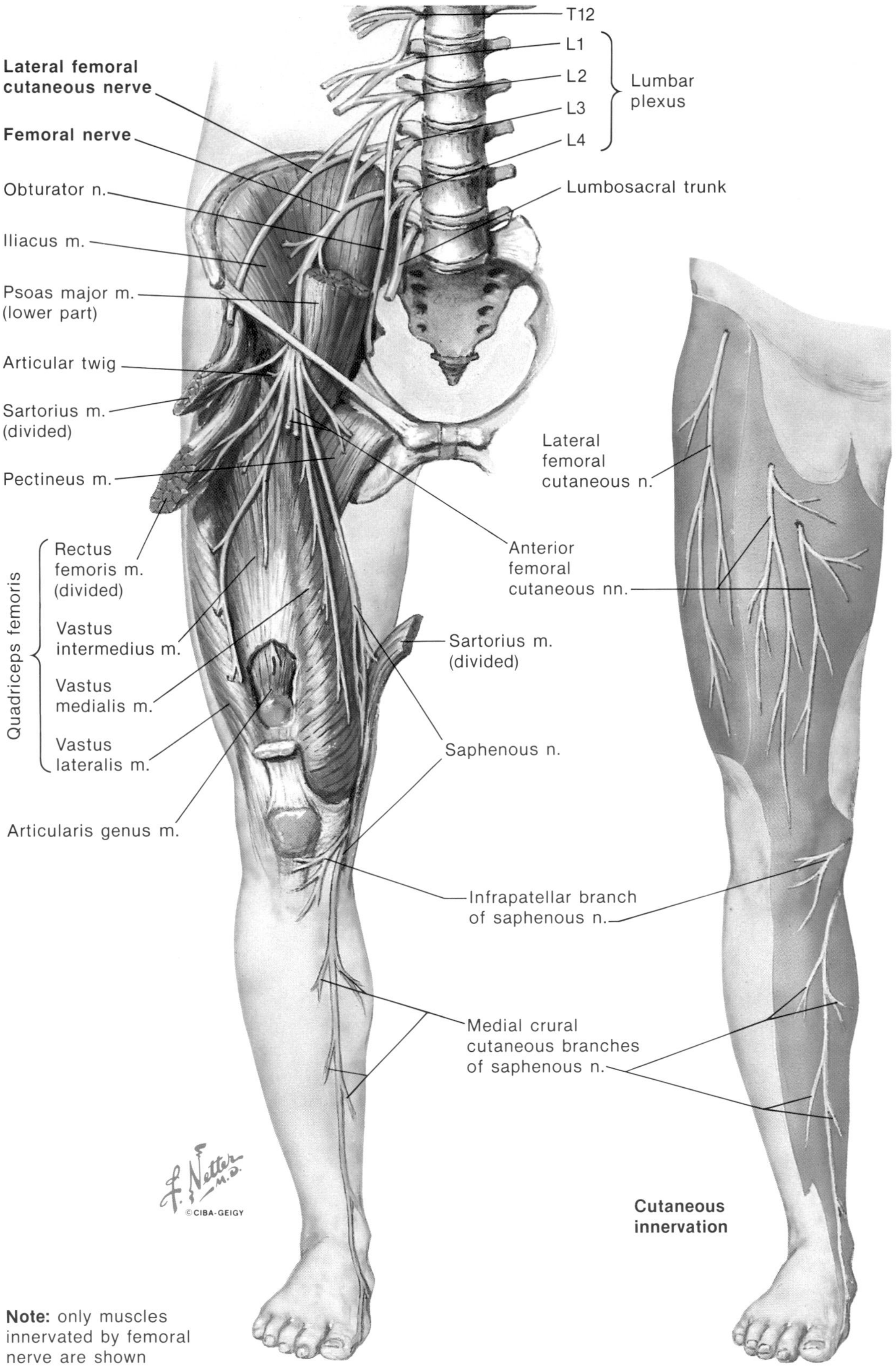

Femoral Nerve

The femoral nerve (L2, 3, 4) is the largest branch of the lumbar plexus (Plate 75). It originates from the posterior divisions of the ventral rami of the second, third, and fourth lumbar nerves, passes inferolaterally through the psoas major muscle, and then runs in a groove between this muscle and the iliacus, which it supplies. It enters the thigh behind the inguinal ligament to lie lateral to the femoral vascular sheath in the femoral triangle. Twigs are given off to the hip and knee joints and adjacent vessels, and cutaneous branches are given off to anteromedial aspects of the lower limb.

Muscular branches supply the pectineus, sartorius, and quadriceps femoris muscles. The nerve to the pectineus muscle arises at the level of the inguinal ligament, while the branches to the sartorius muscle enter the upper two-thirds of the muscle, several arising in common with the anterior femoral cutaneous nerves. The branches to the quadriceps femoris muscle are arranged as illustrated and those to the rectus femoris and vastus lateralis muscles enter the deep surfaces of the muscles. The branch to the vastus intermedius muscle enters its superficial surface and pierces the muscle to supply the underlying articularis genus muscle. The branch to the vastus medialis muscle runs in the adductor canal for a variable distance, on the lateral side of the femoral vessels and saphenous nerve, giving off successive branches to this muscle, some of which end in the vastus intermedius and articularis genus muscles.

The *anterior femoral cutaneous nerves* arise in the femoral triangle. All these branches pierce the fascia lata 8 to 10 cm distal to the inguinal ligament and descend to knee level, supplying the skin and fascia over the front and medial sides of the thigh.

The *saphenous nerve* is the largest and longest of the femoral branches. It arises at the femoral triangle and descends through it on the lateral side of the femoral vessels to enter the adductor canal. Here, it crosses the vessels obliquely to lie on their medial side in front of the lower end of the adductor magnus muscle. In the canal, the saphenous nerve communicates with branches of the anterior femoral cutaneous and obturator nerves to form the *subsartorial plexus*. At the lower end of the canal, it leaves the femoral vessels and gives off its *infrapatellar branch*, which curves around the posterior border of the sartorius muscle, pierces the fascia lata, and runs onward to supply the skin over the medial side and front of the knee and the patellar ligament. This branch assists offshoots from the anterior and lateral femoral cutaneous nerves in forming the *patellar plexus*.

The saphenous nerve continues its descent on the medial side of the knee, pierces the fascia lata between the tendons of the sartorius and gracilis muscles, courses downward on the medial side of the leg close to the greater saphenous vein, and gives off its *medial crural cutaneous branches*. In the lower leg, it subdivides terminally—the smaller branch follows the medial tibial border to the level of the ankle, the larger passes anterior to the medial malleolus to distribute to the skin and fascia on the medial side and dorsum of the foot.

Articular branches arising from the nerve to the rectus femoris muscle accompany the corresponding branches of the lateral femoral circumflex artery to the hip joint. Twigs from the branches to the vastus muscles and from the saphenous nerve supply the knee joint.

Lateral Femoral Cutaneous Nerve

The lateral femoral cutaneous nerve (L2, 3) emerges from the lateral border of the psoas major muscle, passes obliquely over the iliacus muscle behind the parietal peritoneum and iliac fascia (which it supplies) toward the anterior superior iliac spine, and enters the thigh by passing under or through the lateral end of the inguinal ligament. The nerve then passes over or through the proximal part of the sartorius muscle and descends deep to the fascia lata. It gives off a number of small branches to the overlying skin before piercing the fascia about 10 cm below the inguinal ligament. The terminal branches of the lateral femoral cutaneous nerve supply the skin and fascia on the anterolateral surfaces of the thigh between the levels of the greater trochanter of the femur and the knee.

Lower Limb

Nerves of Thigh

(Continued)

Obturator Nerve (L2, 3, 4)

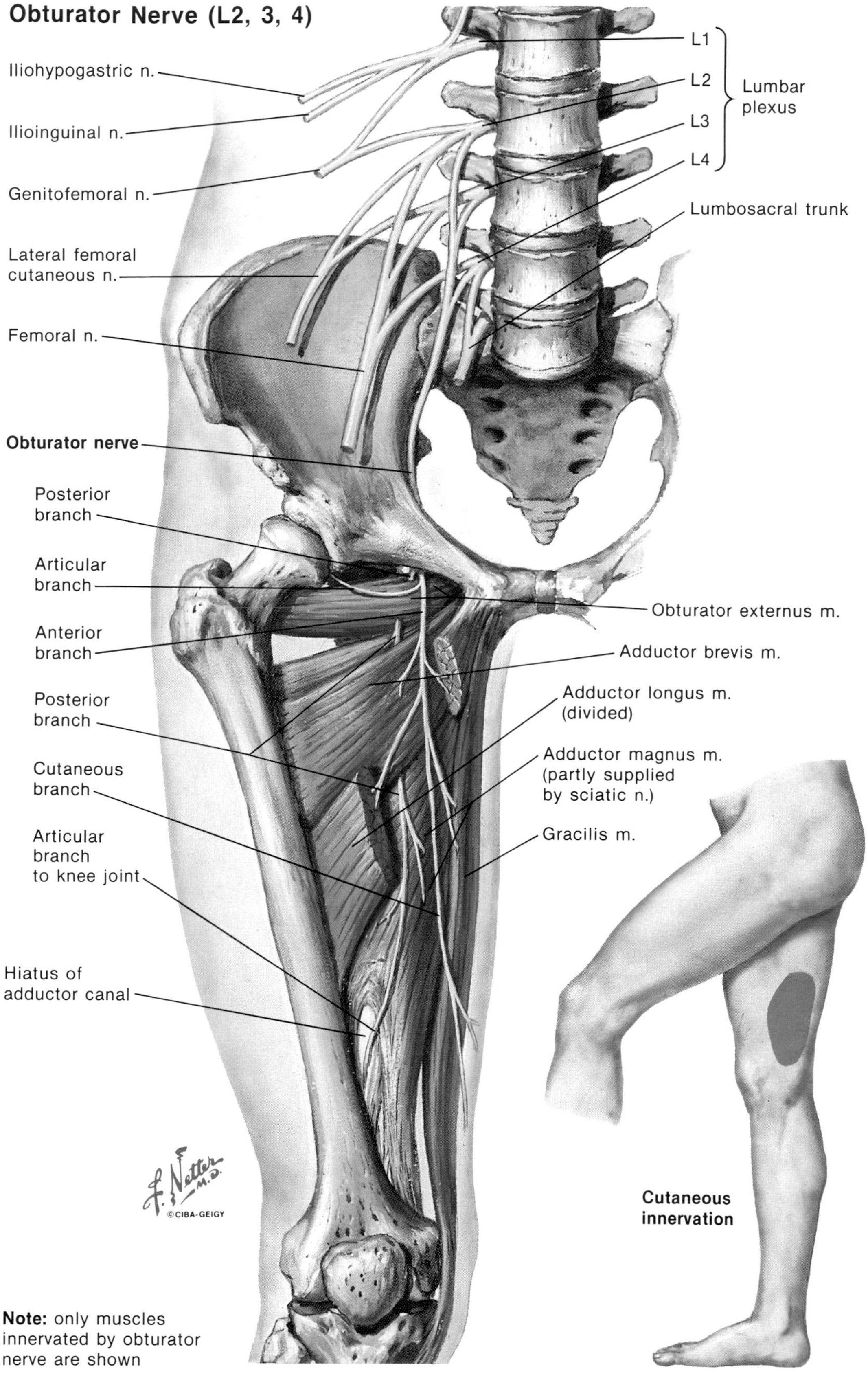
Iliohypogastric n.
Ilioinguinal n.
Genitofemoral n.
Lateral femoral cutaneous n.
Femoral n.
Obturator nerve
Posterior branch
Articular branch
Anterior branch
Posterior branch
Cutaneous branch
Articular branch to knee joint
Hiatus of adductor canal
L1
L2
L3
L4
Lumbar plexus
Lumbosacral trunk
Obturator externus m.
Adductor brevis m.
Adductor longus m. (divided)
Adductor magnus m. (partly supplied by sciatic n.)
Gracilis m.
Cutaneous innervation
F. Netter M.D.
©CIBA-GEIGY

Note: only muscles innervated by obturator nerve are shown

Obturator Nerve

The obturator nerve (L2, 3, 4) supplies the obturator externus and adductor muscles of the thigh, gives filaments to the hip and knee joints, and has a variable cutaneous distribution to the medial sides of the thigh and leg (Plate 76).

The obturator nerve arises from the anterior divisions of the ventral rami of the second, third, and fourth lumbar nerves. The contribution from L2 is commonly the smallest and is sometimes absent. These roots unite within the posterior part of the psoas major muscle, forming a nerve that descends through the muscle to emerge from its medial border opposite the upper end of the sacroiliac joint. The obturator nerve runs outward and downward over the sacral ala and pelvic brim into the lesser pelvis, lying lateral to the ureter and internal iliac vessels. It then bends anteroinferiorly to follow the curvature of the lateral pelvic wall (anterior to the obturator vessels and lying on the obturator internus muscle) to reach the obturator groove at the upper part of the obturator foramen. The nerve passes through this groove and foramen to enter the thigh and divides into *anterior* and *posterior branches* shortly thereafter.

The *anterior branch* runs in front of the obturator externus and adductor brevis muscles and behind the pectineus and adductor longus muscles. Near its origin, it gives off an articular twig that enters the hip joint through the acetabular notch. Rarely, it supplies a branch to the pectineus muscle and sends muscular branches to the adductor longus, adductor gracilis, and adductor brevis muscles. The anterior branch finally divides into cutaneous, vascular, and communicating branches.

The *cutaneous branch* is inconstant. When present, it unites with branches of the saphenous and anterior femoral cutaneous nerves in the adductor canal to form the *subsartorial plexus* and assists in the innervation of the skin and fascia over the distal two-thirds of the medial side of the thigh. Infrequently, this branch is larger and passes between the adductor longus and gracilis muscles to descend behind the sartorius to the medial side of the knee and the adjacent part of the leg, where it assists the saphenous nerve in the cutaneous supply of those areas.

The *vascular branches* end in the femoral artery. Other fine *communicating branches* may link the obturator nerve with the anterior and posterior femoral cutaneous nerves and the inconstant accessory obturator nerve.

The *posterior branch* pierces the anterior part of the obturator externus muscle and supplies it. Thereafter, the nerve runs downward between the adductor brevis and adductor magnus muscles and splits into branches that are distributed to the upper (adductor) part of the adductor magnus and sometimes to the adductor brevis (especially if the latter does not receive a supply from the anterior branch of the obturator nerve). A slender branch emerges from the lower part of the adductor magnus, passes through the hiatus of the adductor canal together with the femoral artery, and then continues to the knee joint. The posterior branch contributes filaments to the femoral and popliteal vessels and ends by perforating the oblique popliteal ligament to supply the articular capsule, cruciate ligaments, and synovial membrane of the knee joint. The fibers to the capsule and ligaments are mostly of somatic origin, while those to the synovial membrane are mainly sympathetic.

The *accessory obturator nerve* (L3, 4) is inconstant, small, and is derived from the anterior divisions of the ventral rami of L3 and L4. It descends on the medial border of the psoas muscle and then crosses the superior pubic ramus to lie behind the pectineus muscle. It ends by helping to supply the pectineus but may also supply one twig to the hip joint and another twig that joins the anterior branch of the obturator nerve.

Lower Limb

Nerves of Thigh

(Continued)

Sciatic Nerve (L4, 5; S1, 2, 3) and Posterior Femoral Cutaneous Nerve (S1, 2, 3)

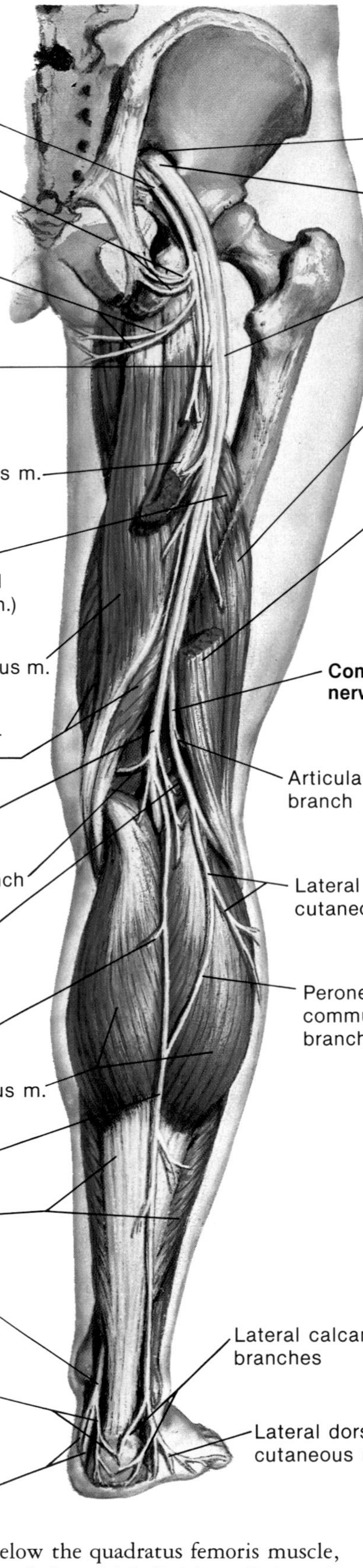

Cutaneous innervation

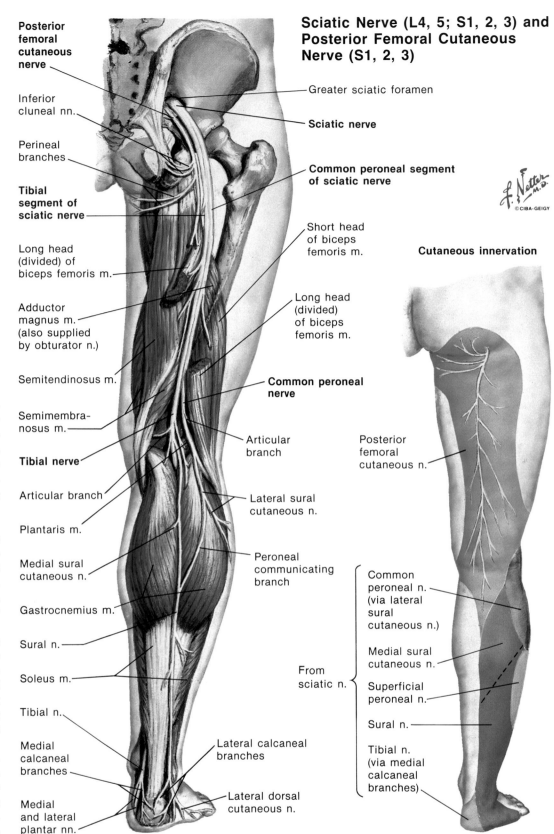

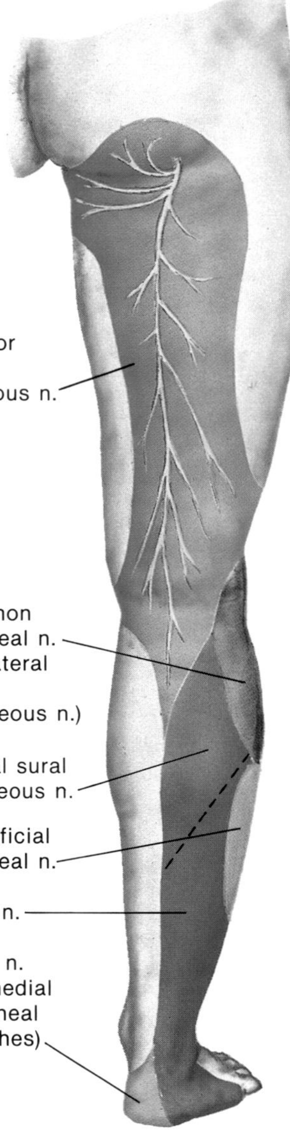

Sciatic Nerve

The roots of the sciatic nerve (L4, 5; S1, 2, 3) arise from the ventral rami of the fourth lumbar to third sacral nerves and unite to form a single trunk that is ovoid in cross section and 16 to 20 mm wide in adults (Plate 77). In the lesser pelvis, the nerve lies anterior to the piriformis muscle, below which it enters the buttock through the greater sciatic foramen (in about 2% of individuals, the nerve pierces the piriformis). Next, the nerve inclines laterally beneath the gluteus maximus muscle, where it rests on the posterior surface of the ischium and the nerve to the quadratus femoris muscle. On its medial side, it is accompanied by the posterior femoral cutaneous nerve and by the inferior gluteal artery and its special branch to the nerve.

On reaching a point about midway between the ischial tuberosity and the greater trochanter of the femur, the nerve turns downward over the gemellus muscles, the obturator internus tendon, and the quadratus femoris muscle (which separate it from the hip joint) and leaves the buttock to enter the thigh beneath the lower border of the gluteus maximus muscle.

The sciatic nerve then descends near the middle of the back of the thigh, lying on the adductor magnus muscle and being crossed obliquely by the long head of the biceps femoris muscle. Just above the apex of the popliteal fossa, it is overlapped by the contiguous margins of the biceps femoris and semimembranosus muscles. In about 90% of cases, the sciatic nerve divides into its terminal *tibial* and *common peroneal branches* near the apex of the popliteal fossa, while in 10% of cases, the division occurs at higher levels. Rarely, the tibial and common peroneal nerves arise independently from the sacral plexus, but pursue closely related courses until they reach the apex of the popliteal fossa.

Branches. In the buttock, the sciatic nerve supplies an *articular branch* to the hip, which perforates the posterior part of the joint capsule (Plate 74). It may also supply *vascular filaments* to the inferior gluteal artery. (The entrance of the sciatic nerve and its variable relationship to the piriformis muscle is described on page 79.)

At levels below the quadratus femoris muscle, two branches of the tibial division of the sciatic nerve spring from its medial side to supply the so-called hamstring muscles of the thigh. The upper branch passes to the long head of the biceps femoris muscle and the upper portion of the semitendinosus; the lower branch innervates the lower portion of the semitendinosus and the semimembranosus muscles and the ischiocondylar portion of the adductor magnus muscle. The nerve to the short head of the biceps femoris muscle arises from the lateral side of the sciatic nerve (common peroneal division of the sciatic nerve) in the middle third of the thigh and enters the superficial surface of the muscle. From this nerve, an articular branch continues to the knee, providing proximal and distal branches that accompany the lateral superior genicular and lateral inferior genicular arteries to the knee joint.

The tibial and common peroneal nerves arise by a division of the sciatic nerve, usually at the upper limit of the popliteal fossa. The tibial nerve continues the vertical course of the sciatic nerve at the back of the knee and into the leg (Plate 97). The common peroneal nerve follows the tendon of the biceps femoris muscle along the upper lateral margin of the popliteal space and into the leg, curving forward around the neck of the fibula (Plate 98). (The posterior femoral cutaneous nerve is described on pages 76 and 79.) □

Muscles of Front of Hip and Thigh

Lower Limb

Fasciae and Muscles of Hip and Thigh

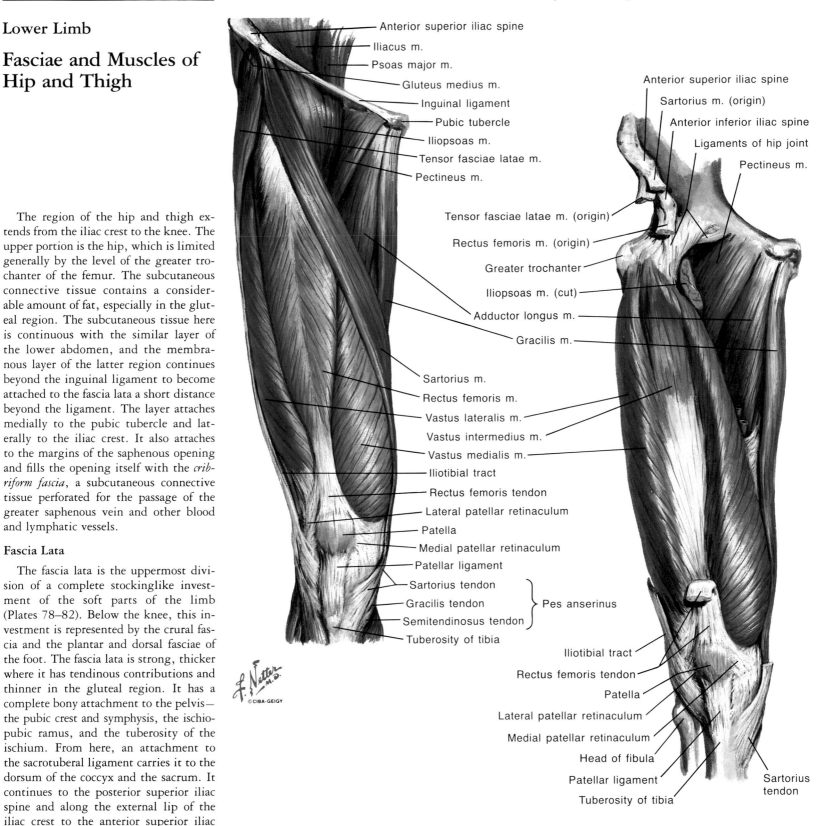

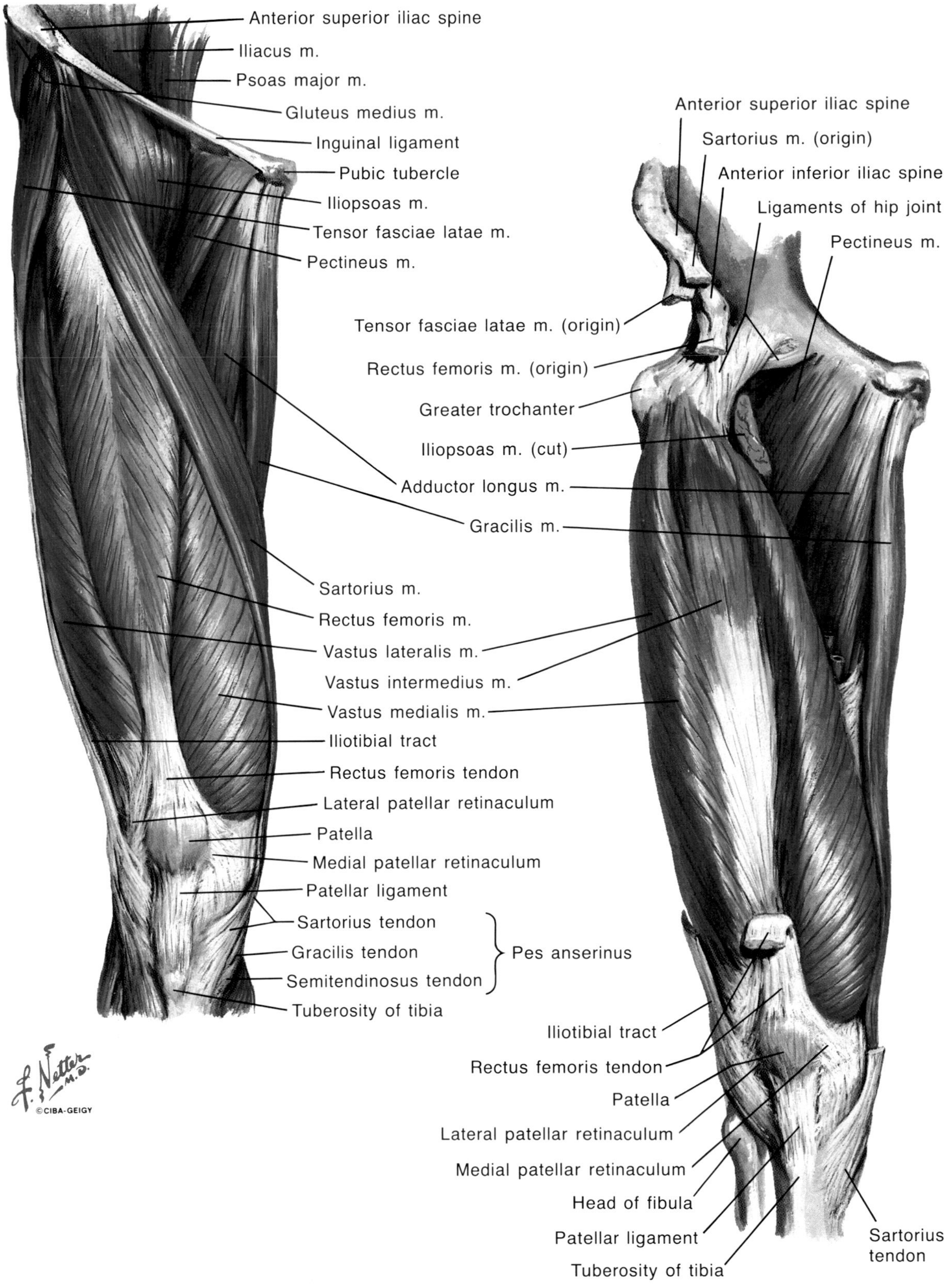

The region of the hip and thigh extends from the iliac crest to the knee. The upper portion is the hip, which is limited generally by the level of the greater trochanter of the femur. The subcutaneous connective tissue contains a considerable amount of fat, especially in the gluteal region. The subcutaneous tissue here is continuous with the similar layer of the lower abdomen, and the membranous layer of the latter region continues beyond the inguinal ligament to become attached to the fascia lata a short distance beyond the ligament. The layer attaches medially to the pubic tubercle and laterally to the iliac crest. It also attaches to the margins of the saphenous opening and fills the opening itself with the *cribriform fascia*, a subcutaneous connective tissue perforated for the passage of the greater saphenous vein and other blood and lymphatic vessels.

Fascia Lata

The fascia lata is the uppermost division of a complete stockinglike investment of the soft parts of the limb (Plates 78–82). Below the knee, this investment is represented by the crural fascia and the plantar and dorsal fasciae of the foot. The fascia lata is strong, thicker where it has tendinous contributions and thinner in the gluteal region. It has a complete bony attachment to the pelvis—the pubic crest and symphysis, the ischiopubic ramus, and the tuberosity of the ischium. From here, an attachment to the sacrotuberal ligament carries it to the dorsum of the coccyx and the sacrum. It continues to the posterior superior iliac spine and along the external lip of the iliac crest to the anterior superior iliac spine, the inguinal ligament, and the pubic tubercle. Here, a deep lamina follows the pecten of the pubis behind the femoral vein.

The *gluteal aponeurosis* lies between the iliac crest and the superior border of the gluteus maximus muscle, which provides part of the origin of the gluteus medius. A strong, lateral band, the *iliotibial tract*, arises from the tubercle of the iliac crest and serves as a tendon of the tensor fasciae latae muscle and as part of the tendon of the gluteus maximus muscle. The tract ends at the knee, where it reinforces the capsule of the knee joint and attaches to the condyle of the tibia.

The lateral and medial intermuscular septa unite the fascia lata to the periosteum of the femur and separate the muscles of the posterior and anterior compartments and the muscles of the medial and anterior groups. The medial intermuscular septum splits to enclose the sartorius muscle and helps to form the adductor canal (Plate 82). The iliotibial tract attaches deeply to the lateral intermuscular septum.

At the *saphenous opening*, a superficial sheet of fascia lata continues along the inguinal ligament to the pubic tubercle, while a deep lamina attaches to the pecten of the pubis.

Muscles

The muscles of the hip and thigh are divided into four groups: anterior, medial, posterior, and lateral femoral muscles (Plates 78–82). Additionally, the psoas major and iliacus muscles, although located for the most part and arising within the lower abdomen, insert into the thigh (lesser trochanter of femur) and have their principal action as flexors of the thigh. (These muscles are described in CIBA COLLECTION, Volume 3/II, page 20.) In systematizing the femoral muscles, it is useful to note that muscles arising from the pubis and ischium are preaxial and are innervated by nerves derived from the anterior branches of the lumbosacral plexus, namely, the obturator or tibial nerves; however, muscles arising from the ilium or femur are postaxial and are innervated by either the femoral or common peroneal nerves of

Muscles of Hip and Thigh

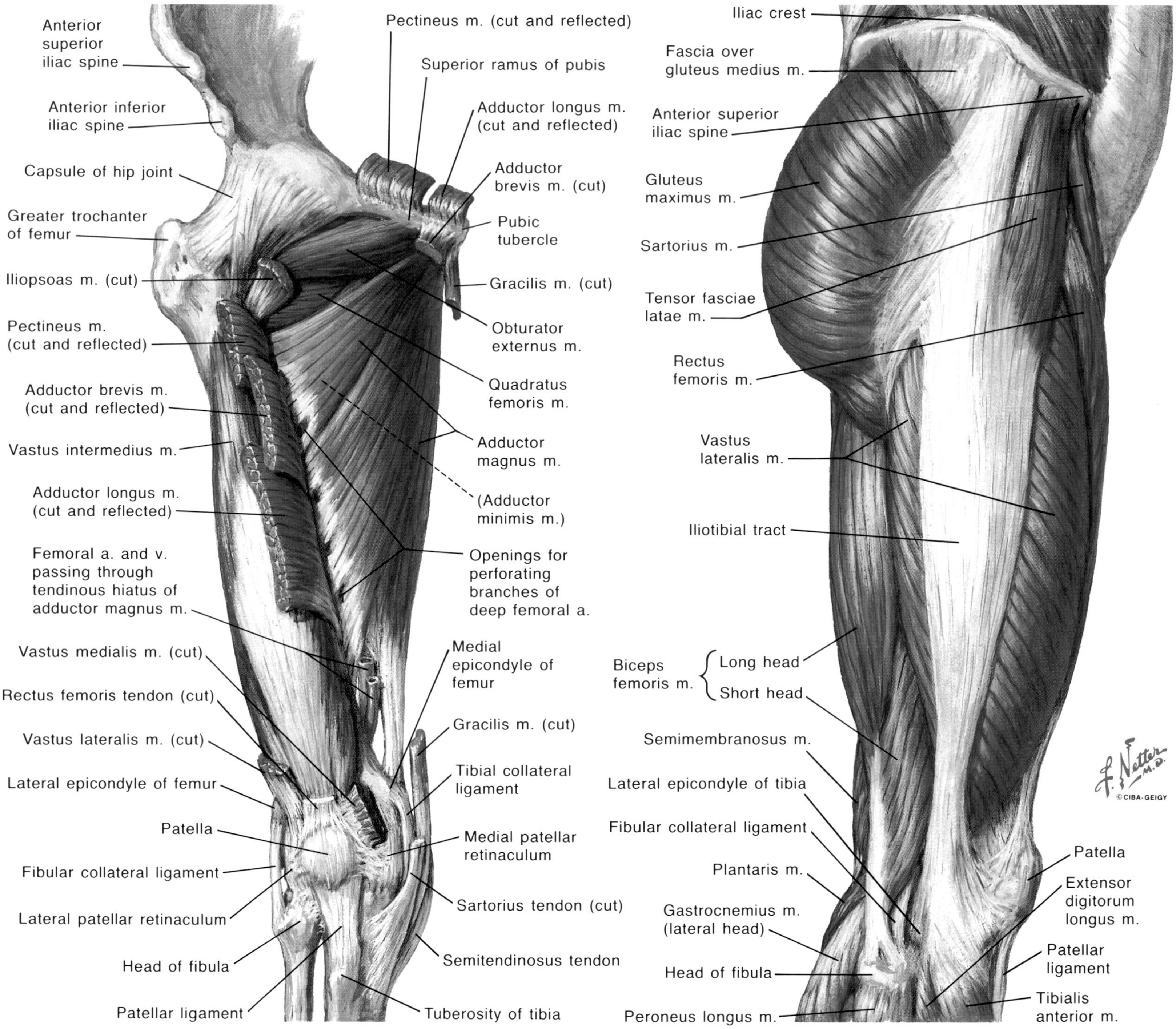

Anterior view: deep dissection
Lateral view: superficial dissection
Anterior superior iliac spine
Anterior inferior iliac spine
Capsule of hip joint
Greater trochanter of femur
Iliopsoas m. (cut)
Pectineus m. (cut and reflected)
Adductor brevis m. (cut and reflected)
Vastus intermedius m.
Adductor longus m. (cut and reflected)
Femoral a. and v. passing through tendinous hiatus of adductor magnus m.
Vastus medialis m. (cut)
Rectus femoris tendon (cut)
Vastus lateralis m. (cut)
Lateral epicondyle of femur
Patella
Fibular collateral ligament
Lateral patellar retinaculum
Head of fibula
Patellar ligament
Pectineus m. (cut and reflected)
Superior ramus of pubis
Adductor longus m. (cut and reflected)
Adductor brevis m. (cut)
Pubic tubercle
Gracilis m. (cut)
Obturator externus m.
Quadratus femoris m.
Adductor magnus m.
(Adductor minimis m.)
Openings for perforating branches of deep femoral a.
Medial epicondyle of femur
Gracilis m. (cut)
Tibial collateral ligament
Medial patellar retinaculum
Sartorius tendon (cut)
Semitendinosus tendon
Tuberosity of tibia
Iliac crest
Fascia over gluteus medius m.
Anterior superior iliac spine
Gluteus maximus m.
Sartorius m.
Tensor fasciae latae m.
Rectus femoris m.
Vastus lateralis m.
Iliotibial tract
Biceps femoris m.
Long head
Short head
Semimembranosus m.
Lateral epicondyle of tibia
Fibular collateral ligament
Plantaris m.
Gastrocnemius m. (lateral head)
Head of fibula
Peroneus longus m.
Patella
Extensor digitorum longus m.
Patellar ligament
Tibialis anterior m.
F. Netter M.D.
©CIBA-GEIGY

SECTION I PLATE 79 Slide 3644

Lower Limb

Fasciae and Muscles of Hip and Thigh

(Continued)

posterior branch derivation. The anterior group muscles are all postaxial in classification and innervation, and the medial group muscles are preaxial. The posterior and lateral groups contain muscles of both types.

Anterior Femoral Muscles

The anterior femoral muscles are the sartorius, quadriceps femoris (combined rectus femoris and vastus muscles), and the articularis genus muscles (Plates 78–79, 82).

The *sartorius* is the longest muscle in the body. Ribbonlike in form, it arises from the anterior superior spine of the ilium and from the notch just below the spine. It is diagonally placed, end ing on the medial side of the leg. Its insertion is into the medial surface of the tibia, below the tuberosity and nearly as far forward as the crest. In this insertion, it is associated with the tendons of the gracilis and semitendinosus muscles in the *pes anserinus*, which is separated from the tibia by a bursa. Contraction of the sartorius produces flexion, abduction, and lateral rotation of the thigh. It also flexes the leg, for its tendon passes behind the transverse axis of the knee. The sartorius forms the lateral border of the femoral triangle in the upper third of the thigh; in the middle third, it forms the roof of the adductor canal. The femoral nerve innervates the sartorius muscle by two branches, with nerve fibers from L2 and L3.

The four parts of the *quadriceps femoris muscle* arise separately but end in closely related parts of the tibia—the tuberosity and condyles. The *rectus femoris muscle*, as its name implies, runs straight down the thigh. Somewhat fusiform in shape, its superficial fibers have a bipennate arrangement. The muscle arises by two tendons. The straight head takes origin from the anterior inferior spine of the ilium; the reflected head arises from the groove above the acetabulum. These tendons unite at an acute angle and continue into a central aponeurosis that is expanded downward into the muscle. From this, the muscle fibers arise, turn around its margin, and end in the tendon of insertion. The latter broadens to attach to the proximal border of the patella and spreads over its surface to the tuberosity of the tibia (patellar ligament). Two branches of the femoral nerve with fibers from L3 and L4 innervate the muscle.

The *vastus lateralis muscle* is the largest component of the quadriceps femoris muscle. It arises from the femur by a broad aponeurosis attached to the upper part of the intertrochanteric line,

Lower Limb

Fasciae and Muscles of Hip and Thigh

(Continued)

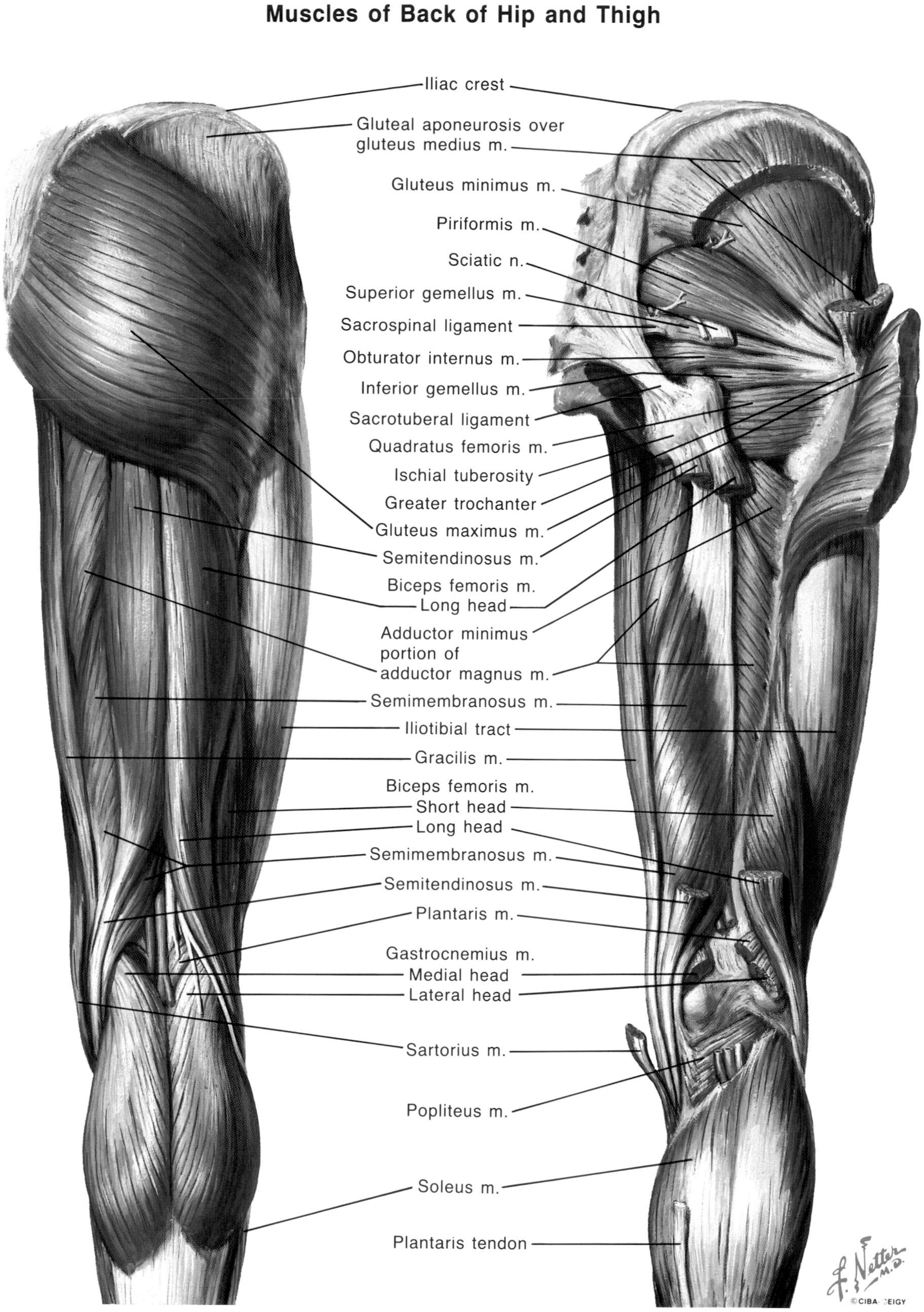
Muscles of Back of Hip and Thigh
Iliac crest
Gluteal aponeurosis over gluteus medius m.
Gluteus minimus m.
Piriformis m.
Sciatic n.
Superior gemellus m.
Sacrospinal ligament
Obturator internus m.
Inferior gemellus m.
Sacrotuberal ligament
Quadratus femoris m.
Ischial tuberosity
Greater trochanter
Gluteus maximus m.
Semitendinosus m.
Biceps femoris m.
Long head
Adductor minimus portion of adductor magnus m.
Semimembranosus m.
Iliotibial tract
Gracilis m.
Biceps femoris m.
Short head
Long head
Semimembranosus m.
Semitendinosus m.
Plantaris m.
Gastrocnemius m.
Medial head
Lateral head
Sartorius m.
Popliteus m.
Soleus m.
Plantaris tendon
F. Netter M.D.
©CIBA-GEIGY

the anterior and inferior borders of the greater trochanter, the gluteal tuberosity, the immediately adjacent portion of the lateral lip of the linea aspera, and the lateral intermuscular septum throughout its length. This aponeurosis covers the superior portion of the muscle, and from its deep surface, many muscle fibers take origin. The tendon of the vastus lateralis muscle inserts into the superolateral border of the patella and the lateral condyle of the tibia. The large branch of the femoral nerve (L3, 4) to the vastus lateralis muscle accompanies the descending branch of the lateral circumflex femoral artery.

The *vastus medialis muscle* arises from the whole extent of the medial lip of the linea aspera, the distal half of the intertrochanteric line, and the medial intermuscular septum. The aponeurotic fibers of origin adhere to the tendons of insertion of the adductor longus and adductor magnus muscles. The fibers are directed downward and forward toward the knee. The aponeurotic tendon inserts into the tendon of the rectus femoris, the superomedial border of the patella, and the medial condyle of the tibia. Two branches of the femoral nerve (L3, 4) supply this muscle.

The *vastus intermedius muscle* arises from the shaft of the femur, from the lower half of the lateral lip of the linea aspera, and from the lateral intermuscular septum. Its fibers end in a superficial aponeurosis, which blends with the deep surface of the tendons of the rectus femoris and with the vastus medialis and vastus lateralis muscles. The vastus intermedius is innervated superficially by a branch of the femoral nerve (L3, 4) and also by the upper nerve to the vastus medialis.

The four parts of the quadriceps femoris muscle converge on the patella, which is regarded as a sesamoid developed in its tendon. The *patellar ligament* is the terminal part of the tendinous insertion of the quadriceps femoris muscle. It ends in the tuberosity of the tibia. The *suprapatellar bursa* intervenes between the quadriceps femoris tendon and the lower end of the femur and communicates freely with the cavity of the knee joint. A deep *infrapatellar bursa* lies between the patellar ligament (just above its insertion) and the tibia. *Medial* and *lateral patellar retinacula* insert into the condyles of the tibia.

The quadriceps femoris muscle is the great extensor of the leg at the knee, all parts of the muscle contributing to this action. The line of traction, which is along the axis of the femur, is not directly in line with the tibia. Thus, there is a tendency to displace the patella lateralward as the muscle contracts. The numerous low almost horizontal fibers of the vastus medialis muscle counter this displacing force. The rectus femoris muscle also acts in flexion of the thigh at the hip. Its two almost right-angled tendons of origin combine to serve the full range of flexion action. Initially in line with the rest of the muscle, the straight tendon becomes less effective as the thigh is flexed, but as this tendon loses its effectiveness, the flexed thigh becomes more and more in line with the reflected tendon, and thus its attachment becomes the optimal site for traction. The quadriceps muscles are generally uncontracted in relaxed standing.

The *articularis genus muscle* consists of a number of small muscular bundles that arise from the lower fourth of the front of the femur. Lying deep to the vastus intermedius muscle, these bundles insert into the upper part of the synovial membrane of the knee joint.

Medial Femoral Muscles

The medial femoral muscles are the gracilis, pectineus, adductor longus, adductor brevis, adductor magnus, and obturator externus muscles (Plates 78–79 and 81–82).

Bony Attachments of Buttock and Thigh Muscles

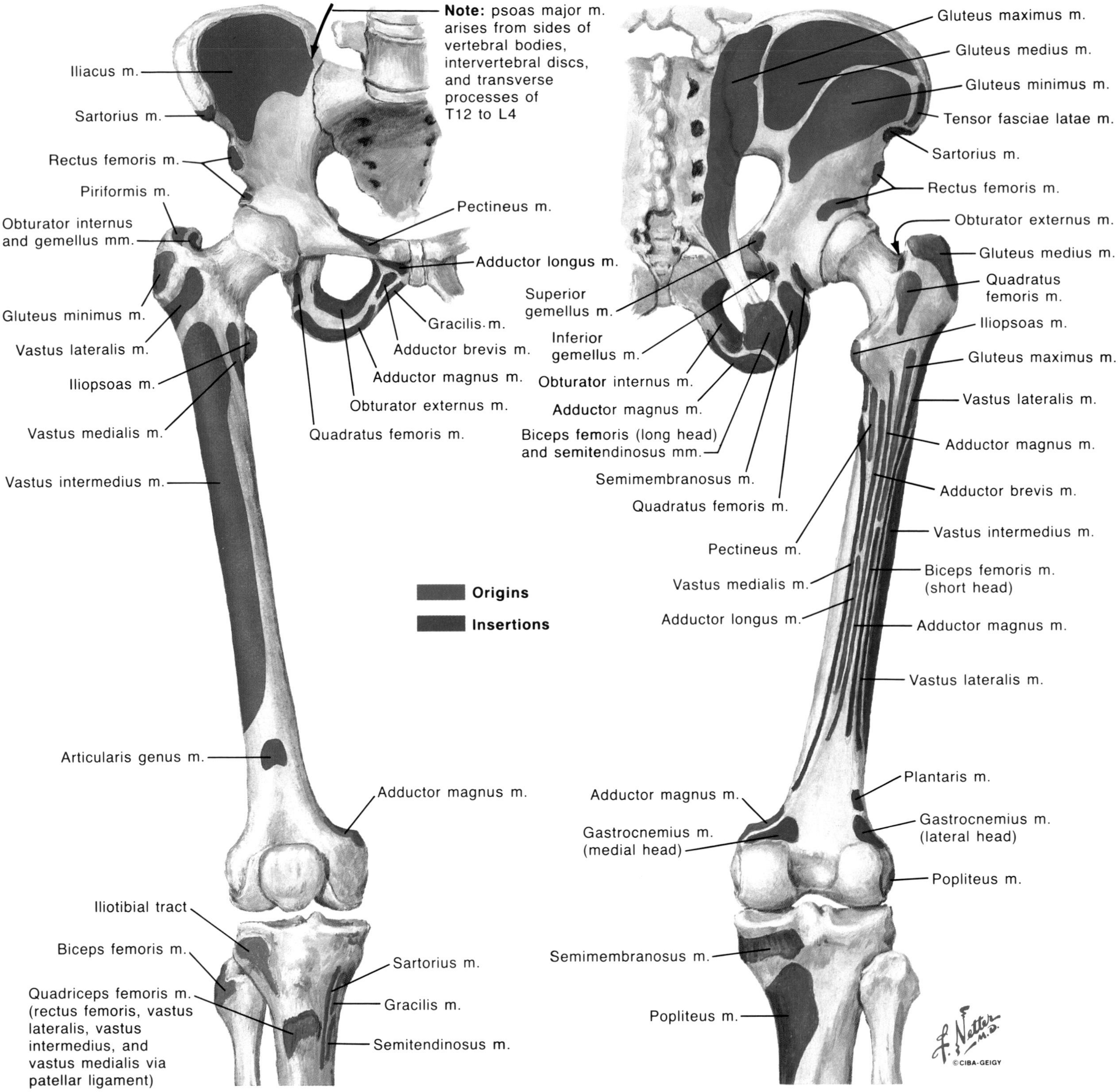

Note: psoas major m. arises from sides of vertebral bodies, intervertebral discs, and transverse processes of T12 to L4
Iliacus m.
Sartorius m.
Rectus femoris m.
Piriformis m.
Obturator internus and gemellus mm.
Gluteus minimus m.
Vastus lateralis m.
Iliopsoas m.
Vastus medialis m.
Vastus intermedius m.
Pectineus m.
Adductor longus m.
Gracilis m.
Adductor brevis m.
Adductor magnus m.
Obturator externus m.
Quadratus femoris m.
Origins
Insertions
Articularis genus m.
Adductor magnus m.
Iliotibial tract
Biceps femoris m.
Quadriceps femoris m. (rectus femoris, vastus lateralis, vastus intermedius, and vastus medialis via patellar ligament)
Sartorius m.
Gracilis m.
Semitendinosus m.
Gluteus maximus m.
Gluteus medius m.
Gluteus minimus m.
Tensor fasciae latae m.
Sartorius m.
Rectus femoris m.
Obturator externus m.
Gluteus medius m.
Quadratus femoris m.
Iliopsoas m.
Gluteus maximus m.
Vastus lateralis m.
Adductor magnus m.
Adductor brevis m.
Vastus intermedius m.
Biceps femoris m. (short head)
Adductor magnus m.
Vastus lateralis m.
Superior gemellus m.
Inferior gemellus m.
Obturator internus m.
Adductor magnus m.
Biceps femoris (long head) and semitendinosus mm.
Semimembranosus m.
Quadratus femoris m.
Pectineus m.
Vastus medialis m.
Adductor longus m.
Adductor magnus m.
Gastrocnemius m. (medial head)
Plantaris m.
Gastrocnemius m. (lateral head)
Popliteus m.
Semimembranosus m.
Popliteus m.
F. Netter M.D.
©CIBA-GEIGY

Lower Limb

Fasciae and Muscles of Hip and Thigh

(Continued)

The *gracilis muscle* is long and slender and is superficially placed on the medial aspect of the thigh. Its thin tendon arises along the pubic symphysis and the inferior ramus of the pubis. Its tapered tendon inserts into the upper part of the shaft of the tibia as part of the pes anserinus, lying between the tendons of the sartorius and semitendinosus muscles. A bursa deep to the tendon separates it from the tibial collateral ligament. The gracilis muscle adducts the thigh and assists in flexion of the leg at the knee. It also participates in flexion and medial rotation of the thigh at the hip. The gracilis muscle is innervated by a branch of the anterior division of the obturator nerve (L2, 3).

The *pectineus muscle* is flat and quadrangular and forms the medial part of the floor of the femoral triangle. It arises from the pecten of the pubis and the surface of the bone below the pecten, between the iliopubic eminence laterally and the pubic tubercle medially. The fibers of the muscle pass downward, backward, and lateralward and insert by a 5-cm-wide tendon into the pectineal line of the femur. The muscle adducts, rotates medially, and assists in flexion of the thigh. It is supplied by a branch of the femoral nerve, which enters the lateral portion of the muscle, and also by the accessory obturator nerve, when this is present. There are variable divisions of the muscle into ventromedial and dorsolateral portions.

The *adductor longus muscle* lies in the same plane as the pectineus and forms the medial boundary of the femoral triangle. It arises in a flat, narrow tendon from the medial portion of the superior

Lower Limb

Fasciae and Muscles of Hip and Thigh

(Continued)

ramus of the pubis. It expands into a broad, triangular muscular belly and inserts by a thin tendon into the middle third of the medial lip of the linea aspera of the femur, between the tendons of the vastus medialis and the adductor magnus muscles. The adductor longus muscle adducts the thigh and assists in its flexion and medial rotation. Its nerve, a branch of the anterior division of the obturator nerve (L2, 3), reaches it on the deep surface of its middle third.

The *adductor brevis muscle* is deep to the pectineus and adductor longus muscles. It is an adductor of the thigh and, to a lesser degree, assists in its flexion and medial rotation. The adductor brevis has a narrow origin from the inferior pubic ramus, between the origins of the gracilis and obturator externus muscles. Its muscular fibers fan out to end in an aponeurosis that inserts into the lower two-thirds of the pectineal line of the femur and the upper half of the medial lip of the linea aspera. Branches of the anterior division of the obturator nerve (L2, 3) enter the middle third of the muscle near its proximal border. Its tendon is pierced by perforating branches of the deep femoral artery and their accompanying veins.

The *adductor magnus muscle* is the largest muscle of the medial femoral group. It is triangular in shape and actually consists of a combination of two muscles that have different innervations. The muscle arises from the lower part of the inferior pubic ramus, the ramus of the ischium, and the ischial tuberosity. Its muscular fibers fan out to the whole length of the linea aspera of the femur; the upper fibers are horizontal and the lower fibers are vertical. The upper horizontal fibers, sometimes designated as the adductor minimus muscle, insert into the medial side of the gluteal ridge and the uppermost part of the linea aspera of the femur. Below this, the aponeurosis of insertion ends in the whole length of the medial lip of the linea aspera and the supracondylar line of the femur.

The most medial and posterior portion of the muscle, its ischiocondylar portion, arises from the ischial tuberosity and forms a round tendon that ends in the adductor tubercle of the medial epicondyle of the femur. The upper anterior portion of the muscle is a strong adductor and assists in flexion and medial rotation of the thigh. It is innervated by the posterior division of the obturator nerve (L3, 4). The *ischiocondylar portion* of the muscle is one of the hamstring muscles of the back of the thigh. Its particular action is to extend the thigh and rotate it medialward. It is innervated on its dorsal surface by a branch from the tibial division of the sciatic nerve (L4; S1).

Cross-Sectional Anatomy of Thigh

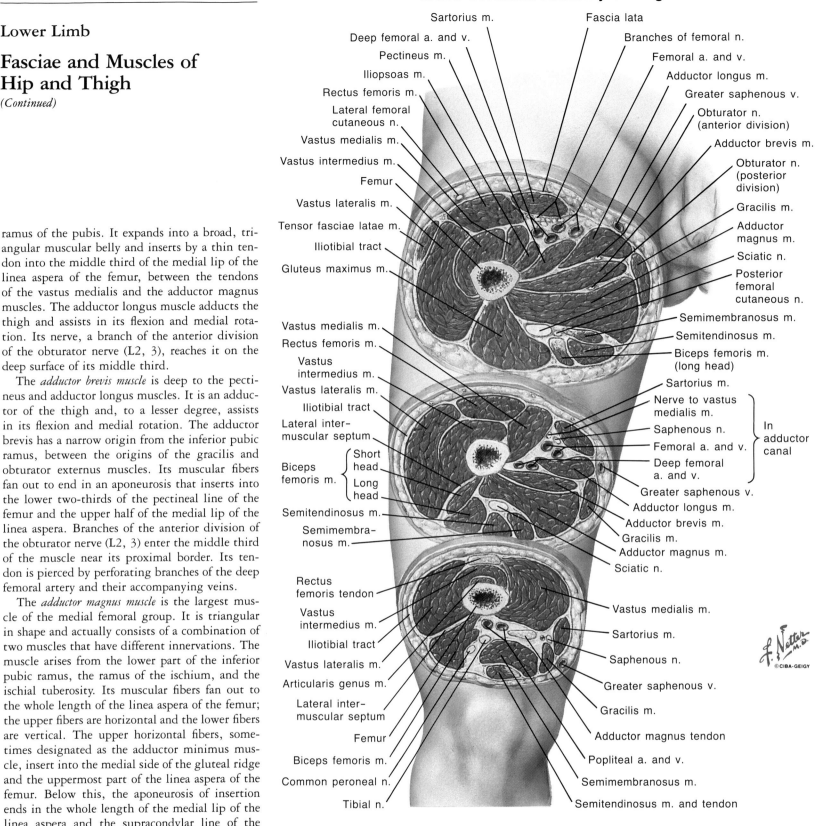

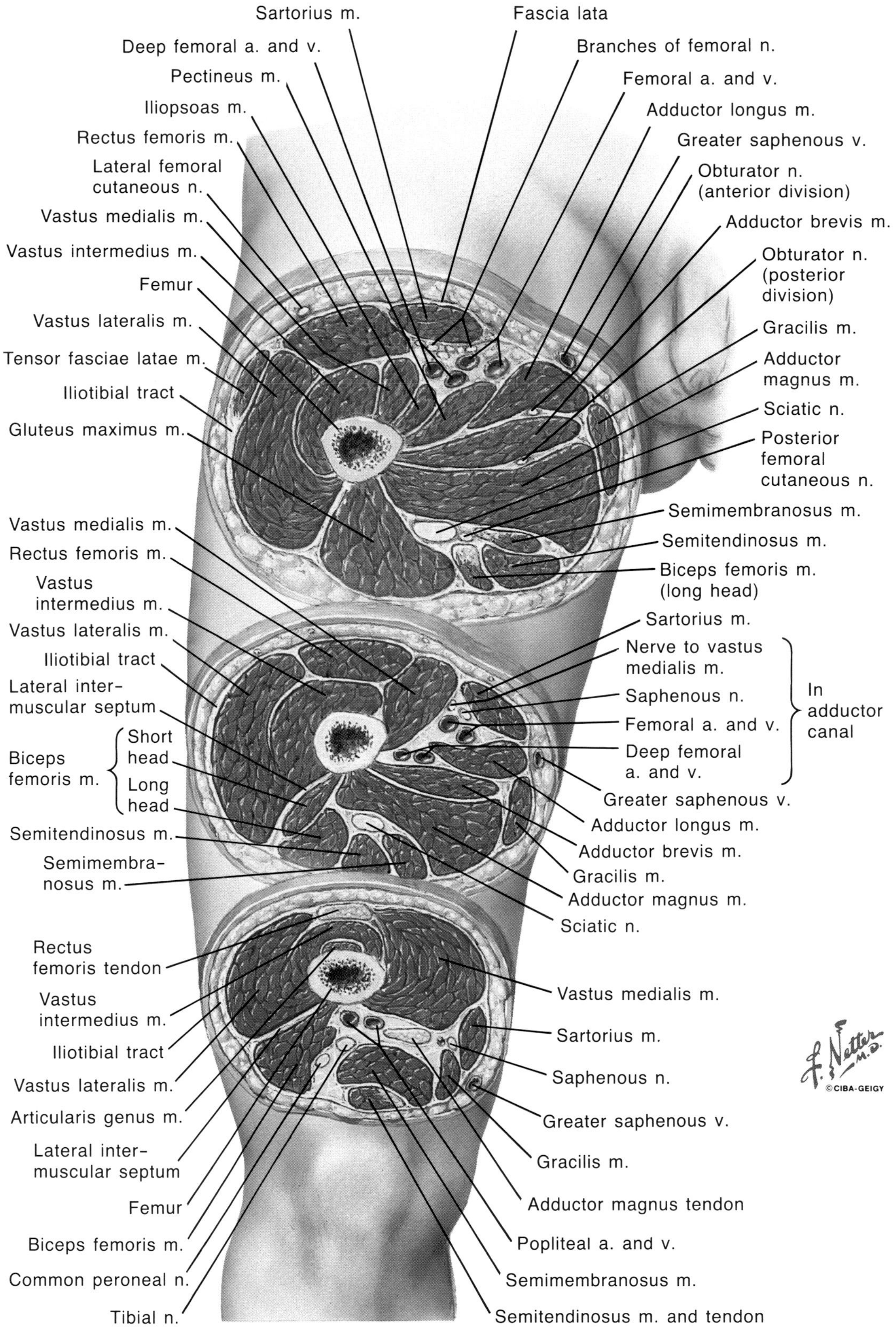

In the lower third of the thigh, aponeurotic fibers spread lateralward from the rounded tendon of the adductor magnus muscle toward the vastus medialis muscle and end in the medial intermuscular septum. This strong *vastoadductor membrane* covers the distal end of the adductor canal and may be pierced for the passage of the saphenous nerve and the descending genicular artery and vein. Further, the aponeurosis of insertion of the adductor magnus muscle is pierced adjacent to the femur by four openings for the passage of the perforating branches of the deep femoral artery and accompanying veins. Finally, a large gap exists between the lower end of the aponeurosis and the tendon to the adductor tubercle. Through this *adductor hiatus*, the femoral vessels pass back and down into the popliteal space, where they become the popliteal vessels.

The *obturator externus muscle* arises from the external aspect of the superior and inferior rami of the pubis and the ramus of the ischium and from the external surface of the obturator membrane. Its tendon passes across the back of the neck of the femur and the capsule of the hip joint to insert in the trochanteric fossa of the femur. The synovial membrane of the hip joint acts as a bursa separating the tendon from the neck of the femur. This muscle is a lateral rotator of the thigh. It is supplied by a branch of the obturator nerve (L3, 4).

Lower Limb

Fasciae and Muscles of Hip and Thigh

(Continued)

Posterior Femoral Muscles

The posterior femoral muscles compose the hamstring group, which includes the semitendinosus, semimembranosus, and biceps femoris muscles, and the ischiocondylar portion of the adductor magnus (Plates 80–82).

The *semitendinosus muscle* is aptly named, for about half its length is tendinous. It arises from the lower and medial impression on the tuberosity of the ischium in common with the long head of the biceps femoris muscle. The semitendinosus tendon forms the medial margin of the popliteal space at the knee; it then curves around the medial condyle of the tibia and inserts as part of the pes anserinus into the upper part of the medial surface of the tibia. It is separated from the tibial collateral ligament by a bursa. Two branches of the tibial division of the sciatic nerve (L4, 5; S1, 2) usually reach this muscle.

The *semimembranosus muscle* arises by a long, flat tendon (or membrane) from the upper and outer impression on the tuberosity of the ischium. The muscular belly begins about halfway down the thigh. The insertion of the muscle at the knee is rather complex. It ends mainly in the horizontal groove on the posteromedial aspect of the medial condyle of the tibia, but a prominent reflection from here forms the oblique popliteal ligament of the knee joint capsule. Other fibers extend from the tendon to the tibial collateral ligament and onto the fascia of the popliteus muscle. The semimembranosus muscle is innervated by a branch of the tibial division of the sciatic nerve arising in common with the lower nerve to the semitendinosus muscle.

The *biceps femoris muscle* is a secondary combination of one preaxial muscle, the long head, and one postaxial muscle, the short head. The long head arises in combination with the semitendinosus muscle from the lower and medial impression of the ischial tuberosity and the lower part of the sacrotuberal ligament. The short head arises from the lateral lip of the linea aspera of the femur, the proximal two-thirds of the supracondylar line, and the lateral intermuscular septum. The muscle fibers of the short head join the tendon of the long head to form the heavy round tendon that forms the lateral margin of the popliteal fossa.

At the knee, the tendon divides around the fibular collateral ligament and ends on the lateral aspect of the head of the fibula, the lateral condyle of the tibia, and in the deep fascia of the lateral aspect of the leg. As a combination of two muscles of differing origins, the two heads are differently innervated. The long head usually receives two branches of the tibial division of the sciatic nerve (S1, 2, 3), one to the upper third and one to the middle third of the muscle. The nerve to the short head is a branch of the common peroneal division of the sciatic nerve (L5; S1, 2), which enters the superficial surface of the muscle.

The hamstring muscles flex the leg and extend the thigh. Their ligamentous, or protective, action at the hip joint is important. In the usual movement, flexion and extension are carried out together, and maximal excursion at one joint carries the limitation of less than maximal excursion at the other. There is also minimal rotary action, the "semi" muscles rotating the flexed leg medially and the biceps femoris muscle rotating it laterally.

Lateral Femoral Muscles

The lateral femoral muscles lie largely in the hip region. They are the gluteus maximus, gluteus medius, gluteus minimus, tensor fasciae latae, piriformis, obturator internus, superior gemellus, inferior gemellus, and quadratus femoris muscles (Plates 79–82).

The *gluteus maximus muscle* is a heavy, coarsely fasciculated muscle, superficially situated in the buttock. It is quadrilateral in form, and its fasciculi are directed downward and outward. The muscle arises from the posterior gluteal line of the ilium and the area of the bone above and behind it, the posterior surface of the sacrum and coccyx, the sacrotuberal ligament, and the gluteal aponeurosis overlying the gluteus medius muscle. The larger upper portion and the superficial fibers of the lower portion insert into the iliotibial tract of the fascia lata; the deeper fibers of the lower portion reach the gluteal tuberosity of the femur and the lateral intermuscular septum.

The upper portion of the muscle is separated from the greater trochanter of the femur by the large trochanteric bursa. Other bursae separate the tendon of the muscle from the origin of the vastus lateralis muscle (of the quadriceps femoris muscle) and the lower portion of the muscle from the ischial tuberosity. The gluteus maximus is distinctly a muscle of the erect posture but becomes active only under conditions of effort. It is a powerful extensor of the thigh and, acting from its insertion, equally an extensor of the trunk. It is a strong lateral rotator, and its superior fibers come into play in forcible abduction of the thigh. Its insertion into the iliotibial tract may promote stability of the femur on the tibia, but the extensive blending of this tract with the lateral intermuscular septum would prevent significant action of the muscle on the tibia. The inferior gluteal nerve (L5; S1, 2) is the sole supply of this muscle.

The *gluteus medius muscle* lies largely anterior to the gluteus maximus under the strong vertical fibers of the gluteal aponeurosis but also partly underlies the gluteus maximus. It arises from the external surface of the ilium between the anterior and posterior gluteal lines and from the gluteal aponeurosis. Its flattened tendon inserts into the posterosuperior angle of the greater trochanter of the femur and into a diagonal ridge on its lateral surface. A bursa separates the tendon from the trochanter proximal to its insertion.

The *gluteus minimus muscle* underlies the gluteus medius, arising from the ilium between its anterior gluteal and inferior gluteal lines. Its insertion is onto the anterosuperior angle of the greater trochanter, a bursa intervening between the tendon and the medial part of the anterior surface of the trochanter. The gluteus medius and minimus muscles abduct the femur and rotate the thigh medialward. These are important functions in walking, for when one foot is raised off the ground, the gluteus muscles, holding the pelvic bone of the other side down toward the greater trochanter, prevent the collapse of the pelvis on its unsupported side. The same muscles, by their rotary action, swing the pelvis forward as the step is taken. Both muscles are innervated by the superior gluteal nerve (L4, 5; S1), accompanied by branches of the superior gluteal vessels between the two muscles.

The *tensor fasciae latae muscle* is fusiform in shape and is enclosed between two layers of the fascia lata. It arises from the anterior part of the external lip of the iliac crest, the outer surface of the anterior superior spine of the ilium, and the notch below the spine. The muscle inserts into the iliotibial tract, with which blend both layers of its investing fascia. The tensor fasciae latae assists in flexion, abduction, and medial rotation (weakly) of the thigh. It probably aids in stabilizing the femur on the tibia, although the long, strong connection of the iliotibial tract with the lateral intermuscular septum must make direct action at the knee minimal. The superior gluteal nerve ends in the tensor fasciae latae, together with an inferior branch of the superior gluteal artery.

The *piriformis muscle* arises within the pelvis from the front of the sacrum between the first to fourth sacral foramina. It passes through the greater sciatic foramen to insert onto the upper border of the greater trochanter of the femur. It is a lateral rotator of the thigh and assists in abduction. Its nerves are one or two branches from S1 and S2.

The *obturator internus muscle* arises from the entire bony margin of the obturator foramen (except the obturator groove), the inner surface of the obturator membrane, and the pelvic surface of the coxal bone behind and above the obturator foramen. The fibers of the muscle converge to pass through the lesser sciatic foramen, where it is separated from the bone of the lesser sciatic notch by a large bursa. The tendon turns a 90° angle at the notch and then, running horizontally, ends in the medial surface of the greater trochanter of the femur above the trochanteric fossa. The muscle is a lateral rotator of the thigh and has some abduction capability. Its nerve (L5; S1, 2) also supplies the superior gemellus muscle. The tendon of the obturator internus muscle receives the superior gemellus tendon along its superior border and superficial surface and the inferior gemellus tendon along its inferior margin.

The *superior* and *inferior gemellus muscles* are small tapered muscles that lie parallel to the tendon of the internal obturator; the superior gemellus muscle above the tendon arises from the ischial spine, and the inferior gemellus muscle below the tendon arises from the ischial tuberosity. Accessories of the internal obturator muscle, the gemellus muscles have the same action. The nerve to the superior gemellus muscle comes from that to the internal obturator muscle; the nerve to the inferior gemellus is in common with the nerve to the quadratus femoris muscle.

The *quadratus femoris muscle* is thick and quadrilateral and is located below the inferior gemellus muscle. It arises from the upper part of the lateral border of the ischial tuberosity and inserts on the quadrate line of the femur, which extends downward from the intertrochanteric crest. This muscle is a strong lateral rotator of the thigh. Its nerve supply is from L4, 5 and S1. □

Arteries and Nerves of Thigh (anterior view)

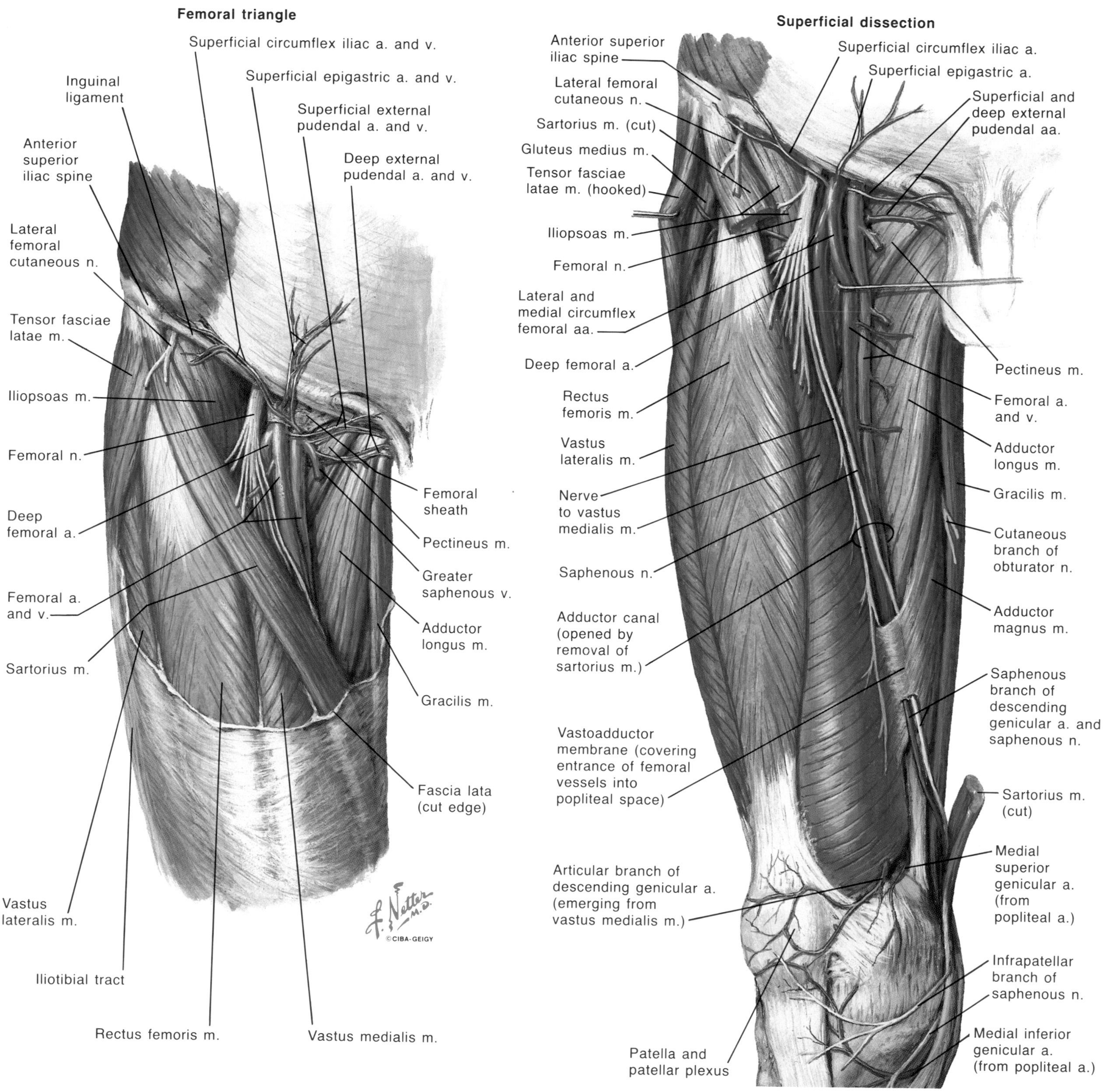

Femoral triangle
Superficial circumflex iliac a. and v.
Superficial epigastric a. and v.
Inguinal ligament
Superficial external pudendal a. and v.
Anterior superior iliac spine
Deep external pudendal a. and v.
Lateral femoral cutaneous n.
Tensor fasciae latae m.
Iliopsoas m.
Femoral n.
Deep femoral a.
Femoral a. and v.
Sartorius m.
Femoral sheath
Pectineus m.
Greater saphenous v.
Adductor longus m.
Gracilis m.
Fascia lata (cut edge)
Vastus lateralis m.
Iliotibial tract
Rectus femoris m.
Vastus medialis m.
F. Netter M.D.
©CIBA-GEIGY
Superficial dissection
Anterior superior iliac spine
Lateral femoral cutaneous n.
Sartorius m. (cut)
Gluteus medius m.
Tensor fasciae latae m. (hooked)
Iliopsoas m.
Femoral n.
Lateral and medial circumflex femoral aa.
Deep femoral a.
Rectus femoris m.
Vastus lateralis m.
Nerve to vastus medialis m.
Saphenous n.
Adductor canal (opened by removal of sartorius m.)
Vastoadductor membrane (covering entrance of femoral vessels into popliteal space)
Articular branch of descending genicular a. (emerging from vastus medialis m.)
Patella and patellar plexus
Superficial circumflex iliac a.
Superficial epigastric a.
Superficial and deep external pudendal aa.
Pectineus m.
Femoral a. and v.
Adductor longus m.
Gracilis m.
Cutaneous branch of obturator n.
Adductor magnus m.
Saphenous branch of descending genicular a. and saphenous n.
Sartorius m. (cut)
Medial superior genicular a. (from popliteal a.)
Infrapatellar branch of saphenous n.
Medial inferior genicular a. (from popliteal a.)

SECTION I PLATE 83 Slide 3648

Lower Limb

Blood Supply of Thigh

Arteries

The femoral, obturator, superior gluteal, and inferior gluteal arteries supply the thigh. The former two distribute principally anteriorly; the latter two distribute in the hip region. The femoral artery is the continuation of the external iliac artery. It distributes largely in the femoral triangle and descends through the midregions of the thigh in the adductor canal.

The *femoral triangle* is a subfascial space in the upper third of the thigh. It is bounded by the inguinal ligament above, and by the sartorius muscle laterally and the adductor longus muscle medially. The apex of the triangle lies downward; it is formed by the crossing of the sartorius muscle over the adductor longus. The floor of the triangle is also muscular. The borders of the iliopsoas and pectineus muscles bound a deep groove in the floor, and here, the medial circumflex femoral artery passes to the back of the thigh. The femoral artery enters the adductor canal at the apex of the triangle. The femoral vein lies on the pectineus muscle, medial to the femoral artery; here, it receives the greater saphenous vein. The femoral nerve descends under the inguinal ligament in the groove between the iliacus and psoas major muscles. In the triangle, it divides into most of its muscular and cutaneous branches; only the saphenous nerve and one of the nerves to the vastus medialis muscle continue into the adductor canal.

The femoral artery and vein are covered by the *femoral sheath* for about 3 cm beyond the inguinal ligament. Here, the extraperitoneal connective tissue of the abdomen extends between the vessels and forms three compartments—a lateral

Lower Limb

Blood Supply of Thigh

(Continued)

one for the artery, a middle one for the vein, and a medial one for one or more deep inguinal lymph nodes and fat. The medial compartment is known as the *femoral canal*, and its abdominal opening is the *femoral ring*.

The *adductor canal* conducts the femoral vessels and one or two nerves through the middle third of the thigh. It begins about 15 cm below the inguinal ligament at the crossing of the sartorius muscle over the adductor longus muscle and ends at the upper limit of the adductor hiatus, a separation in the tendinous insertion of the adductor magnus muscle that allows the femoral vessels to reach the back of the knee. The canal occupies the middle third of the thigh, and its termination is marked medially by a strong fascial band from the vastus medialis to the adductor magnus muscles—the vastoadductor membrane. As the femoral vessels pass behind the femur to become the popliteal artery and vein, this membrane is perforated by the saphenous nerve and the descending genicular artery.

Branches of the femoral artery are the superficial epigastric, superficial circumflex iliac, superficial external pudendal, deep external pudendal, deep femoral, and descending genicular arteries. The first four branches are primarily related to the lower abdominal wall and the perineum and are described in CIBA COLLECTION, Volume 3/II, page 36.

The *deep femoral artery*, the largest branch, arises from the lateral side of the femoral artery about 5 cm below the inguinal ligament. It sinks deeply into the thigh as it descends, lying behind the femoral artery and vein on the medial side of the femur. It crosses the tendon of the adductor brevis muscle and, at its lower border, passes deep to the tendon of the adductor longus muscle. In the lower third of the thigh, the deep femoral artery ends as the fourth perforating artery. In the femoral triangle, this artery gives rise to the medial and lateral circumflex femoral arteries and muscular branches; in the adductor canal, it provides three perforating branches.

The *medial circumflex femoral artery* springs from the medial and posterior aspect of the deep femoral artery. Its course is deep into the femoral triangle, between the pectineus and iliopsoas muscles and under the neck of the femur to the back of the thigh. Deep to the adductor brevis muscle, an acetabular branch enters the hip joint beneath the transverse ligament of the acetabulum. Several muscular branches supply the adductor brevis and adductor magnus muscles, one distributing with the obturator nerve. Anterior to the quadratus femoris muscle, the artery divides into an ascending branch to the trochanteric fossa of the femur and a descending branch to the hamstring muscles beyond the ischial tuberosity.

The *lateral circumflex femoral artery* arises from the lateral side of the deep femoral artery, passes lateralward deep to the sartorius and the rectus femoris muscles, and divides into anterior, transverse, and descending branches. The ascending branch passes upward beneath the tensor fasciae latae muscle and anastomoses with terminals of the superior gluteal artery. The small transverse branch enters the vastus lateralis muscle, winds around the femur below its greater trochanter, and anastomoses on the back of the thigh with the medial circumflex femoral, inferior gluteal, and first perforating arteries (cruciate anastomosis). An articular artery for the hip joint may arise from either branch. The descending branch passes on the vastus lateralis muscle, accompanied by a branch of the femoral nerve to this muscle, and anastomoses with the descending genicular branch of the femoral artery and the lateral superior genicular branch of the popliteal artery.

The *perforating arteries* are usually three in number and arise from the posterior surface of the deep femoral artery. They pass directly against the linea aspera of the femur and pierce the tendons of the adductor muscles to reach the muscles of the posterior compartment of the thigh. The first perforating artery passes immediately below the pectineus and through the middle of the tendon of the adductor brevis muscle; the second, after giving off a nutrient artery to the femur, through its lower 3 or 4 cm; and the third, just below the lowest part of the adductor brevis. The deep femoral artery ends as the fourth perforating

Arteries and Nerves of Thigh: Deep Dissection (anterior view)

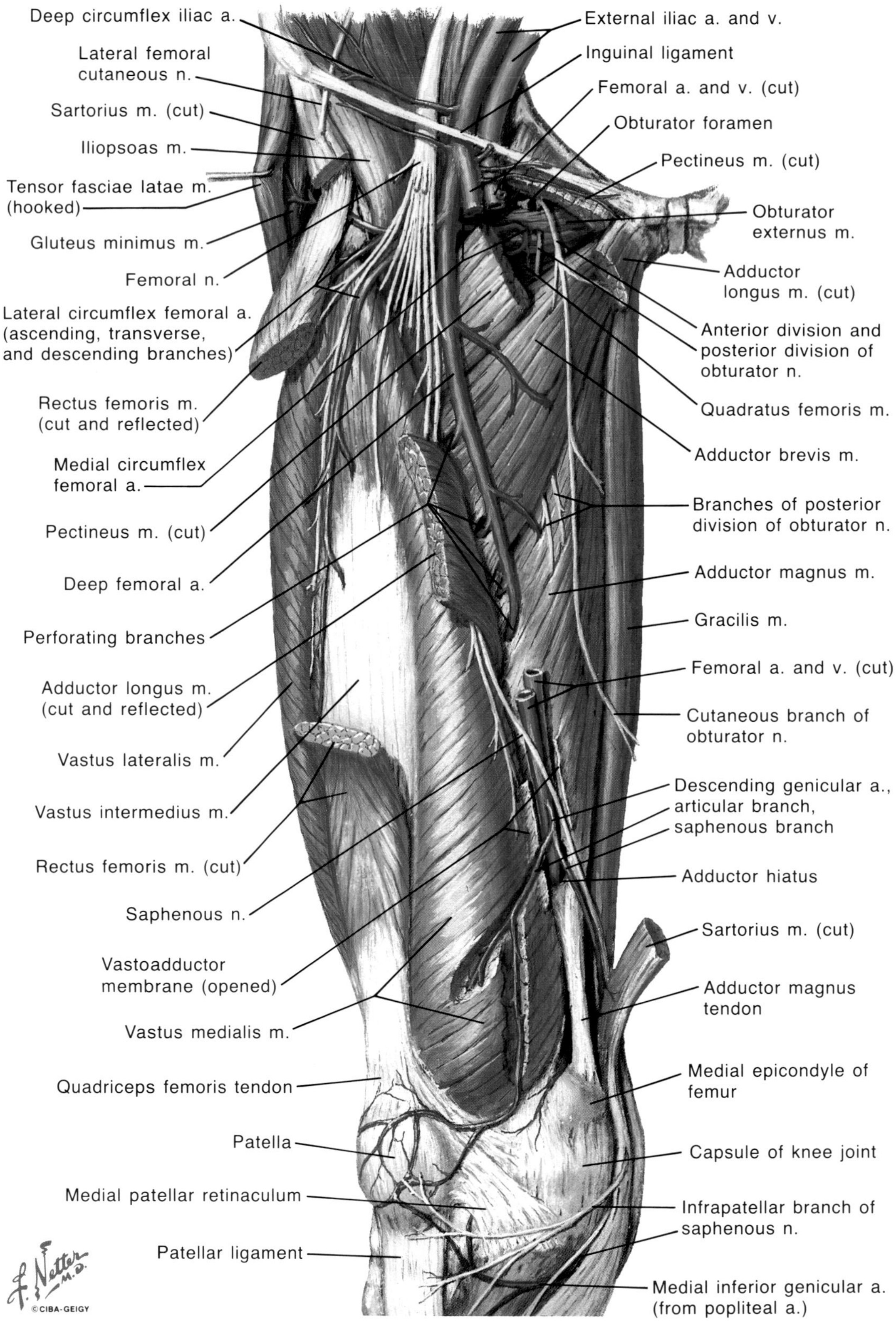

Lower Limb

Blood Supply of Thigh

(Continued)

artery, which pierces the adductor magnus muscle and largely ends in the short head of the biceps femoris muscle. These vessels anastomose with all the vessels of the back of the thigh.

The *descending genicular artery* arises from the femoral artery just before the latter passes through the adductor hiatus and immediately divides into saphenous and articular branches. The *saphenous branch* pierces the vastoadductor membrane and descends between the tendons of the gracilis and sartorius muscles in company with the saphenous nerve, supplying the skin and superficial tissues in the upper medial part of the leg. The *articular branch* descends in the substance of the vastus medialis muscle to the medial side of the knee. It supplies the vastus medialis muscle, and a branch passes lateralward over the patellar surface to anastomose with all other arteries at the knee joint.

The *obturator artery* is a branch of the internal iliac artery. It leaves the pelvis via the obturator canal and immediately divides into anterior and posterior branches. These supply the obturator externus and obturator internus muscles and give branches to the adductor brevis and adductor magnus muscles. The posterior branch gives off an *acetabular artery*, which enters the hip joint through the acetabular notch and provides the small artery of the capitis femoris ligament.

The *superior* and *inferior gluteal arteries* are branches of the internal iliac artery. Although arising within the pelvis, they exit immediately into the gluteal region of the thigh. The *superior gluteal artery* emerges above the piriformis muscle and there divides into a superficial branch to the gluteus maximus muscle and a deep branch to the intermuscular plane between the gluteus medius and gluteus minimus muscles. An upper radicle of this branch supplies the gluteus medius, gluteus minimus, and tensor faciae latae muscles and reaches as far as the anterior superior iliac spine. A lower radicle is directed toward the greater trochanter of the femur and supplies the gluteal muscles and the hip joint.

The *inferior gluteal artery* leaves the pelvis below the piriformis muscle and provides large muscular branches to the gluteus maximus muscle and muscles arising from the ischial tuberosity. Small branches pass medially as far as the skin over the coccyx. An anastomotic branch descends across the short lateral rotator muscles of the hip and contributes to the cruciate anastomosis of the back of the thigh. Cutaneous branches accompany radicles of the posterior femoral cutaneous nerve, and an accompanying artery of the sciatic nerve descends on the surface of that nerve. The inferior gluteal artery is long and slender and may reach as far as the lower part of the thigh. It is developmentally the major artery of the lower limb.

Veins

In general, the veins of the hip and thigh are venae comitantes of the arteries. The superior and inferior gluteal veins and the obturator vein enter

Arteries and Nerves of Thigh: Deep Dissection (posterior view)

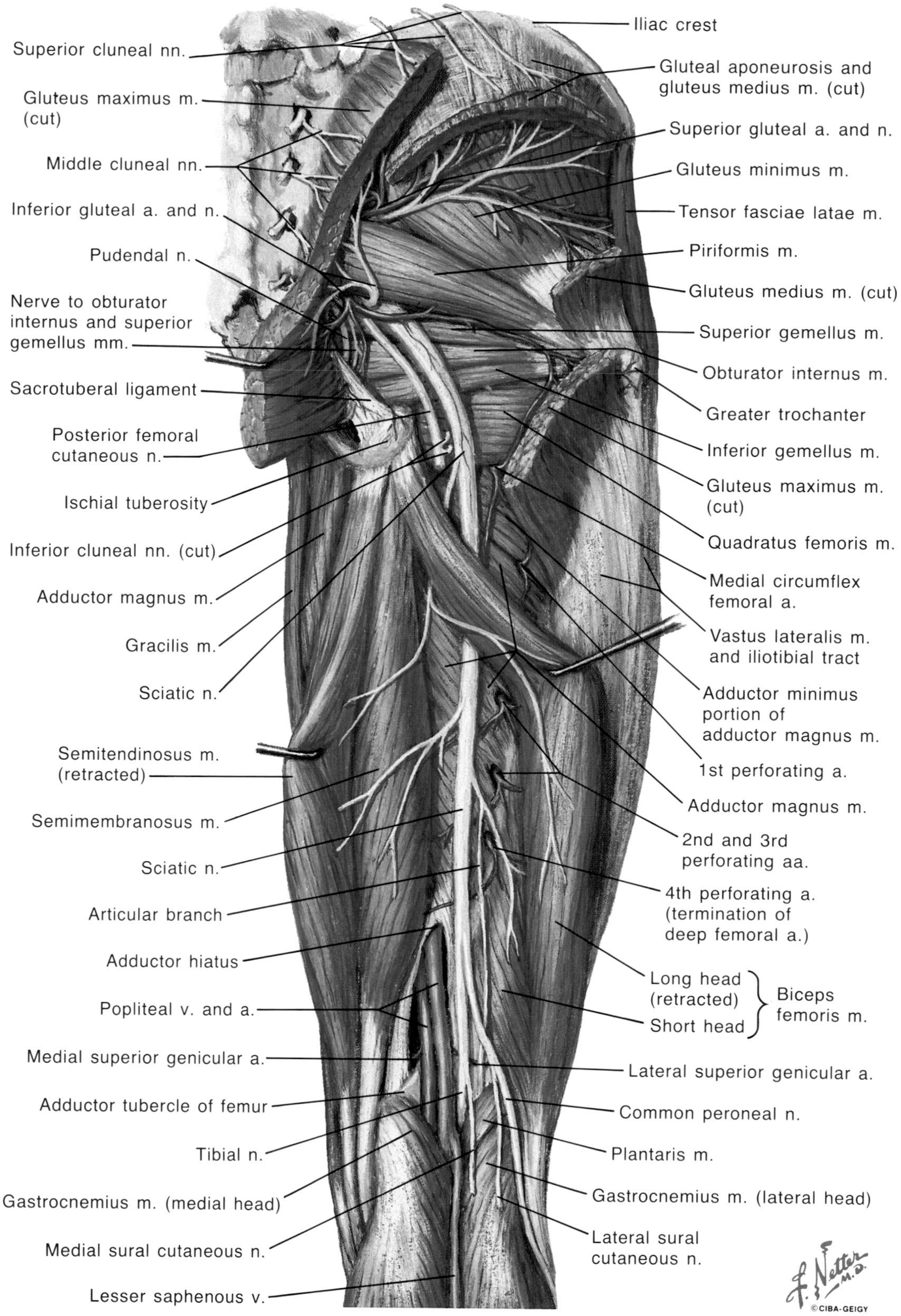

the pelvis with their corresponding arteries; they are tributary to the internal iliac vein. The femoral vein is posterior to the femoral artery at the adductor hiatus but medial to it as it passes under the inguinal ligament to become the external iliac vein. The femoral vein contains two or three bicuspid valves; one is just inferior to the junction with the deep femoral vein, and one is at the level of the inguinal ligament. The superficial circumflex iliac, superficial epigastric, and superficial external pudendal veins enter the greater saphenous and not the femoral vein, and the deep external pudendal vein is usually the highest tributary of the femoral vein. The venae comitantes of the descending genicular artery may enter the lower end of the femoral vein or may end in this vessel at the femoral triangle.

The deep femoral vein enters the femoral vein at a much less regular level than that characterizing the origin of the deep femoral artery (about 8 cm below the inguinal ligament). The perforating veins on the back of the thigh help to form a long anastomotic channel, which is formed by interconnections and connections with tributaries of the popliteal vein below and the inferior gluteal vein above. In three-fourths of cases, the medial and lateral circumflex femoral veins empty into the femoral vein rather than into the deep femoral vein, and their widespread radicles anastomose with many other veins of the thigh. □

Lower Limb

Bones and Ligaments at Hip

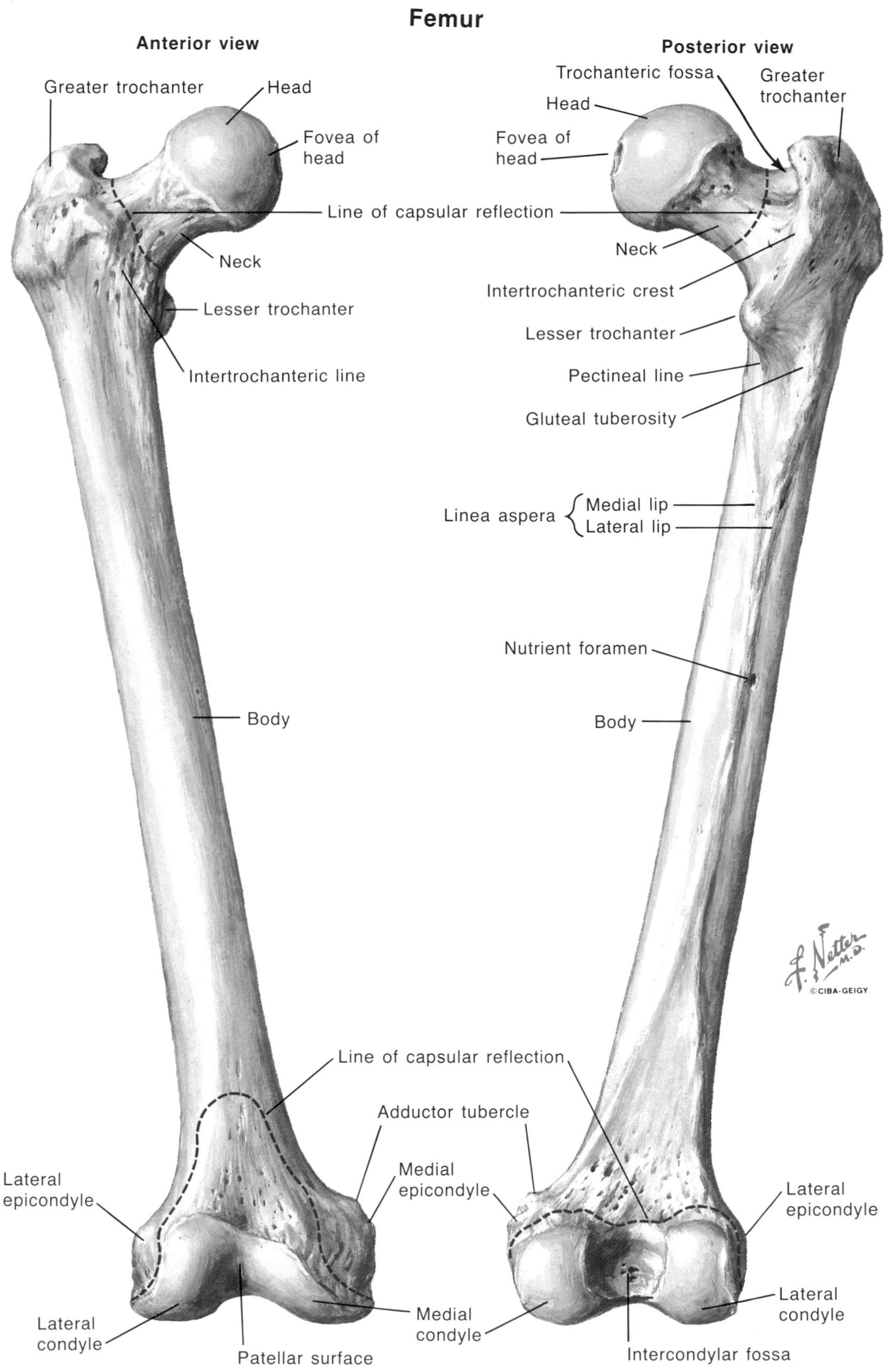
Femur
Anterior view
Greater trochanter
Head
Fovea of head
Line of capsular reflection
Neck
Lesser trochanter
Intertrochanteric line
Body
Line of capsular reflection
Adductor tubercle
Lateral epicondyle
Medial epicondyle
Lateral condyle
Patellar surface
Medial condyle
Posterior view
Trochanteric fossa
Greater trochanter
Head
Fovea of head
Neck
Intertrochanteric crest
Lesser trochanter
Pectineal line
Gluteal tuberosity
Linea aspera
Medial lip
Lateral lip
Nutrient foramen
Body
Lateral epicondyle
Lateral condyle
Intercondylar fossa
F. Netter M.D.
©CIBA-GEIGY

Femur

The femur is the longest and strongest bone in the body, comprising a shaft and two irregular extremities that articulate at the hip and knee joints (Plate 86).

The superior extremity of the bone has a nearly spheric *head* mounted on an angulated neck, and prominent trochanters provide for muscular attachments. The head is smooth, with an articular surface that is largest above and anteriorly; this is interrupted medially by a depression, the fovea capitis femoris, into which attaches the capitis femoris ligament.

The *neck* is about 5 cm long and forms an angle with the shaft, which varies in the normal person from 115° to 140°. It is compressed anteroposteriorly and contains a large number of prominent pits for the entrance of blood vessels.

The *greater trochanter* is the bony prominence of the hip. It is palpable 12 to 14 cm below the iliac crest, is large and square, and marks the upper end of the shaft of the femur. Its large quadrilateral surface is divided by an oblique ridge running from the posterosuperior to its anteroinferior angles. In front and above the ridge, there is a triangular surface (which may be smooth) for a bursa. Below and behind the ridge, the bone is also smooth. Its posterior rounded border bounds the trochanteric fossa and continues downward as the intertrochanteric crest. The trochanteric fossa is a deep pit on the internal aspect of the trochanter.

The *lesser trochanter* is a blunt, conical projection at the junction of the inferior border of the neck with the shaft of the femur. The trochanters are joined behind by the intertrochanteric crest. On the anterior surface of the femur, the junction of the neck and shaft is also ridged. This is the intertrochanteric line, which provides attachment for the capsule of the hip joint across the front of the bone and continues as a spiral line, winding backward to blend into the medial lip of the linea aspera.

The *shaft* of the bone is fairly uniform in caliber but broadens slightly at its extremities. It is bowed forward and its surface is smooth, except for the thickened ridge along its posterior surface, the linea aspera. This is especially prominent in the middle third of the bone, where lateral and medial lips are developed. Superiorly, the lateral lip blends with the prominent gluteal tuberosity; an intermediate lip extends as the pectineal line to the posterior border of the lesser trochanter; and the medial lip continues as the spiral line. The nutrient foramen of the femur, directed upward, is located on the linea aspera.

The *inferior extremity* of the femur is broadened about threefold for the knee joint. Its surfaces, except at the sides, are articular—two oblong *condyles* for articulation with the tibia are separated by an intercondylar fossa and united anteriorly by the patellar surface. The wheellike condyles are also curved from side to side. The intercondylar fossa is especially deep posteriorly and is separated by a ridge from the popliteal surface of the femur above. The medial condyle is longer than the lateral condyle. The condyles rest on the horizontal condyles of the tibia, and the shaft of the femur inclines downward and inward.

The *epicondyles* bulge above and within the curvatures of the condyles. The medial epicondyle is the more prominent, giving attachment to the tibial collateral ligament of the knee joint. It bears on its upper surface a pointed projection, the adductor tubercle. The lateral epicondyle gives rise to the fibular collateral ligament. A groove below the epicondyle borders the articular margin.

The femur is *ossified* from five centers: one for the shaft, one each for the head and inferior extremity, and one for each trochanter. The shaft is ossified at birth; ossification extends into the neck after birth. The center for the inferior extremity of the bone appears during the ninth month of fetal life; that for the head, during the first year. The center in the greater trochanter appears during ages 3 to 5; that for the lesser trochanter, at about age 9 or 10. The epiphyses for the head and trochanters fuse with the shaft at ages 14 to 17; those at the knee fuse with the shaft at about age 17½ in the male but by age 15 in the female.

Lower Limb

Bones and Ligaments at Hip

(Continued)

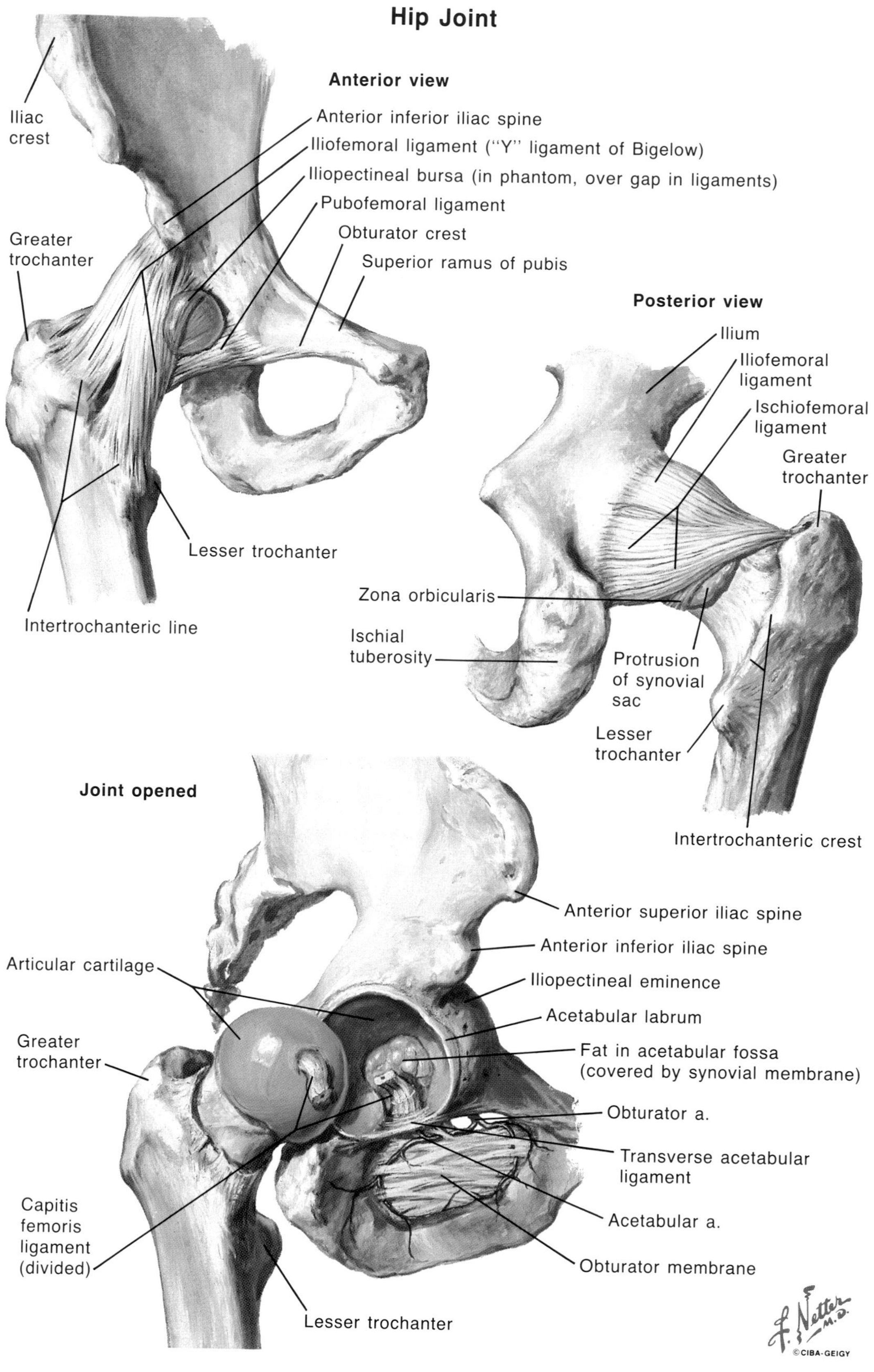
Hip Joint
Anterior view
Iliac crest
Anterior inferior iliac spine
Iliofemoral ligament ("Y" ligament of Bigelow)
Iliopectineal bursa (in phantom, over gap in ligaments)
Pubofemoral ligament
Obturator crest
Superior ramus of pubis
Greater trochanter
Lesser trochanter
Intertrochanteric line
Posterior view
Ilium
Iliofemoral ligament
Ischiofemoral ligament
Greater trochanter
Zona orbicularis
Ischial tuberosity
Protrusion of synovial sac
Lesser trochanter
Intertrochanteric crest
Joint opened
Anterior superior iliac spine
Anterior inferior iliac spine
Iliopectineal eminence
Acetabular labrum
Articular cartilage
Greater trochanter
Fat in acetabular fossa (covered by synovial membrane)
Obturator a.
Transverse acetabular ligament
Capitis femoris ligament (divided)
Acetabular a.
Obturator membrane
Lesser trochanter
F. Netter M.D.
©CIBA-GEIGY

Hip Joint

Movements of the hip joint are flexion-extension, abduction-adduction, and medial and lateral rotation. Circumduction is also allowed.

The hip joint, a synovial ball-and-socket joint, consists of the articulation of the globular head of the femur in the cuplike acetabulum of the coxal bone (Plate 87). Compared with the shoulder joint, it has greater stability and some decrease in freedom of movement. The head forms about two-thirds of a sphere and is covered by articular cartilage, thickest above and thinning to an irregular line of termination at the junction of the head and neck. The acetabulum of the coxal bone exhibits a horseshoe-shaped articular surface arching around the acetabular fossa. The articular fossa lodges a mass of fat covered by synovial membrane; the transverse ligament of the acetabulum closes the fossa below. An acetabular labrum attaches to the bony rim and to the ligament. Its thin, free edge cups around the head of the femur and holds it firmly.

The *articular capsule* of the joint is strong. It is attached to the bony rim of the acetabulum above, and to the transverse ligament of the acetabulum inferiorly. On the femur, it is attached anteriorly to the intertrochanteric line and to the junction of the neck of the femur and its trochanters. Behind, the capsule has an arched free border, covering only two-thirds of the neck of the femur distally. Most of the fibers of the capsule are longitudinal, running from the coxal bone to the femur, but some deeper fibers run circularly. These zona orbicularis fibers are most marked in the posterior part of the capsule; they help to hold the head of the femur in the acetabulum.

Three ligaments, as thickenings of the capsule, add strength. The very strong *iliofemoral ligament* lies on the anterior surface of the capsule, in the form of an inverted Y. Its stem is attached to the lower part of the anterior inferior iliac spine, with the diverging bands attaching below to the whole length of the intertrochanteric line. The iliofemoral ligament becomes taut in full extension of the femur and thus helps to maintain erect posture, for in this position, the body's weight tends to roll the pelvis backward on the femoral heads. The *pubofemoral ligament* is applied to the medial and inferior part of the capsule. Arising from the pubic part of the acetabulum and the obturator crest of the superior ramus of the pubis, it reaches the underside of the neck of the femur and the iliofemoral ligament. The ligament becomes tight in extension and also limits abduction. The articular capsule is thinnest between the iliofemoral and pubofemoral ligaments but is crossed here by the robust iliopsoas tendon. The *iliopectineal bursa* lies between this tendon and the capsule. The *ischiofemoral ligament* forms the posterior margin of the capsule. It arises from the ischial portion of the acetabulum and spirals lateralward and upward, ending in the superior part of the femoral neck. The *capitis femoris ligament*, about 3.5 cm long, is intracapsular, arising from the two margins of the acetabular notch and the lower border of the transverse acetabular ligament and ending in the fossa of the head of the femur. It becomes taut in adduction of the femur.

The *synovial membrane* of the hip joint lines the articular capsule, covers the acetabular labrum, and is extended, sleevelike, over the ligament of the head of the femur. The membrane covers the fat of the acetabular notch and is reflected back along the femoral neck at the femoral attachment of the capsule. Blood vessels to the head and neck of the femur course under these synovial reflections.

The *arteries* of the hip joint are branches of the medial and lateral circumflex femoral arteries, the deep branch of the superior gluteal artery, and the inferior gluteal artery. The posterior branch of the obturator artery provides a significant portion of the blood supply of the femoral head. *Nerve supply* to the hip joint is derived from the nerves supplying the quadratus femoris and rectus femoris muscles, the anterior division of the obturator nerve (rarely also from the accessory obturator nerve), and the superior gluteal nerve. □

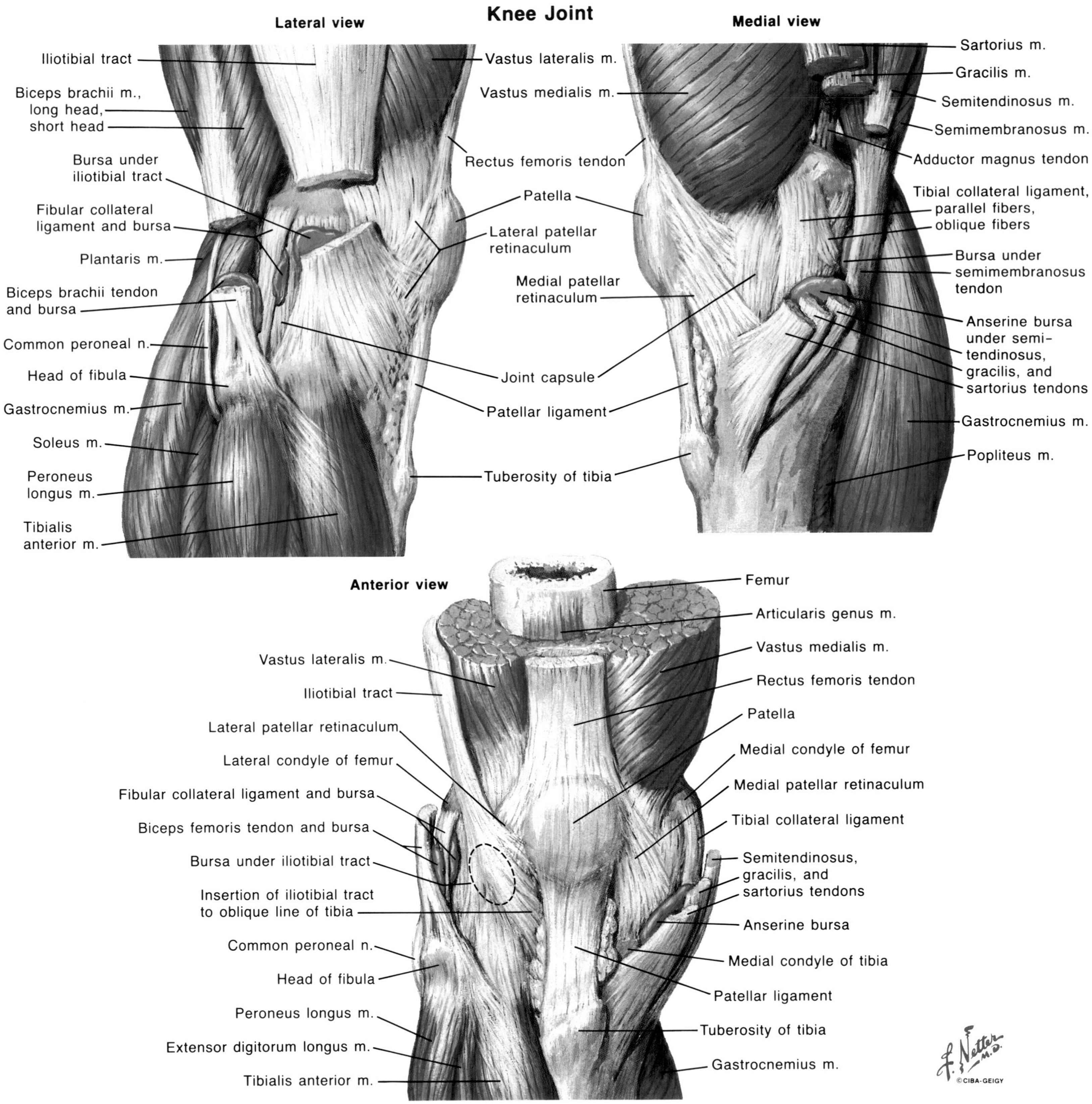
Knee Joint
Lateral view
Medial view
Anterior view
Iliotibial tract
Biceps brachii m.,
long head,
short head
Bursa under
iliotibial tract
Fibular collateral
ligament and bursa
Plantaris m.
Biceps brachii tendon
and bursa
Common peroneal n.
Head of fibula
Gastrocnemius m.
Soleus m.
Peroneus
longus m.
Tibialis
anterior m.
Vastus lateralis m.
Vastus medialis m.
Rectus femoris tendon
Patella
Lateral patellar
retinaculum
Medial patellar
retinaculum
Joint capsule
Patellar ligament
Tuberosity of tibia
Sartorius m.
Gracilis m.
Semitendinosus m.
Semimembranosus m.
Adductor magnus tendon
Tibial collateral ligament,
parallel fibers,
oblique fibers
Bursa under
semimembranosus
tendon
Anserine bursa
under semi-
tendinosus,
gracilis, and
sartorius tendons
Gastrocnemius m.
Popliteus m.
Vastus lateralis m.
Iliotibial tract
Lateral patellar retinaculum
Lateral condyle of femur
Fibular collateral ligament and bursa
Biceps femoris tendon and bursa
Bursa under iliotibial tract
Insertion of iliotibial tract
to oblique line of tibia
Common peroneal n.
Head of fibula
Peroneus longus m.
Extensor digitorum longus m.
Tibialis anterior m.
Femur
Articularis genus m.
Vastus medialis m.
Rectus femoris tendon
Patella
Medial condyle of femur
Medial patellar retinaculum
Tibial collateral ligament
Semitendinosus,
gracilis, and
sartorius tendons
Anserine bursa
Medial condyle of tibia
Patellar ligament
Tuberosity of tibia
Gastrocnemius m.
F. Netter M.D.
©CIBA-GEIGY

SECTION I PLATE 88 Slide 3653

Lower Limb

Bones and Ligaments at Knee

Knee Joint

The knee is primarily a hinge joint that permits flexion and extension (Plates 88–91). In flexion, there is sufficient looseness to allow a small amount of voluntary rotation; in full extension, some terminal medial rotation of the femur (conjunct rotation) achieves the close-packed position. The condyles of the femur provide larger surfaces than those of the tibial condyles, and there is a component of rolling and gliding of the femoral surfaces that uses up this discrepancy. As the extended position is approached, the smaller lateral meniscus is displaced forward on the tibia and becomes firmly seated in a groove on the lateral femoral condyle, which tends to stop extension. However, the medial femoral condyle is still capable of gliding backward, thus bringing its flatter, more anterior surface into full contact with the tibia. These movements of conjunct rotation bring the cruciate ligaments into a taut, or locked, position. The collateral ligaments become maximally tensed, and a full, close-packed, and stable position of extension results. The tension of the ligaments and the close approximation of the flatter parts of the condyles make the erect position relatively easy to maintain.

The sequence of actions in flexion is reversed in extension. Flexion can be carried through about 130° and is finally limited by contact between calf and thigh. The muscles concerned in the movements at the knee are primarily thigh muscles.

There are three articulations in the knee—the femoropatellar articulation and two femorotibial joints. The latter two are separated by the intra-articular cruciate ligaments and the infrapatellar synovial fold. The three joint cavities are connected by restricted openings.

Knee Joint (continued)

Lower Limb

Bones and Ligaments at Knee

(Continued)

Posterior view

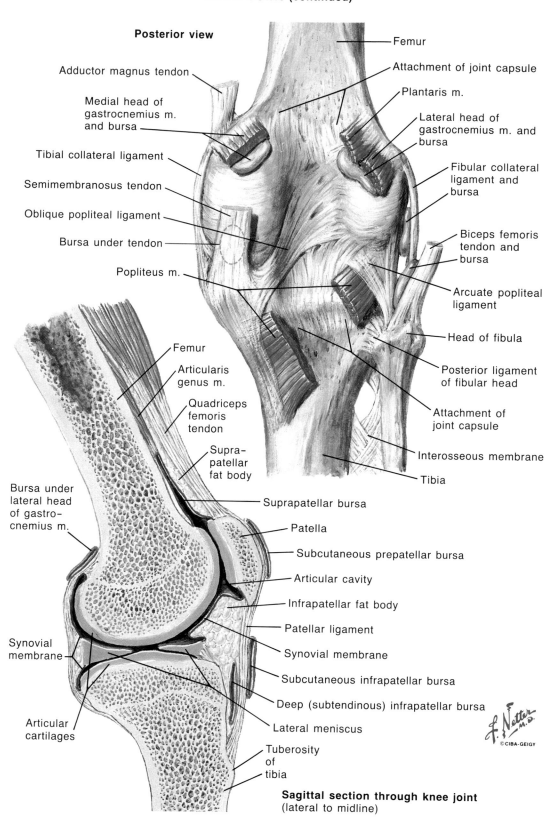

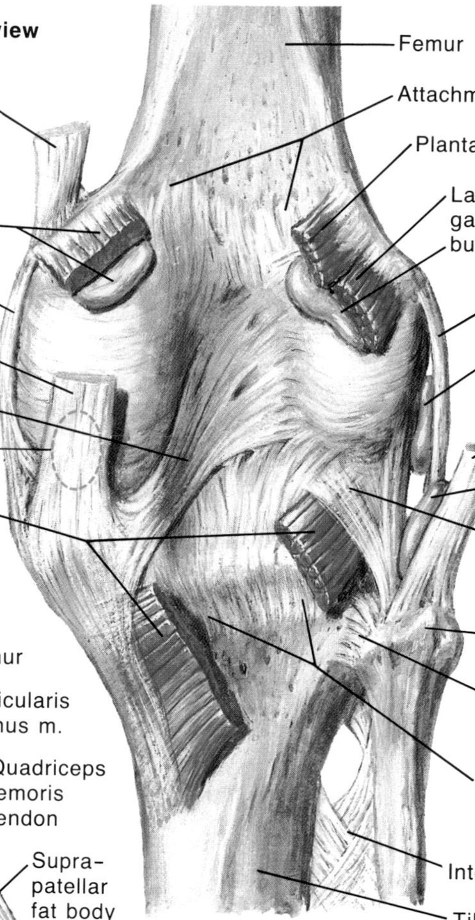
Femur
Attachment of joint capsule
Adductor magnus tendon
Plantaris m.
Medial head of gastrocnemius m. and bursa
Lateral head of gastrocnemius m. and bursa
Tibial collateral ligament
Fibular collateral ligament and bursa
Semimembranosus tendon
Oblique popliteal ligament
Bursa under tendon
Biceps femoris tendon and bursa
Popliteus m.
Arcuate popliteal ligament
Head of fibula
Posterior ligament of fibular head
Attachment of joint capsule
Interosseous membrane
Tibia

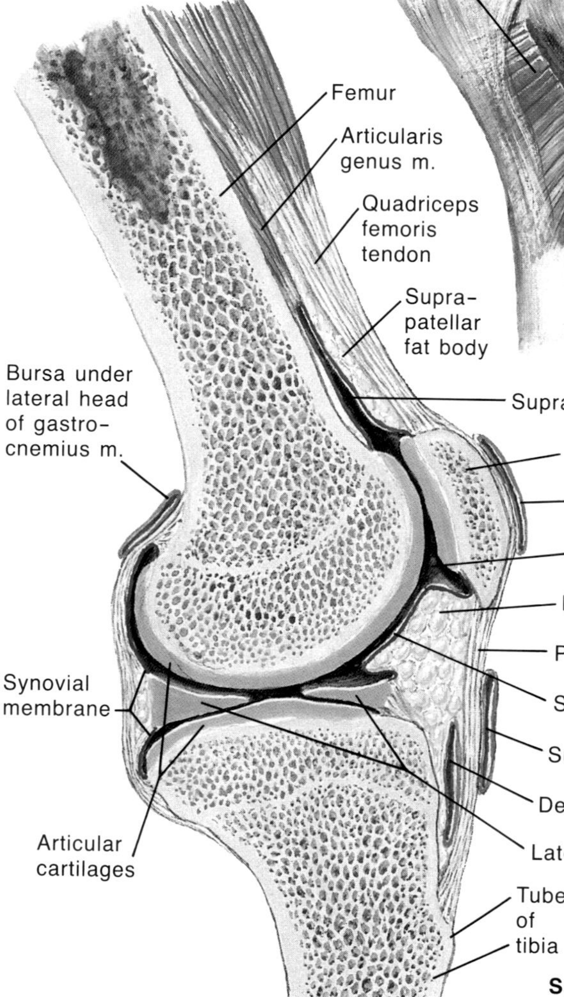
Femur
Articularis genus m.
Quadriceps femoris tendon
Supra-patellar fat body
Bursa under lateral head of gastro-cnemius m.
Suprapatellar bursa
Patella
Subcutaneous prepatellar bursa
Articular cavity
Infrapatellar fat body
Patellar ligament
Synovial membrane
Synovial membrane
Subcutaneous infrapatellar bursa
Deep (subtendinous) infrapatellar bursa
Lateral meniscus
Articular cartilages
Tuberosity of tibia
F. Netter M.D.
©CIBA-GEIGY

Sagittal section through knee joint
(lateral to midline)

The *articular surfaces* of the femur are its medial and lateral condyles and the patellar surface. The condyles are shaped like thick rollers diverging inferiorly and posteriorly. Their surfaces gradually change from a flatter curvature anteriorly to a tighter curvature posteriorly and are separated from the patellar surface by a slight groove.

On the superior surface of the tibia, there are two separate, cartilage-covered areas. The surface of the medial condyle is larger, oval, and slightly concave; that of the lateral condyle is approximately circular, concave from side to side, but concavo-convex from before backward. The fossae of the articular surfaces are deepened by disclike menisci.

The *articular capsule* of the knee joint is scarcely separable from the ligaments and aponeuroses apposed to it. Posteriorly, its vertical fibers arise from the condyles and intercondylar fossa of the femur; inferiorly, these fibers are overlain by the oblique popliteal ligament. The capsule attaches to the tibial condyles and, incompletely, to the menisci. The external ligaments reinforcing the capsule are the fascia lata and the iliotibial tract; the medial patellar and lateral patellar retinacula; and the patellar, oblique popliteal, and arcuate popliteal ligaments. The tibial collateral ligament also closely reinforces the capsule on the medial side.

The aponeurotic tendons of the vastus muscles attach to the sides of the patella and then expand over the front and sides of the capsule as the medial and lateral patellar retinacula. Below, they insert into the front of the tibial condyles and into their oblique lines as far to the sides as the collateral ligaments. Superficially, the fascia lata overlies and blends with the retinacula as it descends to attach to the tibial condyles and their oblique lines. Laterally, the iliotibial tract curves forward over the lateral patellar retinaculum and blends with the capsule anteriorly. Its posterior border is free, and fat tends to be interposed between it and the capsule.

The *patellar ligament* is the continuation of the quadriceps femoris tendon to the tuberosity of the tibia. An extremely strong and relatively flat band, it attaches above the patella and continues over its front with fibers of the tendon, ending somewhat obliquely on the tibial tuberosity. A deep infrapatellar bursa intervenes between the tendon and the bone. A large, subcutaneous infrapatellar bursa is developed in the tissue over the ligament.

The *oblique popliteal ligament* is one of the specializations of the tendon of the semimembranosus muscle, which reinforces the posterior surface of the articular capsule. As this tendon inserts into the groove on the posterior surface of the medial condyle of the tibia, it sends this oblique expansion lateralward and superiorly across the posterior aspect of the capsule.

Collateral Ligaments

These ligaments prevent hyperextension of the joint and any abduction-adduction angulation of the bones (Plates 88–91). The inferior genicular blood vessels pass between them and the capsule of the joint, but only the fibular collateral ligament stands clearly away from the capsule. The *tibial collateral ligament* is a strong, flat band that extends between the medial condyles of the femur and tibia. It is well defined anteriorly, blending with the medial patellar retinaculum. The pes anserinus tendon overlies the ligament below, the two being separated by the anserine bursa. The posterior portion of the ligament is characterized by obliquely running fibers, which converge at the joint level from above and below and give the ligament an attachment into the medial meniscus. The principal inferior attachment of the ligament is about 5 cm below the tibial articular surface immediately posterior to the insertion of the pes anserinus.

The *fibular collateral ligament* is a rounded, pencillike cord, which is entirely separate from the capsule of the knee joint. It is attached to a tubercle on the lateral condyle of the femur above and behind the groove for the popliteus muscle. It ends below on the lateral surface of the head of the fibula, about 1 cm anterior to its apex. The tendon of the popliteus muscle passes deep to the

Cruciate and Collateral Ligaments of Right Knee Joint

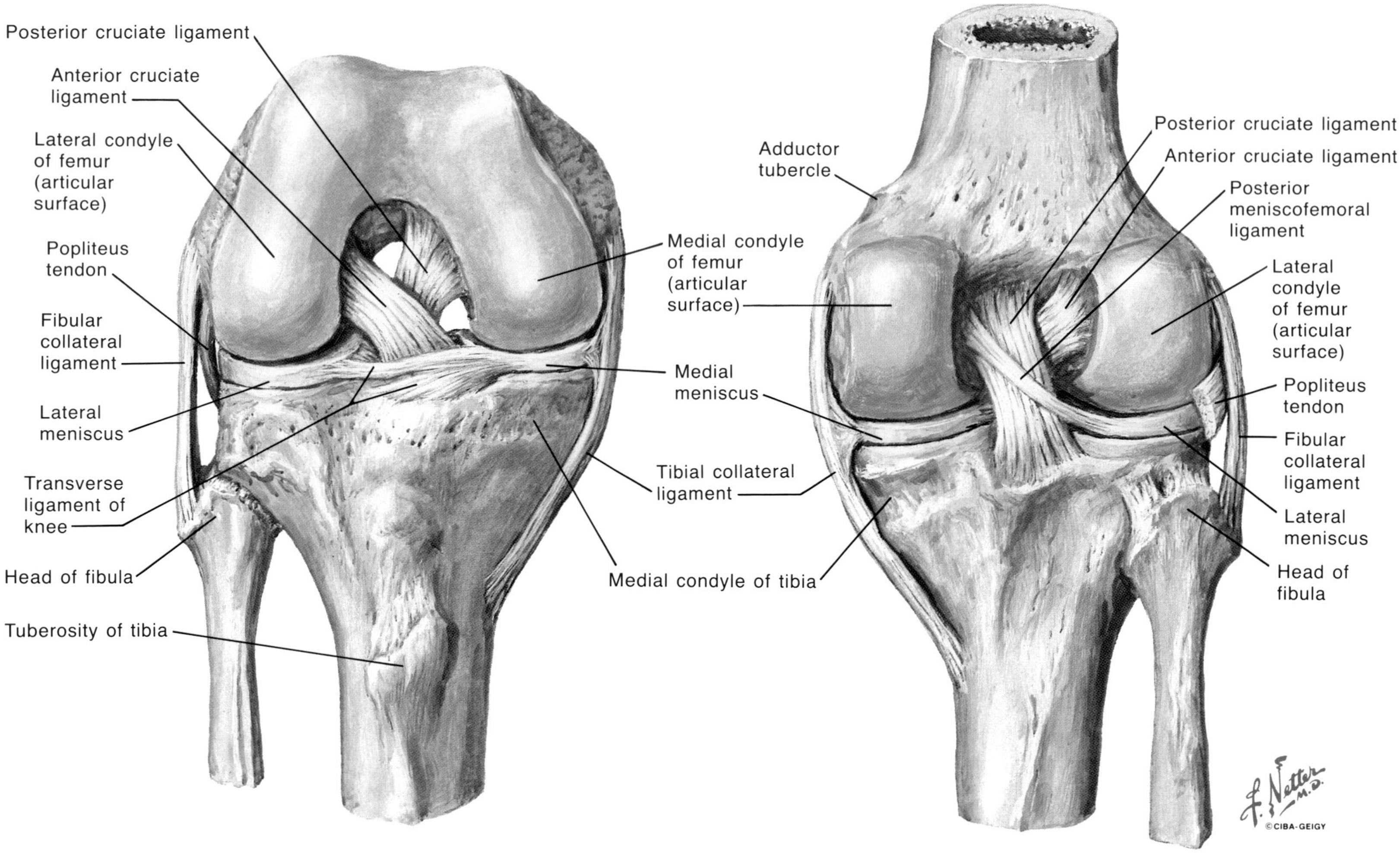
Anterior view: in flexion
Posterior cruciate ligament
Anterior cruciate ligament
Lateral condyle of femur (articular surface)
Popliteus tendon
Fibular collateral ligament
Lateral meniscus
Transverse ligament of knee
Head of fibula
Tuberosity of tibia
Medial condyle of femur (articular surface)
Medial meniscus
Tibial collateral ligament
Medial condyle of tibia
Posterior view: in extension
Adductor tubercle
Posterior cruciate ligament
Anterior cruciate ligament
Posterior meniscofemoral ligament
Lateral condyle of femur (articular surface)
Popliteus tendon
Fibular collateral ligament
Lateral meniscus
Head of fibula
F. Netter M.D.
©CIBA-GEIGY

Lower Limb

Bones and Ligaments at Knee

(Continued)

ligament, and the biceps femoris tendon divides around its fibular attachment, a small inferior subtendinous bursa intervening. Another bursa lies under the upper end of the ligament, separating it from the popliteus tendon. The synovial membrane of the joint, protruding as the subpopliteal recess, separates the popliteus tendon from the lateral meniscus.

Cruciate Ligaments

The cruciate ligaments prevent forward or backward movement of the tibia under the femoral condyles (Plates 90–91). They are somewhat taut in all positions of flexion but become tightest in full extension and full flexion. They lie wholly within the capsule of the knee joint, in the vertical plane between the condyles, but are excluded from the synovial cavity by coverings of synovial membrane. Both ligaments spread linearly at their bony attachments, especially at the femoral condyles. The *anterior cruciate ligament* arises from the rough, nonarticular area in front of the intercondylar eminence of the tibia and extends upward and backward to the posterior part of the medial aspect of the lateral femoral condyle. The *posterior cruciate ligament* passes upward and forward on the medial side of the anterior ligament. It extends from behind the tibial eminence to the lateral side of the medial condyle of the femur.

Menisci. These crescent-shaped wafers of fibrocartilage surmount the peripheral parts of the articular surfaces of the tibia (Plates 89–90). Thicker at their external margins and tapering to thin, unattached edges in the interior of the articulation, they deepen the articular fossae for the reception of the femoral condyles. They are attached to the outer borders of the condyles of the tibia, and at their ends, anterior and posterior to its intercondylar eminence.

The *medial meniscus* is larger and more nearly oval in outline. Broader posteriorly, it narrows anteriorly as it attaches in the intercondylar area of the tibia in front of the origin of the posterior cruciate ligament. The *lateral meniscus* is more nearly circular. Although smaller than the medial meniscus, it covers a somewhat greater proportion of the tibial surface. Anteriorly, it attaches in the anterior intercondylar area, lateral to and behind the end of the anterior cruciate ligament. Posteriorly, it ends in the posterior intercondylar area in front of the end of the medial meniscus. The lateral meniscus is weakly attached around the margin of the lateral tibial condyle and lacks an attachment where it is crossed and notched by the popliteus tendon. At the back of the joint, it gives origin to some of the fibers of the popliteus muscle, and close to its posterior attachment to the tibia, it frequently gives off a collection of fibers, known as the *posterior meniscofemoral ligament*. This may join the posterior cruciate ligament or may insert into the medial femoral condyle behind the attachment of the cruciate ligament. An occasional *anterior meniscofemoral ligament* has a similar but anterior relationship to the posterior cruciate ligament. The *transverse ligament of the knee* connects the anterior convex margin of the lateral meniscus to the anterior end of the medial meniscus.

Synovial Membrane and Joint Cavity. The articular cavity of the knee is the largest joint space of the body. It includes the space between and around the condyles, extends upward behind the patella to include the femoropatellar articulation, and then communicates freely with the suprapatellar bursa between the quadriceps femoris tendon and the femur. The synovial membrane lines the articular capsule and the suprapatellar bursa. Recesses of the joint cavity are also lined by synovial membrane; the subpopliteal recess has been described. Other recesses exist behind the posterior part of each femoral condyle; at the upper end of the medial recess, the bursa under the medial head of the gastrocnemius muscle may open into the joint cavity.

The infrapatellar fat body represents an anterior part of the median septum, which, with the cruciate ligaments, separates the two femorotibial articulations. From the medial and lateral borders of the articular surface of the patella, reduplications of synovial membrane project into the interior of the joint and form two fringelike alar folds, which cover collections of fat.

Lower Limb

Bones and Ligaments at Knee

(Continued)

Blood Vessels and Nerves. In the region of the knee, there is an important *genicular anastomosis.* This consists of a superficial plexus above and below the patella, plus a deep plexus on the capsule of the knee joint and the adjacent bony surfaces. This anastomosis is made up of terminal interconnections of ten vessels. Two of these descend to the joint—the descending branch of the lateral circumflex femoral artery and the descending genicular branch of the femoral artery. Five are branches of the popliteal artery at the level of the knee—the medial superior genicular, lateral superior genicular, middle genicular, medial inferior genicular, and lateral inferior genicular arteries. Three branches of leg arteries ascend to the anastomosis—the posterior tibial recurrent, circumflex fibular, and anterior tibial recurrent arteries. *Veins* of the same names accompany the arteries. The *lymphatics* of the knee joint drain to the popliteal and inguinal node groups.

The *nerves* of the knee joint are numerous. Articular branches of the femoral nerve reach the knee via the nerves to the vastus muscles and the saphenous nerve. The posterior division of the obturator nerve ends in the joint and there are articular branches of the tibial and common peroneal nerves.

Patella. This large sesamoid is developed in the tendon of the quadriceps femoris muscle (Plates 88–89, 91). It bears against the anterior articular surface of the inferior extremity of the femur and, by holding the tendon off the lower end of the femur, improves the angle of approach of the tendon to the tibial tuberosity. The convex anterior surface of the patella is striated vertically by the tendon fibers. The superior border is thick, giving attachment to the tendinous fibers of the rectus femoris and vastus intermedius muscles. The lateral and medial borders are thinner; they receive the fibers of the vastus lateralis and vastus medialis muscles. These borders converge to the pointed apex of the patella, which gives attachment to the *patellar ligament.* The articular surface is a smooth oval area, divided by a vertical ridge into two facets. The ridge occupies the groove on the patellar surface of the femur, the medial and lateral facets corresponding to facing surfaces of the femur. The lateral facet is broader and deeper than the medial. Inferior to the faceted area is a rough nonarticular portion from which the lower half of the patellar ligament arises.

The patella maintains a shifting contact with the femur in all positions of the knee. As the knee shifts from a fully flexed to a fully extended position, first the superior, then the middle, and lastly the inferior parts of the articular surface of the patella are brought into contact with the patellar surfaces of the femur.

Ossification develops from a single center, which appears early in the third year of life. Complete ossification occurs by age 13 in the male and at about age 10 in the female. □

Interior of Right Knee Joint

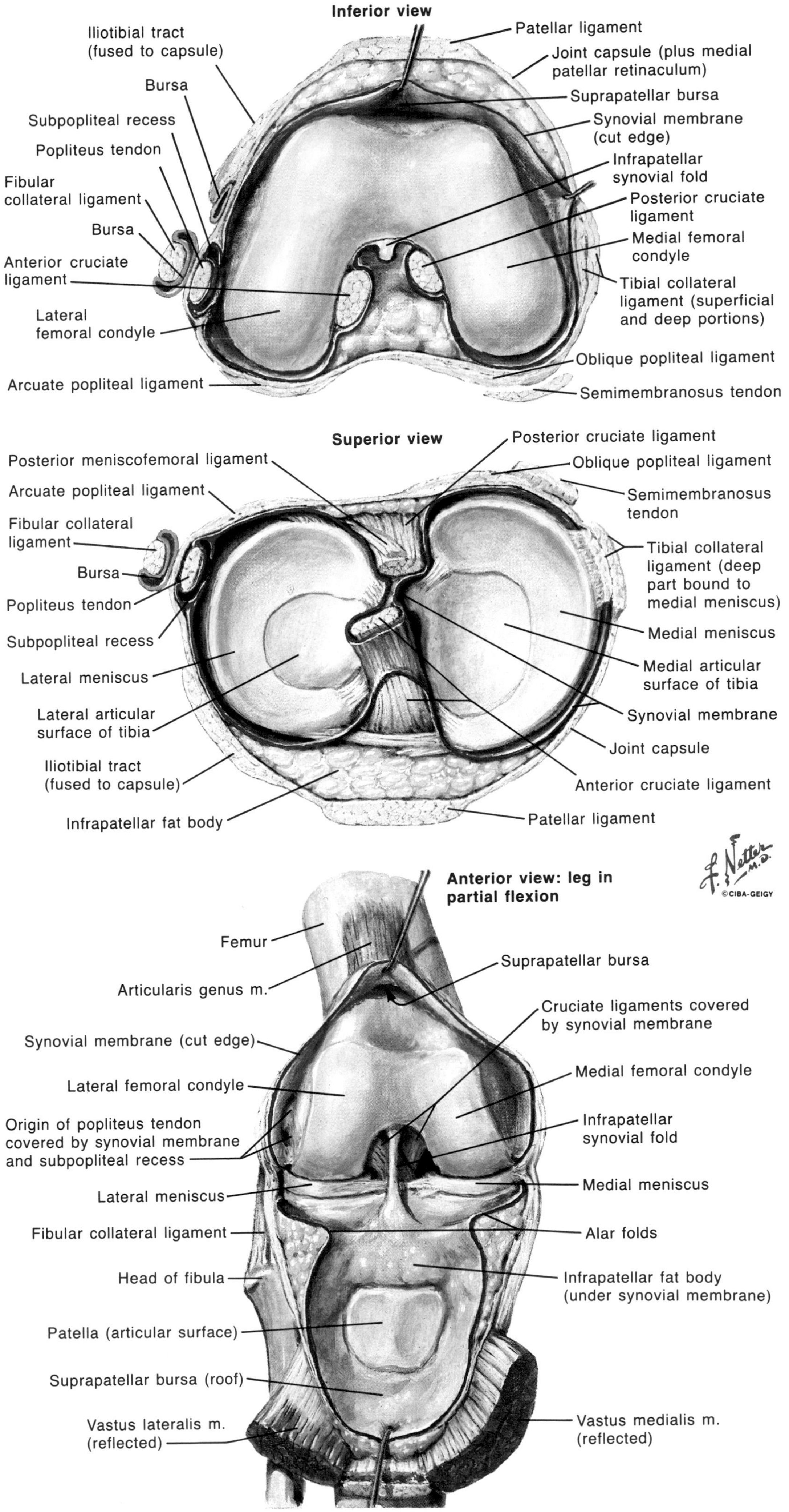
Inferior view
Iliotibial tract (fused to capsule)
Bursa
Subpopliteal recess
Popliteus tendon
Fibular collateral ligament
Bursa
Anterior cruciate ligament
Lateral femoral condyle
Arcuate popliteal ligament
Patellar ligament
Joint capsule (plus medial patellar retinaculum)
Suprapatellar bursa
Synovial membrane (cut edge)
Infrapatellar synovial fold
Posterior cruciate ligament
Medial femoral condyle
Tibial collateral ligament (superficial and deep portions)
Oblique popliteal ligament
Semimembranosus tendon
Superior view
Posterior meniscofemoral ligament
Arcuate popliteal ligament
Fibular collateral ligament
Bursa
Popliteus tendon
Subpopliteal recess
Lateral meniscus
Lateral articular surface of tibia
Iliotibial tract (fused to capsule)
Infrapatellar fat body
Posterior cruciate ligament
Oblique popliteal ligament
Semimembranosus tendon
Tibial collateral ligament (deep part bound to medial meniscus)
Medial meniscus
Medial articular surface of tibia
Synovial membrane
Joint capsule
Anterior cruciate ligament
Patellar ligament
F. Netter M.D.
©CIBA-GEIGY
Anterior view: leg in partial flexion
Femur
Articularis genus m.
Synovial membrane (cut edge)
Lateral femoral condyle
Origin of popliteus tendon covered by synovial membrane and subpopliteal recess
Lateral meniscus
Fibular collateral ligament
Head of fibula
Patella (articular surface)
Suprapatellar bursa (roof)
Vastus lateralis m. (reflected)
Suprapatellar bursa
Cruciate ligaments covered by synovial membrane
Medial femoral condyle
Infrapatellar synovial fold
Medial meniscus
Alar folds
Infrapatellar fat body (under synovial membrane)
Vastus medialis m. (reflected)

Lower Limb

Compartments of Leg

Fasciae and Compartments

The fascia lata of the thigh continues into the leg, where it is designated as the *crural fascia* (Plate 92). At the knee, the fascia has many attachments—the patella, patellar ligament, tibial tuberosity, condyles of the tibia, and head of the fibula, reinforcing the medial and lateral patellar retinacula. The fascia is strengthened by expansions of the tendons of the sartorius, gracilis, semitendinosus, and biceps femoris muscles. Covering the soft parts of the leg, the crural fascia gives origin to superficial muscle fibers and blends with the periosteum of the subcutaneous surface of the tibia. Above the ankle, it blends with the periosteum of the lower part of the shaft of the fibula. Distally, it attaches to the medial and lateral malleoli and the calcaneus.

Deep extensions of the crural fascia separate certain *compartments of the leg*. The *anterior* and *posterior intermuscular septa* dip deeply to attach to the anterior and posterior borders of the fibula, respectively. They define the lateral compartment of the leg, separating it from the anterior and posterior compartments. From the posterior intermuscular septum adjacent to the fibula, the *transverse intermuscular septum* extends medialward, ending on the tibia and in the crural fascia behind it. This sheet is thick over the popliteus muscle and, in the lower part of the leg, lies anterior to the calcaneal tendon. It separates the superficial and deep posterior compartments of the leg. At the ankle, further thickenings of the crural fascia enclose the soft parts and prevent the tendons from bowstringing. One of these is the *superior extensor retinaculum*, which spans across anteriorly from the tibia to the fibula, covers all the muscles of the anterior compartment, and restrains their tendons.

The three intermuscular septa of the leg and the interosseous membrane between the tibia and fibula and the crural fascia define the anterior, lateral, superficial posterior, and deep posterior compartments of the leg. These compartments contain groups of related muscles and vessels and nerves appropriate to them. The two posterior compartments contain preaxial muscles and are served by the tibial nerve; the anterior and lateral compartments contain postaxial muscles and are served by the common peroneal nerve (Plates 97–98).

Muscles of Anterior Compartment

The anterior compartment contains the tibialis anterior, extensor hallucis longus, extensor digitorum longus, and peroneus tertius muscles, which extend the toes and dorsiflex the foot. In addition, the tibialis anterior muscle inverts the sole of the foot (Plates 92–94, 100).

The *tibialis anterior muscle* lies against the lateral surface of the tibia. It arises from the lateral condyle of the tibia and from the upper half of its lateral surface. Fibers also take origin from the interosseous membrane, the overlying crural fascia, and the intermuscular septum between it and the extensor digitorum longus muscle. The tendon appears in the lower third of the leg and passes under the superior and inferior extensor retinacula surrounded by a synovial sheath. It inserts into the medial surface of the medial cuneiform and the base of the first metatarsal (Plate 102).

The *extensor hallucis longus muscle* is a rather thin muscle that outcrops between the tibialis anterior and extensor digitorum longus muscles in the lower half of the leg. It arises from the middle two-fourths of the anterior surface of the fibula and from the interosseous membrane for the same distance. Its tendon develops on the superficial surface, passes under the two extensor retinacula, and inserts into the base of the distal phalanx of the great toe (Plate 103).

The *extensor digitorum longus muscle* occupies the fibular part of the anterior compartment. This pennate muscle arises from the lateral condyle of the tibia, the anterior surface of the fibula, the intermuscular septa (between it and the tibialis anterior muscle and between it and the peroneus longus), and the crural fascia near the tibia. The tendon begins at about the middle of the leg and receives fibers nearly to the ankle. Below the superior extensor retinaculum, the tendon divides into

Fascial Compartments of Leg

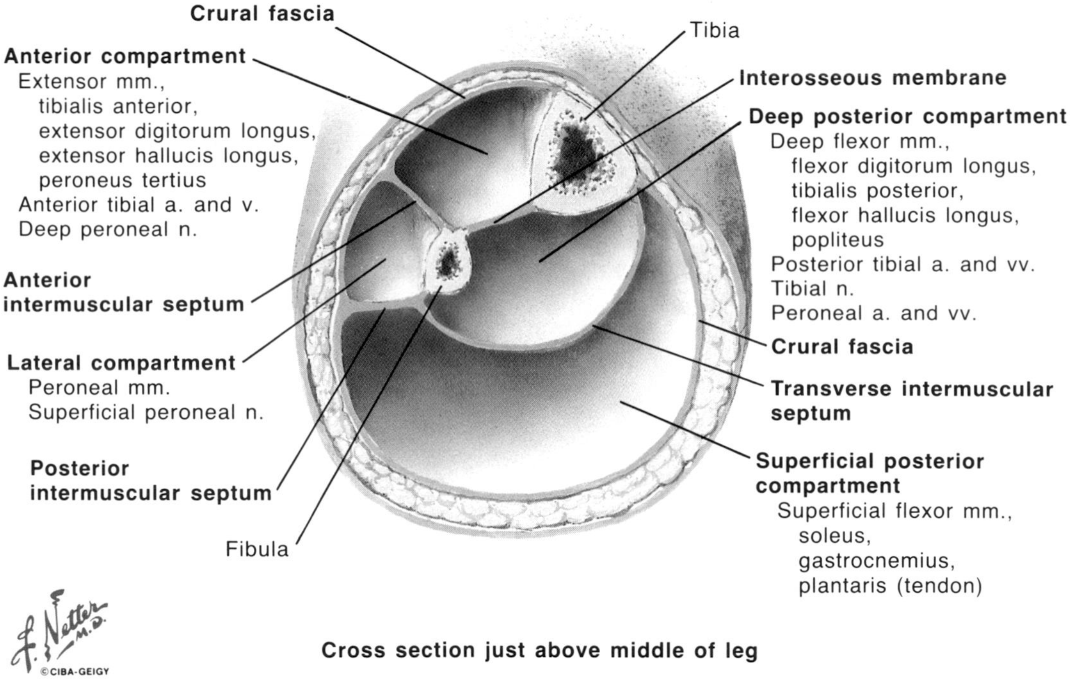

Cross section just above middle of leg

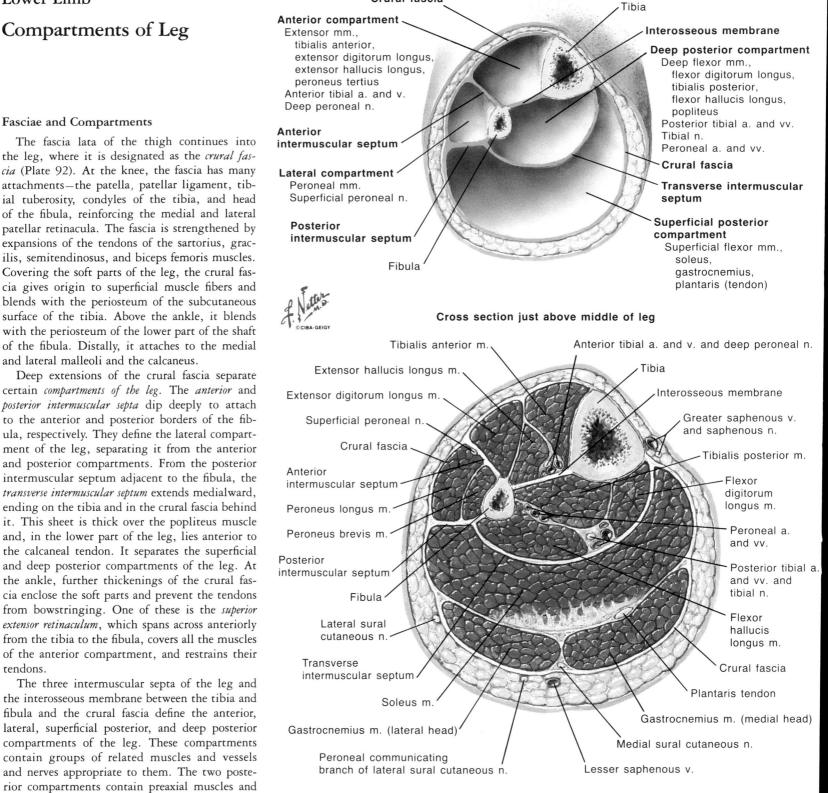

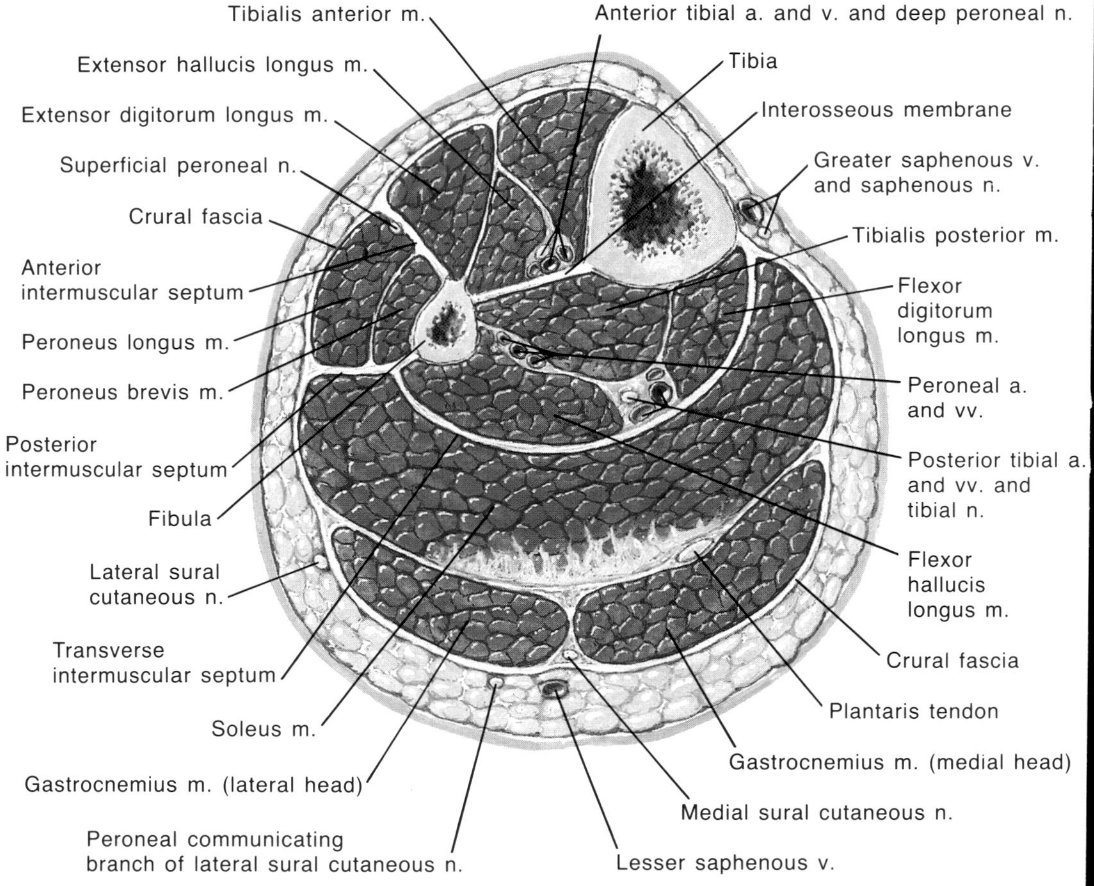

Muscles, Arteries, and Nerves of Leg: Superficial Dissection

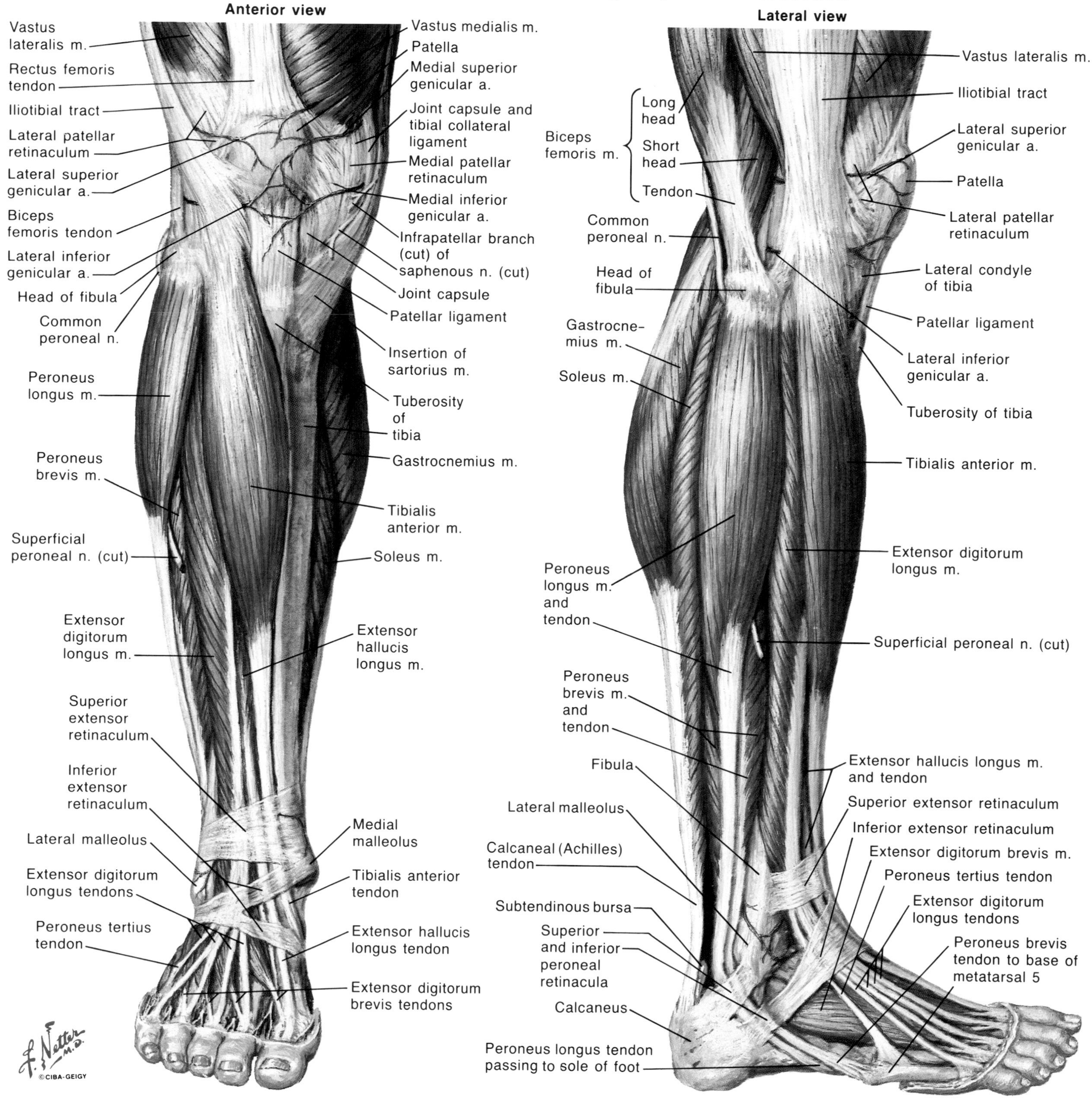

Anterior view
Vastus lateralis m.
Rectus femoris tendon
Iliotibial tract
Lateral patellar retinaculum
Lateral superior genicular a.
Biceps femoris tendon
Lateral inferior genicular a.
Head of fibula
Common peroneal n.
Peroneus longus m.
Peroneus brevis m.
Superficial peroneal n. (cut)
Extensor digitorum longus m.
Superior extensor retinaculum
Inferior extensor retinaculum
Lateral malleolus
Extensor digitorum longus tendons
Peroneus tertius tendon
Vastus medialis m.
Patella
Medial superior genicular a.
Joint capsule and tibial collateral ligament
Medial patellar retinaculum
Medial inferior genicular a.
Infrapatellar branch (cut) of saphenous n. (cut)
Joint capsule
Patellar ligament
Insertion of sartorius m.
Tuberosity of tibia
Gastrocnemius m.
Tibialis anterior m.
Soleus m.
Extensor hallucis longus m.
Medial malleolus
Tibialis anterior tendon
Extensor hallucis longus tendon
Extensor digitorum brevis tendons
F. Netter M.D.
©CIBA-GEIGY
Lateral view
Biceps femoris m.
Long head
Short head
Tendon
Common peroneal n.
Head of fibula
Gastrocnemius m.
Soleus m.
Peroneus longus m. and tendon
Peroneus brevis m. and tendon
Fibula
Lateral malleolus
Calcaneal (Achilles) tendon
Subtendinous bursa
Superior and inferior peroneal retinacula
Calcaneus
Peroneus longus tendon passing to sole of foot
Vastus lateralis m.
Iliotibial tract
Lateral superior genicular a.
Patella
Lateral patellar retinaculum
Lateral condyle of tibia
Patellar ligament
Lateral inferior genicular a.
Tuberosity of tibia
Tibialis anterior m.
Extensor digitorum longus m.
Superficial peroneal n. (cut)
Extensor hallucis longus m. and tendon
Superior extensor retinaculum
Inferior extensor retinaculum
Extensor digitorum brevis m.
Peroneus tertius tendon
Extensor digitorum longus tendons
Peroneus brevis tendon to base of metatarsal 5

SECTION I PLATE 93 Slide 3658

Lower Limb

Compartments of Leg

(Continued)

two parts, which pass under the inferior extensor retinaculum. Redividing, four tendons cross the dorsum of the foot to the lateral four toes.

The tendons of the extensor digitorum longus muscle mimic the divisions and expansions characteristic of the tendons of the extensor digitorum muscle in the hand. They divide into two lateral slips and one central slip; the central slip ends on the dorsum of the middle phalanx, and the lateral slips converge distally to end on the base of the distal phalanx. Extensor expansions are formed over the metatarsophalangeal joints, and distally, the tendons of the interosseous and lumbrical muscles join the lateral slips of the tendon. The tendons of the extensor digitorum brevis muscle join the tendons of the extensor digitorum longus muscle on their fibular sides and help form the extensor expansions of the second, third, and fourth toes.

The *peroneus tertius muscle* is essentially a lateral slip of the extensor digitorum longus muscle; it is seldom completely separate, except at its insertion. It arises from the distal third of the anterior surface of the fibula, the adjacent interosseous membrane, and the anterior intermuscular septum. Descending under the extensor retinacula in the compartment for the extensor digitorum longus it turns lateralward to end on the dorsum of the shaft of the fifth metatarsal.

Muscles of Lateral Compartment

The lateral compartment contains the peroneus longus and the peroneus brevis muscles (Plates 92–94, 100).

The *peroneus longus muscle*, bipennate in form, arises higher in the leg and is more superficial. It

Lower Limb

Compartments of Leg

(Continued)

Muscles, Arteries, and Nerves of Leg: Deep Dissection (anterior view)

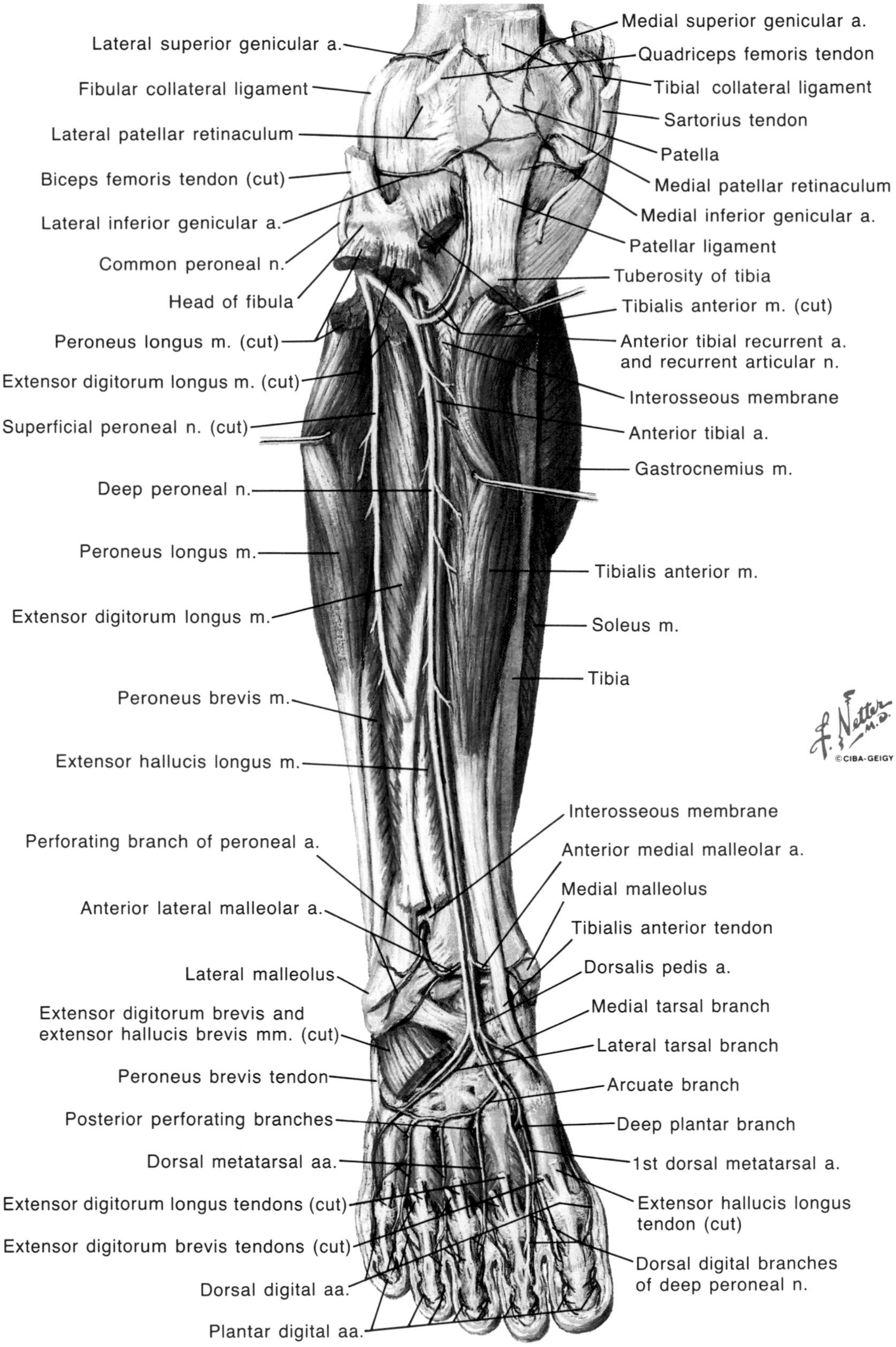
Lateral superior genicular a.
Fibular collateral ligament
Lateral patellar retinaculum
Biceps femoris tendon (cut)
Lateral inferior genicular a.
Common peroneal n.
Head of fibula
Peroneus longus m. (cut)
Extensor digitorum longus m. (cut)
Superficial peroneal n. (cut)
Deep peroneal n.
Peroneus longus m.
Extensor digitorum longus m.
Peroneus brevis m.
Extensor hallucis longus m.
Perforating branch of peroneal a.
Anterior lateral malleolar a.
Lateral malleolus
Extensor digitorum brevis and
extensor hallucis brevis mm. (cut)
Peroneus brevis tendon
Posterior perforating branches
Dorsal metatarsal aa.
Extensor digitorum longus tendons (cut)
Extensor digitorum brevis tendons (cut)
Dorsal digital aa.
Plantar digital aa.
Medial superior genicular a.
Quadriceps femoris tendon
Tibial collateral ligament
Sartorius tendon
Patella
Medial patellar retinaculum
Medial inferior genicular a.
Patellar ligament
Tuberosity of tibia
Tibialis anterior m. (cut)
Anterior tibial recurrent a.
and recurrent articular n.
Interosseous membrane
Anterior tibial a.
Gastrocnemius m.
Tibialis anterior m.
Soleus m.
Tibia
Interosseous membrane
Anterior medial malleolar a.
Medial malleolus
Tibialis anterior tendon
Dorsalis pedis a.
Medial tarsal branch
Lateral tarsal branch
Arcuate branch
Deep plantar branch
1st dorsal metatarsal a.
Extensor hallucis longus
tendon (cut)
Dorsal digital branches
of deep peroneal n.
F. Netter M.D.
©CIBA-GEIGY

takes its origin from the head and upper two-thirds of the lateral surface of the body of the fibula, the anterior and posterior intermuscular septa, and the crural fascia. Its tendon begins high on the superficial surface of the muscle and receives fibers (posteriorly) to almost the lateral malleolus. Behind the lateral malleolus, the tendon is posterior to the tendon of the peroneus brevis; both are enclosed within a common synovial sheath and pass under the superior peroneal retinaculum. The tendon of the peroneus longus muscle passes diagonally forward, inferior to the tendon of the peroneus brevis muscle, and turns into the foot against the anterior slope of the tuberosity of the cuboid. A sesamoid in the tendon protects it over the tuberosity. Crossing the sole of the foot deep to its intrinsic muscles, the peroneus longus tendon ends on the inferolateral surface of the medial cuneiform and on the base and inferolateral surface of the first metatarsal.

The *peroneus brevis muscle* lies deep to the peroneus longus muscle and is smaller and shorter. It arises from the lower two-thirds of the lateral surface of the fibula and from the anterior and posterior intermuscular septa. Its tendon grooves the back of the lateral malleolus, lying in a common synovial sheath with the tendon of the peroneus longus. Turning forward under the superior peroneal retinaculum, it passes through the inferior peroneal retinaculum to insert into the tuberosity on the base of the fifth metatarsal.

Muscles of Superficial Posterior Compartment

The gastrocnemius and soleus muscles are supplemented by the almost vestigial plantaris muscle to form a triceps surae complex in the superior posterior compartment (Plates 92–93, 95, 100). The more superficial *gastrocnemius muscle* arises by two heads. The larger medial head takes origin from the popliteal surface of the femur immediately above the medial femoral condyle. The lateral head arises from the upper posterior portion of the lateral surface of the lateral femoral condyle and from the end of the supracondylar line. Bursae separate both heads from the capsule of the knee joint. The fibers of both heads converge toward the midline of the leg and unite at about its midlength into a tendinous raphe, which broadens into an aponeurosis on the anterior surface of the muscle. This aponeurosis fuses below with the tendon of the soleus muscle and, with it, forms the calcaneal tendon. The medial head is usually somewhat broader and thicker and its muscular fibers descend a bit farther toward the heel.

The *soleus muscle* is broad and fleshy, lying immediately anterior to the gastrocnemius muscle but arising entirely below the knee. It has a triple origin from the posterior surfaces of the head of the fibula and the upper third of its shaft, a tendinous arch between the tibia and fibula that represents the upper part of the transverse intermuscular septum, and the soleal line of the tibia and its medial border along its middle third. The muscle fibers converge into a broad aponeurosis, which fuses below with that of the gastrocnemius muscle to form the calcaneal tendon, the thickest and strongest tendon in the body.

The calcaneal tendon is about 15 cm long, begins at midleg, and receives muscular fibers almost to its termination. Narrowed below, it inserts into the middle part of the posterior surface of the calcaneus; a bursa lies deep to the tendon and separates it from the upper part of the posterior surface of the bone.

The *plantaris muscle* arises from the lower lateral supracondylar line of the femur immediately above the lateral head of the gastrocnemius muscle and from the oblique popliteal ligament. Its short (10 cm) belly ends in a long slender tendon, which descends between the gastrocnemius and soleus muscles, and then along the medial border of the calcaneal tendon to insert into the calcaneus.

Muscles of Deep Posterior Compartment

The deep posterior compartment contains the popliteus, flexor hallucis longus, flexor digitorum longus, and tibialis posterior muscles (Plates 92, 96, 100).

Muscles, Arteries, and Nerves of Leg (posterior view)

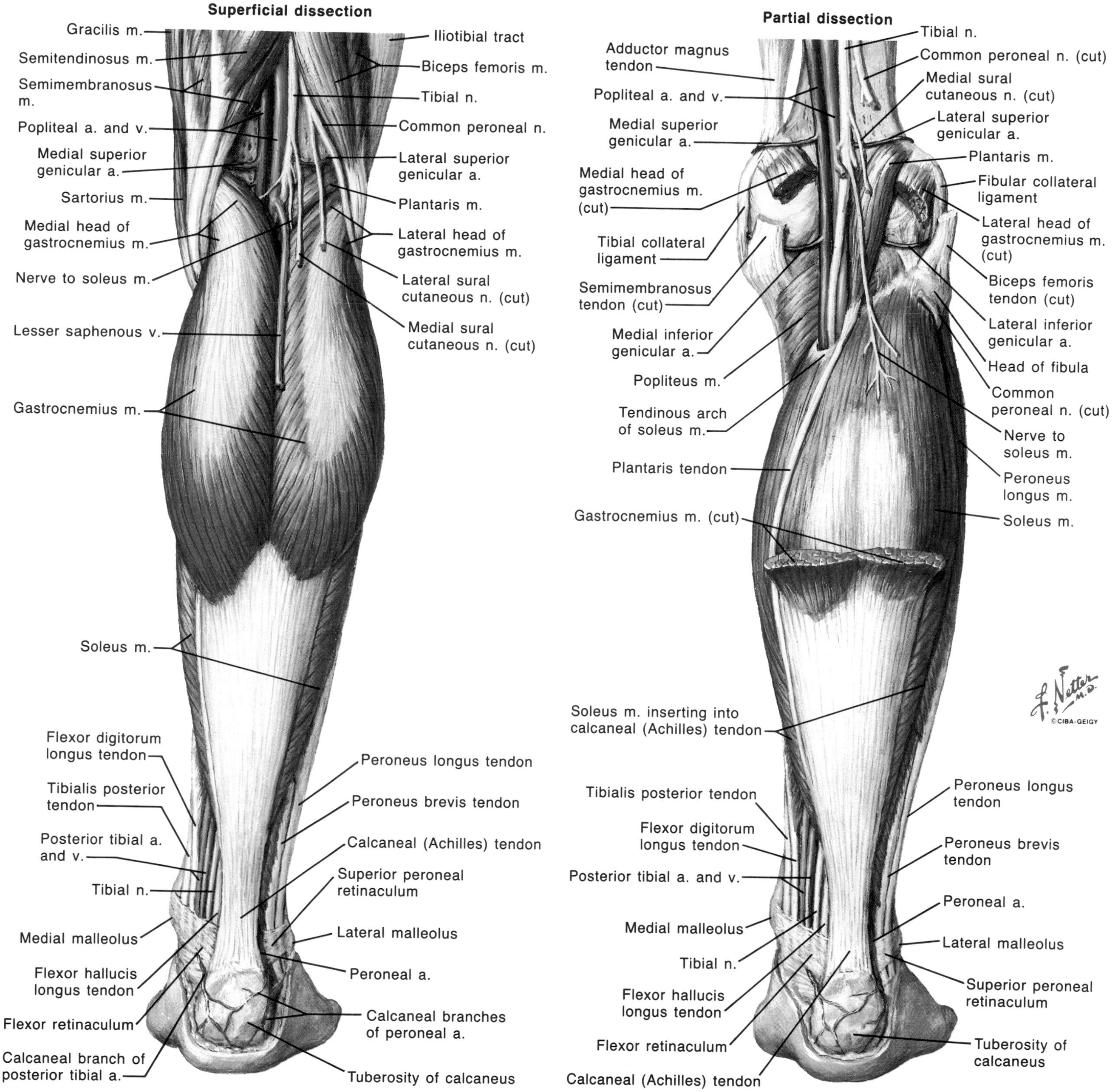

Superficial dissection
Gracilis m.
Semitendinosus m.
Semimembranosus m.
Popliteal a. and v.
Medial superior genicular a.
Sartorius m.
Medial head of gastrocnemius m.
Nerve to soleus m.
Lesser saphenous v.
Gastrocnemius m.
Soleus m.
Flexor digitorum longus tendon
Tibialis posterior tendon
Posterior tibial a. and v.
Tibial n.
Medial malleolus
Flexor hallucis longus tendon
Flexor retinaculum
Calcaneal branch of posterior tibial a.
Iliotibial tract
Biceps femoris m.
Tibial n.
Common peroneal n.
Lateral superior genicular a.
Plantaris m.
Lateral head of gastrocnemius m.
Lateral sural cutaneous n. (cut)
Medial sural cutaneous n. (cut)
Peroneus longus tendon
Peroneus brevis tendon
Calcaneal (Achilles) tendon
Superior peroneal retinaculum
Lateral malleolus
Peroneal a.
Calcaneal branches of peroneal a.
Tuberosity of calcaneus
Partial dissection
Adductor magnus tendon
Popliteal a. and v.
Medial superior genicular a.
Medial head of gastrocnemius m. (cut)
Tibial collateral ligament
Semimembranosus tendon (cut)
Medial inferior genicular a.
Popliteus m.
Tendinous arch of soleus m.
Plantaris tendon
Gastrocnemius m. (cut)
Soleus m. inserting into calcaneal (Achilles) tendon
Tibialis posterior tendon
Flexor digitorum longus tendon
Posterior tibial a. and v.
Medial malleolus
Tibial n.
Flexor hallucis longus tendon
Flexor retinaculum
Calcaneal (Achilles) tendon
Tibial n.
Common peroneal n. (cut)
Medial sural cutaneous n. (cut)
Lateral superior genicular a.
Plantaris m.
Fibular collateral ligament
Lateral head of gastrocnemius m. (cut)
Biceps femoris tendon (cut)
Lateral inferior genicular a.
Head of fibula
Common peroneal n. (cut)
Nerve to soleus m.
Peroneus longus m.
Soleus m.
F. Netter M.D.
©CIBA-GEIGY
Peroneus longus tendon
Peroneus brevis tendon
Peroneal a.
Lateral malleolus
Superior peroneal retinaculum
Tuberosity of calcaneus

SECTION I PLATE 95 Slide 3660

Lower Limb

Compartments of Leg

(Continued)

The *popliteus muscle*, a thin, flat muscle of triangular outline, occupies the floor of the lower part of the popliteal fossa. It arises mainly by a stout cord from the anterior end of the groove on the lateral face of the lateral femoral condyle, close to the articular margin. The tendon passes between the lateral meniscus and the capsule of the knee joint, the subpopliteal recess of synovial membrane lying between them. As much as one half of its fibers may come from the arcuate popliteal ligament, with others arising in the lateral meniscus. The muscle inserts by fleshy fibers into the triangular area of the back of the tibia.

The *flexor hallucis longus muscle* lies on the fibular rather than the tibial side of the leg. It arises from the lower two-thirds of the posterior surface of the shaft of the fibula and from the intermuscular septa separating it from the tibialis posterior and peroneus brevis muscles. Its tendon grooves the posterior surface of the talus and the undersurface of the sustentaculum tali of the calcaneus. It passes forward in the sole of the foot, giving a slip to the overlying flexor digitorum longus muscle, and then passes between the two heads of the flexor hallucis brevis muscle to insert into the base of the distal phalanx of the great toe (Plate 106).

The *flexor digitorum longus muscle* is found on the tibial side of the leg. It arises from the medial side of the posterior surface of the middle three-fifths of the tibia and from the intermuscular septum between it and the tibialis posterior muscle. Its pennate fibers converge to a tendon that lies along the medial margin of the muscle and receives fibers almost to the medial malleolus. At the ankle, the tendon of the tibialis posterior muscle is anterior to that of the flexor digitorum longus muscle. Both tendons enter the foot through a

Lower Limb

Compartments of Leg

(Continued)

groove on the back of the medial malleolus, where they are enclosed in separate synovial sheaths, that on the tendon of flexor digitorum longus muscle passing well into the foot. The tendon of the flexor digitorum longus muscle passes diagonally into the sole of the foot, crossing the deltoid ligament of the ankle joint and running superficial to the tendon of the flexor hallucis longus muscle from which it receives a slip.

In the middle of the sole, the tendon of the flexor digitorum longus muscle receives the insertion of the quadratus plantae muscle and then divides into four tendons; these insert into the bases of the distal phalanges of the second, third, fourth, and fifth toes. Like the tendons of the flexor digitorum profundus muscle in the hand, the tendons of the flexor digitorum longus muscle give origin to lumbrical muscles in the foot. Also, the relationships of these tendons to those of the flexor digitorum brevis muscle and to synovial sheaths and the fibrous digital sheaths are similar to those in the hand.

The *tibialis posterior muscle* is the most deeply situated muscle in the posterior compartment and lies between the flexor digitorum longus and flexor hallucis longus muscles. Beginning above in two pointed processes, the muscle arises from all except the lowest part of the posterior surface of the interosseous membrane, the lateral part of the posterior surface of the tibia in its upper two-thirds, the upper two-thirds of the medial surface of the fibula, the intermuscular septa on either side of it, and the transverse intermuscular septum of the leg. Its tendon emerges from the medial side of the muscle at about the middle of the leg and continues to receive fibers almost to the medial malleolus. The tendon lies behind the medial malleolus anterior to that of the flexor digitorum longus muscle and enters the foot under the flexor retinaculum within its own synovial sheath. It crosses the deltoid ligament of the ankle, passes under the plantar calcaneonavicular ligament, and inserts into the tuberosity of the navicular and the underside of the medial cuneiform. Expansions continue forward and lateralward to the intermediate and lateral cuneiforms and to the plantar surfaces of the bases of the second to fourth metatarsals.

Muscle Actions

The muscles of the superficial posterior compartment all insert into the tuberosity of the calcaneus; they act together to produce plantar flexion of the foot, accompanied by some inversion. Thus, they raise the heel against the weight of the body in locomotion. In standing, these muscles draw back on the leg, stabilizing the ankle joint. The gastrocnemius arises above the knee and flexes it as well. It shows intermittent activity in relaxed standing, since the ankle joint is not then in its close-packed position.

In the deep posterior compartment, the popliteus muscle is a weak flexor of the knee and can rotate the leg medially. With the foot firmly on the ground, the contraction of the muscle leads to lateral rotation of the femur with retraction of the lateral meniscus and starts flexion at the knee. The other muscles of this compartment assist in plantar flexion of the foot and are strong invertors. The flexor hallucis longus and flexor digitorum longus muscles flex the phalanges. The flexor hallucis longus muscle shows its greatest activity at heel-off. The tibialis posterior muscle distributes body weight among the heads of the metatarsals, serving to reduce flatfoot and shifting weight toward the lateral side of the foot.

The muscles of the anterior compartment of the leg dorsiflex and invert the foot at the ankle. The tibialis anterior muscle is powerful in this action; it draws the tibia forward, as in walking. Additionally, the extensor hallucis longus muscle dorsiflexes the great toe, and the extensor digitorum longus muscle has a like action on the second to fifth toes through the insertion of its tendons into the proximal and middle phalanges of these digits. The peroneus tertius muscle produces an eversion rather than an inversion effect. The peroneus longus and peroneus brevis muscles of the lateral compartment evert and abduct the foot and assist in its plantar flexion. With respect to inversion and eversion, the insertions of the tibialis anterior and peroneus longus muscles on the same bones place the foot in a sling that is controlled immediately by the pull of these muscles.

An important fact is that a normally strong foot does not depend on muscles for ordinary static support; the muscles of the leg are generally quiescent in relaxed standing. Static support is provided by the many supporting ligaments of the joint capsules, which hold the bones in the normal arched architecture. However, in a flatfooted person or in one with other serious foot impairment, leg muscles come into play for support.

Blood Vessels

Arteries. The *popliteal artery* is the direct continuation of the femoral artery at the adductor hiatus. Passing to the back of the knee at the hiatus, the artery descends through the popliteal space with a slight mediolateral inclination and ends at the lower border of the popliteus muscle by dividing into the anterior tibial and posterior tibial arteries. The artery is deeper in the popliteal space than the corresponding vein and the tibial nerve, lying in the intercondylar fossa of the femur and against the posterior capsule of the knee joint. Lower down, it crosses the popliteus muscle. The popliteal artery gives off five genicular branches and several large sural arteries.

The *lateral superior genicular artery* passes against the plantaris muscle and above the lateral femoral condyle. Deep to the tendon of the biceps femoris muscle, it winds around the femur and supplies a superficial branch, which enters the vastus lateralis muscle and anastomoses with the descending branch of the lateral circumflex femoral artery and the lateral inferior genicular artery. It also gives rise to a deep branch supplying the knee joint and anastomosing across the front of the femur with the descending genicular and medial superior genicular arteries.

The *medial superior genicular artery* swings across the origin of the medial head of the gastrocnemius muscle anterior to the tendons of the semitendinosus and semimembranosus muscles. One branch supplies the vastus medialis muscle and anastomoses with the descending genicular and medial inferior genicular arteries. A second branch supplies the knee joint and anastomoses with the lateral superior genicular artery by the anastomotic arch across the femur.

The *middle genicular artery* is a small single vessel arising from the anterior surface of the popliteal artery at the back of the knee. It pierces the oblique popliteal ligament to supply the cruciate ligaments and the synovial membrane within the joint cavity.

The *sural arteries* are usually two large muscular branches that enter the heads of the gastrocnemius muscles and send branches to the plantaris muscle and the upper part of the soleus. More proximal unnamed muscular branches supply the lower ends of the hamstring muscles. A cutaneous branch descends along the middle of the back of the calf with the lesser saphenous vein.

The *lateral inferior genicular artery* crosses the upper portion of the popliteus muscle anterior to the lateral head of the gastrocnemius muscle. Turning forward internal to the fibular collateral ligament, it divides into branches, which anastomose with the lateral superior genicular, medial inferior genicular, anterior and posterior tibial recurrent, and circumflex fibular arteries.

The *medial inferior genicular artery* passes medialward along the upper borders of the popliteus muscle and on the medial side of the knee is deep to the tibial collateral ligament. At the anterior border of the ligament, various branches spread to anastomose with the descending genicular and medial superior genicular arteries and with the lateral inferior genicular and anterior tibial recurrent arteries.

The *anterior tibial artery* is one product of the division of the popliteal artery. It passes directly forward above the upper end of the interosseous membrane of the leg and thus enters the anterior compartment, lying against the medial side of the neck of the fibula. The artery descends on the interosseous membrane, first lying between the tibialis anterior and extensor digitorum longus muscles and then between the tibialis anterior and the extensor hallucis longus muscles. It is joined on its lateral side by the deep peroneal nerve. The tendon of the extensor hallucis longus muscle crosses the vessels and nerve, and they then lie between the foregoing and the tendon of the extensor digitorum longus muscle at the ankle. The named branches of the anterior tibial, posterior tibial recurrent, circumflex fibular, anterior tibial recurrent, anterior medial malleolar, and anterior lateral malleolar arteries are concentrated at the knee and ankle.

The *posterior tibial recurrent artery*, sometimes a branch of the posterior tibial artery, usually arises from the anterior tibial in the posterior compartment of the leg. It ascends between the popliteus muscle and the back of the knee, supplies the popliteus and the tibiofibular joint, and anastomoses with the lateral inferior genicular artery. The small *circumflex fibular artery* also arises from the anterior tibial artery before that vessel leaves the posterior compartment of the leg; at times, it may be a branch of the posterior tibial artery. It passes around the neck of the fibula through the soleus muscle, supplying it, and anastomoses with the lateral inferior genicular artery. The *anterior tibial recurrent artery* arises as

Lower Limb

Compartments of Leg

(Continued)

Muscles, Arteries, and Nerves of Leg: Deep Dissection (posterior view)

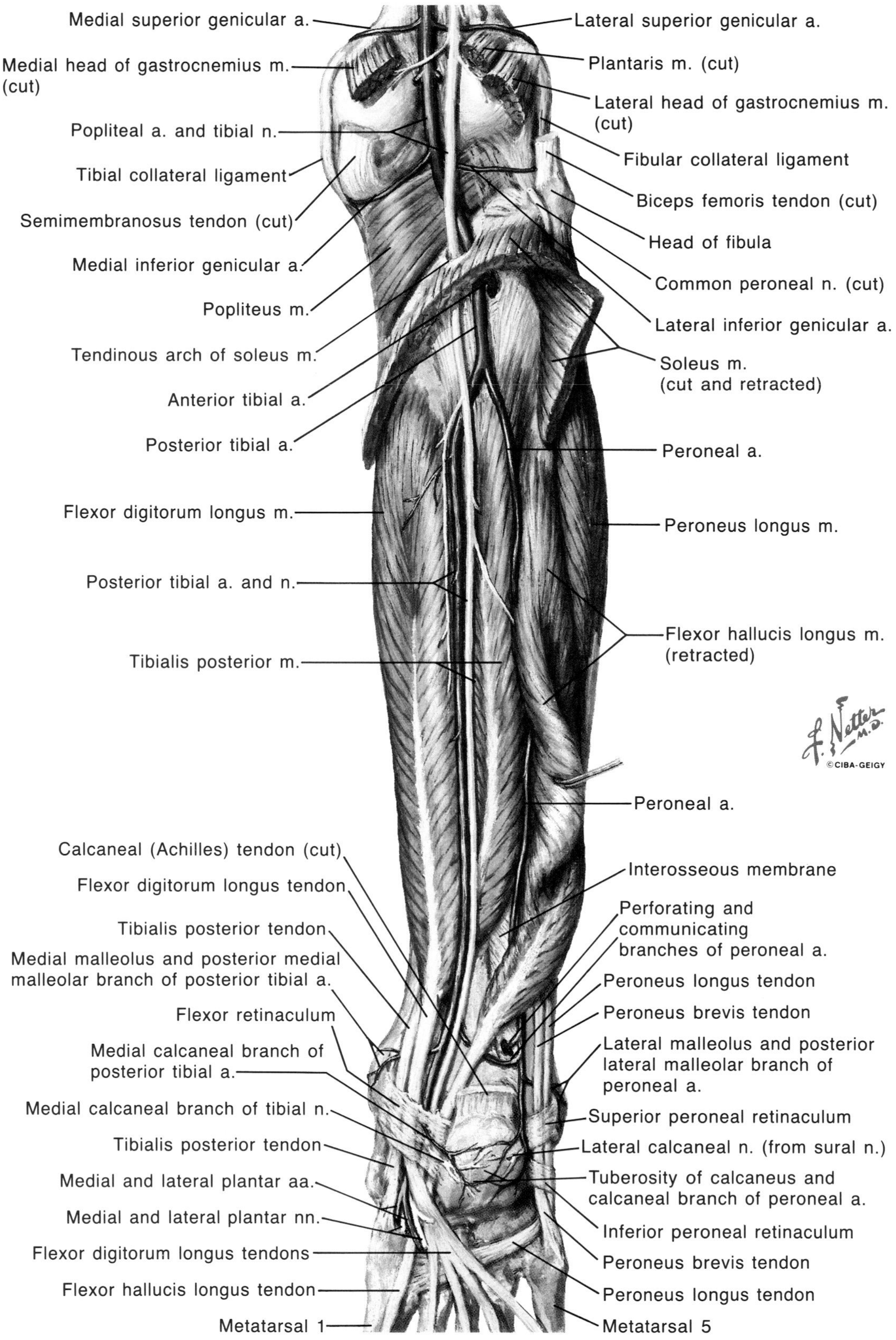
Medial superior genicular a.
Medial head of gastrocnemius m. (cut)
Popliteal a. and tibial n.
Tibial collateral ligament
Semimembranosus tendon (cut)
Medial inferior genicular a.
Popliteus m.
Tendinous arch of soleus m.
Anterior tibial a.
Posterior tibial a.
Flexor digitorum longus m.
Posterior tibial a. and n.
Tibialis posterior m.
Calcaneal (Achilles) tendon (cut)
Flexor digitorum longus tendon
Tibialis posterior tendon
Medial malleolus and posterior medial malleolar branch of posterior tibial a.
Flexor retinaculum
Medial calcaneal branch of posterior tibial a.
Medial calcaneal branch of tibial n.
Tibialis posterior tendon
Medial and lateral plantar aa.
Medial and lateral plantar nn.
Flexor digitorum longus tendons
Flexor hallucis longus tendon
Metatarsal 1
Lateral superior genicular a.
Plantaris m. (cut)
Lateral head of gastrocnemius m. (cut)
Fibular collateral ligament
Biceps femoris tendon (cut)
Head of fibula
Common peroneal n. (cut)
Lateral inferior genicular a.
Soleus m. (cut and retracted)
Peroneal a.
Peroneus longus m.
Flexor hallucis longus m. (retracted)
©CIBA-GEIGY
Peroneal a.
Interosseous membrane
Perforating and communicating branches of peroneal a.
Peroneus longus tendon
Peroneus brevis tendon
Lateral malleolus and posterior lateral malleolar branch of peroneal a.
Superior peroneal retinaculum
Lateral calcaneal n. (from sural n.)
Tuberosity of calcaneus and calcaneal branch of peroneal a.
Inferior peroneal retinaculum
Peroneus brevis tendon
Peroneus longus tendon
Metatarsal 5

soon as the anterior tibial artery enters the anterior compartment of the leg. It ascends among the fibers of the tibialis anterior muscle and then branches over the front and sides of the knee joint. It anastomoses with the genicular branches of the popliteal artery and with the descending genicular and the lateral circumflex femoral branches of the femoral artery. The *anterior medial malleolar artery* takes origin at the ankle. It passes medialward, deep to the tendons of the tibialis anterior and extensor hallucis longus muscles. It supplies the skin and the ankle joint medially, anastomosing with the malleolar branches of the posterior tibial artery. The *anterior lateral malleolar artery* arises opposite the anterior medial malleolar artery and passes lateralward, deep to the tendons of the extensor digitorum longus muscle. It supplies the lateral surface of the ankle and the joint and anastomoses with the perforating branch of the peroneal artery and with ascending branches of the lateral tarsal branch of the dorsalis pedis artery.

The *posterior tibial artery* begins at the lower border of the popliteus muscle and is the direct continuation of the popliteal artery. Accompanied by its veins and by the tibial nerve, it descends in the deep posterior compartment. At first, it inclines toward the fibula; then, after giving off the peroneal artery, swings medially again and passes behind the medial malleolus at the ankle. It ends deep to the origin of the abductor hallucis muscle by dividing into the medial plantar and lateral plantar arteries. Apart from the plantar artery, which is described separately, and muscular arteries throughout the leg, the branches of the posterior tibial artery are not large. A *nutrient artery* arises in the upper part of the leg to enter the tibia posteriorly. A *communicating branch* passes lateralward, just above the tibiofibular syndesmosis, to join a similar branch of the peroneal artery. The *posterior medial malleolar branch* passes onto the medial malleolus and anastomoses with the anterior medial malleolar branch of the anterior tibial artery, and *medial calcaneal branches* arise just proximal to the artery's division. They reach the skin and areolar tissues of the medial side and back of the heel.

The *peroneal artery* is the largest branch of the posterior tibial artery. It supplies the muscles of the lateral side of the leg and is an important longitudinal collateral vessel through its communicating branch to the posterior tibial artery and its perforating branch to the anterior tibial artery. The peroneal artery arises 2 to 3 cm beyond the origin of the posterior tibial artery and descends near the fibula within the substance of the flexor hallucis longus muscle or between it and the tibialis posterior muscle. A *nutrient branch* enters the nutrient foramen of the fibula. The *perforating branch* passes forward at the distal border of the interosseous membrane to enter the anterior compartment of the leg. It supplies the joints at the ankle and anastomoses with the anterior lateral malleolar branch of the anterior tibial artery and with the lateral tarsal and arcuate branches of the dorsalis pedis artery. The *communicating branch* arises just below the perforating branch, runs medialward deep to the tendon of the flexor hallucis longus muscle, and joins the communicating branch of the posterior tibial artery. The *posterior lateral malleolar branch* supplies twigs to the lateral malleolus and anastomoses with the anterior lateral malleolar branch of the anterior tibial artery. It also gives off *lateral calcaneal branches*.

Veins. The veins of the leg are paired accompanying vessels of the arteries. They are supplied with numerous valves and receive many perforating communications from the superficial veins. The unions of the venae comitantes of the anterior tibial, posterior tibial, and peroneal veins are made at various levels; they form the popliteal vein. (A single popliteal vein is typically expected at about 5 cm above the knee joint.) The *popliteal vein* is typically a large single vein, ascending through the popliteal space superficial to its artery and between it and the tibial nerve. It is somewhat medial to the artery inferiorly but against its lateral side above the knee joint. Three or four bicuspid valves prevent descending flow in the vein, and one of these valves is rather constantly located just distal to the adductor hiatus. Other tributaries of the popliteal vein are genicular and muscular, as well as the lesser saphenous vein. □

Lower Limb

Nerves of Leg

Common Peroneal Nerve (L4, 5; S1, 2)

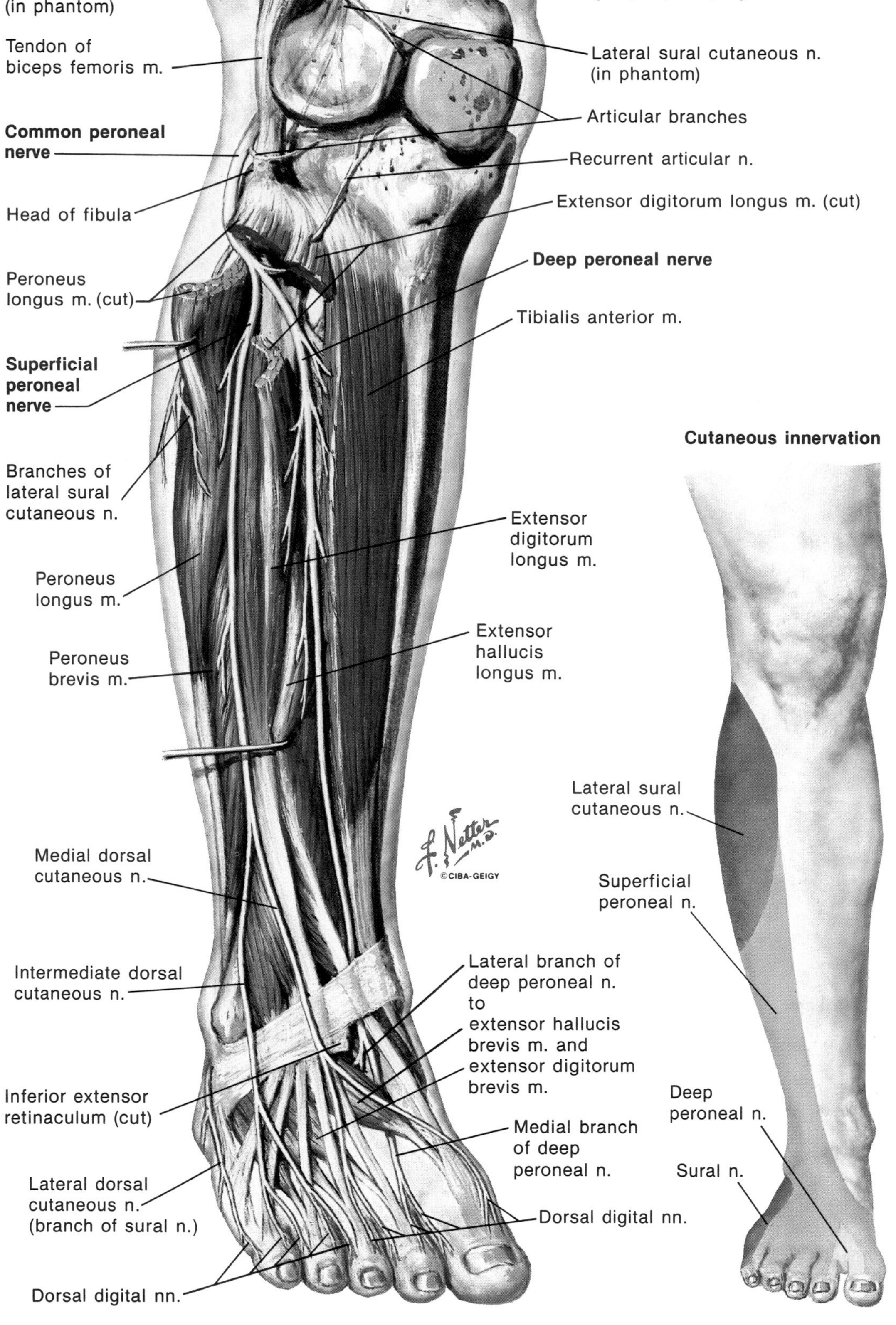
Common peroneal n. (in phantom)
Tendon of biceps femoris m.
Common peroneal nerve
Head of fibula
Peroneus longus m. (cut)
Superficial peroneal nerve
Branches of lateral sural cutaneous n.
Peroneus longus m.
Peroneus brevis m.
Medial dorsal cutaneous n.
Intermediate dorsal cutaneous n.
Inferior extensor retinaculum (cut)
Lateral dorsal cutaneous n. (branch of sural n.)
Dorsal digital nn.
Lateral sural cutaneous n. (in phantom)
Articular branches
Recurrent articular n.
Extensor digitorum longus m. (cut)
Deep peroneal nerve
Tibialis anterior m.
Extensor digitorum longus m.
Extensor hallucis longus m.
Lateral branch of deep peroneal n. to extensor hallucis brevis m. and extensor digitorum brevis m.
Medial branch of deep peroneal n.
Dorsal digital nn.
F. Netter M.D.
©CIBA-GEIGY
Cutaneous innervation
Lateral sural cutaneous n.
Superficial peroneal n.
Deep peroneal n.
Sural n.

Common Peroneal Nerve

The common peroneal nerve (L4, 5; S1, 2) is the smaller, lateral, terminal branch of the sciatic nerve (Plate 97). Its fibers are derived from the posterior divisions of the ventral rami of L4 and L5 and S1 and S2. From its origin, the nerve descends first along the lateral side of the popliteal fossa, overlapped by the medial margin of the biceps femoris; then it passes between the biceps tendon and the lateral head of the gastrocnemius muscle to reach the back of the fibular head; and finally, it winds around the back and outer side of the neck of the fibula between the two heads of the peroneus longus muscle and divides into the *superficial* and *deep peroneal nerves*. At this point, the nerve can easily be compressed against the underlying bone.

Branches. Before dividing, the common peroneal nerve gives off three *articular branches* to the knee, which accompany the lateral superior and inferior genicular and anterior tibial recurrent arteries. It also gives off the *lateral sural cutaneous nerve*, which supplies the skin and fascia on the lateral and adjacent parts of the posterior and anterior surfaces of the upper part of the leg, and the *peroneal communicating branch*, which joins the sural branch of the tibial nerve and is distributed with it.

The *superficial peroneal nerve* descends between the extensor digitorum longus and peroneal muscles and supplies the peroneus longus and peroneus brevis muscles before piercing the deep fascia at about the junction of the middle and lower thirds of the leg. At this level, the nerve divides into medial and intermediate dorsal cutaneous nerves. The *medial dorsal cutaneous nerve* runs in front of the ankle and onto the dorsum of the foot, supplying twigs to the skin and fascia on the anterior surface of the distal third of the leg and the dorsum of the foot. Near the lower border of the inferior extensor retinaculum, it splits into two *dorsal digital nerves*; one of these nerves supplies the medial and dorsal aspects of the dorsum of the foot and the great toe, while the other supplies the adjacent sides of the second and third toes. The *intermediate dorsal cutaneous nerve* runs along the lateral part of the dorsum of the foot, supplying the nearby skin and fascia and providing the dorsal digital nerves for the third and fourth and fourth and fifth toes. It communicates with the lateral dorsal cutaneous nerve, the termination of the sural nerve.

The *deep peroneal nerve* passes obliquely forward and downward around the fibular neck between the peroneus longus and the extensor digitorum longus muscles to the front of the interosseous membrane. It descends lateral to the tibialis anterior muscle and is a medial relation first to the extensor digitorum longus muscle and then to the extensor hallucis longus muscle, the tendon of which crosses the nerve obliquely above the ankle. In its downward course, the nerve first lies lateral to the anterior tibial vessels, then anterior to them, and finally lateral to them again in front of the lower end of the tibia and ankle, where the nerve divides into medial and lateral terminal branches. In the leg, the nerve sends branches to the tibialis anterior, extensor digitorum longus, extensor hallucis longus, and peroneus tertius muscles, an articular branch to the ankle, and filaments to the anterior tibial vessels. The *medial terminal branch* gives rise to a *dorsal digital nerve*, which splits to supply the contiguous sides of the first and second toes. It also supplies filaments to the dorsalis pedis artery and nearby metatarsophalangeal and interphalangeal joints and, occasionally, a twig to the first dorsal interosseous muscle. The *lateral terminal branch* curves outward beneath the extensor digitorum brevis muscle, becomes slightly expanded, and gives off several slender offshoots to supply the extensor digitorum brevis muscle and its medial part (the extensor hallucis brevis muscle), the adjacent tarsal and tarsometatarsal joints and, rarely, the second and third dorsal interosseous muscles.

Lower Limb

Nerves of Leg

(Continued)

Tibial Nerve (L4, 5; S1, 2, 3)

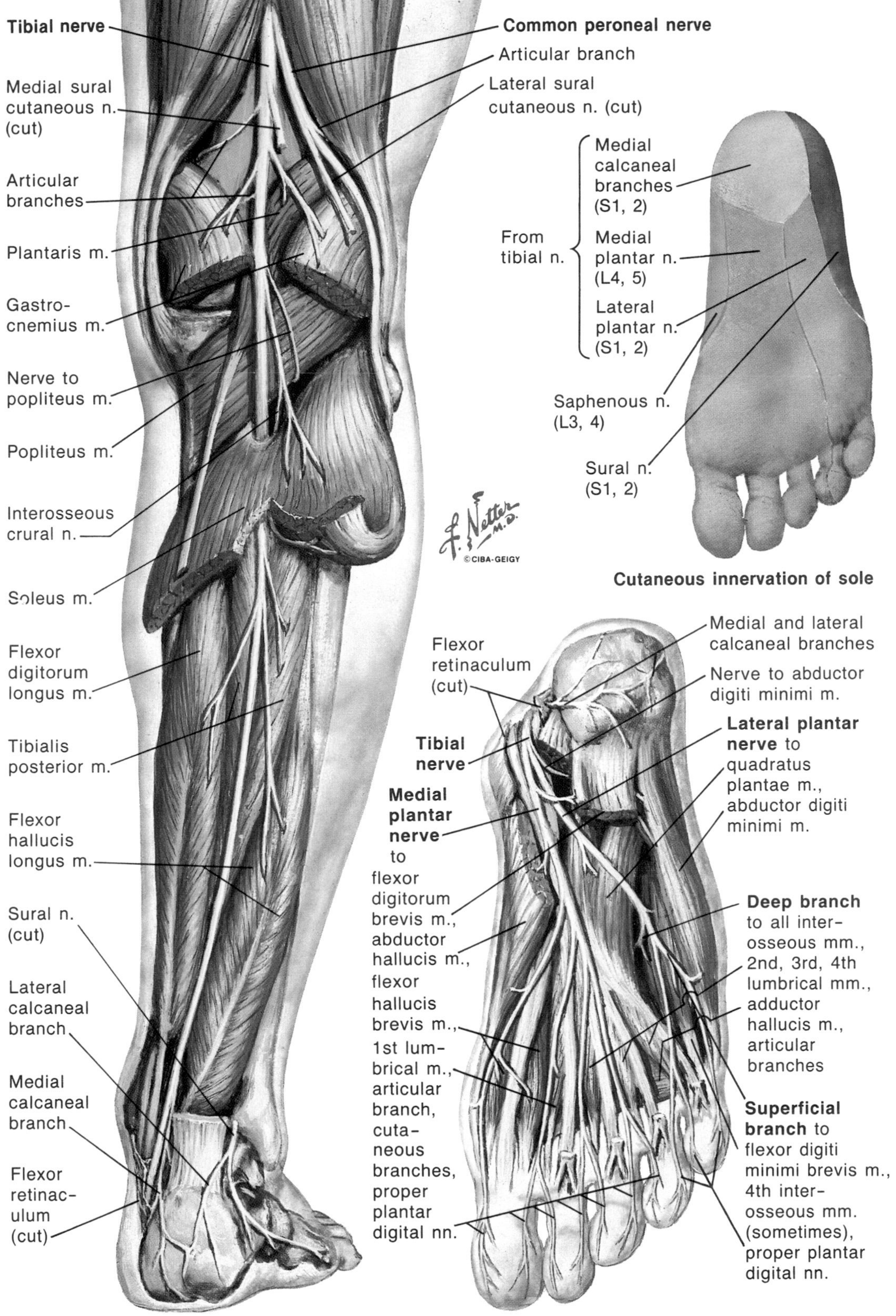
Tibial nerve
Medial sural cutaneous n. (cut)
Articular branches
Plantaris m.
Gastrocnemius m.
Nerve to popliteus m.
Popliteus m.
Interosseous crural n.
Soleus m.
Flexor digitorum longus m.
Tibialis posterior m.
Flexor hallucis longus m.
Sural n. (cut)
Lateral calcaneal branch
Medial calcaneal branch
Flexor retinaculum (cut)
Common peroneal nerve
Articular branch
Lateral sural cutaneous n. (cut)
From tibial n.
Medial calcaneal branches (S1, 2)
Medial plantar n. (L4, 5)
Lateral plantar n. (S1, 2)
Saphenous n. (L3, 4)
Sural n. (S1, 2)
F. Netter M.D.
©CIBA-GEIGY
Cutaneous innervation of sole
Flexor retinaculum (cut)
Tibial nerve
Medial plantar nerve to flexor digitorum brevis m., abductor hallucis m., flexor hallucis brevis m., 1st lumbrical m., articular branch, cutaneous branches, proper plantar digital nn.
Medial and lateral calcaneal branches
Nerve to abductor digiti minimi m.
Lateral plantar nerve to quadratus plantae m., abductor digiti minimi m.
Deep branch to all interosseous mm., 2nd, 3rd, 4th lumbrical mm., adductor hallucis m., articular branches
Superficial branch to flexor digiti minimi brevis m., 4th interosseous mm. (sometimes), proper plantar digital nn.

Tibial Nerve

The *tibial nerve* (L4, 5; S1, 2, 3) is the larger, medial, and terminal branch of the sciatic nerve (Plate 98, see also Plates 73, 77). Its fibers are derived from the anterior divisions of the ventral rami of L4 and L5 and S1, 2, 3.

The tibial nerve continues the line of the sciatic nerve through the popliteal fossa and into the leg. At its origin, the nerve is overlapped by the adjoining margins of the semimembranosus and biceps femoris muscles. In the popliteal fossa, the tibial nerve becomes more superficial, first lying lateral to the popliteal vessels and then crossing obliquely to their medial sides before disappearing into the leg between and beneath the heads of the gastrocnemius and plantaris muscles. Passing over the popliteus muscle and under the tendinous arch of the soleus muscle on the medial side of the posterior tibial vessels, the tibial nerve next enters the space between the gastrocnemius and soleus muscles behind, and the upper part of the tibialis posterior muscle in front. Continuing downward, it crosses over the posterior tibial vessels to reach their lateral sides, so as to lie between the contiguous margins of the flexor digitorum longus and flexor hallucis longus muscles. In the distal third of the leg, the nerve is covered only by skin and fascia as it descends toward the ankle region, where it curves anteroinferiorly into the sole of the foot behind the medial malleolus, deep to the flexor retinaculum and between the tendons of the flexor hallucis longus and the flexor digitorum longus muscles. The nerve ends at this level by dividing into the *medial* and *lateral plantar nerves*.

The tibial nerve consists of muscular, articular, sural, calcaneal, and medial and lateral plantar main branches; it also gives off smaller osseous (medullary) and vascular twigs.

The *muscular branches* supply both heads of the gastrocnemius muscle and the plantaris, popliteus, soleus, tibialis posterior, flexor digitorum longus, and flexor hallucis longus muscles. Branches to the gastrocnemius, plantaris, and popliteus muscles and a few that enter the posterior surface of the soleus muscle arise in the popliteal fossa. The branch to the popliteus muscle descends over the posterior surface of the muscle, hooks around its inferior border, and ascends to enter its anterior surface. Branches to the deep surface of the soleus and to the tibialis posterior, flexor digitorum longus, and flexor hallucis longus muscles are given off in the upper third of the leg. Vasomotor filaments to the popliteal vessels arise from the main tibial nerve or from its branches in the popliteal fossa.

The *articular branches* help to supply the knee, ankle, and superior and inferior tibiofibular joints and may arise in common with twigs supplying adjacent muscles, bones, and vessels. □

Tibia and Fibula

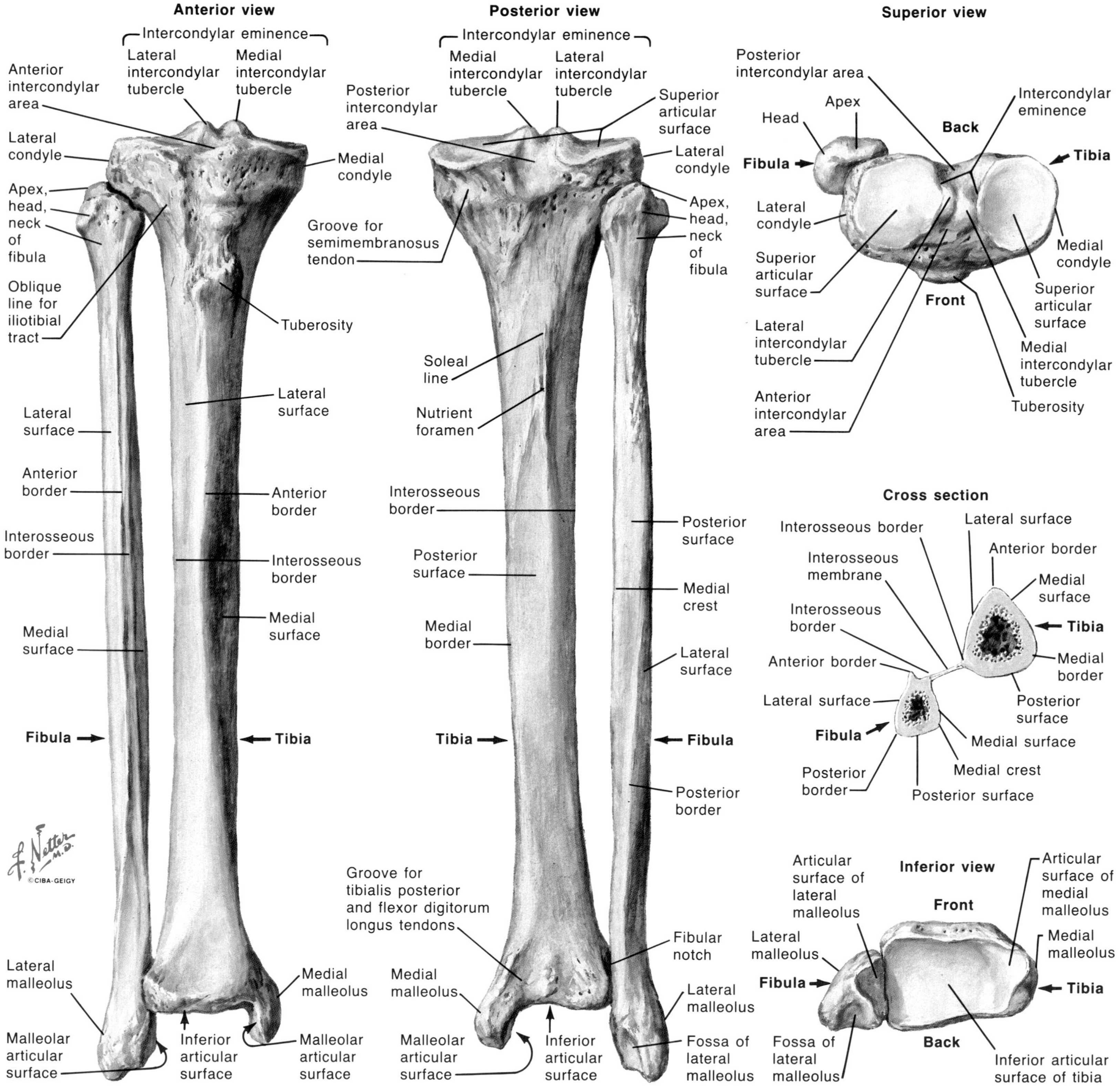

Anterior view
Intercondylar eminence
Lateral intercondylar tubercle
Medial intercondylar tubercle
Anterior intercondylar area
Lateral condyle
Medial condyle
Apex, head, neck of fibula
Oblique line for iliotibial tract
Tuberosity
Lateral surface
Lateral surface
Anterior border
Anterior border
Interosseous border
Interosseous border
Medial surface
Medial surface
Fibula
Tibia
Lateral malleolus
Medial malleolus
Malleolar articular surface
Inferior articular surface
Malleolar articular surface
©CIBA-GEIGY
Posterior view
Intercondylar eminence
Medial intercondylar tubercle
Lateral intercondylar tubercle
Posterior intercondylar area
Superior articular surface
Lateral condyle
Apex, head, neck of fibula
Groove for semimembranosus tendon
Soleal line
Nutrient foramen
Interosseous border
Posterior surface
Posterior surface
Medial crest
Medial border
Lateral surface
Tibia
Fibula
Posterior border
Groove for tibialis posterior and flexor digitorum longus tendons
Fibular notch
Medial malleolus
Lateral malleolus
Malleolar articular surface
Inferior articular surface
Fossa of lateral malleolus
Superior view
Posterior intercondylar area
Apex
Head
Back
Intercondylar eminence
Fibula
Tibia
Lateral condyle
Superior articular surface
Medial condyle
Front
Superior articular surface
Lateral intercondylar tubercle
Medial intercondylar tubercle
Anterior intercondylar area
Tuberosity
Cross section
Interosseous border
Lateral surface
Anterior border
Interosseous membrane
Medial surface
Interosseous border
Tibia
Anterior border
Medial border
Lateral surface
Posterior surface
Fibula
Medial surface
Medial crest
Posterior border
Posterior surface
Inferior view
Articular surface of lateral malleolus
Front
Articular surface of medial malleolus
Lateral malleolus
Medial malleolus
Fibula
Tibia
Fossa of lateral malleolus
Back
Inferior articular surface of tibia

SECTION I PLATE 99 Slide 3664

Lower Limb

Bones of Leg

Bones

Tibia. The tibia is the weight-bearing bone of the leg, the fibula serving for muscular attachments and completing the ankle joint on the lateral side (Plates 99–100). The tibia is a long bone with expanded extremities, especially above, where it is widened to receive the condyles of the femur. Here, there are medial and lateral buttresses, which form the medial and lateral condyles. The *superior articular surface* has two facets. The medial facet is oval and slightly concave. The lateral facet is nearly round. It is concave from side to side, but convex from in front backward. The central parts of these facets receive the condyles of the femur; the rims give attachment to the menisci of the knee joint.

An *intercondylar eminence* is elevated between the two articular facets and is marked by a medial and a lateral intercondylar tubercle. The anterior intercondylar area in front of the eminence provides attachment for the anterior ends of the medial and lateral menisci and for the anterior cruciate ligament. The posterior intercondylar area exhibits a broad groove that lodges the posterior cruciate ligament; in front of this, the area gives attachment to the two menisci. Anteriorly, the two condyles pass over into a triangular area marked by vascular foramina and sloping to the tibial tuberosity below. The triangle is bounded by the *oblique lines*.

The *tibial tuberosity* has a smooth upper portion for the attachment of the patellar ligament; its roughened lower portion is only separated from skin by a subcutaneous infrapatellar bursa. The rough medial surface of the medial condyle gives attachment to the fascia lata; posteriorly, this condyle shows a transverse groove for the insertion of

Lower Limb

Bones of Leg

(Continued)

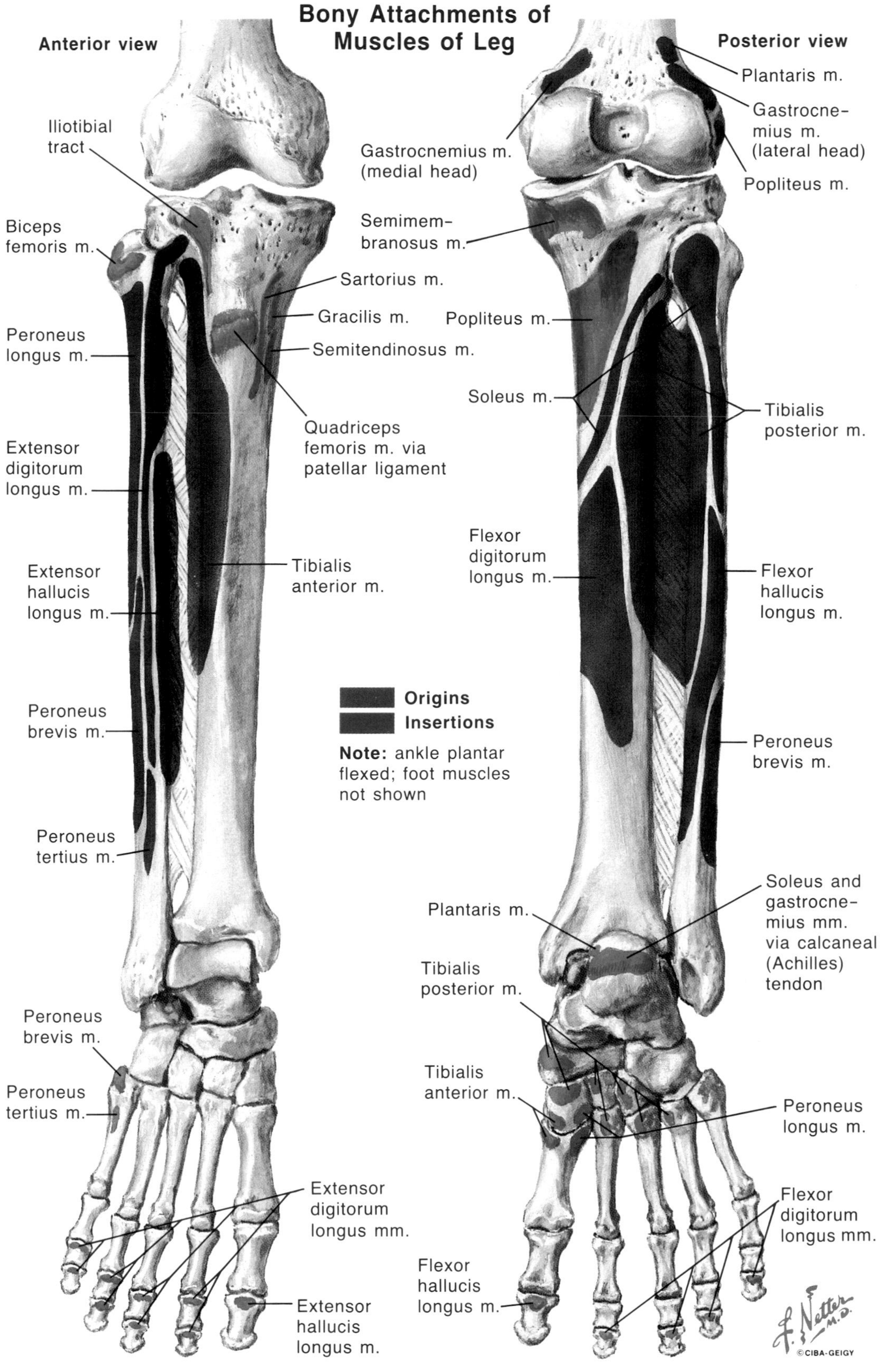
Bony Attachments of Muscles of Leg
Anterior view
Posterior view
Iliotibial tract
Biceps femoris m.
Peroneus longus m.
Extensor digitorum longus m.
Extensor hallucis longus m.
Peroneus brevis m.
Peroneus tertius m.
Peroneus brevis m.
Peroneus tertius m.
Gastrocnemius m. (medial head)
Semimembranosus m.
Sartorius m.
Gracilis m.
Semitendinosus m.
Quadriceps femoris m. via patellar ligament
Tibialis anterior m.
Extensor digitorum longus mm.
Extensor hallucis longus m.
Plantaris m.
Gastrocnemius m. (lateral head)
Popliteus m.
Popliteus m.
Soleus m.
Tibialis posterior m.
Flexor digitorum longus m.
Flexor hallucis longus m.
Origins
Insertions
Note: ankle plantar flexed; foot muscles not shown
Peroneus brevis m.
Soleus and gastrocnemius mm. via calcaneal (Achilles) tendon
Plantaris m.
Tibialis posterior m.
Tibialis anterior m.
Peroneus longus m.
Flexor digitorum longus mm.
Flexor hallucis longus m.
F. Netter M.D.
© CIBA-GEIGY

the tendon of the semimembranosus muscle. On its posteroinferior surface the *lateral condyle* has a nearly circular facet that articulates with the head of the fibula. Anterior to this facet, an oblique line provides attachment for the iliotibial tract.

The *shaft* of the tibia is fairly uniform in size. It is triangular in cross section and has medial, lateral, and posterior surfaces and anterior, medial, and interosseous borders. The anterior border, slightly sinuous, begins at the lateral margin of the tuberosity above and ends on the medial malleolus below. It is subcutaneous and prominent and is sharpest in its middle third. The medial border, sharper in its lower half, extends from the posterior aspect of the medial condyle to the posterior border of the medial malleolus. The interosseous border is on the fibular side of the bone and is sharp throughout its length. Just above the ankle, the interosseous border bifurcates and encloses a triangular area for the attachment of the ligamentous tissue representing the tibiofibular syndesmosis. The medial surface of the shaft of the bone is smooth; it provides for the insertion of certain thigh muscles, such as the pes anserinus muscle, in its upper third. The rest of its surface is subcutaneous. The lateral surface of the shaft is hollowed in its upper two-thirds for the origin of the tibialis anterior muscle. Its lower third is smooth; it spirals anteriorly and is covered by tendons of the muscles of the anterior compartment of the leg.

The *soleal line* is a prominent marking of the posterior surface. It begins behind the facet for the head of the fibula and runs obliquely downward to the medial border of the bone at the junction of the upper and middle thirds of the shaft. The triangular area above the soleal line gives insertion for the popliteus muscle. The soleal line gives origin for the soleus muscle. The nutrient foramen of the tibia is below this line. Also below the soleal line arise the flexor digitorum longus and tibialis posterior muscles, separated by a sometimes distinct vertical ridge of bone.

The *inferior extremity* of the tibia projects medialward and downward as the *medial malleolus*; this forms a subcutaneous prominence at the ankle. The malleolus is grooved posteriorly for the tendons of the tibialis posterior and flexor digitorum longus muscles; there may be a groove more laterally for the tendon of the flexor hallucis longus muscle. The lateral surface of the inferior extremity forms a triangular *fibular notch*, roughened by the ligamentous tissue uniting the bones. The borders of the notch are sharp for the attachments of the anterior tibiofibular and posterior tibiofibular ligaments. The distal end forms a quadrilateral *inferior articular surface* for articulation with the body of the talus. This surface is wider anteriorly than posteriorly and concave anteroposteriorly. It is continuous with the *malleolar articular surface* on the internal aspect of the medial malleolus. The malleolar articular surface lies almost at right angles to the inferior articular surface of the shaft and extends about 1.5 cm beyond it. From the lower edge of the medial malleolus, the deltoid ligament passes to the bones of the foot.

The tibia is *ossified* from three centers, one for the body and one for each extremity. They appear in the seventh week of intrauterine life for the body; in the upper epiphysis, shortly before birth; and in the lower end, between ages 1 and 2. The lower epiphysis joins the body of the bone at about age 16½ in the male and age 14½ in the female. The upper epiphysis joins at approximately age 17½ in the male and at age 15 in the female.

Fibula. The fibula lies parallel to the tibia and is long and slender (Plates 99–100). It is not weight bearing but gives attachment to muscles and aids in forming the ankle joint. It has a slender body and two somewhat expanded extremities.

Lower Limb

Bones of Leg

(Continued)

The *head* of the fibula is knoblike and bears on its slanted superior aspect the almost circular articular surface of the head. The apex of the head projects upward at the posterolateral limit of the articulation; it gives attachment to the fibular collateral ligament of the knee joint and, on its lateral aspect, to the tendon of the biceps femoris muscle. Upper fibers of the tibialis anterior and soleus muscles also arise from the head of the fibula. Like the tibia, the shaft of the fibula has anterior, interosseous, and posterior borders and medial, lateral, and posterior surfaces.

The sharp *anterior border* begins just below the head and distally, immediately above the lateral malleolus, it divides to enclose a triangular subcutaneous surface. This border gives rise to the anterior intermuscular septum. The *interosseous border* is close to the anterior border. It begins in front of the head above, lies anteromedially along the shaft, and divides below into the borders of the tibiofibular syndesmosis. The interosseous border of the fibula is joined in the lower third of the shaft by the *medial crest*. The nutrient foramen of the fibula is located posterior to the distal end of the medial crest. A relatively indistinct posterior border gives attachment to the posterior intermuscular septum. The surfaces of the fibula give origin to the muscles of the adjacent compartments of the leg (Plate 100).

The *lateral malleolus* is the pointed distal extremity of the fibula. Its tip descends about 1.5 cm below the tip of the medial malleolus of the tibia. Its lateral surface, convex and subcutaneous, is continuous with the subcutaneous area of the shaft. The medial surface of the malleolus consists of the triangular malleolar articular surface and the fossa of the lateral malleolus. The *malleolar articular surface*, angulated outward below, makes contact with the lateral side of the talus. The *fossa of the lateral malleolus* provides the area of attachment of the transverse tibiofibular and posterior talofibular ligaments. The borders of the malleolus give attachment to ligaments of the ankle joint, and the posterior border is grooved for the tendons of the peroneus longus and peroneus brevis muscles.

The fibula has *three centers of ossification*. One center for the shaft appears in its middle during the eighth week of uterine life; the ends of the bone are still cartilaginous at birth. Ossification in the lower extremity begins at about the end of the first year; in the upper extremity, it begins in the fourth year in the male and early in the third year in the female. Fusion of the epiphysis of the lower extremity with the shaft takes place at approximately age 14½ in the female and at age 16½ in the male. Fusion of the upper extremity epiphysis takes place between ages 17 and 18 in the male and at age 15½ in the female. Individual differences are common.

Joints Between Tibia and Fibula

The functional union of the tibia and the fibula is in three parts: the tibiofibular articulation proximally, the interosseous membrane, and the distal tibiofibular syndesmosis (Plates 89–90, 99–100, 102).

The *tibiofibular articulation* is a plane joint between the circular facet on the head of the fibula and a matching surface on the underside of the lateral condyle of the tibia. The articular capsule, attached at the margins of the tibial and fibular facets, is strengthened by accessory ligaments. The *anterior ligament of the head of the fibula* consists of fibrous bands, which pass obliquely from the front of the head of the fibula to the front of the lateral condyle of the tibia. The *posterior ligament of the head of the fibula* is a single broad band running obliquely between the head and the back of the lateral tibial condyle. It is crossed by the tendon of the popliteus muscle, and the subpopliteal recess of the knee joint cavity is sometimes in communication with the cavity of this joint.

The tibiofibular articulation receives arteries from the lateral inferior genicular and anterior tibial recurrent arteries, and its lymphatics drain to the popliteal lymph nodes. Nerves to the joint come from the common peroneal nerve, the nerve to the popliteus muscle, and the recurrent articular nerve. Movement is slight at this joint, but it imparts a certain flexibility in the relations of the two bones during ankle movement and in response to the action of the muscles attached to the fibula.

The *interosseous membrane* of the leg extends between the interosseous borders of the two bones. It consists largely of fibers that pass from the tibia lateralward and downward to the fibula. The upper margin does not reach the tibiofibular articulation, and the anterior tibial vessels pass across the upper edge of the membrane to the anterior compartment of the leg. The membrane is continuous below with the interosseous ligament of the tibiofibular syndesmosis and is perforated by a number of vessels, especially the perforating branch of the peroneal artery. It separates the anterior and posterior compartments of the leg and gives origin to muscles of both groups.

The *tibiofibular syndesmosis* is a fibrous joint in which the rough convex surface of the medial aspect of the lower end of the fibula is attached through the interosseous ligament to the corresponding rough, concave surface on the lateral side of the tibia. The *interosseous ligament*, continuous with the interosseous membrane above, consists of short, strong, fibrous bands uniting the two bones. The *anterior tibiofibular ligament* and the *posterior tibiofibular ligament* pass, respectively, from the borders of the fibular notch of the tibia to the anterior and posterior surfaces of the lateral malleolus of the fibula. Each is inclined downward and laterally. The *transverse tibiofibular ligament* is largely deep to the posterior tibiofibular ligament. It arises from nearly the whole inferior border of the tibia posteriorly and attaches to the upper portion of the malleolar fossa of the fibula. This is a thick and strong ligament, and it projects below the bony margin to form part of the articulating fossa for the talus. A recess of the articular cavity of the ankle joint and its enclosing synovial membrane extend upward between the tibia and fibula to the lower end of the interosseous ligament.

The blood supply of the tibiofibular syndesmosis is principally from the perforating branch of the peroneal artery and the malleolar arteries; its nerve supply is from the deep peroneal and tibial nerves. This articulation forms the firm union of the tibia and fibula required in the boxlike mortise of the ankle. Its fibrous tissue and ligaments allow a slight yielding of the bones for the accommodation of the talus in the movements of the ankle. □

Lower Limb

Ankle and Foot

Synovial Tendon Sheaths at Ankle

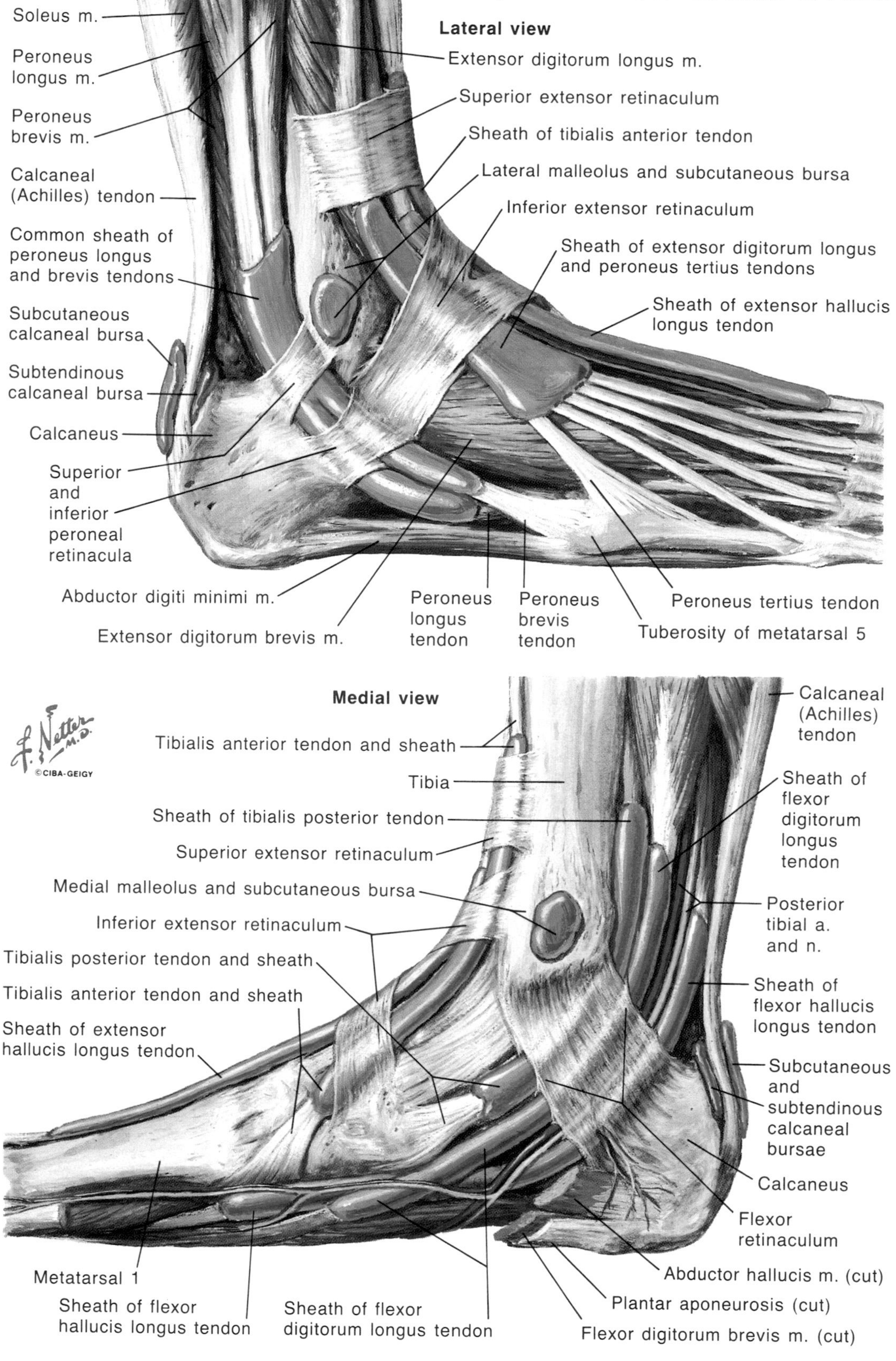
Lateral view
Soleus m.
Peroneus longus m.
Peroneus brevis m.
Calcaneal (Achilles) tendon
Common sheath of peroneus longus and brevis tendons
Subcutaneous calcaneal bursa
Subtendinous calcaneal bursa
Calcaneus
Superior and inferior peroneal retinacula
Abductor digiti minimi m.
Extensor digitorum brevis m.
Peroneus longus tendon
Peroneus brevis tendon
Tuberosity of metatarsal 5
Peroneus tertius tendon
Sheath of extensor hallucis longus tendon
Sheath of extensor digitorum longus and peroneus tertius tendons
Inferior extensor retinaculum
Lateral malleolus and subcutaneous bursa
Sheath of tibialis anterior tendon
Superior extensor retinaculum
Extensor digitorum longus m.
Medial view
Tibialis anterior tendon and sheath
Tibia
Sheath of tibialis posterior tendon
Superior extensor retinaculum
Medial malleolus and subcutaneous bursa
Inferior extensor retinaculum
Tibialis posterior tendon and sheath
Tibialis anterior tendon and sheath
Sheath of extensor hallucis longus tendon
Metatarsal 1
Sheath of flexor hallucis longus tendon
Sheath of flexor digitorum longus tendon
Flexor digitorum brevis m. (cut)
Plantar aponeurosis (cut)
Abductor hallucis m. (cut)
Flexor retinaculum
Calcaneus
Subcutaneous and subtendinous calcaneal bursae
Sheath of flexor hallucis longus tendon
Posterior tibial a. and n.
Sheath of flexor digitorum longus tendon
Calcaneal (Achilles) tendon
F. Netter M.D.
©CIBA-GEIGY

Tendon Sheaths at Ankle

The shift from the vertical organization of the leg to the horizontal orientation of the foot entails the turning forward of all tendons, vessels, and nerves that enter the foot. Provision is made by the various retinacula for holding such structures close to the bones at the ankle and for preventing bowstringing by the tendons (Plates 99, 101–103).

The *superior extensor retinaculum* is a reinforcement of the crural fascia just above the ankle. It is attached laterally to the lower end of the fibula and medially to the tibia, and it covers the structure of the anterior compartment of the leg. A strong septum runs from its deep surface to the tibia, separating a medial compartment for the tendon of the tibialis anterior muscle from a lateral compartment for the tendons of the long extensor muscles.

The *inferior extensor retinaculum* is a well-defined, Y-shaped band overlying the dorsum of the foot and the front of the ankle. The stem of the Y arises from the upper surface of the calcaneus and is in the form of two laminae, one superficial and one deep to the tendons of the peroneus tertius and extensor digitorum longus muscles. At the medial border of the latter tendon, the two laminae merge, and the limbs of the Y begin to diverge. One limb is directed upward and medialward to attach to the medial malleolus. It passes over the tendon of the extensor hallucis longus muscle, the dorsalis pedis vessels, and the deep peroneal nerve, but splits to form a separate canal for the tendon of the tibialis anterior muscle. The lower limb of the Y passes medialward across the medial border of the foot and is lost in the deep fascia of the sole.

The *flexor retinaculum* stretches from the medial malleolus to the medial tubercle of the calcaneus. From its deep surface, septa pass to the back of the lower end of the tibia and the capsule of the ankle joint. The four canals defined by these septa transmit, beginning medially, the tendon of the tibialis posterior muscle, that of the flexor digitorum longus muscle, the posterior tibial vessels and the tibial nerve, and the tendon of the flexor hallucis longus muscle. The upper border of the flexor retinaculum is continuous with the transverse intermuscular septum. Its lower border is continuous with the deep fascia of the sole and gives origin to the fibers of the abductor hallucis muscle.

The peroneal retinacula are thickenings of the fascia on the lateral side of the ankle. The *superior peroneal retinaculum* extends from the lateral malleolus into the fascia of the back of the leg and to the lateral surface of the calcaneus. The *inferior peroneal retinaculum* is a thickening of fascia, both ends of which attach to the lateral surface of the calcaneus. It is continuous superiorly with the stem of the Y of the inferior extensor retinaculum. Deep to the peroneal retinacula pass the tendons of the peroneus longus and peroneus brevis muscles; the peroneus brevis tendon is the anterior of the two behind the medial malleolus and superior to the tendon of the peroneus longus muscle beneath the inferior peroneal retinaculum.

Ankle Joint

This talocrural articulation is a synovial joint of the hinge (ginglymus) type (Plates 99 and 102). Its form is that of a mortise and tenon, the boxlike mortise being formed by the ends of the leg bones. The cartilage-covered articular areas of the end of the tibia, the lateral surface of the medial malleolus, and the triangular facet of the medial surface of the lateral malleolus form the mortise for the trochlea of the body of the talus, which is the tenon. The mortise is deepened behind by the transverse tibiofibular ligament.

The trochlea of the talus is convex from before backward and slightly concave from side to side. Medially, it is straight anteroposteriorly, but its lateral margin is oblique; thus, the trochlea is broader in front than behind. A small articular surface on the anteromedial surface of the trochlea articulates with the medial malleolus. The lateral side of the trochlea is wholly articular; it is triangular in shape and articulates with the lateral malleolus.

Lower Limb

Ankle and Foot

(Continued)

The *articular capsule*, conforming to the requirements of free movement in flexion and extension at the ankle, is weak anteriorly and posteriorly. However, the joint has exceedingly strong collateral ligaments. The thin anterior and posterior parts of the capsule are attached above to the margins of the tibia and fibula and below to the talus both in front and behind the superior surface of its trochlea. The articular capsules at the sides blend with the deltoid ligament on the medial side of the ankle and with the anterior and posterior talofibular ligaments on the lateral side. The *deltoid ligament* is a strong triangular ligament, attached at its anterior and posterior borders and the tip of the medial malleolus. The ligament broadens inferiorly to form a continuous attachment to the bones of the foot; its four parts are designated by their separate distal attachments. The most anterior fibers compose the *anterior tibiotalar ligament*. These are adjacent to, and partly overlain by, the superficial *tibionavicular ligament* to the upper and medial part of the navicular. Below, this ligament blends with the medial margin of the plantar calcaneonavicular ligament. Next, the fibers of the *tibiocalcaneal ligament* descend almost vertically to the whole length of the sustentaculum tali of the calcaneus. The posterior and thickest part of the deltoid is the *posterior tibiotalar ligament*; its fibers run lateralward and backward to the medial side of the talus and to the medial tubercle of its posterior process.

The lateral collateral ligament is made up of three separate bands that do not constitute so strong a ligamentous investment as does the deltoid ligament medially. The *anterior talofibular ligament* passes from the anterior border and tip of the lateral malleolus to the neck of the talus. The *calcaneofibular ligament* is a narrow, rounded cord that descends from the tip of the lateral malleolus to a tubercle at the middle of the lateral surface of the calcaneus. The almost horizontal *posterior talofibular ligament* is strong and thick. It arises in the malleolar fossa of the lateral malleolus and passes medialward and backward to the upper surface of the posterior process of the talus.

The *synovial membrane* of the ankle joint is loose and capacious; the synovial cavity extends upward, between the apposed surfaces of the ends of the tibia and fibula, as far as the interosseous ligament of the tibiofibular syndesmosis. The ankle joint receives its *blood supply* from the four malleolar branches of the anterior tibial, posterior tibial, and peroneal arteries. Its *nerve supply* is provided by twigs from the tibial nerve and the lateral branch of the deep peroneal nerve.

Movements

Primarily dorsiflexion and plantar flexion through a range of 90° are movements permitted at the ankle. In dorsiflexion, the broader anterior portion of the trochlea occupies and completely fills the mortise of the joint, and stability of the foot is greatest in this position. This is the close-

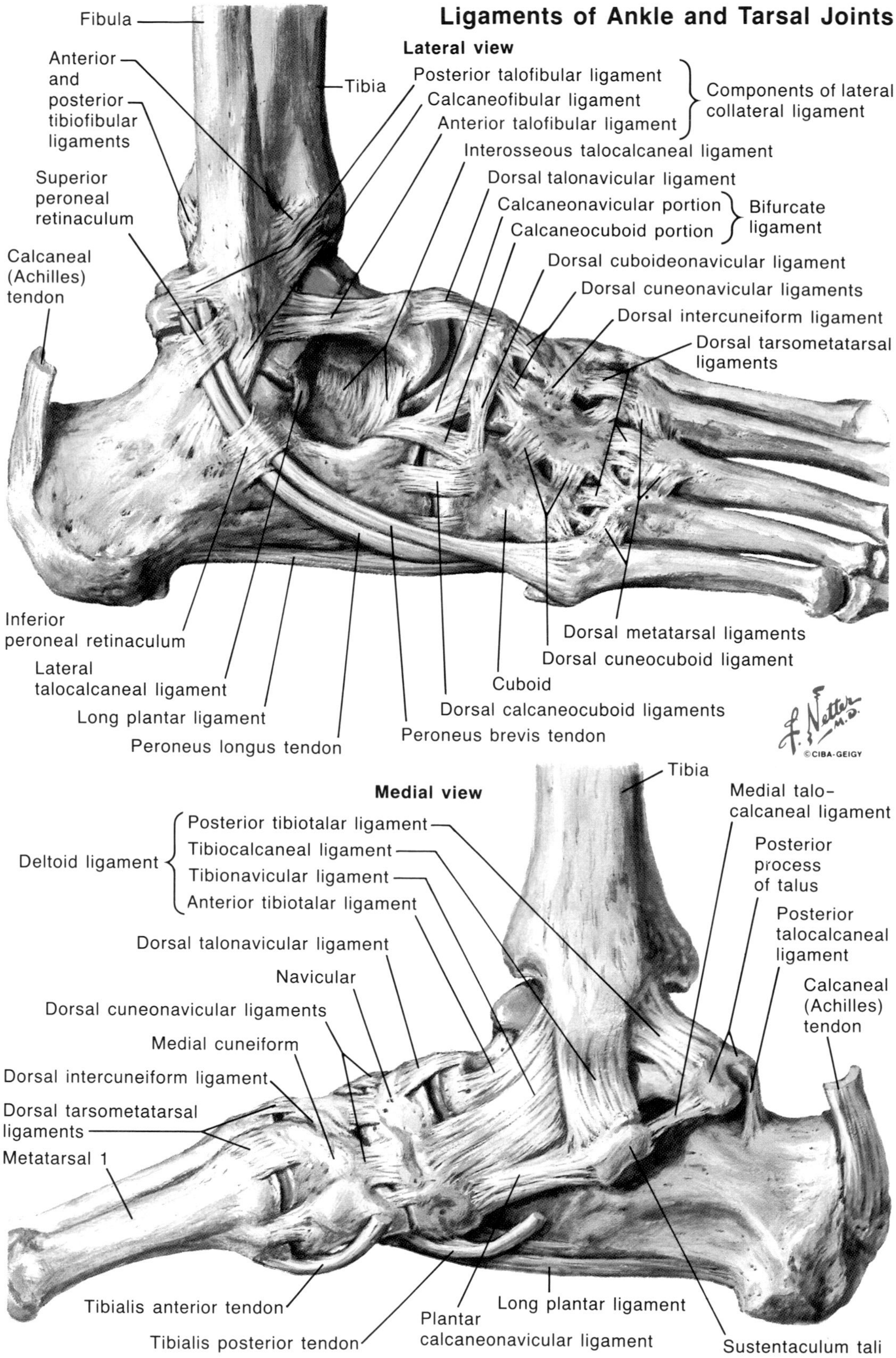
Ligaments of Ankle and Tarsal Joints
Lateral view
Fibula
Anterior and posterior tibiofibular ligaments
Tibia
Superior peroneal retinaculum
Calcaneal (Achilles) tendon
Posterior talofibular ligament
Calcaneofibular ligament
Anterior talofibular ligament
Components of lateral collateral ligament
Interosseous talocalcaneal ligament
Dorsal talonavicular ligament
Calcaneonavicular portion
Calcaneocuboid portion
Bifurcate ligament
Dorsal cuboideonavicular ligament
Dorsal cuneonavicular ligaments
Dorsal intercuneiform ligament
Dorsal tarsometatarsal ligaments
Inferior peroneal retinaculum
Lateral talocalcaneal ligament
Long plantar ligament
Peroneus longus tendon
Peroneus brevis tendon
Dorsal calcaneocuboid ligaments
Cuboid
Dorsal cuneocuboid ligament
Dorsal metatarsal ligaments
F. Netter M.D.
© CIBA-GEIGY
Medial view
Tibia
Deltoid ligament
Posterior tibiotalar ligament
Tibiocalcaneal ligament
Tibionavicular ligament
Anterior tibiotalar ligament
Medial talocalcaneal ligament
Posterior process of talus
Posterior talocalcaneal ligament
Calcaneal (Achilles) tendon
Dorsal talonavicular ligament
Navicular
Dorsal cuneonavicular ligaments
Medial cuneiform
Dorsal intercuneiform ligament
Dorsal tarsometatarsal ligaments
Metatarsal 1
Tibialis anterior tendon
Tibialis posterior tendon
Plantar calcaneonavicular ligament
Long plantar ligament
Sustentaculum tali

packed position of the joint, with maximal congruence of articular surfaces and maximal tension of its ligaments. It is the position from which all thrusting movements of high activity develop. Conversely, in full plantar flexion, the narrowest part of the trochlea engages in the mortise, and stability is markedly decreased; small amounts of side-to-side gliding movements, rotation, and abduction-adduction are permitted. The muscles entering the foot behind the malleoli—those of the lateral and both posterior compartments of the leg—produce plantar flexion at the ankle. Dorsiflexion follows contraction of the muscles of the anterior compartment of the leg. In the erect position, the line of gravity of the body passes in front of the transverse axis of the ankle joint and, as a result, the body tends to fall forward at this joint. In a normally strong foot, the ligaments take most of the strain of the erect posture, but some contraction of the soleus muscle (and perhaps of the gastrocnemius as well) may be needed for easy standing and will certainly come into play with activity. The tendons of the deep posterior and lateral compartments of the leg turn sharply forward at the ankle, and part of their traction is such as to draw the foot backward under the ankle. This posteriorly displacing force on the foot is resisted by the collateral ligaments at the ankle, the strongest of them running posteriorly between the malleoli and the tarsals.

Lower Limb

Ankle and Foot

(Continued)

Muscles, Arteries, and Nerves of Front of Ankle and Dorsum of Foot: Superficial Dissection

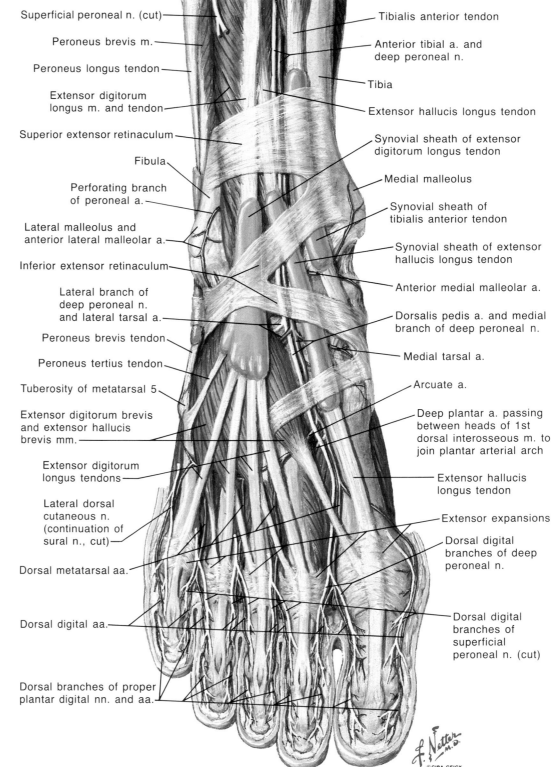

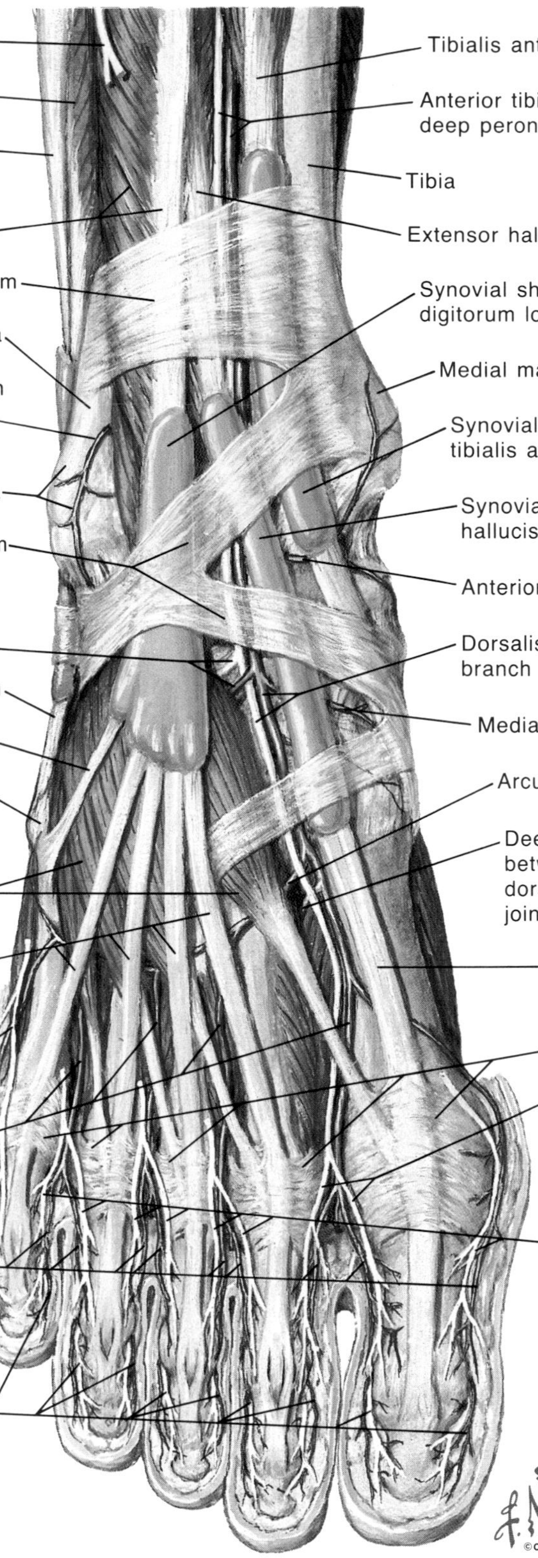
Superficial peroneal n. (cut)
Peroneus brevis m.
Peroneus longus tendon
Extensor digitorum longus m. and tendon
Superior extensor retinaculum
Fibula
Perforating branch of peroneal a.
Lateral malleolus and anterior lateral malleolar a.
Inferior extensor retinaculum
Lateral branch of deep peroneal n. and lateral tarsal a.
Peroneus brevis tendon
Peroneus tertius tendon
Tuberosity of metatarsal 5
Extensor digitorum brevis and extensor hallucis brevis mm.
Extensor digitorum longus tendons
Lateral dorsal cutaneous n. (continuation of sural n., cut)
Dorsal metatarsal aa.
Dorsal digital aa.
Dorsal branches of proper plantar digital nn. and aa.
Tibialis anterior tendon
Anterior tibial a. and deep peroneal n.
Tibia
Extensor hallucis longus tendon
Synovial sheath of extensor digitorum longus tendon
Medial malleolus
Synovial sheath of tibialis anterior tendon
Synovial sheath of extensor hallucis longus tendon
Anterior medial malleolar a.
Dorsalis pedis a. and medial branch of deep peroneal n.
Medial tarsal a.
Arcuate a.
Deep plantar a. passing between heads of 1st dorsal interosseous m. to join plantar arterial arch
Extensor hallucis longus tendon
Extensor expansions
Dorsal digital branches of deep peroneal n.
Dorsal digital branches of superficial peroneal n. (cut)
©CIBA-GEIGY

Dorsum of Foot

The structures of the dorsum of the foot are continuations of the anterior compartment of the leg (Plates 103–104). The skin here is thin, and there is relatively little subcutaneous fat. The deep fascia is thin; it is continuous with the extensor retinacula and curves over the margins of the foot to become the fascia of the sole. Anteriorly, the fascia of the dorsum encloses the extensor tendons. There is one muscle on the dorsum of the foot. It underlies the tendons of the extensor digitorum longus muscle and largely covers the dorsalis pedis artery and its branches and the deep peroneal nerve.

Muscles. The *extensor digitorum brevis muscle* is broad and thin. It arises from the distal part of the superior and lateral surfaces of the calcaneus and the stem of the inferior extensor retinaculum. The muscle divides into four tendons for the medial four toes. The largest and most medial tendon, together with its belly, is often separately designated as the *extensor hallucis brevis muscle.* It inserts into the base of the first phalanx of the great toe. The other three tendons join the lateral sides of the tendons of the extensor digitorum longus muscle to the second, third, and fourth toes and assist in forming the extensor expansions on these digits. The muscle assists the long extensor muscle in extending the proximal phalanges of the medial four toes.

Arteries. The *dorsalis pedis artery* is the continuation of the anterior tibial artery at the ankle joint; it is directed forward across the dorsum of the foot to the proximal end of the first metatarsal space. Here, it divides into the deep plantar and first dorsal metatarsal arteries. The dorsalis pedis artery lies against the bones and ligaments of the dorsum, with the medial branch of the deep peroneal nerve lateral to it. It is accompanied by two venae comitantes. Its branches are the lateral tarsal, medial tarsal, arcuate, first dorsal metatarsal, and deep plantar arteries.

The *lateral tarsal artery* arises over the navicular bone and passes lateralward and distalward. It supplies the extensor digitorum brevis muscle and the tarsal articulations and anastomoses with branches of the arcuate, anterior lateral malleolar, and the perforating branch of the peroneal arteries. The dorsalis pedis artery is absent or greatly reduced in about 5% of cases, and this lateral

Lower Limb

Ankle and Foot

(Continued)

Muscles, Arteries, and Nerves of Front of Ankle and Dorsum of Foot: Deeper Dissection

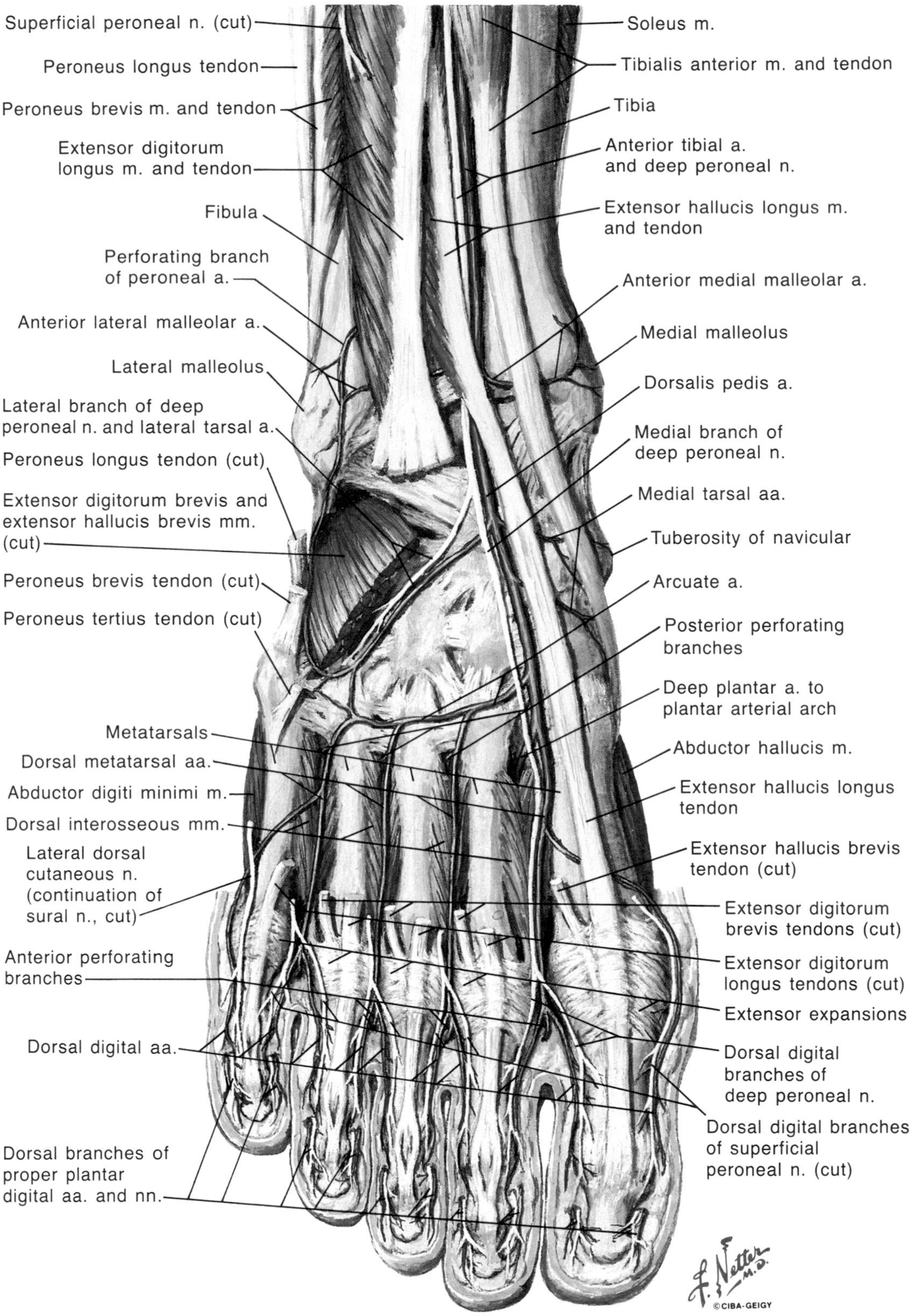
Superficial peroneal n. (cut)
Peroneus longus tendon
Peroneus brevis m. and tendon
Extensor digitorum longus m. and tendon
Fibula
Perforating branch of peroneal a.
Anterior lateral malleolar a.
Lateral malleolus
Lateral branch of deep peroneal n. and lateral tarsal a.
Peroneus longus tendon (cut)
Extensor digitorum brevis and extensor hallucis brevis mm. (cut)
Peroneus brevis tendon (cut)
Peroneus tertius tendon (cut)
Metatarsals
Dorsal metatarsal aa.
Abductor digiti minimi m.
Dorsal interosseous mm.
Lateral dorsal cutaneous n. (continuation of sural n., cut)
Anterior perforating branches
Dorsal digital aa.
Dorsal branches of proper plantar digital aa. and nn.
Soleus m.
Tibialis anterior m. and tendon
Tibia
Anterior tibial a. and deep peroneal n.
Extensor hallucis longus m. and tendon
Anterior medial malleolar a.
Medial malleolus
Dorsalis pedis a.
Medial branch of deep peroneal n.
Medial tarsal aa.
Tuberosity of navicular
Arcuate a.
Posterior perforating branches
Deep plantar a. to plantar arterial arch
Abductor hallucis m.
Extensor hallucis longus tendon
Extensor hallucis brevis tendon (cut)
Extensor digitorum brevis tendons (cut)
Extensor digitorum longus tendons (cut)
Extensor expansions
Dorsal digital branches of deep peroneal n.
Dorsal digital branches of superficial peroneal n. (cut)
F. Netter M.D.
©CIBA-GEIGY

anastomosis is greatly enlarged, taking over laterally the main supply that usually comes down medially on the dorsum of the foot. In such variations, the peroneal artery is the principal source of blood supply. The *medial tarsal arteries* are two or three small branches that ramify over the medial border of the foot, anastomosing with the medial malleolar arteries. The *arcuate artery* arises at the level of the bases of the metatarsals and runs lateralward across the proximal ends of these bones beneath the extensor tendons. Lateralward, it ends by anastomosing with the lateral tarsal and lateral plantar arteries. Three *dorsal metatarsal arteries* arise from the arcuate artery and pass distalward over the dorsal interosseous muscle to the clefts of the toes. Here, each divides into two *dorsal digital arteries* for the adjacent sides of the toes on either side of the cleft. Like the dorsal digital arteries of the fingers, they do not reach the distal phalanx. The fourth dorsal metatarsal artery has an additional dorsal digital branch for the lateral side of the small toe.

The metatarsal arteries have posterior and anterior perforating arteries proximally and distally in the interosseous space, which perforate to anastomose with corresponding plantar metatarsal arteries. The *first dorsal metatarsal artery* is like the other dorsal metatarsal arteries in its course and division into two dorsal digital arteries to the adjacent sides of the first and second toes. It also gives off an extra dorsal digital artery for the medial side of the great toe. The *deep plantar artery* is the much-enlarged posterior perforating branch of the first dorsal metatarsal artery. Passing proximally between the two heads of origin of the first dorsal interosseous muscle of the sole of the foot, it unites with the lateral plantar artery to form the plantar arterial arch (Plate 104). It behaves like the radial artery in the hand as that artery completes the deep palmar arterial arch.

Nerves. The *deep peroneal nerve* (S1, 2) divides into its medial and lateral terminal branches at the lower border of the inferior extensor retinaculum. As noted, the lateral branch supplies the extensor digitorum brevis muscle and the tarsal joints. The medial branch passes distalward lateral to the dorsalis pedis artery. It divides into two dorsal digital nerves to the adjacent sides of the great and second toes. Other small twigs are supplied to the metatarsophalangeal and interphalangeal articulations of the great toe, and one supplies the first dorsal interosseous muscle.

Muscles, Arteries, and Nerves of Sole of Foot

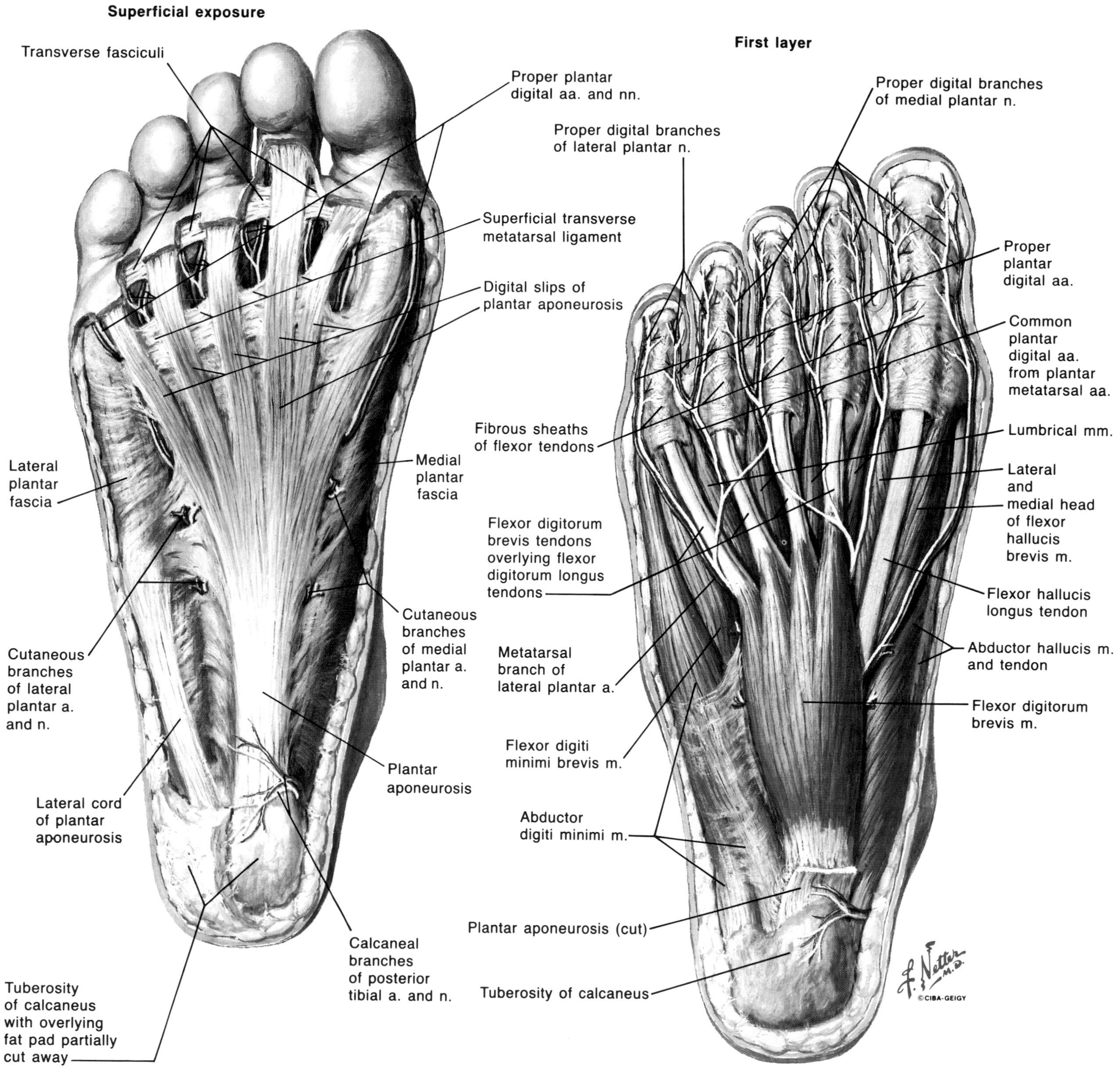

Superficial exposure
Transverse fasciculi
Proper plantar digital aa. and nn.
Superficial transverse metatarsal ligament
Digital slips of plantar aponeurosis
Lateral plantar fascia
Medial plantar fascia
Cutaneous branches of lateral plantar a. and n.
Cutaneous branches of medial plantar a. and n.
Lateral cord of plantar aponeurosis
Plantar aponeurosis
Calcaneal branches of posterior tibial a. and n.
Tuberosity of calcaneus with overlying fat pad partially cut away
First layer
Proper digital branches of medial plantar n.
Proper digital branches of lateral plantar n.
Proper plantar digital aa.
Common plantar digital aa. from plantar metatarsal aa.
Lumbrical mm.
Fibrous sheaths of flexor tendons
Lateral and medial head of flexor hallucis brevis m.
Flexor digitorum brevis tendons overlying flexor digitorum longus tendons
Flexor hallucis longus tendon
Abductor hallucis m. and tendon
Metatarsal branch of lateral plantar a.
Flexor digitorum brevis m.
Flexor digiti minimi brevis m.
Abductor digiti minimi m.
Plantar aponeurosis (cut)
Tuberosity of calcaneus
F. Netter M.D.
©CIBA-GEIGY

SECTION I PLATE 105 Slide 3670

Lower Limb

Ankle and Foot

(Continued)

Sole of Foot

The structures of the plantar region are shown in Plates 105–107. The *skin* of the foot is thin on the toes and instep. It is thickened over the heel and the heads of the metatarsals in response to friction and weight bearing. There is much fat in the subcutaneous tissue of the sole and plantar aspects of the toes. Intermingled with fibrous connective tissue, the firmly supported fat over the sole and heel forms a cushioning pad for the weight-bearing parts of the foot. The *plantar fascia* is continuous with the deep fascia of the dorsum of the foot after attachments to the periosteum of the sides of the first and fifth metatarsals. Thinner membranous sheets medially and laterally enclose the compartments of the great and little toes, and a thickened plantar aponeurosis covers the central compartment. This *plantar aponeurosis* (comparable to the palmar aponeurosis in the hand) consists of longitudinally arranged bands of fibrous connective tissue, which diverge toward the toes from the medial process of the tuberosity of the calcaneus. Five digital slips pass to the plantar surface of the toes. Deeper-lying transverse fibers form a reinforcing band, the *superficial transverse metatarsal ligament*, over the heads of the metatarsals. *Transverse fasciculi* reinforce the webs of the toes.

Marginal fibers of the digital slips pass deeply to blend with the proximal ends of the fibrous sheaths of the flexor tendons and attach to the deep transverse metatarsal and plantar ligaments. The superficial central fibers of the digital slips end largely in the skin of the flexion creases between the toes and the sole. At the lateral and medial margins of the plantar aponeurosis, fibers

Lower Limb

Ankle and Foot

(Continued)

radiate onto the thinner membranous fascia of the side compartments, and deep intermuscular septa penetrate the soft parts of the sole of the foot and separate the central compartment from the others. These septa reach the plantar interosseous fascia and the bones and ligaments deep in the foot. The *lateral plantar fascia* is thick and well developed near the heel and thinner toward the small toe. A *calcaneometatarsal ligament* extends within it from the lateral process of the tuberosity of the calcaneus to the tuberosity of the fifth metatarsal. The thinner *medial plantar fascia* covers the intrinsic muscles of the great toe.

Compartments

The compartments of the sole of the foot are similar to those of the palm of the hand (Plates 105–107). There is a compartment for the great toe and its associated soft parts; another for the small toe; a central compartment; and a deep interosseous-adductor compartment. Each side compartment contains an abductor and a flexor for the toe concerned. The opponens muscles are only minimally represented in the foot. The central compartment accommodates the flexor digitorum brevis muscle and the tendons of the flexor digitorum longus and its associated muscles—the quadratus plantae and the four lumbrical muscles. The interosseous-adductor compartment contains the dorsal and plantar interosseous muscles and the adductor hallucis muscle.

The compartments contain branches of the medial and lateral plantar nerves and blood vessels. These are branches of the tibial nerve and the posterior tibial artery. The muscular nerves correspond in their distribution to the median and ulnar nerves of the hand.

The *compartment of the great toe* contains the abductor hallucis and flexor hallucis brevis muscles, the tendon of the flexor hallucis longus muscle, the medial plantar nerve and vessels, and the first metatarsal. The *abductor hallucis muscle* arises superficially from the medial process of the tuberosity of the calcaneus, the flexor retinaculum, and the intermuscular septum separating it from the flexor hallucis brevis muscle. It inserts into the medial side of the base of the proximal phalanx of the great toe, partly blending with the medial head of the flexor hallucis brevis muscle.

The *flexor hallucis brevis* is a two-bellied muscle; between its bellies, the tendon of the flexor hallucis longus muscle passes to its insertion on the base of the distal phalanx of the great toe. The muscle arises from the plantar aspect of the cuboid and the adjacent part of the lateral cuneiform, the tibialis posterior tendon, and the medial side of the first metatarsal. The two bellies have an intermediate tendinous raphe, and the medial belly blends with the abductor hallucis muscle to insert into the medial side of the base of the proximal phalanx of the great toe. The lateral belly, combining with the tendon of the adductor hallucis muscle, inserts into the lateral side of the base of the same phalanx. A sesamoid in each tendon of the short flexor muscle acts as a bearing and plays against the underside of the head of the first metatarsal.

The *medial plantar nerve* is the larger of the two plantar nerves. It arises from the division of the tibial nerve beneath the posterior part of the abductor hallucis muscle and passes forward, accompanied by the small medial plantar artery, in the medial intermuscular septum between the abductor hallucis and flexor digitorum brevis muscles. The muscular branches to the abductor hallucis and flexor digitorum brevis muscles arise here and enter the deep surfaces of the muscles, and articular branches supply the joints of the tarsals and metatarsals. The proper digital nerve for the great toe supplies the flexor hallucis brevis muscle, and a branch of the first common digital nerve supplies the first lumbrical muscle.

The *medial plantar artery* does not usually form an arch like the superficial palmar arch of the palm. The artery accompanies the medial plantar nerve and, like it, provides three digital branches. The branches anastomose with the three plantar metatarsal arteries of the plantar arch at the base of the interdigital clefts. These vessels are small and may be partially or completely absent.

The *compartment of the small toe* includes an abductor and a short flexor for this toe and the fifth metatarsal. The *abductor digiti minimi muscle* arises from both the medial and lateral processes of the tuberosity of the calcaneus, from the lateral plantar fascia, and from the intermuscular septum between it and the flexor digitorum brevis muscle. Its tendon inserts on the lateral side of the base of the proximal phalanx of the fifth toe. The *flexor digiti minimi brevis muscle* underlies and is medial to the abductor digiti minimi tendon. Its origin is the base of the fifth metatarsal and the sheath of the peroneus longus tendon. Its tendon inserts on the lateral side of the base of the proximal phalanx of the fifth digit.

The *central compartment* of the sole of the foot lies deep to the plantar aponeurosis, separated from the side compartments by deep extensions of the aponeurosis. It contains the flexor digitorum brevis muscle, the tendon of the flexor digitorum longus muscle with its associated quadratus plantae and four lumbrical muscles, a portion of the tendon of the flexor hallucis longus muscle, and the lateral plantar nerve and vessels. The *flexor digitorum brevis muscle* immediately underlies the plantar aponeurosis. It arises from the medial process of the tuberosity of the calcaneus, the posterior third of the plantar aponeurosis, and the intermuscular septa on either side of it. The muscle divides into four tendons. Opposite each proximal phalanx, each tendon divides into two slips, between which passes the tendon of the flexor digitorum longus muscle. Turning under the tendon of the flexor digitorum longus, the two slips of each tendon of the flexor digitorum brevis unite and insert into the base of the middle phalanx.

Digital fibrous sheaths begin over the heads of the metatarsals and extend to the bases of the distal phalanges. They arch over the tendons, attaching to the capsules of the joints and the margins of the proximal and middle phalanges. Over the shafts of these bones, the fibers of the sheath are transverse and strong, but over the joints, the sheaths are much thinner and most of their fiber bundles run obliquely from side to side. Slender transverse bands cross the joint intervals. The marginal fibers of the digital slips of the plantar aponeurosis terminate in the fibrous sheaths.

Synovial sheaths occupy the digital sheaths of the toes. They enclose the tendon of the flexor hallucis longus muscle and the tendons of the flexor digitorum longus and flexor digitorum brevis muscles of the lateral four toes. The synovial sheaths invest the tendons from just proximal to the openings of the digital fibrous sheaths to the bases of the distal phalanges. The *tendon of the flexor digitorum longus muscle* passes the ankle in the second compartment under the flexor retinaculum and enters the foot deep to the abductor hallucis muscle. Passing diagonally toward the center of the sole, it expands and receives from behind the broad insertion of the quadratus plantae muscle. The tendon now divides into four slips, which enter the digital fibrous sheaths surrounded by separate synovial sheaths, pass through the separation of the tendons of the flexor digitorum brevis muscle, and terminate in the bases of the distal phalanges.

The *tendon of the flexor hallucis longus muscle* passes through the lateralmost compartment under the flexor retinaculum and enters the foot under the sustentaculum tali of the calcaneus. Directed toward the great toe, it passes superior to the tendon of the flexor digitorum longus muscle (to which it contributes a tendinous slip) and then lies in the groove between the two bellies of the flexor hallucis brevis muscle. It inserts on the base of the distal phalanx.

The *quadratus plantae muscle*, accessory to the long digital flexor, arises by two heads separated from one another by the long plantar ligament. Its tendinous lateral head arises from the lateral border of the plantar surface of the calcaneus and from the long plantar ligament. Its fleshy medial head takes origin from the medial surface of the calcaneus and the medial border of the long plantar ligament. The two parts join to form a flattened, muscular band inserting into the lateral margin and both surfaces of the tendon of the flexor digitorum longus muscle.

The *lumbrical muscles*, as in the hand, are four small, cylindric muscles arising from the four tendons of the flexor digitorum longus muscle. Except for the first one, each arises from the two adjacent tendons of the flexor digitorum longus muscle; the first muscle springs from the medial side of the first tendon alone. The tendons cross on the plantar side of the deep transverse metatarsal ligaments and end in the medial surface of the extensor expansion over the lateral four toes.

The *lateral plantar nerve*, smaller than the medial plantar, passes diagonally in the sole of the foot, between the flexor digitorum brevis and quadratus plantae muscles, and provides muscular branches to the abductor digiti minimi and quadratus plantae muscles and articular branches to the calcaneocuboid joint. At the lateral margin of the quadratus plantae, the nerve divides and a deep branch sinks into the interosseous-adductor compartment of the sole. The remaining superficial branch splits into one common digital branch for adjacent sides of the fourth and fifth toes, and into a nerve that supplies a proper digital branch for the lateral side of the fifth toe and muscular branches to the flexor digiti minimi brevis muscle and (sometimes) the two interosseous muscles of the fourth interosseous space.

Muscles, Arteries, and Nerves of Sole of Foot (continued)

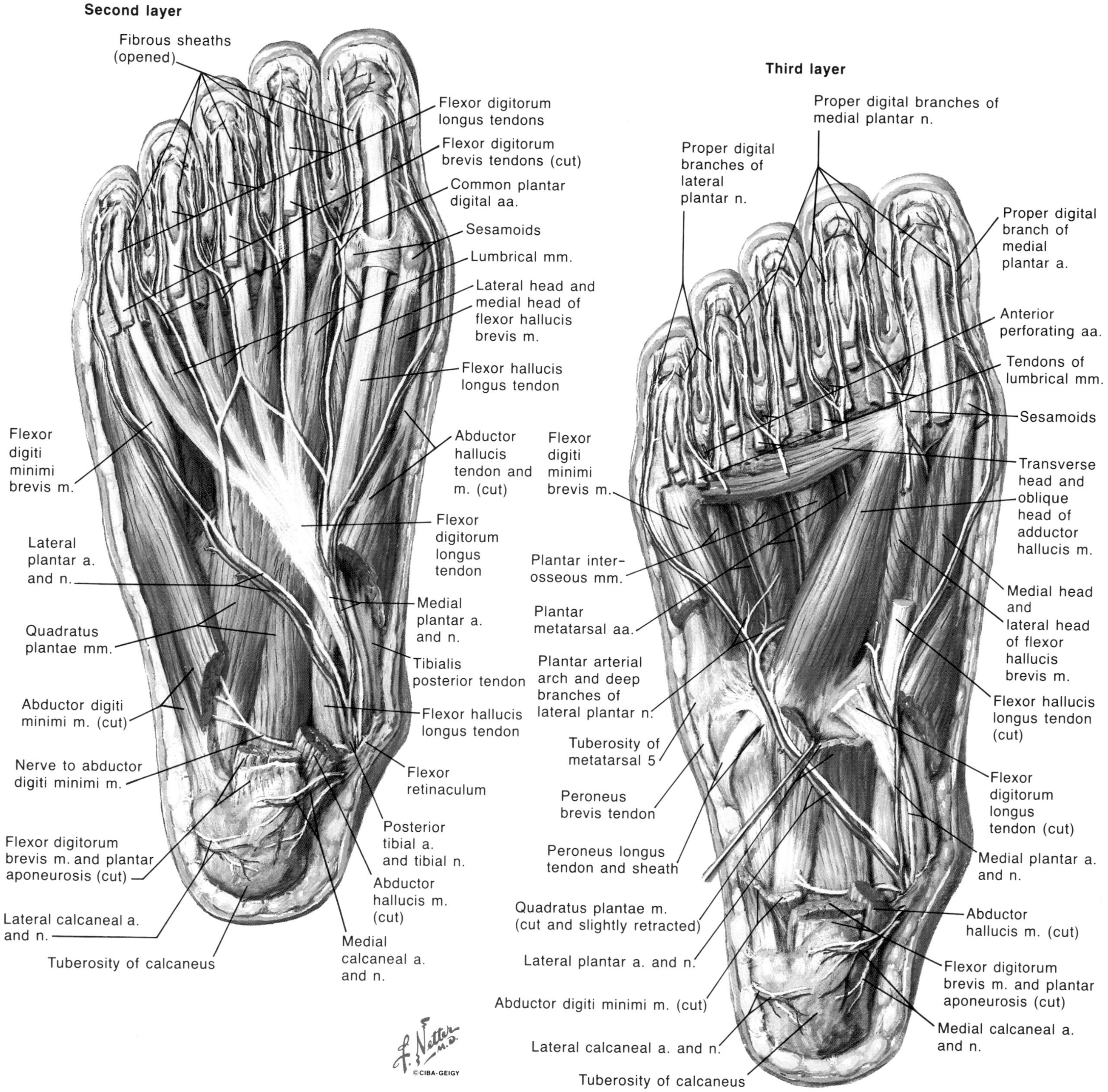

Second layer
Fibrous sheaths (opened)
Flexor digitorum longus tendons
Flexor digitorum brevis tendons (cut)
Common plantar digital aa.
Sesamoids
Lumbrical mm.
Lateral head and medial head of flexor hallucis brevis m.
Flexor hallucis longus tendon
Abductor hallucis tendon and m. (cut)
Flexor digitorum longus tendon
Medial plantar a. and n.
Tibialis posterior tendon
Flexor hallucis longus tendon
Flexor retinaculum
Posterior tibial a. and tibial n.
Abductor hallucis m. (cut)
Medial calcaneal a. and n.
Flexor digiti minimi brevis m.
Lateral plantar a. and n.
Quadratus plantae mm.
Abductor digiti minimi m. (cut)
Nerve to abductor digiti minimi m.
Flexor digitorum brevis m. and plantar aponeurosis (cut)
Lateral calcaneal a. and n.
Tuberosity of calcaneus
Third layer
Proper digital branches of medial plantar n.
Proper digital branches of lateral plantar n.
Proper digital branch of medial plantar a.
Anterior perforating aa.
Tendons of lumbrical mm.
Sesamoids
Transverse head and oblique head of adductor hallucis m.
Medial head and lateral head of flexor hallucis brevis m.
Flexor hallucis longus tendon (cut)
Flexor digitorum longus tendon (cut)
Medial plantar a. and n.
Abductor hallucis m. (cut)
Flexor digitorum brevis m. and plantar aponeurosis (cut)
Medial calcaneal a. and n.
Flexor digiti minimi brevis m.
Plantar inter-osseous mm.
Plantar metatarsal aa.
Plantar arterial arch and deep branches of lateral plantar n.
Tuberosity of metatarsal 5
Peroneus brevis tendon
Peroneus longus tendon and sheath
Quadratus plantae m. (cut and slightly retracted)
Lateral plantar a. and n.
Abductor digiti minimi m. (cut)
Lateral calcaneal a. and n.
Tuberosity of calcaneus
F. Netter M.D.
©CIBA-GEIGY

Lower Limb

Ankle and Foot

(Continued)

The *lateral plantar artery* accompanies the lateral plantar nerve diagonally across the sole of the foot. Accompanied by venae comitantes, it turns around the margin of the quadratus plantae muscle and sinks into the deeper plane of the foot. Perforating the plantar interosseous fascia, it passes medialward across the fourth to second metatarsals and interosseous muscles as the *plantar arterial arch* (Plate 107). In the proximal part of its course, the artery gives off calcaneal branches to the heel, muscular branches to the small toe, and cutaneous branches to the lateral side of the foot. The lateral plantar veins mirror the artery.

The *interosseous-adductor compartment* is enclosed dorsally by the dorsal interosseous fascia and its attachments to the periosteum of the metatarsals (Plates 106–107). It is limited on its plantar aspect by the plantar interosseous fascia covering the adductor hallucis muscle. The interosseous and adductor hallucis muscles occupy this compartment; in it are included the plantar arterial arch, the deep branch of the lateral plantar nerve, and the dorsal metatarsal branches of the dorsalis pedis artery.

The *adductor hallucis muscle* arises by oblique and transverse heads. Occupying the hollow on the plantar surface of the metatarsals, the oblique head arises from the bases of the second, third, and fourth metatarsals, and from the sheath of the peroneus longus muscle. The transverse head takes origin from the plantar metatarsophalangeal ligaments of the third, fourth, and fifth digits, and from the deep transverse metatarsal ligament. The tendons of both heads and that of the lateral head of the flexor hallucis brevis muscle insert into the lateral side of the base of the proximal phalanx of the great toe.

Lower Limb

Ankle and Foot

(Continued)

Interosseous Muscles and Plantar Arterial Arch

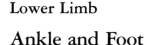

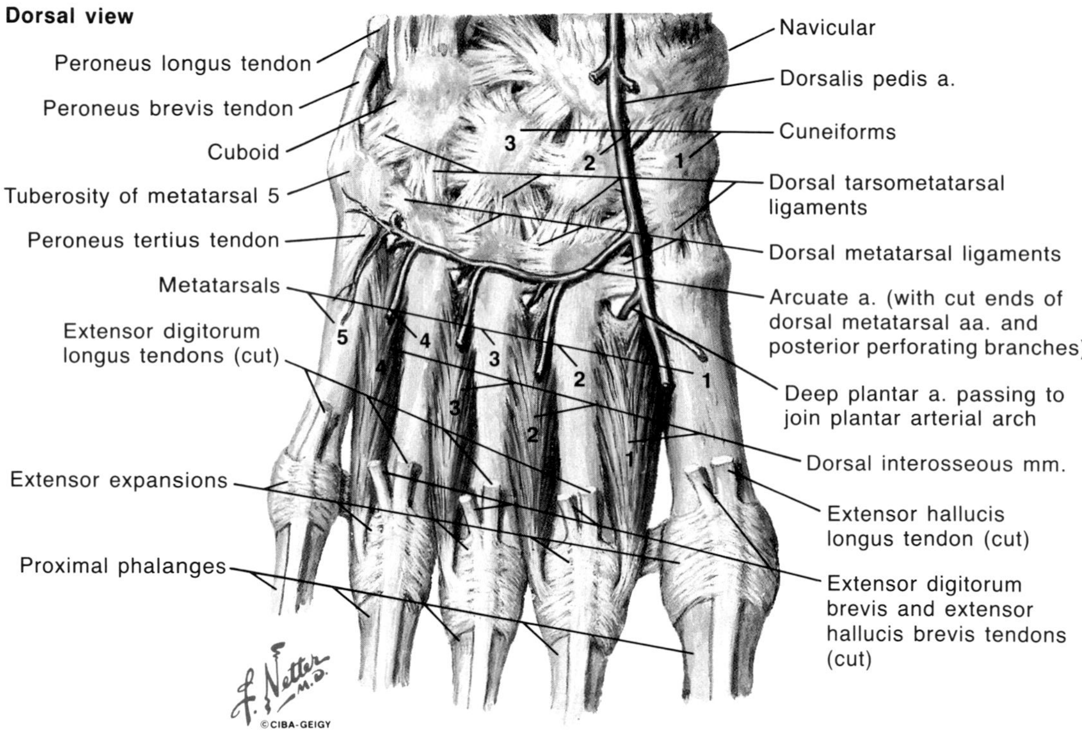
Dorsal view
Peroneus longus tendon
Peroneus brevis tendon
Cuboid
Tuberosity of metatarsal 5
Peroneus tertius tendon
Metatarsals
Extensor digitorum longus tendons (cut)
Extensor expansions
Proximal phalanges
Navicular
Dorsalis pedis a.
Cuneiforms
Dorsal tarsometatarsal ligaments
Dorsal metatarsal ligaments
Arcuate a. (with cut ends of dorsal metatarsal aa. and posterior perforating branches)
Deep plantar a. passing to join plantar arterial arch
Dorsal interosseous mm.
Extensor hallucis longus tendon (cut)
Extensor digitorum brevis and extensor hallucis brevis tendons (cut)
3
2
1
5
4
3
2
1
©CIBA-GEIGY

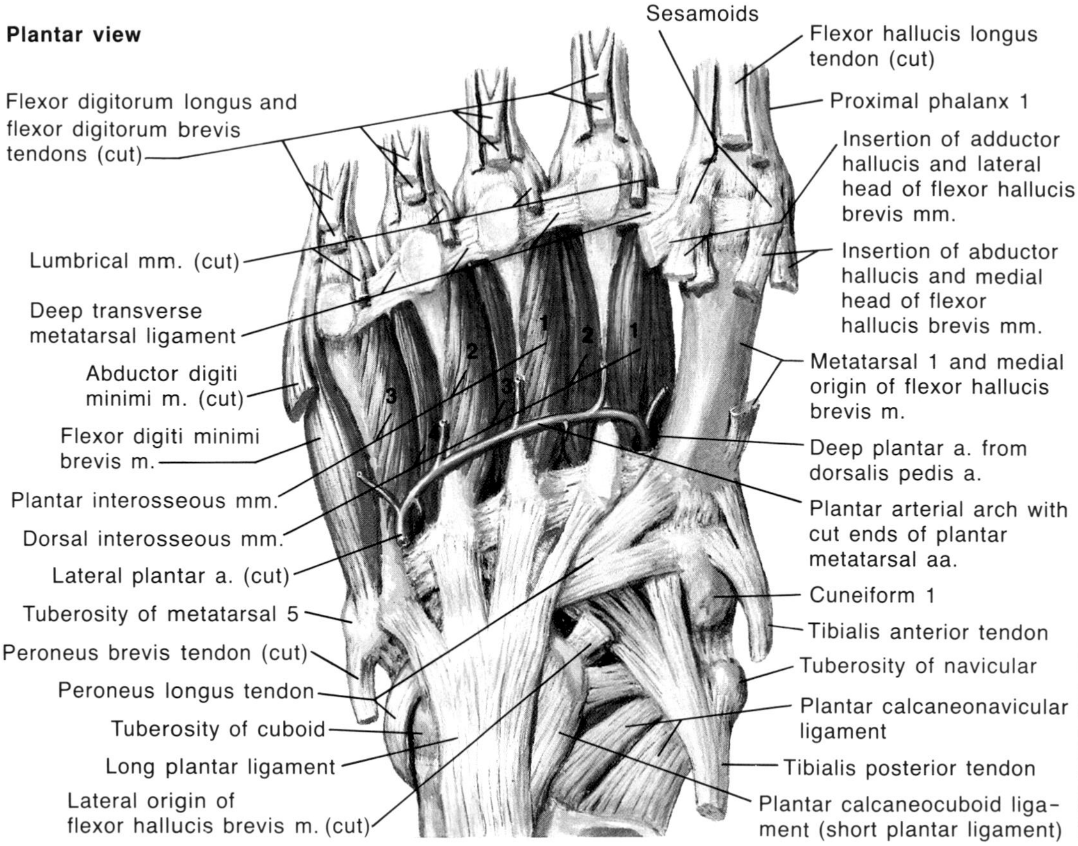
Plantar view
Flexor digitorum longus and flexor digitorum brevis tendons (cut)
Lumbrical mm. (cut)
Deep transverse metatarsal ligament
Abductor digiti minimi m. (cut)
Flexor digiti minimi brevis m.
Plantar interosseous mm.
Dorsal interosseous mm.
Lateral plantar a. (cut)
Tuberosity of metatarsal 5
Peroneus brevis tendon (cut)
Peroneus longus tendon
Tuberosity of cuboid
Long plantar ligament
Lateral origin of flexor hallucis brevis m. (cut)
Sesamoids
Flexor hallucis longus tendon (cut)
Proximal phalanx 1
Insertion of adductor hallucis and lateral head of flexor hallucis brevis mm.
Insertion of abductor hallucis and medial head of flexor hallucis brevis mm.
Metatarsal 1 and medial origin of flexor hallucis brevis m.
Deep plantar a. from dorsalis pedis a.
Plantar arterial arch with cut ends of plantar metatarsal aa.
Cuneiform 1
Tibialis anterior tendon
Tuberosity of navicular
Plantar calcaneonavicular ligament
Tibialis posterior tendon
Plantar calcaneocuboid ligament (short plantar ligament)

The *plantar arterial arch* crosses the sole medialward under the plantar interosseous fascia. Its formation as an arch is due to the free anastomosis of the lateral plantar artery and the deep plantar branch of the dorsalis pedis artery. The plantar arch gives off four plantar metatarsal arteries, three perforating branches, and twigs to the tarsal joints and muscles of the compartment.

The *plantar metatarsal arteries* run forward from the arch on the plantar surface of the interosseous muscles. Each artery divides into a pair of *proper plantar digital arteries*, which supply the adjacent sides of the toes. Each plantar metatarsal artery gives off, near its point of division, an anterior perforating branch that passes through the interosseous space to anastomose with a corresponding branch of a dorsal metatarsal artery. The proper plantar digital arteries, as in the fingers, provide terminal branches, which pass dorsally to supply the nail beds and skin of the distal phalanges. The perforating branches of the plantar arch anastomose with posterior perforating branches of the dorsal metatarsal arteries. The *deep branch of the lateral plantar nerve* passes into the interosseous-adductor compartment directly behind the plantar arterial arch. It supplies muscular branches to the lateral three lumbrical muscles, the interosseous muscles of each space (except the fourth space in some cases), and both heads of the adductor hallucis muscle. Articular branches reach the intertarsal and tarsometatarsal joints.

The *interosseous muscles of the foot* are very similar in structure and placement to the comparable muscles of the hand, except that here, the plane of reference for abduction and adduction of the toes is through the second digit rather than the third. The four *dorsal interosseous muscles* are bipennate in form and arise from the adjacent sides of both metatarsals of the space in which they lie. They have a longer origin from the metatarsal of the digit into which they insert. The first and second dorsal interosseous muscles lie on the medial and lateral sides of the second metatarsal and insert into the same sides of the bases of the proximal phalanx. The third and fourth dorsal interosseous muscles lie on the lateral surfaces of the third and fourth metatarsals, respectively, and insert into the lateral sides of the bases of their proximal phalanges. Minor insertions occur into the dorsal extensor expansions. These muscles are abductors of the digits with reference to the midplane of the second digit. The three *plantar interosseous muscles* (adductors of digits III, IV, and V) are unipennate. They arise from the bases and medial sides of metatarsals 3, 4, and 5, and insert into the medial sides of the proximal phalanges.

Muscle Actions

Electromyographic evidence indicates that the muscles of the foot have little if any action in support of the body weight in the normally strong foot in the erect balanced position. In cases of flatfoot, some muscular action is usually required for relaxed standing and, of course, any movement of the body is immediately accompanied by muscular activity. The great and small toe compartments each lodge abductor and flexor muscles. The abductors abduct and flex the proximal phalanx of the digit and the flexors assist in flexion of the same digit. The flexor digitorum brevis muscle flexes the middle phalanges of the lateral four toes and assists in metatarsophalangeal flexion of the same digits. The quadratus plantae muscle assists the tendons of the flexor digitorum longus muscle in flexing the toes and helps to bring the line of traction of those tendons more nearly parallel with the long axis of the foot. The adductor hallucis muscle adducts the great toe and aids in maintaining the transverse arch of the foot. The lumbrical muscles flex the proximal phalanges at the metatarsophalangeal joints and extend the middle and distal phalanges. The interosseous muscles, as in the hand, are abductors and adductors of the digits and also serve in flexion of the metatarsophalangeal joints. Any action of the interosseous muscles in the extension of the middle and distal phalanges (as in the hand) is uncertain in the foot. The extensor digitorum brevis muscle of the dorsum of the foot assists in extension of the medial four toes.

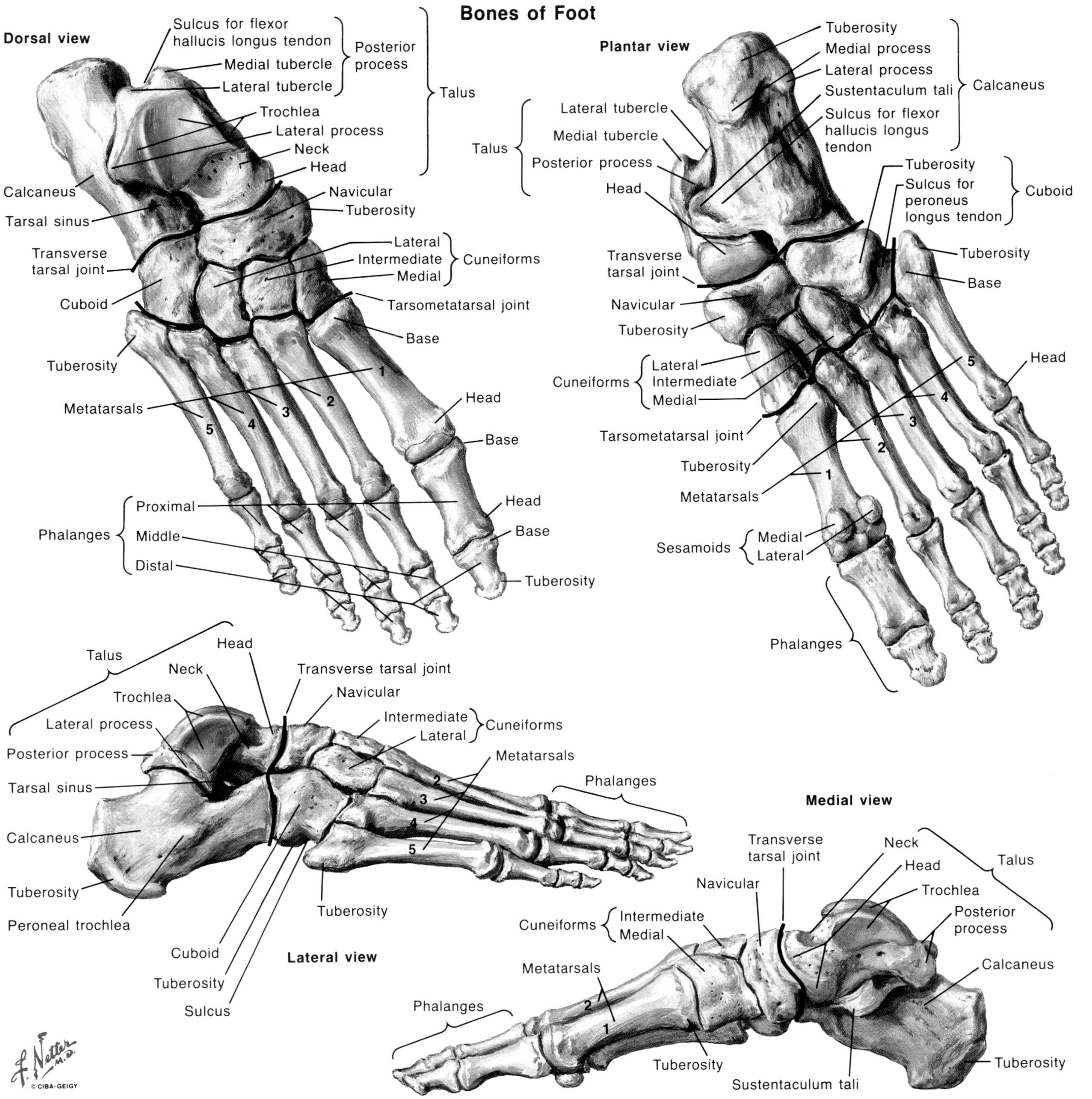
Bones of Foot
Dorsal view
Sulcus for flexor hallucis longus tendon
Medial tubercle
Lateral tubercle
Posterior process
Talus
Trochlea
Lateral process
Neck
Head
Calcaneus
Navicular
Tuberosity
Tarsal sinus
Transverse tarsal joint
Lateral
Intermediate
Medial
Cuneiforms
Cuboid
Tarsometatarsal joint
Base
Tuberosity
1
2
3
4
5
Metatarsals
Head
Base
Proximal
Middle
Distal
Phalanges
Head
Base
Tuberosity
Plantar view
Tuberosity
Medial process
Lateral process
Sustentaculum tali
Calcaneus
Sulcus for flexor hallucis longus tendon
Lateral tubercle
Medial tubercle
Posterior process
Talus
Head
Tuberosity
Sulcus for peroneus longus tendon
Cuboid
Transverse tarsal joint
Tuberosity
Base
Navicular
Tuberosity
Lateral
Intermediate
Medial
Cuneiforms
Head
5
4
3
2
1
Tarsometatarsal joint
Tuberosity
Metatarsals
Sesamoids
Medial
Lateral
Phalanges
Talus
Head
Neck
Trochlea
Lateral process
Posterior process
Tarsal sinus
Calcaneus
Tuberosity
Peroneal trochlea
Transverse tarsal joint
Navicular
Intermediate
Lateral
Cuneiforms
Metatarsals
2
3
4
5
Phalanges
Tuberosity
Cuboid
Tuberosity
Sulcus
Lateral view
Medial view
Transverse tarsal joint
Neck
Head
Talus
Navicular
Trochlea
Posterior process
Cuneiforms
Intermediate
Medial
Calcaneus
Metatarsals
2
1
Phalanges
Tuberosity
Sustentaculum tali
Tuberosity
F. Netter M.D.
©CIBA-GEIGY

SECTION I PLATE 108 Slide 3673

Lower Limb

Ankle and Foot

(Continued)

Bones

The bones of the foot are similar in arrangement to those of the wrist and hand (Plates 102 and 108–109). There are seven tarsals, five metatarsals, and fourteen phalanges. The tarsals are the talus, calcaneus, navicular, three cuneiforms (medial, intermediate, and lateral), and cuboid. The arrangement of the bones indicates a limited independence between the bones forming the medial three digits and those forming the lateral two digits. The talus receives the weight of the body at the ankle and constitutes the summit of the bony longitudinal arch of the foot.

The *talus* articulates with many bones—the tibia and the fibula above and on its sides, the calcaneus below, and the navicular in front. No muscles insert into it, but it receives a number of ligaments. It has a head, a body, and a neck. The *head* of the talus is its rounded anterior extension; it is directed forward and medialward and has three articular surfaces. The large *navicular surface* is rounded, convex, and oval; followed to its underside, it passes into a flat, triangular *anterior calcaneal articular surface*. Through this latter surface, the head of the talus bears on the anterior facet of the calcaneus and on the plantar calcaneonavicular ligament. The third and most posterior facet, the oval *middle calcaneal articular surface*, bears on the upper surface of the sustentaculum tali of the calcaneus. The *neck* of the talus is its somewhat restricted part between the head and the body. On the upper side, it is rough for ligaments and shows a number of vascular foramina. Inferiorly, there is a deep groove, the *talar sulcus*, which forms the roof of the tarsal sinus, occupied in life by the interosseous talocalcaneal

Lower Limb

Ankle and Foot

(Continued)

ligament. The *body* of the talus is roughly quadrilateral, its upper portion, the trochlea, entering into the formation of the ankle joint. The region of the talus inferior to its small medial articular surface is rough for the deltoid ligament and has numerous vascular foramina. Prominent on the underside of the body is an oblong articular facet, deeply concave from side to side. This is the *posterior calcaneal articular surface*. The posterior process of the talus is grooved by the tendon of the flexor hallucis longus muscle. Medial to this groove is the medial tubercle for the medial talocalcaneal and posterior tibiotalar ligaments; on the lateral aspect of the groove is a lateral tubercle for the attachment of the posterior talofibular ligament of the ankle joint.

The *calcaneus* is the largest and strongest bone of the foot. It is long, flattened from side to side, and bulbous posteriorly where it forms the heel. The superior surface of the calcaneus exhibits, anteriorly, three articular facets for the talus. The largest is the *posterior talar articular surface*, triangular in form and convex from behind forward. Anterior to it is a deep depression, which leads onto the sustentaculum tali as the *calcaneal sulcus*, the floor of the tarsal sinus. The upper surface of the sustentaculum tali carries the *middle talar articular surface* for the posterior facet of the head of the talus. A small oval *anterior talar articular surface* characterizes the anterior end of the superior surface of the bone.

Toward the posterior extremity of the calcaneus, its superior surface is rough and is in relation with the fat deep to the calcaneal tendon. The inferior surface of the calcaneus is narrow and uneven. The long plantar ligament attaches here. The *tuberosity* is the posteroinferior part of the bone. It is rough and striated for the attachment of the calcaneal tendon; toward its superior surface, it is smooth for the subtendinous bursa of the calcaneal tendon. Two processes characterize the tuberosity inferiorly—the larger medial process gives origin to the abductor hallucis, flexor digitorum brevis, and abductor digiti minimi muscles; the narrow lateral process gives origin to the abductor digiti minimi muscle. The *medial surface of the calcaneus* is smoothly concave and overhung by the prominent *sustentaculum tali*; the latter has a sulcus on its underside for the flexor hallucis longus muscle. The *lateral surface of the calcaneus* is rough and at about its middle has a small swelling for attachment of the calcaneofibular ligament; inferior to this is a *peroneal trochlea* that separates the tendons of the peroneus longus and peroneus brevis muscles, the peroneus longus tendon inferior to it. The anterior extremity of the calcaneus is the *cuboidal articular surface*. It is roughly triangular in form and carries a saddle-shaped articulation for the cuboid.

The *navicular* is a flattened oval bone located between the head of the talus and the three cuneiforms. It is characterized by a large, oval, concave articular facet on its posterior surface for the head of the talus, and a rounded eminence at its medial plantar extremity, the *tuberosity*, for the primary attachment of the tibialis posterior muscle. Three triangular facets separated by two vertical ridges occupy its anterior surface; they articulate with the three cuneiforms. The superior and inferior surfaces of the bone are rough for ligaments, and the lateral surface frequently exhibits a small articular facet for the cuboid.

The *cuneiforms* are all wedge shaped, but the broader side of the wedge faces plantarward on the medial cuneiform and dorsalward on the other two. The posterior extremity of each bone is concave and articulates with one of the facets of the navicular. The anterior extremity of each bone enters into the tarsometatarsal joint of the first, second, or third digit; there are articular surfaces between the adjacent cuneiforms and one between the lateral cuneiform and the cuboids. The dorsal and plantar surfaces of these bones are rough for the attachment of the ligaments and tendons. The medial cuneiform is the largest, and the middle bone is the shortest. This shortness forms a recess into which the second metatarsal is received. The articulation of the lateral cuneiform with the cuboid is by a large triangular or oval facet situated toward the posterosuperior aspect of its lateral surface.

The *cuboid* is interposed on the lateral side of the foot between the calcaneus and the fourth and fifth metatarsals. Its dorsal surface is rough and nonarticular; its plantar surface has a prominent ridge, which receives the long plantar ligament (Plate 109) and ends laterally in the *tuberosity* of the bone. The tuberosity has a convex cartilage-covered facet over which the tendon of the peroneus longus muscle plays as it enters the foot. This tendon is not usually seen in the cadaver in the groove anterior to the tuberosity. Laterally, the cuboid is short and concave for the peroneus longus tendon; its longer medial side bears, posteriorly, either a large triangular or an oval facet for articulation with the lateral cuneiform. The posterior surface of the cuboid is entirely articular; it is saddle shaped for participation in the calcaneocuboid joint. The distal surface of the cuboid has a small medial and a larger lateral facet. These slightly concave facets articulate with the bases of the fourth and fifth metatarsals, respectively.

The *metatarsals*, like the metacarpals of the hand, are long bones; each consists of a base, a body, and a head. They are 6 to 8 cm long and are relatively flat dorsally but concave longitudinally on their plantar sides. The bases carry smooth articular surfaces for articulation with the cuneiforms and cuboid (and in most cases with each other) and show pits for ligaments on their sides. The bodies are narrow and tend to be triangular in cross section. The heads present convex articular surfaces, somewhat flattened from side to side, for articulation with the proximal phalanges; dorsally on their sides they exhibit tubercles for the attachment of the collateral ligaments of the metatarsophalangeal joints.

The *first metatarsal* is the shortest, broadest, and most massive of the series. At its base, a tuberosity projects downward and lateralward to receive the tendon of the peroneus longus muscle. The head of the bone is broad, and its plantar surface has two deep grooves separated by a ridge; in these grooves play the sesamoids in the tendons of the flexor hallucis longus muscle. The *second metatarsal* is the longest, and its base fits into the recess formed by the three cuneiforms. Thus, the base of this bone has articular facets for all the cuneiforms and, in addition, for the base of the third metatarsal. The *third metatarsal* articulates with the end of the lateral cuneiform and, by facets on the sides of its base, with the adjacent sides of the second and fourth metatarsals. The *fourth metatarsal* articulates with the medial of the two facets of the cuboid and with the adjacent third and fifth metatarsals. There may also be a facet for contact with the lateral cuneiform. The base of the *fifth metatarsal* is expanded laterally into a rough tuberosity for the insertion of the peroneus brevis tendon. It articulates with the lateral facet of the cuboid and, on its medial side, with the fourth metatarsal. Its shaft is compressed dorsoplantarward rather than from side to side.

The *phalanges* of the toes, like the fingers, are fourteen in number—three for each digit except the great toe, which has two. Except for that of the great toe, which is broad and thick, the *proximal phalanges* are broadened at their extremities and narrow throughout their bodies. The bases have single, round or oval, cuplike facets for reception of the heads of the corresponding metatarsals. The heads of the proximal phalanges present rounded pulleylike surfaces, grooved in the middle and raised at the edges, for articulation with the bases of the middle phalanges. The *middle phalanges* are short, but their bodies are proportionately broader than those of the proximal phalanges. Both ends of the middle phalanges have trochlear surfaces. The *distal phalanges* are also short. They exhibit broadened bases with trochlear surfaces and rough, broadened distal tuberosities for support of the nails and pulp of the toes.

Ossification

The *tarsals* are ossified from a single center for each bone, except the calcaneus, which has a separate epiphysis for its tuberosity. The principal ossification center for the calcaneus appears at the sixth fetal month; that for the talus, during the seventh fetal month; that for the cuboid, at the time of birth; that for the lateral cuneiform, during the first year; and that for the medial cuneiform, during the third year. Ossification centers for the intermediate cuneiform and naviculars appear in the fourth year. The epiphysis for the tuberosity of the calcaneus appears between ages 8 to 10 and is united with the rest of the bone at puberty. Ossification of all the tarsals is complete shortly after puberty.

In each of the *metatarsals*, a primary center of ossification for the body and the base (except the body and head for the first metatarsal) appears about the ninth week of fetal life, and these bones are well ossified at birth. A secondary center for each of the heads (base of first metatarsal) appears in the third year and fuses to the shaft between ages 14 to 17. The *phalanges* are each ossified from two centers, one for the body and head, and one

Lower Limb

Ankle and Foot

(Continued)

for the base. Those for the bodies and heads appear from between the tenth fetal week to the time of birth—in the distal phalanges first, and in the middle phalanges last. The secondary ossification centers for the bases appear during the third year of life and unite with the shafts from ages 14 to 17.

Joints

The intertarsal joints are the subtalar, talocalcaneonavicular, calcaneocuboid, transverse tarsal, cuneonavicular, intercuneiform, and the plantar ligaments (Plates 102 and 107). In order to maintain the foot against the weight of the body, the plantar ligaments are stronger and more extensive than the dorsal ligaments. Blood vessels are supplied from adjacent branches of the dorsalis pedis, medial plantar, and lateral plantar arteries; the nerve supply comes from the deep peroneal and medial and lateral plantar nerves.

The *subtalar joint* is formed between the large concave facet on the underside of the body of the talus and the convex posterior articular surface of the superior aspect of the calcaneus. A loose, thin-walled articular capsule unites the bones, attaching to the margins of the articular surfaces. Somewhat stronger portions are designated as the posterior, medial, and lateral talocalcaneal ligaments. The *medial talocalcaneal ligament* connects the medial tubercle of the posterior process of the talus with the posterior margin of the sustentaculum tali; the *lateral talocalcaneal ligament* is parallel to, and deeper than, the calcaneofibular ligament. The *posterior talocalcaneal ligament* is a short band, its fibers radiating from a narrow attachment on the lateral tubercle of the talus to the upper and medial parts of the calcaneus. The *interosseous talocalcaneal ligament* is located in the tarsal sinus. It is a strong band, composed of several layers of fibers interspersed with fatty tissue, which connects the adjacent surfaces of the talus and calcaneus along their oblique grooves. Support for the subtalar joint is also derived from those ligaments of the ankle joint that, passing from the tibia and fibula to the calcaneus, span the talus.

Tendon Insertions and Ligaments of Sole of Foot

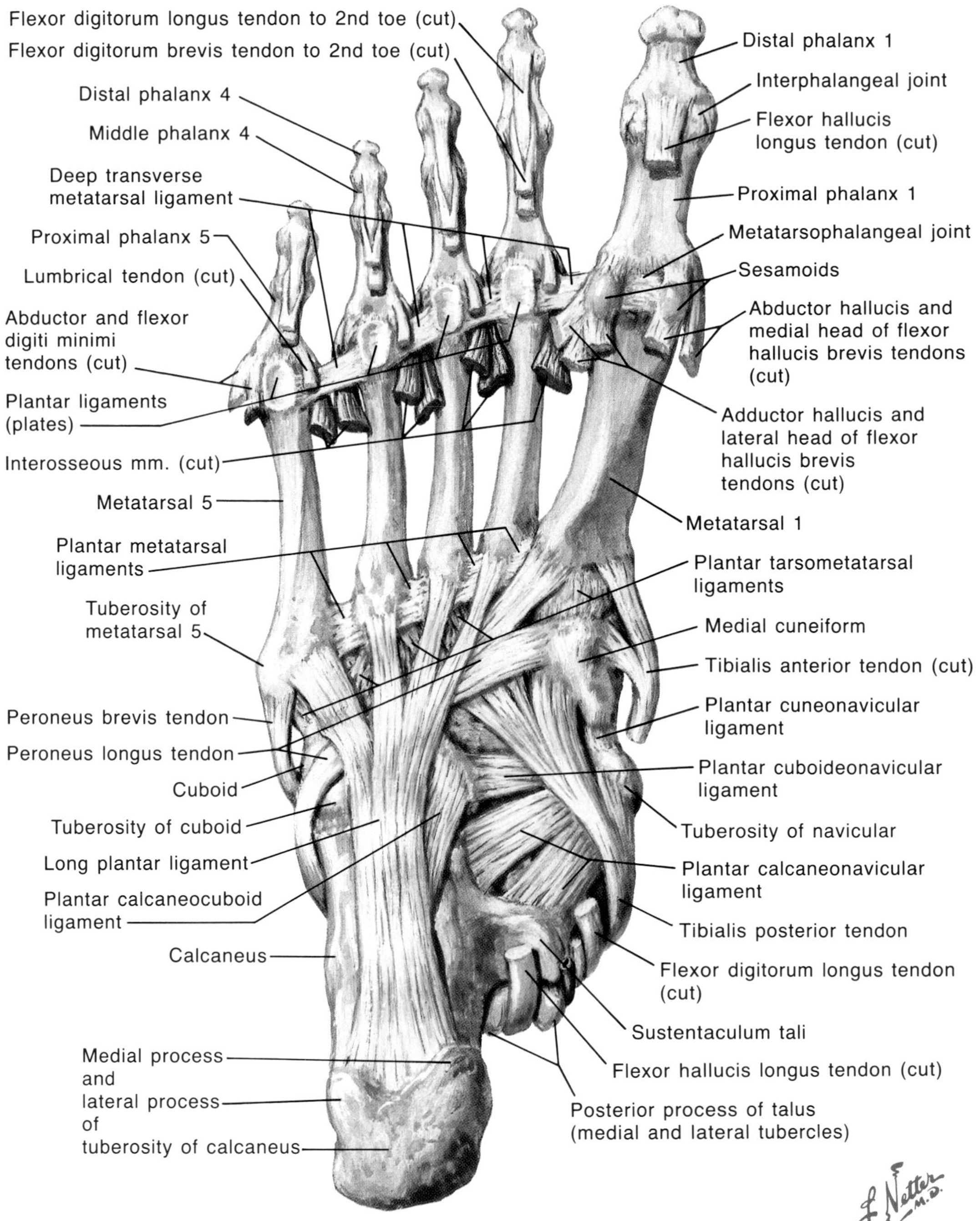
Flexor digitorum longus tendon to 2nd toe (cut)
Flexor digitorum brevis tendon to 2nd toe (cut)
Distal phalanx 4
Middle phalanx 4
Deep transverse metatarsal ligament
Proximal phalanx 5
Lumbrical tendon (cut)
Abductor and flexor digiti minimi tendons (cut)
Plantar ligaments (plates)
Interosseous mm. (cut)
Metatarsal 5
Plantar metatarsal ligaments
Tuberosity of metatarsal 5
Peroneus brevis tendon
Peroneus longus tendon
Cuboid
Tuberosity of cuboid
Long plantar ligament
Plantar calcaneocuboid ligament
Calcaneus
Medial process and lateral process of tuberosity of calcaneus
Distal phalanx 1
Interphalangeal joint
Flexor hallucis longus tendon (cut)
Proximal phalanx 1
Metatarsophalangeal joint
Sesamoids
Abductor hallucis and medial head of flexor hallucis brevis tendons (cut)
Adductor hallucis and lateral head of flexor hallucis brevis tendons (cut)
Metatarsal 1
Plantar tarsometatarsal ligaments
Medial cuneiform
Tibialis anterior tendon (cut)
Plantar cuneonavicular ligament
Plantar cuboideonavicular ligament
Tuberosity of navicular
Plantar calcaneonavicular ligament
Tibialis posterior tendon
Flexor digitorum longus tendon (cut)
Sustentaculum tali
Flexor hallucis longus tendon (cut)
Posterior process of talus (medial and lateral tubercles)
F. Netter M.D.
©CIBA-GEIGY

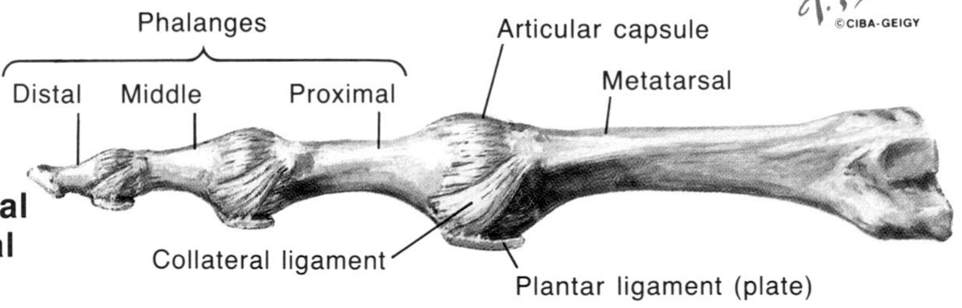
Phalanges
Distal
Middle
Proximal
Articular capsule
Metatarsal
Collateral ligament
Plantar ligament (plate)

Capsules and Ligaments of Metatarsophalangeal and Interphalangeal Joints (lateral view)

The *talocalcaneonavicular joint* is formed between the articular surfaces of the head of the talus and the navicular, the plantar calcaneonavicular ligament, the sustentaculum tali, and the adjacent part of the anterior articular surface of the calcaneus. The thin articular capsule encloses this common articular cavity. The capsule is reinforced between the neck of the talus and the dorsal surface of the navicular by the broad *dorsal talonavicular ligament*. Supporting the joint below is the thick, dense, fibroelastic *plantar calcaneonavicular ligament*. This ligament extends from the sustentaculum tali and the distal surface of the calcaneus to the entire width of the inferior surface of the navicular and to its medial surface behind its tuberosity. Medially, it blends with the deltoid ligament and laterally, with the lower border of the calcaneonavicular portion of the bifurcate ligament. Its upper surface is smooth and contains a fibrocartilaginous plate on which the head of the talus bears. The *calcaneonavicular portion of the bifurcate ligament* completes the socket on the lateral side. Its short fibers pass from the upper surface of the anterior end of the calcaneus to the adjacent lateral surface of the navicular.

The *calcaneocuboid joint* unites the saddle-shaped articular surfaces of the calcaneus and the cuboids. Its joint cavity is separate from adjacent cavities, and an articular capsule encloses it. The thin, broad *dorsal calcaneocuboid ligament* reinforces the

Lower Limb

Ankle and Foot

(Continued)

capsule dorsally. The *bifurcate ligament*, also concerned in the talocalcaneonavicular joint through its calcaneonavicular portion, has a calcaneocuboid band that ends on the dorsomedial angle of the cuboid and is one of the main connections between the first and second rows of tarsals. The calcaneocuboid joint takes much of the thrust of the body weight onto the lateral side of the foot and the lateral part of its longitudinal arch. It is therefore supported by strong plantar ligaments—the plantar calcaneocuboid and the long plantar ligaments.

The *plantar calcaneocuboid ligament* is attached to the rounded eminence at the anterior end of the inferior surface of the calcaneus and to the plantar surface of the cuboid behind its tuberosity and oblique ridge. The fibers of this wide ligament are short and strong, and the ligament is partially overlain by the long plantar ligament. The *long plantar ligament* stretches from the plantar surface of the calcaneus in front of its tuberosity to the tuberosity of the cuboid, its more superficial fibers spreading forward to the bases of the third, fourth, and fifth metatarsals.

The *transverse tarsal joint* is a name given to the irregular articular plane crossing the foot from side to side and composed of the talonavicular articulation medially and the calcaneocuboid joint laterally. These separate joints combine functionally to contribute primarily to the inversion-eversion action of the foot. With inversion is combined adduction and flexion; with eversion, abduction and dorsiflexion.

Contributory to allied actions of the foot are the subtalar and talocalcaneonavicular joints. These provide movement around an axis passing through the tarsal sinus and thus allow the foot to be placed firmly on slanting and irregular surfaces. Additional ligaments unite the navicular and the cuboid. The *dorsal cuboideonavicular* and *plantar cuboideonavicular ligaments* unite adjacent surfaces of the two bones, and a strong *interosseous cuboideonavicular ligament* connects the rough nonarticular portions of their adjacent surfaces. There may be a small joint cavity between the posterior medial angle of the cuboid and the lateral margin of the navicular; it will be continuous with the cuneonavicular joint in front of it.

Distal intertarsal joints are the *cuneonavicular*, *intercuneiform*, and *cuneocuboid articulations*. These are united by a common articular capsule enclosing a common articular cavity, which also extends downward to include the tarsometatarsal joint between the intermediate cuneiform and the metatarsals 2 and 3, and the intermetatarsal joints between metatarsals 2 and 3 and 3 and 4. Adjacent bones are united by weak *dorsal cuneonavicular ligaments* for each of the cuneiforms, *dorsal intercuneiform ligaments*, and a *dorsal cuneocuboid ligament*. *Plantar ligaments* correspond in names to the dorsal, and there are also *intercuneiform and cuneocuboid interosseous ligaments*. Slight gliding motions at these distal joints contribute to the adaptability and flexibility of the foot.

The *tarsometatarsal joints* are plane joints between the distal row of tarsals and the bases of the metatarsals; the medial three metatarsal bases with the cuneiforms and the lateral two with the cuboid. There are three tarsometatarsal joint cavities. Medially, there is a separate joint cavity between the first metatarsal and the medial cuneiform. The intermediate articulation includes articulations of the second and third metatarsals and is an extension of the distal intertarsal joint space. The lateral articulation includes the contact of the fourth and fifth metatarsals with the cuboid. Weak *dorsal* and somewhat stronger *plantar tarsometatarsal ligaments* connect adjacent borders of the cuneiforms and the second and third metatarsals.

Intermetatarsal joints are interposed between the bases of the lateral four metatarsals, mostly as forward extensions of the tarsometatarsal joints. *Dorsal, plantar*, and *interosseous metatarsal ligaments* close these joint spaces; these interosseous ligaments are strong and help maintain the transverse arch of the foot. These joints provide slight gliding movements contributing to the flexibility of the foot, although the first joint also permits slight rotary movements of the great toe.

The *metatarsophalangeal joints* are condyloid joints between the rounded heads of the metatarsals and the cupped proximal extremities of the proximal phalanges. They are very similar to the metacarpophalangeal joints of the fingers; each joint is enclosed by an articular capsule, reinforced by plantar and collateral ligaments. The articular capsule is loose and reinforced dorsally by fibers from the extensor tendon expansions. The *plantar ligament*, like its palmar counterpart, is a dense, fibrocartilaginous plate that is firmly attached to the proximal plantar border of the phalanx and serves as part of the bearing surface for the head of the metatarsal. At the sides, it is attached to the collateral ligaments and the deep transverse metatarsal ligaments. For the great toe, the sesamoids and their interconnecting ligaments replace the plantar ligament. The strong *collateral ligaments* pass from the tubercles on each side of the head of the metatarsal to the sides of the proximal end of the phalanx and the plantar ligament. The plantar ligaments are interconnected by the *deep transverse metatarsal ligament*, which connects the heads and the joint capsules of all the metatarsal heads. The movements allowed at the metatarsophalangeal joints are dorsiflexion, plantar flexion, abduction, adduction, and circumduction.

The *interphalangeal joints* are similar to the metatarsophalangeal joints, but their trochlear surfaces permit only dorsiflexion and plantar flexion. Each joint has an articular capsule and plantar and collateral ligaments. Blood vessels and nerves to these and to the metatarsophalangeal joints are branches of digital vessels and nerves.

Foot Dynamics

The foot is strong enough to support the weight of the body but it is also flexible and resilient enough to absorb the shocks it receives and to provide spring and lift for body activities. Skeletally, it has an arched structure composed of a number of bones linked together by joints and ligaments. The bones of the foot are arranged in longitudinal and transverse arches. The longitudinal arch is supported posteriorly on the tuberosity of the calcaneus; anteriorly, it rests on the heads of the five metatarsals. The transverse arch results from the shape of the distal tarsals and the bases of the metatarsals. These are generally broader dorsally so that, as they fit against one another, a domed configuration results. The talus is at the summit of the foot and is primarily related in its thrust to the navicular, the three cuneiforms, and the medial three metatarsals. These bones, then, constitute the medial segment of the longitudinal arch.

Laterally, the calcaneus relates forward to the cuboid and to the lateral two metatarsals, forming the lateral segment of the longitudinal arch. Conformably, the medial segment shows a much higher arch and considerable elasticity, whereas the lateral segment is flatter, more rigid, and makes the initial contact with the ground in weight bearing.

A recent alternative suggestion would relate the skeletal and joint structure of the foot to a twisted plate spanning from the metatarsal heads to the calcaneus, with the elevation of the instep medially contrasted to the flattened border of the plate laterally. In either case, this is a body part of considerable stability and resistance to deformation coupled with the capacity for elastic recoil and the development of strong dynamic responses. In standing, the weight of the body is distributed equally between the heel and the ball of the foot and is shared between the feet, depending on posture. The ligaments are primary in relaxed standing, and the action of muscles is not normally induced except in flatfoot, imbalance, or the initiation of movement.

The plantar ligaments of the foot are the strongest ligaments, and their support function is enhanced by robust interosseous ligaments that keep the bones from spreading apart. Notable on the sole of the foot are the long plantar and plantar calcaneonavicular and calcaneocuboid (short plantar) ligaments. The elasticity of the plantar calcaneonavicular ligament, and its reception of the head of the talus, have led to its being called the *spring ligament*. The plantar aponeurosis may be thought of as a *tie rod* for the longitudinal arch, resisting spread of its two ends. The toes add to the *grasp* of the foot on the ground, and the great toe is of special importance. The foot is raised against the contact of the great toe with the ground, and its bones and muscles contribute much to the *pushoff*. The stresses of standing are borne at the ball of the foot between the head of the first metatarsal and those of the second to fifth metatarsals in a ratio of 1:2.

Electromyographic research is increasingly adding to the understanding of the staged participation of muscles in standing, locomotion, and other activities, and the movement patterns contributed to by the specific muscles of the limb reported in this account reflect such studies. □

Lower Limb

Lymphatic Drainage

Lymph Vessels and Nodes of Lower Limb

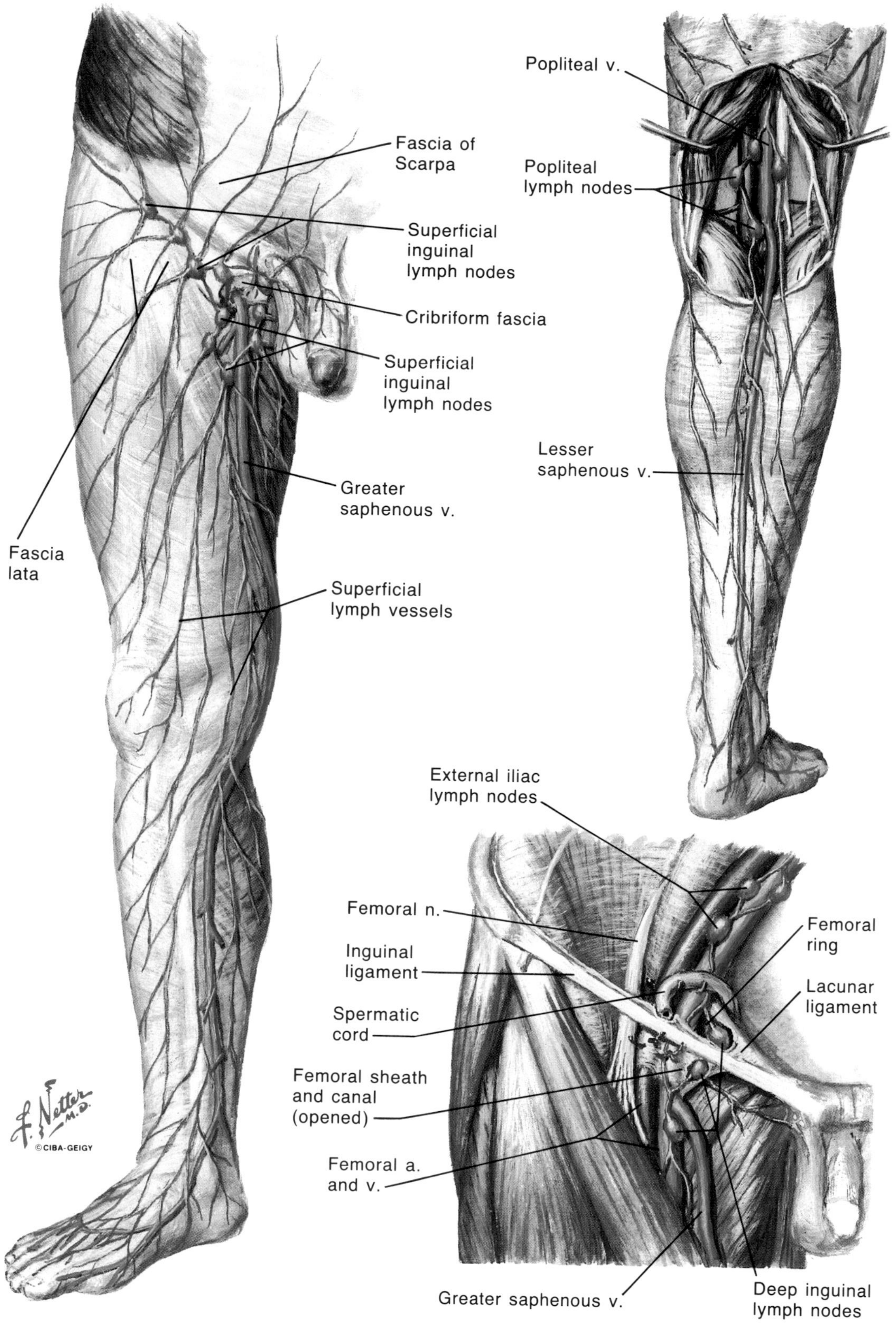
Fascia of Scarpa
Superficial inguinal lymph nodes
Cribriform fascia
Superficial inguinal lymph nodes
Greater saphenous v.
Fascia lata
Superficial lymph vessels
Popliteal v.
Popliteal lymph nodes
Lesser saphenous v.
External iliac lymph nodes
Femoral n.
Inguinal ligament
Spermatic cord
Femoral sheath and canal (opened)
Femoral a. and v.
Femoral ring
Lacunar ligament
Greater saphenous v.
Deep inguinal lymph nodes
©CIBA-GEIGY

Superficial Lymphatic Vessels. These vessels, which have a similar arrangement to the superficial lymphatic vessels of the hand, arise in plexuses on the plantar side of the toes and foot. Collecting vessels pass through interdigital clefts to the dorsum of the foot, to join there with collecting vessels from the dorsum of the toes. Collecting vessels from the medial and dorsal parts of the foot accompany the greater saphenous vein those from the lateral part of the foot accompanying the lesser saphenous vein.

The larger stream of ascending vessels is with the greater saphenous vein, toward which also converge vessels from the lateral and medial borders and the front and back of the leg and thigh. These ascending vessels end above in the superficial inguinal lymph nodes, to which also pass collecting vessels from the lower abdomen and perineum, scrotum and penis in the male (or the vulvar region in the female), and the gluteal region. The area of drainage of the lesser saphenous vein provides lymph vessels that accompany that vein, pierce the popliteal fascia with it, and end in the popliteal lymph nodes.

The *superficial inguinal lymph nodes*, 12 to 20 in number, are arranged in the form of a T in the subcutaneous tissue of the groin. Most of the nodes lie in the horizontal part of the T in a chain parallel to, and about 1 cm below, the inguinal ligament. They receive lymph from the lower abdominal wall, the buttocks, the penis and scrotum (or vulvar region in the female), and the perineum. The fewer, larger nodes of the vertical limb of the T lie along the termination of the greater saphenous vein. These nodes principally receive afferent vessels superficially from the limb below them, but also from the penis and scrotum, perineum, and buttocks. The superficial inguinal lymph nodes send their efferent channels through the femoral sheath to the external iliac nodes. Only a few of these channels end in the deep inguinal nodes.

Deep Lymphatic Vessels. These vessels accompany the deep blood vessels of the limb. In the leg, they follow the anterior and posterior tibial and peroneal vessels to the popliteal nodes. Certain lymph vessels of the gluteal region follow the superior and inferior gluteal vessels to the internal iliac nodes. The *popliteal nodes* are usually small, six to seven in number, and lie in the fat of the popliteal fossa. One lies at the termination of the lesser saphenous vein and receives the lymph channels accompanying that vein. Another node usually lies between the popliteal artery and the capsule of the knee joint and is especially concerned with the lymphatic drainage of the knee. Other nodes of the popliteal group receive the channels that follow the deep blood vessels of the leg. The efferent vessels of the popliteal nodes follow the femoral vessels to the deep inguinal nodes.

The *deep inguinal nodes* are from one to three in number, lying on the medial side of the femoral vein. If three are present, one is usually located in the femoral canal, one at its upper end (femoral ring), and one below the junction of the greater saphenous and femoral veins. These nodes receive the deep lymphatic drainage of the lower limb, some channels from the penis (or clitoris), and a few of the efferent channels from the superficial inguinal nodes. They discharge to the external iliac nodes. Drainage from the external iliac nodes is through the common iliac and lateral lumbar lymph node groups to the thoracic duct. □

Section II

Embryology

Frank H. Netter, M.D.

in collaboration with

Edmund S. Crelin, Ph.D., D.Sc.
Plates 1–21

Development of Musculoskeletal System

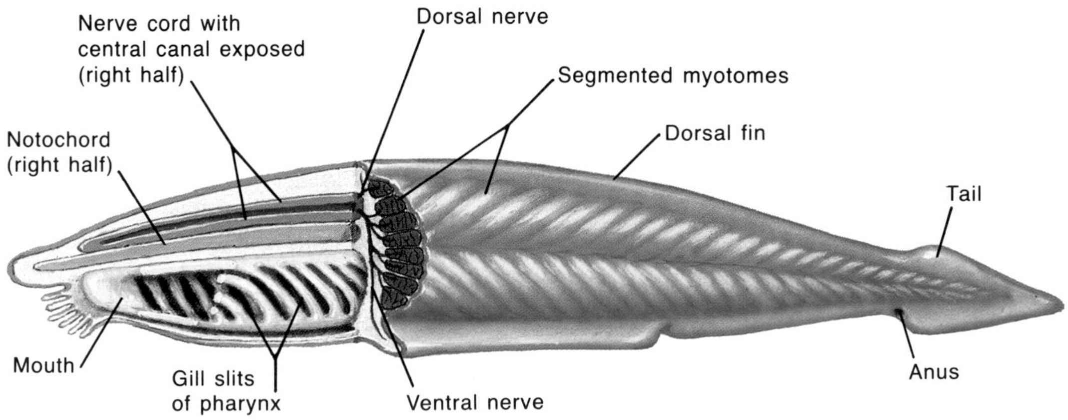

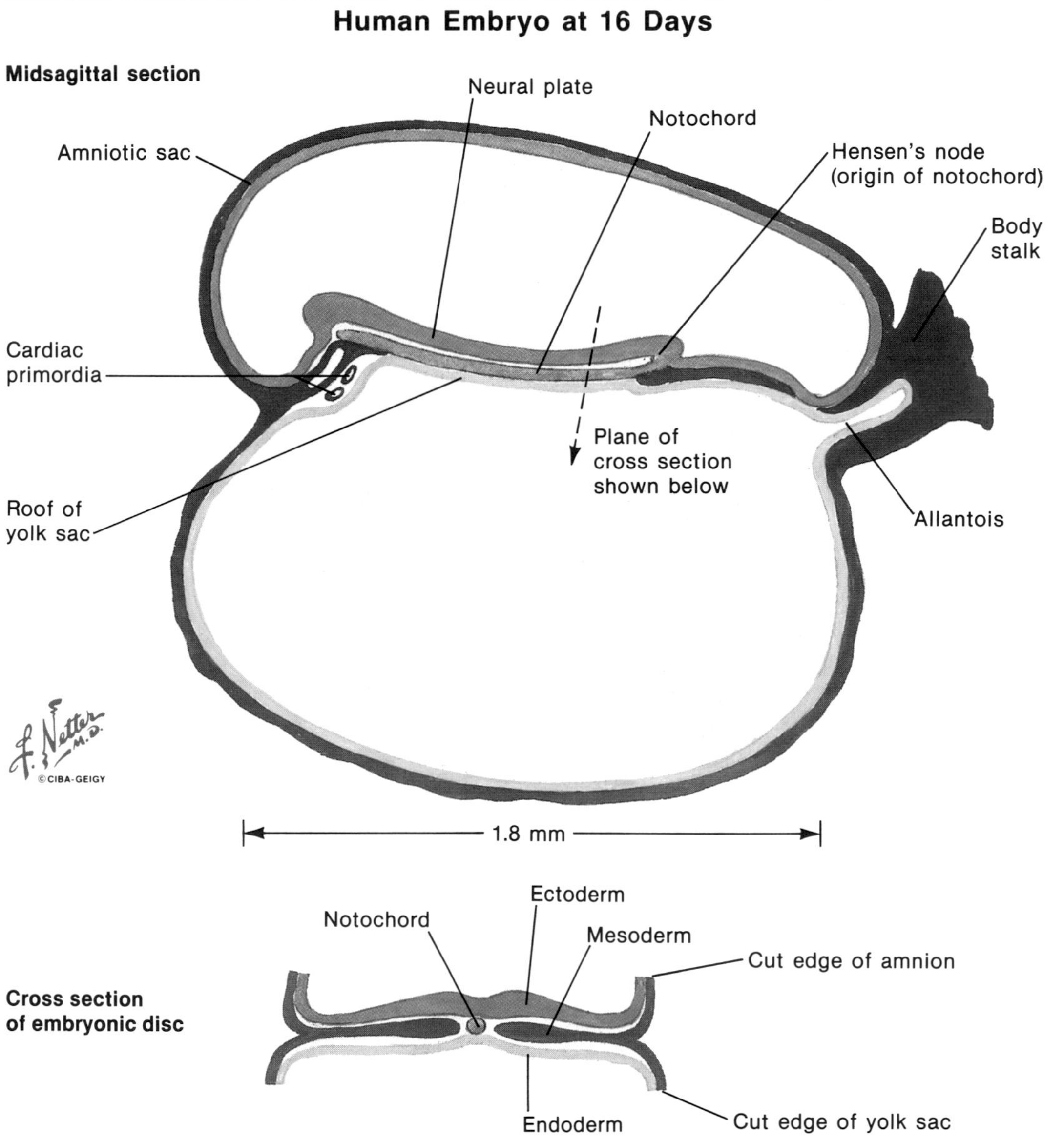

Evolution

The development of the musculoskeletal system in man is an interesting demonstration of ontogeny recapitulating phylogeny. The genetic code that guides the continually changing body plan of the developing human results in a résumé of body plans of the various forms of man's vertebrate ancestors from which fish, amphibians, reptiles, and mammals evolved. In their adult state, a number of living animals resemble some of the ancient ancestors of the central stem line. The knowledge of the fossil record of extinct forms and the comparative anatomy and physiology of living animals makes rational so many aspects of human development that would otherwise have to be regarded as completely wasteful and nonsensical, or both.

Amphioxus

The extant adult amphioxus, or lancelet, is considered to resemble an ancient ancestor of the vertebrates (Plate 1). It is a fishlike animal, about 2 in. long, that has the basic body plan of the early human embryo. The central nervous system consists of a nerve cord resembling the portion of the embryonic neural tube in man that becomes the spinal cord. The digestive, respiratory, excretory, and circulatory systems of the amphioxus also closely resemble those of the early human embryo. As in the early human embryo, the skeleton of the amphioxus consists of a notochord, a slender rod of turgid cells that runs the length of the body directly beneath the nerve cord, or neural tube. The muscular system of the amphioxus consists of individual muscle segments on each side of the body, known as myotomes or myomeres, which are similar in appearance to the myotomes of the early human embryo. The nerve cord of the amphioxus gives off a pair of nerves to each myotome, and the striated muscle fibers of the myotomes contract to produce the lateral bending movements of swimming.

Axial Skeleton

The axial skeleton includes the vertebrae, ribs, sternum, and skull. The first structure of the future axial skeleton to form is the notochord (Plate 1). It appears in the midline of the embryonic disc at 15 days of development as a cord of cells budding off from a mass of ectoderm known as Hensen's node. The notochordal cells become temporarily intercalated in the endoderm, which forms the roof of the yolk sac. After separating from the endoderm, the notochord becomes a slender rod of cells running the length of the embryo between the neural tube and the developing gut.

The dorsal mesoderm on either side of the notochord becomes thickened and arranged into 42 to 44 pairs of cell masses known as somites (4 occipital, 8 cervical, 12 thoracic, 5 lumbar, 5 sacral, 8–10 coccygeal) between the nineteenth and thirty-second day of development. The formation of these primitive segments, or somites, reflects the serial repetition of homologous parts known as metamerism, which is retained in many adult prevertebrates. The vertebrate embryo is fundamentally metameric, even though much of its segmentation is lost as development proceeds to the adult form. The first significant change in the somite of the human embryo is the formation of a cluster of mesenchymal cells, the sclerotome, on the ventromedial border of the somite (Plate 2). The sclerotomal cells migrate from the somites and become aggregated about the notochord to ultimately give rise to the vertebral column and ribs (Plate 3).

Vertebral Column and Ribs

During the fourth week of development, a clustering of sclerotomal cells derived from two adjacent somites on either side of the notochord becomes the primordium of the body, or centrum, of a vertebra. Soon after the body takes shape, paired concentrations of mesenchymal cells extend dorsally and laterally from the body to form the primordia of the neural arches and the costal processes. The costal process becomes a rib that articulates with the body and transverse process of

Development of Musculoskeletal System

(Continued)

Differentiation of Somites Into Myotomes, Sclerotomes, and Dermatomes

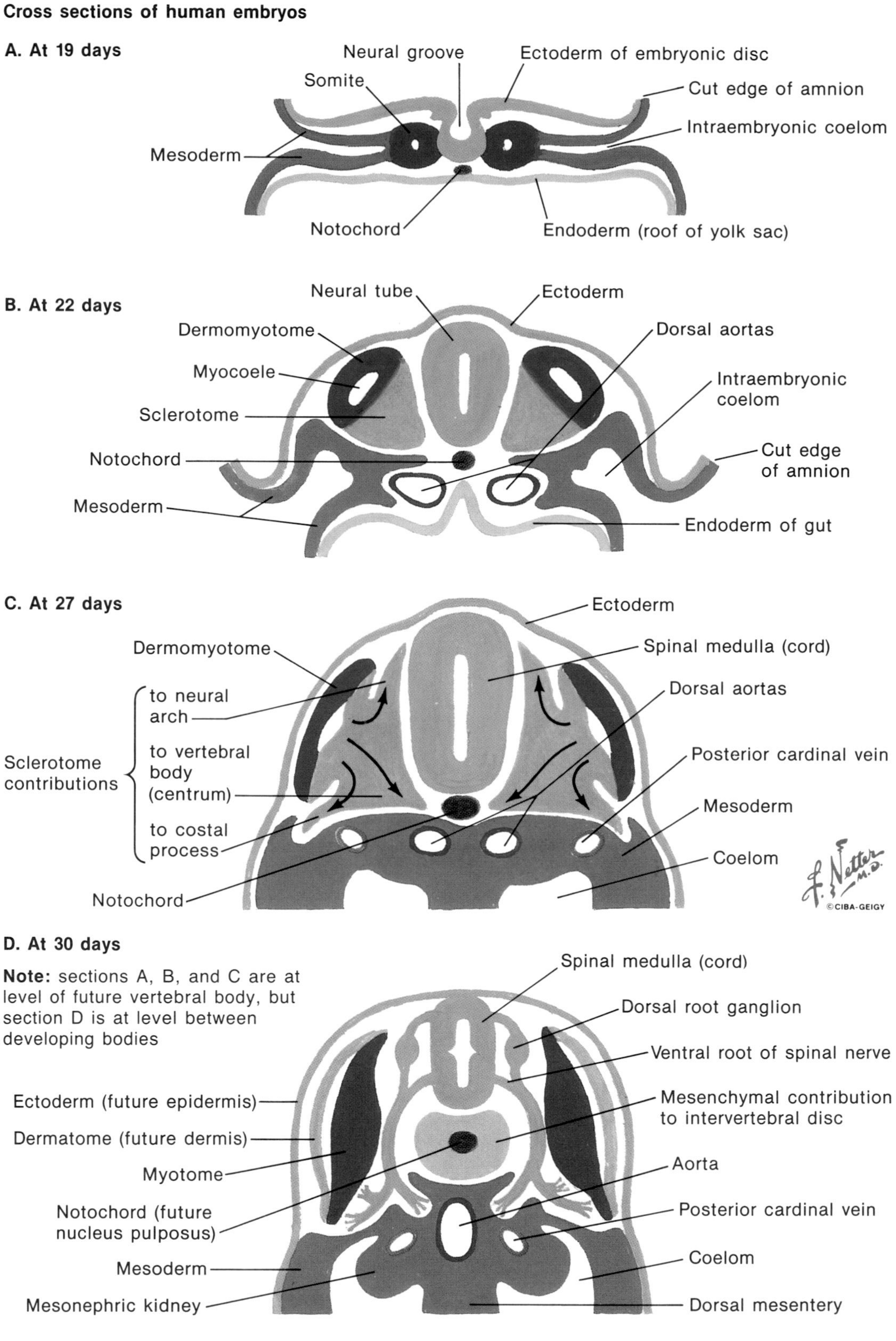
Cross sections of human embryos
A. At 19 days
Neural groove
Ectoderm of embryonic disc
Somite
Cut edge of amnion
Intraembryonic coelom
Mesoderm
Notochord
Endoderm (roof of yolk sac)
B. At 22 days
Neural tube
Ectoderm
Dermomyotome
Dorsal aortas
Myocoele
Intraembryonic coelom
Sclerotome
Notochord
Cut edge of amnion
Mesoderm
Endoderm of gut
C. At 27 days
Ectoderm
Dermomyotome
Spinal medulla (cord)
to neural arch
Dorsal aortas
Sclerotome contributions
to vertebral body (centrum)
Posterior cardinal vein
Mesoderm
to costal process
Coelom
Notochord
F. Netter M.D.
©CIBA-GEIGY
D. At 30 days
Note: sections A, B, and C are at level of future vertebral body, but section D is at level between developing bodies
Spinal medulla (cord)
Dorsal root ganglion
Ventral root of spinal nerve
Ectoderm (future epidermis)
Mesenchymal contribution to intervertebral disc
Dermatome (future dermis)
Aorta
Myotome
Notochord (future nucleus pulposus)
Posterior cardinal vein
Coelom
Mesoderm
Mesonephric kidney
Dorsal mesentery

the neural arch of the thoracic vertebrae (Plate 4). The costal process becomes the anterior part of the transverse foramen of the cervical vertebrae, the transverse process of the lumbar vertebrae, and the lateral part of the sacrum. Occasionally, the costal process of the seventh cervical or the first lumbar vertebra becomes a supernumerary rib. Failure of fusion of the neural folds results in various types of spina bifida.

The vertebrae and ribs in the mesenchymal, or blastemal, stage are one continuous mass of cells. This stage is quickly followed by the cartilage stage, when the mesenchymal cells become chondrocytes and produce cartilage matrix during the seventh week, beginning in the upper vertebrae. By the time ossification begins at 9 weeks, the rib cartilages have become separated from the vertebrae.

The clustering of sclerotomal cells to form the bodies of the vertebrae establishes intervertebral fissures that fill with mesenchymal cells to become the intervertebral discs (Plate 3). The notochord in the center of the developing intervertebral disc expands as its cells produce a large amount of mucoid semifluid matrix to form the nucleus pulposus. The mesenchymal cells surrounding the nucleus pulposus produce cartilage matrix and collagen fibers to become the fibrocartilage annulus fibrosus of the intervertebral disc. At birth, the nucleus pulposus makes up the bulk of an intervertebral disc. From birth to adulthood, it serves as a shock-absorbing mechanism, but by 10 years of age, the notochordal cells have disappeared and the surrounding fibrocartilage begins to gradually replace the mucoid matrix. The water-binding capacity and elasticity of the matrix are also gradually reduced.

The portion of the notochord surrounded by the developing body of a vertebra usually disappears completely before maturity. This is also true of the portions that become incorporated into the body of the sphenoid and the basilar part of the occipital bone. However, the portion of the notochord that normally becomes the nucleus pulposus in the intervertebral discs becomes the apical dental ligament, connecting the dens of the axis with the occipital bone. The dens evolved as an addition to the body of the first cervical vertebra, the atlas, in those reptiles that gave rise to mammals. The most primitive of mammals, the duck-billed platypus and the spiny anteater, have a large atlas body and a dens. In the human embryo, the atlas body and dens become dissociated as a unit from the rest of the atlas and fuse with the body of the second cervical vertebra, the axis (Plate 5). This fusion results in a mature ring-shaped atlas with an anterior arch lacking a body.

At 5 weeks, a prominent tail containing coccygeal vertebrae is present in the human embryo (Plate 3). A free-moving tail is characteristic of most adult vertebrates. However, the human tail is concealed by the growing buttocks and actually regresses to become the coccyx, which consists of four or five rudimentary vertebrae fused together.

Sternum

At 6 weeks, a pair of bands of mesenchymal cells, the sternal bars, appear ventrolaterally in the body wall (Plate 5). They have no connection with the ribs or with each other, and their formation is independent of any sclerotomal derivatives. Following the attachment of the upper ribs to the sternal bars, they fuse together progressively in a craniocaudal direction. At 9 weeks, the union of the bars, which have become cartilaginous, is complete. At the cranial end of the sternal bars, two suprasternal masses form and fuse with the future manubrium to serve as sites where the clavicles articulate. Influenced by the ribs, the cartilaginous body of the sternum becomes secondarily segmented into six sternebrae. Faulty fusion of the sternal bars in the midline results either in a cleft or perforated sternum or in a bifid xiphoid process.

Development of Musculoskeletal System

(Continued)

Skull

The skeleton of the head consists of three primary components: (1) the capsular investments of the sense organs, (2) the brain case, and (3) the branchial arch skeleton (Plate 6). Other than some exceptions of the branchial arch skeleton, these three primary components unite into a composite mammalian skull.

The notochord originally extends into the head of the embryo as far as the oropharyngeal membrane. Its termination later shifts to the caudal border of the hypophyseal fossa of the sphenoid bone. (The replacement of the notochord in the head region during evolution involved the formation of a cartilaginous cranium similar to that in the primitive fish of the shark type, which had a skeleton composed of only cartilage.) The earliest indication of skull formation in the human embryo is the concentration of mesenchyme about the notochord at the level of the hindbrain during the fifth and sixth weeks (Plate 3). This mesenchymal skull formation extends forward to form a floor for the developing brain. By the seventh week, the skull begins to become cartilaginous as it completely or incompletely encapsulates the organs of olfaction (nasal capsule), vision (orbitosphenoid), and audition and equilibrium (otic capsule). This chondrocranium is essentially roofless.

As the evolving brain increased in size, additional rudiments were acquired to form a top to the braincase—the calvaria (skullcap). In bony fish, these were derived from the enlarged scales of the head region, which sank into the head and sheathed the chondrocranium to become the bones of the top and sides of the skull and the jaws. These encasing bones derived from the skin are known as dermal, or membrane, bones. In the human embryo, the mesenchymal membrane bone rudiments form the top and sides of the skull and the bones of the face and jaws. They never transform into cartilage; therefore, bone forms directly within the membranous tissue. Most of the membrane bone rudiments become independent bones, but a few become parts of bones formed in the chondrocranium.

The branchial arch skeleton is derived from the embryonic counterparts of the gill arches that support the mouth and pharynx of present-day adult fish and tailed amphibians. The most primitive skeletal rudiments of the branchial arches develop from neural crest cells that migrate into the arches, not from the mesoderm of the arches. The neural crest rudiments become cartilaginous and are retained as cartilage in present-day adult cartilaginous fish, such as the shark, to support the jaw and aqueous respiratory system. In the

Progressive Stages in Formation of Vertebral Column, Dermatomes, and Myotomes

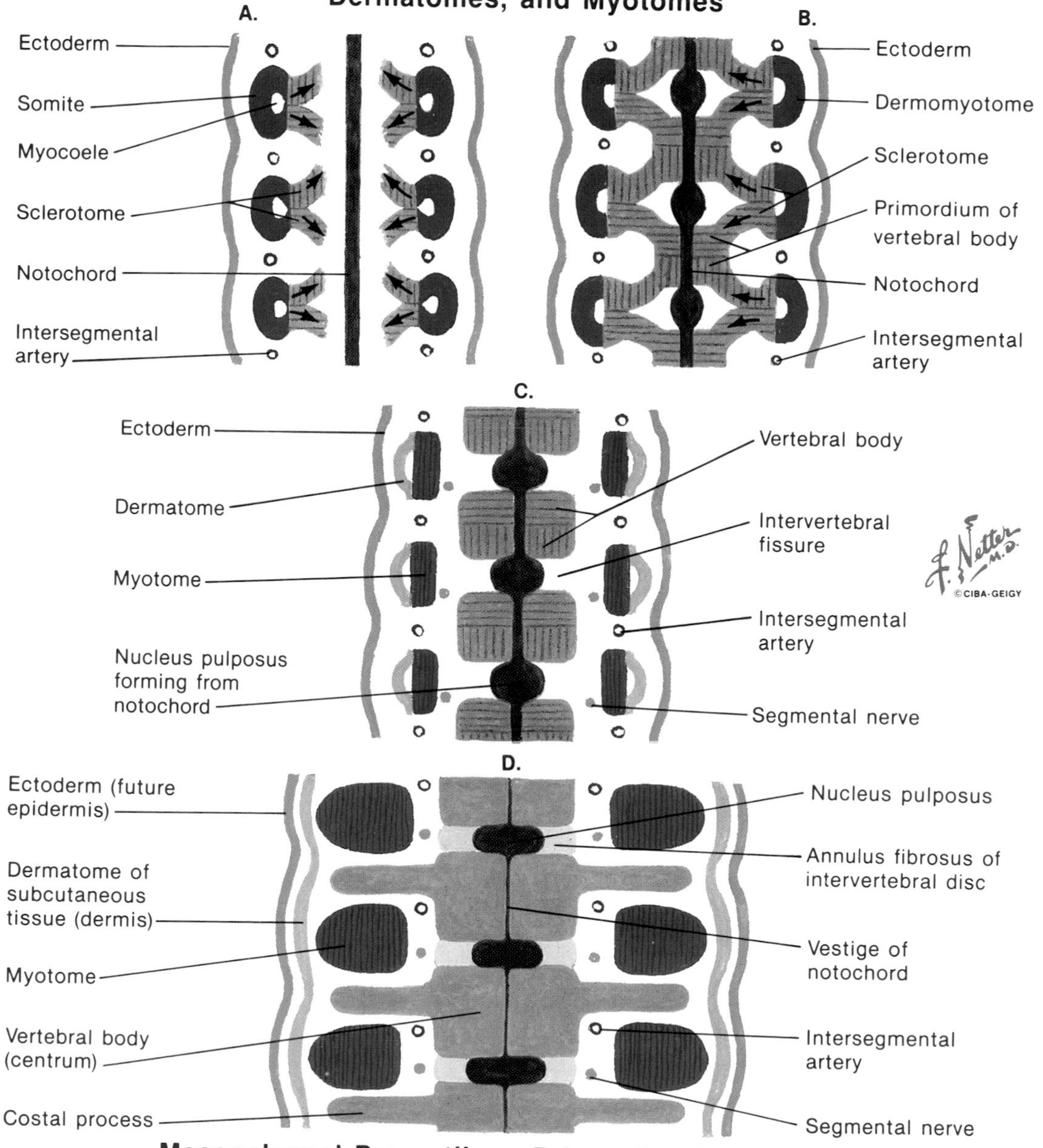
A.
B.
Ectoderm
Somite
Myocoele
Sclerotome
Notochord
Intersegmental artery
Ectoderm
Dermomyotome
Sclerotome
Primordium of vertebral body
Notochord
Intersegmental artery
C.
Ectoderm
Dermatome
Myotome
Nucleus pulposus forming from notochord
Vertebral body
Intervertebral fissure
Intersegmental artery
Segmental nerve
F. Netter M.D.
©CIBA-GEIGY
D.
Ectoderm (future epidermis)
Dermatome of subcutaneous tissue (dermis)
Myotome
Vertebral body (centrum)
Costal process
Nucleus pulposus
Annulus fibrosus of intervertebral disc
Vestige of notochord
Intersegmental artery
Segmental nerve

Mesenchymal Precartilage Primordia of Axial and Appendicular Skeletons at 5 Weeks

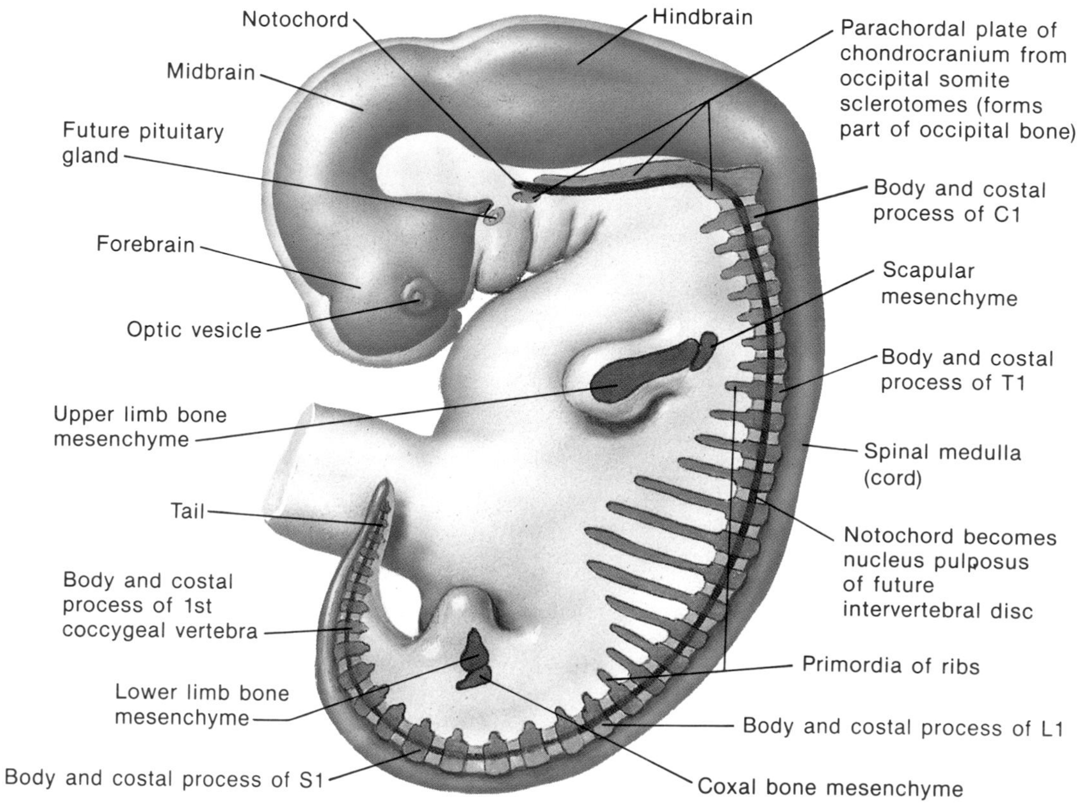
Notochord
Hindbrain
Midbrain
Future pituitary gland
Forebrain
Optic vesicle
Upper limb bone mesenchyme
Tail
Body and costal process of 1st coccygeal vertebra
Lower limb bone mesenchyme
Body and costal process of S1
Parachordal plate of chondrocranium from occipital somite sclerotomes (forms part of occipital bone)
Body and costal process of C1
Scapular mesenchyme
Body and costal process of T1
Spinal medulla (cord)
Notochord becomes nucleus pulposus of future intervertebral disc
Primordia of ribs
Body and costal process of L1
Coxal bone mesenchyme

Development of Musculoskeletal System

(Continued)

Fate of Body, Costal Process, and Neural Arch Components of Vertebral Column, With Sites and Time of Appearance of Ossification Centers

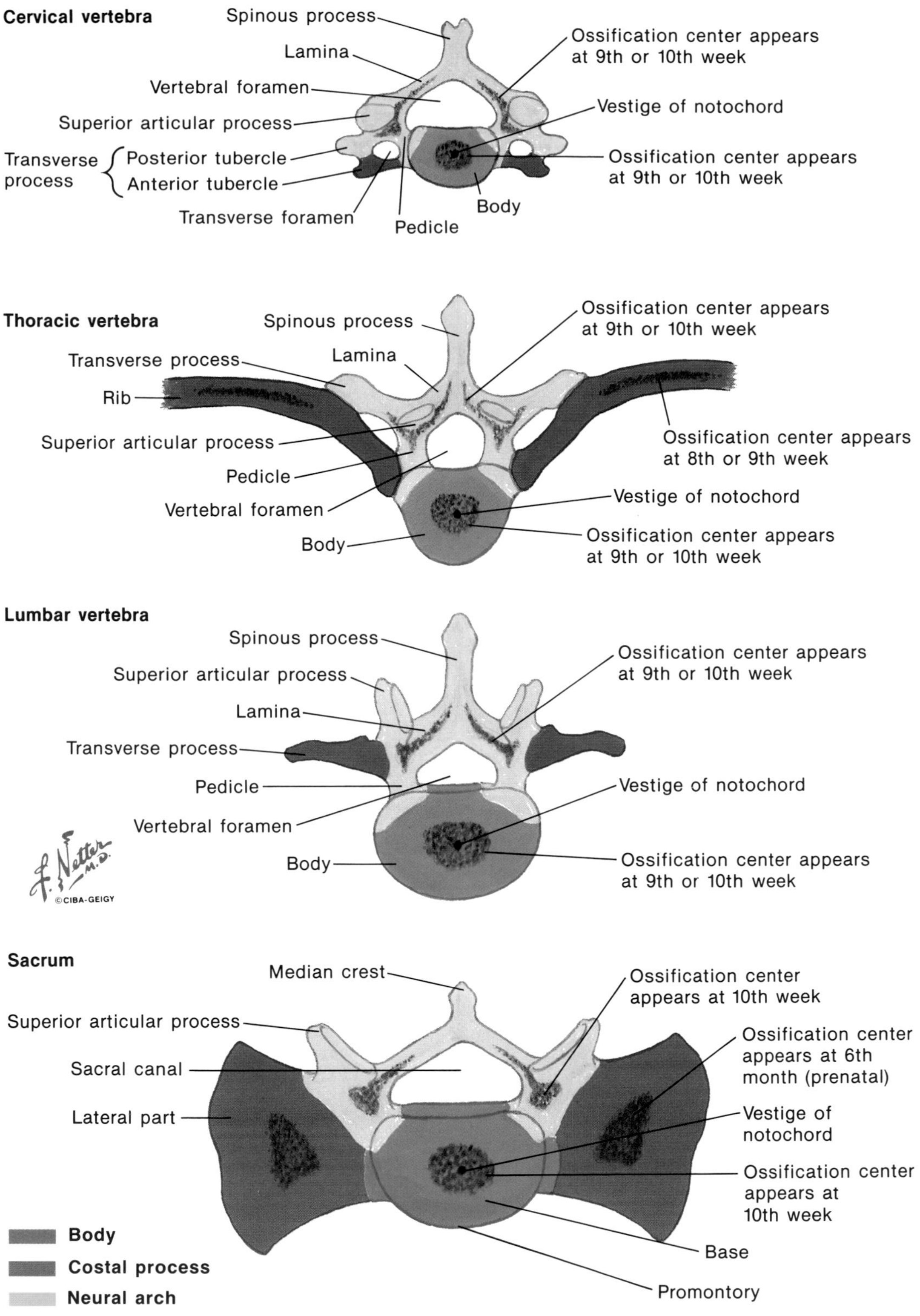
Cervical vertebra
Spinous process
Lamina
Vertebral foramen
Superior articular process
Transverse process
Posterior tubercle
Anterior tubercle
Transverse foramen
Pedicle
Body
Ossification center appears at 9th or 10th week
Vestige of notochord
Ossification center appears at 9th or 10th week
Thoracic vertebra
Spinous process
Lamina
Transverse process
Rib
Superior articular process
Pedicle
Vertebral foramen
Body
Ossification center appears at 9th or 10th week
Ossification center appears at 8th or 9th week
Vestige of notochord
Ossification center appears at 9th or 10th week
Lumbar vertebra
Spinous process
Superior articular process
Lamina
Transverse process
Pedicle
Vertebral foramen
Body
Ossification center appears at 9th or 10th week
Vestige of notochord
Ossification center appears at 9th or 10th week
F. Netter M.D.
©CIBA-GEIGY
Sacrum
Median crest
Superior articular process
Sacral canal
Lateral part
Ossification center appears at 10th week
Ossification center appears at 6th month (prenatal)
Vestige of notochord
Ossification center appears at 10th week
Base
Promontory
Body
Costal process
Neural arch

evolutionary transformation from water breathing to air breathing, much of the skeleton of the aqueous respiratory system was modified to become parts of the air respiratory system, as well as of the modified acoustic apparatus. The human embryo goes through the essential structural stages of this evolutionary water-breathing to air-breathing transformation. Some of the cartilages remain in the adult human (laryngeal cartilages), whereas others become bone (hyoid, styloid process, and ossicles of the middle ear). The branchial arch components originally subserved the function of mastication as well as that of respiration. Although the primitive cartilages of the first branchial arches become the skeletons of the upper and lower jaws in cartilaginous fish, they do not do so in man, in whom the maxillae and mandible are derived from membrane bones.

Because the brain grows large before birth, the calvaria is much larger than the facial skeleton in the newborn infant—a ratio of 8:1, compared with a ratio of 2:1 in the adult (Plate 7).

Appendicular Skeleton

The appendicular skeleton consists of the pectoral and pelvic girdles and the bones of the free appendages attached to them. The paired appendages of land vertebrates evolved from the paired fins of fish. The development of the human limbs is a résumé of their evolution.

The upper limb buds appear first, differentiate sooner, and attain their final relative size earlier than the lower limbs (Plates 3, 8–9). Not until birth do the lower limbs equal the upper limbs in length (Plate 7). However, throughout childhood, the lower limbs elongate faster than the upper limbs. In essence, an upper limb was never a lower limb, and vice versa; each has its own unique evolutionary and developmental history. Even so, it is interesting that the structures of the mature upper and lower limbs have a number of similarities. They are most similar during the earliest stages of development, when both sets of finlike appendages point caudally. They then become paddlelike and project outward almost at right angles to the body wall. Following this, they bend at the elbow and knee directly anteriorly, so that the elbow and knee point laterally, or outward, and the palm and sole face the trunk. Then a series of major changes occurs that causes the upper and lower limbs to differ markedly both structurally and functionally (Plate 8). By the seventh week, both undergo a 90° torsion about their long axes, but in opposite directions, so that the elbow points caudally and the knee points cranially. Accompanying this torsion is a permanent twisting of the entire lower limb, which results in its cutaneous innervation assuming a twisted, "barber pole" arrangement (Plate 9). This would be similar to twisting the upper limb so that the forearm and hand become fully and permanently pronated.

The limb buds appear during the fourth week and consist of a core of condensed mesenchyme covered with an epidermal cap, the apical ectodermal ridge. They are functionally related in a two-way process of induction: the mesenchyme induces the development and maintenance of the ridge, which in turn gives the mesenchymal cells the "competence" to form the skeletal rudiments. Any genetic breakdown of differentiating cells or the presence of a teratogenetic substance that interferes with this two-way process of induction results in various limb malformations, such as amelia (total failure of limb development), hemimelia (failure of development of distal parts of limbs), or phocomelia (failure of development of the bulk of the limb but not of its distal part).

Once the appendicular skeleton starts to develop, the progress is rapid. Early in the sixth week, only vague concentrations of mesenchyme represent the primordia of future bones. By the end of the sixth week, these cellular concentrations are sufficiently molded so that some of the larger future bones can be detected. During the seventh week, the primordia of many of the smaller bones of the hand and foot are present.

Development of Musculoskeletal System

(Continued)

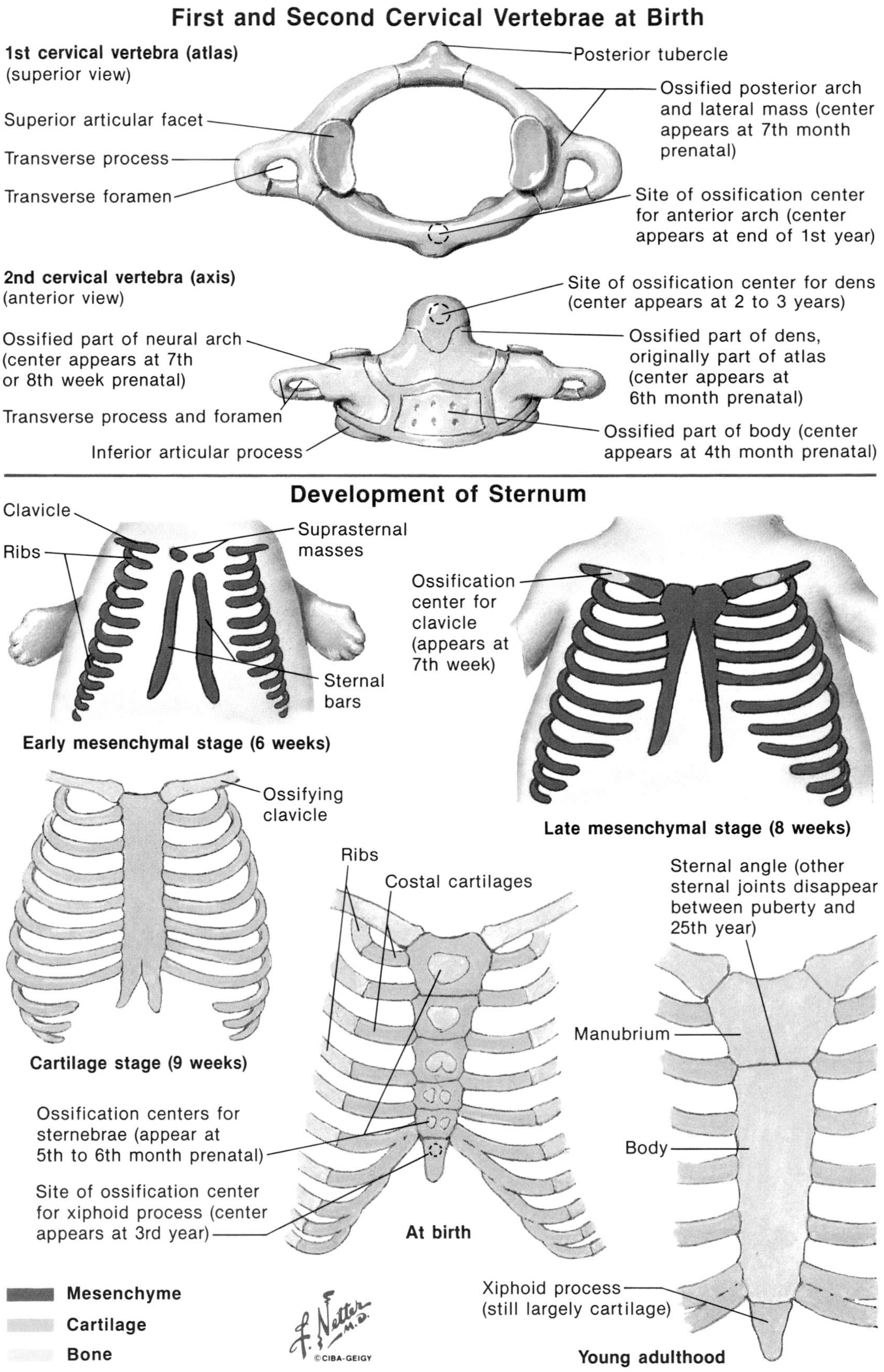
First and Second Cervical Vertebrae at Birth
1st cervical vertebra (atlas)
(superior view)
Superior articular facet
Transverse process
Transverse foramen
Posterior tubercle
Ossified posterior arch and lateral mass (center appears at 7th month prenatal)
Site of ossification center for anterior arch (center appears at end of 1st year)
2nd cervical vertebra (axis)
(anterior view)
Ossified part of neural arch (center appears at 7th or 8th week prenatal)
Transverse process and foramen
Inferior articular process
Site of ossification center for dens (center appears at 2 to 3 years)
Ossified part of dens, originally part of atlas (center appears at 6th month prenatal)
Ossified part of body (center appears at 4th month prenatal)
Development of Sternum
Clavicle
Ribs
Suprasternal masses
Sternal bars
Early mesenchymal stage (6 weeks)
Ossification center for clavicle (appears at 7th week)
Late mesenchymal stage (8 weeks)
Ossifying clavicle
Cartilage stage (9 weeks)
Ribs
Costal cartilages
Ossification centers for sternebrae (appear at 5th to 6th month prenatal)
Site of ossification center for xiphoid process (center appears at 3rd year)
At birth
Sternal angle (other sternal joints disappear between puberty and 25th year)
Manubrium
Body
Xiphoid process (still largely cartilage)
Young adulthood
Mesenchyme
Cartilage
Bone
F. Netter M.D.
©CIBA-GEIGY

By the eighth week, well-molded cartilage rudiments represent all the major future bones of the appendicular skeleton.

Bone Formation

Bone forms in areas occupied by either connective tissue or cartilage. Bone formed in connective tissue is of intramembranous origin and is called membrane bone. Most of the bones of the calvaria, the facial bones, and, in part, the clavicle and mandible, are membrane bones. All the other bones of the body form in areas occupied by cartilage, which they gradually replace. These bones are of endochondral origin and are called cartilage bones. The terms "membrane bone" and "cartilage bone" merely describe the environment in which a bone forms, not the microscopic structure once the bone is completely developed.

Membrane Bone. The cells of the mesenchymal rudiment of a membrane bone begin to produce a mucoprotein matrix in which collagen fibers are embedded (Plate 10). Within this organic matrix, which is known as osteoid, inorganic crystals of calcium phosphate are deposited between, on, and within the collagen fibers. This mineralization of the osteoid is known as ossification. The calcium-to-phosphate ratio increases in the bone matrix as ossification proceeds before birth, chiefly in the form of a series of minerals known as apatites. As development proceeds to the time of birth, hydroxyapatite emerges as the dominant component of bone mineral. Hydroxyapatite is the basic inorganic constituent of mature bone, and its hydroxyl groups are partially substituted by other chemical elements and radicals, such as fluoride or carbonate.

The mesenchymal cells involved in bone formation become known as osteoblasts. As bone formation proceeds, the osteoblasts divide and some become completely surrounded by osteoid. The trapped osteoblasts, then known as osteocytes, send out long, thin extensions of their cell bodies in all directions, which make contact with the cellular extensions of adjacent osteocytes also laying down osteoid (Plate 10). When bone mineral is deposited in the osteoid, the space in the matrix housing the portion of the osteocyte containing its nucleus is known as a lacuna, and the tiny, tubular spaces radiating out from the lacuna containing the extensions of the osteocyte are known as canaliculi (Plate 11).

Once the matrix is ossified, diffusion of nutrients to sustain the osteocytes and transport of ions through it cannot occur. Therefore, the canaliculi are the transport channels that interconnect the bone spaces containing blood capillaries and the lacunae surrounding the part of the osteocyte in which the nucleus is located. Since the extensions of the osteocytes fill the canaliculi, the passage of material through the canaliculi is via cell transport.

In the formation of membrane bone, individual shafts of bone, known as trabeculae, are laid down (Plate 10). Trabeculae increase in length and thickness and join each other at various points to produce a lattice framework of primary trabecular bone. At the outer surface of the bone rudiment, the dense sheath of connective tissue acquires an inner layer of osteoblasts to become the periosteum. The osteoblastic layer lays down bone in the form of subperiosteal layers, or lamellae. The coalescing trabeculae in the deeper parts of the rudiment surround capillaries and nerves. Bone is laid down in layers on these trabeculae to constitute the lamellae of primary trabecular bone. Up to the time of birth, the bones of the fetal skeleton are made up chiefly of this type of bone, but near the time of birth, this primary trabecular bone begins to transform into compact bone (Plate 11).

The transformation from trabecular to compact bone is essentially the reduction in the size of the marrow spaces containing mesenchymal cells, capillaries, and nerve fibers. The relatively large marrow spaces with their surrounding bony trabeculae are known as primary osteons. The osteoblasts lining the trabeculae surrounding a marrow space (which contains one or two capillaries, some

Development of Musculoskeletal System

(Continued)

Early Development of Skull

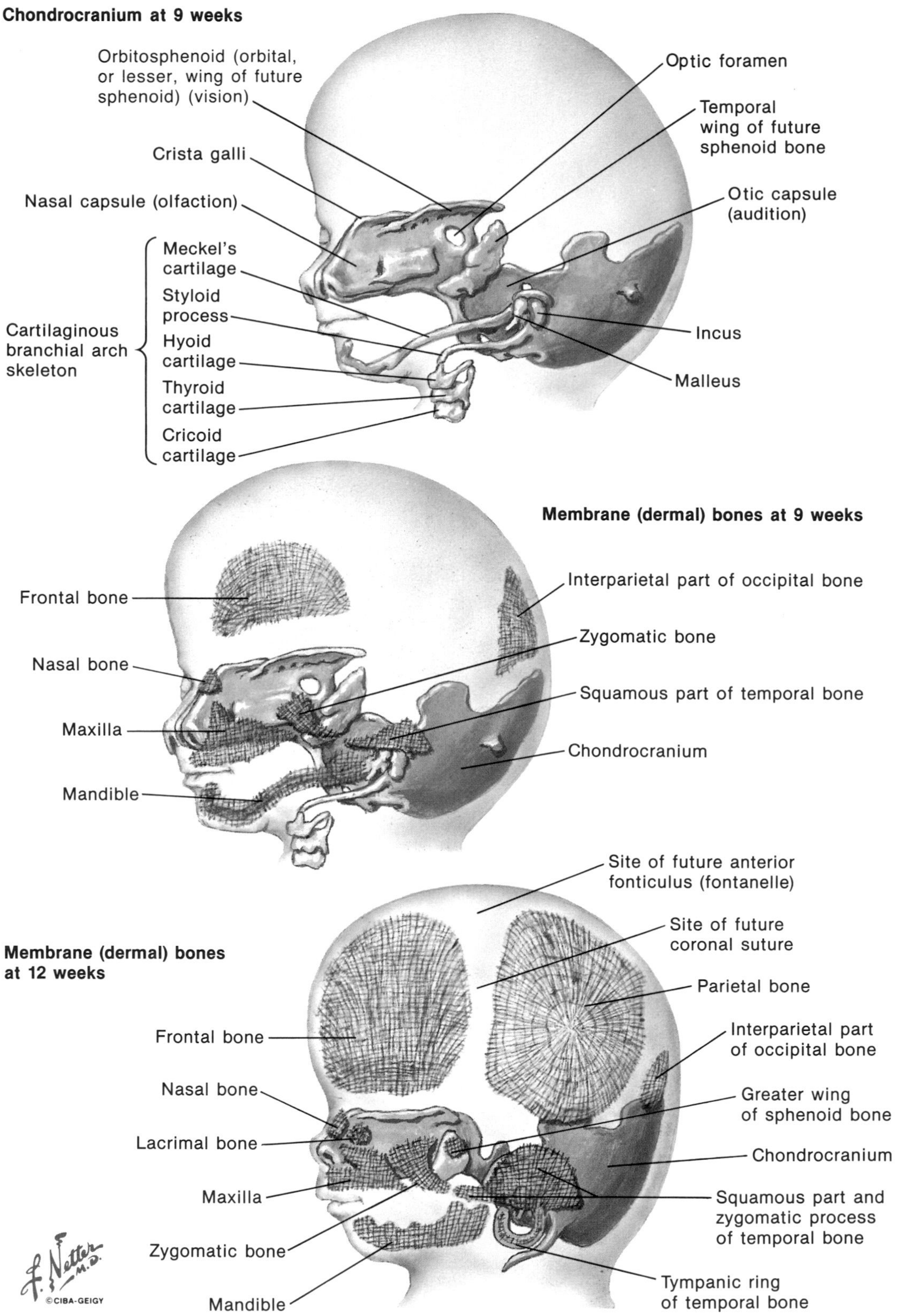

Chondrocranium at 9 weeks
Orbitosphenoid (orbital, or lesser, wing of future sphenoid) (vision)
Crista galli
Nasal capsule (olfaction)
Optic foramen
Temporal wing of future sphenoid bone
Otic capsule (audition)
Cartilaginous branchial arch skeleton
Meckel's cartilage
Styloid process
Hyoid cartilage
Thyroid cartilage
Cricoid cartilage
Incus
Malleus
Membrane (dermal) bones at 9 weeks
Frontal bone
Nasal bone
Maxilla
Mandible
Interparietal part of occipital bone
Zygomatic bone
Squamous part of temporal bone
Chondrocranium
Membrane (dermal) bones at 12 weeks
Site of future anterior fonticulus (fontanelle)
Site of future coronal suture
Parietal bone
Interparietal part of occipital bone
Greater wing of sphenoid bone
Chondrocranium
Squamous part and zygomatic process of temporal bone
Tympanic ring of temporal bone
Frontal bone
Nasal bone
Lacrimal bone
Maxilla
Zygomatic bone
Mandible
F. Netter M.D.
©CIBA-GEIGY

perivascular cells, and a nonmyelinated and occasionally a myelinated nerve fiber) lay down bone in concentric layers, or lamellae. This process continues until the marrow space is nearly obliterated, leaving a small central osteonal, or haversian, canal. The canal is about 50 μm in diameter and usually contains a single capillary and nerve fiber and some perivascular cells in the center of what is known as a secondary osteon (haversian system). There are from 4 to 20 (usually 6 or less) concentric lamellae that are each 3 to 7 μm thick. The formation of many such adjacent secondary osteons converts what was originally trabecular bone into compact bone. In the central core of a membrane bone, the marrow cavities persist, and their mesenchymal tissue develops into hematopoietic red bone marrow. Thus, in a fully formed, flat bone of the calvaria, there is an inner and outer table of compact bone, between which is trabecular bone surrounding a marrow cavity, the diploë.

The secondary osteons of compact bone usually run the length of a bone. In cross section, the outer limit of each osteon is clearly demarcated by a narrow refractile ring known as a cement line, which lacks collagen fibrils and is highly mineralized. The central haversian canals are connected to one another and communicate with the periosteal surface as well as with the marrow cavity via transverse and oblique channels known as Volkmann's canals. The blood flows through the compact bone from the inner marrow cavity via vessels in Volkmann's and haversian canals until it emerges at the periosteal surface (see Section III, Plate 22).

Cartilage Bone. The cartilage rudiments of bones of endochondral origin are temporary miniatures of the future adult bone. With the exception of the clavicle, the long bones are of endochondral origin. The first of two or more ossification centers of a long bone appears in the shaft, or diaphysis (Plate 12). Diaphyseal ossification is actually a form of intramembranous ossification, because bone is laid down by the connective tissue outer sheath of the cartilage rudiment known as the perichondrium. The perichondrium becomes known as the periosteum once it starts to lay down bone in the form of a delicate collar surrounding the center of the diaphysis of the cartilage rudiment. Deep to this collar of bone, the cartilage matrix becomes calcified and the chondrocytes hypertrophy.

Irruption canals appear in the bony collar through which vascular buds of capillaries and mesenchymal cells pass from the periosteum to the calcified cartilage, which undergoes a breakdown. Some of the chondrocytes die and others modulate or transform into chondroclasts and osteoblasts. This process brings into existence primordial marrow cavities, which contain osteoblasts, and vascular marrow tissue, which is derived from the irruption canal cells. The osteoblasts initially lay down bone along the remaining spicules of calcified cartilage matrix. As a result, the endochondral bone becomes trabecular. As the periosteal and the endochondral bone formation occurring at the center of the diaphysis extends toward each end of the long bone, a large central medullary (marrow) cavity arises in the trabecular bone of the diaphysis. Toward the end of fetal life and continuing into puberty, ossification centers appear in the two cartilaginous ends, or epiphyses, of the long bone (Plate 12). Between the bone formed in the diaphysis and that formed in the epiphysis is the epiphyseal plate, a circular mass of cartilage in a region of the long bone known as the metaphysis. It is at the epiphyseal plate that the diaphysis continues to grow in length.

Bone Growth

Cartilage grows continually on the side of the epiphyseal plate facing the epiphysis of a long bone, while on the opposite side of the plate facing the diaphysis, cartilage breaks down continually and is replaced by bone (Plate 12). These epiphyseal growth plates persist during the entire postnatal growth period. The plates are finally resorbed and replaced by bone that joins the

Development of Musculoskeletal System

(Continued)

Skeleton of Full-Term Newborn

Time of appearance of ossification centers (primary unless otherwise indicated)

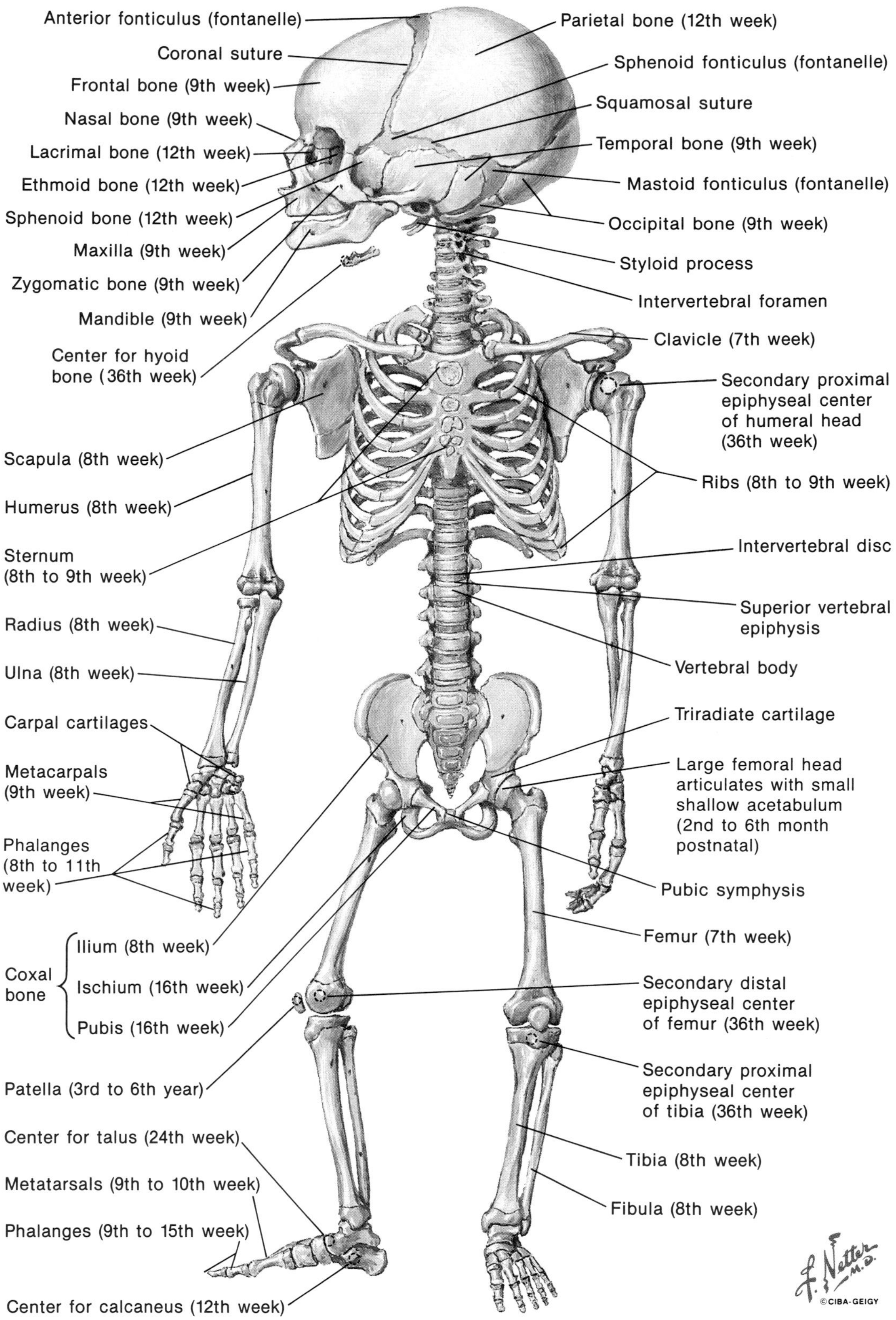

Anterior fonticulus (fontanelle)
Coronal suture
Frontal bone (9th week)
Nasal bone (9th week)
Lacrimal bone (12th week)
Ethmoid bone (12th week)
Sphenoid bone (12th week)
Maxilla (9th week)
Zygomatic bone (9th week)
Mandible (9th week)
Center for hyoid bone (36th week)
Scapula (8th week)
Humerus (8th week)
Sternum (8th to 9th week)
Radius (8th week)
Ulna (8th week)
Carpal cartilages
Metacarpals (9th week)
Phalanges (8th to 11th week)
Coxal bone
Ilium (8th week)
Ischium (16th week)
Pubis (16th week)
Patella (3rd to 6th year)
Center for talus (24th week)
Metatarsals (9th to 10th week)
Phalanges (9th to 15th week)
Center for calcaneus (12th week)
Parietal bone (12th week)
Sphenoid fonticulus (fontanelle)
Squamosal suture
Temporal bone (9th week)
Mastoid fonticulus (fontanelle)
Occipital bone (9th week)
Styloid process
Intervertebral foramen
Clavicle (7th week)
Secondary proximal epiphyseal center of humeral head (36th week)
Ribs (8th to 9th week)
Intervertebral disc
Superior vertebral epiphysis
Vertebral body
Triradiate cartilage
Large femoral head articulates with small shallow acetabulum (2nd to 6th month postnatal)
Pubic symphysis
Femur (7th week)
Secondary distal epiphyseal center of femur (36th week)
Secondary proximal epiphyseal center of tibia (36th week)
Tibia (8th week)
Fibula (8th week)
F. Netter M.D.
©CIBA-GEIGY

epiphyses permanently to the diaphysis when the skeleton has acquired its adult size. The epiphyses unite with the diaphysis sooner in females than in males, so that growth in length ceases about 2 years earlier in females. In males, most fusions of the epiphyses with the diaphyses end at about age 20. Interference with the normal growth occurring at the epiphyseal plates of the appendicular skeleton results in abnormally short limbs, such as those of an achondroplastic dwarf who may have a head and body of normal length. Achondroplasia is usually a genetic abnormality.

Peripheral growth of a typical flat bone of membrane origin occurs at the margins that articulate via connective tissue with other flat bones. At first, these articulations are broad. At certain intervals between the growing skull bones, even wider gaps known as fonticuli, or fontanelles, occur (Plate 7). Of these large, soft spots, the two sphenoid fonticuli may become nearly obliterated as early as 6 months after birth, whereas the two mastoid fonticuli and the single anterior fonticulus are nearly obliterated by age 2. Obliteration of the narrow intervals between the bones of the calvaria, the sutures, does not begin until about age 30.

Growth in width of a flat membrane bone and a long endochondral bone is similar. The osteoblasts of the periosteum of both the outer and inner tables of a flat bone and of the surface of a long bone lay down bone in the form of subperiosteal circumferential layers, or lamellae, that are parallel to the bone surface (Plate 13). To prevent an overly thick mass of compact bone from forming as the bone grows in width, bone is resorbed concomitantly at the endosteal surface bordering the marrow cavity. This laying down of bone at the surface involves a peripheral shift of osteons that retains the necessary distance between the intrinsic blood supply and the osteocytes of the bone. There is an eccentric resorption of osteons on the side facing the outer surface of the widening bone. The bone resorption is the result of progenitor cells within the central haversian canal modulating into osteoclasts, as well as osteoclastic activity of the osteocytes within the lacunae of the circular lamellae in the path of bone erosion. The dissolution of their surrounding matrix by individual osteocytes is known as osteocytic osteolysis. When these osteoclastic osteocytes are released from their lacunae, they may fuse with each other to form multinucleated osteoclasts. Plate 13 shows the sequence of events in this destruction of osteons and the formation of new ones.

Bone Remodeling

For a bone to maintain its proper form and proportions while it lengthens and thickens, the growth process must involve more than merely bone formation at the periosteal surface and concomitant bone resorption at the endosteal surface (Plate 14). Progressive remodeling, with formation and resorption (or a reversal of this process), must occur at all parts of the bone as its dimensions alter. At times, all activity ceases.

Although bone remodeling begins during the fetal period, it is not very active before birth but accelerates during the first year after birth. The annual rate of bone renewal during the first 2 years after birth is 50%, compared with a rate of 5% in the adult. During the first 2 years after birth, the infant progresses from an essentially helpless state to an erect walking individual.

At birth, the ossification centers present in the skeleton are, with few exceptions, primary centers (Plate 7). The exceptions are the secondary, or epiphyseal, centers in the distal condyle of the femur, in the proximal condyle of the tibia, and possibly in the head of the humerus; numerous primary centers do not form until a number of years after birth. The mechanical stresses on the skeleton, as the infant begins to acquire increasing voluntary neuromuscular function during its first 2 years, serve to stimulate skeletal growth, ossification, and especially remodeling. During bone remodeling, the attachments of muscles and ligaments are also shifted and modified. Bone remodeling is most active during the growing period but

Development of Musculoskeletal System

(Continued)

Changes in Position of Limbs Before Birth

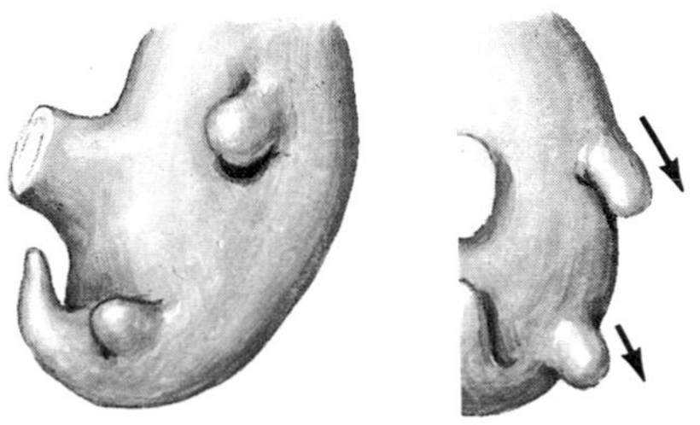

At 5 weeks. Upper and lower limbs have formed as finlike appendages pointing laterally and caudally

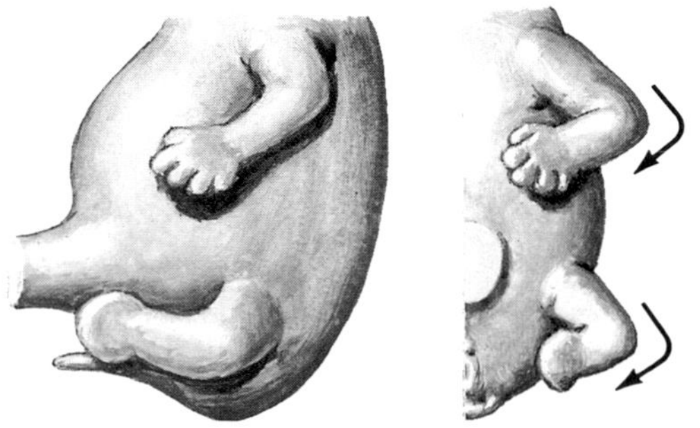

At 6 weeks. Limbs bend anteriorly, so that elbows and knees point laterally, palms and soles face trunk

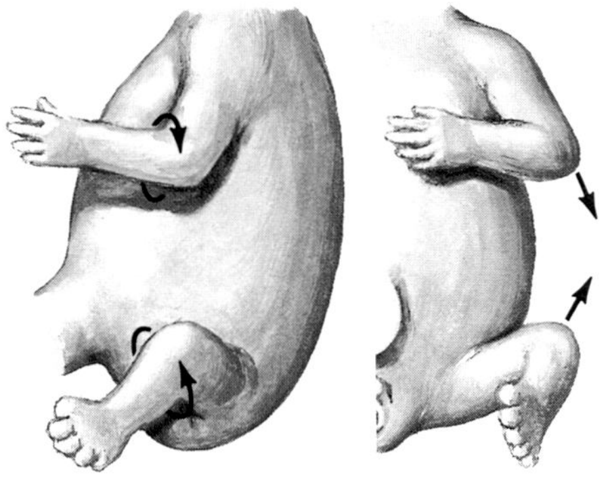

At 7 weeks. Upper and lower limbs have undergone 90° torsion about their long axes, but in opposite direction, so that elbows point caudally and knees cranially

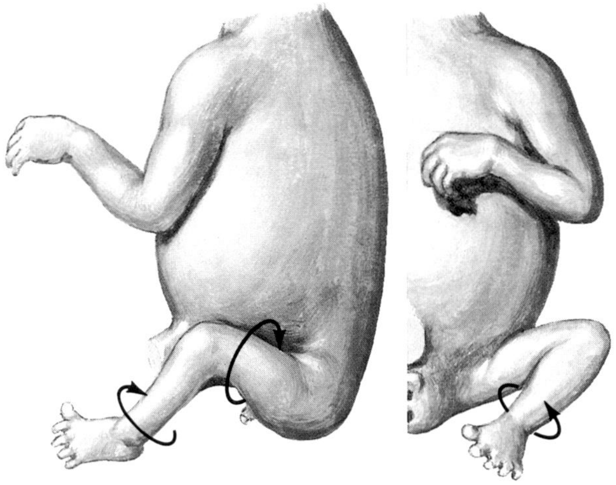

At 8 weeks. Torsion of lower limbs results in twisted, or "barber pole," arrangement of their cutaneous innervation

Precartilage Mesenchymal Cell Concentrations of Appendicular Skeleton at 6th Week

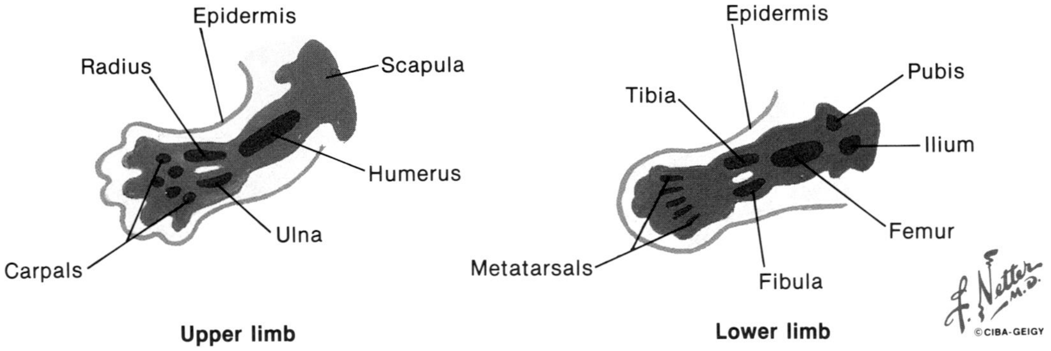

continues throughout life in response to stresses created by an individual's ever-changing types of physical activity.

Developmental History of Bones

Each of the more than 200 bones of the skeleton has its own developmental history. Some bones have a simple history, whereas others have quite a complicated one. The history of the clavicle and mandible is unique. The clavicle is the first bone in the entire skeleton to ossify (during the 7th week), followed shortly thereafter by the mandible (Plates 5–6). Both the clavicle and the mandible are originally membrane bones that secondarily develop growth cartilage. The temporal bone is a good example of a bone with a complicated developmental history. It is a composite bone that forms initially as the otic capsule enclosing the organ for audition and equilibrium (inner ear) of the primitive chondrocranium, which then acquires secondary additions. Its squamous part, zygomatic process, and tympanic ring are derived from membrane bones, whereas its styloid process and ear ossicles are derived from the branchial arch skeleton. Although the overall size of the temporal bone is less than half its adult size at birth, the bony labyrinth of the inner ear, the middle ear cavity, the ear ossicles, and the eardrum have attained their adult size at birth. In contrast, the articular tubercle and mastoid process are absent at birth (Plates 6–7).

Homeostasis

Homeostasis is the maintenance of constant conditions in the internal environment of the body. There is a constant turnover of bone mineral throughout life in response to mechanical stresses exerted on the skeleton. The bones of athletes become considerably heavier than those of nonathletes. Due to the atrophy of disuse, the bones of a limb immobilized in a cast become thin and demineralized. In astronauts, a general demineralization of the entire skeleton occurs in response to the weightlessness caused by the lack of gravity in outer space. These alterations in the mineral content of bones allow the skeleton to serve as a dynamic structural support of the body. However, this support function is not really significant until the end of the first year after birth when the child starts to walk. Long before that time, the alterations in the mineral content of the bones are a part of another function of the skeleton related to the homeostasis of the body.

About 56% of the adult human body consists of fluid. There is intracellular fluid within the 75 trillion cells of the body and extracellular fluid outside the cells. The cells are capable of living, growing, and providing their special functions as long as proper concentrations of oxygen, glucose, ions, amino acids, and fatty substances are available in the internal environment. The skeleton plays a vital role in the regulation of calcium metabolism, which is fully described in Section III.

Blood Supply

Hematopoiesis, or the formation of blood cells, begins before birth. The first hematopoietic cells to appear are erythrocytes, or red blood cells. They are derived from the extraembryonic mesoderm of the yolk sac during the third week. During the fifth week, the erythrocytes are derived primarily from the liver and secondarily from the spleen. The myeloid, or bone marrow, period of hematopoiesis begins during the fourth month. Chiefly, granulocytes, or white blood cells, are initially derived from the bone marrow, while the liver and spleen continue to give rise to only erythrocytes. The marrow tissue also gives rise to the lymphoid stem cells that migrate both to the thymus to induce differentiation of T cells involved in cellular immunity and to the intestinal walls to induce differentiation of B cells involved in antibody production. During the fifth month, the liver erythropoiesis begins to diminish, while the bone marrow, in addition to granulocytes, begins to give rise to erythrocytes. The bone marrow is the principal site of all blood cell formation during the last three months before birth. At birth, hematopoiesis occurs almost exclusively in the bone marrow, because only residual hematopoiesis occurs in the liver and spleen.

During the first 3 or 4 years after birth, almost all the bones of the body contain hematopoietic marrow, although regression of hematopoiesis begins in the distal phalanges of the digits before birth, and the red marrow of the phalanges of the toes is completely replaced by yellow, fatty

Development of Musculoskeletal System

(Continued)

Changes in Ventral Dermatome Pattern (Cutaneous Sensory Nerve Distribution) During Limb Development

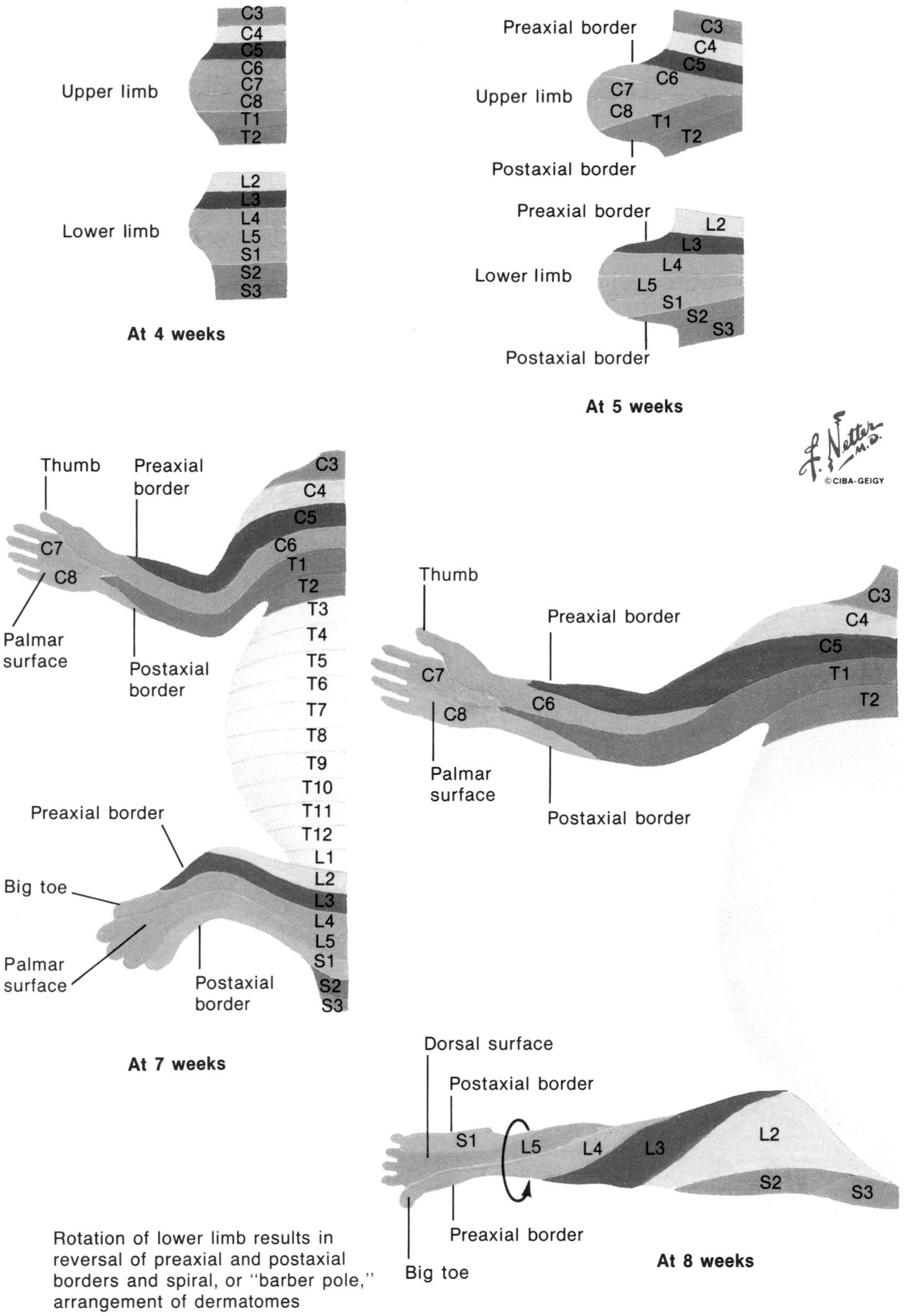

marrow by 1 year of age. Shortly before puberty, yellow marrow appears in the distal ends of the long bones of the forearm, arm, leg, and thigh and gradually extends proximally until 20 years of age, by which age only the upper end of the humerus and femur still contain red marrow.

The other bones in which hematopoiesis occurs in the skeleton of the young adult are the vertebrae, ribs, sternum, clavicles, scapulae, coxal (hip) bones, and skull.

Blood reaches the marrow cavity of the diaphysis of a long bone via one or two relatively large diaphyseal nutrient arteries (see Section III, Plate 22). The nutrient artery passes obliquely through the nutrient foramen of the bone, without branching and in a direction that usually points away from the end of the bone, where the greatest amount of growth is occurring at the epiphyseal plate. Once the nutrient artery enters the marrow cavity, it sends off branches that pass toward the two ends of the bone to anastomose with a number of branches of small metaphyseal arteries that pass directly through the bone into the marrow cavity at the two metaphyses. The arteries of the metaphysis supply the metaphyseal side of the epiphyseal growth plate of cartilage.

Numerous small epiphyseal arteries pass directly through the bone into the marrow cavity of the epiphyses at each end of the bone. The epiphyseal arteries supply the deep part of the articular cartilage and the epiphyseal side of the epiphyseal growth plate. In a growing bone with a relatively thick growth plate, there are few, if any, anastomoses between the epiphyseal and metaphyseal vessels. The growth plate also receives a blood supply from a collar of periosteal arteries adjacent to the periphery of the plate.

The branches of the diaphyseal nutrient arteries, which pass to each end of the bone to anastomose with the metaphyseal arteries, give off two sets of branches along the way, one peripheral and one central. The peripheral set passes directly to the bone as arterioles that give off the capillaries that enter Volkmann's canals and branch to supply the central haversian canals, ultimately emerging at the outer surface of the bone and anastomosing with the periosteal vessels. The direction of the blood flow in these capillaries is from within the bone outward; thus, the blood flow through the canal system of the bony wall is relatively slow and at a low pressure.

The central set of branches given off by the diaphyseal nutrient arteries become arterioles that join plexuses of large irregularly shaped capillaries known as sinusoids. In a young child, sinusoids, which are the sites of hematopoiesis, are found throughout the marrow cavity. An extensive, delicate meshwork of reticular fibers containing hematopoietic cells, fibroblasts, and occasional fat cells surrounds the single-celled endothelial wall of the sinusoids; this constitutes red marrow. The newly formed blood cells eventually pass out of the sinusoids into large veins that directly pierce the diaphyseal bony shaft, without branching, as the venae comitantes of the nutrient diaphyseal arteries. Others pass directly through the bony wall, without branching, as independent emissary veins.

The myeloid, or bone marrow, period of hematopoiesis begins during the fourth month. The bone marrow is the principal site of all blood cell formation during the last 3 months before birth, at which time only residual hematopoiesis occurs in the liver and spleen.

Symphyseal Joints

In the first vertebrates, the skeleton evolved as an axial skeleton, the vertebral column. The segmentation that evolved in the increasingly substantial column allowed the necessary swimming movements that the flexible notochord afforded the prevertebrates. Intervening regions between the firmer segments of the column became pliable cartilage that allowed very limited and yet every possible type of motion between the firmer segments. Thus, in man, intervertebral discs

Development of Musculoskeletal System

(Continued)

Initial Bone Formation in Mesenchyme

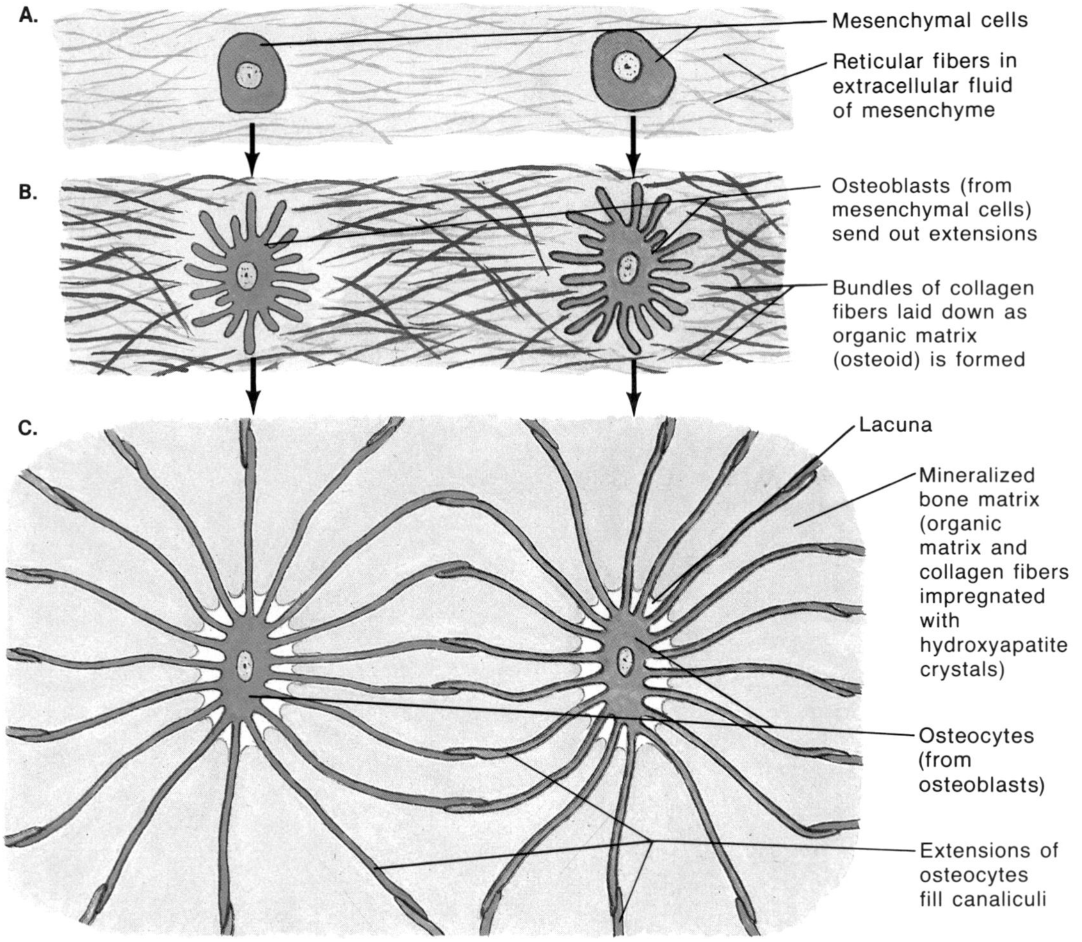
A.
Mesenchymal cells
Reticular fibers in extracellular fluid of mesenchyme
B.
Osteoblasts (from mesenchymal cells) send out extensions
Bundles of collagen fibers laid down as organic matrix (osteoid) is formed
C.
Lacuna
Mineralized bone matrix (organic matrix and collagen fibers impregnated with hydroxyapatite crystals)
Osteocytes (from osteoblasts)
Extensions of osteocytes fill canaliculi

Early Stages of Flat (Membrane or Dermal) Bone Formation

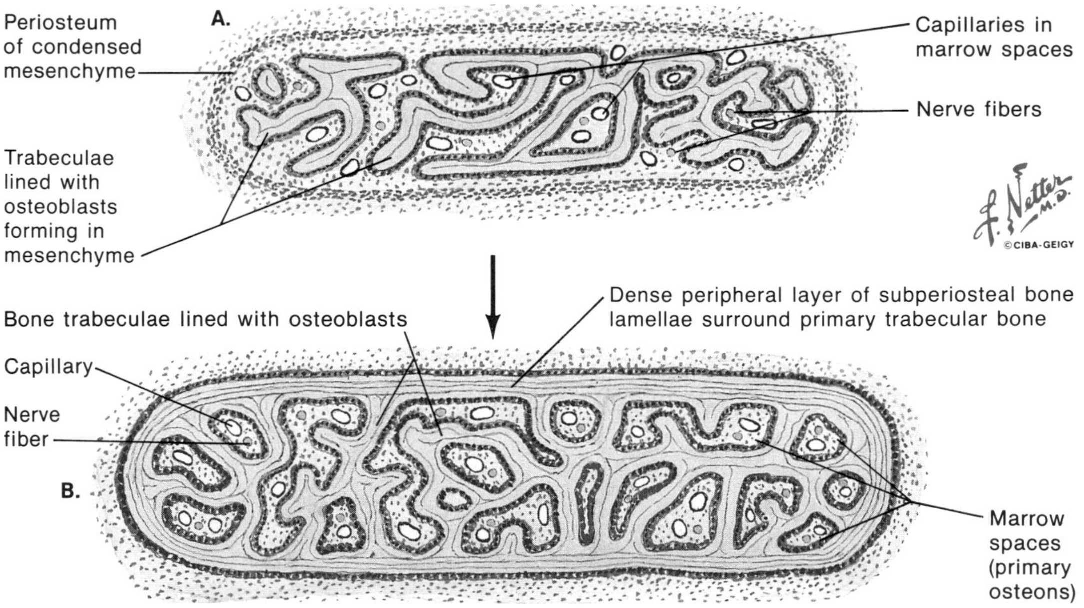
A.
Periosteum of condensed mesenchyme
Capillaries in marrow spaces
Nerve fibers
Trabeculae lined with osteoblasts forming in mesenchyme
F. Netter M.D.
©CIBA-GEIGY
Bone trabeculae lined with osteoblasts
Dense peripheral layer of subperiosteal bone lamellae surround primary trabecular bone
Capillary
Nerve fiber
B.
Marrow spaces (primary osteons)

between the vertebral bodies allow a limited degree of twisting and bending in all directions. However, the sum total of a given motion occurring between the vertebral bodies throughout the column is considerable.

The multiaxial joint between the vertebral bodies is known as a symphysis because of its structure. A central portion of fibrocartilage, including the nucleus pulposus, blends with a layer of hyaline cartilage lining the surface of each of the two vertebral bodies bordering the joint. The only symphysis of the appendicular skeleton is the pubic symphysis (Plate 7).

Since a symphyseal joint has limited motion, it is an amphiarthrosis. A central cleft containing fluid occurs in some symphyses, such as the pubic and manubriosternal (sternal angle) joints, but true gliding surfaces do not develop (Plate 5). This is an intermediate phase in the evolution of synovial joints.

Although the majority of articulations of the appendicular skeleton are synovial joints, many of the articulations of the axial skeleton are also typical synovial joints. For example, the numerous joints between the articular processes of the vertebral arches are synovial joints of the plane variety in that their apposed articular surfaces are fairly flat (Plate 4).

Synovial Joints

Synovial, or diarthrodial, joints have a wide range of motion; they link cartilaginous bones with one another and with certain membrane bones, such as the mandible and clavicle.

The earliest mesenchymal rudiments of long bones are essentially continuous. As the rudiments pass into the precartilage stage, the sites of the future joints can be discerned as intervals of less concentrated mesenchyme (Plate 15). When the mesenchymal rudiments transform into cartilage, the mesenchymal cells in the future joint region become flattened in the center. At the periphery of the future joint, these flattened cells are continuous with the investing perichondrium; this perichondral investment becomes the joint capsule.

During the third month, the joint cavity arises from a cleft that appears in the circumferential part of the mesenchyme. The mesenchymal cells in the center of the developing joint disappear, allowing the cartilage rudiments to come into direct contact with each other, and for a time, a transitory fusion may result in a small area of direct cartilaginous union. Soon, all the remaining mesenchymal cells undergo dissolution, and a distinct joint cavity is formed. The surrounding joint capsule maintains its continuity with the perichondrium when it becomes transformed into periosteum as the cartilage rudiments become bones. The original cartilage of the rudiment forming the joint surface is retained as the articular hyaline cartilage.

Deep to the articular cartilage, epiphyseal bone is laid down. Since the articular cartilage was never actually lined with perichondrium, it grows in thickness by intrinsic, or interstitial, growth. Some perichondrium is retained at the periphery of the articular cartilage, which continues to form cartilage until the articular surface of the joint attains adult size. Once full growth is attained, the chondrocytes normally do not undergo division.

Articular cartilage, especially that found in weight-bearing joints, is uniquely structured to withstand tremendous abuse. It can resist crushing by static loads considerably greater than those required to break a bone. No painful sensations are elicited in traumatized cartilage because it lacks nerves. The chondrocytes in weight-bearing joints are genetically programmed to tolerate crushing forces without overreacting, such as by inducing their surrounding matrix to undergo extensive dissolution or by laying down excessive amounts of matrix. Such responses would markedly alter the surface contour of the cartilage in a manner that would interfere with the normal joint motion.

Development of Musculoskeletal System

(Continued)

Secondary Osteon (Haversian System)

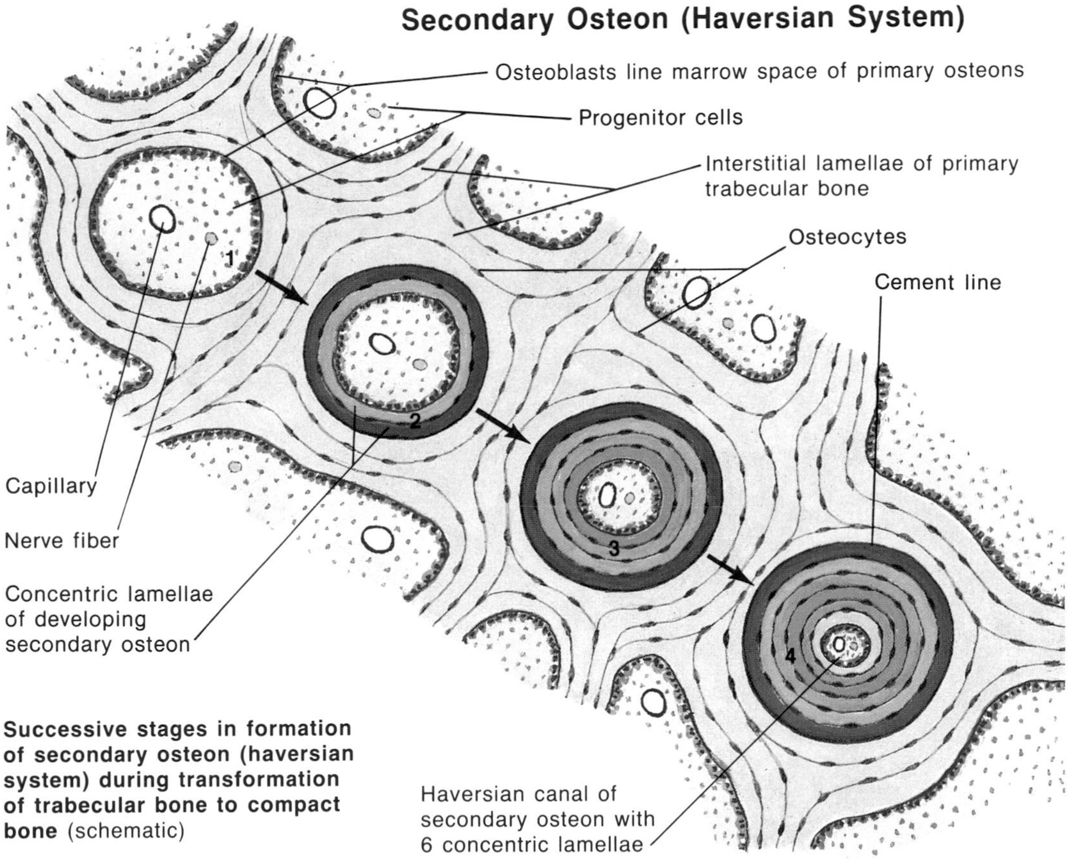
Osteoblasts line marrow space of primary osteons
Progenitor cells
Interstitial lamellae of primary trabecular bone
Osteocytes
Cement line
Capillary
Nerve fiber
Concentric lamellae of developing secondary osteon
Haversian canal of secondary osteon with 6 concentric lamellae
1
2
3
4

Successive stages in formation of secondary osteon (haversian system) during transformation of trabecular bone to compact bone (schematic)

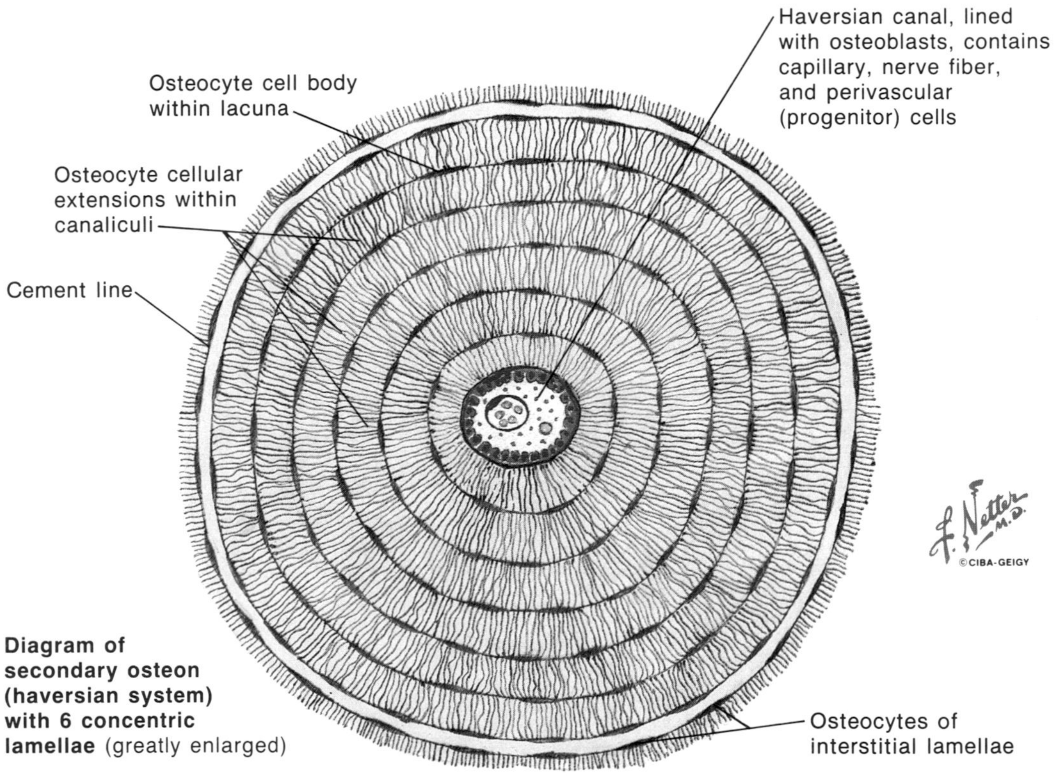
Haversian canal, lined with osteoblasts, contains capillary, nerve fiber, and perivascular (progenitor) cells
Osteocyte cell body within lacuna
Osteocyte cellular extensions within canaliculi
Cement line
Osteocytes of interstitial lamellae
F. Netter M.D.
©CIBA-GEIGY

Diagram of secondary osteon (haversian system) with 6 concentric lamellae (greatly enlarged)

As soon as the joint cavity appears during development, it contains watery fluid. The joint capsule develops an outer fibrous portion that is lined with an inner, more highly vascularized synovial membrane. Although this membrane lines the fibrous capsule as well as any bony surfaces, ligaments, and tendons within the joint, it does not line the surfaces of the joint discs, menisci, or articular cartilage.

The synovial membrane is the site of formation of the synovial fluid that fills the joint cavity. This fluid is similar to that found in bursae and tendon sheaths. Before birth, it is sticky, viscous, and much like egg white in consistency. Only a small amount of the fluid is normally present in a joint cavity, where it forms a sticky film that lines all the surfaces of the joint cavity (for example, the adult knee joint contains only a little more than 1 ml of synovial fluid). Even so, before birth and thereafter, the fluid is the chief source of nourishment of the chondrocytes of the articular cartilage, which lacks blood and lymphatic vessels.

The articular cartilage is never very thick, averaging 1 to 2 mm in thickness in the adult and reaching a maximum of 5 to 7 mm in the larger joints of young individuals. However, compared with cells in the vascularized tissue of the body, which are not more than 25 to 50 μm from a capillary, the chondrocytes are at an enormous distance from their source of nourishment. Joint activity enhances both the diffusion of nutrients through the cartilage matrix to the chondrocytes and the diffusion of metabolic waste products away from them. The alternating compression and decompression of the cartilage during joint activity produce a pumping action that enhances the exchange of nutrients and waste products between the cartilage matrix and the synovial fluid.

In some developing joints, the mesenchymal tissue between the cartilage rudiments, instead of disappearing, gives rise to a fibrous sheet that completely divides the joint into two separate compartments. The sheet develops into an intra-articular disc, which is made up of fibrous connective tissue and, possibly, a small amount of fibrocartilage. A separate synovial cavity develops on each side of the disc, as found in the temporomandibular joint.

In other developing joints, the mesenchymal tissue between the cartilage rudiments gives rise to a fibrous sheet that is incomplete centrally. This crescentic sheet, which projects from the joint capsule into a single joint cavity, gives rise to articular menisci consisting of fibrous tissue and possibly a small amount of fibrocartilage, such as found in the knee joint.

After the synovial joint cavity is established during the third month, the muscles that move the joint begin to undergo contractions. This movement is essential for the normal development of the synovial joints, because it not only enhances the nutrition of the articular cartilage, but also prevents fusion between the apposed articular cartilages.

Restriction of joint motion by permanent paralysis early in development can result in the loss of the joint cavity by having a permanent fusion occur between the apposed surfaces of the articular cartilage. If the restriction of joint movement occurs later in development, the joint space may be present but the associated soft tissues of the joint are abnormal. An example is the nongenetic form of clubfoot (talipes varus) caused by the severe restriction of movement of the ankle joint before birth. The normal positioning of the fetus in the uterus allows a fair degree of movement of the upper limbs, but the lower limbs are folded together and pressed firmly against the body. The hip and knee joints are flexed and the feet are inverted in the pigeon-toed position. The ankle joint may become fixed in this inverted position because of the abnormal shortening of the muscles that invert the foot and the lengthening of their antagonists. Also, the ligaments on the medial side of the ankle joint may become abnormally shortened.

Development of Musculoskeletal System

(Continued)

Growth and Ossification of Long Bones (humerus, midfrontal sections)

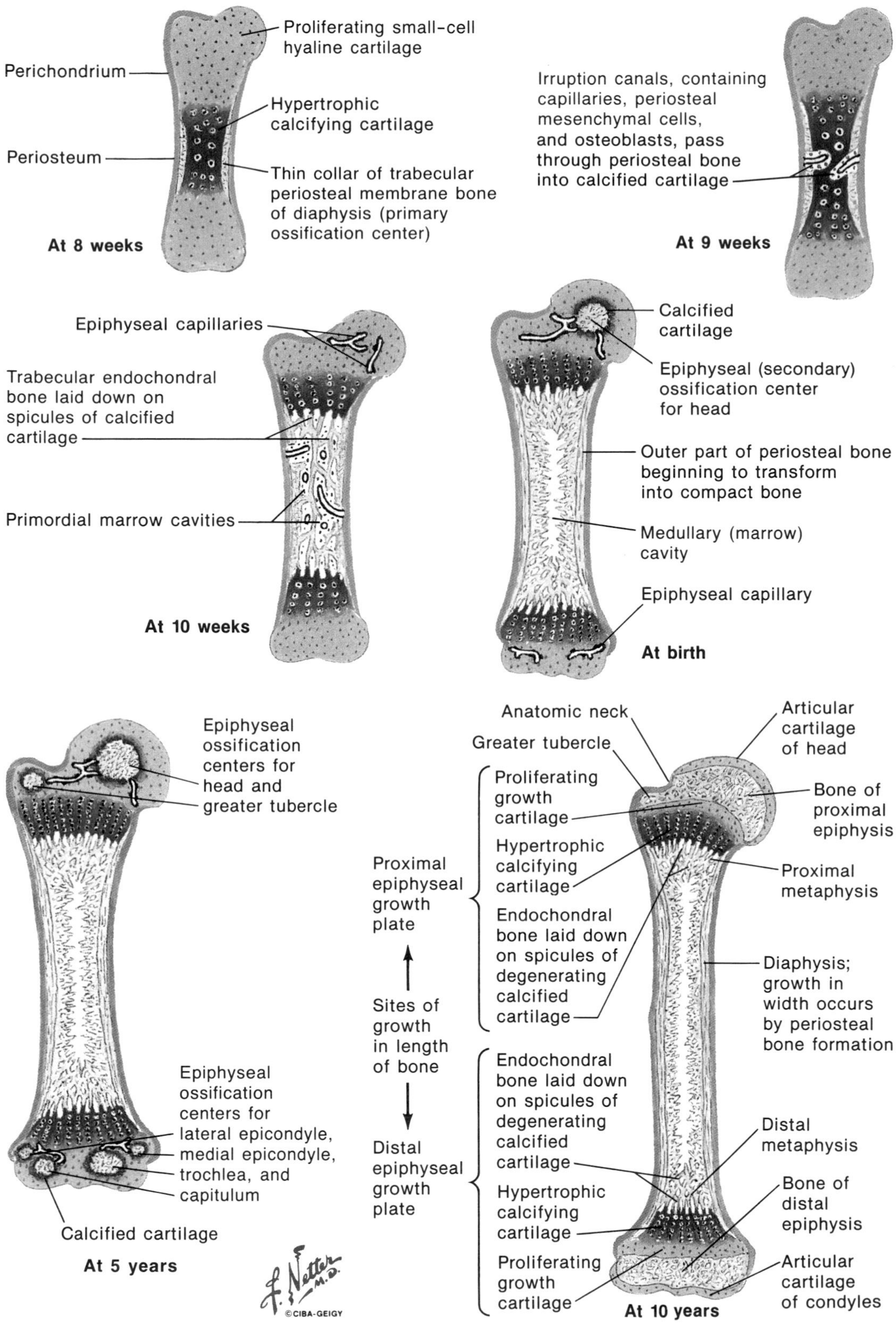
Proliferating small-cell hyaline cartilage
Perichondrium
Hypertrophic calcifying cartilage
Periosteum
Thin collar of trabecular periosteal membrane bone of diaphysis (primary ossification center)
At 8 weeks
Irruption canals, containing capillaries, periosteal mesenchymal cells, and osteoblasts, pass through periosteal bone into calcified cartilage
At 9 weeks
Epiphyseal capillaries
Trabecular endochondral bone laid down on spicules of calcified cartilage
Primordial marrow cavities
At 10 weeks
Calcified cartilage
Epiphyseal (secondary) ossification center for head
Outer part of periosteal bone beginning to transform into compact bone
Medullary (marrow) cavity
Epiphyseal capillary
At birth
Epiphyseal ossification centers for head and greater tubercle
Epiphyseal ossification centers for lateral epicondyle, medial epicondyle, trochlea, and capitulum
Calcified cartilage
At 5 years
Anatomic neck
Greater tubercle
Articular cartilage of head
Proliferating growth cartilage
Bone of proximal epiphysis
Hypertrophic calcifying cartilage
Proximal metaphysis
Proximal epiphyseal growth plate
Endochondral bone laid down on spicules of degenerating calcified cartilage
Diaphysis; growth in width occurs by periosteal bone formation
Sites of growth in length of bone
Endochondral bone laid down on spicules of degenerating calcified cartilage
Distal metaphysis
Distal epiphyseal growth plate
Hypertrophic calcifying cartilage
Bone of distal epiphysis
Proliferating growth cartilage
Articular cartilage of condyles
At 10 years
F. Netter M.D.
©CIBA-GEIGY

Hip Joint

The upper limbs are far more functionally advanced at birth than are the lower limbs. The newborn infant can reflexly grasp objects firmly with the hands. In contrast, the infant reflexly maintains the underdeveloped lower limbs in the position they were held in before birth and in fact strongly resists their straightening. Relative to this, the very underdeveloped hip joint is prone to dislocation when the limbs are shortened. The hip socket, or acetabulum, is normally very small compared to the relatively large head of the femur (Plate 7). When the lower limbs are in the fetal position, the firm ligament of the head of the femur, by virtue of its attachments, strongly prevents the hip joint from becoming dislocated posterosuperiorly. However, if the ligament is abnormally long, it will not prevent a posterosuperior dislocation.

Normally, the ligament does not function to prevent hip dislocation in any limb position other than the fetal one. The thin, flimsy joint capsule is the chief resistance to dislocation when the limbs are not held in the fetal position. Once the infant tends to maintain the lower limbs in extension in the months after birth, the hip joint becomes secure, and the ligament of the head of the femur serves no further useful function.

Erect Posture

During the evolution of the human erect posture, the lumbar joints and especially the lumbosacral joint acquired the ability to undergo a pronounced extension that allows a marked lumbar curvature, or lordosis, of the vertebral column. Except for the fixed sacral curve, the vertebral column at birth has no curves. The thoracic part of the spine gradually develops a relatively fixed curve in the young child. A flexible cervical curve appears when the infant is able to raise the head, and a flexible lumbar curve appears at the end of the first year when the child starts to walk. The lumbar curve is necessary to attain the erect posture, because the pelvis remains essentially in the same position as that in a standing quadruped.

The fact that the pelvis did not shift from its quadruped position during evolution of the erect posture also necessitated placing the hip and knee joints into full extension. In addition, the arch of the foot evolved so that the bones were structurally arranged to bear the body weight with a minimum of muscular activity. Therefore, in the human, the passive ligaments of the foot bones and those of the fully extended hip and knee joints bear the brunt of the forces involved in standing erect.

Only man stands perfectly erect. Quadrupeds, including the knuckle-walking apes, can only mimic the erect human posture. They do it with a great expenditure of muscular energy because their hip and knee joints cannot be fully extended so that the passive ligaments of the joints can withstand the brunt of the forces involved in standing erect. This same expenditure of energy is made when a child first starts to stand with the hip and knee joints partially flexed. The erect posture of man may appear to be a most awkward position compared with the normal standing posture of quadrupeds, but it is the most efficient and economical posture that ever evolved. Once man rises by muscular activity to the fully erect position, only occasional brief contractions of postural muscles are required to keep the head, trunk, and limbs aligned with the vertical line of the center of gravity. The upper limbs are included in the economics of the erect posture because the passive ligaments of the joints, not the muscles of the upper limbs, bear the brunt of supporting the limbs as they hang at the sides of the body.

Development of Musculoskeletal System

(Continued)

Peripheral Shift of Osteons of Compact Diaphyseal Bone With Growth in Width

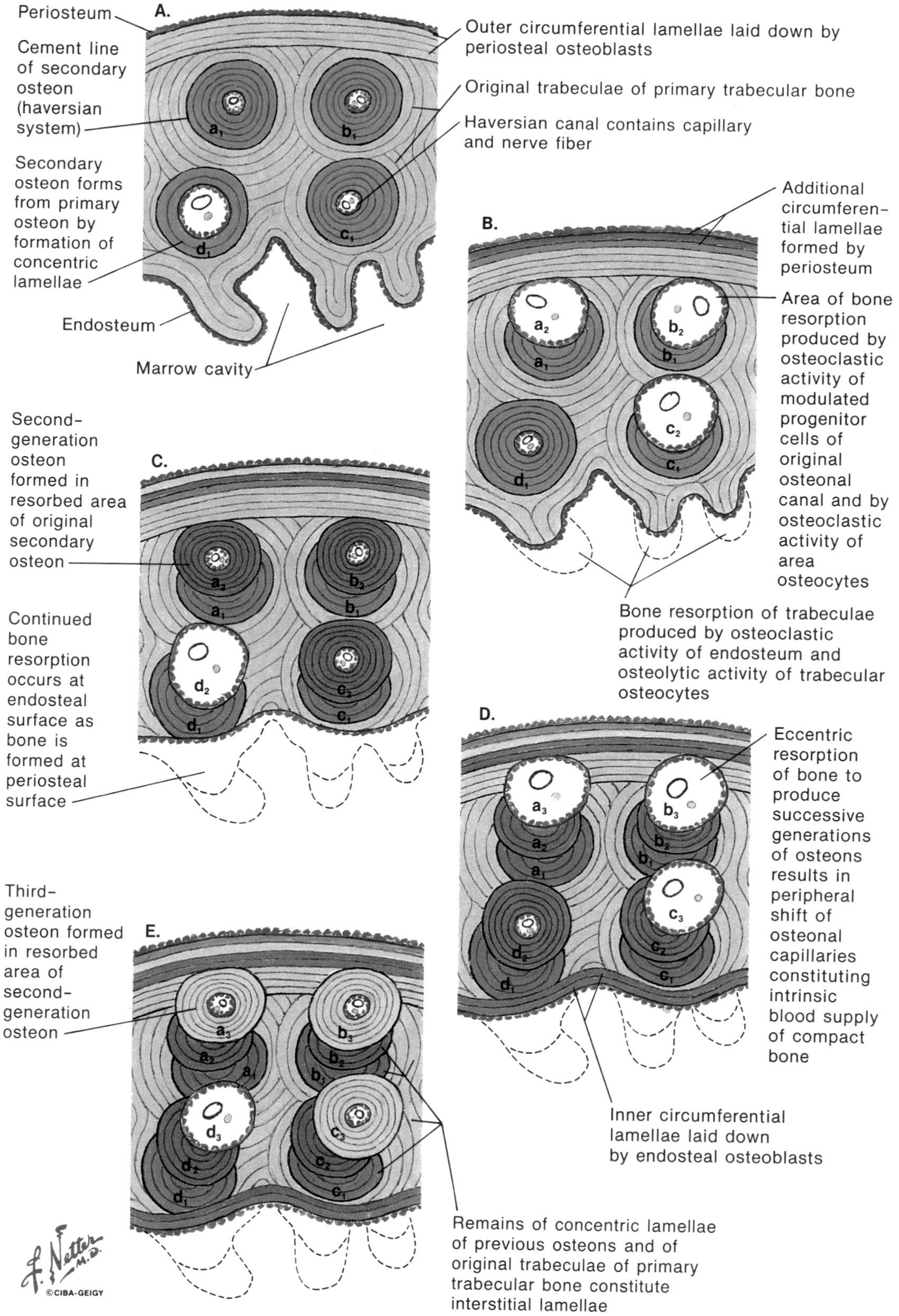
A.
Periosteum
Cement line of secondary osteon (haversian system)
Secondary osteon forms from primary osteon by formation of concentric lamellae
Endosteum
Marrow cavity
Outer circumferential lamellae laid down by periosteal osteoblasts
Original trabeculae of primary trabecular bone
Haversian canal contains capillary and nerve fiber
B.
Additional circumferential lamellae formed by periosteum
Area of bone resorption produced by osteoclastic activity of modulated progenitor cells of original osteonal canal and by osteoclastic activity of area osteocytes
Bone resorption of trabeculae produced by osteoclastic activity of endosteum and osteolytic activity of trabecular osteocytes
C.
Second-generation osteon formed in resorbed area of original secondary osteon
Continued bone resorption occurs at endosteal surface as bone is formed at periosteal surface
D.
Eccentric resorption of bone to produce successive generations of osteons results in peripheral shift of osteonal capillaries constituting intrinsic blood supply of compact bone
Inner circumferential lamellae laid down by endosteal osteoblasts
E.
Third-generation osteon formed in resorbed area of second-generation osteon
Remains of concentric lamellae of previous osteons and of original trabeculae of primary trabecular bone constitute interstitial lamellae
F. Netter M.D.
©CIBA-GEIGY

Muscles

Characteristically, all living cells, including protozoa and slime molds, contain the contractile proteins actin and myosin. Thus, actin and myosin are present in all the cells of the human body—from the most highly differentiated nerve cells to the shed fragments of megakaryocyte cytoplasm, the platelets, which are important in the formation of blood clots. Actin and myosin are arranged in the cytoplasm of a cell to interact and slide in relationship to one another to produce contraction of the cell when driven by the energy supplied by the hydrolysis of adenosine triphosphate (ATP).

During the evolution of single-celled protozoa into metazoa, or multicellular organisms, cells became specialized to perform specific functions. Certain cells accumulated larger than usual amounts of actin and myosin in their cytoplasm to become muscle cells scattered throughout the body of the primitive metazoan. As the higher forms developed distinct organ systems, the muscle cells grouped together to become the smooth (involuntary, visceral, nonsegmental) muscles of the viscera and blood vessels.

Smooth and Cardiac Muscle

All the smooth and cardiac muscle cells in the human embryo arise from mesoderm, except the sphincter and dilator smooth muscles of the iris of the eye and the myoepithelial cells of the sweat and mammary glands, which arise from ectoderm. Both smooth and cardiac muscle cells have a centrally placed nucleus. During development, numerous smooth muscle cells become elongated in the same direction and form layers, such as the circular and longitudinal smooth muscle layers of the small intestine.

For a time during evolution, a simple layer of smooth muscle surrounding the vessels of the circulatory system was also sufficient for the demands of function. However, as organisms became larger and increasingly complex, the need arose for the system to have a strong pump, the heart. In the human embryo, two endothelial tubes fuse to become one vessel, which then becomes surrounded with mesenchyme that differentiates into cardiac muscle (see CIBA COLLECTION, Volume 5, pages 114–126). The muscle cells surrounding the developing heart accumulated a larger amount of more compactly and more orderly arranged actin and myosin molecules than did simple smooth muscle cells. Despite undergoing repeated mitotic divisions, they remained attached to one another in such a manner that they formed long tubes of cells known as fibers.

Within each cell of the fibers, the myosin formed thick myofilaments and the actin formed thin myofilaments that ran parallel to the longitudinal axis of the cell. The myofilaments became identically aligned and organized within the cell into larger longitudinal bundles, the myofibrils, which in turn became aligned with the adjacent myofibrils. Mitochondria were interspersed between the myofibrils. This identical, side-by-side alignment coincided with that of the cells of adjacent fibers, resulting in the cross-banded, or striated, appearance of longitudinally sectioned cardiac muscle at the microscopic level.

The dense concentration in cardiac muscle of orderly arrangements of interdigitating actin and myosin molecules, which could synchronously slide across each other throughout the atrial or ventricular muscle, resulted in an organ that could make strong, quick contractions of short duration. And so, between the third and fourth week, the cardiac muscle of the single-tube heart begins to contract. The bundles, nodes, and Purkinje fibers, which are the components of the conducting system of the heart, are merely modified cardiac muscle fibers.

If damaged, smooth muscle is able to regenerate to a limited degree by division of preexisting muscle cells, and by division and differentiation of nearby connective tissue cells of the mesenchymal type. However, there is no regeneration of damaged cardiac muscle; repair of damaged myocardium is by means of fibrous scar tissue.

Development of Musculoskeletal System
(Continued)

Remodeling: Maintenance of Basic Form and Proportions of Bone During Growth

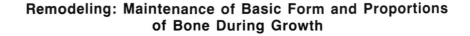

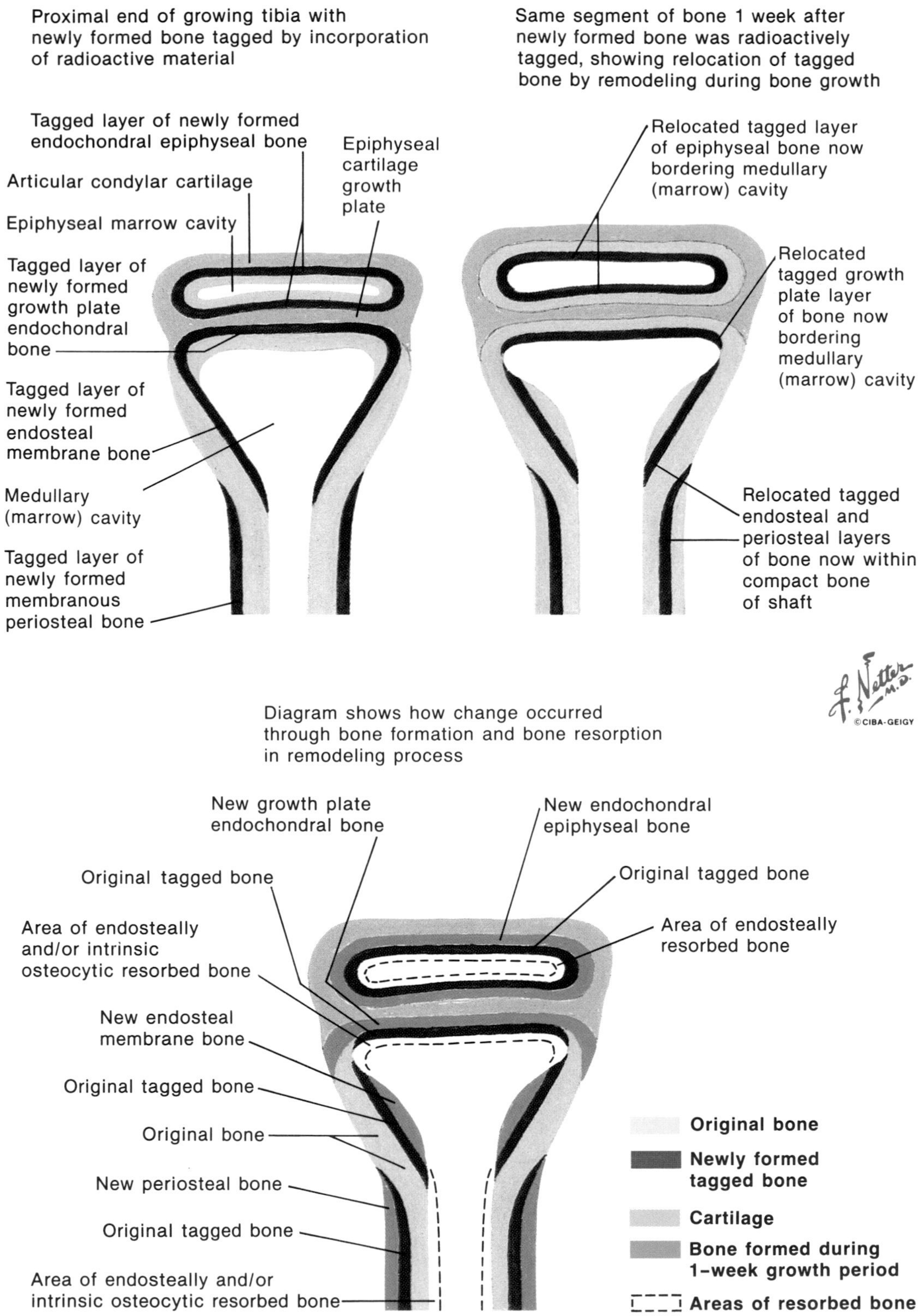
Proximal end of growing tibia with newly formed bone tagged by incorporation of radioactive material
Same segment of bone 1 week after newly formed bone was radioactively tagged, showing relocation of tagged bone by remodeling during bone growth
Tagged layer of newly formed endochondral epiphyseal bone
Epiphyseal cartilage growth plate
Articular condylar cartilage
Epiphyseal marrow cavity
Tagged layer of newly formed growth plate endochondral bone
Tagged layer of newly formed endosteal membrane bone
Medullary (marrow) cavity
Tagged layer of newly formed membranous periosteal bone
Relocated tagged layer of epiphyseal bone now bordering medullary (marrow) cavity
Relocated tagged growth plate layer of bone now bordering medullary (marrow) cavity
Relocated tagged endosteal and periosteal layers of bone now within compact bone of shaft
F. Netter M.D.
© CIBA-GEIGY
Diagram shows how change occurred through bone formation and bone resorption in remodeling process
New growth plate endochondral bone
New endochondral epiphyseal bone
Original tagged bone
Original tagged bone
Area of endosteally and/or intrinsic osteocytic resorbed bone
Area of endosteally resorbed bone
New endosteal membrane bone
Original tagged bone
Original bone
New periosteal bone
Original tagged bone
Area of endosteally and/or intrinsic osteocytic resorbed bone
Original bone
Newly formed tagged bone
Cartilage
Bone formed during 1-week growth period
Areas of resorbed bone

Skeletal Muscle

Skeletal muscle is also known as voluntary, striated, striped, or segmental muscle. The last term refers to the origin of most of the skeletal muscles of the vertebrate body from the segmented paraxial mesoderm, the somites.

In the adult prevertebrate amphioxus, there are, according to the species, from 50 to 85 muscle segments known as myotomes, or myomeres (Plate 1). The V-shaped myotomes are dovetailed into one another along the length of the body. The individual striated muscle fibers of each myotome run parallel to the long axis of the body and each myotome receives a pair of nerves from the dorsal nerve cord. The original myotomic segments are retained in a similar fashion throughout the trunk of adult fish. However, each myotome is divided into a dorsal, or epaxial, and a ventral, or hypaxial, portion, which are separated in fish by the transverse processes of the vertebral column and a fibrous septum extending from these processes to the lateral body line. Each myotome is supplied by a spinal nerve, with a dorsal ramus innervating the epaxial portion and a ventral ramus innervating the hypaxial portion.

In the human embryo, the maximum number of 42 to 44 somites is attained during the fifth week, after which the first of the four occipital and the last seven or eight coccygeal somites regress and disappear. In addition to the somites, there are three masses of mesenchyme on each side of the embryonic head that are anterior to the otic vesicles—the future membranous labyrinths of the inner ears—which represent the three pairs of preotic somites found in primitive vertebrate embryos that give rise to the striated extrinsic muscles of the eye. The three preotic mesenchymal masses in the human embryo aggregate into one mass around the developing eyeball during the fifth week, giving rise to the extrinsic ocular muscles that become innervated by the initially nearby oculomotor (III), trochlear (IV), and abducens (VI) nerves (Plate 16).

In the human embryo, the early differentiation of all the persisting *somites* (the second occipital to the third or fourth coccygeal) is similar: the ventromedial portion of the somite becomes the *sclerotome*; the sclerotomal cells migrate toward the notochord to give rise to the vertebral column and ribs, and the remaining portion of the somite is then called the *dermomyotome* (a fluid-filled cavity, the myocoele, appears in the somite but is soon obliterated); the cells of the dermomyotome then proliferate to form a medial mass, the *myotome*, which can be distinguished from the less proliferative lateral portion, the *dermatome* (Plates 2–3). Finally, the cells of the dermatome spread beneath the overlying ectoderm to give rise to the subcutaneous fascia and the dermis of the skin. The segmental dermatome distribution of the embryo is reflected in the innervation of the skin of the trunk and limbs of the adult. The area of skin supplied by a single spinal nerve in the adult constitutes a dermatome.

In fish, the myotome stays in place and occupies the equivalent position of its parent somite, giving rise to a segmental muscle that attaches to the vertebral column. This prevents the sclerotome portion of the somite from also retaining its original position and giving rise to only a single vertebra. If this had happened, each muscle would attach to only a single vertebra, and then the vertebral column could not move when the muscle contracted. The process of establishing an overlapping arrangement between myotomes and vertebrae is recapitulated in the human embryo.

The cells of the myotome, the mononucleated myoblasts, elongate in a direction parallel to the long axis of the embryo (Plate 17) and undergo repeated mitotic divisions, subsequently fusing with each other to form syncytia. Each syncytium becomes a tube with continuous cytoplasm, and the numerous nuclei within it are centrally located. The process is similar to the formation of the tubular cardiac muscle fiber except that in the latter, each centrally located nucleus is within a separate cell.

Development of Musculoskeletal System

(Continued)

Development of Three Types of Diarthrodial (Synovial) Joints

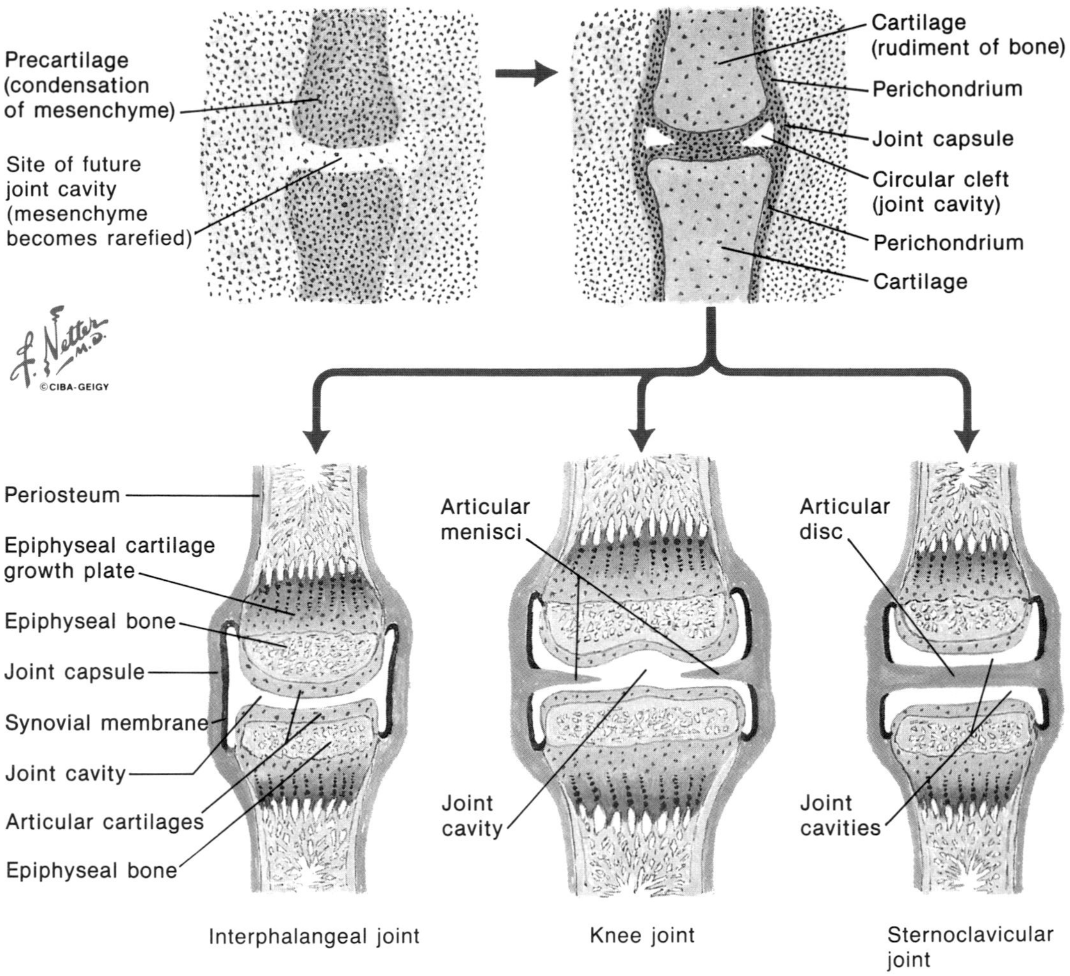
Precartilage
(condensation
of mesenchyme)
Site of future
joint cavity
(mesenchyme
becomes rarefied)
Cartilage
(rudiment of bone)
Perichondrium
Joint capsule
Circular cleft
(joint cavity)
Perichondrium
Cartilage
©CIBA-GEIGY
Periosteum
Epiphyseal cartilage
growth plate
Epiphyseal bone
Joint capsule
Synovial membrane
Joint cavity
Articular cartilages
Epiphyseal bone
Articular
menisci
Joint
cavity
Articular
disc
Joint
cavities
Interphalangeal joint
Knee joint
Sternoclavicular
joint

The syncytial myotubes of skeletal muscle become muscle fibers as myofilaments of actin and myosin are laid down within the cytoplasm. The thin actin and the thick myosin polypeptide myofilaments become strung out parallel to the long axis of the fiber and are arranged in a side-by-side, interdigitating relationship so that they can slide past each other to cause muscle contraction. The cross-banded, or striated, appearance of skeletal muscle at the microscopic level reflects this relationship between the two types of submicroscopic filaments. The myofilaments group together into numerous longitudinal bundles known as myofibrils, which occupy the bulk of the fiber; the nuclei and nearly all the mitochondria are relocated to the periphery, where they are in contact with the outer membrane of the fiber, the sarcolemma (Plate 17).

The myotube stage of fiber formation begins at about the fifth week. Subsequent generations of myotubes develop from the persisting population of myoblasts found in close relationship to muscle fibers. The nuclei of the muscle fibers themselves, once in place, do not divide mitotically or amitotically; consequently, in order to increase their number, the incorporation of new myoblasts into the syncytia is required, especially when the fibers grow in length.

The growth of skeletal muscles is the result of an increase in both the number of muscle fibers and the size of the individual fibers. The greatest increase in the number of fibers occurs before birth, after which time both the number and size of the fibers increase. In the male, there is a fourteenfold increase in fiber number from 2 months to age 16, with a rapid spurt at age 2, and a maximum rate of increase from ages 10 to 16, during which time the fibers double in number. There is also a steady linear increase in the size of muscle fibers from infancy to adolescence and beyond in the male. In the female, the increase in fiber number is more linear than in the male, with an overall tenfold postnatal increase. However, in the female, the increase in fiber size is more rapid than in the male after age 3½, reaching a plateau at age 10½. After age 14½, fiber size in males exceeds that in females. Fiber numbers increase steadily in both sexes up to about age 50, after which there is a steady decline.

Muscle fibers are very fine threads, up to 30 cm in length but less than 0.1 mm in width, which contract to about 57% of their resting length. Only the largest muscle fibers in an adult would be visible to the naked eye if they could be individually excised. Muscles will develop completely in the absence of an innervation that is due to a congenital nervous system abnormality. Thus, nerves do not supply a necessary organizing stimulus, and gross muscle morphogenesis will go to completion with function never having occurred. However, a muscle that never had a nerve supply does not attain its full differentiation at the fiber level and disappears with time.

Skeletal muscles make up the bulk of the adult body and comprise about 45% of its total weight. There are over 650 named muscles, and nearly all are paired. Each has a characteristic shape that is circumscribed by a connective tissue sheath.

During vertebrate evolution, the head underwent changes related to the development of the special senses. The anterior end of the nerve cord became a brain, and the nerves passing to and from the brain became the cranial nerves. In the prevertebrate amphioxus, which has no brain, muscle is present in the region of the mouth of the digestive system (Plate 1). In the vertebrate fish, the gills have a branchial arch musculature that arises from the mesoderm associated with the developing pharyngeal region of the foregut. Therefore, this musculature can properly be called visceral musculature, even though it is voluntary and striated. A better term is branchial, or branchiomeric, musculature because it represents a serial division, or metamerism, of the lateral (gill or branchial) mesoderm that does not segment in its counterpart in the trunk.

In the human embryo, the branchial arches and their contained structures initially develop as though the aqueous gill-slit type of breathing apparatus were going to be retained. Instead of disappearing, most of the branchial arch structures are gradually modified and incorporated into the permanent acoustic and air-breathing respiratory systems. The branchiomeric musculature that develops from the mesoderm of the series of branchial arches on each side of the embryonic head becomes innervated by cranial nerves. Most of these muscles ultimately attach to the skull.

In addition to the skeletal muscles derived from the myotome and branchial arch, there are those that arise, in situ, directly from the local mesenchyme. Some of these locally derived muscles are the result of the slurring over of the sequence of evolutionary events during development, so that their derivation from myotome or branchial arch mesenchyme is obscured. Others, such as the limb muscles, appear relatively late in evolution and development. In the human embryo, the muscles of the limbs that evolved from fins appear after the myotomic and branchial arch musculature formation is well under way. The muscles of the pelvic diaphragm, perineum, and external genitalia also appear relatively late in development.

A developing skeletal muscle normally provides attractive forces that serve to guide a nerve to it. With only a few exceptions, the muscles retain their original innervation throughout life, no matter how far they may migrate from their site of origin during development; this is true whether a muscle is of myotomic origin and innervated by a spinal nerve or of branchial arch origin and innervated by a cranial nerve. Therefore, the innervation of adult muscles can be used as a clue to determine their embryonic origin. Embryonic muscle masses receive their motor innervation very early at or near their midpoint.

Development of Musculoskeletal System

(Continued)

Segmental Distribution of Myotomes in Fetus of 6 Weeks

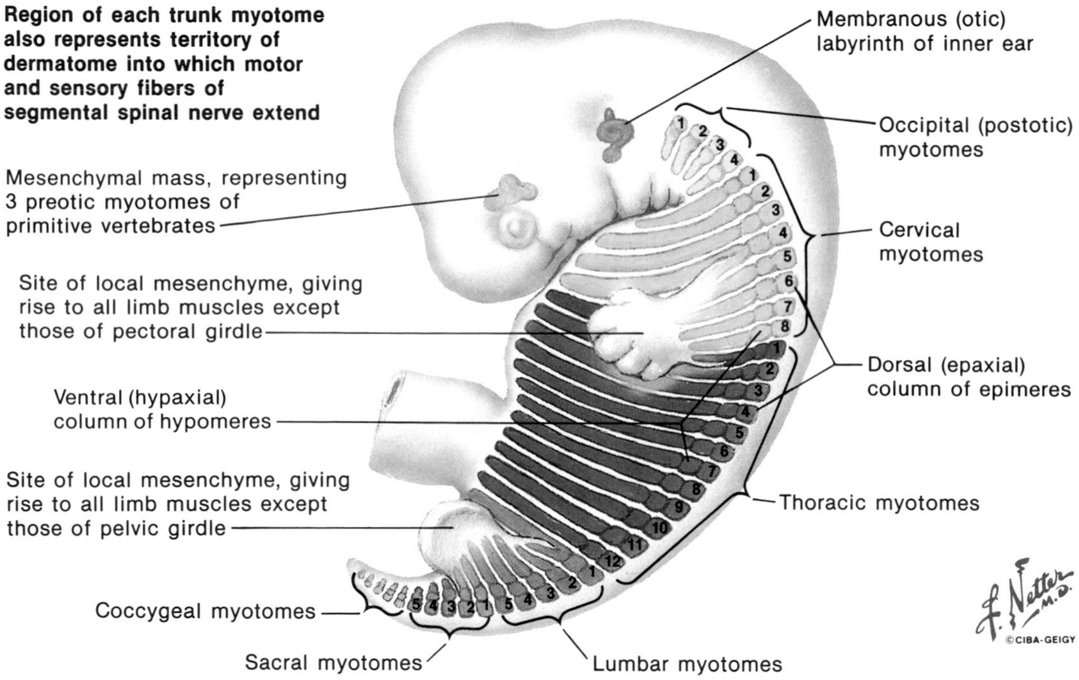
Region of each trunk myotome also represents territory of dermatome into which motor and sensory fibers of segmental spinal nerve extend
Mesenchymal mass, representing 3 preotic myotomes of primitive vertebrates
Site of local mesenchyme, giving rise to all limb muscles except those of pectoral girdle
Ventral (hypaxial) column of hypomeres
Site of local mesenchyme, giving rise to all limb muscles except those of pelvic girdle
Coccygeal myotomes
Sacral myotomes
Lumbar myotomes
Thoracic myotomes
Dorsal (epaxial) column of epimeres
Cervical myotomes
Occipital (postotic) myotomes
Membranous (otic) labyrinth of inner ear
©CIBA-GEIGY

Developing Skeletal Muscles at 8 Weeks (superficial dissection)

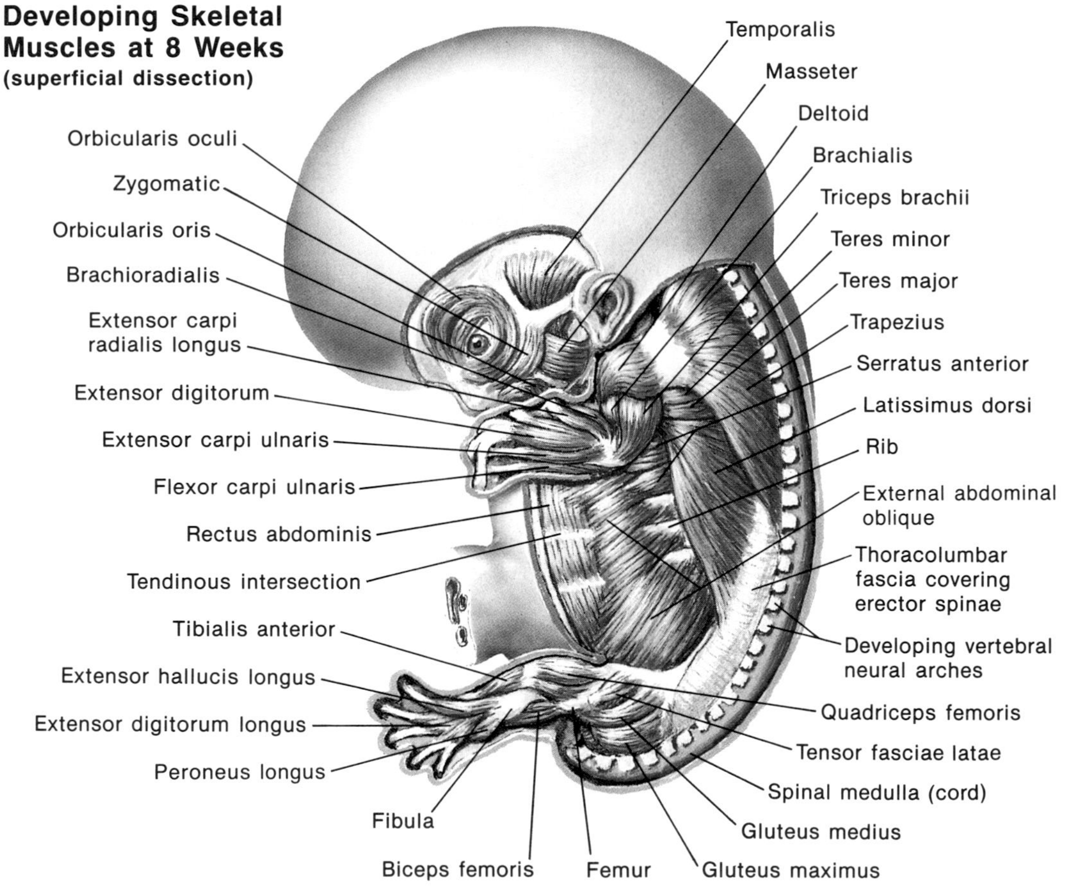
Orbicularis oculi
Zygomatic
Orbicularis oris
Brachioradialis
Extensor carpi radialis longus
Extensor digitorum
Extensor carpi ulnaris
Flexor carpi ulnaris
Rectus abdominis
Tendinous intersection
Tibialis anterior
Extensor hallucis longus
Extensor digitorum longus
Peroneus longus
Fibula
Biceps femoris
Femur
Gluteus maximus
Gluteus medius
Spinal medulla (cord)
Tensor fasciae latae
Quadriceps femoris
Developing vertebral neural arches
Thoracolumbar fascia covering erector spinae
External abdominal oblique
Rib
Latissimus dorsi
Serratus anterior
Trapezius
Teres major
Teres minor
Triceps brachii
Brachialis
Deltoid
Masseter
Temporalis

If a nerve supplies more than one muscle, it can be assumed that the muscles are subdivisions of an original myotome. Thus, the developmental histories of adult muscles formed by early fusion, splitting, migration, or other modifications can be reconstructed with considerable certainty.

Nearly all the skeletal muscles are present and, in essence, have their mature form in a fetus of 8 weeks with a crown-to-rump length of about 30 mm (Plate 16). From the time the first myotomes begin to differentiate into skeletal muscles early in the fifth week, six fundamental processes that occur up to the eighth week are involved in the gross development of the muscles. Frequently, the formation of a muscle is the result of more than one of these processes.

1. The direction of the muscle fibers may change from the original craniocaudal orientation in the myotome. Only a few muscles retain their initial fiber orientation parallel to the long axis of the body (the rectus abdominis, erector spinae, and some small vertebral column muscles). Good examples of muscles that undergo a directional change are the flat muscles of the abdominal wall—the external and internal abdominal oblique muscles and especially the transverse abdominal muscle.

2. Portions of successive myotomes commonly fuse to form a composite single muscle (the erector spinae and rectus abdominis muscles). The latter is formed by the fusion of the ventral portions of the last six or seven thoracic myotomes. Only a few muscles are derivatives of single myotomes (the intercostals and some deep, short vertebral column muscles).

3. A myotome, or branchial arch muscle primordium, may split longitudinally into two or more parts that become separate muscles (the sternohyoid and omohyoid and trapezius and sternocleidomastoid muscles).

4. The original myotome masses may split tangentially into two or more layers (the external and internal intercostal and abdominal oblique and transverse abdominal muscles).

5. A portion or all of a muscle segment may degenerate. The degenerated muscle leaves connective tissue that becomes a sheet known as an aponeurosis (the epicranial aponeurosis [galea aponeurotica], which connects the frontal and occipital portions of the occipitofrontalis muscle).

6. Finally, muscle primordia may migrate, wholly or in part, to regions more or less remote from their original site of formation. An example is the formation of certain muscles of the upper limb that arise from cervical myotomes. The serratus anterior muscle migrates to the thoracic region, to attach ultimately to the scapula and the upper eight or nine ribs, taking along its fifth, sixth, and seventh cervical spinal nerve innervation. The trapezius muscle, along with the upper five cervical spinal nerves, migrates to attach ultimately to the skull, the nuchal ligament, and the spinous processes of the seventh cervical to twelfth thoracic vertebrae. The migration of the latissimus dorsi muscle is even more extensive; it carries with it its seventh and eighth cervical spinal nerve innervation to attach ultimately to the humerus, the lower thoracic and lumbar vertebrae, the last three or four ribs, and the iliac crest of the pelvis.

As these migrating upper limb muscles acquire their attachments to the trunk, they are all superficial to the underlying muscles of the body wall. The muscles of facial expression are also good examples of muscle migration. They arise from the mesenchyme of the second or hyoid branchial arch of the future neck and migrate with their facial (VII) nerve innervation to their final positions around the mouth, nose, and eyes.

A wide range of normal variations in skeletal muscle morphology result from one or more of the six fundamental processes going awry. Usually, the variations do not interfere with an individual's normal functional ability, except when a greater part or all of a muscle is absent due to an initial failure to form, or when the usual amount of degeneration of a muscle segment is excessive.

Development of Musculoskeletal System

(Continued)

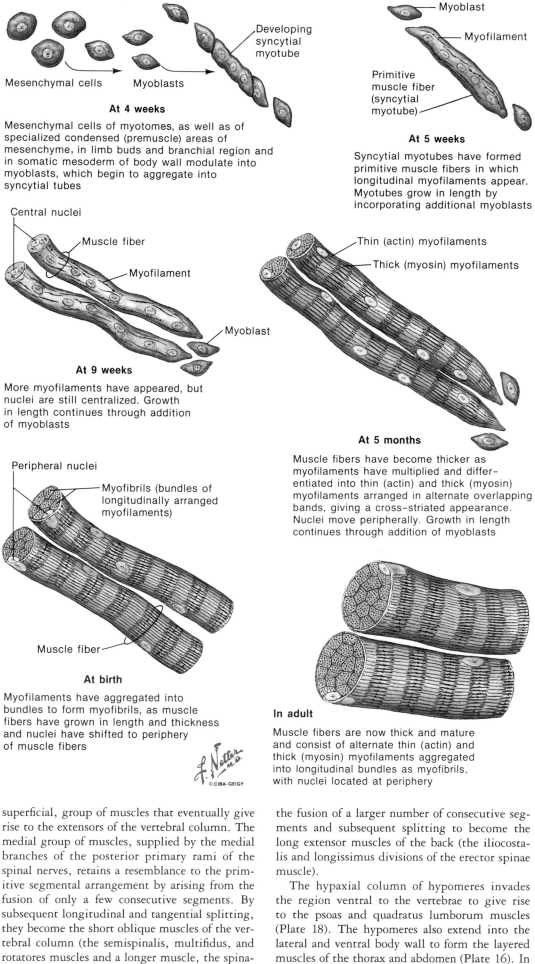

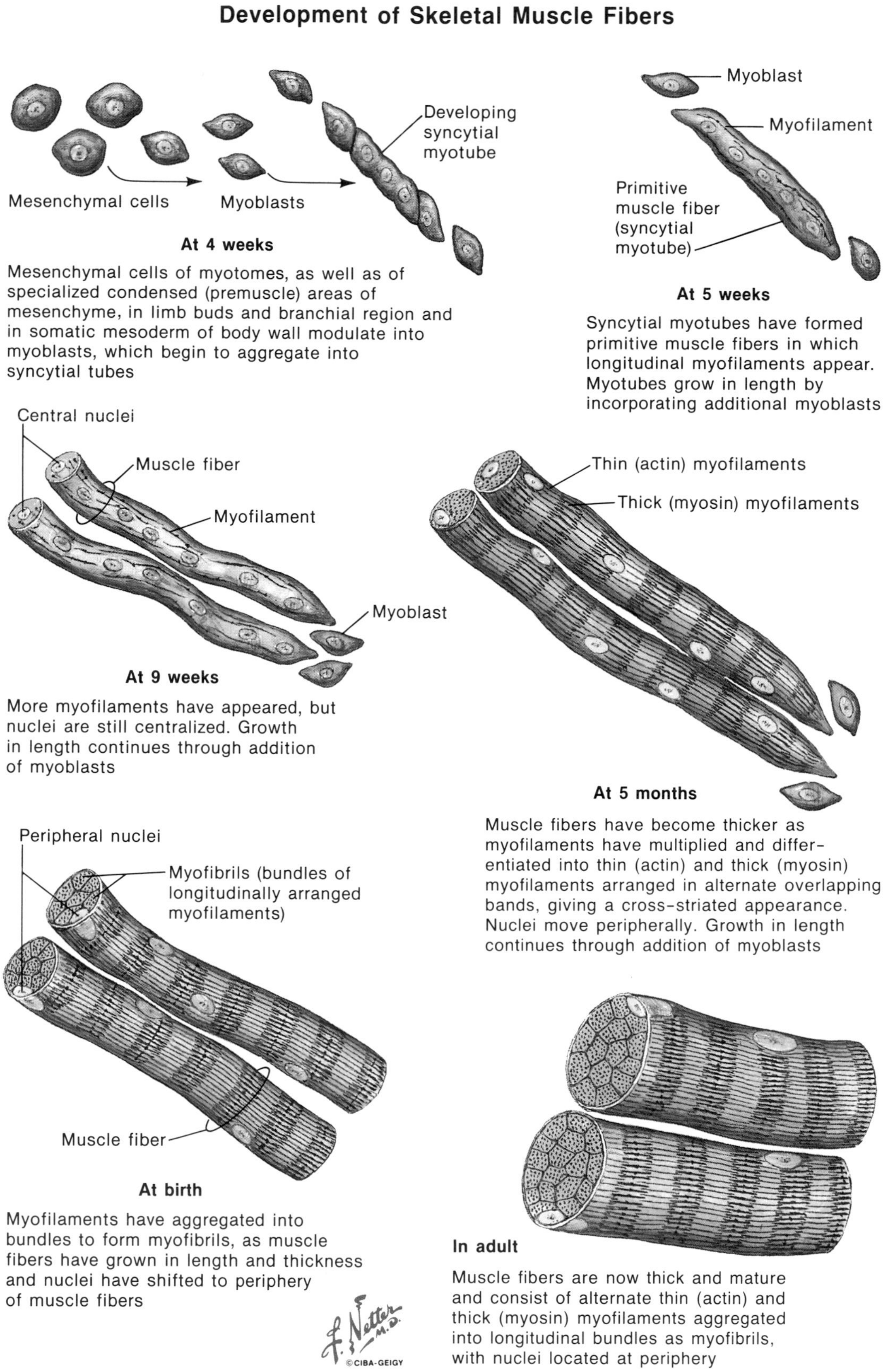
Development of Skeletal Muscle Fibers
Mesenchymal cells
Myoblasts
Developing syncytial myotube
At 4 weeks
Mesenchymal cells of myotomes, as well as of specialized condensed (premuscle) areas of mesenchyme, in limb buds and branchial region and in somatic mesoderm of body wall modulate into myoblasts, which begin to aggregate into syncytial tubes
Myoblast
Myofilament
Primitive muscle fiber (syncytial myotube)
At 5 weeks
Syncytial myotubes have formed primitive muscle fibers in which longitudinal myofilaments appear. Myotubes grow in length by incorporating additional myoblasts
Central nuclei
Muscle fiber
Myofilament
Myoblast
At 9 weeks
More myofilaments have appeared, but nuclei are still centralized. Growth in length continues through addition of myoblasts
Thin (actin) myofilaments
Thick (myosin) myofilaments
At 5 months
Muscle fibers have become thicker as myofilaments have multiplied and differentiated into thin (actin) and thick (myosin) myofilaments arranged in alternate overlapping bands, giving a cross-striated appearance. Nuclei move peripherally. Growth in length continues through addition of myoblasts
Peripheral nuclei
Myofibrils (bundles of longitudinally arranged myofilaments)
Muscle fiber
At birth
Myofilaments have aggregated into bundles to form myofibrils, as muscle fibers have grown in length and thickness and nuclei have shifted to periphery of muscle fibers
In adult
Muscle fibers are now thick and mature and consist of alternate thin (actin) and thick (myosin) myofilaments aggregated into longitudinal bundles as myofibrils, with nuclei located at periphery
F. Netter M.D.
©CIBA-GEIGY

Some unusual muscle variations can be explained as genetic atavisms or muscles that were typical in one of the vertebrate ancestors of man.

Skeletal muscle can undergo limited regeneration. When damaged, macrophages enter the necrotic area and remove the dead material. The damaged muscle fibers on each side of the necrotic area, which are actually open-ended syncytial tubes, form growth buds on their ends that grow toward each other, meet, and fuse. This reestablishes muscle fiber continuity across the damaged area and may be sufficient for the repair of a small muscle injury. When there is more extensive damage, the repair process is similar to the embryologic process of muscle fiber formation. Undifferentiated mononucleated cells normally present within the damaged muscle become myoblasts that divide and then fuse together to become new multinucleated syncytial myotubes replacing the damaged segment. The myotubes go on to differentiate into typical muscle fibers. Even so, when large areas are damaged, the muscle regeneration may be so limited that the missing muscle is replaced chiefly with connective tissue.

Trunk Muscles

Between the fifth and sixth weeks, the myotomes of the trunk of the human embryo become divided by a slight longitudinal constriction into a dorsal epaxial column of *epimeres* and a more ventral hypaxial column of *hypomeres* (Plates 16 and 18). The original spinal nerve to the myotome that gives rise to an epimere and a hypomere also divides into dorsal and ventral rami. Thus, the epimeres and hypomeres are innervated, respectively, by the dorsal and ventral rami of the serially repeated spinal nerves, just as in adult primitive fish. In addition, the developing transverse processes of the vertebrae serve to help separate the epaxial and hypaxial columns. The mesenchyme between the two columns attaches to the transverse processes and becomes a connective tissue sheet or intermuscular septum, the rudiment of the thoracolumbar fascia, which permanently separates the two columns.

After the transverse processes appear, the ribs form in the sclerotomal tissue that extends by differentiation into the ventral portions of the original clefts between the somites. The maximum development of the ribs is in the thoracic region; consequently, of all the muscles in the adult, the intercostal muscles retain to the greatest degree the original segmental pattern of the hypaxial musculature.

The epaxial column of epimeres divides further into a medial, or deep, and a lateral, or superficial, group of muscles that eventually give rise to the extensors of the vertebral column. The medial group of muscles, supplied by the medial branches of the posterior primary rami of the spinal nerves, retains a resemblance to the primitive segmental arrangement by arising from the fusion of only a few consecutive segments. By subsequent longitudinal and tangential splitting, they become the short oblique muscles of the vertebral column (the semispinalis, multifidus, and rotatores muscles and a longer muscle, the spinalis division of the erector spinae muscle). The lateral, more superficial group of muscles, which is supplied by the lateral branches of the posterior primary rami of the spinal nerves, arises by the fusion of a larger number of consecutive segments and subsequent splitting to become the long extensor muscles of the back (the iliocostalis and longissimus divisions of the erector spinae muscle).

The hypaxial column of hypomeres invades the region ventral to the vertebrae to give rise to the psoas and quadratus lumborum muscles (Plate 18). The hypomeres also extend into the lateral and ventral body wall to form the layered muscles of the thorax and abdomen (Plate 16). In the thorax, they are the intercostals; in the abdomen, they are the external and internal oblique, transverse abdominal, and rectus abdominis muscles (Plate 18). The rectus abdominis muscle

Development of Musculoskeletal System

(Continued)

develops from the most ventral extension of the lower thoracic and first abdominal hypomeres that fuse in a cephalocaudal direction to become a single longitudinal muscle on either side of the midline of the body, which is separated in the abdomen by the linea alba of dense connective tissue. The tendinous intersections (inscriptions) are indicative of the original segmental character of the rectus abdominis muscle (Plate 16). Also, the fibers of this muscle retain the cephalocaudal orientation of the original myotomic fibers. In the upper thoracic region, there is also a longitudinal muscle sheet that is continuous with the sheet giving rise to the rectus abdominis muscle. It normally disappears but is occasionally retained as the sternalis muscle. All muscles derived from the hypomeres are primarily flexors of the vertebral column.

Perineal Muscles

The formation of the muscles derived from both the epimeres and the hypomeres is well advanced by the seventh week, except for the muscles of the pelvic diaphragm, perineum, and external genitalia (Plate 19). These muscles develop later because of the late division of the single cloacal opening into a urethral and anal opening in the male and female, and the acquisition of an additional opening in the female—the vagina.

This late development is a reflection of the more recent changes occurring in the evolution of the urogenital system. A single cloacal opening is characteristic of all adult fish, amphibians, reptiles, birds, and the primitive egg-laying mammals. In all mammals higher up the phylogenetic ladder than egg layers, there are separate anal and urogenital openings; however, it is only in female primates that the urethra and vagina are completely separate and have separate openings to the exterior.

In man, a striated cloacal sphincter muscle and levator ani muscle (pelvic diaphragm) arise from the third sacral to the first coccygeal myotomic hypomeres and are well developed by the eighth week. The striated external anal sphincter, perineal, and external genital muscles arise from the cloacal sphincter muscle by its rearrangements and additions during the establishment of the urogenital and anal openings. The deep, or inner, fibers of the cloacal sphincter muscle give rise to the urethral sphincter muscle. Although the muscles of the external genitalia are the same in both sexes, they, of necessity, must undergo a different arrangement in each sex. The mature pelvic muscle arrangement in the two sexes is present by the sixteenth week of development. However, not until sometime during the second year after birth do the urethral and external anal sphincter muscles come under voluntary control.

Cross Sections of Body at 6 to 7 Weeks

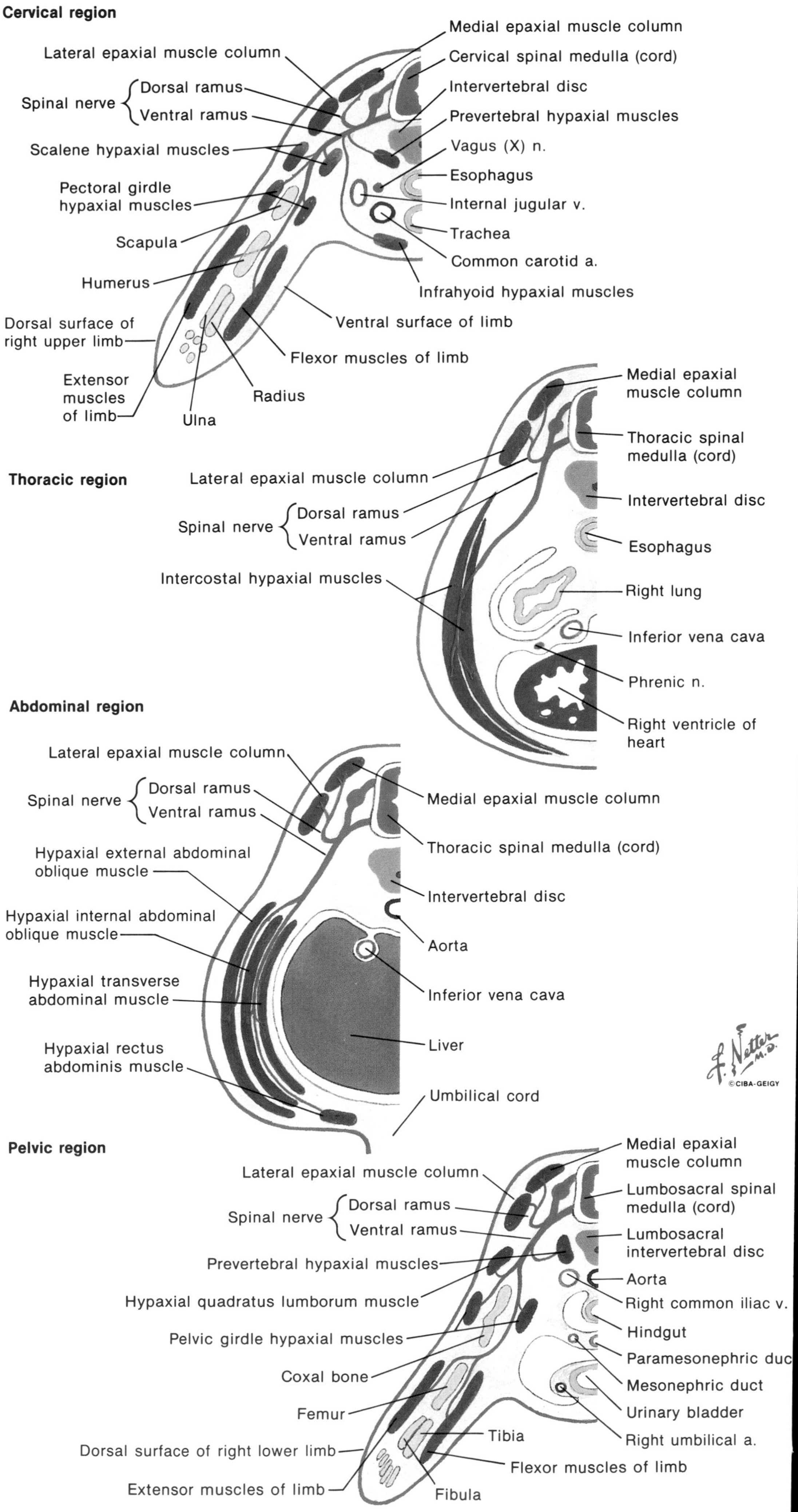

Development of Musculoskeletal System

(Continued)

Prenatal Development of Perineal Musculature

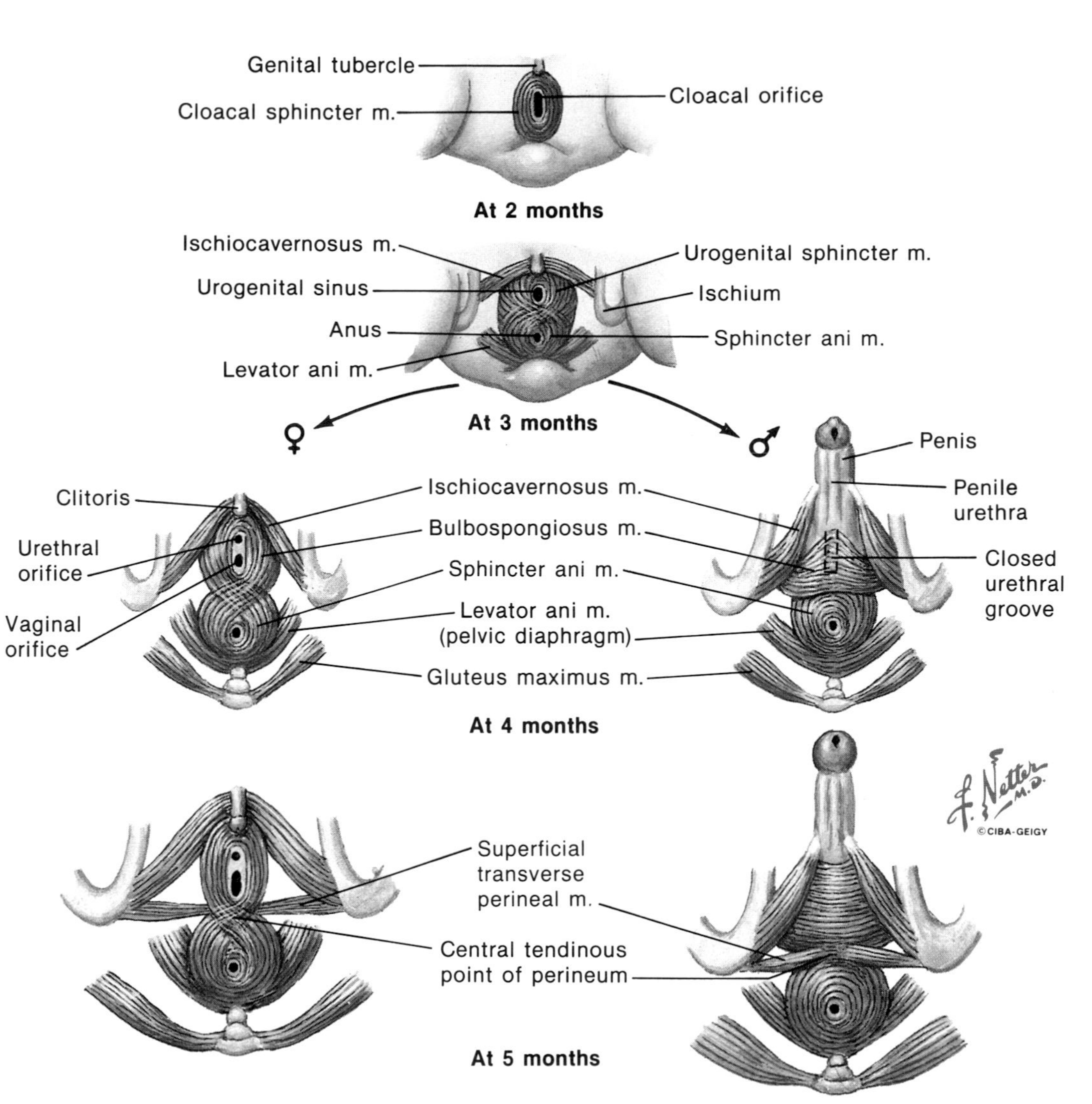
Genital tubercle
Cloacal sphincter m.
Cloacal orifice
At 2 months
Ischiocavernosus m.
Urogenital sinus
Anus
Levator ani m.
Urogenital sphincter m.
Ischium
Sphincter ani m.
At 3 months
♀
♂
Penis
Clitoris
Urethral orifice
Vaginal orifice
Ischiocavernosus m.
Bulbospongiosus m.
Sphincter ani m.
Levator ani m. (pelvic diaphragm)
Gluteus maximus m.
Penile urethra
Closed urethral groove
At 4 months
Superficial transverse perineal m.
Central tendinous point of perineum
At 5 months
F. Netter M.D.
©CIBA-GEIGY

Limb Muscles

During their early development, the limbs are literally ectodermal sacs that become stuffed with mesenchyme. As the limb buds grow, the proliferating local somatic mesenchyme eventually gives rise to all skeletal rudiments. Myotome cells from the adjacent somites invade the limb buds to give rise to all the skeletal muscles. When the ingrowth of myotome cells, nerve fibers, neurilemmal cells, pigment cells, and, possibly, the endothelium of the blood and lymphatic systems are excluded, the limb buds would still have the capacity for self-differentiation to become limbs containing all the normal skeletal rudiments. The muscles of the pectoral and pelvic girdles are also of myotomic origin.

Early in the seventh week, the mesenchymal premuscle masses of the girdle musculature are formed in the human embryo. As the rudiments of the appendicular skeleton become differentiated within the developing limb, the mesenchyme from which the limb muscles arise is aggregated into masses grouped dorsal and ventral to the developing skeletal parts. The progressive formation of distinct muscles reaches the level of the hand and foot during the seventh week. The muscles of the upper limb develop slightly ahead of those of the lower limb.

The early limbs are flattened dorsoventrally and look like paddles projecting straight out from the body. They each have a cephalic (preaxial) border and a caudal (postaxial) border, as well as a craniocaudal attachment to the body opposite a number of myotomes (Plate 9). Each upper limb bud lies opposite the lower five cervical and the first thoracic myotomes. Each lower limb bud is opposite the second and fifth lumbar and the upper three sacral myotomes. The branches of the spinal nerves supplying these myotomes reach the base of their respective limb bud. As the bud elongates to form a limb, the nerves grow into it in such a manner that the group of limb muscles along the preaxial border of the upper limb becomes innervated by the fourth to the seventh cervical nerves, and those of the postaxial border, by the eighth cervical and the first thoracic nerves. In the lower limb, the group of muscles along the preaxial border receives innervation from the second to the fifth lumbar nerves, and the group of muscles along the postaxial border, from the first to the third sacral nerves.

The preaxial and postaxial groups of developing muscles become split and rearranged. In so doing, they both contribute to the formation of the ventral, or anterior, limb-flexor group of muscles and a dorsal, or posterior, limb-extensor group (Plate 18). The original preaxial and postaxial nerves of the limbs are similarly divided into anterior and posterior divisions, supplying the flexors and extensors, respectively. Thus, the ulnar and median nerves in the upper limb, which contain both preaxial and postaxial nerve fibers, are branches of the anterior divisions of the trunks of the brachial plexus and innervate flexor muscles. Likewise, the radial nerve, containing both preaxial and postaxial nerve fibers, is derived from the posterior divisions of the trunks of the brachial plexus and innervates extensor muscles.

In the lower limb, the tibial part of the sciatic nerve, which contains both preaxial and postaxial nerve fibers, arises from the anterior divisions of the sacral plexus and innervates flexor muscles via branches of the sciatic and tibial nerves. The femoral nerve, containing only preaxial nerve fibers, arises from the posterior divisions of the lumbar plexus and innervates the extensor muscles. The peroneal part of the sciatic nerve and its peroneal branch, which contains both preaxial and postaxial nerve fibers, arise from the posterior divisions of the sacral plexus and also innervate the extensor muscles.

At 6 weeks, the flexed limbs have not yet rotated out of their primary position (Plate 8). Because the upper and lower limbs later undergo opposite rotations to reach their definitive positions, the eventual anterior, or ventral, flexor muscle compartment of the mature arm corresponds to the posterior, or dorsal, flexor muscle compartment of the mature thigh. Also, the eventual anterior, or ventral, flexor muscle compartment of the mature forearm corresponds to the posterior, or dorsal, flexor muscle compartment (calf) of the mature leg. Because of the twist of the lower limb during development that results in permanent pronation of the foot, extension of the mature wrist corresponds to the so-called dorsiflexion of the ankle that is actually its extension.

Head and Neck Myotomic Muscles

The formation of the three preotic somites is slurred over in the human embryo. What would have been their myotomes appear as three closely apposed aggregations of mesenchyme in the region of the developing eye that give rise to the extrinsic ocular muscles (Plate 20). The three surviving postotic occipital somites of the original four give rise to typical myotomes.

Comparative anatomy indicates that during evolution, the tongue muscles first appeared in amphibian forms because, in fish, the tongue is a membranous sac lacking muscle. In ancestral forms, the tongue muscles are derived from the occipital myotomes that are innervated exclusively by the hypoglossal (XII) nerves.

In the human embryo, the origin of the tongue muscles is abbreviated and slurred over. The muscles arise directly from an ill-defined mass of mesenchyme located adjacent to the pharynx in the region of the branchial arch mesenchyme from which the branchiomeric skeletal muscles arise (Plate 20). However, because of the close relationship of the hypoglossal nerves to the occipital somites when they first form in the human

Development of Musculoskeletal System

(Continued)

Origins and Innervations of Branchiomeric and Adjacent Myotomic Muscles

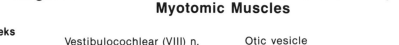

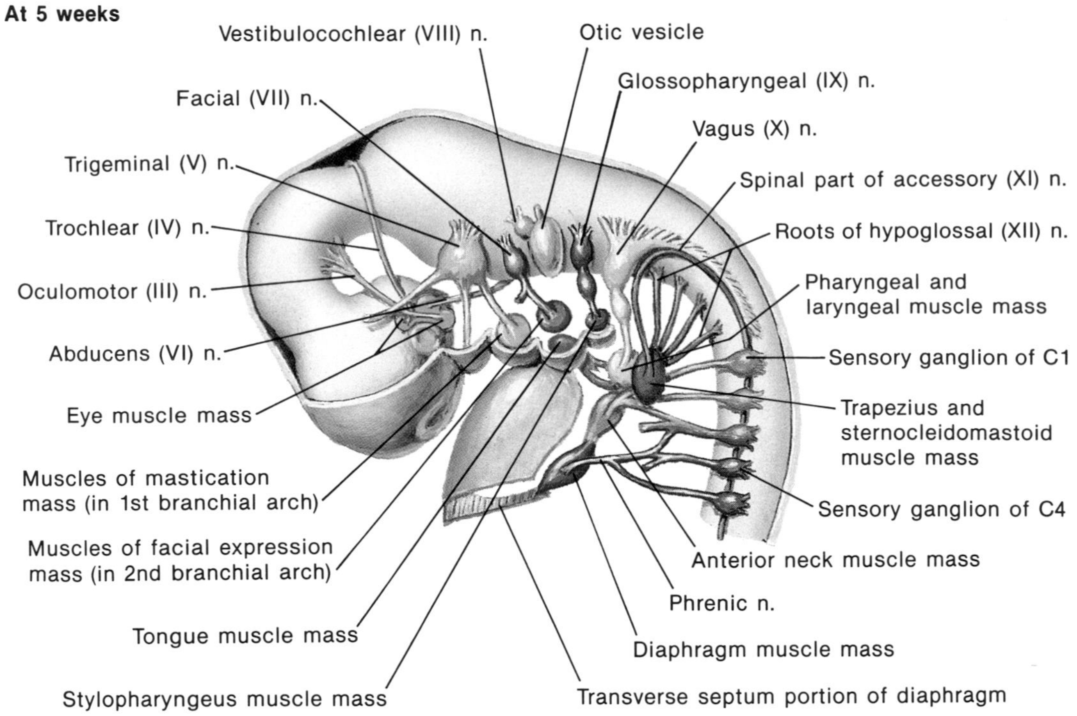

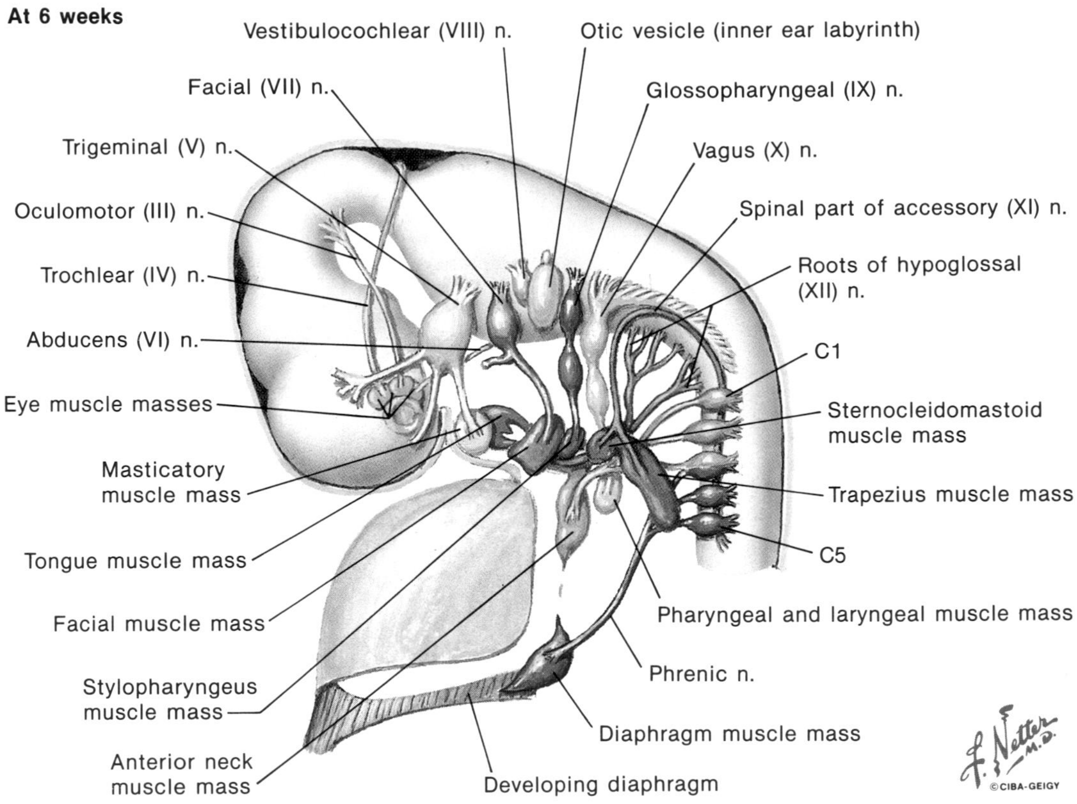

embryo, the tongue muscles are regarded as being derived from occipital myotomes even though they appear to arise directly from mesenchyme in the region of the tongue rudiment.

Another muscle mass that has slurred-over development gives rise to the trapezius and sternocleidomastoid muscles. It forms in mesenchyme situated between the occipital myotomes and the branchiomeric mesenchyme of the most caudal branchial arch. The innervation of the muscle mass is unique because it arises as a number of motor roots from the side of the upper five segments of the cervical spinal medulla (cord) between the dorsal and ventral roots of the cervical spinal nerves, which eventually become the spinal part of the accessory (XI) nerve.

The epaxial column of epimeres derived from the cervical myotomes becomes the extensor musculature of the neck in the same manner as that in the trunk. However, the neck musculature is more elaborately developed than that of the thorax. The medial, or deep, group of muscles derived from the epaxial column are the short oblique muscles of the vertebral column—the multifidus and rotatores muscles, and some longer muscles—the spinalis and semispinalis muscles that also attach to the skull. The lateral, or superficial, group of muscles derived from the epaxial column are the long extensor muscles of the vertebral column—the iliocostalis cervicis, longissimus cervicis, and capitis divisions of the erector spinae muscle, and the splenius capitis muscle.

The formation of muscles from the cervical hypaxial column of hypomeres, however, is quite different from what happens in the thorax; this is due to the development of the adjacent upper limbs, to the caudal recession of the coelomic, or body, cavity that originally extended into the head region, and to the presence of the branchial arches. It is interesting that the muscle mass giving rise to the infrahyoid muscles is continuous with the mass giving rise to the tongue muscles, and that the infrahyoid muscle mass is also continuous caudally with the muscle mass that becomes the diaphragmatic striated muscle.

The diaphragm is originally located in the neck region. Because of its caudal migration, mainly due to differential growth, its cervical spinal innervation via the phrenic nerves has to elongate markedly.

Head and Neck Branchiomeric Muscles

During evolution, the switch from water breathing to air breathing resulted in the loss of the branchial arch, gill slit, and aqueous respiratory apparatus and the acquisition of a definitive face and neck. Many of the branchial arch structures, especially the skeleton, underwent modification and were retained in the resulting air-breathing upper respiratory system and acoustic system. A résumé of these modifications is recapitulated in the human embryo. Of the six branchial arches of primitive vertebrates, the fifth and sixth arches are completely rudimentary in man. Even so, the deep tissue in the territory of the fifth and sixth arches gives rise to certain primitive structures that undergo modifications and are retained in the adult. The four definitive arches are present by the fifth week.

During the fifth week, condensations of mesoderm appear in the dorsal end of each of the four branchial arches, including the territories of the fifth and sixth arches. In the development of primitive vertebrates, there is continuity between the mesodermal condensations of each arch and one of the head somites, indicating that the condensations represent the hypaxial portion of the head somites. However, in the human embryo, this phase of development is slurred over because no such continuity occurs between the condensations and the somites. Therefore, the myoblasts that differentiate directly from the mesodermal condensations of the arches give rise to skeletal striated muscles that are regarded as branchiomeric in origin. The voluntary motor part of a special visceral cranial nerve grows into each of the muscle rudiments of the arches, including those of the territories of the fifth and sixth arches.

The muscles of branchiomeric origin retain their original cranial nerve innervation as they migrate to their final destinations (Plates 20–21).

Development of Musculoskeletal System

(Continued)

Branchiomeric and Adjacent Myotomic Muscles at Birth

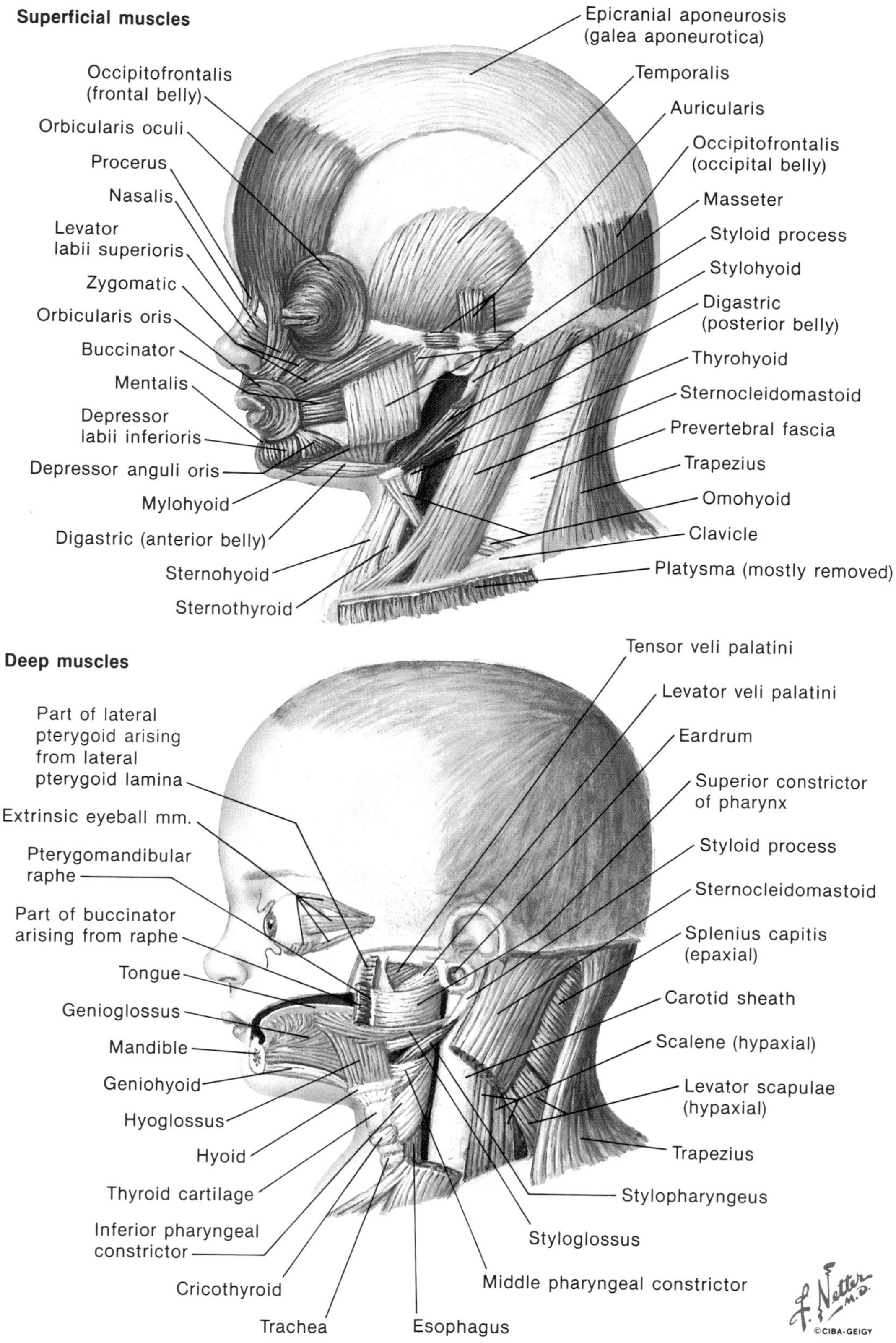
Superficial muscles
Epicranial aponeurosis (galea aponeurotica)
Occipitofrontalis (frontal belly)
Temporalis
Auricularis
Orbicularis oculi
Occipitofrontalis (occipital belly)
Procerus
Nasalis
Masseter
Levator labii superioris
Styloid process
Stylohyoid
Zygomatic
Digastric (posterior belly)
Orbicularis oris
Buccinator
Thyrohyoid
Mentalis
Sternocleidomastoid
Depressor labii inferioris
Prevertebral fascia
Trapezius
Depressor anguli oris
Omohyoid
Mylohyoid
Clavicle
Digastric (anterior belly)
Platysma (mostly removed)
Sternohyoid
Sternothyroid
Deep muscles
Tensor veli palatini
Levator veli palatini
Part of lateral pterygoid arising from lateral pterygoid lamina
Eardrum
Superior constrictor of pharynx
Extrinsic eyeball mm.
Styloid process
Pterygomandibular raphe
Sternocleidomastoid
Part of buccinator arising from raphe
Splenius capitis (epaxial)
Tongue
Carotid sheath
Genioglossus
Scalene (hypaxial)
Mandible
Geniohyoid
Levator scapulae (hypaxial)
Hyoglossus
Hyoid
Trapezius
Thyroid cartilage
Stylopharyngeus
Inferior pharyngeal constrictor
Styloglossus
Middle pharyngeal constrictor
Cricothyroid
Trachea
Esophagus
F. Netter M.D.
©CIBA-GEIGY

The muscles that arise from the primordial mesenchymal mass of the first or mandibular branchial arch become innervated by the motor neurons of the trigeminal (V) nerve. These muscles become the masticatory muscles (the temporal, masseter, and pterygoid muscles) as well as the mylohyoid, anterior belly of the digastric, tensor veli palatini, and tensor tympani muscles. The muscles arising in the region of the second, or hyoid, branchial arch become the muscles of facial expression and receive their motor innervation from the facial (VII) nerve. Other muscles arising from the second arch mesenchyme and innervated by the facial nerve are the posterior belly of the digastric, stylohyoid, and stapedius muscles. The glossopharyngeal (IX) nerve supplies motor innervation to the muscle mass of the third branchial arch, which becomes the stylopharyngeus muscle.

The muscles arising in the fourth arch and in the territories of the fifth and sixth branchial arches become those of the soft palate (the levator veli palatini, uvulae, and palatoglossus muscles); those of the pharynx (the pharyngeal constrictor, palatopharyngeus, and salpingopharyngeus muscles); and all the intrinsic muscles of the larynx. The innervation of all these muscles derived from the fourth arch and the fifth and sixth arch territories is actually from the vagus (X) nerve. However, the rootlets containing the axons of the motor neurons leave the side of the medulla oblongata portion of the brainstem to become what is named the cranial part of the accessory (XI) nerve. The cranial part, after being attached by connective tissue to the spinal part of the accessory nerve as they pass through the jugular foramen of the skull, separates from the spinal part in the neck to join the main trunk of the vagus nerve. Its motor neurons to the striated muscles of the soft palate and pharynx pass via the pharyngeal branches of the vagus, while those to the intrinsic muscles of the larynx pass via the superior and recurrent laryngeal branches.

Skeletal Muscle Innervation

The establishment of neural contacts with developing skeletal muscle fibers is a critical developmental stage. The contacts enhance muscle development and are important for the complete differentiation and function of the fibers. The motor nerve axons make contact with the masses of myoblasts constituting the developing muscles as early as between the fifth and sixth week if they are trunk muscles. However, it is between this time and the tenth week that the branches of the large somatic (alpha) motor neurons begin to ramify among the developing motor fibers of the muscles and to establish the formation of neuromuscular junctions. Muscle spindles (proprioceptors) can be distinguished at about the twelfth week. They become innervated by the small gamma motor nerves.

Movements of the mother, and especially of the uterus, serve as stimuli to induce muscular activity to occur in the fetus before the fourth month, although the mother is not aware of it until the "quickening" at about the fourth month. Long before birth, the diaphragm contracts periodically in response to phrenic nerve activity (hiccups). The fetus begins to swallow amniotic fluid at 12½ weeks; before birth, it may at times suck the fingers. Therefore, the phrenic nerves and the muscular diaphragm used for breathing, and the sensory nerves of the lips, mouth, and throat, as well as the striated muscles with their motor nerves of the lips, tongue, jaws, and throat used for the complicated reflex functions of suckling and swallowing, are functionally well developed at birth. In contrast, the trunk and limb muscles at birth are uniformly slow in contracting.

Voluntary control of the skeletal muscles cannot occur in the newborn infant because of the lack of dendritic development of the cerebral neurons, especially those of the motor cortex, and the fact that the fibers of the upper motor neurons of the corticobulbar and corticospinal tracts have only begun to be myelinated. It is not until the end of the first year after birth that the myelination of the nerve fibers of the corticospinal tract is nearly completed. This is about the time when the child has sufficient voluntary control over the skeletal muscles to be able to stand and walk. □

Section III

Physiology

Frank H. Netter, M.D.

in collaboration with

Jonathan Black, Ph.D. *Plates 22–23*

Carl T. Brighton, M.D., Ph.D. *Plates 16–18, 34, 36–38*

Joseph A. Buckwalter, M.D. *Plate 25*

Bruce M. Carlson, M.D., Ph.D. *Plates 2–3, 7, 21, 27*

Charles C. Clark, Ph.D. *Plates 24–25*

Joseph P. Iannotti, M.D. *Plate 19*

Frederick S. Kaplan, M.D. *Plates 16, 19–25, 28–31, 33–36, 39*

Henry J. Mankin, M.D. *Plate 32*

Richard G. Schmidt, M.D. *Plate 39*

H. Ralph Schumacher, Jr., M.D. *Plate 26*

Michael E. Selzer, M.D., Ph.D. *Plates 1–15*

Microscopic Appearance of Skeletal Muscle Fibers

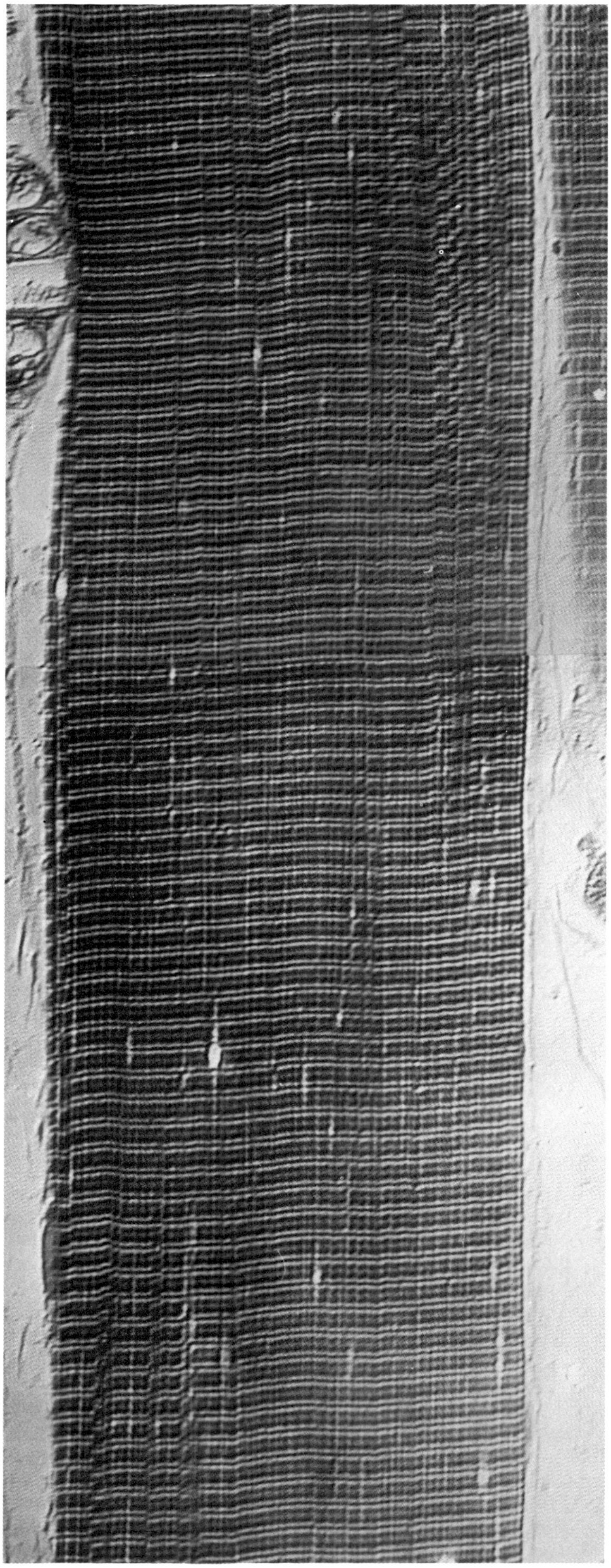

On light microscopy skeletal muscle fibers have a strikingly regular banding pattern (longitudinal section of frog skeletal muscle under differential interference microscopy, x1,280)

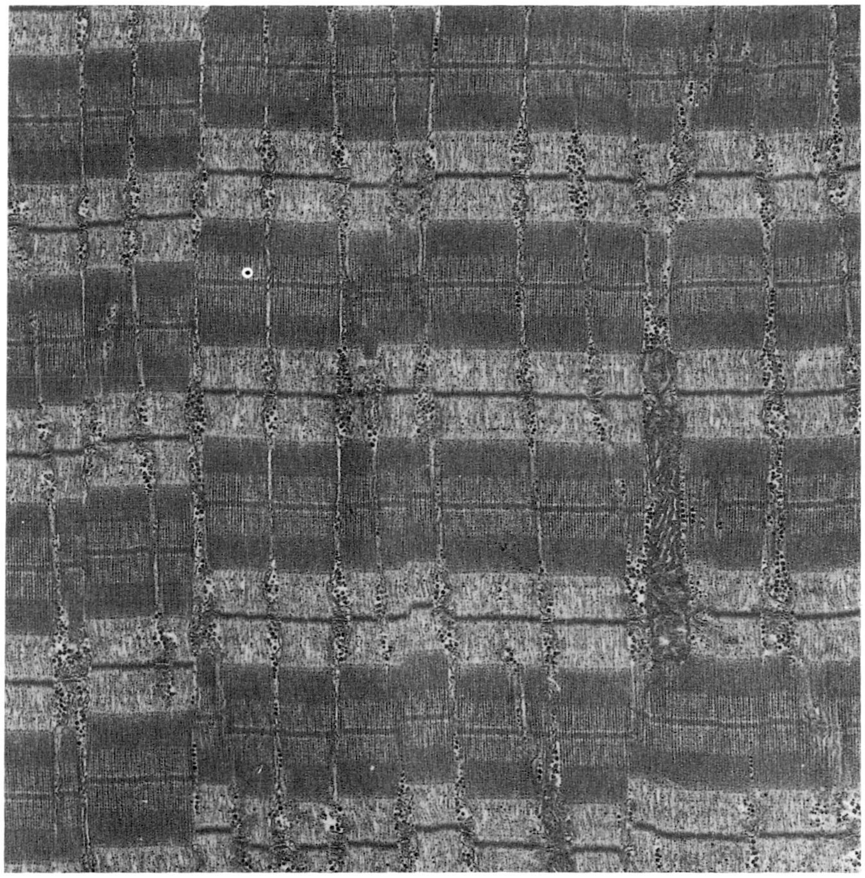

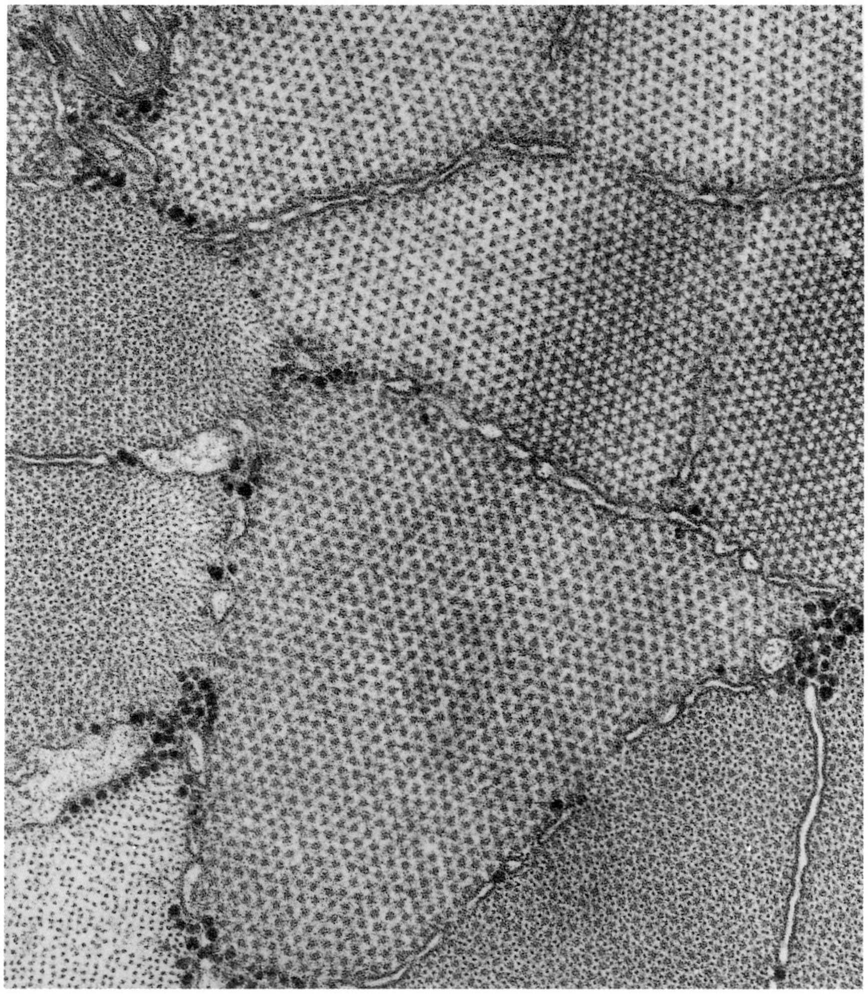

On electron microscopy banding pattern is seen to result from overlap of regularly arranged thick and thin filaments. Above: longitudinal section stained with lead, x9,800. Below: transverse section stained with lead, x66,000

Organization of Skeletal Muscle

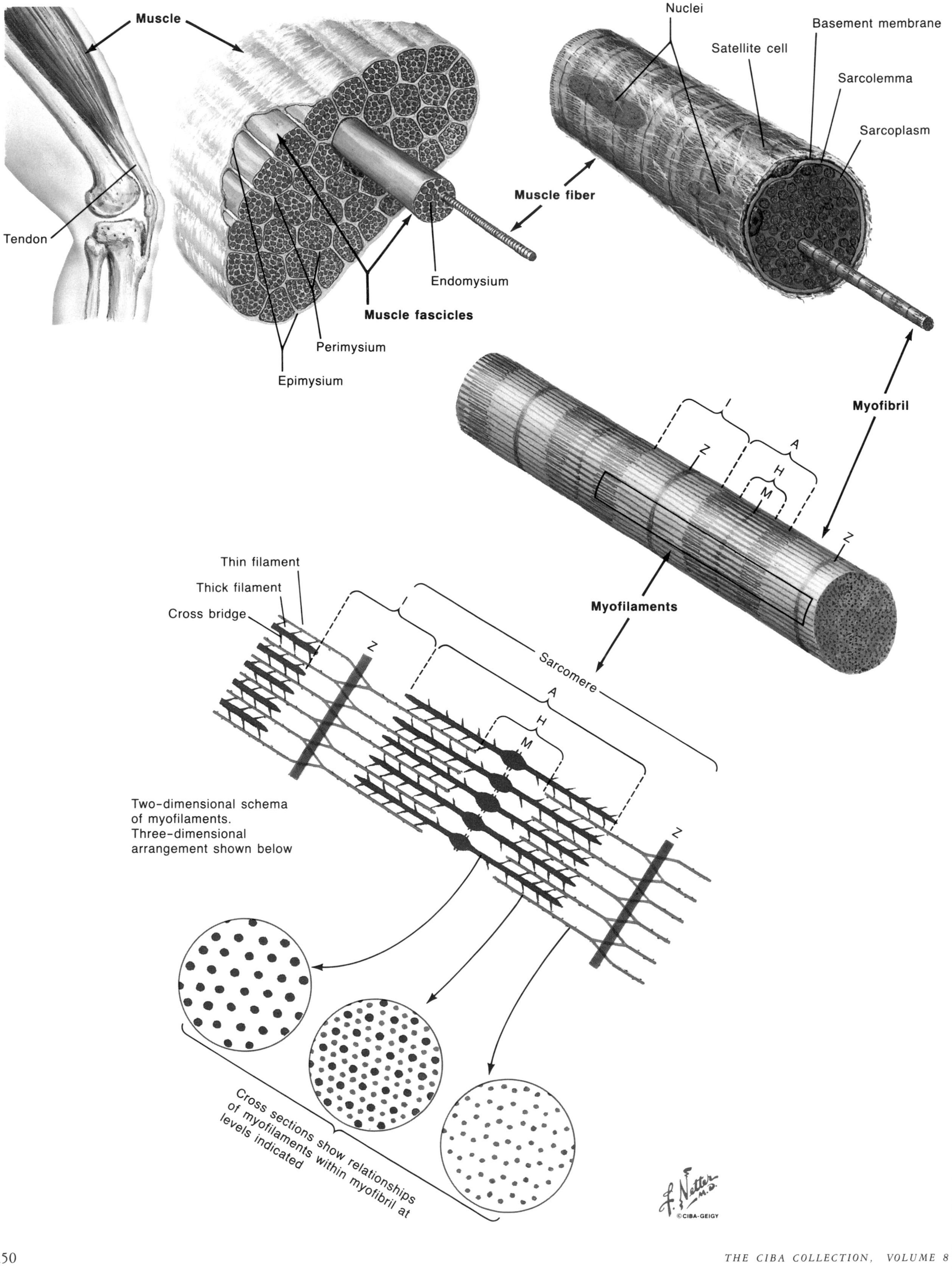
Muscle
Tendon
Muscle fascicles
Endomysium
Perimysium
Epimysium
Muscle fiber
Nuclei
Satellite cell
Basement membrane
Sarcolemma
Sarcoplasm
Myofibril
I
A
H
M
Z
Myofilaments
Thin filament
Thick filament
Cross bridge
Sarcomere
Two-dimensional schema of myofilaments. Three-dimensional arrangement shown below
Cross sections show relationships of myofilaments within myofibril at levels indicated
F. Netter M.D.
©CIBA-GEIGY

SECTION III PLATE 2 Slide 3698

Structural Organization of Skeletal Muscle

The principal function of skeletal muscles is to move the limbs, trunk, head, respiratory apparatus, and eyes. Most skeletal muscles are under voluntary control. They are made up of long multinucleated cells called muscle fibers, which are derived by end-to-end fusion of many embryonic cells called *myoblasts* to form *myotubes* during development. The ends of the muscle fibers insert into tendons that in turn attach to bones across the joints. The entire muscle is surrounded by a connective tissue sheath, the *epimysium*. The connective tissue extends into the muscle as the *perimysium*, which divides the muscle into a number of fascicles, each containing several muscle fibers. Within the fascicle, muscle fibers are separated from one another by the *endomysium*.

Each muscle fiber is invested by a thin layer of connective tissue called the basal lamina, or *basement membrane*. It is now believed that the basement membrane contains molecules important to the development and differentiation of the neuromuscular apparatus. Satellite cells, enclosed within the basement membrane, are believed to derive from undifferentiated myoblasts and are also thought to be capable of fusing with damaged muscle fibers in a regenerative process.

A muscle fiber exerts force by contracting. The microscopic structure of the muscle fiber gives a great deal of information about the way it functions. The contractile apparatus of each muscle fiber is subdivided into *myofibrils*, longitudinally oriented bundles of thick and thin filaments. The thick and thin filaments provide the mechanical force of contraction by sliding past one another. A myofibril measures about 1 μm in diameter and extends the entire length of the fiber. The thin filaments of the myofibril are anchored at one end to a meshlike structure made up largely of protein and oriented at right angles to the filaments. Seen from the side, this lattice appears narrow and dense. The resulting image in a longitudinal section observed on light microscopy is called the *Z band* (Zwischenscheibe). Z bands occur at very regular intervals along the length of the myofibril. The stretch of myofibril between two adjacent Z bands is called a *sarcomere*, which can be considered the unit of contractile action. Thus, myofibrils are made up of many sarcomeres linked end to end. The thick filaments are disposed in the center of the sarcomere. Because they strongly rotate polarized light, the thick filaments are responsible for the appearance of the anisotropic bands, or *A bands*, on longitudinal section.

The contractile filaments slide past one another by a grappling action. The thick filaments are linked to the thin filaments by *cross bridges*, which are part of the structure of the thick filaments (Plate 4). Electron microscopy reveals that, except at the middle portion, the cross bridges are located along the length of the thick filament. The cross bridges slant away from the middle portions of the filament toward the Z band closest to them. Thick filaments widen slightly at their middle portions, and the widened middle portions of adjacent thick filaments are in register, thus creating the appearance of the *M band*.

Most of the time, the sarcomere is in a state of relaxation. Because it is longer than a thick filament, there is a region at either end of the sarcomere that contains only thin filaments. The thin filaments rotate polarized light very little; therefore, the region of the sarcomere on either side of the Z band where thin filaments are not overlapped by thick filaments is called the isotropic band, or *I band*. In the relaxed state, the thin filaments of a single sarcomere that are attached to adjacent Z bands point toward each other but do not touch. Thus, there is a region in the middle of the sarcomere where thick filaments are not overlapped by thin filaments, which is called Hensen's disk, or the *H zone*.

The three-dimensional structure of the sarcomere is very regular. On cross section, each thick filament is surrounded by six thin filaments, and each thin filament is equidistant to three thick filaments. □

SECTION III PLATE 3 Slide 3699

Intrinsic Blood Supply of Skeletal Muscle

Muscles use a great deal of energy and therefore require a rich blood supply. Arteries and veins usually enter the muscle together with the nerve. This grouping, called a *neurovascular bundle*, is a common anatomic organization in many organs of the body. The main arteries supplying muscles run longitudinally within the connective tissue perimysium. They give rise to smaller branches, or *arterioles*, which penetrate the endomysium of the fascicle. These endomysial arterioles give rise to capillaries that nourish the muscle fibers. Other branches of the main arteries are transversely oriented and remain within the epimysium and perimysium. Because these branches give rise to only a few capillaries, they do not serve to nourish the muscle fibers; instead, they form nonnutritive connections, or anastomoses, with other arteries, or they form shunts directly to the veins.

During muscle contraction, the shortened muscle fibers bulge, squeezing against the surrounding connective tissue and one another. During very vigorous contraction, the blood vessels within the endomysium can be choked off completely. Arterial blood would back up and the blood pressure would rise excessively were it not for the fact that the anastomotic channels permit blood to bypass the nutritive circulation. Muscles that need to generate a lot of force—for example, muscles used during sprinting—work much of the time without a supply of oxygen, because their nutritive blood supply is closed off while they are contracting. These muscles are specialized to function anaerobically, but in so doing, they rapidly use up their energy stores. The so-called oxygen debt is repaid when the muscle stops working and the nutritive arterioles open once again. □

Intrinsic Blood Supply of Skeletal Muscle

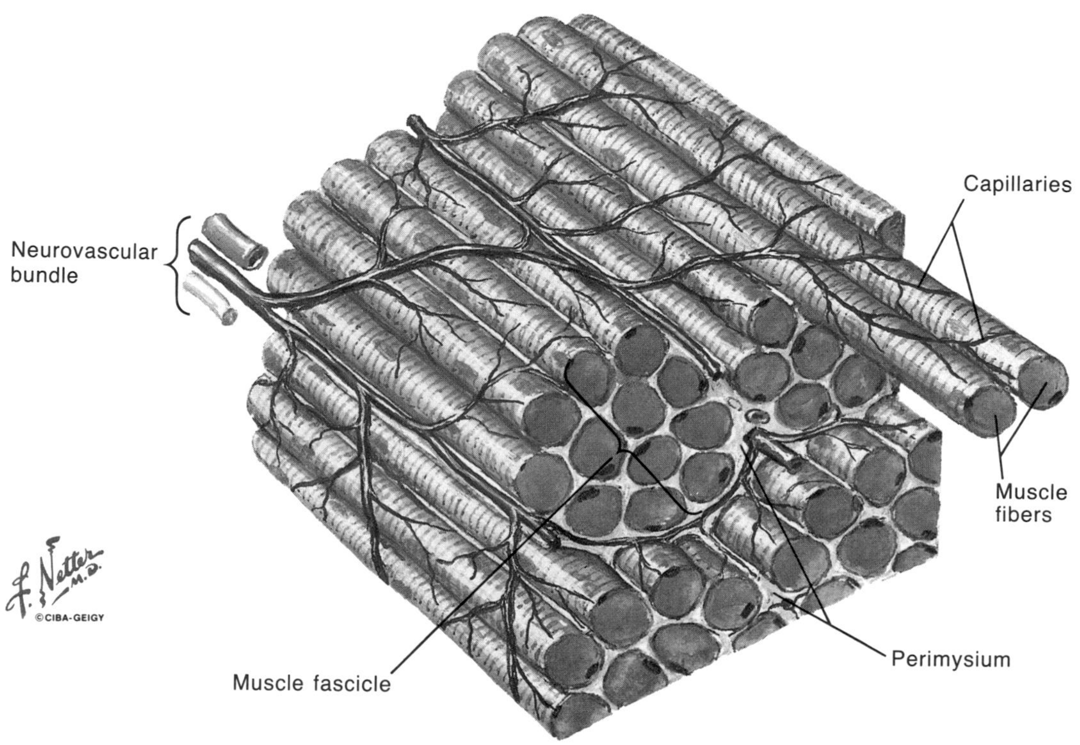

Composition and Structure of Myofilaments

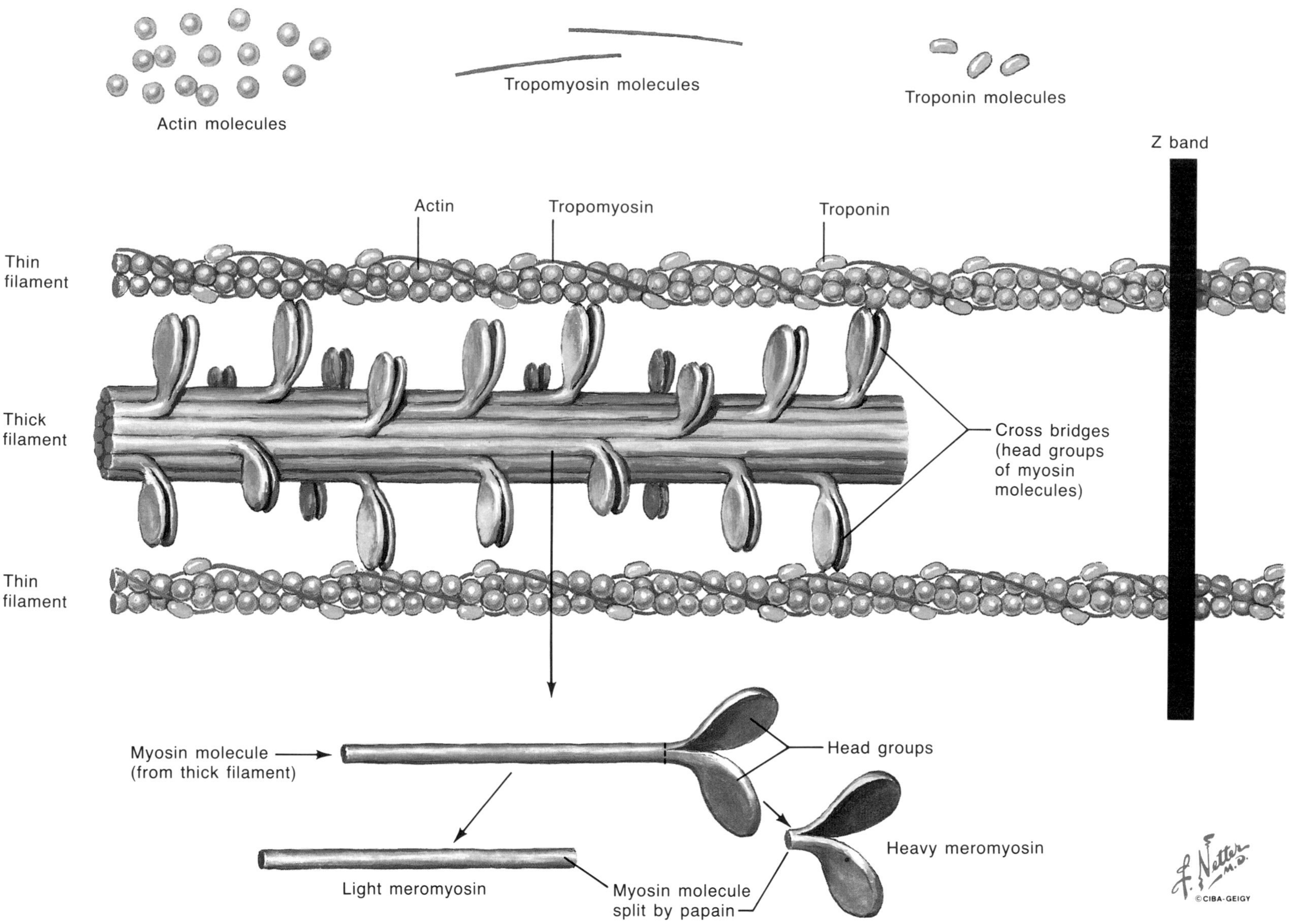

Actin molecules
Tropomyosin molecules
Troponin molecules
Z band
Actin
Tropomyosin
Troponin
Thin filament
Thick filament
Thin filament
Cross bridges (head groups of myosin molecules)
Myosin molecule (from thick filament)
Head groups
Heavy meromyosin
Light meromyosin
Myosin molecule split by papain
F. Netter M.D.
©CIBA-GEIGY

SECTION III PLATE 4 Slide 3700

Composition and Structure of Myofilaments

Thick filaments are composed primarily of a protein called myosin, which can be extracted from muscle by treating it with concentrated salt solutions. *Myosin* is a large protein with a molecular weight of approximately 500,000. On electron microscopy, a myosin molecule looks like a long rod with two paddles attached to one end. Actually, a myosin molecule consists of a pair of long filaments, each coiled in a configuration called an α-helix, a pattern of protein folding frequently seen in nature. Although wound around each other, the two filaments can be separated by treatment with high concentrations of urea or detergent. This procedure reveals that each filament has a globular enlargement, or *head group*, at one end; that is, each paddle is associated with one filament. These paddles form the *cross bridges* between the thick and thin filaments. The angle between the cross bridges and the rod portion of the myosin molecule becomes more acute during muscle contraction. This change of angle occurs when the end of the paddle is bound to a nearby thin filament, which provides the mechanical force for pulling the thin filaments past the thick filaments. This in turn results in a shortening of the sarcomere and therefore in muscle contraction.

The structure of myosin has also been studied by breaking it down into smaller pieces with enzymatic digestion. For example, the enzyme papain splits off the head groups and a small portion of the rod from the rest of the myosin molecule. The portion with the head groups is called *heavy meromyosin*, whereas the rod portion is called *light meromyosin*. With further digestion, the two head groups can be separated from each other. As far as is known, the head groups are identical, each weighing about 120,000 daltons. In the muscle, the myosin molecules are arranged with the head groups slanting away from the middle of the thick filament. In the middle of the thick filament, the tails of the myosin molecules overlap one another end to end, creating a region devoid of head groups and with a smooth appearance on electron microscopy.

Thin filaments consist chiefly of a protein called fibrous actin, or *F-actin*, which is in the form of a double helix. In very dilute salt solutions, F-actin breaks down into globular protein molecules called globular actin, or *G-actin*. These molecules are much smaller than myosin, with molecular weights of about 42,000. If the concentration of salt in the solution is increased, the G-actin molecules repolymerize end to end into their normal chainlike configuration. Thus, the actin filament is like a double string of G-actin "pearls" wound around each other. One turn of the helix contains 13.5 molecules of G-actin.

Although G-actin is the largest constituent of thin filaments, two other proteins form part of the structure and play important roles in muscle contraction. Along the notches between the two strands of actin subunits lie molecules of a globular protein, *troponin*. (Actually, this is a complex of three polypeptide subunits—troponin I, troponin C, and troponin T—each of which plays an important role in muscle contraction.) Attached to each troponin (at the T subunit) is a molecule of a thin, fibrous protein, *tropomyosin*, which lies along the grooves in the double helix. The precise disposition of tropomyosin along the F-actin chain probably varies importantly during the contraction-relaxation cycle. □

Muscle Contraction and Relaxation

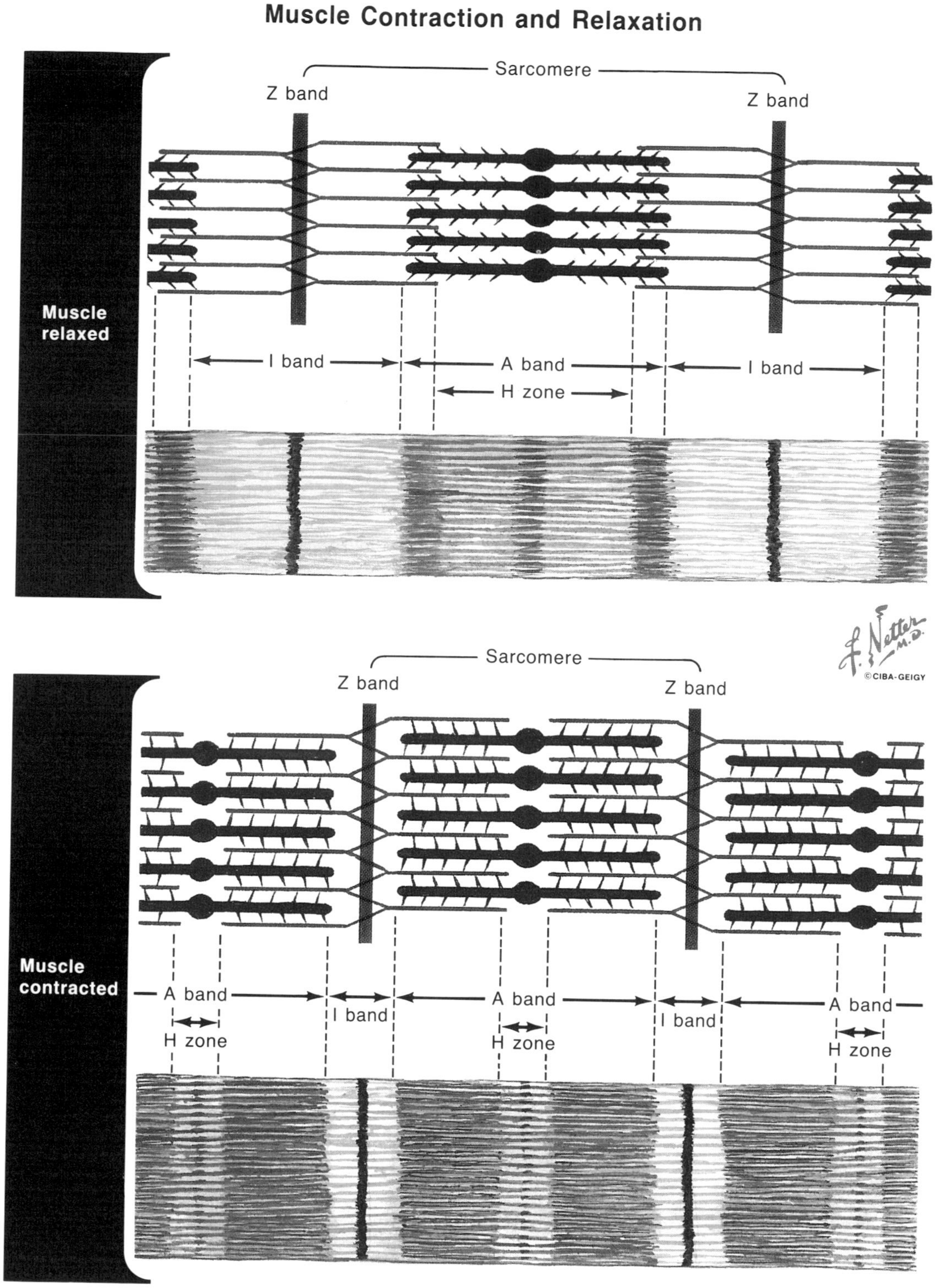

During muscle contraction, thin filaments of each myofibril slide deeply between thick filaments, bringing Z bands closer together and shortening sarcomeres. A bands remain same width, but I bands narrowed. H zones also narrowed or disappear as thin filaments encroach upon them. Myofibrils, and consequently muscle fibers (muscle cells), fascicles, and muscle as whole grow thicker. During relaxation, reverse occurs

Under normal conditions, the arrival of a nerve impulse at the neuromuscular junction causes muscle fibers to contract. Usually, the amount of the transmitter substance *acetylcholine (ACh)* released at the nerve terminal is sufficient to evoke a rapidly conducting electric impulse, or *action potential*, in the muscle fiber. This impulse is transmitted into the depth of the fiber and triggers the mechanical contraction. A single impulse in the motor nerve results in contraction of the muscle fiber in an all-or-nothing fashion. This is because the muscle action potential is propagated along the entire length of the fiber and thus activates the entire contractile machinery almost simultaneously.

The contraction of a muscle fiber in response to a single nerve impulse is called a *twitch*. Under a given set of starting conditions, the force of a single fiber's twitch is fixed, and the strength of a muscular contraction is therefore determined by the number of muscle fibers contracting at the same time. Contraction is under voluntary control of the central nervous system.

Muscle contraction thus results from the simultaneous shortening of all the sarcomeres in all the activated muscle fibers. It is brought about by the increase in overlap between the thick and thin filaments within each sarcomere. The increase in overlap is accomplished by a cycle of making and breaking cross-bridge linkages between the thick and thin filaments.

The head groups of the myosin molecules alternately flex and extend to interact with successive actin subunits on the thin filaments, which are brought progressively closer to the opposite Z band.

This "rowing" action slides the thin filaments past the thick filaments, narrowing the I band. As the ends of the actin filaments get closer to the M band, the I band appears denser and the H zone becomes narrower. The force of the contraction depends on the number of cross bridges linking the thick and the thin filaments at the same time.

Muscle relaxation occurs when the cross-bridge linkages are broken, allowing the thick and thin filaments to slide in the reverse direction. The elastic properties of the muscle and the tension on the ends of the muscle (for example, due to the weight of the limb) determine the muscle length during relaxation. □

Biochemical Mechanics of Muscle Contraction

Biochemical Mechanics of Muscle Contraction

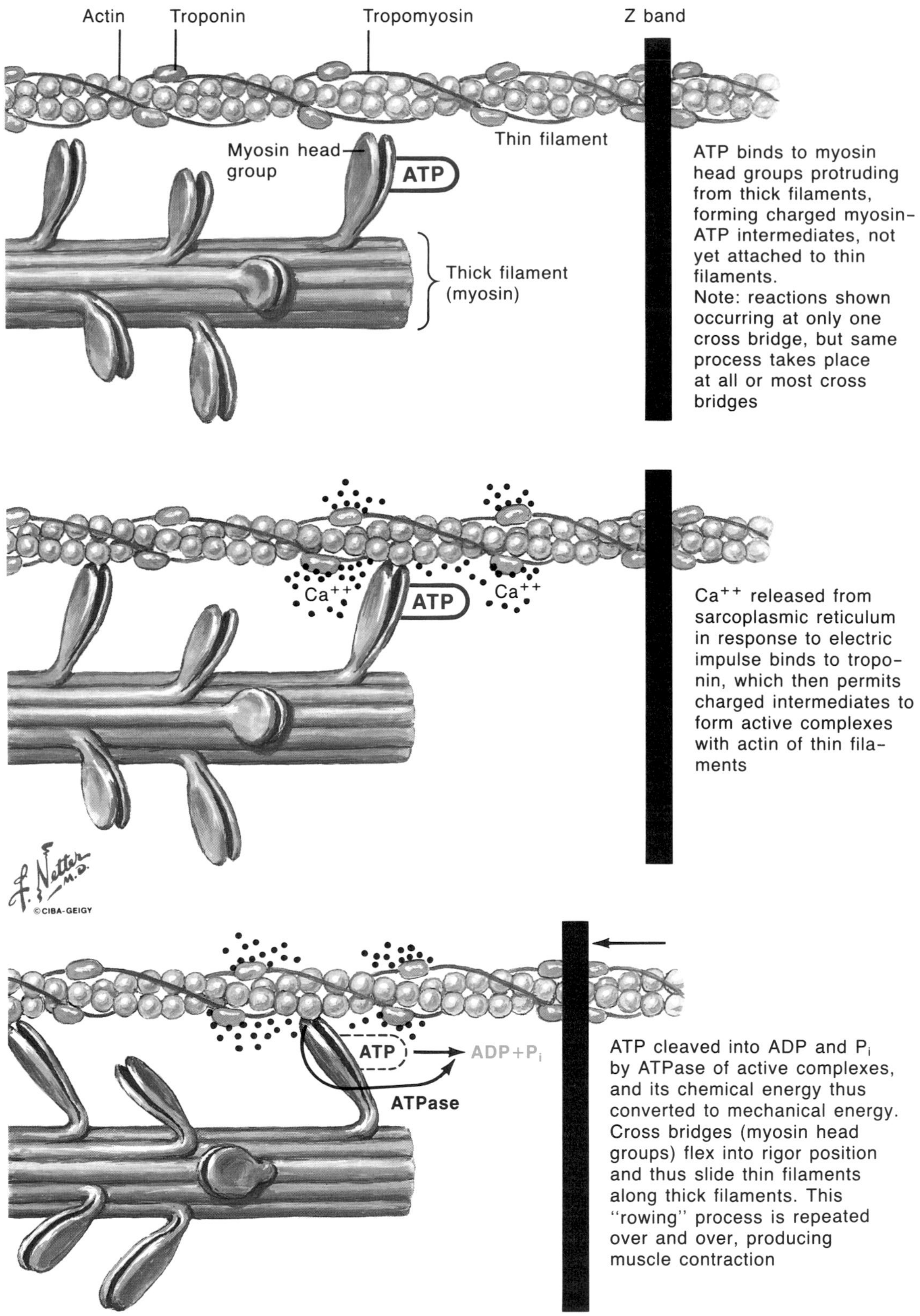
Actin
Troponin
Tropomyosin
Z band
Thin filament
Myosin head group
ATP
Thick filament (myosin)
ATP binds to myosin head groups protruding from thick filaments, forming charged myosin-ATP intermediates, not yet attached to thin filaments.
Note: reactions shown occurring at only one cross bridge, but same process takes place at all or most cross bridges
Ca++
ATP
Ca++
Ca++ released from sarcoplasmic reticulum in response to electric impulse binds to troponin, which then permits charged intermediates to form active complexes with actin of thin filaments
F. Netter M.D.
©CIBA-GEIGY
ATP
ADP+Pi
ATPase
ATP cleaved into ADP and Pi by ATPase of active complexes, and its chemical energy thus converted to mechanical energy. Cross bridges (myosin head groups) flex into rigor position and thus slide thin filaments along thick filaments. This "rowing" process is repeated over and over, producing muscle contraction

In the process of making cross-bridge linkages, the thick filaments "grip" the thin filaments, producing the force for muscle contraction. Cross bridges are formed by the globular head groups of the myosin molecules of the thick filament. In order for the cross bridges to occur, adenosine triphosphate (ATP) must bind to the myosin head groups, forming a charged myosin-ATP intermediate. This charged intermediate is capable of binding to an appropriate site on the actin subunit. When the tropomyosin molecules are in the resting configuration, the intermediate does not appear to have access to the actin subunit site. When the muscle fiber is electrically excited, calcium ions are released from the sarcoplasmic reticulum and bind to the troponin C subunit of the troponin molecules on the actin filaments, with four calcium ions binding to each troponin molecule. Consequently, there appears to be a change in the configuration of the tropomyosin, which is attached to the troponin on the T subunit, so that binding sites for the cross bridges are exposed. These sites are then bound by the closest myosin head groups. At this point, the thick and thin filaments are mechanically connected, but no movement has occurred. Movement requires the head groups to change their angle and drag the thick and thin filaments past one another. Energy is needed for this process and is provided in the form of ATP.

The head groups possess adenosine triphosphatase (ATPase) enzymatic sites that are active only when the heads are complexed with actin. The active complex hydrolyzes ATP into inorganic phosphate and adenosine diphosphate (ADP). Since the head group has a low affinity for ADP, the ADP dissociates from the myosin. Part of the energy released is used to change the position of the head groups from extension to flexion. This process is repeated, and the thin filament is pulled toward the middle of the sarcomere and the sarcomere is shortened.

The flexed position of the myosin head groups bound to the actin of the thin filament is called the *rigor complex*. It is so named after the term "rigor mortis," because after death, muscle fibers run out of ATP and all the myosin and actin molecules are tightly cross-linked in this configuration. However, in healthy muscle, when electric activity ceases, excess calcium is rapidly taken up by the sarcoplasmic reticulum. Without calcium bound to the troponin, the head groups cannot remain bound to actin. The rigor complex is broken, the sarcomeres lengthen, and the muscle once again relaxes. □

Sarcoplasmic Reticulum and Initiation of Muscle Contraction

Sarcoplasmic Reticulum

Segment of muscle fiber greatly enlarged to show endosarcoplasmic structures and inclusions

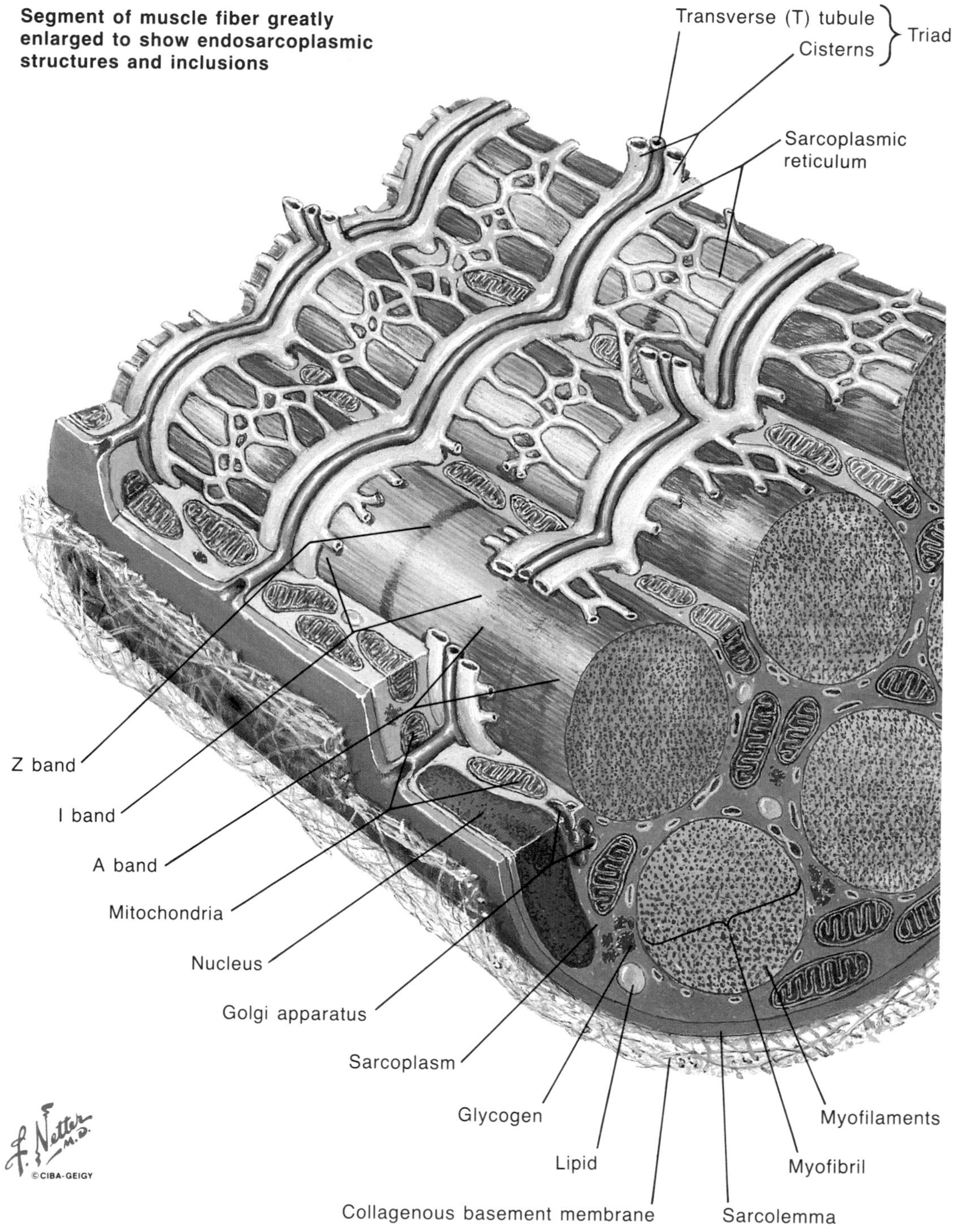
Transverse (T) tubule
Cisterns
Triad
Sarcoplasmic reticulum
Z band
I band
A band
Mitochondria
Nucleus
Golgi apparatus
Sarcoplasm
Glycogen
Lipid
Collagenous basement membrane
Sarcolemma
Myofibril
Myofilaments
F. Netter M.D.
©CIBA-GEIGY

An impulse in the motor nerve releases the neurotransmitter substance acetylcholine (ACh) at the neuromuscular junction (also called the motor end plate). Acetylcholine excites the muscle fiber membrane, or *sarcolemma*, causing an electric impulse to spread over the surface of the muscle. The electric impulse is coupled to the activation of the muscle's contractile mechanism, and some aspects of this process are described below (Plates 7–8).

The immediate trigger for muscular contraction is a sudden increase in the concentration of calcium ions in the cytoplasm of the muscle fiber, the *sarcoplasm*. To prevent the muscle from being in a continual state of contraction, the calcium is stored in a system of intracellular membrane-bound channels. This system, called the *sarcoplasmic reticulum*, permeates the entire muscle fiber, so that each sarcomere is surrounded by it.

The membranes of the sarcoplasmic reticulum contain a calcium pump that uses the energy stored in adenosine triphosphate (ATP) to transport calcium ions from the sarcoplasm, where the calcium concentration is maintained at a very low level, into the sarcoplasmic reticulum, where the calcium concentration is very high. The pump contains an enzyme that catalyzes the splitting of ATP into adenosine diphosphate (ADP) and inorganic phosphate. This converting enzyme requires calcium and magnesium ions for its operation and thus is called calcium-magnesium ATPase. During the cleavage of one ATP molecule, two calcium ions are transported into the sarcoplasmic reticulum. The capacity of the sarcoplasmic reticulum to store calcium is enhanced by the existence of a special calcium-binding protein called calsequestrin, which has been identified in purified preparations of sarcoplasmic reticulum. It is estimated that when the muscle is at rest, the calcium concentration in the sarcoplasmic reticulum is more than 100 mmol/kg of dry weight.

Maintenance of the steep concentration gradient for calcium across the membranes of the sarcoplasmic reticulum and activation of the contractile mechanism use up ATP, which must be replenished quickly. ATP is most efficiently replenished by the oxidative pathway. Because of their high energy requirements, muscle fibers are rich in mitochondria, which contain the enzymatic machinery for oxidative metabolism. Mitochondria are most heavily concentrated near the sarcolemma, close to the capillaries that supply them with oxygen.

The muscle action potential is propagated from the region of the neuromuscular junction along the entire length of the muscle fiber. The electric impulse of muscle is similar to that of most nerve fibers. The sarcolemma contains voltage-dependent sodium channels that open in response to an injection of depolarizing (positive) current into the muscle fiber. Since the action of acetylcholine is to depolarize the sarcolemma at the neuromuscular junction, sodium channels open in the neighboring area of sarcolemma. The concentration of sodium ions inside the muscle fiber is kept very low by a pump consisting of a sodium/potassium–activated ATPase. The cleavage of one ATP molecule into ADP and phosphate is accompanied by the transport of three sodium ions out of the fiber and two potassium ions into the fiber. Because the intracellular sodium concentration is so low (about 10 mmol/L), when sodium channels open, sodium ions move into the muscle fiber from the extracellular fluid, where the concentration is much higher (about 110 mmol/L). The inward movement of these positively charged ions further depolarizes the sarcolemma, opening more sodium channels in a cycle of depolarization and increase in sodium conductance until the membrane potential reaches almost +50 mV.

Sarcoplasmic Reticulum and Initiation of Muscle Contraction

(Continued)

Initiation of Muscle Contraction by Electric Impulse and Calcium Movement

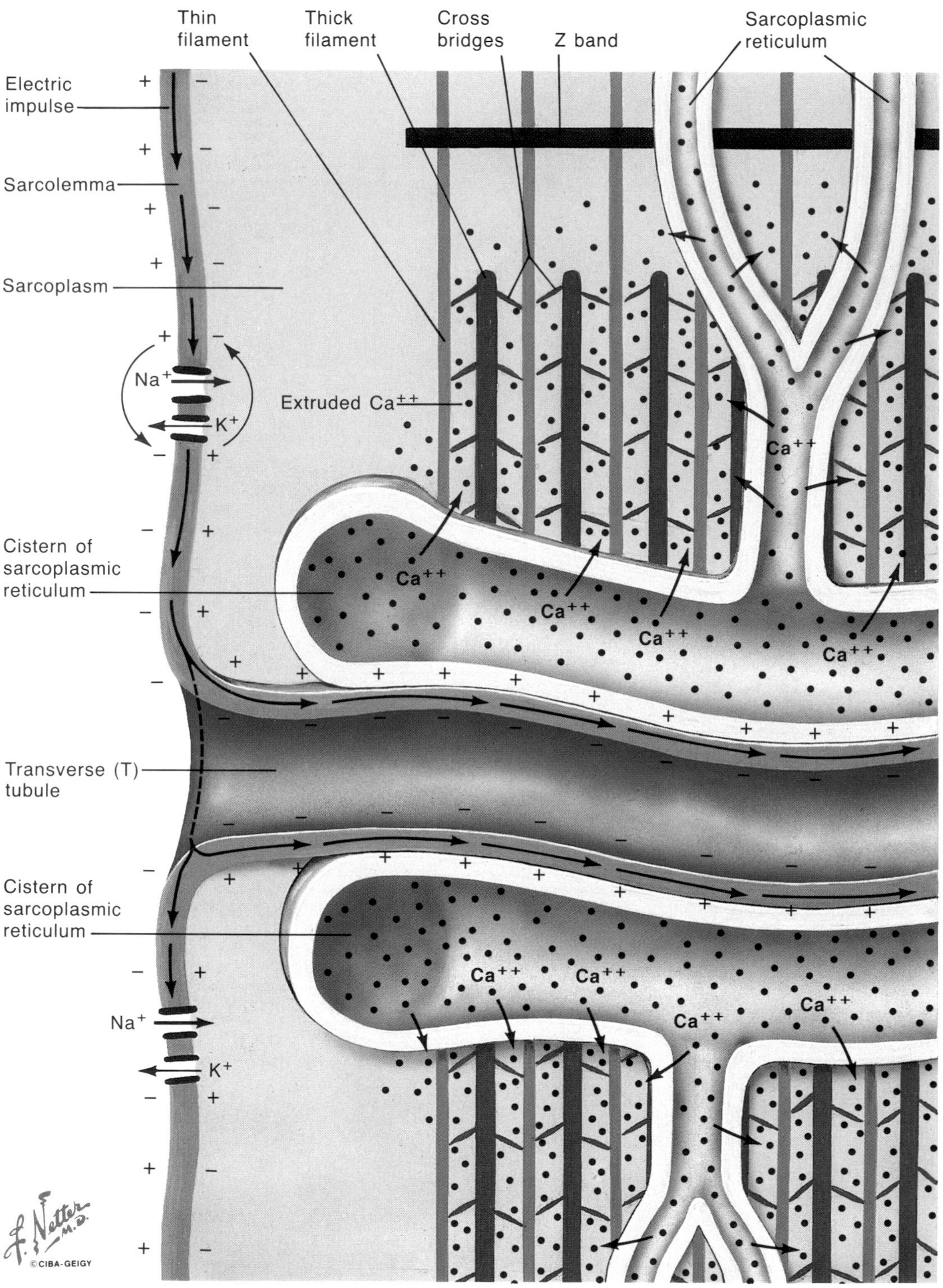

Electric impulse traveling along muscle cell membrane (sarcolemma) from motor end plate (neuromuscular junction) and then along transverse tubules affects sarcoplasmic reticulum, causing extrusion of Ca^{++} to initiate contraction by "rowing" action of cross bridges, sliding filaments past one another

This process turns itself off by two mechanisms. First, the sodium conductance channels are not only voltage dependent, they are also time dependent: they close if the depolarization of the sarcolemma is maintained for longer than a few milliseconds. Second, the sarcolemma also contains voltage-dependent potassium channels. The depolarization associated with the muscle action potential opens these channels, allowing the positively charged potassium ions to escape from the muscle fiber. This causes the sarcolemma to be repolarized (the inside becomes negative), thereby closing the sodium channels. Thus, immediately before contraction, the sarcolemma undergoes a large depolarization lasting only 1 or 2 msec. This electric impulse is in some way responsible for the sudden release of large amounts of calcium from the sarcoplasmic reticulum. This calcium release from a storage site within the muscle fiber triggers the contraction of the muscle fiber.

However, in order for the action potential to affect the sarcoplasmic reticulum deep within the muscle fiber, it must be propagated inward as well as along the surface. This is accomplished through invaginations of the sarcolemma called the transverse tubules, or *T tubules*. In mammalian skeletal muscle, T tubules occur in register with the junction between the A bands and the I bands. Thus, each sarcomere is associated with two systems of T tubules, one at each end of the A band. (This is not true throughout the animal kingdom. In frogs, from which we have derived much of our knowledge about the structure and function of skeletal muscle, the T tubules occur in register with the Z band; thus, there is only one system of T tubules per sarcomere.)

Flanking the T tubules are paired dilatations of the sarcoplasmic reticulum called cisternae, or *cisterns*. On electron microscopy, the characteristic grouping of one T tubule and two cisterns seen in cross sections is called the *triad*. Thus, the action potential is propagated into the depths of the muscle fiber very close to elements of the sarcoplasmic reticulum. Nevertheless, the precise mechanism by which the invasion of the T tubules by the muscle action potential is coupled to the release of calcium from the sarcoplasmic reticulum has not yet been determined. □

Motor Unit

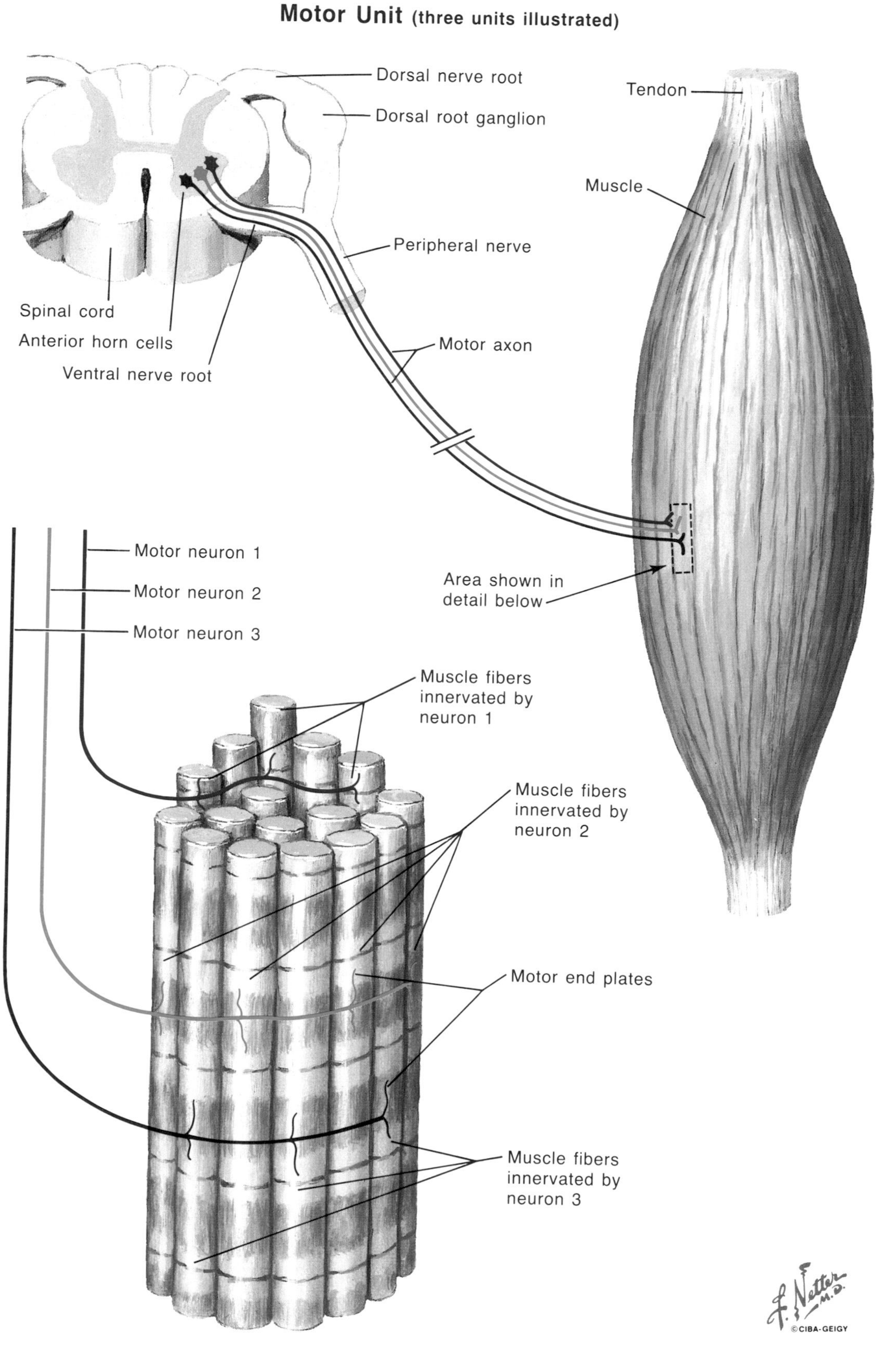
Motor Unit (three units illustrated)
Dorsal nerve root
Dorsal root ganglion
Tendon
Muscle
Peripheral nerve
Spinal cord
Anterior horn cells
Ventral nerve root
Motor axon
Motor neuron 1
Motor neuron 2
Motor neuron 3
Area shown in detail below
Muscle fibers innervated by neuron 1
Muscle fibers innervated by neuron 2
Motor end plates
Muscle fibers innervated by neuron 3
©CIBA-GEIGY

Muscle fibers are innervated by neurons whose cell bodies are located in the anterior (ventral) horn of the spinal cord gray matter and in the motor nuclei of the brainstem. The nerve fibers, or *axons*, of these motor neurons leave the spinal cord via the ventral roots and are distributed to the motor nerves. They enter the muscle at a region called the *end plate zone*. Each motor axon branches several times and innervates many muscle fibers.

In mammalian skeletal muscle, each muscle fiber is innervated by only one motor neuron. The combination of a single motor neuron and all the muscle fibers it innervates is called a *motor unit*. Although the muscle fibers of a given motor unit tend to be located near one another, motor units have overlapping territories.

The strength of muscle contraction depends on the number of muscle fibers active at the same time. However, the central nervous system cannot control each individual muscle fiber. It can only activate the motor neurons and therefore the motor units. The degree of control that can be exerted on the strength of contraction depends on the number of muscle fibers in a motor unit. Motor units of large muscles such as the gastrocnemius, which exert a great deal of power, may contain more than 2,000 muscle fibers. Motor units of small muscles such as the extraocular muscles, which exert very fine control but not much power, may contain as few as six muscle fibers.

Even within a given muscle, the motor units are not equal in size. In general, small motor neurons innervate fewer muscle fibers (they have smaller motor units). Small motor neurons are also more easily activated by synaptic inputs than are large motor neurons. Therefore, when signals from the brain initiate a movement, the smallest motor neurons and motor units are usually activated first. If only a small fine movement is required, the smallest motor units alone can be activated. As more power and less fine control is needed, the larger motor units are progressively recruited. This process is called the size principle of motor control.

The mechanical properties of the muscle fibers are also matched to the size of their motor unit. Because the muscle fibers of the smallest motor units are those most often activated, they must be relatively resistant to fatigue (Plate 15). □

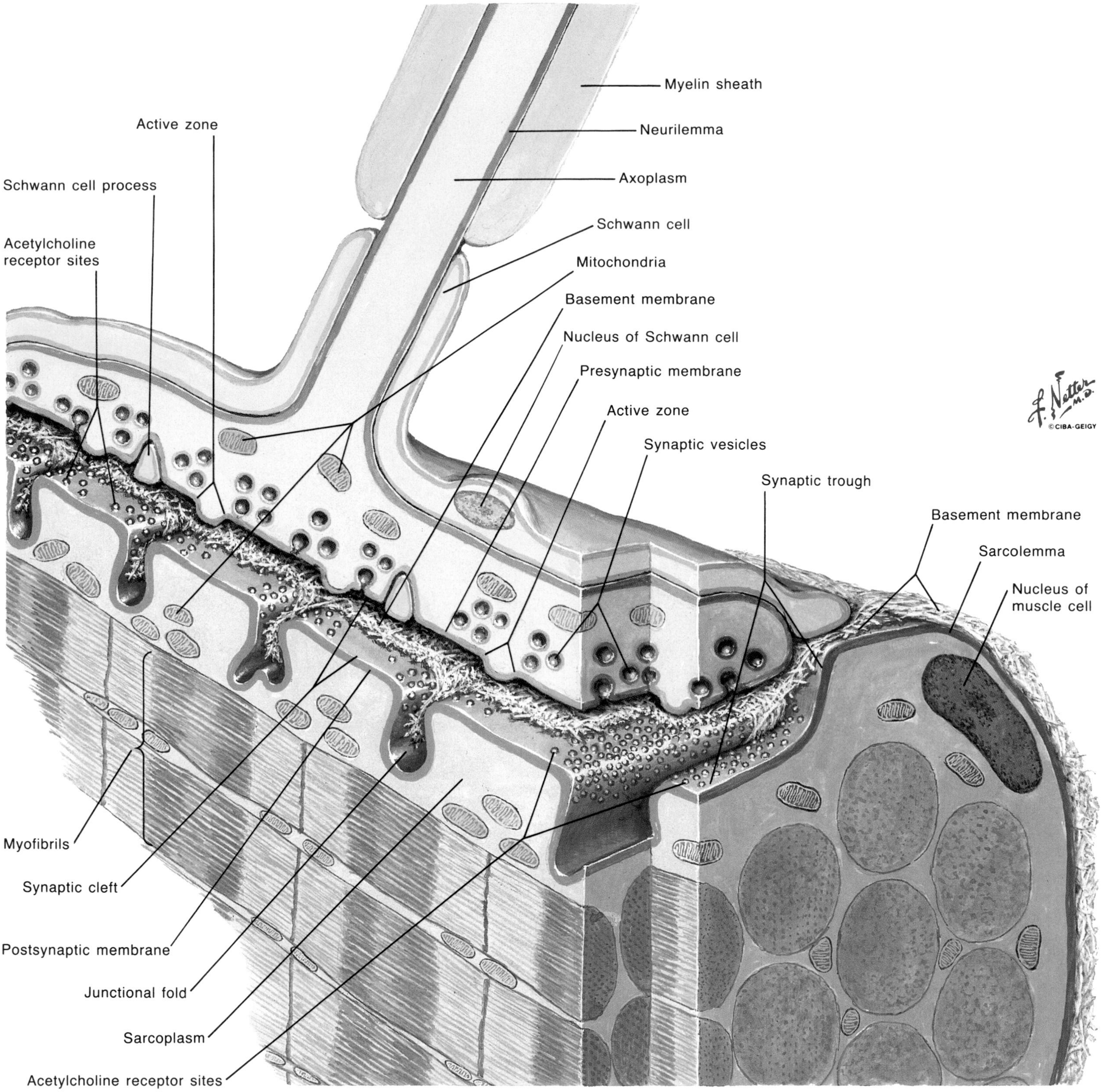

Myelin sheath
Neurilemma
Axoplasm
Schwann cell
Mitochondria
Basement membrane
Nucleus of Schwann cell
Presynaptic membrane
Active zone
Synaptic vesicles
Synaptic trough
Basement membrane
Sarcolemma
Nucleus of muscle cell
Active zone
Schwann cell process
Acetylcholine receptor sites
Myofibrils
Synaptic cleft
Postsynaptic membrane
Junctional fold
Sarcoplasm
Acetylcholine receptor sites
F. Netter M.D.
©CIBA-GEIGY

SECTION III PLATE 10 Slide 3706

Structure of Neuromuscular Junction

Motor axons are generally large myelinated fibers. The terminals, although unmyelinated, are invested by a Schwann cell, which projects fingerlike processes between the membranes of the nerve and the muscle. The nerve terminal lies in a trough within the muscle fiber membrane (sarcolemma); it is rich in mitochondria and contains numerous synaptic vesicles about 50 nm in diameter. These vesicles, which contain the neurotransmitter acetylcholine (ACh), are clustered around nipple-shaped active zones located at regular intervals along the terminal membrane. Acetylcholine is released by exocytosis of vesicles lying adjacent to both sides of the active zone. Opposite each active zone, the sarcolemma is invaginated by junctional folds. The presynaptic and postsynaptic membranes are separated by a space approximately 50 nm wide. Freeze-fracture electron microscopy reveals granular structures embedded in the postsynaptic membrane. These structures, which are concentrated on the banks of the junctional folds opposite the sites of acetylcholine release, are acetylcholine receptors that mediate the action of the transmitter. They are sparse in regions of the sarcolemma not close to the neuromuscular junction.

The muscle fiber is surrounded by a connective tissue basement membrane that continues into the synaptic cleft, sending extensions into the junctional folds. The basement membrane contains a large amount of collagen as well as most of the acetylcholinesterase (AChE) in the neuromuscular junction (Plate 12). It also contains other important molecules that help guide the growth of the nerve terminal during development and regeneration, determine the locations of the presynaptic active zones, and induce the clumping of acetylcholine receptors opposite the synaptic vesicle. □

Physiology of Neuromuscular Junction

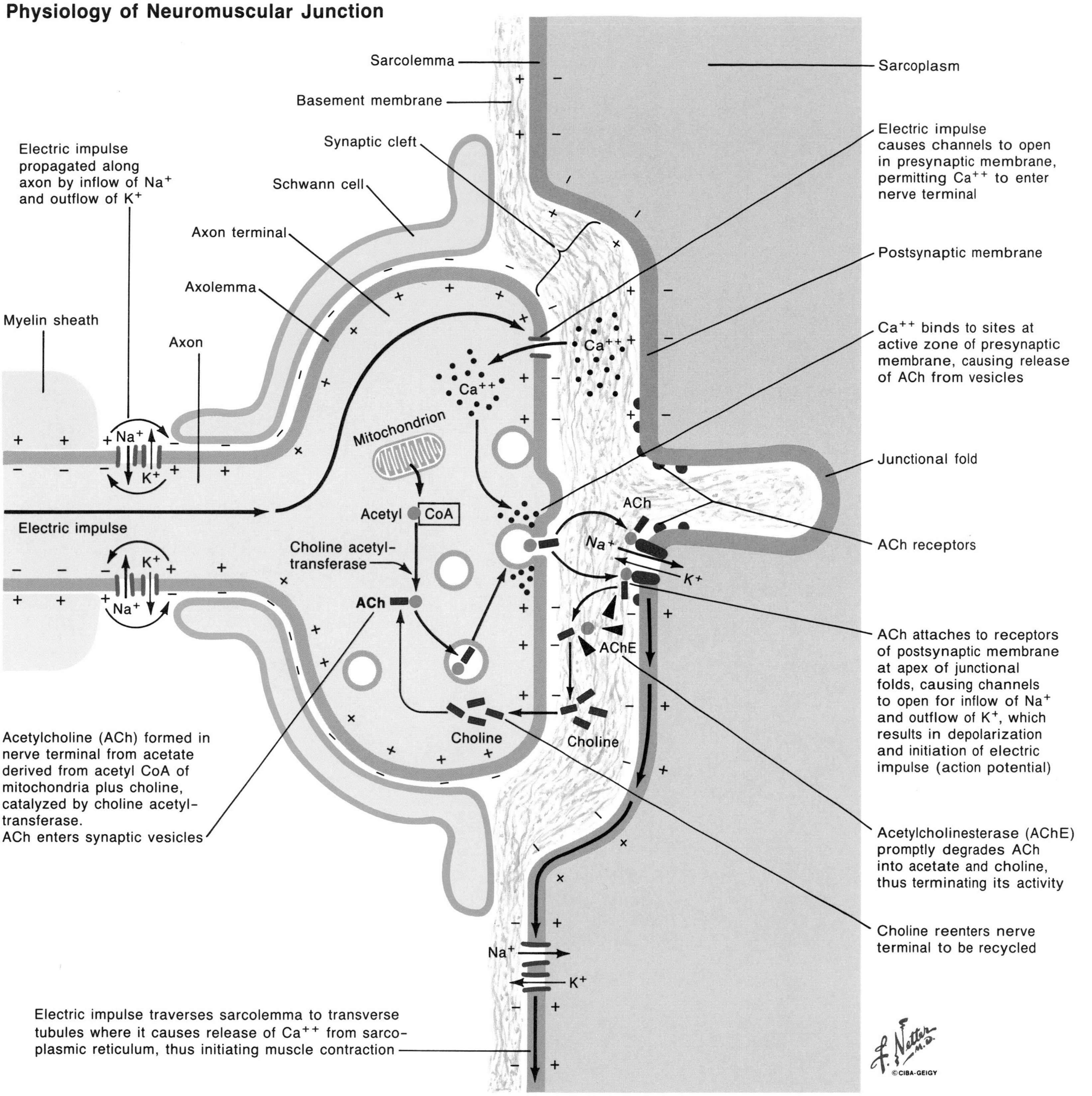
Sarcolemma
Sarcoplasm
Basement membrane
Electric impulse causes channels to open in presynaptic membrane, permitting Ca++ to enter nerve terminal
Synaptic cleft
Electric impulse propagated along axon by inflow of Na+ and outflow of K+
Schwann cell
Axon terminal
Postsynaptic membrane
Axolemma
Myelin sheath
Ca++ binds to sites at active zone of presynaptic membrane, causing release of ACh from vesicles
Axon
Ca++
Mitochondrion
Na+
K+
Junctional fold
ACh
Acetyl
CoA
Electric impulse
ACh receptors
Choline acetyl-transferase
ACh
AChE
ACh attaches to receptors of postsynaptic membrane at apex of junctional folds, causing channels to open for inflow of Na+ and outflow of K+, which results in depolarization and initiation of electric impulse (action potential)
Choline
Choline
Acetylcholine (ACh) formed in nerve terminal from acetate derived from acetyl CoA of mitochondria plus choline, catalyzed by choline acetyl-transferase.
ACh enters synaptic vesicles
Acetylcholinesterase (AChE) promptly degrades ACh into acetate and choline, thus terminating its activity
Choline reenters nerve terminal to be recycled
Electric impulse traverses sarcolemma to transverse tubules where it causes release of Ca++ from sarco-plasmic reticulum, thus initiating muscle contraction
F. Netter M.D.
©CIBA-GEIGY

SECTION III PLATE 11 Slide 3707

Physiology of Neuromuscular Junction

Electric impulses are propagated along the motor axon by the inward movement of positively charged sodium ions through depolarization-activated channels located at the nodes of Ranvier. The insulating properties of myelin prevent leakage of current. Adjacent nodes are thus depolarized, and the nerve impulse is regenerated. The axon is repolarized by the outward movement of potassium ions through other depolarization-activated channels, which are thought to be located in the paranodal membrane.

The membrane of the nerve terminal has a different assortment of ion channels: fewer sodium channels, several types of potassium channels, and most important, voltage-dependent calcium channels. When an action potential arrives at the nerve terminal, it opens the calcium channels and calcium ions move from the extracellular fluid, where the concentration is about 2.5 mmol/L, into the nerve terminal. There, active pumping of calcium ions across the nerve membrane into intracellular organelles (especially the endoplasmic reticulum and the mitochondria) keeps the concentration at about 1 mmol/L. The sudden increase in intraterminal concentration of calcium ions is linked to the release of acetylcholine (ACh) by exocytosis at the synaptic vesicle release sites. The acetylcholine binds to receptor molecules on the postsynaptic membrane, opening channels that permit the influx of sodium ions and efflux of potassium ions. The net effect is a depolarization of the muscle membrane and a triggering of the muscle action potential. The acetylcholine is then rapidly hydrolyzed by acetylcholinesterase (AChE) into choline and acetate. The choline is conserved by active uptake into the nerve terminal, where it is reconverted into acetylcholine by the enzyme choline acetyltransferase. □

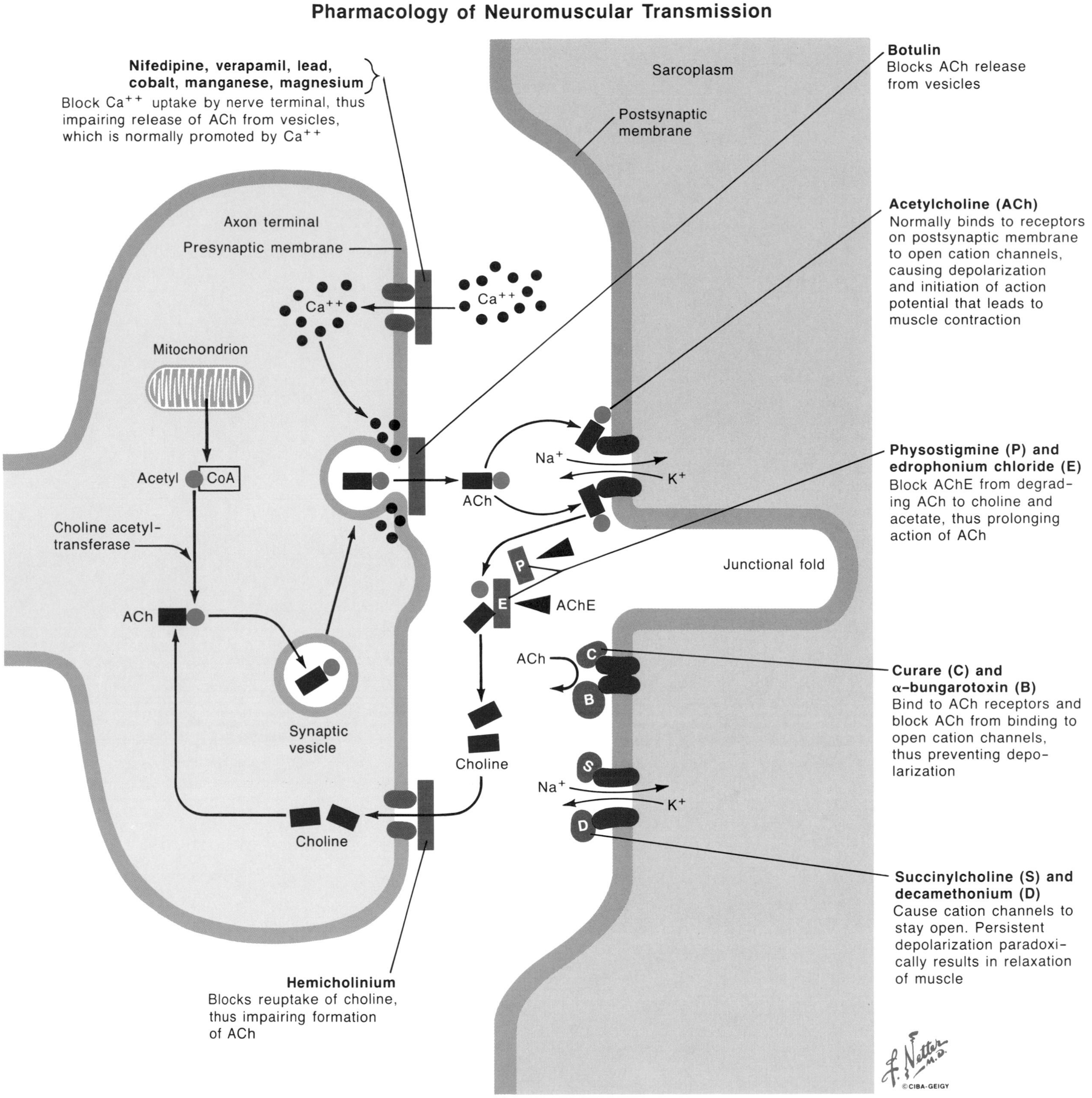
Pharmacology of Neuromuscular Transmission
Nifedipine, verapamil, lead, cobalt, manganese, magnesium
Block Ca++ uptake by nerve terminal, thus impairing release of ACh from vesicles, which is normally promoted by Ca++
Sarcoplasm
Postsynaptic membrane
Botulin
Blocks ACh release from vesicles
Axon terminal
Presynaptic membrane
Ca++
Ca++
Acetylcholine (ACh)
Normally binds to receptors on postsynaptic membrane to open cation channels, causing depolarization and initiation of action potential that leads to muscle contraction
Mitochondrion
Acetyl
CoA
Na+
K+
ACh
Physostigmine (P) and edrophonium chloride (E)
Block AChE from degrading ACh to choline and acetate, thus prolonging action of ACh
Choline acetyltransferase
P
E
AChE
Junctional fold
ACh
ACh
C
B
Curare (C) and α-bungarotoxin (B)
Bind to ACh receptors and block ACh from binding to open cation channels, thus preventing depolarization
Synaptic vesicle
Choline
S
Na+
K+
D
Choline
Succinylcholine (S) and decamethonium (D)
Cause cation channels to stay open. Persistent depolarization paradoxically results in relaxation of muscle
Hemicholinium
Blocks reuptake of choline, thus impairing formation of ACh
F. Netter M.D.
©CIBA-GEIGY

SECTION III PLATE 12 Slide 3708

Pharmacology of Neuromuscular Transmission

Neuromuscular transmission can be interrupted by many drugs. Organic calcium channel blockers, such as verapamil, and divalent cations, such as lead, prevent calcium ions from entering the nerve terminal and thus block release of the neurotransmitter acetylcholine (ACh). The neurotoxin botulin blocks acetylcholine release more directly by a mechanism that is still unknown. Some drugs block the acetylcholine receptors either reversibly (curare) or irreversibly (α-bungarotoxin). Succinylcholine and decamethonium produce muscle relaxation not by preventing the opening of the acetylcholine-activated channels but by keeping them open too long.

The voltage-dependent sodium channel, which mediates the muscle action potential, is also time dependent: prolonged depolarization inactivates the action potential. This can be reversed only by a repolarization of the membrane. By keeping the acetylcholine channel open, these depolarizing blockers keep the muscle membrane depolarized and refractory to impulse initiation. Hemicholinium weakens neuromuscular transmission by blocking the reuptake of choline, thus reducing the synthesis of acetylcholine.

Some drugs strengthen neuromuscular transmission. Acetylcholine agonists, such as nicotine, react directly with the acetylcholine receptor. Others, such as physostigmine, pyridostigmine bromide, and edrophonium chloride, inhibit acetylcholinesterase (AChE) and strengthen transmission by delaying the breakdown of acetylcholine. The action of physostigmine and pyridostigmine bromide persists for hours, and these agents are used to treat the neuromuscular disease myasthenia gravis; edrophonium chloride, on the other hand, is short acting and is used in diagnosing the disease. □

Physiology of Muscle Contraction

Muscle Response to Nerve Stimuli

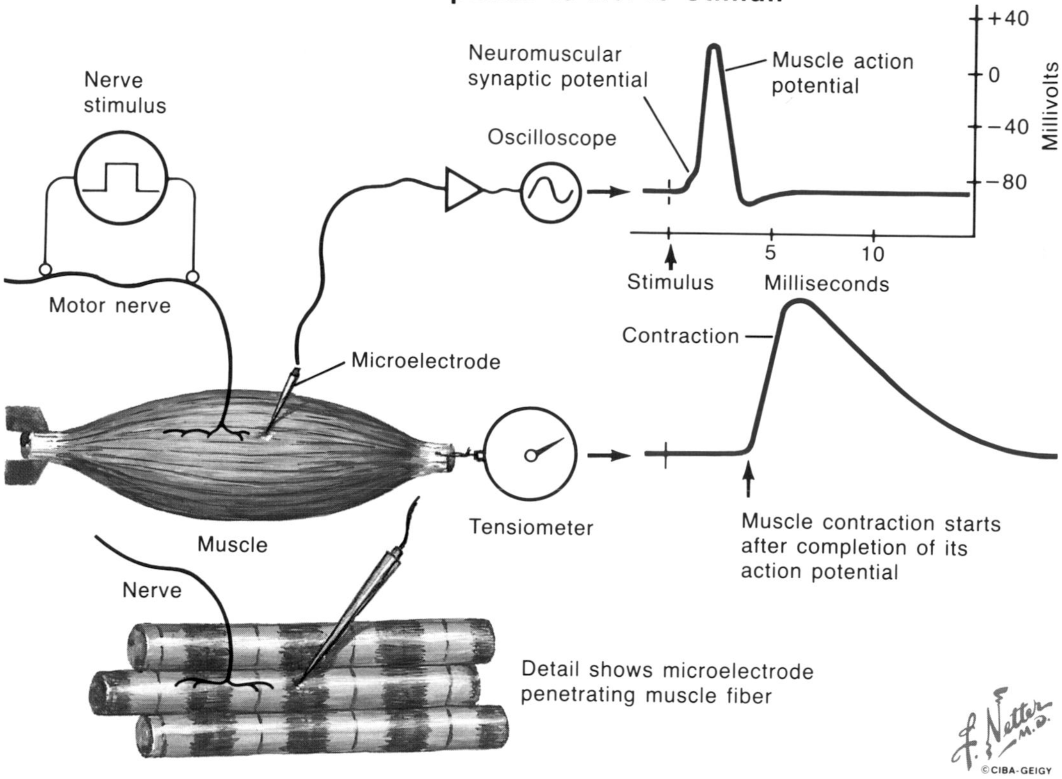

Summation of Muscle Response With Progressive Frequency of Stimulation

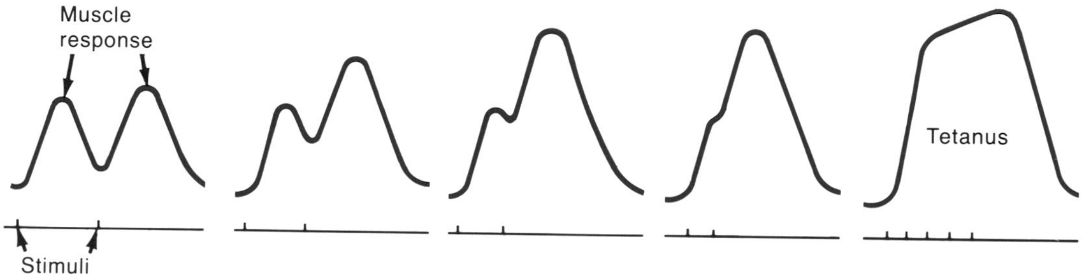

Muscle Length–Muscle Tension Relationships

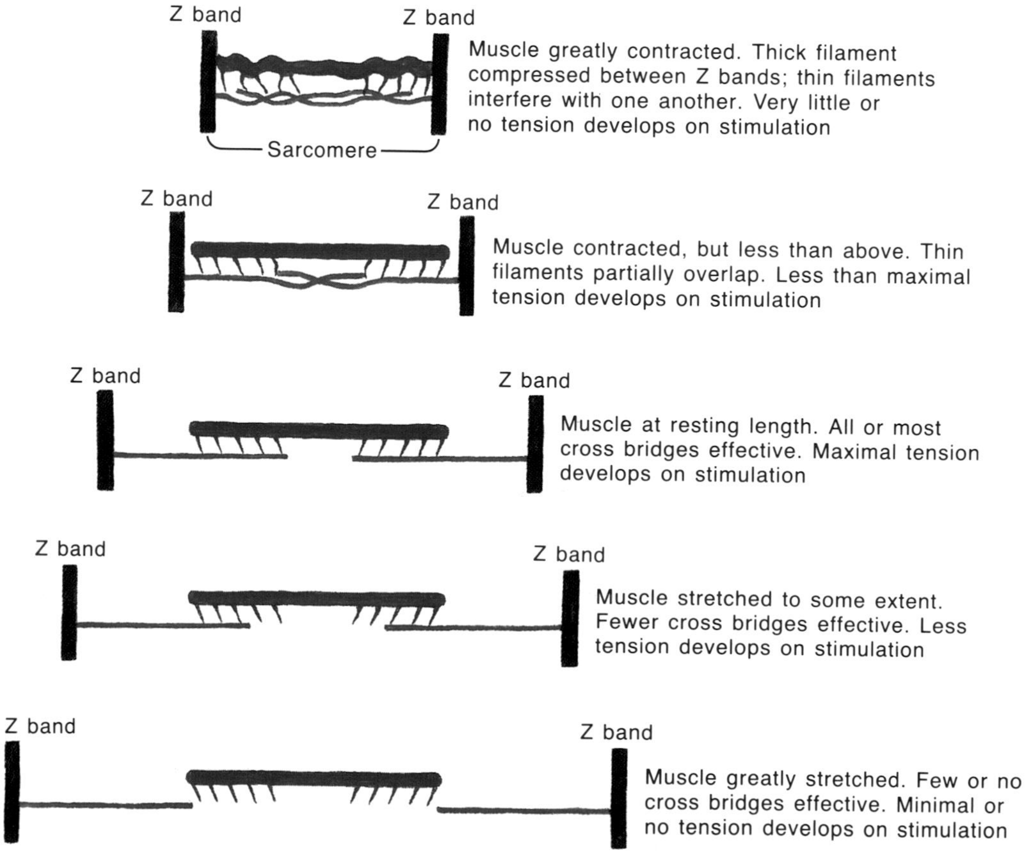

Muscle Response to Nerve Stimuli. The amount of transmitter substance released at the nerve terminal of the skeletal neuromuscular junction is sufficient to produce an end plate potential that always reaches the threshold required to activate a muscle action potential. This is not the case in certain pathologic conditions, such as myasthenia gravis or the Lambert-Eaton (myasthenic) syndrome. The muscle action potential is propagated along the entire length of the muscle fiber and deep into the fiber through the transverse (T) tubules. Calcium ions are released almost synchronously from the entire sarcoplasmic reticular system, which elicits a contraction, or twitch.

The tension of the twitch can be measured under conditions in which the muscle is not allowed to shorten (isometric contraction). Since all the sarcomeres are activated, the single-twitch strength of a fiber tends to be the same each time, providing the muscle remains the same length. If a second twitch is elicited before the first has relaxed, the maximum tension achieved is increased. If the muscle is activated at a high enough frequency, the twitches fuse into a continuous smooth contraction (tetanus) of even greater tension.

Muscle Length-Muscle Tension Relationships. The tension developed by a tetanically stimulated muscle depends on the final length the muscle fibers are permitted to reach. Maximum tension is exerted when the length of the sarcomere allows activation of all the cross bridges between the thick and the thin filaments. This occurs at the normal resting length of the muscle fiber. If the muscle is contracted too far, the thin filaments overlap, which interferes with their interactions with the thick filaments, reducing the maximum attainable tension.

On the other hand, if the muscle is stretched, the thin filaments do not have access to all of the available myosin head groups, and fewer than the maximum number of cross bridges are formed. If the muscle is greatly stretched, the thick and thin filaments may not overlap at all, and no additional tension can develop in response to stimulation. □

Regeneration of ATP for Source of Energy in Muscle Contraction

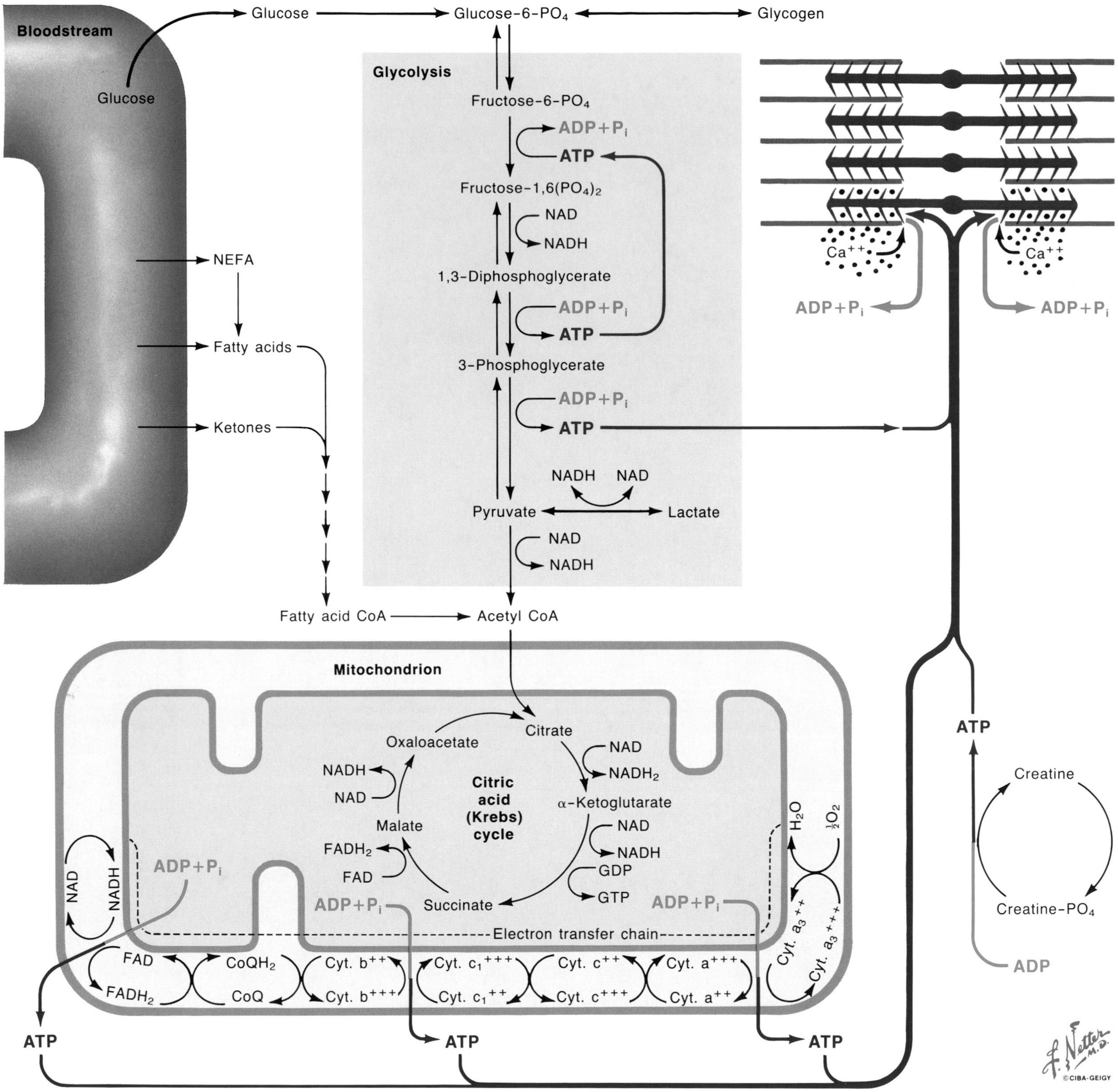

SECTION III PLATE 14 Slide 3710

Energy Metabolism of Muscle

The immediate source of energy for muscle contraction is adenosine triphosphate (ATP) generated in the mitochondria by glycolysis and the oxidative metabolism of carbohydrates and fats. The economy of ATP can be understood by counting all the points of its synthesis and breakdown in these metabolic pathways.

In glycolysis, glucose-6-phosphate, derived either from the degradative phosphorylation of glycogen or from the phosphorylation of glucose in serum, is broken down into two molecules of acetyl coenzyme A (CoA). This process may take place in the absence of oxygen, yet three molecules of ATP are generated for each molecule of glucose used. Acetyl CoA enters the citric acid (Krebs) cycle, which generates the reduced forms of flavin adenine dinucleotide ($FADH_2$) and nicotinamide adenine dinucleotide (NADH). Both of these can fuel the conversion of adenosine diphosphate (ADP) into ATP only through the cytochrome oxidative pathway, which involves the reduction of oxygen to water. (The glycolytic pathway also generates NADH, whereas the citric acid cycle generates guanosine triphosphate [GTP], which can contribute a high-energy phosphate to ADP and thus make ATP.) Thus, while only three molecules of ATP are generated under anaerobic conditions, 35 molecules of ATP are generated by oxygen-requiring steps in the metabolism of one molecule of glucose derived from glycogen.

The level of ATP in muscle must remain high; it does not decrease substantially, even during continuous contraction, because muscle fibers have a built-in ATP-buffering system. Energy is stored as creatine phosphate. If the ATP level falls, a small amount of creatine phosphate transfers a phosphate into ADP, forming creatine and regenerating ATP. □

Muscle Fiber Types

For a short time, muscles are able to function without oxygen by using the glycolytic pathway to generate adenosine triphosphate (ATP). Muscle fibers specialized for a high-power output over a short time (type I fibers) make extensive use of this pathway; however, carbohydrates are utilized rather inefficiently in the production of energy, and the carbohydrate store is depleted rapidly.

Muscle fibers that must remain active over a long time (type II fibers) are rich in mitochondria, whose iron-containing cytochrome oxidase enzymes give the fibers their red appearance. These type II fibers stain darkly for enzymes of the oxidative pathway, such as succinic acid dehydrogenase (SDH), but they do not have to generate high tensions and thus do not stain deeply for myofibrillar adenosine triphosphatase (ATPase) and glycolytic enzymes. Type II fibers, which tend to be small, are used in fine manipulations. They are the first fibers in any muscle to be activated when a low level of power is required. Because of their mechanical properties, they are called *slow-twitch, fatigue-resistant (SR) fibers.* Energy is conserved by SR fibers by a slow rate of relaxation following a twitch, and thus they require a low frequency of stimulation for the twitches to fuse into a sustained contraction (tetanus).

Compared with type II fibers, type I fibers, which must generate high tensions rapidly but need not remain active for prolonged periods, are relatively poor in mitochondrial enzymes and are white in appearance. On the other hand, they are rich in ATPase and glycolytic enzymes. Because of their mechanical properties, they are called *fast-twitch, fatigable (FF) fibers.*

In recent years, it has become clear that some fibers have mechanical properties intermediate between those of SR and FF fibers. They can generate a relatively fast twitch but are still fatigue resistant and are therefore called *fatigue-resistant (FR) fibers.*

Muscles can respond to exercise patterns by an appropriate shift in their metabolic characteristics. Isometric, anaerobic exercise results in an increase in the number of myofibrils and in the amount of contractile protein per fiber. Either other fiber types are converted to FF fibers, or a higher actomyosin ATPase and glycolytic enzyme activity occurs for all fiber types in the muscle. Thus, muscles of weight lifters and sprinters contain a high proportion of FF fibers. In contrast, aerobic exercise such as long-distance running and swimming induces the reverse enzymatic pattern and increases the muscle's ability to use oxygen. The proportion of FR and SR fibers is larger in marathon runners than in other athletes. It is still unclear whether the fiber types actually change as a consequence of training, or whether marathon runners choose this sport because of their muscle fiber composition. □

Muscle Fiber Types

Structural classification

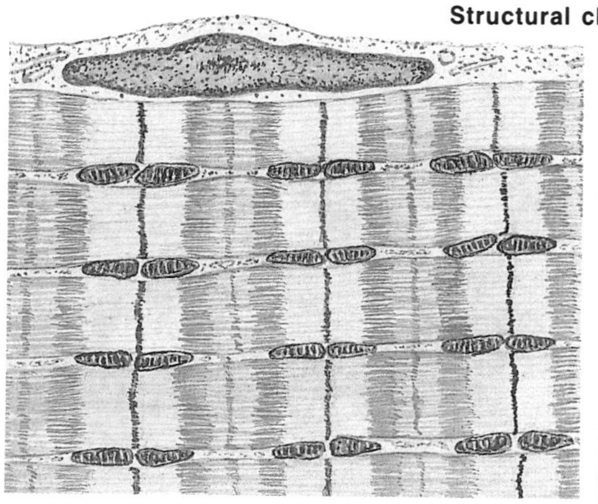

Type I: light or white skeletal muscle fiber in longitudinal section on electron microscopy. Small, relatively sparse mitochondria, chiefly paired in interfibrillar spaces at Z bands

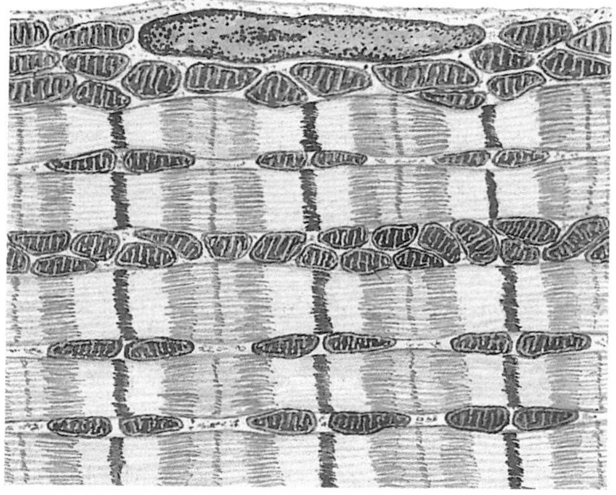

Type II: dark or red fiber. Large, profuse mitochondria beneath sarcolemma and in rows as well as paired in interfibrillar regions. Z bands wider than in type I

Histochemical classification

Fiber type	ATPase stain	SDH stain
1. Fast-twitch, fatigable (FF) Stain deeply for ATPase, poorly for succinic acid dehydrogenase (SDH), a mitochondrial enzyme active in citric acid cycle. Therefore, fibers rapidly release energy from ATP but poorly regenerate it, thus becoming fatigued		
2. Fast-twitch, fatigue-resistant (FR) Stain deeply for both ATPase and SDH. Therefore, fibers rapidly release energy from ATP and also rapidly regenerate ATP in citric acid cycle, thus resisting fatigue		
3. Slow-twitch, fatigue-resistant (SR) Stain poorly for ATPase but deeply for SDH. Therefore, fibers only slowly release energy from ATP but regenerate ATP rapidly, thus resisting fatigue		

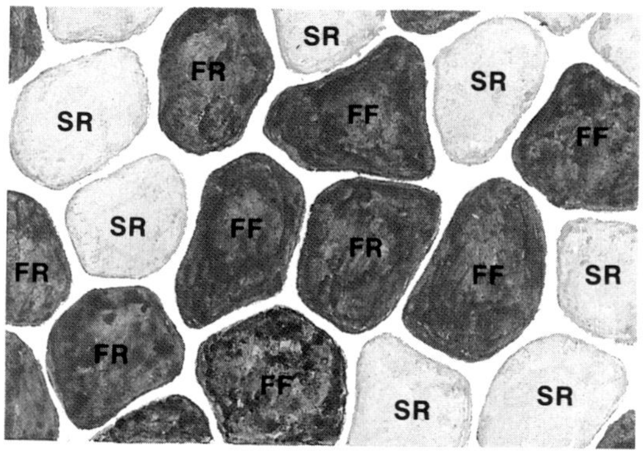

Cross section of skeletal muscle fibers stained for ATPase

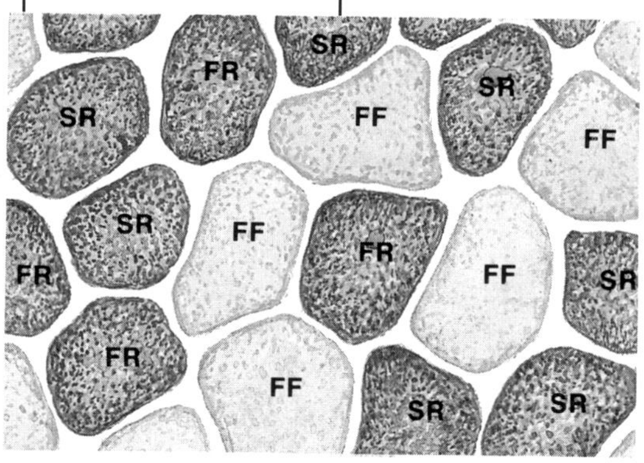

Identical section stained for SDH

Sprinter
Fast-twitch, fatigable fibers predominate

F. Netter M.D.
©CIBA-GEIGY

Marathon runner
Slow-twitch, fatigue-resistant fibers predominate

Zones Structures	Histology	Functions	Blood supply	Po_2	Cell (chondrocyte) health	Cell respiration	Cell glycogen
Secondary bony epiphysis Epiphyseal artery	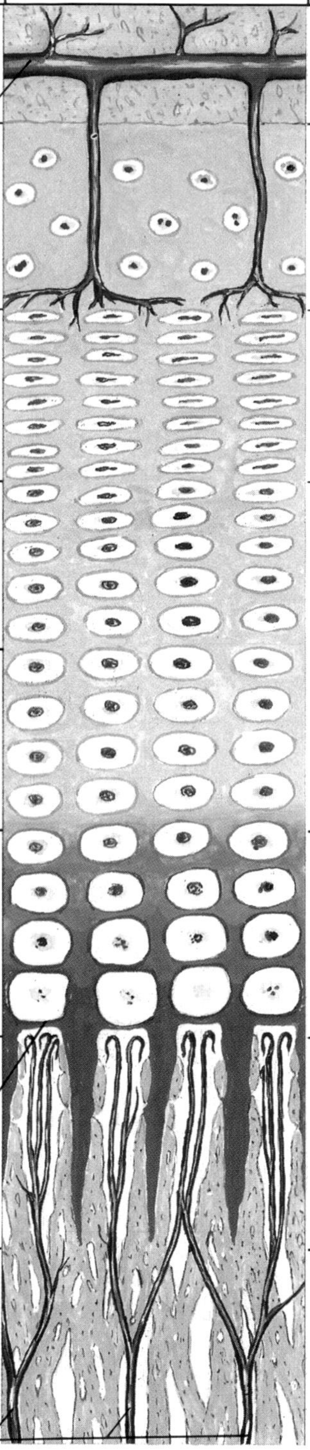						
Reserve zone		Matrix production Storage	Vessels pass through, do not supply this zone	Poor (low)	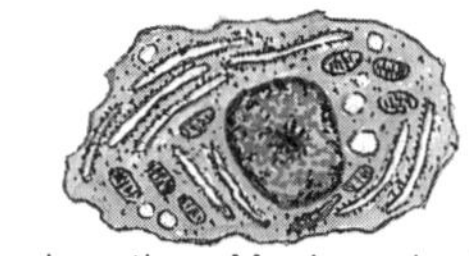 Good, active. Much endoplasmic reticulum, vacuoles, mitochondria	Anaerobic	High concentration
Proliferative zone		Matrix production Cellular proliferation (longitudinal growth)	Excellent	Excellent Fair	 Excellent. Much endoplasmic reticulum, ribosomes, mitochondria. Intact cell membrane	Aerobic ↓ Progressive change to anaerobic	High concentration (less than in above)
Hypertrophic zone: **Maturation zone**		Preparation of matrix for calcification	↓ Progressive decrease	Poor (low) ↓ Progressive decrease	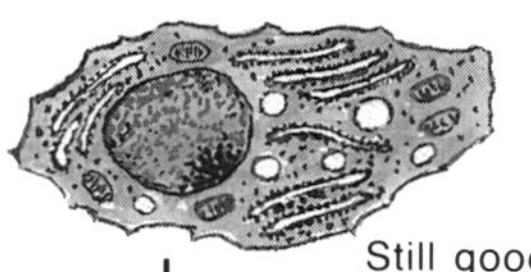 Still good		↓ Glycogen consumed until depleted
Degenerative zone					↓ Progressive deterioration	Anaerobic glycolysis	
Zone of provisional calcification		Calcification of matrix	Nil	Poor (very low)	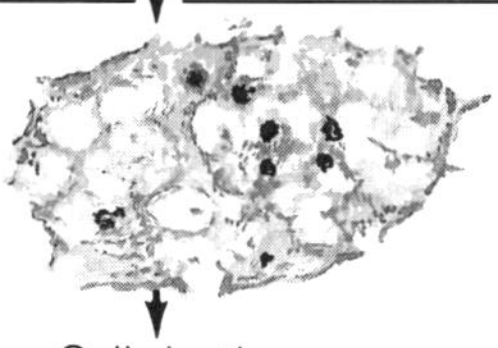 Cell death	Anaerobic glycolysis	Nil
Metaphysis: Last intact transverse septum **Primary spongiosa**		Vascular invasion and resorption of transverse septa Bone formation	Closed capillary loops Good	Poor Good		↓ Progressive reversion to aerobic	?
Secondary spongiosa Branches of metaphyseal and nutrient arteries		Remodeling Internal: removal of cartilage bars, replacement of fiber bone with lamellar bone External: funnelization	Excellent	Excellent		Aerobic	?

SECTION III PLATE 16 Slide 3712

Growth Plate

The growth plate is an organ composed of cartilage, bone, and fibrous components. The two growth plates in a typical long bone are peripheral extensions of the primary center of ossification in the midportion of the fetal cartilaginous anlage of the bone. The primary center of ossification grows and expands centrifugally in all directions until it eventually becomes confined to two platelike structures at each end of the bone (Plates 16–18).

Structure, Blood Supply, and Physiology

The growth plate may be divided into three anatomic components: a cartilaginous component with various histologic zones; a bony component, or metaphysis; and a fibrous component that surrounds the periphery of the plate and consists of the ossification groove of Ranvier and the perichondral ring of La Croix. Each of the three components of the growth plate has its own distinct blood supply. The vascular differences have important implications for metabolic activity (Plate 17).

Reserve Zone. The functions of this zone are storage and matrix production. It lies immediately adjacent to the secondary ossification center and comprises cells that appear to be storing lipid and other materials. The cells are spheric and may exist singly or in pairs. They are relatively few in number, with more extracellular matrix between them than between cells in any other zone.

The matrix shows a positive histochemical reaction for the presence of a neutral polysaccharide or an aggregated proteoglycan. The cytoplasm exhibits a positive stain for glycogen. Electron microscopy reveals that these cells contain abundant endoplasmic reticulum, a clear indication that they are actively synthesizing protein.

Oxygen tension (Po_2) is low, which means that blood vessels passing through the reserve

Pathophysiology of Growth Plate

Proteoglycans in matrix	Mitochondrial activity	Matrix calcification	Matrix vesicles	Exemplary diseases	Defect (if known)
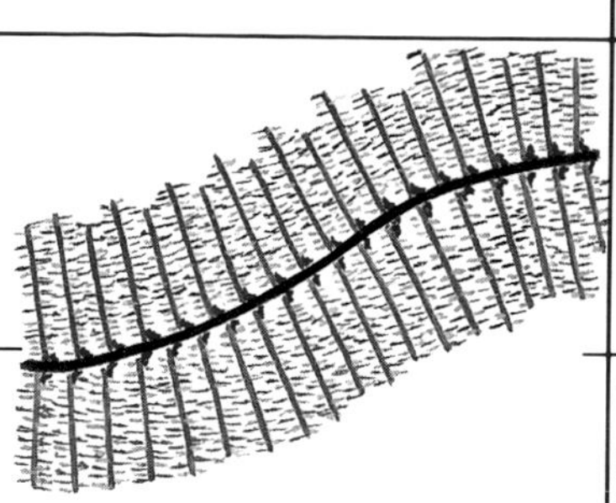	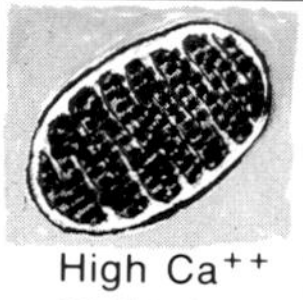 High Ca^{++} content	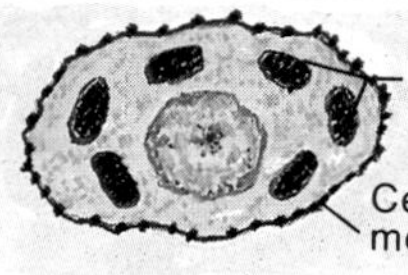 Ca^{++} intracellular	 Few vesicles, contain little Ca^{++}	Diastrophic dwarfism (also, defects in other zones) Pseudoachondroplasia (also, defects in other zones) Kneist syndrome (also, defects in other zones)	Defective type II collagen synthesis Defective processing and transport of proteoglycans Defective processing of proteoglycans
Aggregated proteoglycans (neutral mucopolysaccharides) inhibit calcification	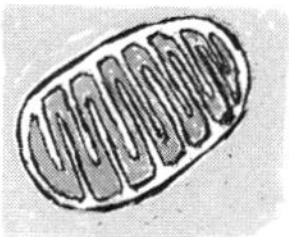 ATP made	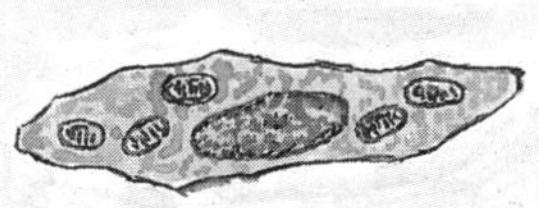 Ca^{++} intracellular	Few vesicles, contain little Ca^{++}	Gigantism Achondroplasia Hypochondroplasia Malnutrition, irradiation injury, glucocorticoid excess	Increased cell proliferation (growth hormone increased) Deficiency of cell proliferation Less severe deficiency of cell proliferation Decreased cell proliferation and/or matrix synthesis
Progressively disaggregated	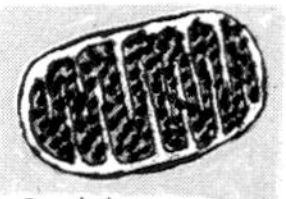 Ca^{++} uptake, no ATP made	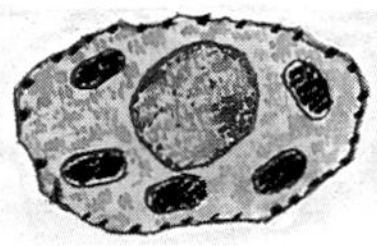 Ca^{++} intracellular	 Contain little Ca^{++}		
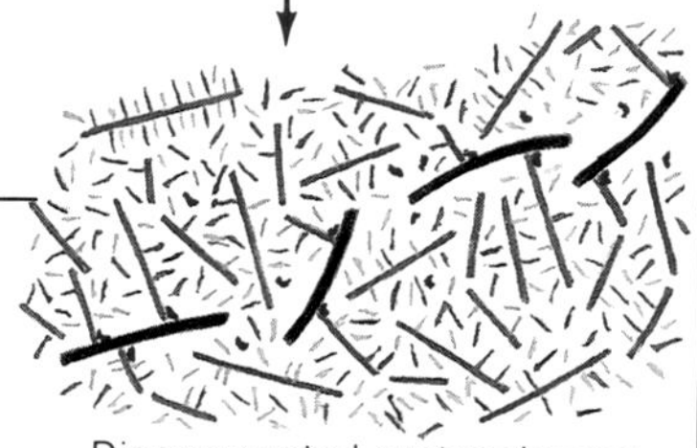	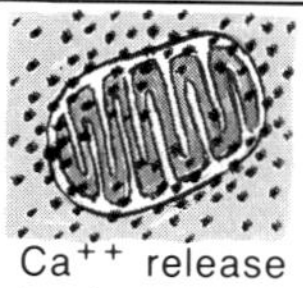 Ca^{++} release begins	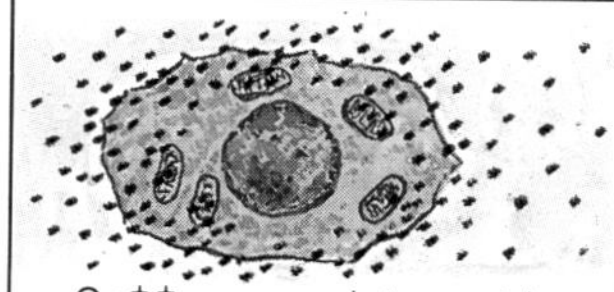 Ca^{++} passes into matrix	 Begin Ca^{++} uptake	Mucopolysaccharidosis (Morquio's syndrome, Hurler's syndrome)	Deficiencies of specific lysosomal acid hydrolases, with lysosomal storage of mucopolysaccharides
Disaggregated proteoglycans (acid mucopolysaccharides) permit calcification	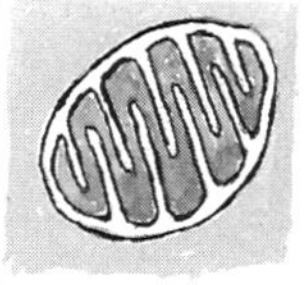 Ca^{++} released	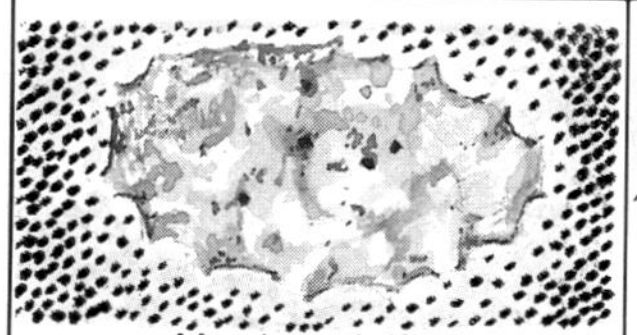 Matrix calcified	 Crystals in and on vesicles	Rickets, osteomalacia (also, defects in metaphysis)	Insufficiency of Ca^{++} and/or P_i for normal calcification of matrix
				Metaphyseal chondro-dysplasia (Jansen and Schmid types) Acute hematogenous osteomyelitis	Extension of hypertrophic cells into metaphysis Flourishing of bacteria due to sluggish circulation, low Po_2, reticuloendothelial deficiency
				Osteopetrosis Osteogenesis imperfecta Scurvy Metaphyseal dysplasia (Pyle disease)	Abnormality of osteoclasts (internal remodeling) Abnormality of osteoblasts and collagen synthesis Inadequate collagen formation Abnormality of funnelization (external remodeling)

F. Netter M.D.

zone in cartilage canals do not actually supply it (Plate 16). Chondrocytes in the reserve zone do not proliferate or do so only sporadically.

Proliferative Zone. The functions of this zone are matrix production and cellular proliferation, which together produce longitudinal growth. The chondrocytes are flattened and aligned in longitudinal columns, with the long axis of the cells lying perpendicular to the long axis of the bone; they are packed with endoplasmic reticulum. The cytoplasm stains positively for glycogen.

With few exceptions, chondrocytes in the proliferative zone are the only cells in the cartilaginous portion of the growth plate that divide. The top cell of each column is the true "mother" cartilage cell for each column, and the top of the proliferative zone is the true germinal layer of the growth plate. Longitudinal growth in the growth plate equals the rate of production of new chondrocytes at the top of the proliferative zone multiplied by the maximum size of the chondrocytes at the bottom of the hypertrophic zone.

Because of the rich vascular supply to the top of the proliferative zone, Po_2 is highest in this region of the growth plate. The high Po_2 coupled with the presence of glycogen in the chondrocytes indicates that aerobic metabolism with glycogen storage is taking place.

Hypertrophic Zone. The functions of this zone are preparation of the matrix for calcification and calcification. Chondrocytes in this zone become spheric and greatly enlarged. By the bottom of the zone, they have enlarged to five times their size in the proliferative zone. The cytoplasm of chondrocytes in the top half of the zone stains positively for glycogen; near the middle of the zone, it abruptly loses all glycogen-staining ability.

On electron microscopy, chondrocytes in the top half of the hypertrophic zone appear normal and contain the full complement of cytoplasmic components, but in the bottom half of the zone, the cytoplasm contains holes that occupy over 85% of the total cytoplasmic volume.

The last cell at the base of each cell column is clearly nonviable and shows extensive fragmentation of the cell membrane and the nuclear envelope, with loss of all cytoplasmic components except a few mitochondria and scattered remnants of endoplasmic reticulum. Mitochondria and cell membranes of chondrocytes in the top half of the hypertrophic zone are loaded with calcium.

Growth Plate

(Continued)

Structure and Blood Supply of Growth Plate

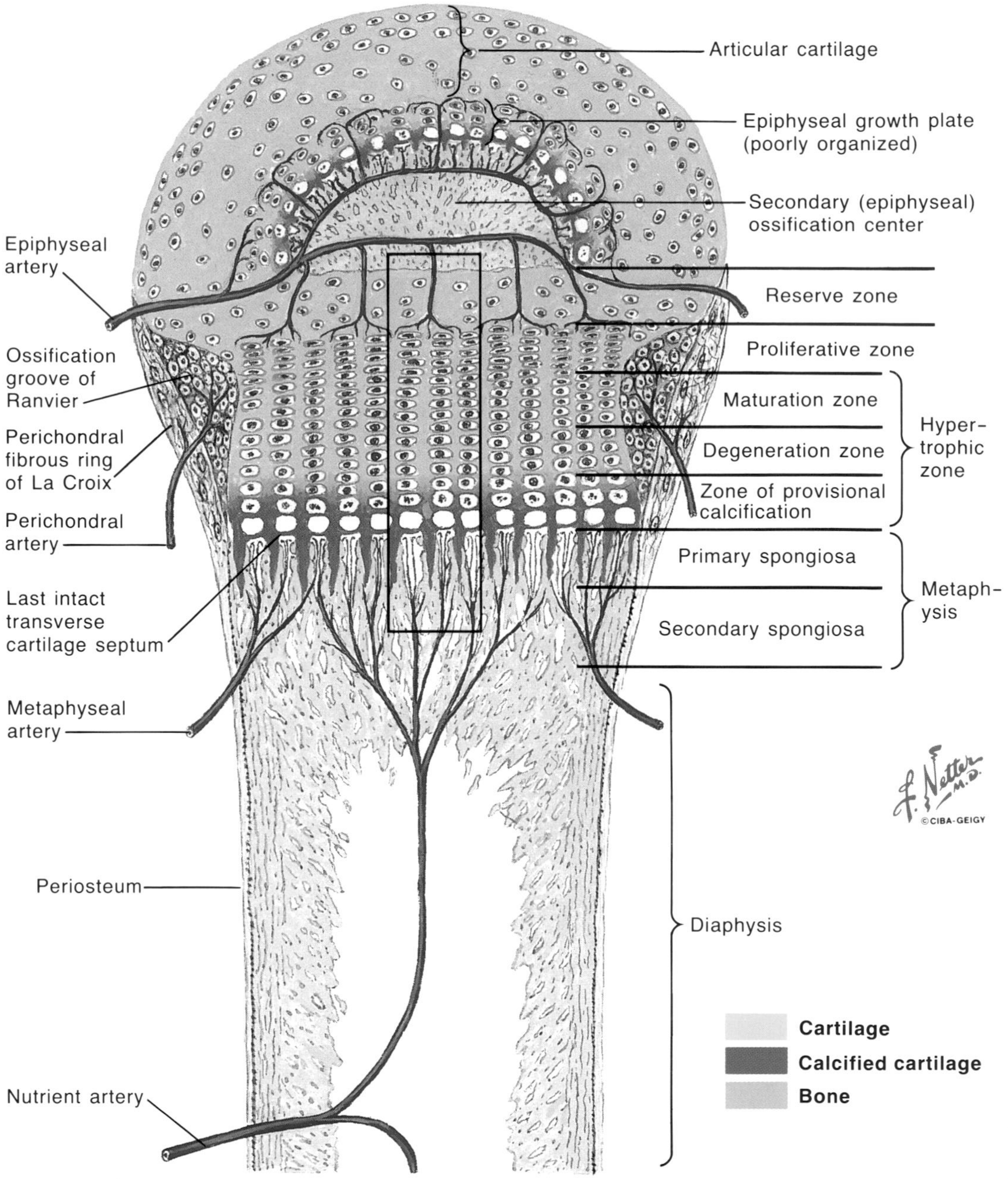

Articular cartilage
Epiphyseal growth plate (poorly organized)
Secondary (epiphyseal) ossification center
Epiphyseal artery
Reserve zone
Proliferative zone
Ossification groove of Ranvier
Maturation zone
Perichondral fibrous ring of La Croix
Degeneration zone
Hyper-trophic zone
Zone of provisional calcification
Perichondral artery
Primary spongiosa
Last intact transverse cartilage septum
Metaph-ysis
Secondary spongiosa
Metaphyseal artery
Periosteum
Diaphysis
Cartilage
Calcified cartilage
Bone
Nutrient artery
©CIBA-GEIGY

Branches of epiphyseal artery pass directly through reserve zone of cartilaginous growth plate without contributing capillaries to it, but on reaching proliferative zone, they arborize into capillaries that supply only top of cell columns. Metaphyseal and nutrient arteries subdivide into numerous small branches, which pass axially to bone–cartilage junction where they form loops or tufts but do not enter cartilage. Thus, metaphysis and bony portion of growth plate are well supplied with blood, but only uppermost portion of cell columns (ie, proliferative zone of cartilage) is vascularized. Hypertrophic zone is avascular; cells are progressively poorly oxygenated and nourished from top downward, and lowermost cells degenerate and die. Peripherally located perichondral ring of La Croix and ossification groove of Ranvier have own distinct blood supply. These vascular phenomena have profound physiologic significance (see Plate 16 for enlarged view of outlined rectangle)

Toward the middle of the zone, mitochondria rapidly lose calcium, and at the bottom of the zone, both mitochondria and cell membranes have no calcium. These findings suggest that mitochondrial calcium may be involved in cartilage calcification.

The hypertrophic zone is avascular, and PO_2 is therefore low. In the bottom half of the zone, glycogen is completely depleted. There is no other source of nutrition to serve as an energy source for the mitochondria. Since calcium uptake and retention require energy, as soon as the chondrocytes' glycogen supplies are exhausted, mitochondria release calcium, a factor that may play a role in matrix calcification.

The metabolic events in the proliferative and hypertrophic zones can be summarized as follows. In the proliferative zone, PO_2 is high, aerobic metabolism occurs, glycogen is stored, and mitochondria form adenosine triphosphate (ATP). In the hypertrophic zone, PO_2 is low, anaerobic metabolism occurs, and glycogen is consumed until near the middle of the zone, where mitochondria switch from forming ATP to accumulating calcium. ATP formation and calcium accumulation cannot take place simultaneously. Both processes require energy, which comes from the respiratory chain in the mitochondria. In addition, ATP formation requires the presence of adenosine diphosphate (ADP), whereas calcium accumulation does not. Possibly, in the hypertrophic zone, there is insufficient ADP for significant ATP formation.

The matrix of the hypertrophic zone shows a positive histochemical reaction for an acid mucopolysaccharide, or a disaggregated proteoglycan (Plate 25). From the reserve zone through the hypertrophic zone, there is a progressive decrease in the length of proteoglycan aggregates and in the number of aggregate subunits in the matrix. The distance between the subunits also increases.

The large proteoglycan aggregates with tightly packed subunits may inhibit mineralization or its spread, whereas smaller aggregates with widely spaced subunits at the bottom of the hypertrophic zone may be less effective in preventing mineral growth. In any event, proteoglycan disaggregation, or degradation, must take place before significant mineralization occurs.

The initial calcification (seeding, or nucleation) in the bottom of the hypertrophic zone, called the *zone of provisional calcification*, occurs within or on vesicles in the longitudinal septa of the matrix (Plate 16). Matrix vesicles are densest in the hypertrophic zone. They are very small structures (1,000–1,500 Å in diameter), enclosed in a trilamellar membrane and therefore produced by chondrocytes. Matrix vesicles are rich in alkaline phosphatase, which may act as a pyrophosphatase to destroy pyrophosphate, another inhibitor of calcium-phosphate precipitation. Matrix vesicles begin to accumulate calcium at the same level in the hypertrophic zone at which mitochondria begin to lose it. This suggests that mitochondrial calcium is involved in the initial calcification in the growth plate. Initial calcification, whether it is within or on matrix vesicles, collagen fibers, or proteoglycan disaggregates, may be in the form of amorphous calcium phosphate, but this rapidly gives way to hydroxyapatite crystal formation. With crystal growth and confluence, the longitudinal septa become calcified.

Matrix calcification in the zone of provisional calcification makes the intracellular matrix relatively impermeable to metabolites. The hypertrophic zone has the lowest diffusion coefficient in

Growth Plate

(Continued)

the entire growth plate, primarily because of its high mineral content.

Metaphysis. The three functions of the metaphysis are vascular invasion of the transverse septa at the bottom of the cartilaginous portion of the growth plate, bone formation, and bone remodeling (Plate 16). The metaphysis begins just distal to the last intact transverse septum at the base of each cell column of the cartilaginous portion of the growth plate and ends at the junction with the diaphysis. In the first part of the metaphysis, PO_2 is low, which, together with the rouleau formation frequently seen just distal to the last intact transverse septum, indicates that this is a region of vascular stasis.

Electron microscopy shows capillary sprouts, which are lined with a layer of endothelial and perivascular cells, invading the base of the cartilaginous portion of the plate. Cytoplasmic processes from these cells push into the transverse septa and, presumably through lysosomal enzyme activity, degrade and remove the nonmineralized transverse septa. This region of the metaphysis is known as the *primary spongiosa*. The longitudinal septa are partially or completely calcified, with osteoblasts lining up along the calcified bars. Between this layer of osteoblasts and the capillary sprouts are osteoprogenitor cells that contain little cytoplasm but have a prominent ovoid-to-spindle–shaped nucleus.

A short distance down the calcified longitudinal septa is the region called the *secondary spongiosa*. Osteoblasts begin laying down bone by a process called *endochondral ossification*, or bone formation within or on cartilage. The amount of bone formed on the cartilage bars increases downward and into the metaphysis. At the same time, the calcified cartilage bars gradually become thinner until they disappear altogether. Still farther down in the metaphysis, the original fiber bone is replaced with lamellar bone. The gradual replacement of the calcified longitudinal septa with newly formed fiber bone, as well as the gradual replacement of fiber bone with lamellar bone, is called *internal*, or *histologic, remodeling*. Large, irregularly shaped osteoclasts are distributed evenly throughout the metaphysis (except in the primary spongiosa), and subperiosteally around the outside of the metaphysis where it narrows to meet the diaphysis. This narrowing of the metaphysis is called *external*, or *anatomic, remodeling*.

Peripheral Fibrocartilaginous Element

Encircling the periphery of the growth plate in a typical long bone are two structures: a wedge-shaped groove of cells, the *ossification groove*, first described by Ranvier, and a band of fibrous tissue and bone called the *perichondral fibrous ring*, studied by La Croix (Plate 18). Although both structures are simply different parts of the peripheral fibrocartilaginous element of the growth plate, they can be considered separate entities because of their different functions.

The function of the ossification groove of Ranvier appears to be the contribution of chondrocytes for the increase in width of the growth plate. The groove of Ranvier contains round-to-ovoid cells that, on light microscopy, appear to "flow" from the groove into the cartilage at the level of the reserve zone. The perichondral fibrous ring of La Croix acts as a limiting membrane that provides mechanical support for the bone-cartilage junction of the growth plate. It is a dense fibrous band encircling the growth plate, in which collagen fibers run vertically, obliquely, and circumferentially. The structure is continuous at one end with the ossification groove, and at the other end with the periosteum and subperiosteal bone of the metaphysis.

Pathophysiology

Certain representative disorders whose pathophysiology exemplifies the known functions of the highly synchronized and interrelated zones in the growth plate have been identified and are shown in Plate 16. □

Peripheral Fibrocartilaginous Element of Growth Plate

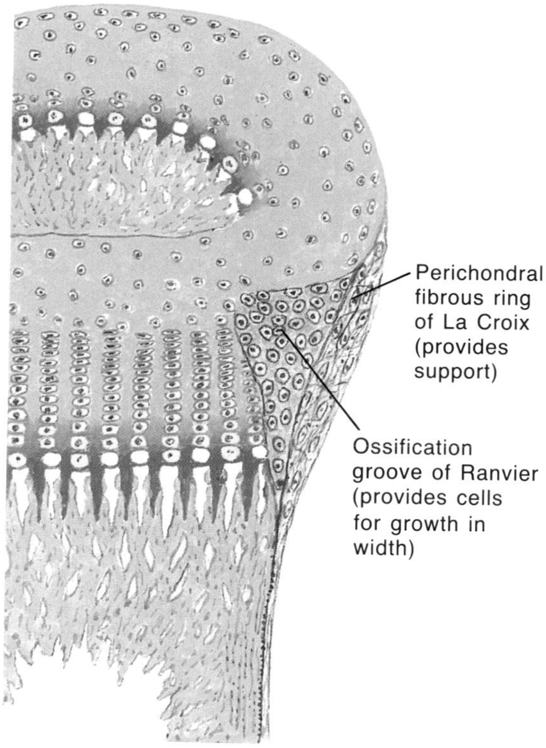
Perichondral fibrous ring of La Croix (provides support)
Ossification groove of Ranvier (provides cells for growth in width)

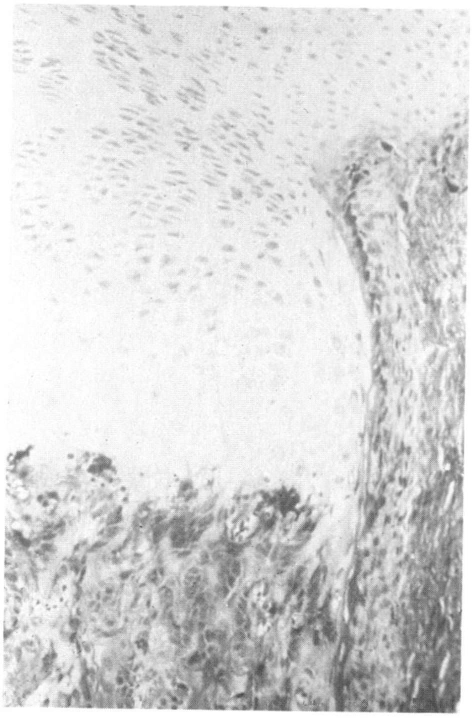

Microscopic section (H and E) corresponds generally to illustration at left

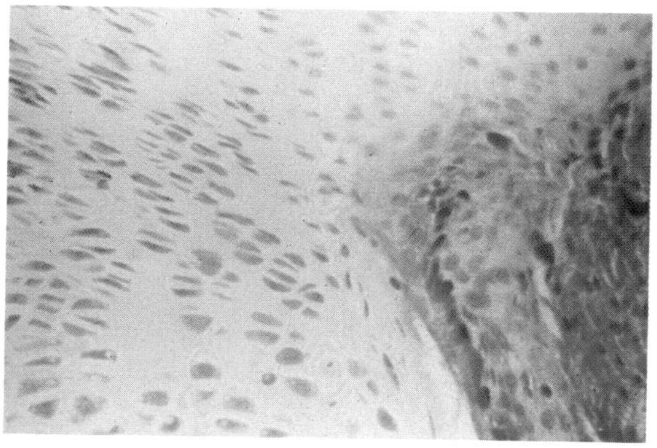

High-power section shows cells of ossification groove of Ranvier apparently "flowing" into cartilage at level of reserve zone, thus contributing to growth in width of growth plate. Note presence of arterioles (cut-in section)

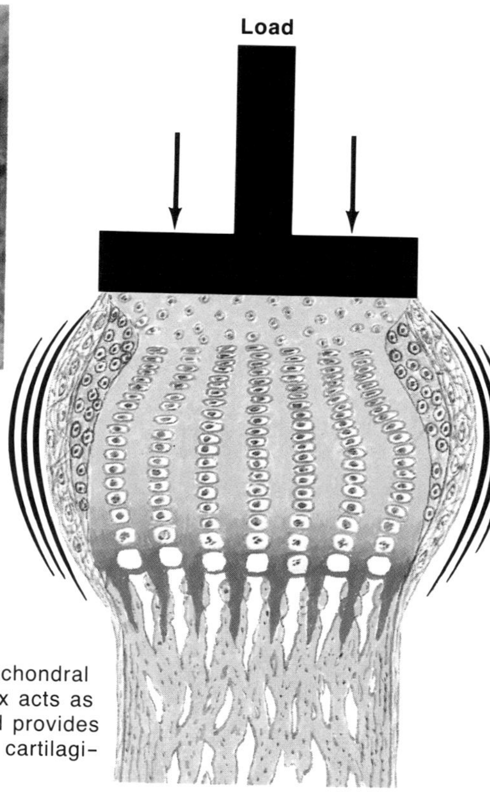
Load

Illustration of how perichondral fibrous ring of La Croix acts as limiting membrane and provides mechanical support to cartilaginous growth plate

F. Netter M.D. ©CIBA-GEIGY

Composition and Structure of Cartilage

Cartilage is a complex and versatile connective tissue. *Growth plate cartilage* is responsible for much of the shape, growth, and development of the skeleton. *Articular cartilage* provides the self-lubricating, low-friction gliding and load-distributing surfaces of the synovial (diarthrodial) joints. *Fibrocartilage* attaches tendons and ligaments to bone. *Elastic cartilage* contributes structural integrity to the auricles, nose, eustachian tubes, epiglottis, and trachea. *Fibroelastic cartilage* is responsible for the load-distributing and shock-absorbing properties of the intervertebral discs and intraarticular menisci.

Regardless of its specialized function, all cartilage consists of cells—chondrocytes and chondroblasts. These cells synthesize and deposit around them an elaborate matrix of macromolecules that are some of the largest in nature. The mechanical properties of cartilage tissue are derived primarily from the properties of the complex extracellular matrix.

On gross examination and on light microscopy, all cartilage appears smooth and homogeneous. However, electron microscopy reveals that its basic fibrillar structure consists of a meshwork of collagen fibers and giant proteoglycans in approximately equal amounts. In addition, water is a major component of cartilage, contributing 65% to 80% of its weight. Type II collagen, the major fibrillar component of cartilage matrix, contributes tensile strength and form to the tissue. Proteoglycans, by their ability to trap and hold large amounts of water (tissue fluid), give cartilage a resiliency and stiffness to compression (Plate 25). The exact mechanisms by which collagen and proteoglycans interact in the various types of cartilage remain unclear. However, another function of collagen is to trap proteoglycans and restrain their swelling pressure.

In addition to properties shared with other types of hyaline cartilage, articular cartilage has a complex internal structure. Electron microscopy and biochemical studies reveal four poorly demarcated zones: a small superficial, or tangential, zone (I); a larger intermediate, or transitional, zone (II); a deep vertical zone (III), which occupies the greatest volume; and a zone of calcified cartilage (IV), which lies adjacent to the subchondral bone. On light microscopy, the boundary between zones III and IV is demarcated by an undulating plate, referred to as a tidemark.

The four zones differ dramatically in cell size, shape, orientation, and number, as well as in the relative composition, proportion, and orientation of macromolecules in the matrix. Even small differences in the composition and organization of the matrix give each zone slightly different mechanical properties. □

Composition and Structure of Cartilage

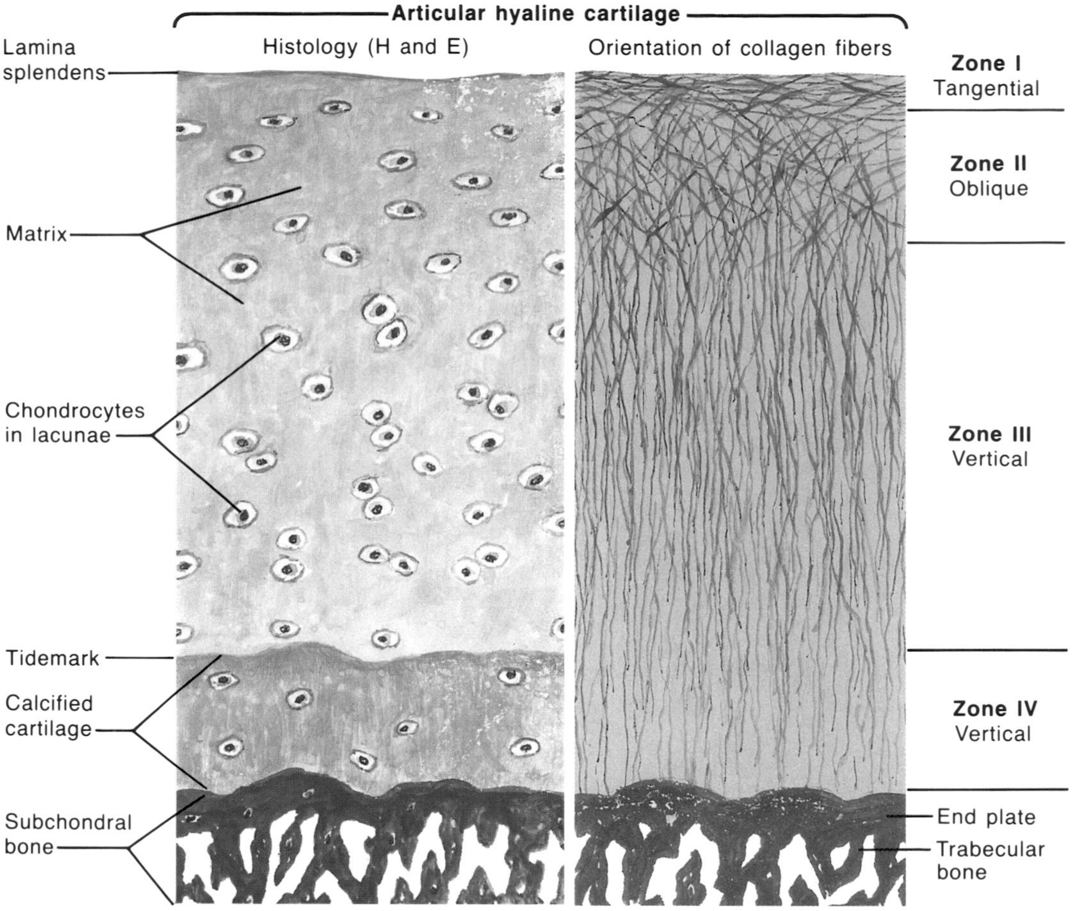
Articular hyaline cartilage
Histology (H and E)
Orientation of collagen fibers
Lamina splendens
Matrix
Chondrocytes in lacunae
Tidemark
Calcified cartilage
Subchondral bone
Zone I
Tangential
Zone II
Oblique
Zone III
Vertical
Zone IV
Vertical
End plate
Trabecular bone

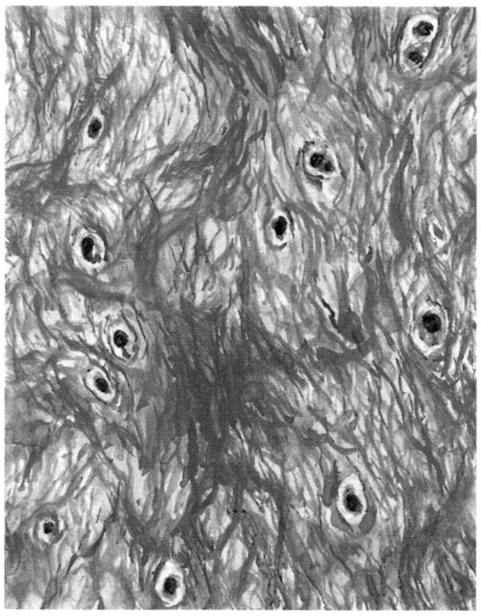

Fibrocartilage
In annulus fibrosus, meniscus, tendon–bone interface

Interlacing strands of fibrous tissue throughout matrix (H and E)

F. Netter M.D.

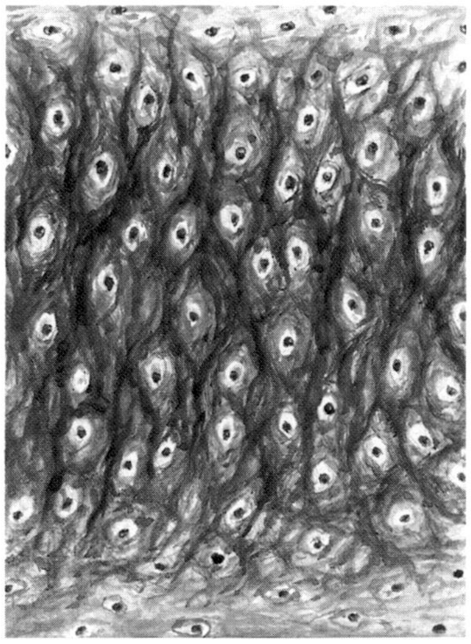

Elastic cartilage
In auricle, eustachian tube, nose, epiglottis

Dark-staining elastic fibers between and around lacunae (H and E)

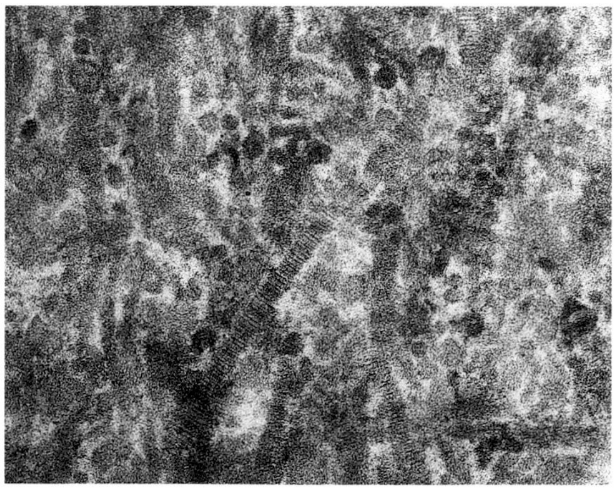

Although hyaline cartilage appears smooth and homogeneous to naked eye, electron microscopy reveals basic structure of network of collagen fibers and proteoglycans (×80,000)

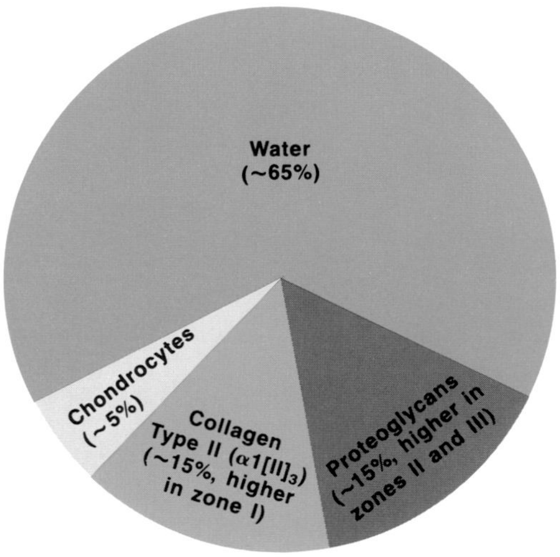
Water
(~65%)
Chondrocytes
(~5%)
Collagen
Type II (α1[II]$_3$)
(~15%, higher
in zone I)
Proteoglycans
(~15%, higher in
zones II and III)

Composition of hyaline cartilage

Composition and Structure of Bone

Composition of Bone

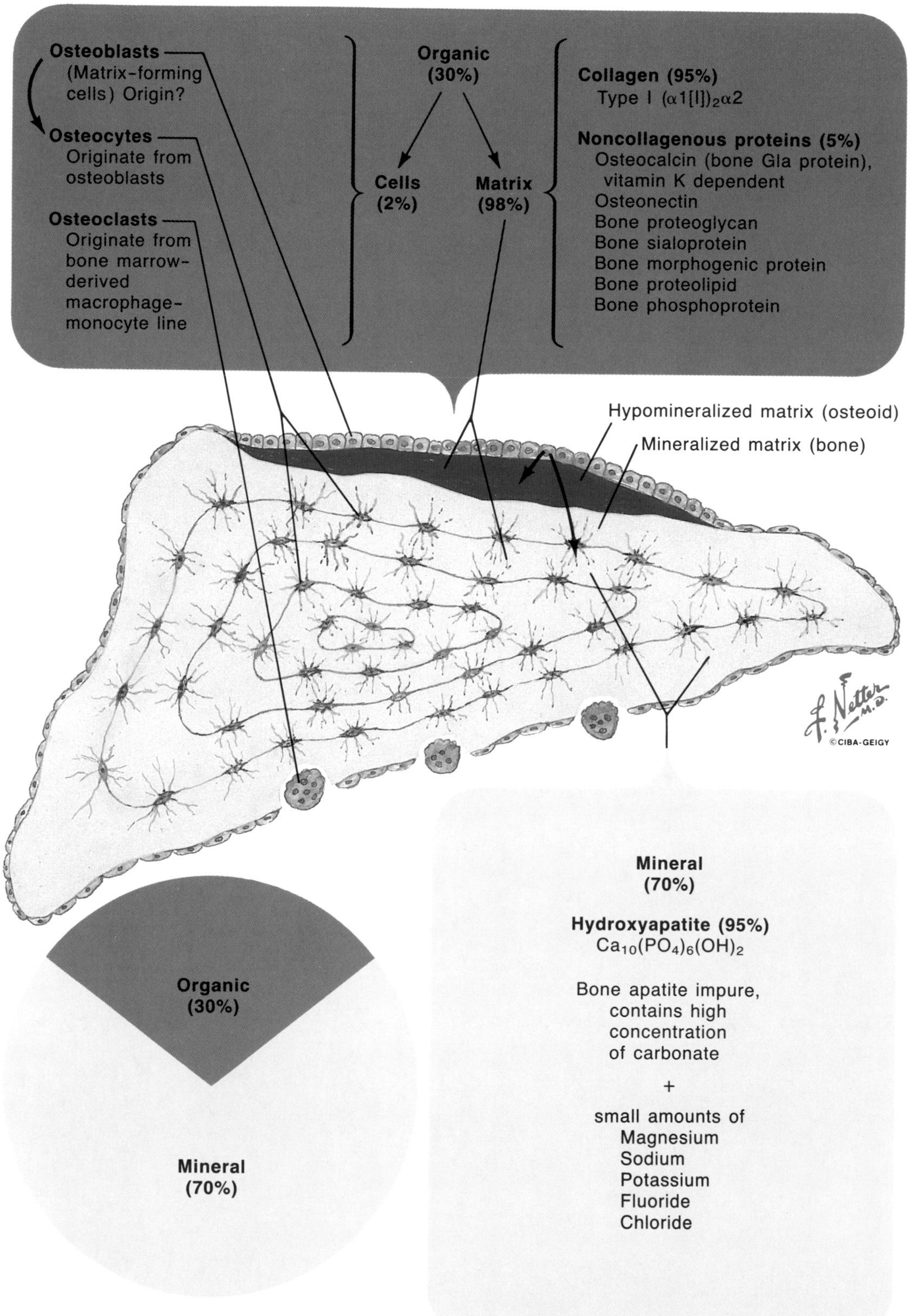
Osteoblasts
(Matrix-forming cells) Origin?
Osteocytes
Originate from osteoblasts
Osteoclasts
Originate from bone marrow-derived macrophage-monocyte line
Organic
(30%)
Cells
(2%)
Matrix
(98%)
Collagen (95%)
Type I $(\alpha 1[I])_2\alpha 2$
Noncollagenous proteins (5%)
Osteocalcin (bone Gla protein), vitamin K dependent
Osteonectin
Bone proteoglycan
Bone sialoprotein
Bone morphogenic protein
Bone proteolipid
Bone phosphoprotein
Hypomineralized matrix (osteoid)
Mineralized matrix (bone)
F. Netter M.D.
©CIBA-GEIGY
Organic
(30%)
Mineral
(70%)
Mineral
(70%)
Hydroxyapatite (95%)
$Ca_{10}(PO_4)_6(OH)_2$
Bone apatite impure, contains high concentration of carbonate
+
small amounts of
Magnesium
Sodium
Potassium
Fluoride
Chloride

Composition of dried bone (by weight)

The skeleton is not only an adaptable and well-articulated frame but also a dynamic mineral reserve bank in which the body stores its calcium and phosphate in a metabolically stable and structurally useful manner. The cells of the bone—the osteoblasts, osteocytes, and osteoclasts—function as both construction workers and metabolic bankers, dual roles that often conflict.

The *osteoblast*, or bone-forming cell, is approximately 20 μm in diameter and contains a single eccentric nucleus. Its precursor is still unknown, although it is thought to be a fixed-tissue osteoprogenitor cell. Osteoblasts require a scaffolding (periosteal, endosteal, trabecular, or haversian surface) on which to synthesize the organic bone matrix, or osteoid. In synthesizing the bone matrix and presiding over its mineral growth, the osteoblast makes large quantities of bone alkaline phosphatase, an important enzyme in preparing the bone matrix for mineralization.

The mature *osteocyte*, derived from an osteoblast, is an oval cell approximately 20 to 60 μm long and buried deep within the mineralized bone matrix in a small cavern called a lacuna. Numerous processes extend from the cell surface and leave the lacuna via a network of canals, or canaliculi. Many osteocyte processes extend into the canalicular system and contact processes from other osteocytes. This extensive osteocyte-canaliculi network is believed to play a vital role in transportation of cell metabolites, communication between cells, and regulation of mineral homeostasis.

The other major type of bone cell, the *osteoclast*, resorbs mineralized bone matrix. The osteoclast is a large cell (as great as 100 μm in diameter) containing as many as 100 nuclei per cell (athough most osteoclasts contain many fewer nuclei). It is rich in lysosomal enzymes (including acid phosphatase) and possesses a specialized cell membrane (the ruffled border) at sites where active bone resorption occurs. Unlike the osteoblast and the osteocyte (which possibly develop from a fixed-tissue osteoprogenitor cell), the osteoclast is thought to derive from circulating marrow cells, or preosteoclasts, in the monocyte-macrophage cell line.

Bone cells account for only a small portion (2%) of the entire organic component of bone, most of which consists of osteoid produced by osteoblasts.

Collagen (predominantly type I) is the major organic component of bone, accounting for as much as 95% of the osteoid. Bone collagen is deposited along lines of mechanical stress according to Wolff's law and provides an important passive template for bone crystal nucleation (Plate 36).

Although constituting only 5% of the weight of the osteoid, noncollagenous bone proteins play an important role in bone metabolism and mineralization of bone matrix.

The major noncollagenous proteins unique to bone include osteonectin, osteocalcin (bone Gla protein), bone proteoglycan, bone proteolipid, bone sialoprotein, and bone morphogenic protein.

The organic component of bone (cells plus organic matrix) makes up approximately 30% of the bone (dry weight). The inorganic, or mineral, component of bone (70% of dry weight) consists mainly of a carbonate-rich hydroxyapatite analogue called bone apatite, which is smaller and less perfect in crystal arrangement than pure hydroxyapatite. Because of its crystalline imperfections, bone apatite is more soluble than pure hydroxyapatite and is therefore more readily available for metabolic activity and for exchange with body fluids. In addition to incorporating carbonate, bone apatite possesses the ability to incorporate magnesium, sodium, potassium, chloride, fluoride, strontium, and other bone-seeking elements.

Mature lamellar bone has the same chemical composition and material properties throughout the skeleton, regardless of its mechanism of formation—intramembranous or endochondral—or its structural organization—cortical (compact) or trabecular bone. □

Composition and Structure of Bone

(Continued)

Composition of Bone (continued)

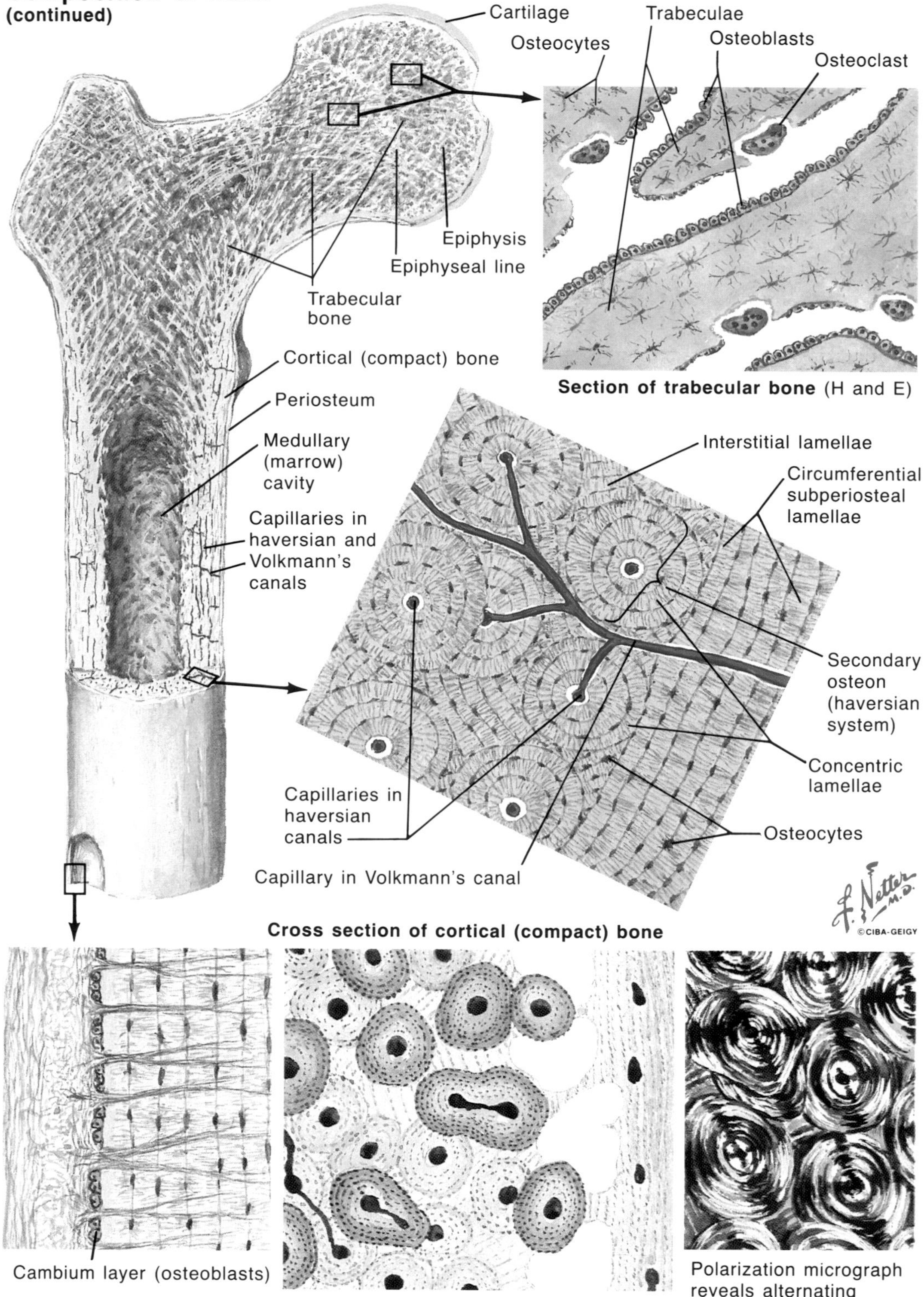

Section of trabecular bone (H and E)

Cross section of cortical (compact) bone

Section of decalcified bone shows attachment of periosteum to bone by perforating (Sharpey's) fibers

Electron micrograph shows different generations of secondary osteons. New bone (dark) replaces old bone (light) and nonhaversian bone

Polarization micrograph reveals alternating light and dark rings of mineral hydroxyapatite crystals embedded in osteoid

Skeletal growth and development begin in utero and continue for nearly 2 decades in a series of well-orchestrated events. These events are determined genetically and regulated by central endocrine and peripheral biophysical and biochemical processes.

Normal bone forms either by intramembranous ossification from mesenchymal osteoblasts or by endochondral ossification from a preexisting cartilage model. Long bones and vertebrae increase in size by a combination of these two processes. For example, ossification of the shaft of a long bone is an intramembranous process: subperiosteal deposition of new bone widens the shaft, while endosteal resorption widens the medullary canal. Long bones increase in length by cartilage proliferation at the growth plate in an elaborate process of endochondral ossification.

Histology

The adult skeleton contains only two types of bone, cortical (compact) bone and trabecular bone.

Both of these histologic types are represented in a typical long bone such as the femur (Plate 21). Cortical bone forms the wall of the shaft, and trabecular bone is concentrated at each end.

The articular surface of the femur is covered with a cap of hyaline cartilage, which is better suited than bone to withstand the friction and relative motion in the joint. The cartilage cap is continuous with the synovial membrane lining the joint cavity (Plate 26). The rest of the outer surface of the bone is lined with *periosteum*, a dense fibrous connective tissue. In a growing bone, the inner surface of the periosteum contains osteogenic cells that are actively laying down sheets of bone matrix. The cell morphology ranges from active cells near the bone itself to inactive-appearing fibroblasts embedded among dense collagen fibers in the outer edge of the periosteum. As the layers of bone are deposited, bundles of collagen fibers become embedded in them, forming perforating, or Sharpey's, fibers.

The inner layer of the shaft of a long bone is lined with *endosteum*, a much less substantial layer. Endosteal cells possess an osteogenic capacity that is expressed during fracture healing.

Structure of Cortical (Compact) Bone

The fundamental functional unit of cortical bone is the *osteon*, or haversian system, a cylindric structure measuring approximately 250 μm by 1 to 5 cm. The osteon consists of concentric layers of bony lamellae, each 2 to 3 μm thick, which surround a central haversian canal (Plates 22–23). The *haversian canal*, first described in 1691 by the English anatomist Clopton Havers, contains the blood and nerve supplies of the bone. Lateral branches, called *Volkmann's canals*, carry blood vessels from one osteon to another. Each cylindric lamella within the osteon is lined with a sparse population of regularly arranged osteocytes, which communicate with one another by fine cell processes projecting into the lamellae through minute channels, or canaliculi. Oxygen and nutrients reach osteocytes in the outer lamellae by diffusion through these canaliculi. In addition to osteons, the compact collar of a long bone contains at its periphery subperiosteal circumferential lamellae, which are deposited by the inner layer of the periosteum.

Composition and Structure of Bone

(Continued)

Structure of Cortical (Compact) Bone

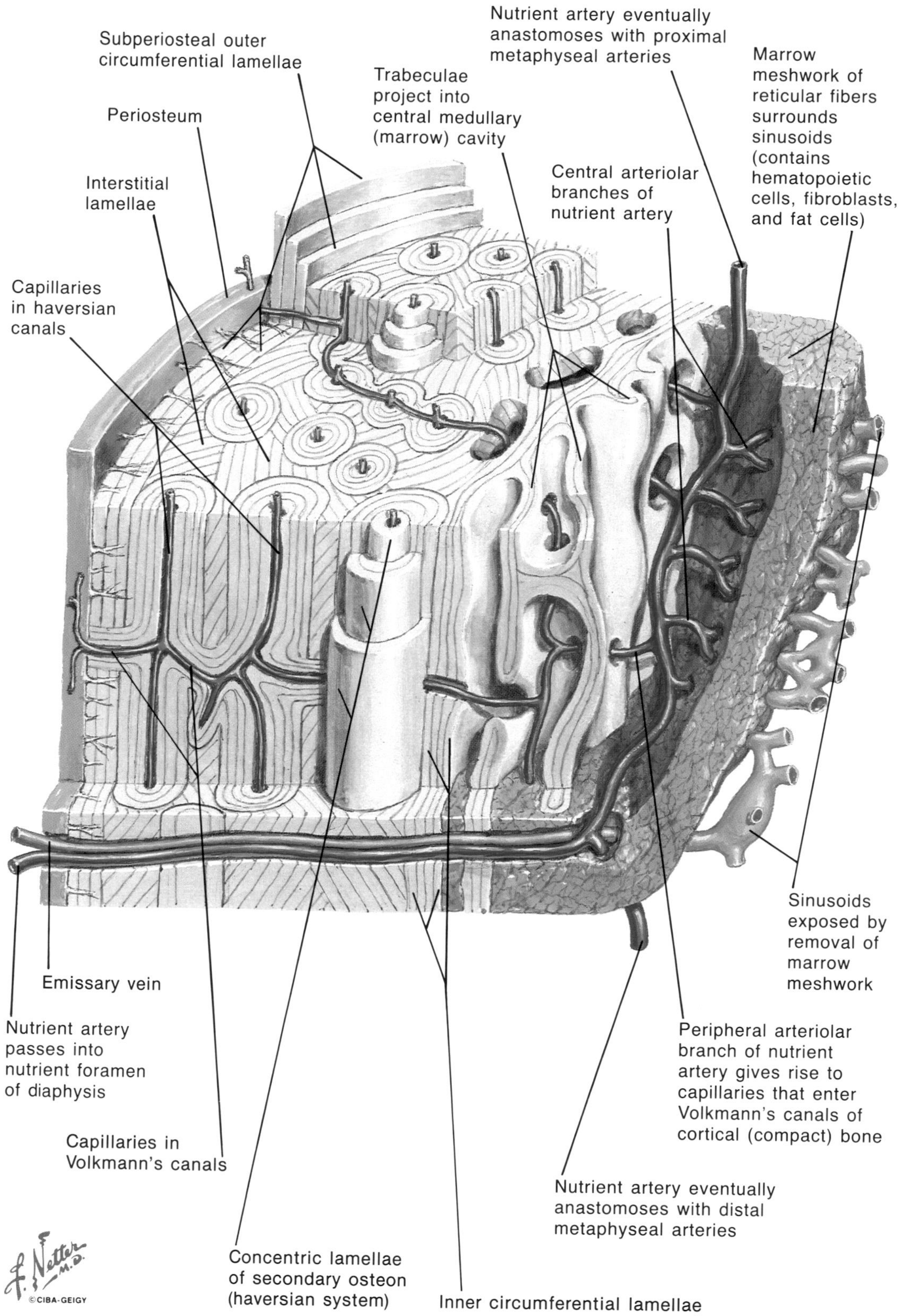
Subperiosteal outer circumferential lamellae
Nutrient artery eventually anastomoses with proximal metaphyseal arteries
Marrow meshwork of reticular fibers surrounds sinusoids (contains hematopoietic cells, fibroblasts, and fat cells)
Trabeculae project into central medullary (marrow) cavity
Periosteum
Central arteriolar branches of nutrient artery
Interstitial lamellae
Capillaries in haversian canals
Sinusoids exposed by removal of marrow meshwork
Emissary vein
Nutrient artery passes into nutrient foramen of diaphysis
Peripheral arteriolar branch of nutrient artery gives rise to capillaries that enter Volkmann's canals of cortical (compact) bone
Capillaries in Volkmann's canals
Nutrient artery eventually anastomoses with distal metaphyseal arteries
Concentric lamellae of secondary osteon (haversian system)
Inner circumferential lamellae
F. Netter M.D.
©CIBA-GEIGY

As the long bone grows in width (or if it is subjected to changing stress patterns), remodeling occurs. The initial step in bone remodeling is the removal of portions of osteons by the activity of osteoclasts. After the bone is removed, new lamellae are deposited in new concentric layers, from outside inward, until a complete new osteon is formed. In mature bone, extensive remodeling causes the destruction and formation of many generations of osteons. The newest ones can be recognized by their complete outer circumference. In some osteons of earlier generations, a portion of the outer border has been removed and occupied by the outer border of a new osteon. Indeed, in some old osteons, these destructive processes have occurred so frequently that only small portions of the original lamellae remain (see Section II, Plate 13). These remnants, which may also include portions of circumferential lamellae, are called interstitial lamellae.

Cortical bone is remodeled by bone cells on the periosteal, endosteal, and haversian canal surfaces. These surfaces are called bone envelopes, or remodeling bays. The *periosteal surface* is responsible for the growth in bone width. The *endosteum*, which lines the medullary cavity of long bones, carries out complex metabolic and structural activities throughout life. These activities include phases of bone formation alternating with phases of bone resorption (Plate 39). Endosteal activity determines the diameter of the medullary canal, while the combined activities of the periosteum and endosteum determine the thickness of the bone cortex. The *haversian canal surface* is important in bone remodeling and is responsible for the density of the cortex.

Arterial blood supply to the bone cortex is predominantly centrifugal and is carried out by nutrient arteries entering from the medullary canal (Plate 22); periosteal arterioles supply approximately one-third of the outer cortex of a long bone. A highly developed anastomotic network connects the centripedal periosteal arterial system with the centrifugal endosteal arterial system. Venous drainage of the cortex is predominantly centripedal via a large plexus of veins in the medullary canal. (See pages 132–133 for a comprehensive discussion of blood supply to bone.)

Structure of Trabecular Bone

In contrast to the compact structure of cortical bone, *trabecular bone* is a complex network of intersecting curved plates and tubes (Plate 23). The bone within each trabecula is mature lamellar bone; the osteocytes are concentrically oriented and have a well-developed canalicular network.

Trabecular bone is typically located at the ends of a bone. Here, the well-defined medullary cavity of the shaft gives way to a different organization: bony trabeculae fill the entire cross section of the bone, occupying approximately 20% of its volume. In the proximal end of the femur, the trabeculae are quite regularly arranged, reflecting the direction of the principal mechanical stresses to which this bone is subjected.

Cortical bone accounts for 80% of skeletal bone mass, whereas trabecular bone constitutes the remaining 20%. However, because of the vast surface area of trabecular bone, its surface-to-volume ratio is approximately 10 times that of cortical bone.

Composition and Structure of Bone

(Continued)

Structure of Trabecular Bone

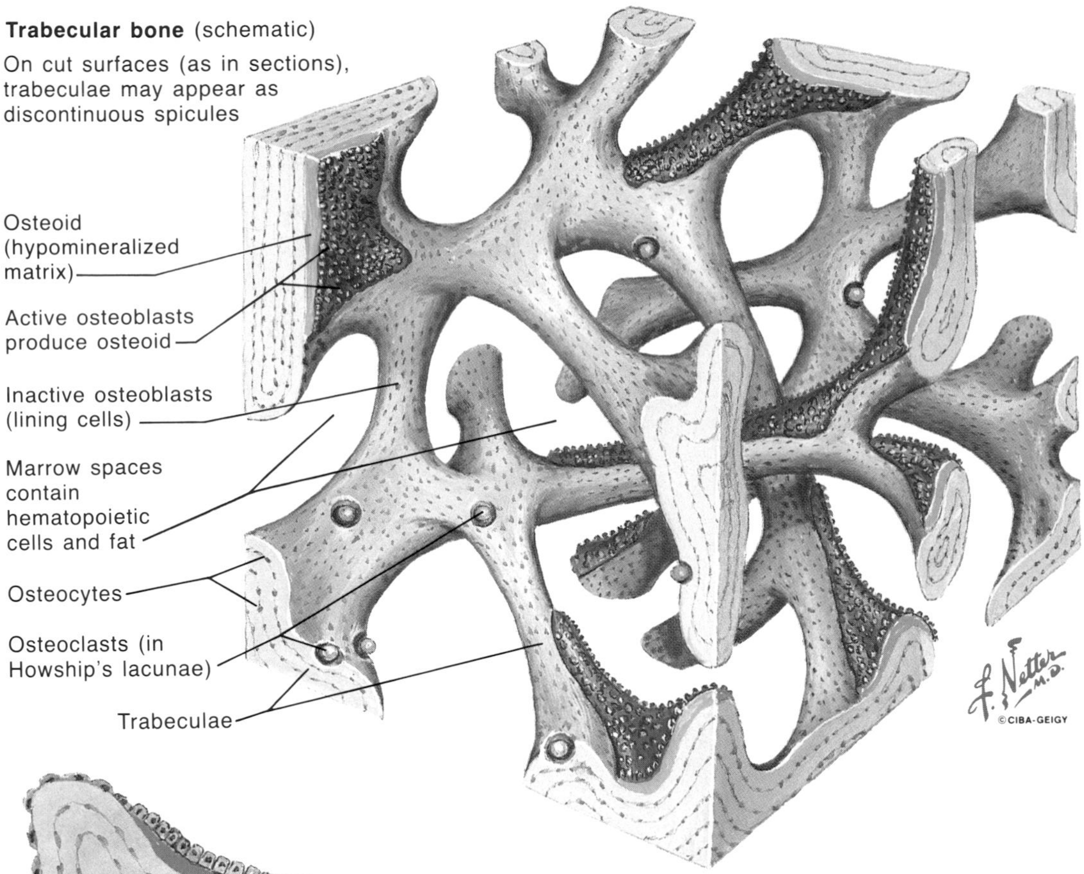

Trabecular bone (schematic)
On cut surfaces (as in sections), trabeculae may appear as discontinuous spicules
Osteoid (hypomineralized matrix)
Active osteoblasts produce osteoid
Inactive osteoblasts (lining cells)
Marrow spaces contain hematopoietic cells and fat
Osteocytes
Osteoclasts (in Howship's lacunae)
Trabeculae
F. Netter M.D.
©CIBA-GEIGY

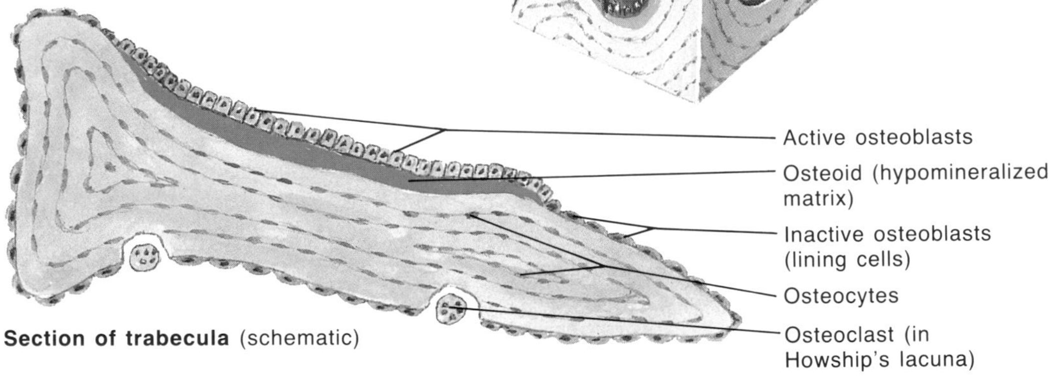

Active osteoblasts
Osteoid (hypomineralized matrix)
Inactive osteoblasts (lining cells)
Osteocytes
Osteoclast (in Howship's lacuna)

Section of trabecula (schematic)

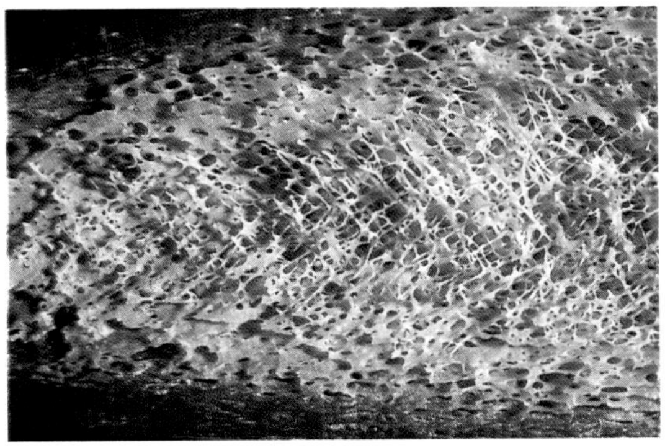

Cross section of trabecular bone (marrow elements removed). Trabecular bone in center; thin cortical (compact) bone at bottom

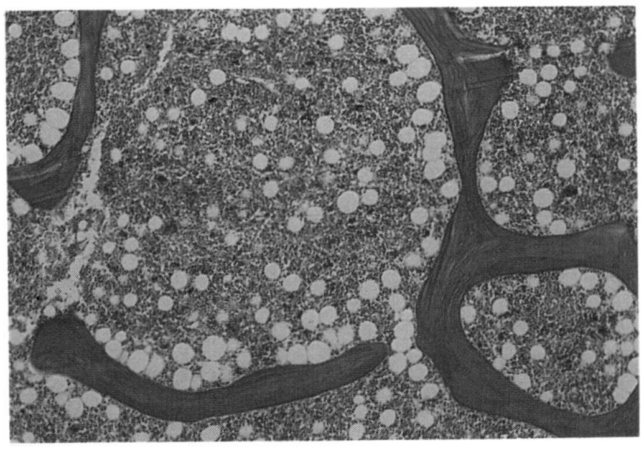

Photomicrograph of decalcified trabecular bone shows relationship of trabeculae to marrow (H and E, ×35)

The metabolic activity of trabecular bone is nearly eight times that of cortical bone, which may help to explain why disorders of skeletal homeostasis (metabolic bone diseases) have a greater effect on trabecular bone than on cortical bone. Recent observations indicate that the rate of remodeling in trabecular bone may vary greatly in different parts of the skeleton. For example, in adults, trabecular bone at the ends of long bones is in contact with a fatty marrow, whereas in the axial skeleton, it is in contact with a highly cellular hematopoietic marrow. These differences may help to account for the axial distribution of trabecular osteopenia.

Despite its apparent porosity and relatively small volume, trabecular bone is well adapted to resist compressive force, a capacity best exemplified by the vertebral body. In contrast, the structural properties of cortical bone are best suited to resist bending and torsional stresses.

In actively growing, or remodeling, trabecular bone, the direction of deposition can be determined by a row of osteoblasts on one border of the trabecula (Plate 23). The deposition of new bone by these osteoblasts is counterbalanced by the removal of bone by osteoclasts from the opposite surface of the trabecula. By this means of coordinated resorption and deposition, the position of a trabecula can shift within a bone.

With age, the balance between the rate of bone formation and the rate of bone resorption changes, leading to a decrease in bone mass. Any number of combinations can cause this effect. Evidence from kinetic studies indicates that after age 40, bone formation remains constant whereas bone resorption increases. Over several decades, through age-related bone loss (men and women) and postmenopausal bone loss (women), the skeletal mass may be reduced to 50% of what it was at age 30. If the bone density becomes so low that the skeleton can no longer withstand the mechanical stresses of everyday life, pathologic fractures may result. In an in vitro study, Carter and Hayes showed that the compressive strength of bone is proportional to the square of its apparent density; thus, if its density decreases by a factor of 2, its compressive strength decreases by a factor of 4. Many variables determine the fracture threshold, not the least of which is peak bone density at the time of skeletal maturity.

The biochemical composition and microscopic physical properties are similar in both cortical and trabecular bone. However, the macroscopic structure of bone produces markedly different physical properties that have broad variations in strength and stiffness to suit local physical requirements. Thus, the thin cortical shell supported by trabecular bone at the ends of long bones is well suited to sustain the concentrated loads in the joints, whereas the tubular cortical midshaft is better suited to support the large torsional and bending loads applied to this area.

All normal adult bone is lamellar bone, whether it has a cortical or a trabecular structure. In adults, immature woven bone, or fiber bone, is seen only in normal fracture healing or in pathologic conditions such as hyperparathyroidism or Paget's disease. □

Formation and Composition of Collagen

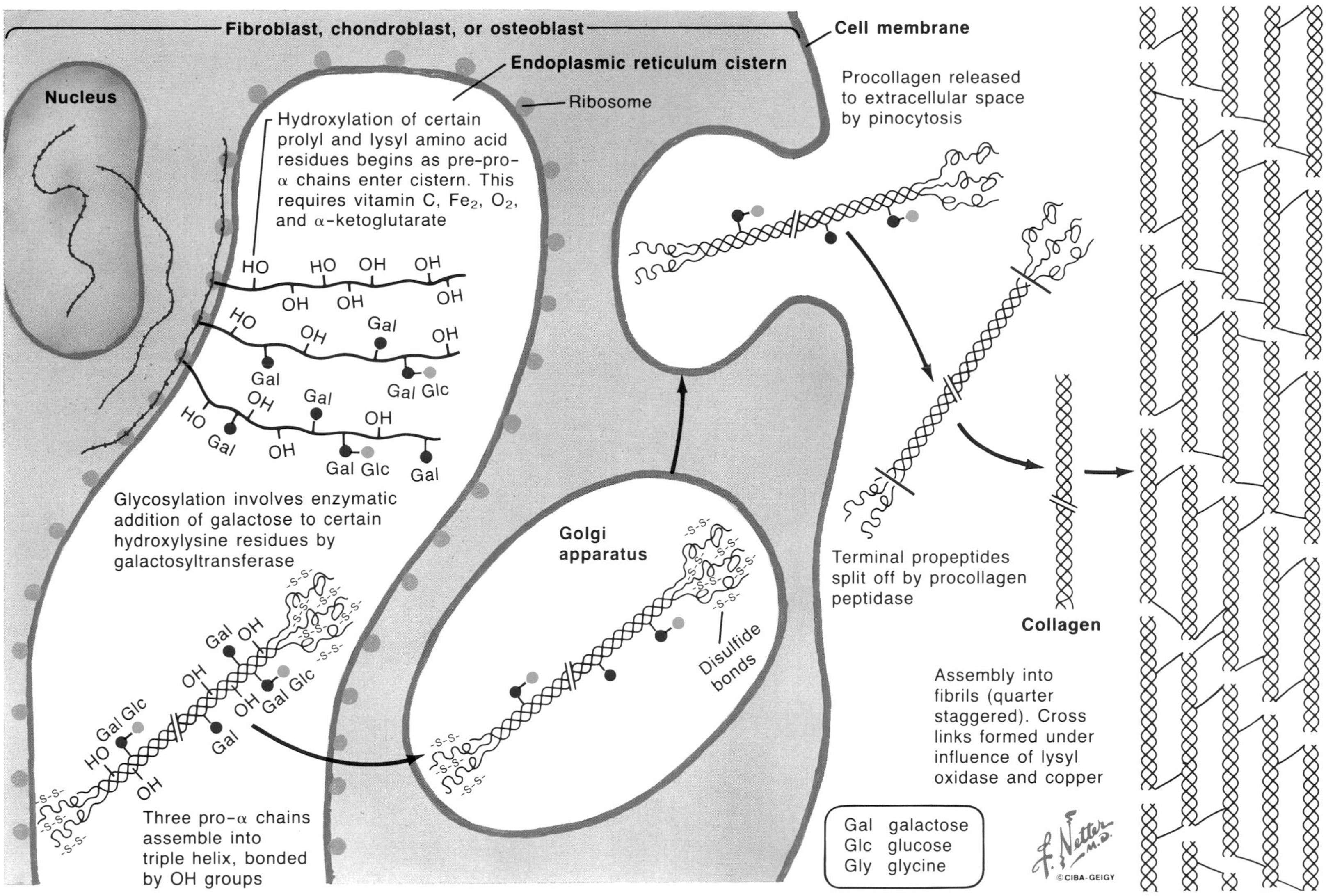

Structure of α chains

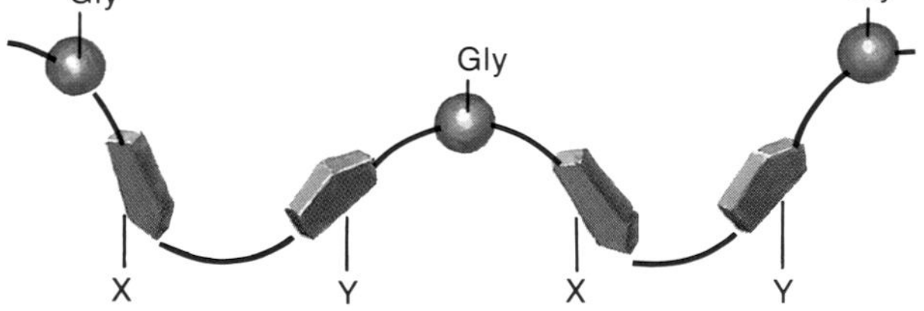

Each α chain comprises about 1,000 amino acids. Every third amino acid in chain is glycine, smallest of amino acids. Glycine has no side chains, which thus permits tight coil. X and Y here indicate other amino acids (X often proline; Y often hydroxyproline). Proline and hydroxyproline, respectively, constitute about 20% and 25% of total amino acids in each α chain

Types of collagen
(based on α chain composition of fibrils)

Type I

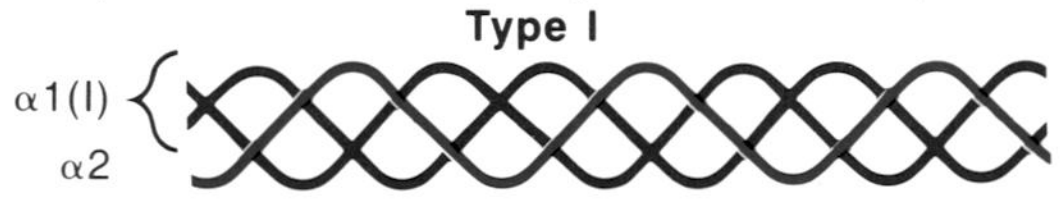

Two α1(I) chains and one α2 chain = $(\alpha 1[I])_2\ \alpha 2$; in bone, tendon, ligament, fascia, skin, artery, uterus

Type II

Three α1(II) chains = $(\alpha 1[II])_3$; in articular cartilage

Type III $(\alpha 1[III])_3$; in skin, artery, uterus, GI tract.
Type IV $(\alpha 1[IV])_3$; in basement membranes, lens capsule.
Type V $(\alpha B)_3$ or $(\alpha B)_2\ \alpha A$; in basement membranes, other tissues. At least 12 different collagen molecules identified

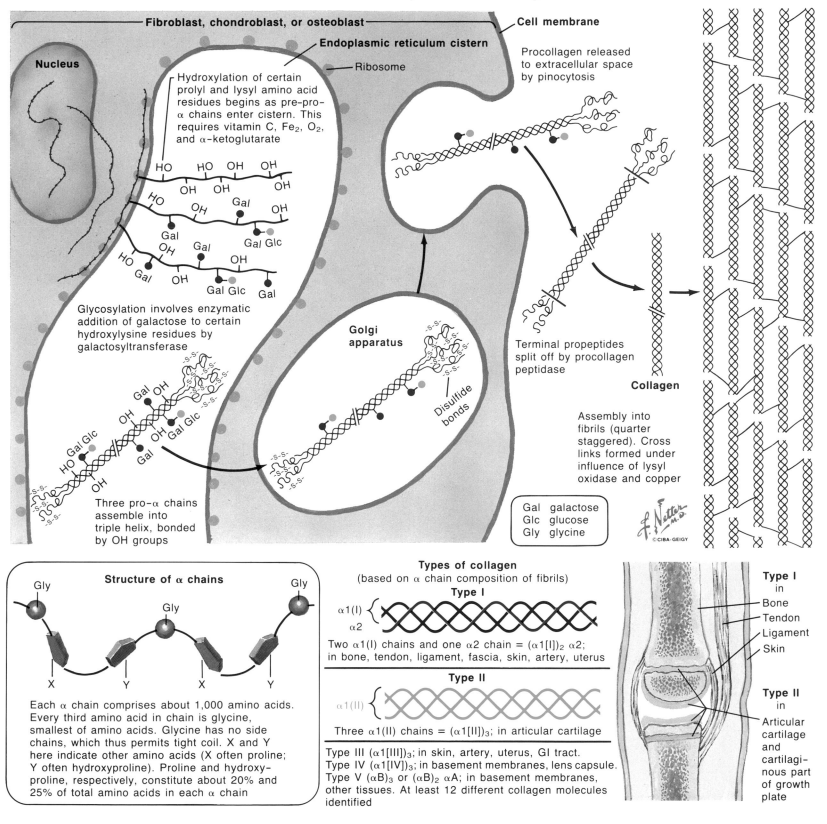

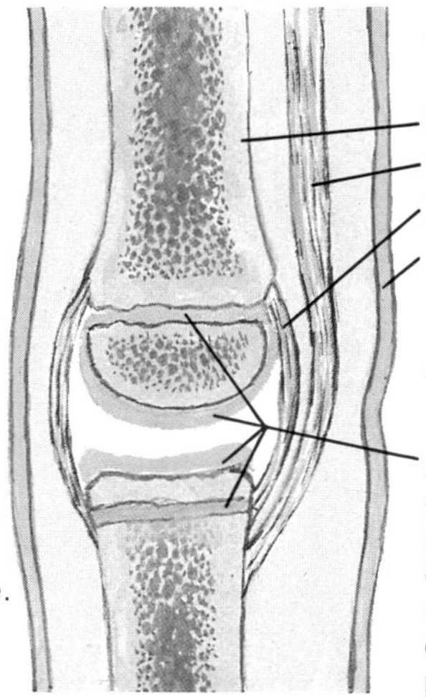

Formation and Composition of Collagen

Collagen is the most abundant and ubiquitous family of proteins in the body, and its members are major constituents of all connective tissues. One of its most amazing biologic properties is the ability to spontaneously self-assemble outside the cell into a variety of fibrillar and nonfibrillar forms. Collagen formation plays a vital role in the process of tissue repair.

Like members of most families, the collagens share certain similarities but also possess characteristic differences. At least 11 types of collagen macromolecules have been identified. The most abundant, *type I* collagen, is found in skin, fasciae, tendons, ligaments, and bones. *Type II* collagen is found in all forms of cartilage (including growth plate and articular cartilage) and in the nucleus pulposus of the intervertebral disc. *Type III* collagen is less abundant but is generally found with type I collagen. *Type IV* collagen, the most abundant nonfibrillar type, is a major constituent of the basement membrane. *Type V* collagen, the least abundant fibrillar collagen, is found in the placenta and blood vessels. In addition, there is a variety of minor collagens whose distribution is still unclear.

All collagen molecules are composed of three polypeptide α chains wrapped around one another like a three-stranded rope.

Although each collagen type is a unique combination of three α chains (in the form of either a homotrimer or a heterotrimer) and although each α chain is encoded by a unique gene and possesses a unique amino acid sequence, there are many similarities among the various types. Each α chain has a primary structure that is relatively simple and highly repetitive; a good example is glycine-X-Y[334]. Glycine, the smallest amino acid, occupies every third amino acid position, and

Formation and Composition of Collagen

(Continued)

X and Y are often proline and hydroxyproline, respectively. This repeating triplet allows the α chains to form a tight helix.

Despite the relatively simple structure of collagen, its biosynthetic pathway is complex and can be divided into intracellular and extracellular events. The intracellular assembly begins with the transcription of messenger ribonucleic acid (mRNA) from a collagen gene. The pro-α chains of procollagen are synthesized on the rough endoplasmic reticulum (RER) by translation of the corresponding mRNA. Subsequently, many posttranslational modifications occur. Hydroxylation of specific proline and lysine residues takes place in the lumen of the RER while the α chains are attached to ribosomes. This process requires the presence of vitamin C, oxygen, ferrous iron, α-ketoglutarate, and the appropriate hydroxylation enzymes—prolyl 4-hydroxylase, prolyl 3-hydroxylase, and lysyl hydroxylase. Deficiencies in cofactors or enzymes can lead to deficits in secretion. Other posttranslational modifications involve glycosylation of hydroxylysine residues, glycosylation of the carboxyl (C)-terminal propeptide, and formation of disulfide bonds among the C-terminal propeptides of the three α chains. The last process initiates the formation of the triple helix in the lumen of the RER. Once the triple helix is formed, procollagen is transported from the RER to the Golgi apparatus and packaged for secretion by exocytosis.

Once outside the cell, the C-terminal and N-terminal propeptides of some collagens are cleaved by procollagen peptidase C and procollagen peptidase N, respectively. Following cleavage of the terminal propeptides, these collagen molecules spontaneously precipitate as fibrils under physiologic conditions. The 68-nm periodic staining of fibrillar collagen results from the staggered structure of the fibrils. Collagen structures are stabilized by intermolecular cross-linking between lysine or hydroxylysine residues in adjacent collagen molecules.

As the major component of the connective tissue matrix, collagen determines the tensile strength of tissues, provides the framework for tissues, limits the movement of other components of tissue and matrix, induces platelet aggregation and clot formation, regulates the deposition of hydroxyapatite crystals in bone, and plays an important role in the regulation and differentiation of various cells and tissues.

Heritable disorders of collagen metabolism include Ehlers-Danlos syndrome, Marfan's syndrome, and osteogenesis imperfecta. Acquired disorders include scurvy, keloid formation, proliferative scar formation, atherosclerosis, pulmonary fibrosis, and cirrhosis. □

SECTION III PLATE 25 Slide 3721

Formation and Composition of Proteoglycan

The normal growth and development of cartilage are critically dependent on the presence of proteoglycans in the cartilage matrix. Cartilage proteoglycans are relatively large, complex macromolecules. They are composed of a core protein to which are attached a variable number of glycosaminoglycan chains consisting of repeating, negatively charged disaccharide units of varying length. The core protein of the major cartilage proteoglycan monomer contains more than 2,000 amino acids and is divided into three regions, a hyaluronic acid–binding region, a keratan sulfate–rich region distal to the binding region, and a most distal chondroitin sulfate–rich region. (Chondroitin sulfate and keratan sulfate are examples of glycosaminoglycans.) The average cartilage proteoglycan may contain more than 80 to 100 chondroitin sulfate chains, 50 to 60 keratan sulfate chains, and a variable number of smaller oligosaccharide chains. Thus, the total molecular weight of a proteoglycan monomer is in the millions.

To make the situation even more complex, most proteoglycans in the cartilage matrix are not monomers but form aggregates with two other matrix components, hyaluronic acid and link proteins. This aggregate consists of a variable number of proteoglycan monomers, which are noncovalently attached to a single hyaluronic acid chain through the hyaluronic acid–binding region and stabilized by the noncovalent association of link proteins to the monomer and the hyaluronic acid. The typical molecular weight of such an aggregate approaches 100 million.

All the components of the proteoglycan aggregate are synthesized by chondroblasts and chondrocytes and are then transported by them for extracellular self-assembly. For example, the core protein is synthesized and some oligosaccharides are added in the rough endoplasmic reticulum (RER), whereas the synthesis of remaining oligosaccharides and glycosaminoglycan chains and their subsequent sulfation occur in the Golgi apparatus. While the proteoglycan monomer is being assembled and secreted, the cartilage cells are also synthesizing and secreting the hyaluronic acid and link proteins required for the final self-assembly of the giant proteoglycan aggregate, one of the largest molecular complexes in nature.

The proteoglycan content of articular cartilage, elastic cartilage, and fibrocartilage gives these tissues many of their characteristic properties. For example, the critical mechanical properties of hyaline cartilage—resiliency and stiffness to compression—exist because the gigantic proteoglycan aggregates possess the ability to sequester water.

In solution, the electronegative charges on the sulfated sugars cause the monomer side chains to repel one another and to attract water. The water molecules in the extracellular fluid act as dipoles: the positively charged hydrogen atoms are attracted to the negatively charged sulfate domain, and the negatively charged oxygen atoms of the water molecules repel one another.

The result is an enormous hydration sphere that can resist compression and reabsorb water after compressive forces are removed. The association of collagen molecules gives this system a tensile strength, restrains movement of the proteoglycan aggregates, and limits the maximal absorption of water. Because cartilage is avascular, diffusion of nutrients and exchange of waste products occur through the tissue fluid. The tissue fluid of articular cartilage exchanges freely with the water and solutes of synovial fluid. □

Formation and Composition of Proteoglycan

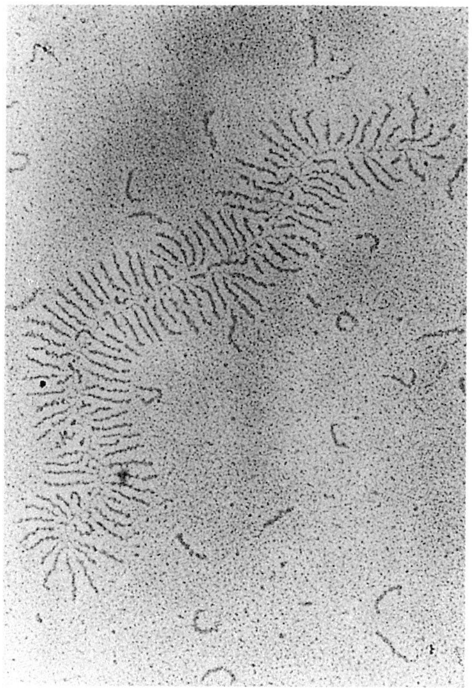

Electron micrograph of large, aggregated proteoglycan molecule from epiphyseal cartilage. Numerous closely spaced monomers bound to central hyaluronic acid filament; free monomers surround aggregate (×50,000)

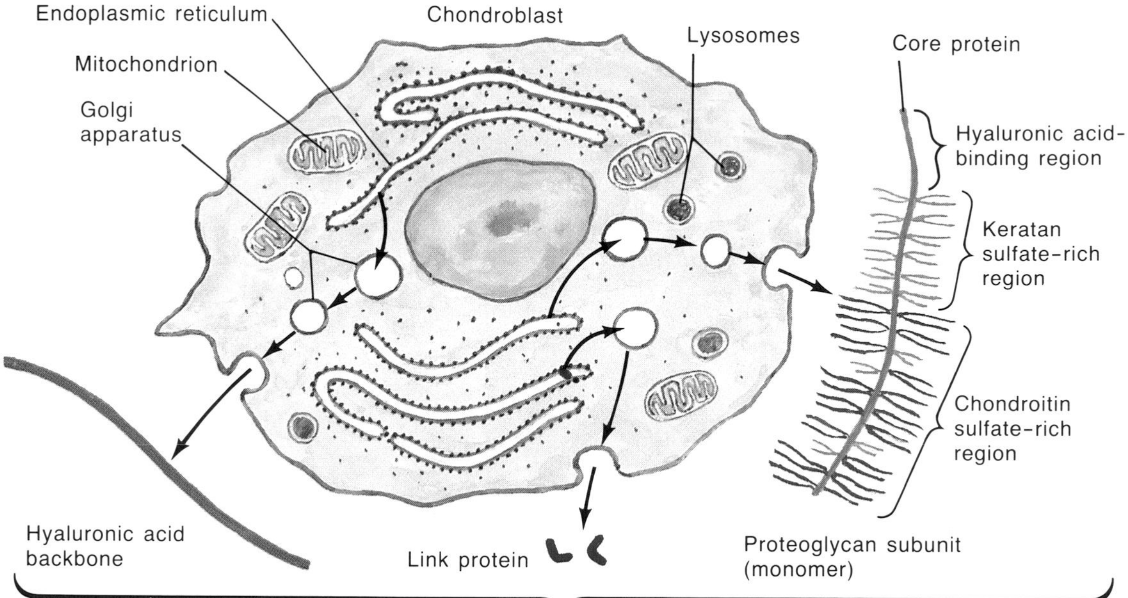

Endoplasmic reticulum
Chondroblast
Lysosomes
Core protein
Mitochondrion
Golgi apparatus
Hyaluronic acid-binding region
Keratan sulfate-rich region
Chondroitin sulfate-rich region
Hyaluronic acid backbone
Link protein
Proteoglycan subunit (monomer)

Each component synthesized separately in endoplasmic reticulum and transported out of chondroblast via Golgi apparatus for assembly into giant aggregated proteoglycan molecule of cartilage matrix

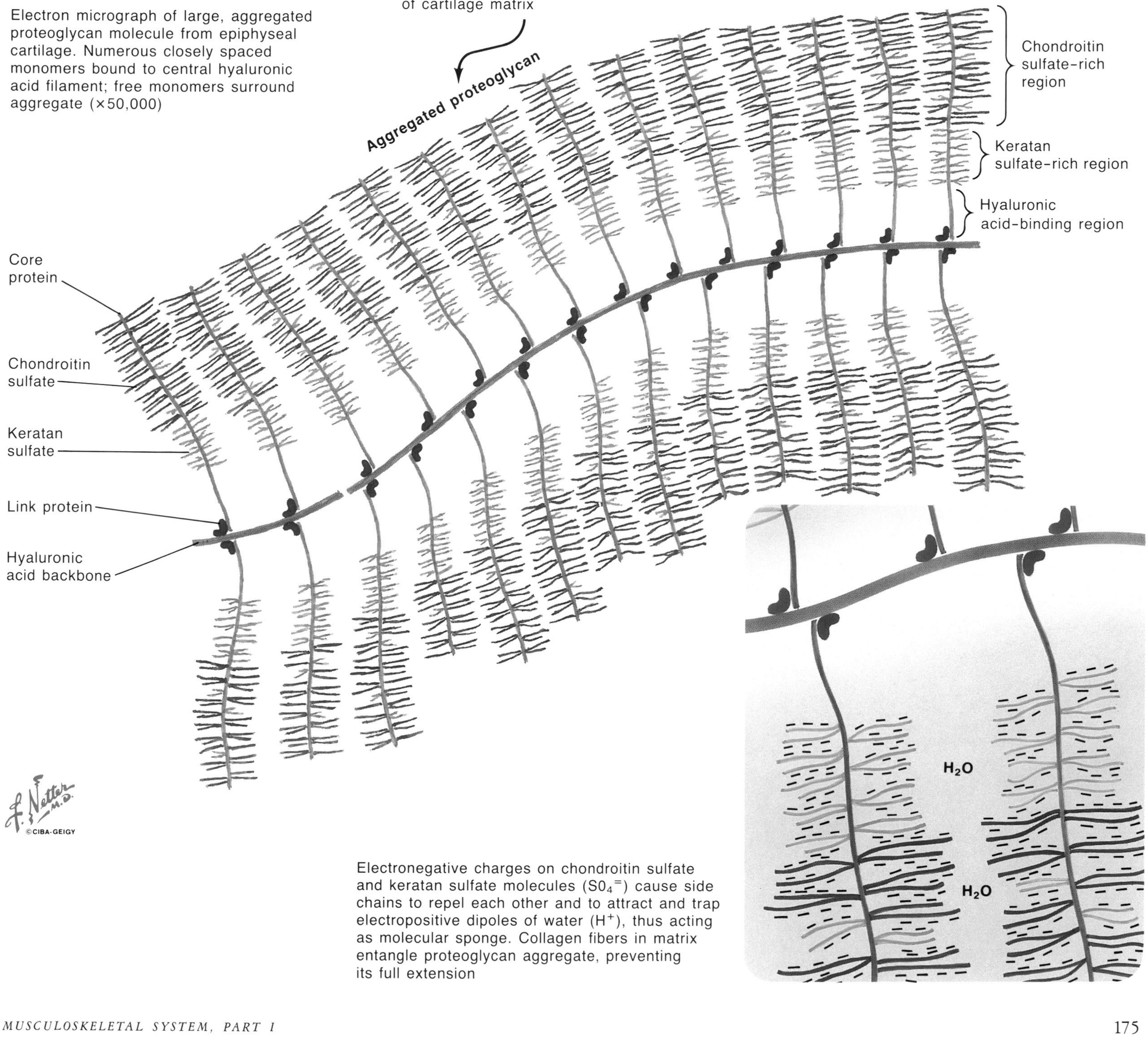

Aggregated proteoglycan
Chondroitin sulfate-rich region
Keratan sulfate-rich region
Hyaluronic acid-binding region
Core protein
Chondroitin sulfate
Keratan sulfate
Link protein
Hyaluronic acid backbone
H_2O
H_2O

Electronegative charges on chondroitin sulfate and keratan sulfate molecules ($SO_4^{=}$) cause side chains to repel each other and to attract and trap electropositive dipoles of water (H^+), thus acting as molecular sponge. Collagen fibers in matrix entangle proteoglycan aggregate, preventing its full extension

Structure and Function of Synovial Membrane

The synovial membrane, or synovium, is the vascular mesenchymal tissue that lines the joint space of all synovial (diarthrodial) joints. Only the cartilage and surfaces of the meniscus are not covered by synovial membrane. In normal joints, this tissue serves primarily to produce the joint fluid with its diverse components and to remove cellular and connective tissue debris from the joint space.

On gross examination, the synovial surface appears pale pink and shiny. Although some folds can be seen with the naked eye, the characteristic villi that increase the effective surface area of the synovial membrane are visible only on microscopic examination. One or two layers of cells, with their long axes generally lying parallel to the surface, line the synovial membrane; these lining cells are not jointed by intracellular junctions. The deeper tissue consists predominantly of loose connective tissue, fibrous tissue, or fat; thus, the associated synovial membrane is described as areolar, fibrous, or adipose. Fibrous synovial membrane is found in areas that need more strength but less flexibility.

Capillaries and venules lie immediately beneath the lining cells. Lymphatics, which are difficult to identify with standard light microscopy, are most abundant in areolar synovial membrane. Nonmyelinated nerve fibers extend from the capsule into the adventitia of the synovial blood vessels.

Ultrastructural and immunopathologic studies have added considerably to the understanding of the synovial membrane. The lining cell layer consists of some cells that are rich in rough endoplasmic reticulum (RER). These cells, called type B, are probably related to fibroblasts. Type B cells are most important because of their ability to secrete prostaglandins, collagenase, hyaluronic acid, and many other components of joint fluid.

Phagocytic cells (type A), which have prominent lysosomes, are now known to originate from monocytes. They often lie superficially to the type B cells. Some cells, which appear to have features of both type A and type B cells, are less well understood. Mast cells in perivascular areas, easily identified on electron microscopy, are a source of important vasoactive substances. Collagen (types I and III), fibronectin, and proteoglycans are present in the matrix.

Electron microscopic examination reveals that the superficial capillaries and venules have a fenestrated endothelium through which fluid, together with small amounts of low-molecular-weight protein, transudes to form the joint fluid. The addition of hyaluronic acid by the lining cells gives the joint fluid its characteristic viscosity. The deeper vessels, which have thicker walls, are the vessels through which most inflammatory cells emigrate. □

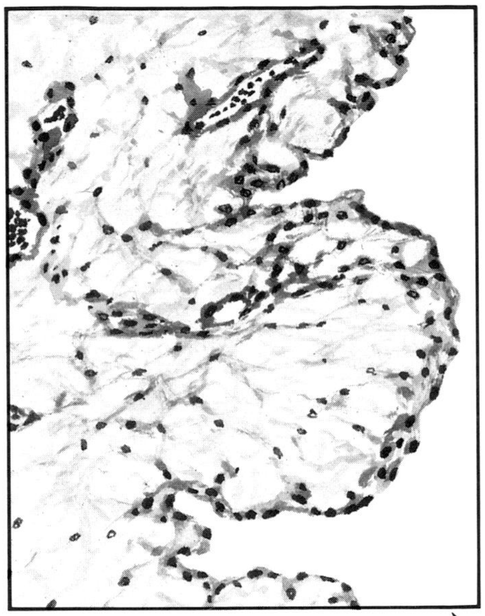
Areolar synovial membrane

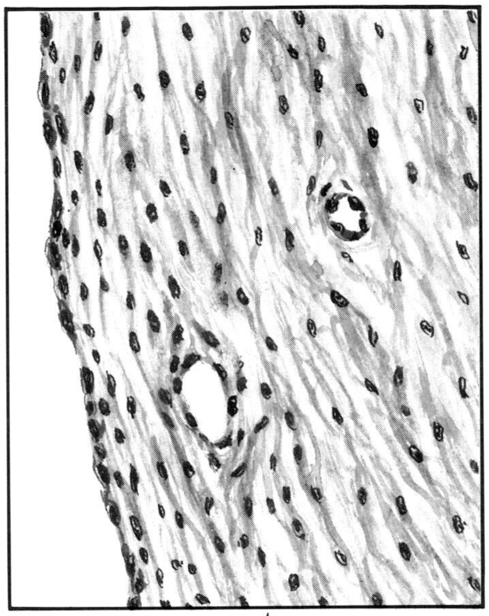
Fibrous synovial membrane

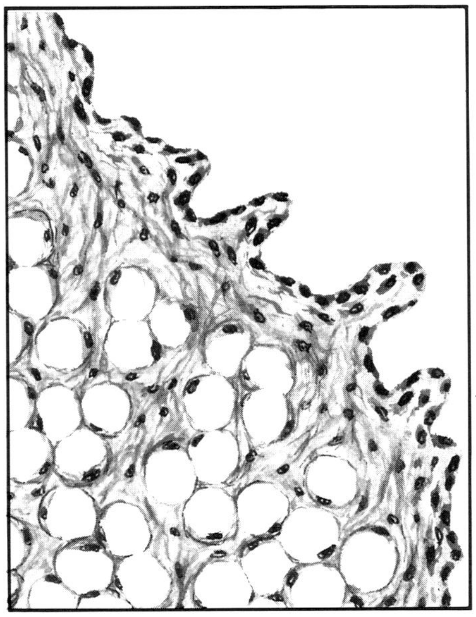

Adipose synovial membrane

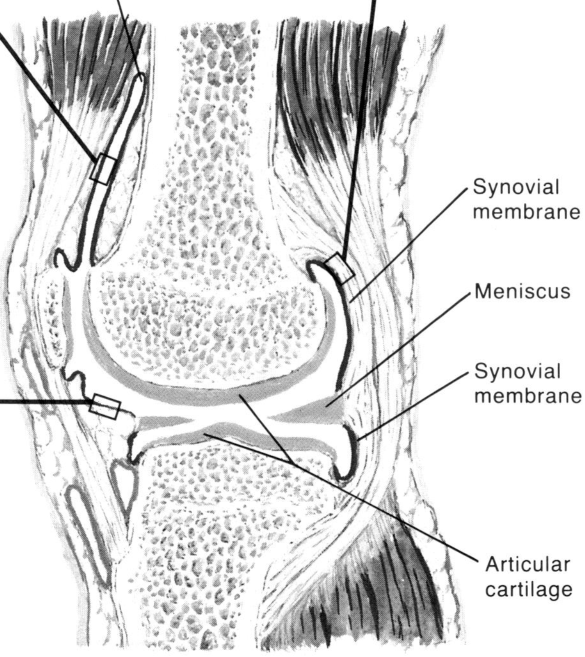
Synovial membrane
Synovial membrane
Meniscus
Synovial membrane
Articular cartilage

Sagittal section of knee joint (lateral to midline)

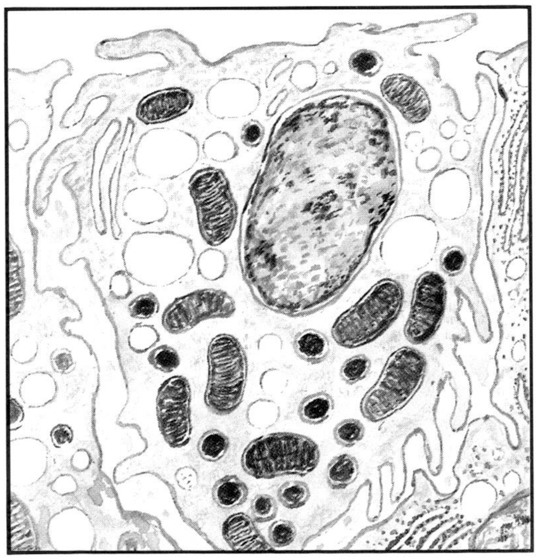

Electron micrograph shows type A cell of synovial lining characterized by many vacuoles, lysosomes (dark bodies), mitochondria, and filopodia but little endoplasmic reticulum, all evidence of macrophage and phagocytic function

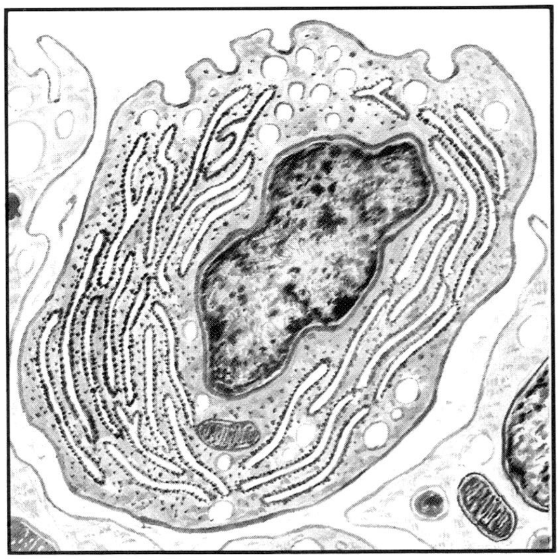

Type B cell of synovial lining shows much rough endoplasmic reticulum and pinocytotic vesicles, related to its presumed function of synthesizing and excreting hyaluronic acid and glycoprotein for synovial fluid

Histology of Connective Tissue

Histology of Connective Tissue

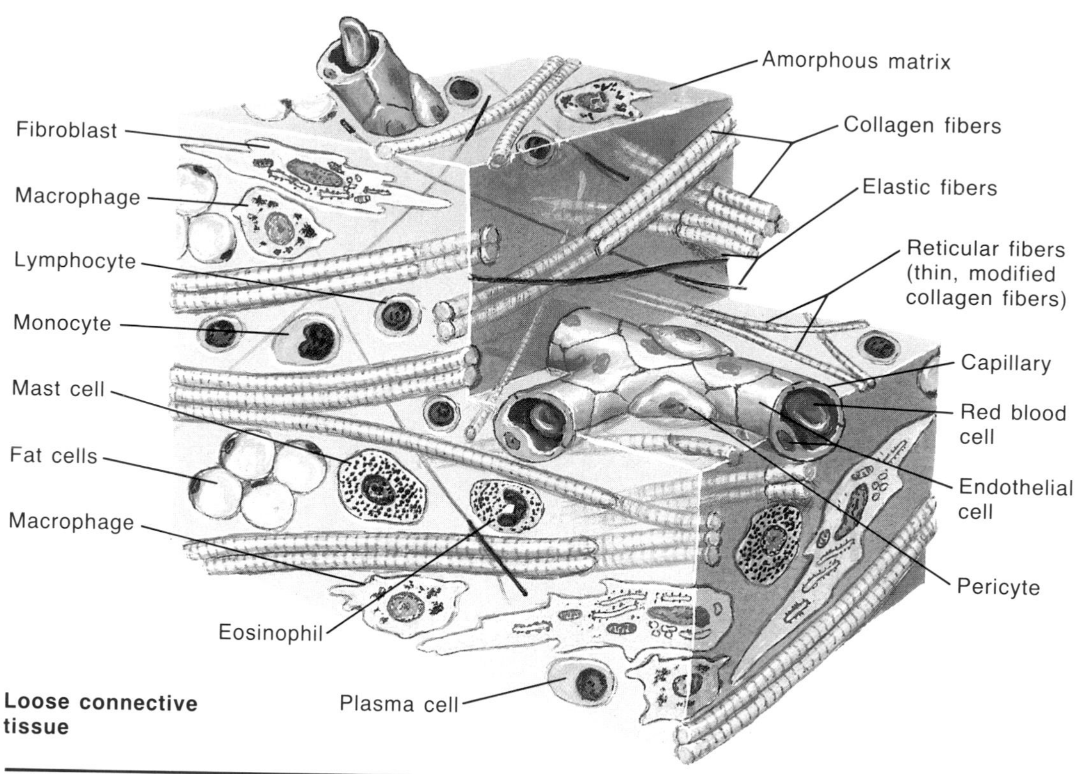
Amorphous matrix
Fibroblast
Collagen fibers
Macrophage
Elastic fibers
Lymphocyte
Reticular fibers (thin, modified collagen fibers)
Monocyte
Capillary
Mast cell
Red blood cell
Fat cells
Endothelial cell
Macrophage
Pericyte
Eosinophil
Loose connective tissue
Plasma cell

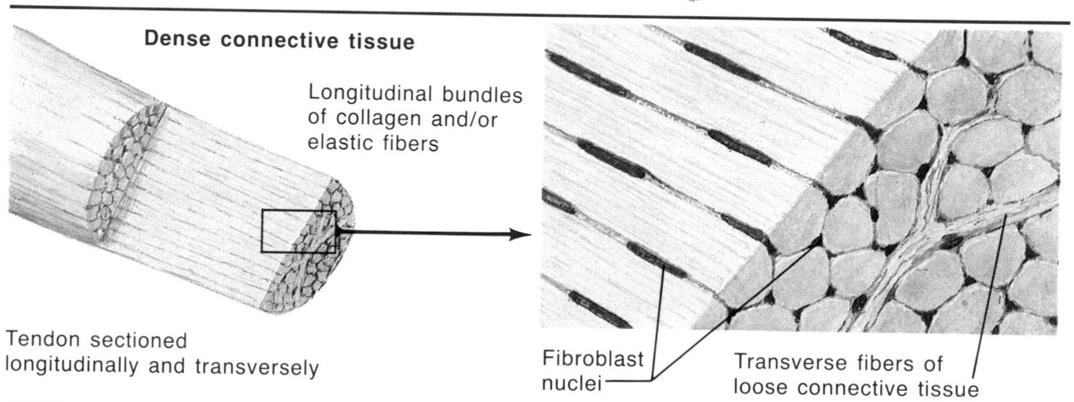
Dense connective tissue
Longitudinal bundles of collagen and/or elastic fibers
Tendon sectioned longitudinally and transversely
Fibroblast nuclei
Transverse fibers of loose connective tissue

Light microscopic longitudinal section of tendon shows fascicles (predominantly collagen) and fibroblasts (H and E)

Adult connective tissue comprises a family of tissue types that includes connective tissue proper, blood, cartilage, and bone. All tissues in this family originate from embryonic mesenchymal tissue. Connective tissue itself includes a range of recognizable histologic types, which are determined by the proportion of the various components in the tissue. The two extremes of the continuum of connective tissue are *loose connective tissue* and *dense connective tissue*. These two classifications are based on the proportion and density of the fibrous component of the tissue. Connective tissue may have a regular arrangement, as in the tendon, or an irregular arrangement, as in the dermis. Often, one component of the connective tissue predominates, such as the fat cell in adipose tissue.

Regardless of its histologic type, connective tissue is made up of two components, cells and extracellular matrix. Fibroblasts (which synthesize collagen), fat cells, fixed macrophages, mast cells, plasma cells, and some leukocytes are the principal cellular elements of connective tissue. Extracellular matrix consists of collagenous, elastic, or reticular fibers, and a variety of mucopolysaccharides (glycosaminoglycans), which are either sulfated or nonsulfated. These are often present as components of gigantic macromolecular assemblies called proteoglycans (Plate 25). Among the important sulfated compounds are chondroitin sulfate, keratan sulfate, and heparin. The principal nonsulfated compounds are chondroitin and hyaluronic acid. In addition, connective tissue also contains blood and lymphatic vessels and nerves that vary in number and size.

The composition and organization of a particular connective tissue largely depends on its function. Loose, or areolar, connective tissue is found throughout the body wherever biologic packing material is needed. It is well vascularized and highly cellular, with a large proportion of matrix. The fibrous component varies in amount and orientation, depending on the mechanical stresses in the region. Adipose tissue is a specialized form of loose connective tissue in which fat cells predominate. Although the fat cell is sometimes thought of as a type of fibroblast, it is increasingly being regarded as a separate cell type that, when it does not contain lipid, resembles a fibroblast.

Dense connective tissue, found in tendons and ligaments, is poorly vascularized. Parallel bundles of densely packed collagen fibers are the predominant elements. Alongside the bundles of collagen fibers are inactive-appearing fibroblasts with densely staining, elongated nuclei. In some species, the fibroblasts of tendons have the tendency to form nodules of ectopic cartilage and bone after injury. Occasionally, nodules of fibrocartilage are seen in tendons. □

Bone Homeostasis

The skeleton acts as a dynamic mineral reserve bank in which the body stores its ionized calcium and phosphorus (in the form of phosphate ions) in a metabolically stable and structurally useful way. Although each bone cell population—the osteoblasts, osteoclasts, and osteocytes—is under the direction of numerous endocrine factors and is influenced by various local biochemical and bioelectric factors, the cells themselves are endowed with genetic instructions that determine their ability to form, resorb, or maintain bone (Plate 28).

Calcium Requirements. The body regulates few functions with greater fidelity than the concentration of calcium in the extracellular fluid. Although extracellular calcium represents less than 1% of the body's calcium stores, it is the metabolically active component that is critically important for numerous life-sustaining processes that include enzymatic reactions, mitochondrial function, cell membrane maintenance, intercellular communication, interneuronal transmission, neuromuscular transmission, muscle contraction, and blood clotting. An elaborate endocrine system maintains the serum calcium concentration within a very narrow physiologic range. When this level falls, even momentarily, it is restored to normal through the parathyroid hormone (PTH)–vitamin D system, which increases calcium absorption in the gastrointestinal tract and reabsorption in the kidneys, and the resorption of bone (Plate 33).

The National Research Council of the National Academy of Sciences has established recommended daily allowances (RDAs) of calcium for all age groups. These values, which may be conservative estimates, reflect the average amount of calcium required to maintain a positive calcium balance and to prevent withdrawal of the mineral stores banked in bone. For young adults, the RDA of calcium is 750 to 1,000 mg. Unfortunately, large-scale dietary surveys of women with osteoporosis show that the average American woman consumes less than 500 mg/day. On such a calcium-deficient diet, the body mobilizes calcium from its skeletal reserve for its daily needs by increasing secretion of PTH and 1,25-dihydroxyvitamin D, or $1,25(OH)_2D$ (the hormonally active metabolite of vitamin D).

Absorption of calcium from the upper gastrointestinal tract becomes less efficient with age because of decreases in baseline levels and secretory reserves of $1,25(OH)_2D$; thus, older persons need more dietary calcium to maintain a calcium balance. Healthy premenopausal women over age 30 may require as much as 1,000 mg/day, and pregnant women and women over age 50 need more than 1,500 mg/day. Lactating women need 2,000 mg/day to prevent untimely catabolism of bone.

Calcium consumption may be inadequate in persons with lactase deficiency who avoid eating dairy products, the primary source of dietary calcium. Increased protein intake accelerates calcium excretion by the kidney. Therefore, the high-protein diet common in western industrialized countries may be a contributing factor to accelerated bone loss in these populations.

Vitamin D Requirements. The vitamin D metabolite $1,25(OH)_2D$ helps to maintain normal serum calcium and phosphate levels by increasing the absorption of these substances from the intestine and the osteoclastic resorption of bone. About half of our vitamin D comes from dietary sources (particularly from vitamin D–enriched milk), and the remainder comes from a reaction in the skin stimulated by ultraviolet radiation. Only a few natural foods, such as fish liver oils, contain vitamin D. Elderly persons frequently have a mild vitamin D deficiency because of their meager exposure to sunlight, decreased intake of milk and other dairy products, and decreased intestinal absorption of vitamin D. The RDA of vitamin D is 400 IU for young adults, but no more than 800 IU is recommended for elderly persons. Larger amounts may cause hypercalcemia. Premature infants may require 500 to 1,000 IU daily.

In addition to age- and sex-related effects on bone loss, endocrine and metabolic changes affect bone homeostasis, leading to osteoporosis. In both osteoporotic and normal elderly women, $1,25(OH)_2D$ levels are normal. However, in the elderly patients, the kidney's production of $1,25(OH)_2D$ in response to PTH infusion is

Dynamics of Bone Homeostasis

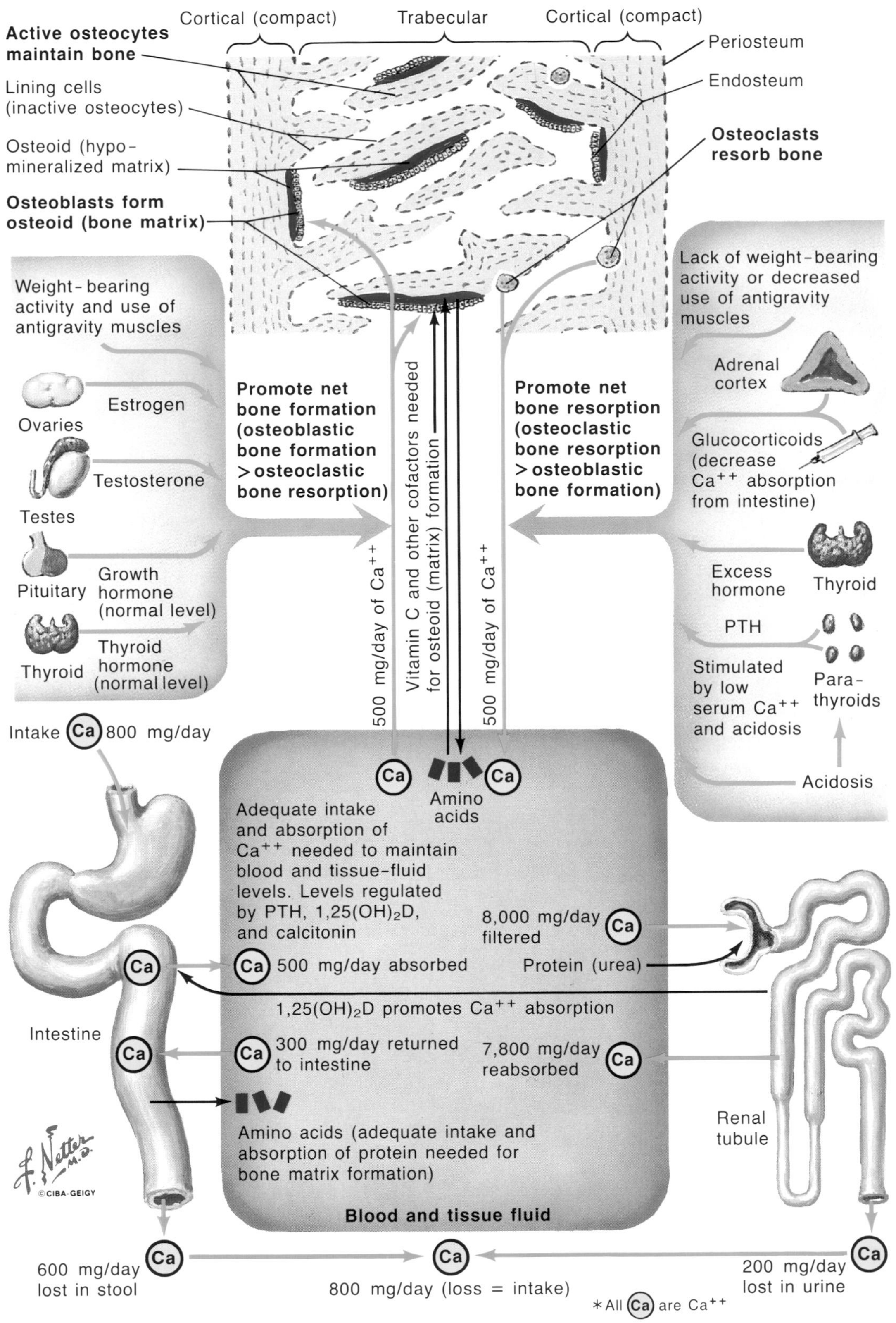

Bone Homeostasis

(Continued)

impaired. Also, in postmenopausal women, bone-resorbing cells (osteoclasts) appear to be excessively responsive to endogenous PTH. Although estrogen receptors have not been found in bone cells, estrogen does decrease the sensitivity of osteoclasts to PTH. Estrogen deficiency is thus the major cause of bone loss in the early postmenopausal period.

Regulation of Calcium and Phosphate Metabolism

Despite considerable daily variations in calcium intake, the body maintains the serum calcium concentration at a remarkably constant level. The primary homeostatic regulation of serum calcium concentration is under the control of the parathyroid gland, which produces PTH; the thyroid gland, which secretes calcitonin; and the kidney, which provides $1,25(OH)_2D$ from less active vitamin D metabolites. Other factors contributing to the regulation of serum calcium levels include hormones (gonadal steroids, thyroid hormone, growth hormone, glucocorticoids, insulin); vitamins C and D; proteins (albumin, calcium-binding protein, vitamin D–binding protein); phosphate; small inhibitors of mineralization such as pyrophosphates; and pH of blood (Plate 29).

The exchange of calcium between extracellular fluid and bone can be considered kinetically (Plate 28). Through a slow phase of bone formation and resorption, approximately 1,000 mg/day of calcium is exchanged between bone and extracellular fluid; this represents approximately one-tenth of 1% of the total calcium reserve (1,000–1,200 g). Most of the calcium in bone does not readily diffuse into the extracellular compartment but must be mobilized by endocrine-regulated, cell-mediated bone resorption. There is no known sustained biologic process by which the body can remove just the mineral component of bone. Thus, when the body needs to withdraw calcium from its mineral reserves, it can do so only by resorbing bone (mineral component plus organic matrix). Through the process of *coupling*, bone formation increases and osteoblasts are stimulated to fill in the resorption defect, although this repair may not be complete in the elderly.

Endocrine-mediated bone formation and bone resorption involve more than the stimulation of existing differentiated bone cells; these processes are dependent on the transformation of undifferentiated stem cells in both osteoblast and osteoclast cell lines. Thus, bone formation and bone resorption are contingent not only on the metabolic activity of each cell but also on the recruitment of cells to the job.

The major function of calcitonin, a hormone secreted by the parafollicular cells of the thyroid gland, is to inhibit osteoclastic bone resorption in response to elevated serum calcium levels. The biologically active vitamin D metabolite $1,25(OH)_2D$ regulates intestinal absorption of calcium and phosphate and activates bone resorption by stimulating the recruitment of osteoclast precursors (preosteoclasts). There is also some evidence that $1,25(OH)_2D$ also stimulates the recruitment of osteoblasts. Conversion of 25-hydroxyvitamin D, or 25(OH)D, to $1,25(OH)_2D$ in the kidney is controlled by the enzyme 25-hydroxyvitamin D-1α-hydroxylase, or 25(OH)D-1α-OH_{ase}, and stimulated maximally by increased serum PTH and decreased serum phosphate levels.

Efficient gastrointestinal absorption of calcium depends primarily on daily calcium intake, vitamin D status, and age. As the major storage form of vitamin D is 25(OH)D, the serum level of 25(OH)D is an excellent indicator of the body's total vitamin D reserves. With the vitamin D axis intact, efficiency of calcium absorption increases if calcium intake decreases. Some calcium secreted into the intestine is absorbed, but much of it passes into the stool, together with any unabsorbed calcium.

The kidney filters about 8,000 mg of calcium daily and, under the influence of PTH, reabsorbs more than 95%. For each additional gram of calcium ingested, only about 50 mg of additional calcium appears in urine. Thus, the urinary calcium level is better determined by the rate of calcium absorption by the intestine than by the amount of calcium ingested. □

Regulation of Calcium and Phosphate Metabolism

	Parathyroid hormone (PTH) (peptide)	$1,25(OH)_2D$ (steroid)	Calcitonin (peptide)
Hormone	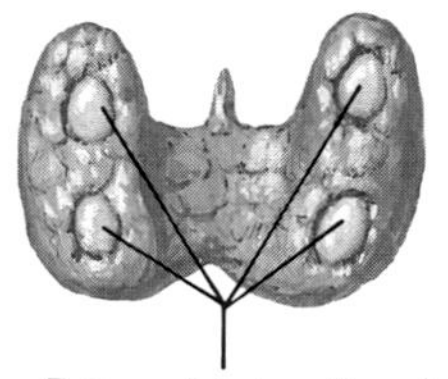 From chief cells of parathyroid glands	 From proximal tubule of kidney	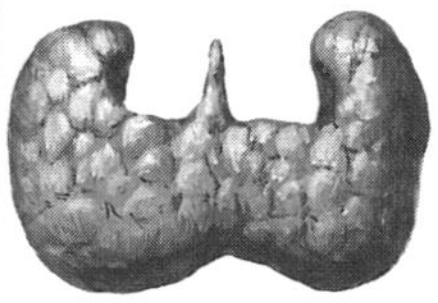 From parafollicular cells of thyroid gland
Factors stimulating production	Decreased serum Ca^{++}	Elevated PTH Decreased serum Ca^{++} Decreased serum P_i	Elevated serum Ca^{++}
Factors inhibiting production	Elevated serum Ca^{++} Elevated $1,25(OH)_2D$	Decreased PTH Elevated serum Ca^{++} Elevated serum P_i	Decreased serum Ca^{++}
End organs for hormone action: Intestine 	No direct effect Acts indirectly on bowel by stimulating production of $1,25(OH)_2D$ in kidney	Strongly stimulates intestinal absorption of Ca^{++} and P_i	?
Kidney 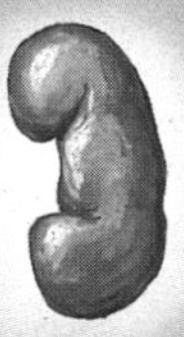	Stimulates 25(OH)D-1α-OH_{ase} in mitochondria of proximal tubular cells to convert 25(OH)D to $1,25(OH)_2D$ Increases fractional reabsorption of filtered Ca^{++} Promotes urinary excretion of P_i	?	?
Bone 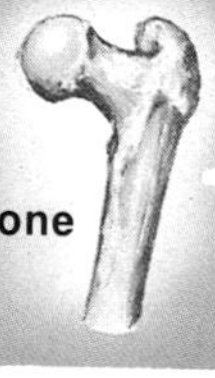	Stimulates osteoclastic resorption of bone Stimulates recruitment of preosteoclasts	Strongly stimulates osteoclastic resorption of bone	Inhibits osteoclastic resorption of bone ? Role in normal human physiology
Net effect on calcium and phosphate concentrations in extracellular fluid and serum	**Increased serum calcium** **Decreased serum phosphate**	**Increased serum calcium** **Increased serum phosphate**	**Decreased serum calcium (transient)**

F. Netter M.D.

Regulation of Bone Mass

Effects of Bone Formation and Bone Resorption on Skeletal Mass

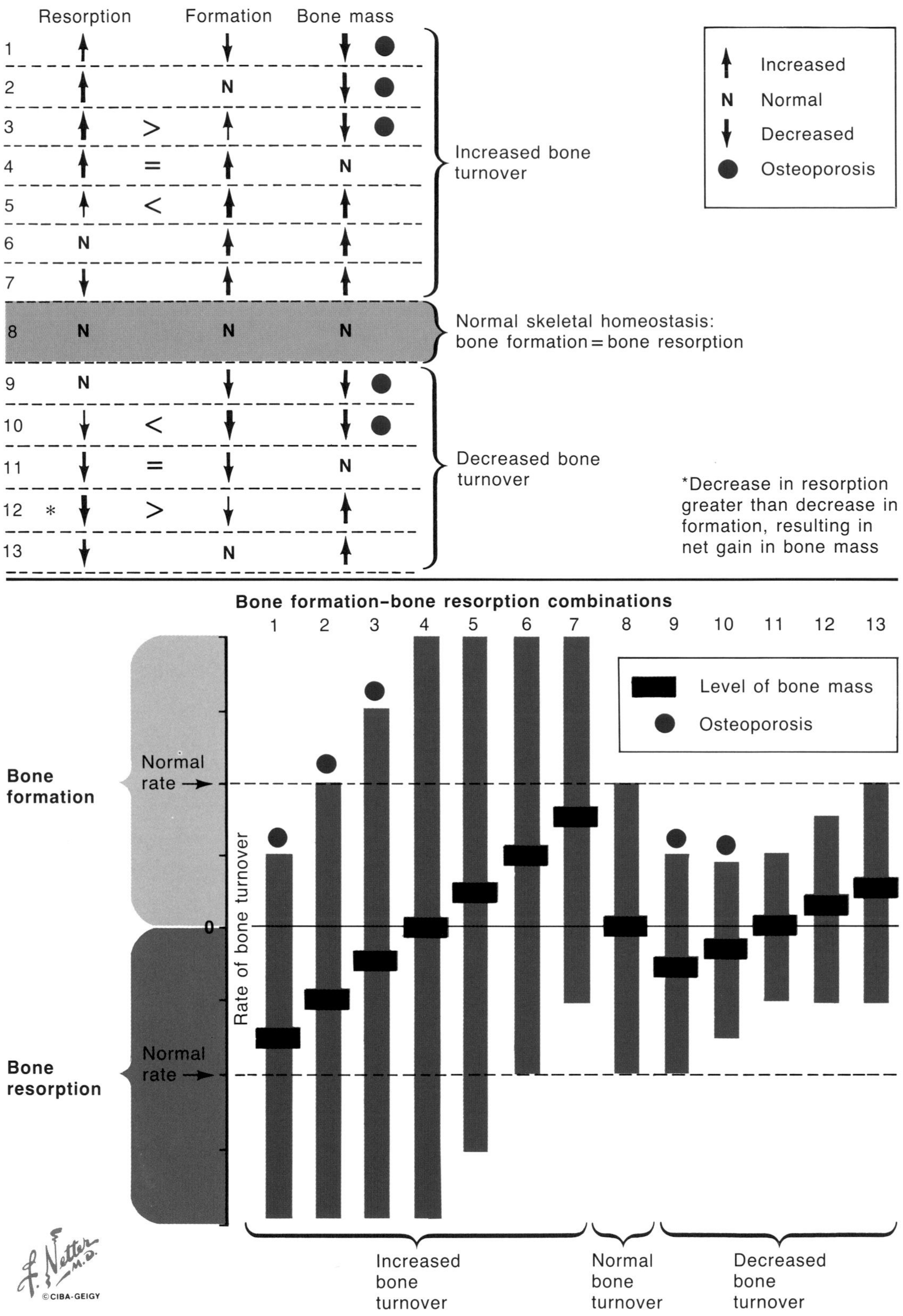

Effects of Bone Formation and Resorption on Skeletal Mass

Once peak bone mass has been achieved by the middle of the fourth decade, net bone mass remains relatively constant throughout early adult life. However, living bone is never metabolically at rest but constantly remodels and reappropriates its mineral stores along lines of mechanical stress. The factors controlling bone formation and resorption are not well understood, but in the normal adult skeleton, the two processes are balanced in a process called *coupling*, so that net bone formation equals net bone resorption.

Unless a significant metabolic insult occurs to bone early in life, the peak adult bone mass will exceed the threshold for spontaneous fracture. With the introduction of noninvasive techniques to measure bone density, a new and useful definition of osteoporosis has emerged: a mass-per-unit volume of normally mineralized bone that falls below a population-defined threshold for spontaneous fracture.

However, if by adulthood peak bone mass far exceeds the threshold for spontaneous fracture, significant osteoporosis can develop only if there is a net loss of bone mass; in other words, an *uncoupling* of bone formation and bone resorption processes.

If net bone resorption exceeds net bone formation, bone mass declines with time (Plate 30). With a rapid rate of bone turnover, the decrease of bone mass is rapid. With a slow rate of bone turnover, the decline in mass is correspondingly slow. Similarly, if net bone formation exceeds net bone resorption (as in normal bone growth and even after longitudinal growth ceases), bone mass increases with time.

Some possible combinations of bone formation/resorption are shown in the table in the upper half of Plate 30 and graphically presented in the lower half. *Example 8* represents normal bone turnover: bone formation and bone resorption are appropriately coupled, leading to a stable bone mass with no net change (black bar on middle line). *Example 4* represents greatly increased bone turnover, but again because formation and resorption are appropriately coupled, there is no net change in bone mass. This state of bone remodeling is seen in the active stage of Paget's disease (see Section IV, Plates 41–43).

Example 11 shows decreased bone turnover, but here, too, bone formation and resorption are appropriately coupled, with no net change in bone mass. This state of bone remodeling might be seen in the inactive phase of normal bone turnover (example 8), or it might reflect a decrease in appropriately coupled bone remodeling that represents a normal variation.

Increased Bone Turnover. *Example 1* depicts a state of severe uncoupling (decreased bone formation and increased bone resorption), which causes a net decline in bone mass over time. Such an imbalance in formation and resorption occurs in chronic glucocorticoid excess, which may be endogenous (Cushing's syndrome) or iatrogenic (see Section IV, Plate 23).

Example 2 illustrates increased bone turnover with a normal rate of bone formation but an increased rate of bone resorption. To use the bank analogy illustrated in Plate 31: with time, net withdrawal from the bone bank exceeds net deposition, resulting in a decreased skeletal reserve (bone mass). This state is represented in the lower half of Plate 30 by the black bar below the zero line and by the red dot above the green column, indicating resultant osteoporosis. (An example of this situation is the rapid increase in bone resorption that occurs after menopause.)

Regulation of Bone Mass

(Continued)

Four Mechanisms of Bone Mass Regulation

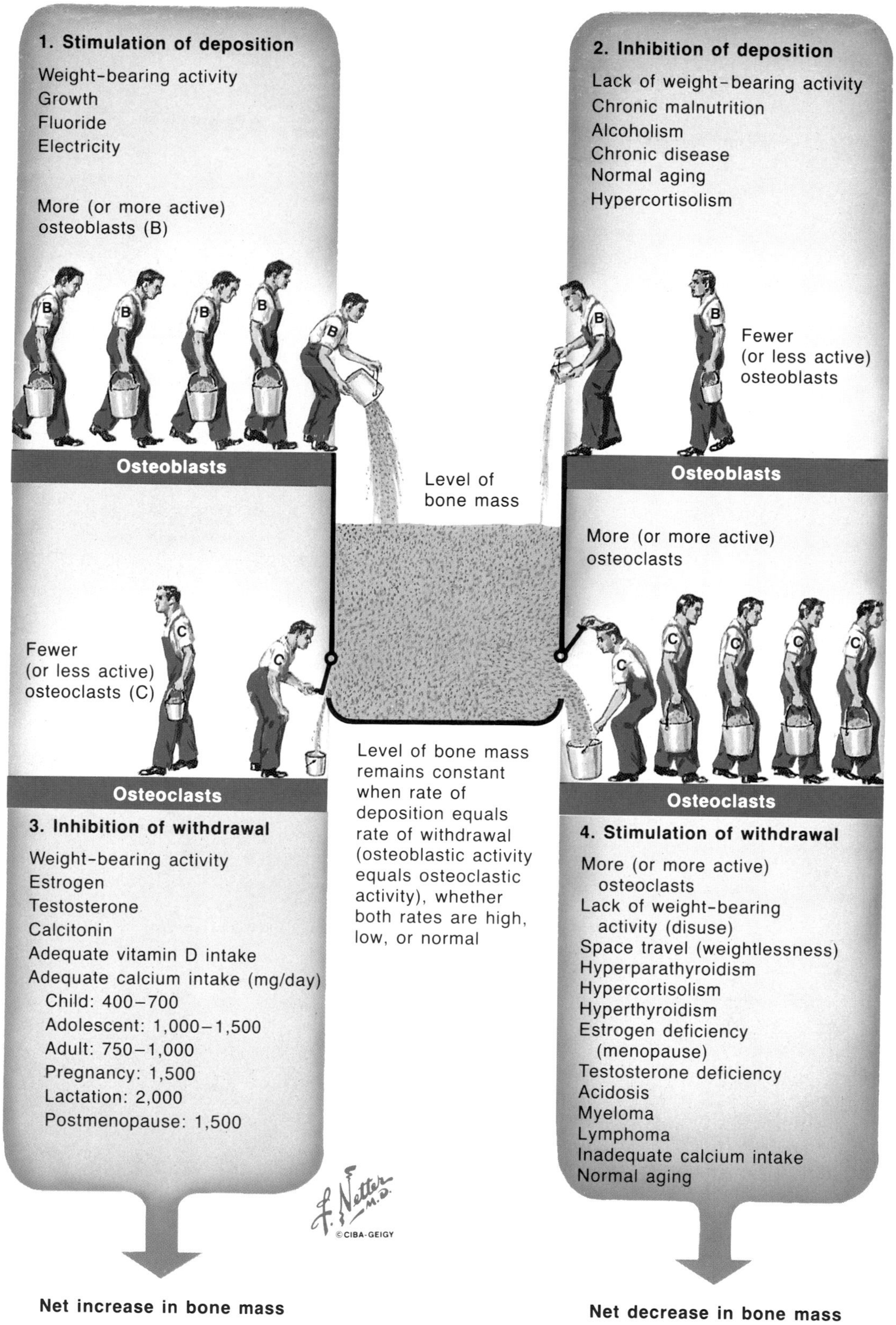

In *example 3*, both bone formation and resorption are increased, but the rate of resorption is greater than the rate of formation. Thus, there is a net loss of bone mass over a period of time, which leads to osteoporosis. Mild hyperthyroidism, mild hyperparathyroidism, and a chronic dietary calcium deficiency can cause increased bone turnover, with resorption exceeding formation. In severe hyperthyroidism and hyperparathyroidism, remodeling rates can be even greater than those shown in example 3, with a correspondingly greater difference between formation and resorption and thus a greater net loss of bone mass.

Example 5 illustrates a state of increased bone formation and resorption, with formation exceeding resorption. In this circumstance, bone mass increases with time. Although such a generalized state of remodeling is unlikely to occur in the adult skeleton, a localized (focal) phenomenon may be seen in the osteoblastic stage of Paget's disease.

Example 6 illustrates normal bone resorption paired with an increase in the rate of bone formation, which leads to a net increase in bone mass with time. This is typical during normal growth and development, especially between late adolescence (when longitudinal growth ceases) and approximately age 35 (when peak adult bone mass is reached).

Example 7 shows the highly uncoupled state of increased bone formation and decreased bone resorption that results in a net increase in bone mass. Such a state of remodeling is seen in the adult, or autosomal dominant, form of osteopetrosis (see Section IV, Plates 39–40). In this disorder, the reduced rate of bone resorption is due to the relative failure of osteoclastic bone resorption; the rate of bone formation may be normal or increased. This form of uncoupling may also be induced by pharmacologic doses of fluoride administered to stimulate bone formation and stabilize bone apatite crystal and render it more resistant to breakdown.

Decreased Bone Turnover. *Example 9* shows a state of normal bone resorption with decreased bone formation, which leads to reduced bone mass and osteoporosis. This can occur after exposure to a poison or toxin that affects the osteoblasts, as with use of certain chemotherapeutic agents and chronic alcohol abuse.

In *example 10*, both bone formation and resorption are decreased but bone resorption still exceeds bone formation. This is seen when calcium supplementation helps to diminish age-related bone loss.

Example 13 shows a positive bone balance resulting from a normal rate of bone formation and a decreased rate of bone resorption. This occurs in women receiving replacement therapy with calcium and estrogen or calcium and calcitonin soon after menopause. These agents, administered independently or in combination, act to decrease bone resorption. The coupling process then causes bone formation rates to adjust downward accordingly (*example 12*); eventually, bone mass stabilizes at a much lower rate of bone remodeling (*example 11*).

Four Mechanisms of Bone Mass Regulation

Plate 31 illustrates four basic mechanisms of bone mass regulation: (1) stimulation and (2) inhibition of bone formation, or deposition, and (3) stimulation and (4) inhibition of bone resorption, or withdrawal. However, in many cases, early changes will be checked by concomitant stimulation of the opposing bone cell population—stimulation of osteoclasts (resorption) leads to stimulation of osteoblasts (formation). □

Normal Calcium and Phosphate Metabolism

Normal Calcium and Phosphate Metabolism

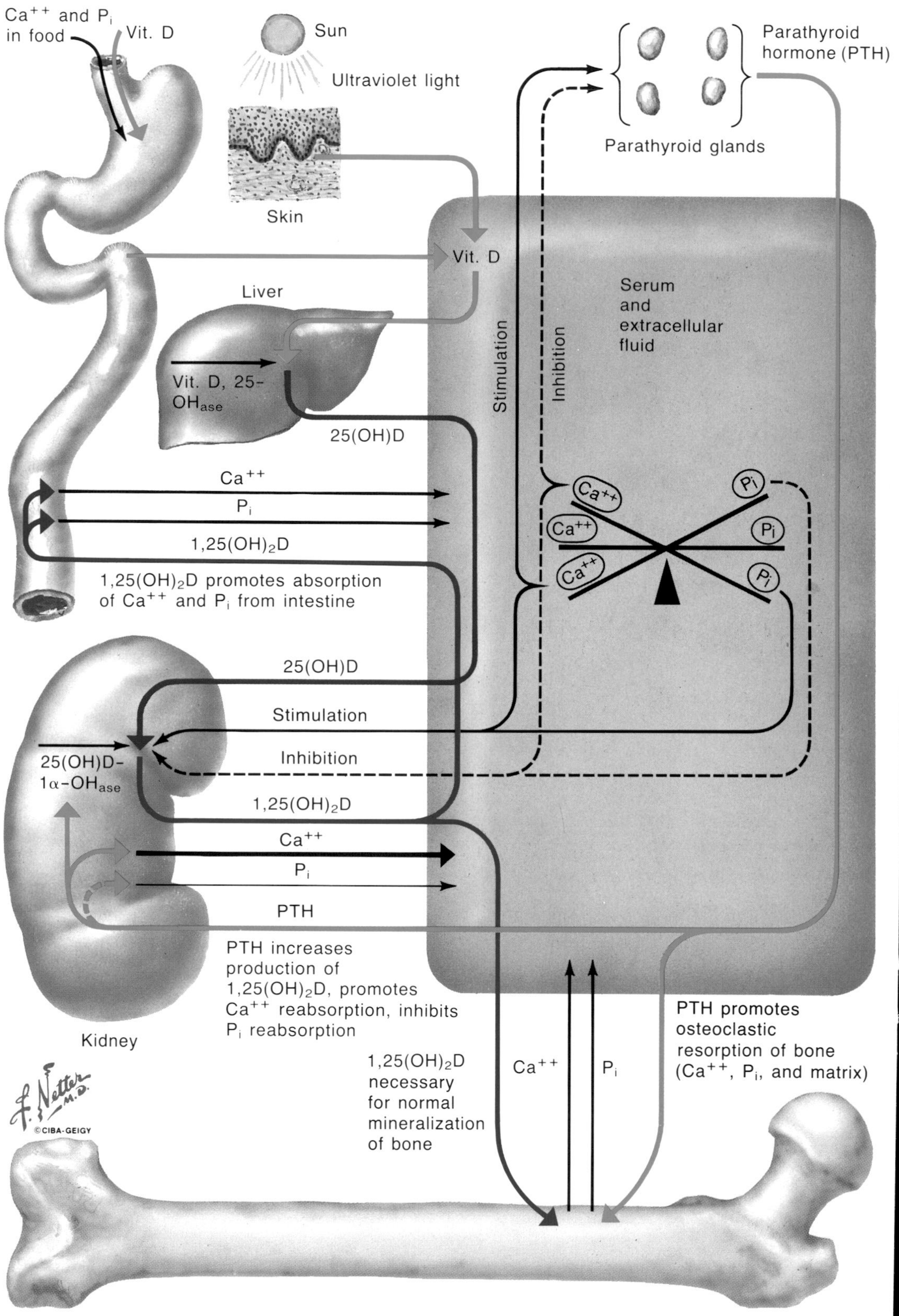

Three major "truths" govern the metabolism of calcium and phosphate and in large measure explain systems that are closely guarded by a complex homeostatic mechanism. The truths are distinct, representing chemical, physiologic, and biologic facts.

1. Calcium phosphate is not freely soluble in water. In fact, at the pH of body fluids and the concentrations of calcium and phosphate ions found in the extracellular space, the critical solubility product is probably exceeded but precipitation is prevented by some not-too-well-defined inhibitor systems. For obvious reasons, the body jealously guards the critical (and "metastable") concentrations of calcium and phosphate. (Otherwise, we would "turn to stone!")

2. The irritability, contractility, and conductivity of skeletal and smooth muscles, and the irritability and conductivity of nerves are exquisitely sensitive and inversely proportional to the concentration of calcium. (The relationship between the calcium level and cardiac muscle is an equally sensitive but direct one.) The body goes to great lengths to protect these important systems from the danger of hypocalcemia or hypercalcemia.

3. Absorption of calcium across the gut cell wall, reabsorption of filtered calcium by the renal tubule, and resorption of calcium released from the apatite crystal of bone cannot take place without a transport system. This system, at least for the gut cell, is a polypeptide of low molecular weight synthesized at the endoplasmic reticulum and named calcium-binding protein (CBP) by Wasserman. The mechanism by which CBP is synthesized and acts to transport calcium from the lumen of the distal duodenum (and proximal jejunum) across the cell wall and into the extracellular fluid is partly dependent on the parathyroid hormone (PTH)—which acts through adenyl cyclase and cyclic adenosine monophosphate (cyclic AMP) to render the membrane more porous for calcium—and on a low serum concentration of cytosol phosphate (a high concentration inhibits calcium absorption). However, synthesis of CBP is principally dependent on 1,25-dihydroxyvitamin D, or $1,25(OH)_2D$, which enhances the transcription of messenger ribonucleic acid (mRNA) for the synthesis of the polypeptide.

Perhaps the greatest advance in the understanding of the homeostatic mechanisms of calcium and phosphate in the last 3 decades has come from studies on the actions of vitamin D by DeLuca and others. Provitamin D_2 (ergosterol) is ingested or provitamin D_3 (7-dehydrocholesterol) is synthesized from cholesterol by the liver, and both are stored in the skin. Sunlight at wavelengths of approximately 315 nm (ultraviolet range) activates the provitamins into vitamin D_2 (calciferol) or vitamin D_3 (cholecalciferol), respectively, which are then transported to the liver. Here, they are acted upon by a vitamin D,25-hydroxylase (vitamin D, 25-OH_{ase}) to form 25-hydroxyvitamin D, or 25(OH)D. (Since both D_2 and D_3 are treated identically and act in the same way, the numeric designator has been omitted.) The 25(OH)D then travels to the renal tubule and, in response to lowered calcium, high PTH, and low phosphate levels in serum, is transformed by 25-hydroxyvitamin D-1α-hydroxylase, or 25(OH)D-1α-OH_{ase}, into the highly potent polar metabolite $1,25(OH)_2D$. If a surplus of either calcium or phosphate is present (indicated by a high

Normal Calcium and Phosphate Metabolism

(Continued)

serum concentration of calcium or phosphate, or both, or a low PTH level), an alternate pathway is selected in which the $25(OH)_2D$ is acted on by $25(OH)D\text{-}24\text{-}OH_{ase}$ and the far less potent 24, $25(OH)_2D$ is synthesized.

In a balanced diet, calcium and phosphate are ingested in adequate amounts. However, calcium is difficult to obtain except from dairy products, especially milk, and many otherwise reasonable diets may be calcium deficient. On the other hand, phosphate is present in almost all foods, and dietary deficiencies are uncommon. Accessory factors promoting absorption of calcium from the gut include an acid pH, a low serum phosphate concentration (to avoid exceeding the critical solubility product—see the first "truth"), and the absence of chelators such as phytate, oxalate, or excessive free fatty acids. Transport across the gut cell wall is controlled principally by the interaction of PTH, which renders the cell more permeable to luminal calcium, and $1,25(OH)_2D$, which activates the transport polypeptide CBP.

Reabsorption of filtered calcium from the proximal tubule obeys the same rules. A diminished PTH level or decreased synthesis of $1,25(OH)_2D$ leads to decreased tubular reabsorption of calcium, whereas a high level of PTH and an increased level of $1,25(OH)_2D$ enhance reabsorption. Although both PTH and $1,25(OH)_2D$ are apparently able to cause resorption of bone, the vitamin D metabolite is at least partly responsible for the normal mineralization of bone.

The mechanism of phosphate absorption is far less selective than that of calcium absorption but also appears to be at least partly dependent on the vitamin D metabolites. Since dietary intake of phosphate varies widely and absorption is almost unrestricted, the first "truth" suggests that humans stand poised on the brink of metastatic calcification and ossification because of a high, uncontrollable intake of phosphate. In fact, the renal excretory mechanisms exert a fine-tuned control over phosphate levels. In addition to both a tubular maximum and tubular secretion, tubular reabsorption of phosphate is exquisitely and inversely responsive to the concentration of PTH. Thus, PTH, which acts to *increase* tubular reabsorption of calcium, *diminishes* tubular reabsorption of phosphate, thus avoiding the potential disaster associated with exceeding the critical solubility product.

PTH plays a critical role in regulating levels of calcium and phosphate in serum and extracellular fluid. If the normal calcium concentration is not maintained, the diminished level signals the parathyroid glands to produce more PTH. The release of PTH, which is almost entirely dependent on the calcium level, has six separate functions, five of which are designed to correct the calcium deficit in serum and extracellular fluid. The six functions of PTH are as follows:

1. Increasing the synthesis of $1,25(OH)_2D$ in the kidney.
2. Acting at the level of the gut cell (with vitamin D) to increase absorption of calcium.
3. Acting at the level of the renal tubule (with vitamin D) to increase tubular reabsorption of filtered calcium.
4. Acting at the level of bone (with vitamin D) to break down crystals of hydroxyapatite, releasing calcium and phosphate.
5. Activating and increasing the population of osteoclasts, which destroy not only the hydroxyapatite crystals but also massive segments of organic and inorganic bone material, thus releasing both calcium and phosphate.
6. After flooding the system with phosphate (see 4 and 5), lowering the tubular reabsorption of phosphate to reduce the potential danger of violating the critical solubility product.

A brief mention of two other homeostatic systems should be included in this discussion, namely, the body's response system to hypercalcemia and hyperphosphatemia.

The standard physiologic mechanism that controls hypercalcemia (see first and second "truths," above) is twofold: (1) "Turnoff" of the vitamin D-PTH-calcium–sparing system, resulting in limited production of $1,25(OH)_2D$, greatly diminished PTH elaboration, diminished gut absorption of calcium, resorption of bone, and greatly diminished tubular reabsorption of calcium. (2) Increased elaboration of calcitonin, a hormone of low molecular weight secreted by the parafollicular cells (C cells) of the thyroid gland, which, at least in theory, acts to lower the serum calcium concentration. This is achieved principally by diminishing the osteoclast population and activity and, to some extent, by reducing gastrointestinal absorption. However, it should be clearly noted that although the second mechanism may be well developed in avian species and although administration of exogenous, nonspecies-specific calcitonin may have a profound effect on the skeleton, the natural mechanism in humans appears to be too limited to protect the body from hypercalcemia.

Hyperphosphatemia, or increased concentration of serum phosphate, may lead to metastatic calcification, particularly in renal failure, since the critical solubility product can be exceeded even if calcium levels are normal. An increase in cytosol phosphate, however, appears to effectively impair the vitamin D-PTH-calcium–sparing system. Thus, with increase of phosphate levels in serum and extracellular fluid, synthesis of $1,25(OH)_2D$ markedly declines. Also, gastrointestinal absorption and tubular reabsorption of calcium, and even bone breakdown, are initially reduced, thus diminishing the concentration of calcium. (If these mechanisms continue for a long period of time, however, they will induce a secondary hyperparathyroidism in response to the lowered serum calcium level.)

The system of calcium and phosphate metabolism is complex, and the variables are multiple. The fundamental "truths" and the interactions of the various hormonal and mineral materials discussed here are important in understanding the principles that govern and control the homeostatic mechanisms and the alterations that lead to the rachitic syndrome (Plates 28–29, and Section IV, Plates 11–20). □

Nutritional Calcium Deficiency

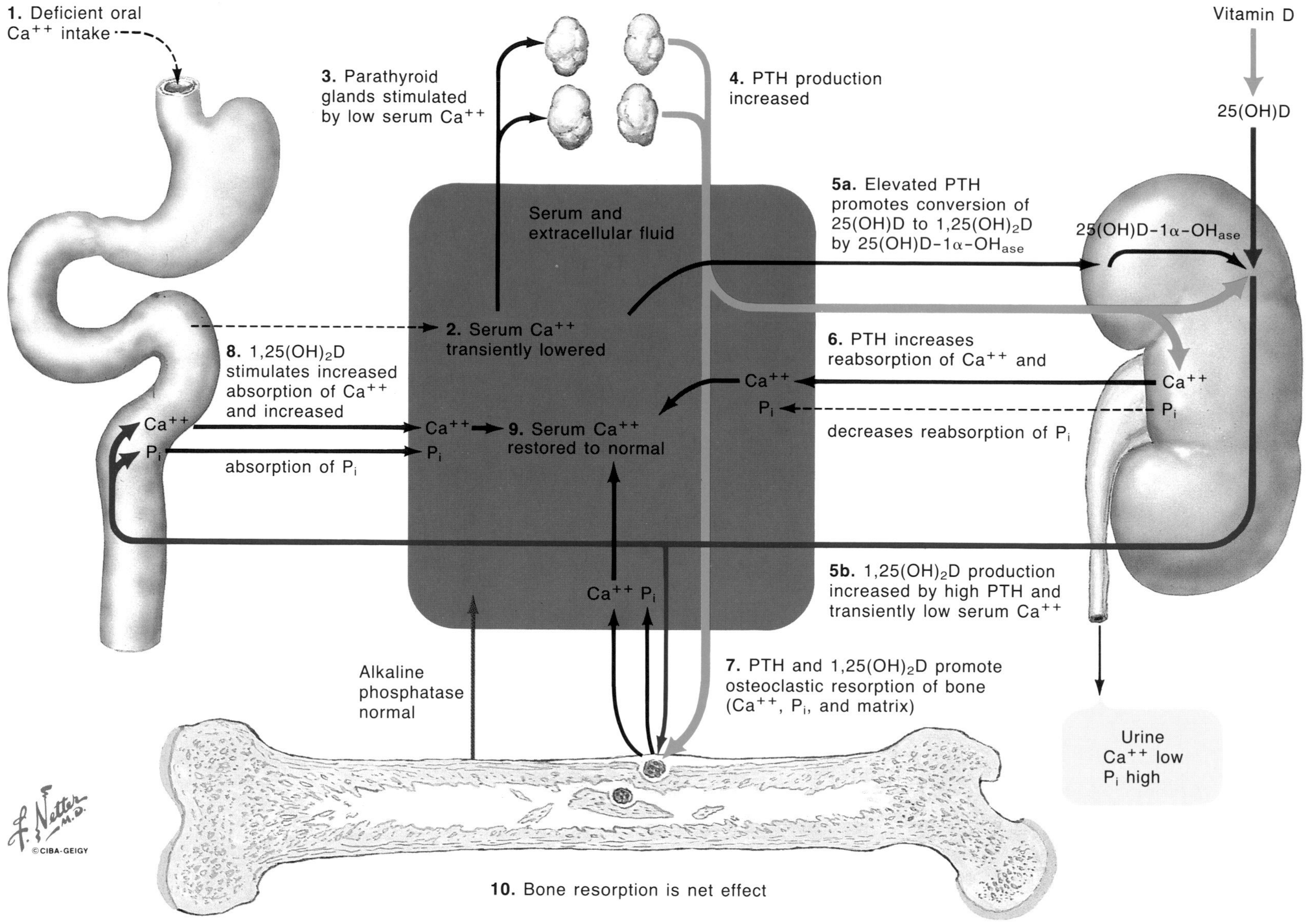

1. Deficient oral Ca++ intake
2. Serum Ca++ transiently lowered
3. Parathyroid glands stimulated by low serum Ca++
4. PTH production increased
Vitamin D
25(OH)D
Serum and extracellular fluid
5a. Elevated PTH promotes conversion of 25(OH)D to 1,25(OH)2D by 25(OH)D-1α-OHase
25(OH)D-1α-OHase
6. PTH increases reabsorption of Ca++ and
decreases reabsorption of Pi
Ca++
Pi
8. 1,25(OH)2D stimulates increased absorption of Ca++ and increased
absorption of Pi
9. Serum Ca++ restored to normal
5b. 1,25(OH)2D production increased by high PTH and transiently low serum Ca++
Alkaline phosphatase normal
7. PTH and 1,25(OH)2D promote osteoclastic resorption of bone (Ca++, Pi, and matrix)
Urine
Ca++ low
Pi high
10. Bone resorption is net effect
©CIBA-GEIGY

SECTION III PLATE 33 Slide 3729

Nutritional Calcium Deficiency

Calcium deficiency poses a constant threat to all life-forms. Although 99% of the body's total calcium is stored in bone in the form of an imperfect hydroxyapatite crystal, it is the 1% remaining in the extracellular fluid (including serum) that the body monitors vigilantly and controls assiduously. Many functions critical to life, such as cell proliferation, differentiation, secretion, coagulation, excitation, and contraction, require a stable and steep gradient of calcium across cell membranes. That gradient is dependent on the concentration of calcium in the extracellular fluid.

An intricate physiologic mechanism has evolved to prevent dangerous hypocalcemia: a reduction in calcium intake stimulates calcium absorption by the gastrointestinal tract, promotes renal calcium conservation, and stimulates net bone resorption (mineral plus matrix), thus restoring the serum calcium level to normal.

Take as an example a 30-year-old woman who begins a diet with a very low calcium intake. Let us explore the mechanisms that maintain a constant calcium level in the extracellular fluid.

In response to reduced calcium intake, net absorption decreases and serum calcium levels decline transiently. As a result, secretion of parathyroid hormone (PTH) increases. The target organs for PTH are the bones and the kidneys. In bone, osteoclastic activity is stimulated via a cyclic adenosine monophosphate (cyclic AMP)–mediated pathway (probably indirectly through osteoblastic activity), leading to net resorption of both bone matrix and bone mineral. In the kidney, PTH promotes the tubular reabsorption of filtered calcium and impairs the reabsorption of inorganic phosphate. In addition, PTH activates the enzyme 25-hydroxyvitamin D-1α-hydroxylase, or 25(OH)D-1α-OH_{ase}, in the mitochondria of the proximal tubular cells, thus promoting the conversion of the 25(OH)D substrate to the potent hormonal metabolite 1,25-dihydroxyvitamin D (1,25[OH]$_2$D, or calcitriol). The activation of cyclic AMP also mediates the action of PTH in the kidney. The major target organs for 1,25(OH)$_2$D are the bones, the duodenum, and the jejunum. The exact mechanism of 1,25(OH)$_2$D-induced bone resorption is not known, but the hormone is thought to act by stimulating osteoclastic precursors. It promotes calcium absorption in the duodenum and jejunum by stimulating numerous events and proteins, including the production of calcium-binding protein (CBP) in the enterocytes. However, the kidney's ability to adapt to low calcium intake by the mechanisms described declines with age. An age-related decline in 1,25(OH)$_2$D production by the kidneys impairs the efficiency of gastrointestinal absorption of calcium, thus an even greater resorption of bone is required to maintain serum calcium at normal levels. This latter mechanism is thought to play a major role in the evolution of type II, or age-related, osteoporosis. □

Effects of Disuse and Stress (Weight Bearing) on Bone Mass

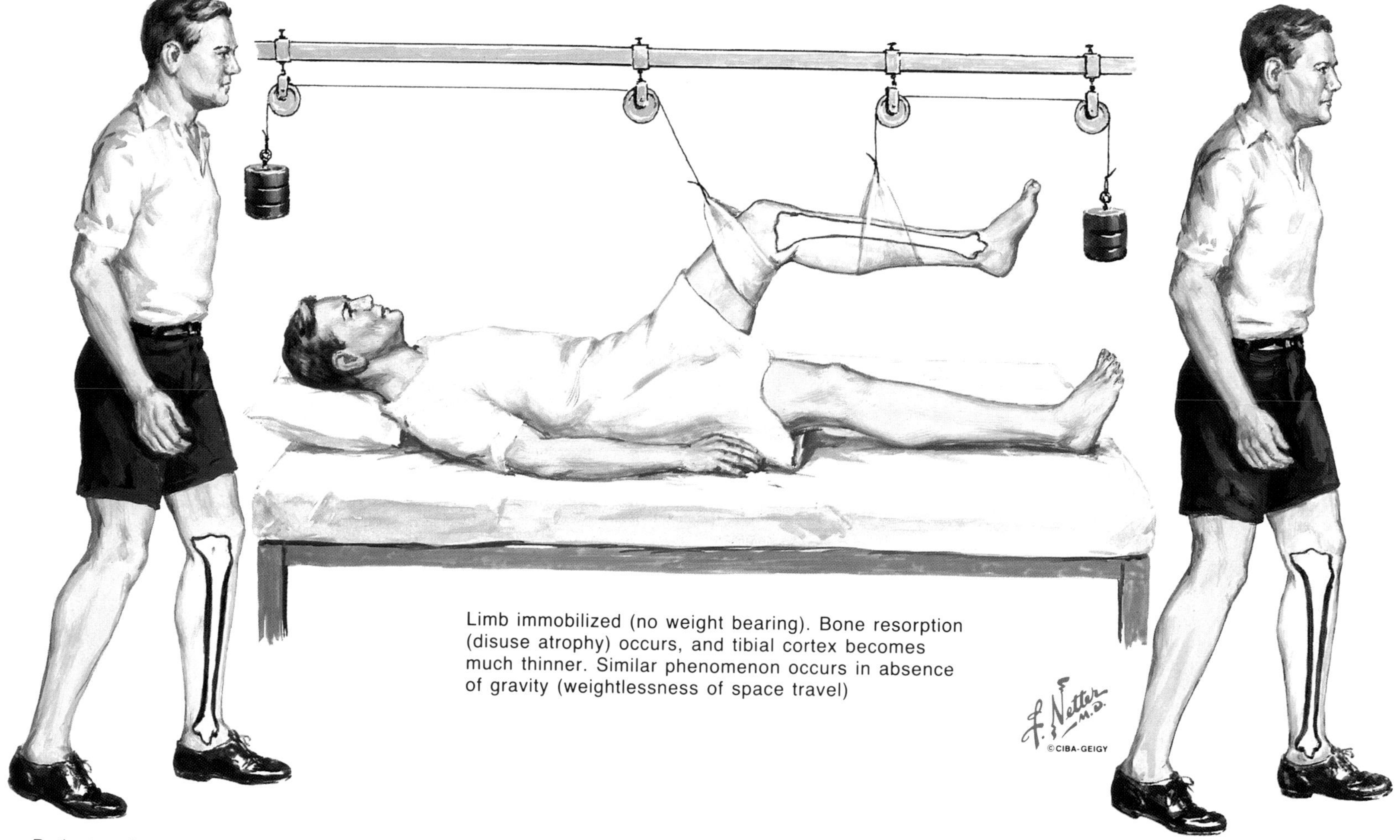

Limb immobilized (no weight bearing). Bone resorption (disuse atrophy) occurs, and tibial cortex becomes much thinner. Similar phenomenon occurs in absence of gravity (weightlessness of space travel)

Patient ambulatory (weight bearing). Cortex of tibial diaphysis has substantial thickness

Patient again ambulatory. Tibial cortex soon restored almost to original thickness

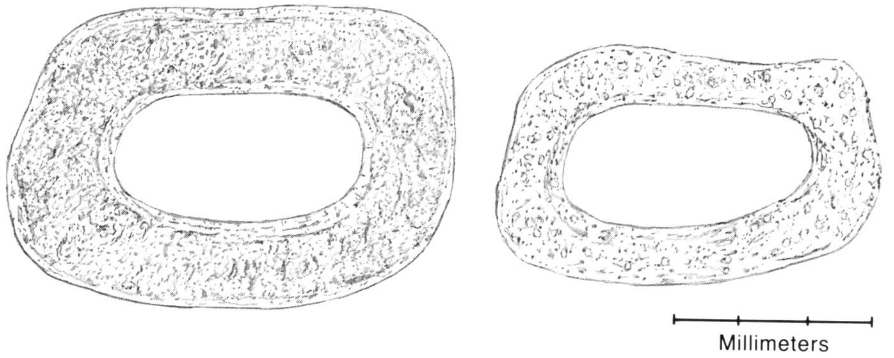

Cross sections of metacarpals of young adult dog (metacarpals weight bearing in dogs). Left from control limb. Right from limb immobilized 40 weeks. Right shows greatly decreased cortical thickness, reduced bone diameter (periosteal resorption), increased bone porosity, and enlarged medullary cavity. (Endosteal resorption predominates in older animals)

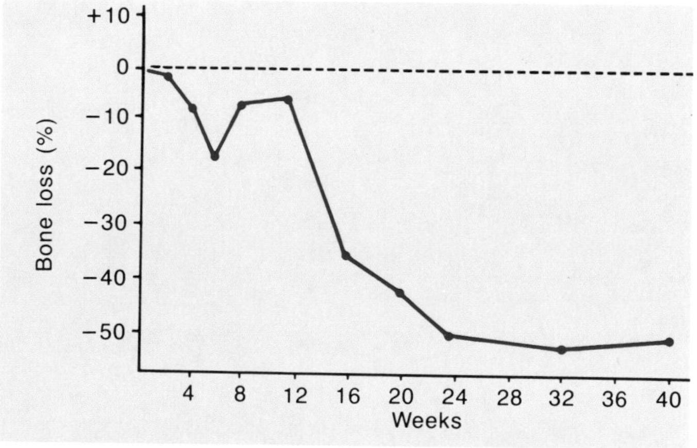

Pattern of bone loss in metacarpal 3 of dog in relation to duration of immobilization (expressed as percentage of control). Partial recovery from 8th to 12th week; then progressive bone loss to lower steady state

Effects of Disuse and Stress (Weight Bearing) on Bone Mass

Disuse osteoporosis is an important example of Wolff's law (Plate 36). Normally, the tibial cortex in the midshaft region is quite thick. If the limb is immobilized, disuse osteoporosis sets in rapidly, and the cortex thins measurably. With resumption of weight bearing, the bone quickly rebuilds and the cortical mass and thickness are restored almost to normal. However, prefracture bone mass may not be fully restored.

Disuse osteoporosis may be focal, involving a single bone or limb (arthritis or immobilization), or it may be generalized (prolonged bed rest or paralysis). With generalized immobilization, bone loss is most profound and rapid in the trabecular bones of the axial skeleton.

Results of studies carried out on dogs show that tubular bones exhibited a temporal and envelope-specific pattern of change in bone mass after long-term immobilization in a cast. However, even with prolonged immobilization, loss of bone mass does not continue indefinitely. The study by Uhthoff and Jaworski demonstrated a rapid initial loss of bone lasting approximately 6 weeks. By the twelfth week of immobilization, a rapid rebound phenomenon occurred, with bone mass approaching control values. A slower, longer-lasting bone loss began at 16 weeks and ended at 32 weeks, with bone mass reaching a plateau at approximately 50% of original values.

In the short first phase, bone loss involved the periosteal, osteonal (haversian), and corticoendosteal envelopes. In the later, sustained-loss phase, the periosteal envelope was affected (except in older dogs), resulting in a smaller bone with a slightly widened medullary canal and a thinner, more porous cortex. □

Musculoskeletal Effects of Weightlessness (Spaceflight)

Musculoskeletal Effects of Weightlessness (Spaceflight)

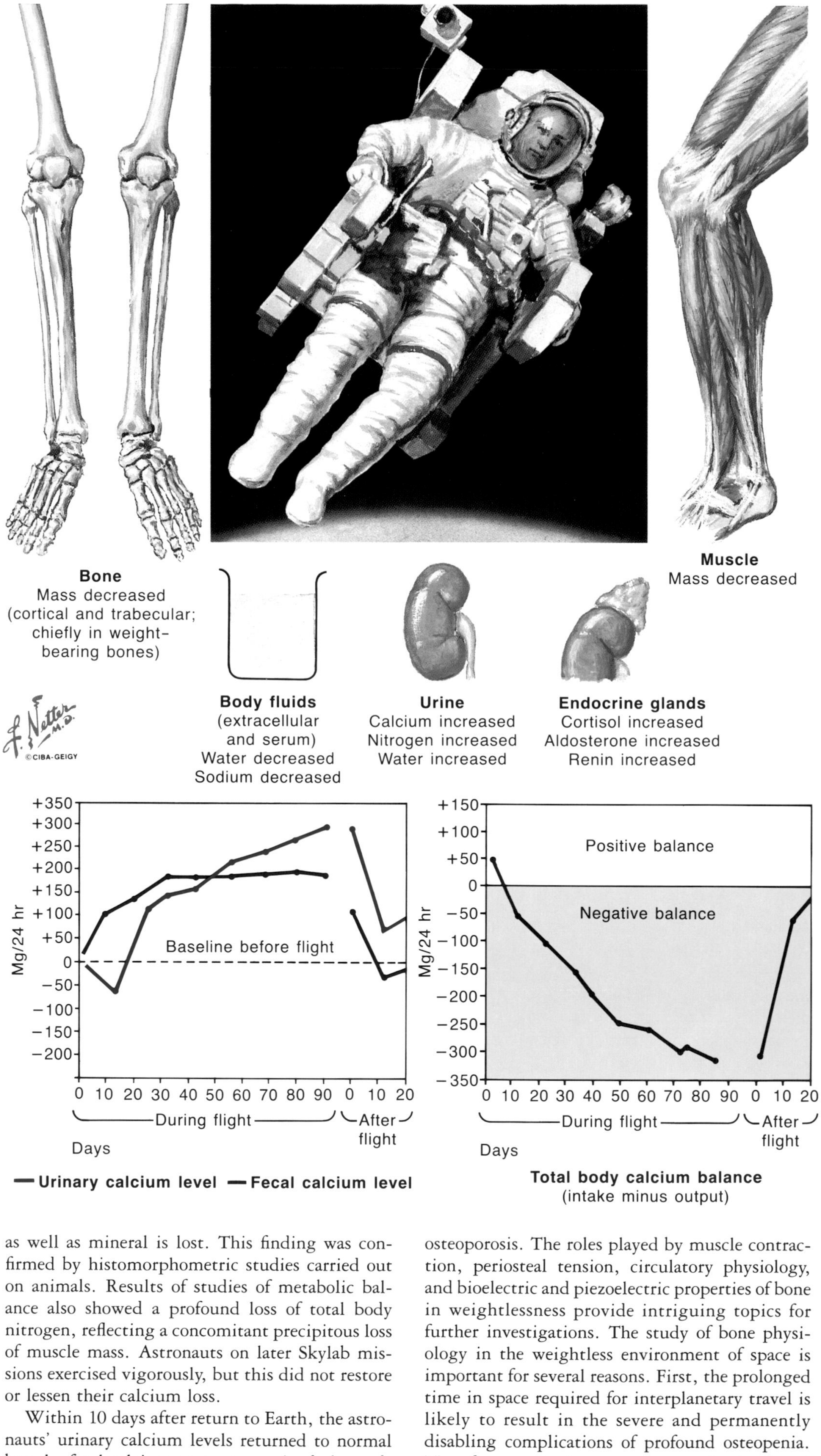

Gravity has played a major role in the evolution of life on Earth. The structure of living organisms and basic physiologic processes such as growth, development, and locomotion evolved in the presence of a gravitational field. Space travel has made it possible for humans to escape the pervasive influence of Earth's gravitational field and to experience the potential benefits and consequences of weightlessness.

One of the most intriguing and potentially hazardous biomedical phenomena discovered from space travel is the continuous loss of bone mass (both matrix and mineral) in a weightless environment. During the early Gemini and Apollo flights, x-ray densitometry studies of the calcaneus of astronauts indicated that bone loss may be extensive and rapid even during a brief period of weightlessness. In later Apollo and Skylab spaceflights, a more precise technique, photon absorptiometry, was employed to assess the preflight and the postflight bone mass of the calcaneus. Findings showed a direct dose-response relationship between time spent in a weightless environment and loss of bone mass, although wide variations were observed. An average decrease of 3.9% in calcaneal density was observed in the crew of Skylab 4, whose flight lasted 84 days. This finding was corroborated by studies of changes in bone mass carried out on Soviet cosmonauts.

In addition, more severe changes occur in the weight-bearing bones of the lower limb, with the greatest changes seen in the high-remodeling trabecular bone of the axial skeleton. After return from space, skeletal mass is gradually restored, but bone mass may not be completely restored, even after a prolonged recovery in Earth's gravitational field. With prolonged weightlessness, full recovery is unlikely because of irreversible loss of the surface scaffolding that is necessary for bone cell activity. Studies of metabolic balance, performed on the Skylab crew during flights lasting 28 to 84 days, revealed a striking increase in urinary calcium levels, which reached a plateau after 28 days in space. Fecal calcium levels, however, continued to increase, with no sign of leveling off even after 84 days in space. After only 10 days of weightlessness, the preflight positive calcium balance was abolished, and a net negative calcium balance prevailed for the remainder of the flight.

This loss of total body calcium was much greater than that predicted from bed-rest studies. These findings led Rambaut and Johnston to postulate that a year of weightlessness could result in a loss of 25% of the body's total calcium reserve, 99% of which is stored in bone apatite. As occurs with immobilization, hypercalciuria is accompanied by hydroxyprolinuria, indicating that matrix as well as mineral is lost. This finding was confirmed by histomorphometric studies carried out on animals. Results of studies of metabolic balance also showed a profound loss of total body nitrogen, reflecting a concomitant precipitous loss of muscle mass. Astronauts on later Skylab missions exercised vigorously, but this did not restore or lessen their calcium loss.

Within 10 days after return to Earth, the astronauts' urinary calcium levels returned to normal but the fecal calcium content remained elevated. Thus, total body calcium balance remained negative even 20 days after the spaceflight.

Osteoporosis caused by weightlessness is more severe and unrelenting than any form of disuse osteoporosis. The roles played by muscle contraction, periosteal tension, circulatory physiology, and bioelectric and piezoelectric properties of bone in weightlessness provide intriguing topics for further investigations. The study of bone physiology in the weightless environment of space is important for several reasons. First, the prolonged time in space required for interplanetary travel is likely to result in the severe and permanently disabling complications of profound osteopenia. Therefore, protective measures must be developed for long spaceflights. Second, the weightless environment provides a natural opportunity for studying the complexities of normal bone physiology as well as a multitude of osteopenic conditions. □

Physical Factors in Bone Remodeling

Bone Architecture in Relation to Physical Stress

Wolff's law. Bony structures orient themselves in form and mass to best resist extrinsic forces (ie, form and mass follow function)

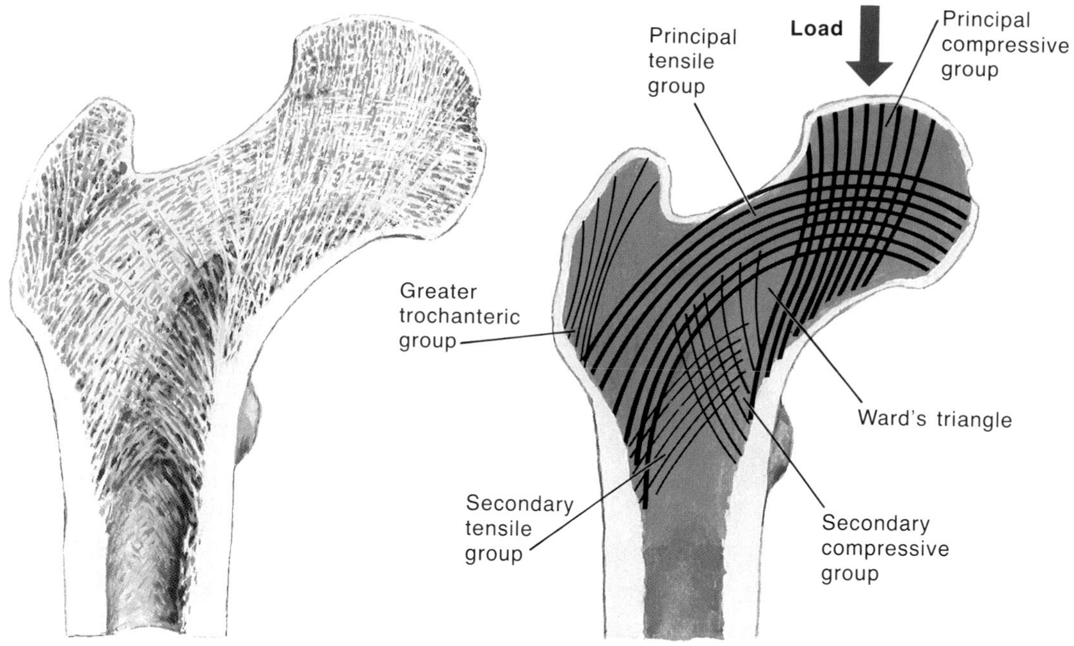

Trabecular configuration in proximal femur

Trabecular groups conform to lines of stress in weight bearing

Bone Remodeling in Response to Stress

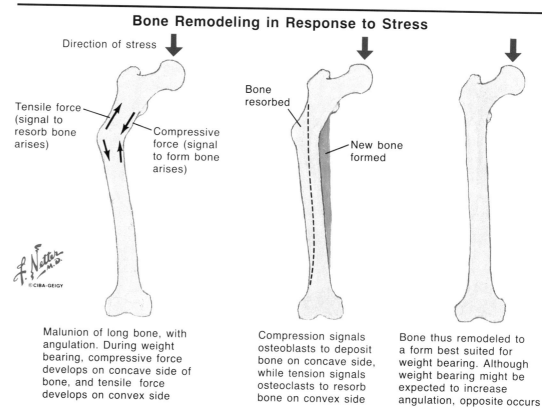

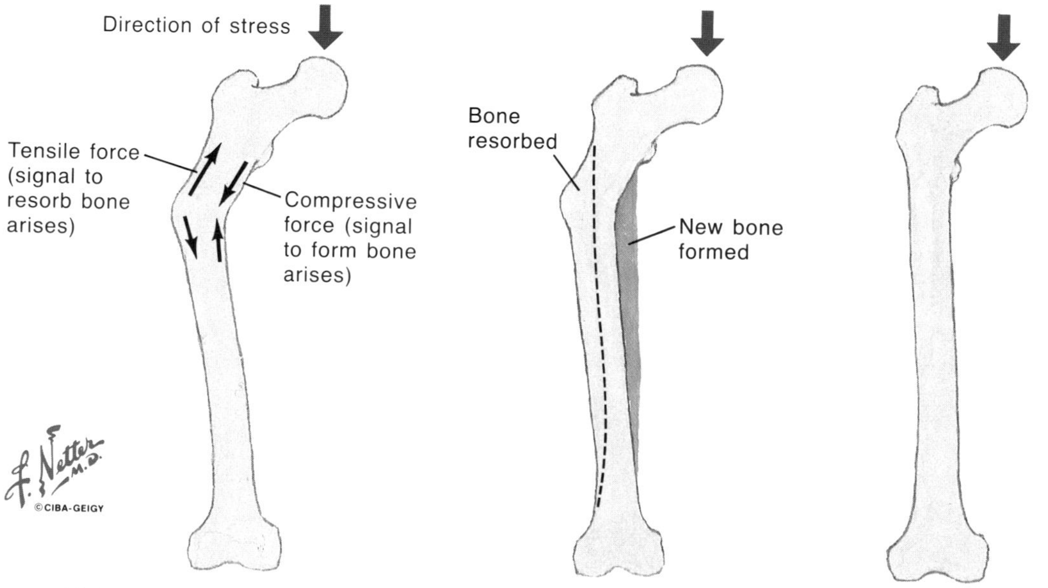

Malunion of long bone, with angulation. During weight bearing, compressive force develops on concave side of bone, and tensile force develops on convex side

Compression signals osteoblasts to deposit bone on concave side, while tension signals osteoclasts to resorb bone on convex side

Bone thus remodeled to a form best suited for weight bearing. Although weight bearing might be expected to increase angulation, opposite occurs

Galileo first recognized the relationship between applied load and bone morphology. In 1683, he noted a direct correlation between body weight and bone size. During the next two centuries, others observed that bone remodels, but Julius Wolff, a German anatomist, was the first to link the two vital concepts. He noted that changes in bone mass accompanied changes in load, through the process of skeletal remodeling. In "The Law of Bone Transformation," published in 1892, Wolff explained: "Every change in the function of a bone is followed by certain definite changes in internal architecture and external conformation in accordance with mathematical laws." Stated more simply, *form follows function*. Although the mechanism by which bone cells transform mechanical or bioelectric signals into a useful biologic response is not fully understood, Wolff's observations are as valid today as they were nearly a century ago.

Bone Architecture

The architecture of the proximal femur beautifully illustrates the general principle that the external form and shape of bone *as an organ* and the internal organization of bone *as a tissue* are well adapted to the forces placed upon them. There are dynamic internal forces as well as static and dynamic external forces on bone. The internal forces are created by muscle contraction; the external forces, by Earth's ubiquitous gravitational field and by the dynamic compressive forces of weight bearing. The upper half of Plate 36 depicts the bony trabeculae of the proximal femur aligned along the lines of stress according to Wolff's law. This intersecting network of trabeculae is the biologic response to the sum of internal and external physical forces on that region of the skeleton. Both tensile and compressive trabeculae are present and correspond to the lines of force. Reduced weight bearing resulting from disuse or immobilization leads to a progressive thinning and eventual loss of trabeculae; those bearing the least weight are resorbed first. A similar pattern is seen in all weight-bearing bones, but the loss of trabeculae is most dramatic in the axial skeleton, especially in the vertebral bodies, which are largely made up of weight-bearing trabecular bone.

Mechanical forces also play a significant role in the external shape of bone. For example, the applied dynamic force of contraction of the gluteal muscles influences both the size and shape of the greater trochanter. If these muscles are paralyzed during skeletal development (as in certain types of poliomyelitis or in meningomyelocele), the greater trochanter does not attain its normal size and shape.

Bone Remodeling

Wolff's law is also demonstrated by the straightening of a malunion of a long bone. With time, growth, and weight bearing, a malunion that has an angulation of as much as 30° will straighten completely, at least in the infant and young child (lower half of Plate 36). This phenomenon runs contrary to the laws of biomechanics, since continued weight bearing should cause an angulated structure to bend further until fatigue occurs. However, the exact opposite happens and the bone straightens with growth.

What is the explanation of this phenomenon? Some biologic or physical signal must arise from the concave side of the bone at the site of the malunion, inducing the osteoblasts there to lay down bone, and a corresponding signal must arise from the convex side of the malunion, stimulating the osteoclasts there to remove bone. What is the nature of this signal?

Four independent research teams began looking at this problem, each adopting the hypothesis that if an important function of bone is physical—namely, to bear load—then the signal that directs bone formation and resorption is perhaps a physical one. In the 1950s and early 1960s, Yasuda and Fukada, Bassett and Becker, and Shamos and Lavine carried out studies to discover the nature of signals in stressed bone. Also in the early 1960s, Friedenberg and Brighton began looking for signals in viable nonstressed bone. These studies determined that two types of electric signals (action potentials) are present in bone: *stress-generated*, or strain-related, potentials and *bioelectric*, or standing, potentials.

Physical Factors in Bone Remodeling

(Continued)

Stress-Generated Electric Potentials in Bone

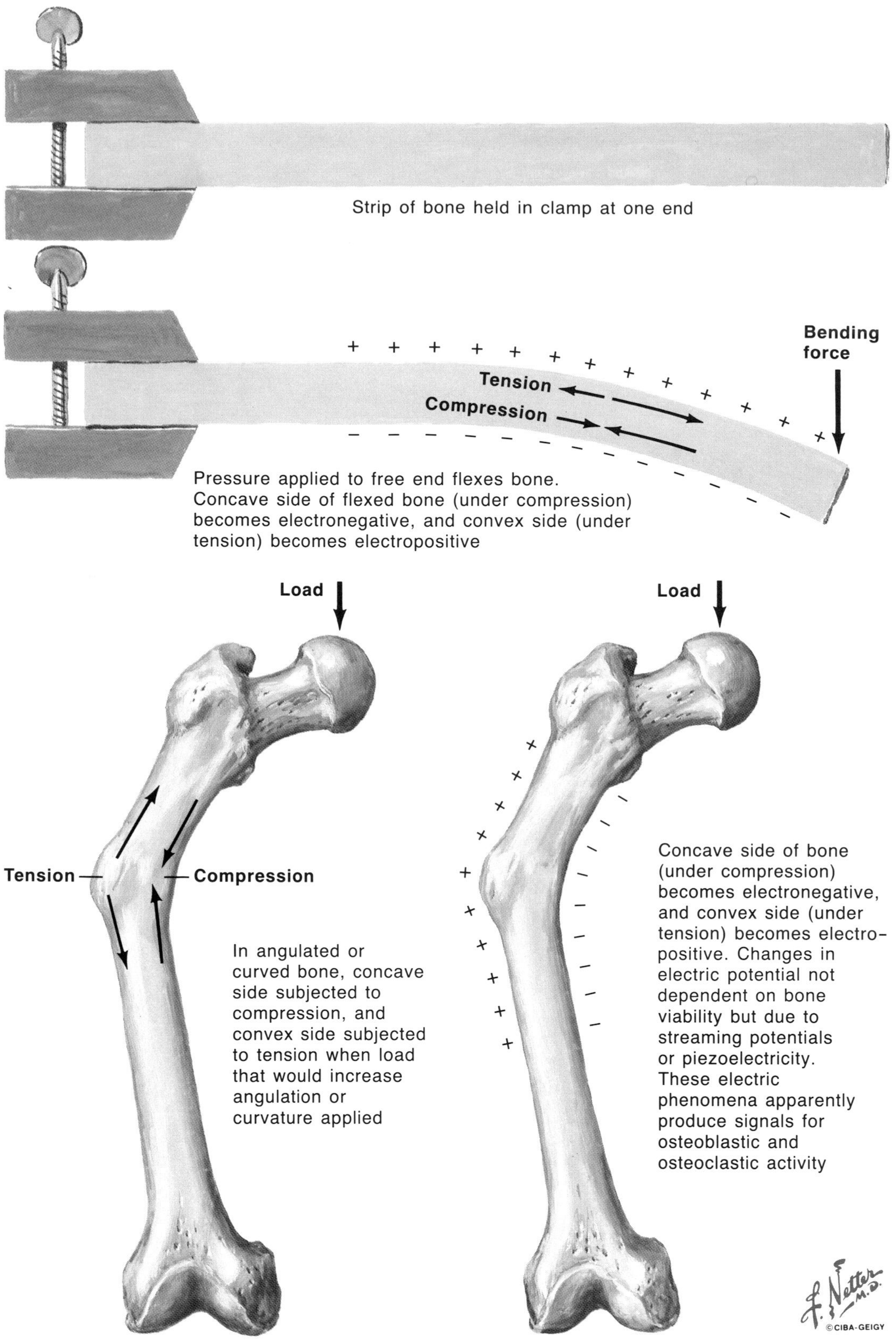
Strip of bone held in clamp at one end
Bending force
Tension
Compression
Pressure applied to free end flexes bone. Concave side of flexed bone (under compression) becomes electronegative, and convex side (under tension) becomes electropositive
Load
Load
Tension
Compression
In angulated or curved bone, concave side subjected to compression, and convex side subjected to tension when load that would increase angulation or curvature applied
Concave side of bone (under compression) becomes electronegative, and convex side (under tension) becomes electropositive. Changes in electric potential not dependent on bone viability but due to streaming potentials or piezoelectricity. These electric phenomena apparently produce signals for osteoblastic and osteoclastic activity
F. Netter M.D.
©CIBA-GEIGY

Stress-Generated Electric Potentials in Bone

When bone is stressed, the concave side of the bone (the area under compression) becomes negatively charged, or electronegative, and the convex side of the bone (the area under tension) becomes positively charged, or electropositive (Plate 37). In the malunion of a long bone, the area of compression, where new bone will be formed, is electronegative, and the area under tension, where bone will be removed, is electropositive. These stress-generated potentials arise when the bone is stressed and are not dependent on cell viability. Research has also shown that the electric signal arises from the organic and not the mineral component of bone. Thus, stress-generated signals arise even if the bone is totally decalcified.

Bioelectric Potentials in Bone

Bioelectric potentials are measured from the surface of nonstressed bone (Plate 38). In the intact tibia, the growth plate–metaphyseal regions are electronegative, whereas the diaphyseal, or midshaft, region is not. When a fracture occurs in the diaphysis, the entire tibial surface becomes electronegative, with a large peak of electronegativity occurring over the fracture site and persisting until the fracture heals. A second peak of electronegativity occurs over the farthest growth plate. This latter finding is fascinating because a fractured extremity in a child frequently exhibits overgrowth not at the fracture site but in the growth plate near the end of the bone. The nature of the signal directing the growth plate to accelerate growth has never been identified, but the peak of electronegativity over the growth plate–metaphyseal area that accompanies a midshaft fracture may be such a signal.

To determine the source of action potentials in nonstressed bone, the following experiments were performed on a rabbit:

1. The vascular supply of the leg was interrupted, yet the electric potential over the proximal 7 cm of the tibia did not change. That is, the peak of electronegativity in the proximal tibia showed no significant change 30 minutes after ligation of the vessels.

2. The same result was found after the leg was denervated.

3. After injection of a cytotoxic drug (dinitrophenol), there was an immediate statistically significant drop in the electronegative potential. This suggested that the potential, which was measured from the surface of bone, was indeed linked to cell viability.

4. An in situ segment of the tibia was subjected to high-energy ultrasound waves, and a small segment of bone was killed. A corresponding statistically significant drop in the electric potential occurred over the nonviable region.

Potentials arising from nonstressed bone are called bioelectric potentials, meaning that they arise from living bone. Such potentials are dependent on cell viability and not on stress. Active areas of growth and repair are electronegative, and less active areas are electrically neutral or electropositive.

Studies have also shown that the application of small electric currents to bone stimulates osteogenesis at the site of the negative electrode (cathode).

Various in vitro and in vivo models have identified the processes and results of electrically induced osteogenesis. (1) Given the proper current and voltage, bone forms only in the vicinity of the cathode, whereas cell necrosis occurs around the anode when stainless steel electrodes are used. (2) Resistance rapidly increases between the electrodes, leading to a concomitant decrease in the current. (If a constant current is to be maintained, an active power supply using a transistorized control current circuit must be provided.)

Physical Factors in Bone Remodeling

(Continued)

Bioelectric Potentials in Bone

Apparatus measures electric potentials from surface of in situ bone in rabbit. Differences in potentials present over length of bone in absence of stress. Potentials dependent on bone cell viability (eliminated by cytotoxic drugs and ultrasound, which kill bone cells). Potentials therefore referred to as bioelectric potentials

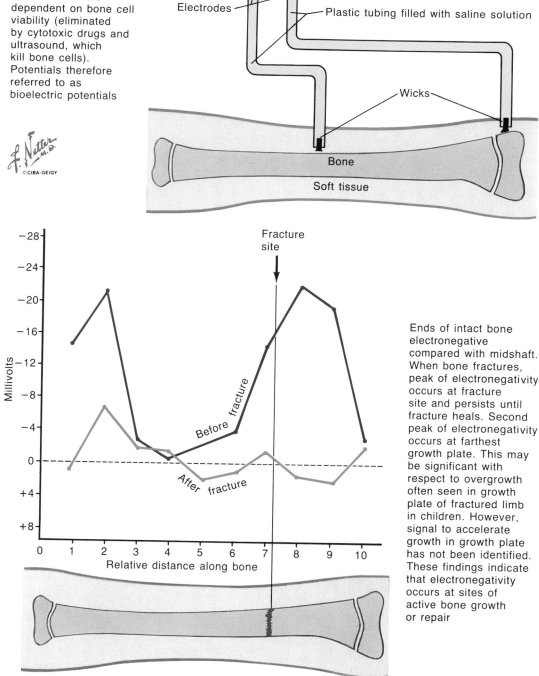

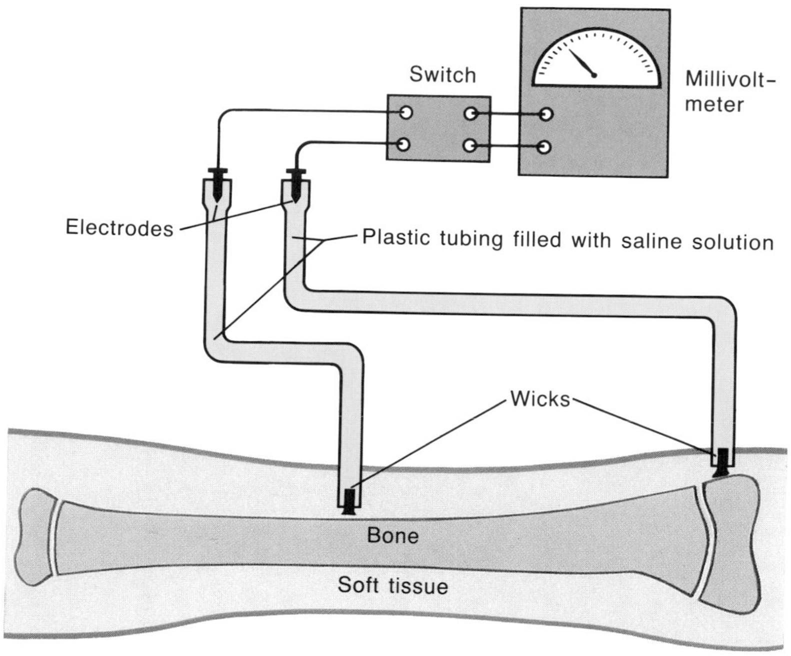

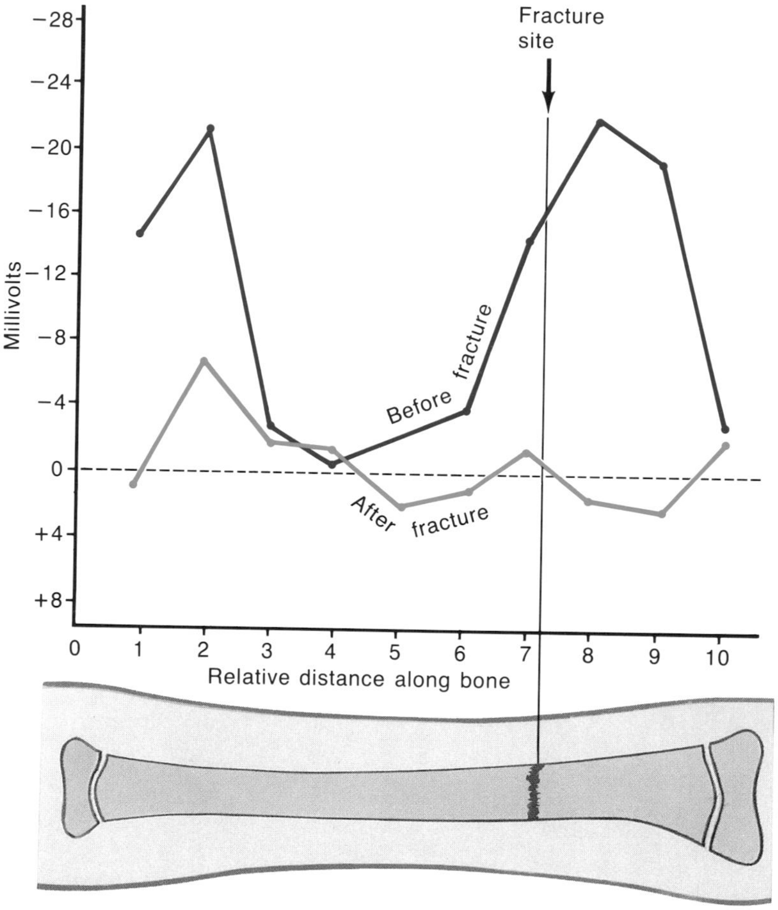

Ends of intact bone electronegative compared with midshaft. When bone fractures, peak of electronegativity occurs at fracture site and persists until fracture heals. Second peak of electronegativity occurs at farthest growth plate. This may be significant with respect to overgrowth often seen in growth plate of fractured limb in children. However, signal to accelerate growth in growth plate has not been identified. These findings indicate that electronegativity occurs at sites of active bone growth or repair

(3) Electrically induced osteogenesis exhibits a dose-response curve; a current of less than 5 μA delivered through a stainless steel cathode does not produce osteogenesis; a current of 5 to 20 μA produces progressively increasing degrees of bone formation; and a current greater than 20 μA induces bone formation that gives way to cellular necrosis. (4) Electricity can favorably influence fracture healing in laboratory animals, but for this to occur, the cathode must be placed directly in the fracture site. (5) With the proper current and voltage, electricity can induce bone formation in the absence of trauma and in areas of inactive bone formation, such as in the medullary canal of an adult animal. (6) The reaction at the cathode results in consumption of oxygen and production of hydroxyl radicals. (7) Pulsed direct current is not as effective as constant direct current in inducing osteogenesis. (8) The electrically active area of the cathode is at the insulation-bare wire junction and measures approximately 0.02 mm; therefore, when stainless steel is used as the cathode, the actual current density is 1×10^{-3} A/mm^2.

In addition, electricity can be induced in bone by means of an electric field with the electric apparatus remaining completely external to the limb. The electric field can be inductively or capacitively coupled to the bone. In inductive coupling, a current varying with time produces a time-varying magnetic field, which in turn induces a time-varying electric field.

In capacitive coupling, an electric field is induced in bone by an external capacitor (two charged metal plates are placed on either side of the limb and attached to a voltage source). Several studies have shown that both constant and pulsed capacitively coupled electric fields can favorably influence fracture repair in experimental rabbits and in the in vitro growth of the epiphyseal plate.

The mechanism by which electricity induces osteogenesis is unclear. It is known that the cathode consumes oxygen and produces hydroxyl radicals according to the equation $2H_2O + O_2 + 4e^- \rightarrow 4OH^-$. Thus, the oxygen tension (Po_2) is lowered in the local tissue, and pH is raised in the vicinity of the cathode. Studies have also shown that low Po_2 in tissue encourages bone formation: (1) low Po_2 has been measured at the bone-cartilage junction in the growth plate and in newly formed bone and cartilage in fracture calluses; (2) optimum in vitro bone growth occurs in a low-oxygen (5%) environment; and (3) cells of the growth plate cartilage, as well as bone cells, follow a predominantly anaerobic metabolic pathway. Howell and associates found that the pH in the growth plate at the calcification front was rather high (7.70 ± 0.05), suggesting that an alkaline environment is favorable to the mineralization of bone.

These local microenvironmental changes in the vicinity of the cathode lead indirectly to cellular changes that ultimately result in osteogenesis. Electricity may also act directly on bone and cartilage. Such a direct effect may be expected to activate the cell's cyclic adenosine monophosphate (cyclic AMP) system—activation of the intracellular, or second, messenger, which in turn activates various enzyme systems within the cell to bring about a specific physiologic response. Physical forces increase production of cyclic AMP at the site of new bone formation. If the cyclic AMP system is indeed activated by electricity, the electron or charge acts like a hormone in being the first messenger to bone or cartilage cells. Studies by Norton and associates support this hypothesis. They discovered a significant increase in cyclic AMP in epiphyseal cartilage cells that are subjected to an oscillating electric field. If this hypothesis is confirmed by further studies, the application of electricity in one or more of its forms may enable the orthopedic surgeon to modulate growth, maintenance, and repair of bone and cartilage. □

Age-Related Changes in Bone Geometry

The long bones of the appendicular skeleton grow in length by endochondral ossification and in width by a process of subperiosteal bone formation and endosteal bone resorption. Even after longitudinal growth ceases at skeletal maturity, bone modeling continues throughout life. Age-related changes in the geometry of long bones reflect the body's ability to rearrange its remaining skeletal assets in the most biomechanically useful way—another example of Wolff's law (Plate 36).

The growth in width of a long bone is due primarily to subperiosteal formation of new bone, a process that begins before birth and continues even into the ninth and tenth decades. In all population samples studied, subperiosteal formation of new bone is greater in males than in females. The growth in width of long bones is particularly accelerated in the first 2 years of life. For example, by age 2, the diameter of the medullary canal at the middiaphysis of the femur is nearly equal to the diameter of the entire middiaphysis at birth. Thus, the rate of cortical modeling approaches 50% per year in the first 2 years of life. The growth in width of long bones continues at a slower rate in childhood, then increases rapidly during the adolescent growth spurt. During this period of rapid longitudinal as well as latitudinal growth, as much as 300 mg of elemental calcium is incorporated into bone apatite every day.

Most traditional views of bone development imply that all growth ceases after skeletal maturity, near the beginning of the third decade. However, results of cross-sectional studies on large samples of the adult population and longitudinal studies on individuals indicate that subperiosteal bone apposition continues throughout adulthood and into old age. The greatest increase in bone width occurs in the femur, but the general process is observed in the entire skeleton, in bones as diverse as the skull, ribs, and vertebrae. Also, subperiosteal bone formation occurs in both men and women and in all population samples studied. Although the total subperiosteal area is greater in men than in women, the percentage of gain is greater in women.

In conjunction with the age-specific subperiosteal bone apposition that continues throughout life, a complex age-related activity, characterized by alternating phases of resorption and apposition, occurs at the endosteal surface. Whereas subperiosteal activity determines the width of the bone, endosteal activity determines the width of the medullary canal. The combination of the relative activities at the two modeling surfaces over a period of time determines the thickness of the cortex, and modeling within the individual osteons of the cortex determines intracortical porosity.

During the first few years of life, a great deal of activity takes place at both the subperiosteal and the endosteal surface of cortical bone, tremendously increasing the width of both the bone and the medullary canal. Then, for the next several years, subperiosteal bone formation continues at a slower rate, accompanied by a large decrease in endosteal resorption and a short period of endosteal apposition. These processes enlarge the diameter of the bone and reduce the width of the medullary canal. Then, from about age 6 until the middle teenage years, endosteal bone resorption resumes, with resultant enlargement of the medullary canal.

The greatest natural uncoupling of bone activity in favor of bone formation occurs during adolescence. (For a more complete discussion of coupling, see Plates 30–31.) The growth in bone length by endochondral bone formation at the growth plates is accompanied by a corresponding burst of subperiosteal bone formation; once again, endosteal activity is reversed, with a new wave of endosteal bone apposition taking place. These processes increase the length and width of the bone, increase the thickness of the cortex, and decrease the width of the medullary canal. The apposition of bone at the endosteal surface begins earlier in females and continues until nearly age 40 in both sexes. After age 40, the activity at the endosteal surface again reverses, with endosteal bone resorption persisting for the remainder of life. Subperiosteal bone formation continues for the rest of life as well, at a slow but steady rate.

As a result of these two activities at the bone surfaces, the width of the bone increases slightly throughout adulthood and into old age; the width of the medullary canal also increases, resulting in a wider but thinner cortex. Although these changes in bone geometry have been noted in all long bones studied, they are most striking in the femur. However, the changes in the metacarpals (the bones most extensively studied and documented) provide an excellent means of assessing the state of cortical bone modeling in the appendicular skeleton (see Section IV, Plate 28). However, changes in the cross-sectional geometry of the bone surfaces do not take into account changes in intracortical bone density.

The age-related subperiosteal expansion of long bones may compensate mechanically for the endosteal resorption and resultant cortical thinning that occurs with aging. This radial outward displacement of the cortical mass serves to protect the long bones from bending and torsional stresses. This can best be understood by envisioning a solid rod of a certain cross-sectional area. If the material in the solid rod were displaced radially from the central axis of the rod to create a hollow tube, the result would be a structure that was stronger in both bending and torsion and thus better able to resist fracture. Thus, rearranging the same amount of material into a hollow tube improves the structural properties. It is no coincidence that the long bones are hollow and that the osteon is also essentially a hollow tube with a central haversian canal. By a process of natural selection, a structure has evolved that best accommodates the local biomechanical requirements of a system whose metabolic resources are under strict systemic control. □

Age-Related Changes in Bone Geometry

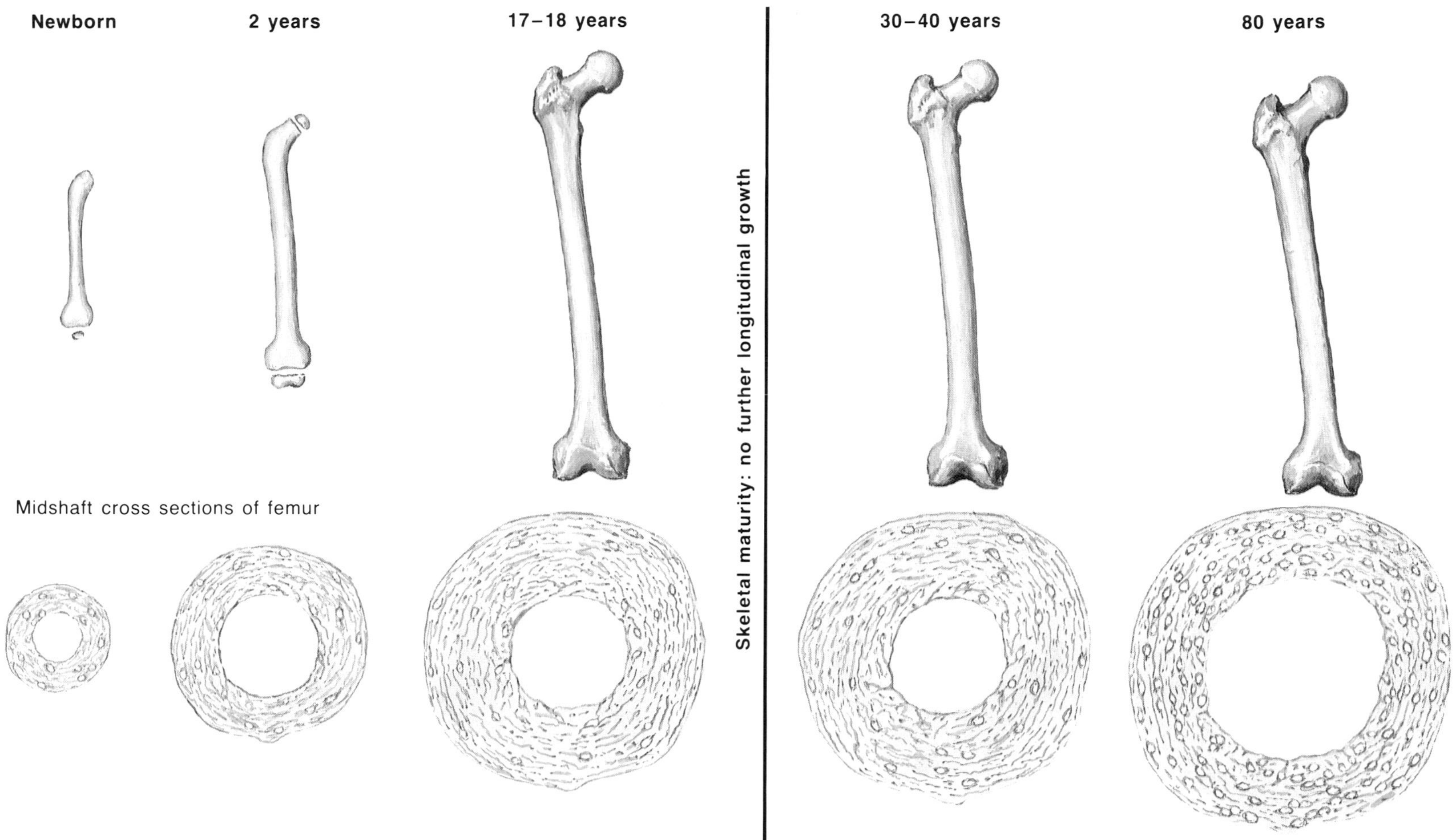
Newborn
2 years
17–18 years
Skeletal maturity: no further longitudinal growth
30–40 years
80 years
Midshaft cross sections of femur

At age 2, diameter of medullary canal nearly equal to diameter of entire shaft at birth. Thus, in first 2 years, rate of cortical modeling approaches 50%/yr

Subperiosteal diameter greatly increased, but medullary canal not enlarged. Thus, total bone mass greatly increased

Subperiosteal diameter only slightly increased, but medullary canal almost unchanged. Thus, total bone mass only slightly increased

Subperiosteal diameter further increased, but medullary canal considerably enlarged. Porosity also increased. Thus, total bone mass decreased

Radiographs of postmortem cross sections at midfemoral region

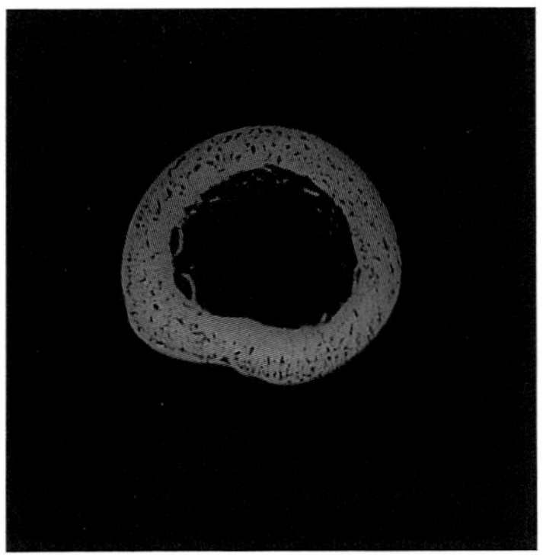

6-year-old female

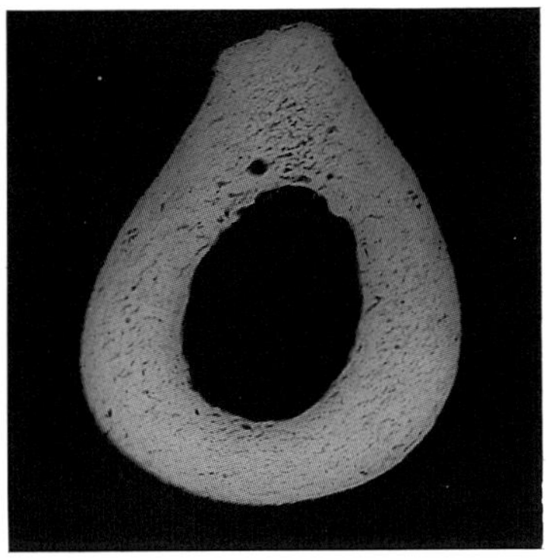

33-year-old female

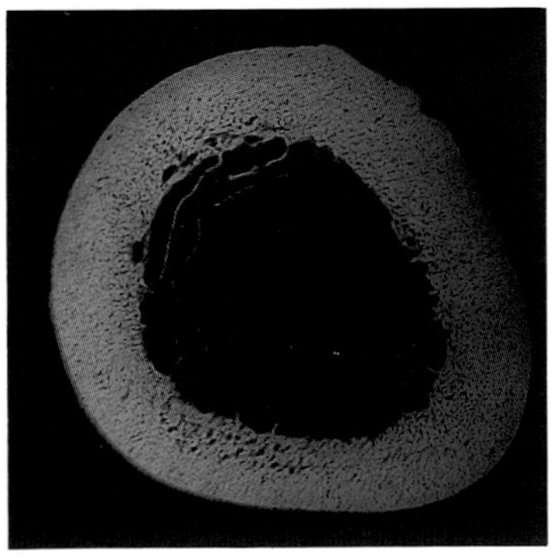

62-year-old male

Graph shows changes in subperiosteal and medullary cavity cross-sectional areas of metacarpals that occur with age. After age 60, subperiosteal area slowly increases but medullary cavity enlarges faster, resulting in net decrease of cortical thickness and mass

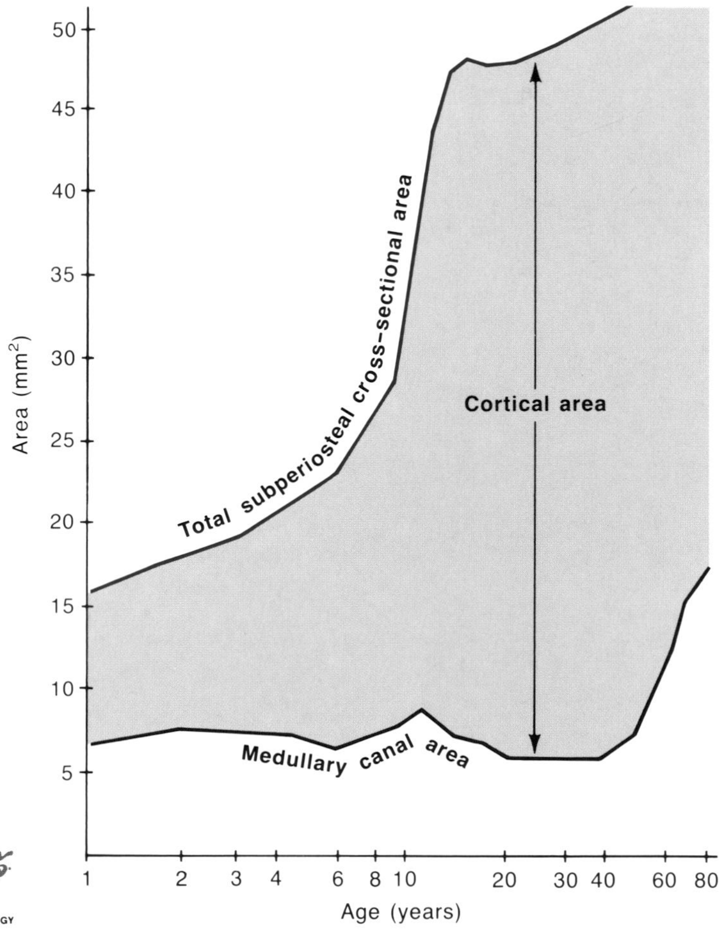

F. Netter M.D.
©CIBA-GEIGY

Section IV

Metabolic Disorders

Frank H. Netter, M.D.

in collaboration with

Abass Alavi, M.D. *Plate 28*

Maurice F. Attie, M.D. *Plates 1–10*

Charles S. August, M.D. *Plates 39–40*

Murray K. Dalinka, M.D. *Plates 25–27, 29*

Michael D. Fallon, M.D. *Plates 21, 30*

John G. Haddad, M.D. *Plates 32, 41–43*

Frederick S. Kaplan, M.D. *Plates 1–4, 22–33, 37–44*

Joel S. Karp, Ph.D. *Plate 28*

Henry J. Mankin, M.D. *Plates 11–20*

Edward A. Millar, M.D. *Plates 34–36*

Synthesis, Secretion, and Function of Parathyroid Hormone (PTH)

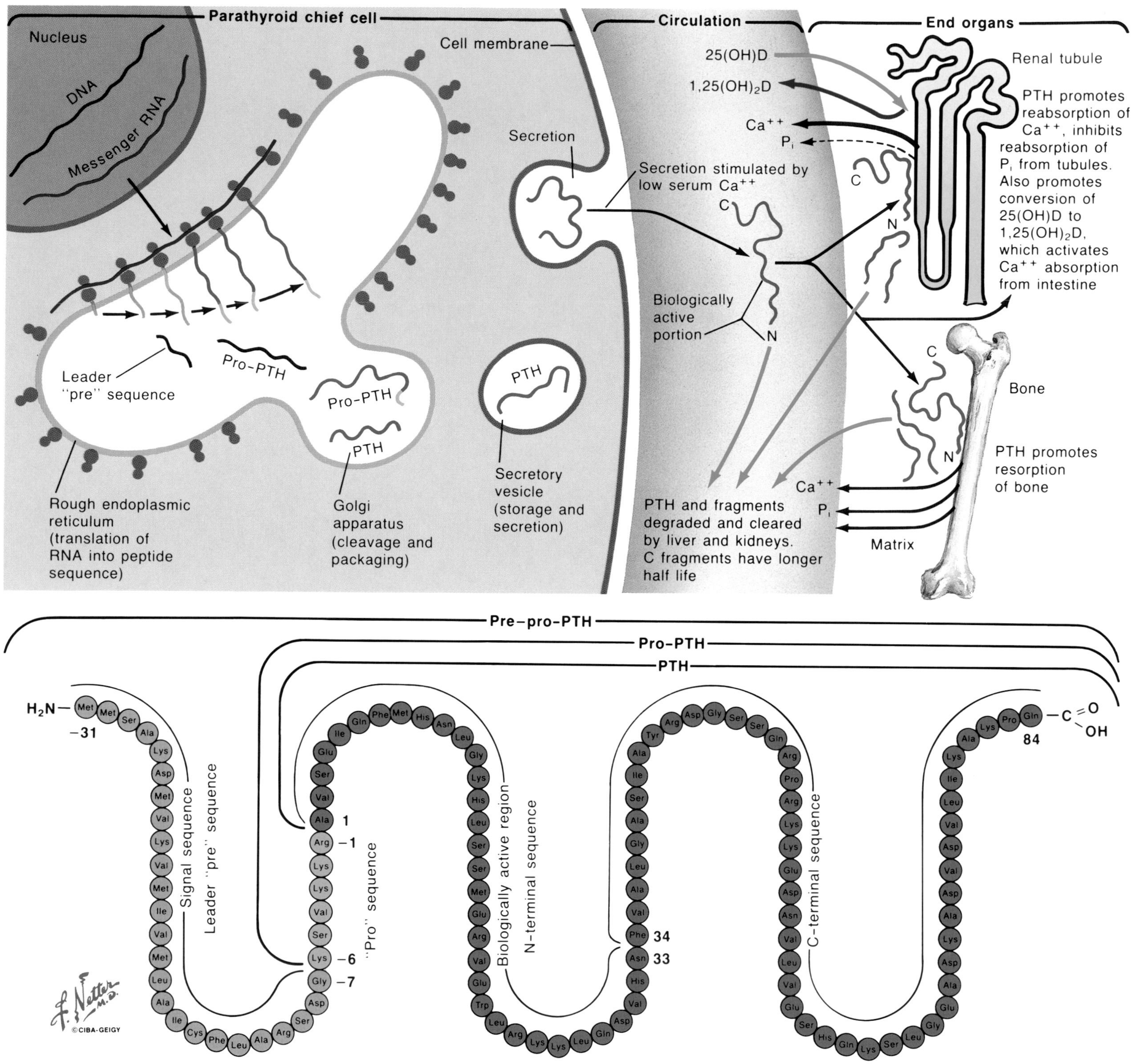

SECTION IV PLATE 1 Slide 3736

Parathyroid Hormone

The parathyroid gland regulates the calcium level in the extracellular fluid by sensing small changes in this level and rapidly modifying the secretion of parathyroid hormone (PTH). With a fall in the calcium level, PTH secretion is increased and in turn leads to increased calcium concentration and suppressed PTH secretion, thus completing a feedback loop. PTH raises the calcium level by promoting the entry of calcium from bone, renal tubule, and intestine. In bone, PTH increases resorption of formed bone, which contains calcium and phosphate. In the kidney, PTH enhances renal tubular reabsorption of calcium (which decreases urinary calcium excretion) and decreases tubular reabsorption of phosphate (which eliminates the phosphate released from bone). PTH indirectly increases calcium absorption from the gut by increasing the synthesis of the active form of 1,25-dihydroxyvitamin D, or 1,25$(OH)_2$D, from its precursor 25-hydroxyvitamin D, or 25(OH)D. This metabolite of vitamin D mediates the gut effect of PTH by directly activating calcium absorption in the small intestine.

PTH is initially synthesized as pre–pro-PTH, a large precursor peptide of 115 amino acids. During translation from its messenger ribonucleic acid (mRNA), the initial "pre" sequence directs the growing peptide across the membrane of the endoplasmic reticulum and is then cleaved before synthesis of the entire hormone is completed. The 90-amino-acid pro-PTH is then transported through membrane channels to the Golgi apparatus where the hexapeptide extension of pro-PTH is removed. PTH is stored in the secretory vesicles until secretion occurs.

After secretion, PTH is rapidly cleaved into fragments devoid of bioactivity. Fragments containing the bioactive region of PTH—the first third of the peptide, the amino-terminal (N-terminal)—are potentially bioactive, but are also rapidly removed from plasma. In contrast, the inert carboxyl-terminal (C-terminal) has a longer half-life and is the predominant species in plasma. It is also the form of PTH measured in most radioimmunoassays. □

Primary Hyperparathyroidism

Pathologic Physiology of Primary Hyperparathyroidism

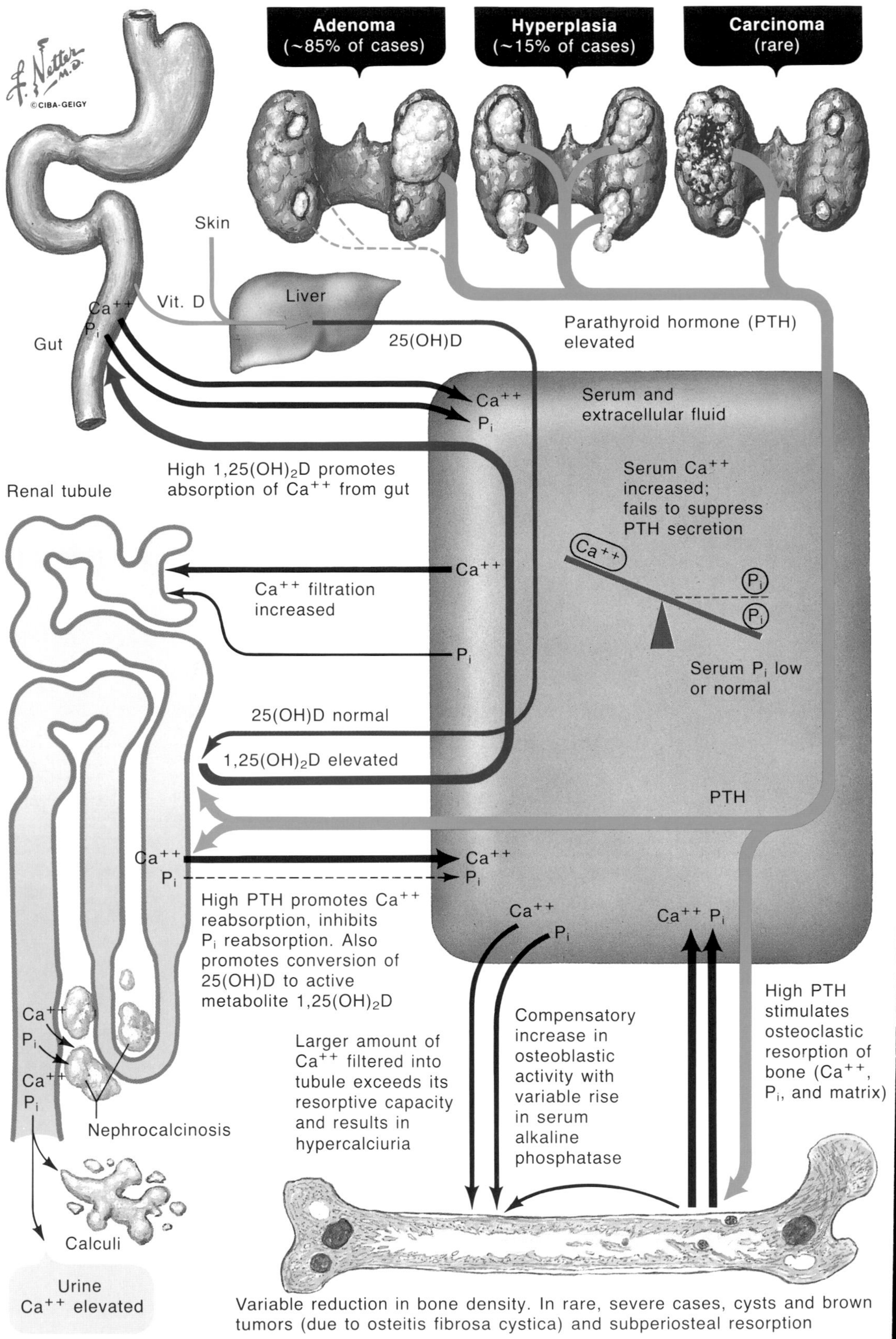
F. Netter M.D.
©CIBA-GEIGY
Adenoma
(~85% of cases)
Hyperplasia
(~15% of cases)
Carcinoma
(rare)
Skin
Vit. D
Liver
25(OH)D
Gut
Ca^{++}
P_i
Parathyroid hormone (PTH) elevated
Serum and extracellular fluid
Ca^{++}
P_i
High $1,25(OH)_2D$ promotes absorption of Ca^{++} from gut
Renal tubule
Serum Ca^{++} increased; fails to suppress PTH secretion
Ca^{++}
P_i
Ca^{++} filtration increased
Serum P_i low or normal
25(OH)D normal
$1,25(OH)_2D$ elevated
PTH
Ca^{++}
P_i
High PTH promotes Ca^{++} reabsorption, inhibits P_i reabsorption. Also promotes conversion of 25(OH)D to active metabolite $1,25(OH)_2D$
Larger amount of Ca^{++} filtered into tubule exceeds its resorptive capacity and results in hypercalciuria
Compensatory increase in osteoblastic activity with variable rise in serum alkaline phosphatase
High PTH stimulates osteoclastic resorption of bone (Ca^{++}, P_i, and matrix)
Nephrocalcinosis
Calculi
Urine
Ca^{++} elevated
Variable reduction in bone density. In rare, severe cases, cysts and brown tumors (due to osteitis fibrosa cystica) and subperiosteal resorption

Pathologic Physiology

Primary hyperparathyroidism is caused by excessive production of parathyroid hormone (PTH) by enlarged parathyroid glands (Plate 2). Factors causing the gland enlargement and the loss of the suppressive effects of calcium are not known. In 85% of cases, only a single parathyroid gland is enlarged (adenoma); in 15% of cases, all four glands are enlarged (hyperplasia). These two pathologic types of hyperparathyroidism have identical clinical manifestations and can only be distinguished at surgery. Parathyroid carcinoma occurs in less than 1% of patients, and severe hypercalcemia is often the initial symptom.

Hypercalcemia results from increased resorption of bone, reabsorption of calcium in the renal tubule, and absorption of calcium from the gut. PTH also decreases reabsorption of phosphate in the renal tubule, and in moderate-to-severe hyperparathyroidism, the serum phosphate level is reduced; in the mild form of the disease, the serum phosphate level is often normal.

Clinical Manifestations

Pathologic manifestations of hyperparathyroidism result from either a high PTH level or a high serum calcium level. In bone, PTH activates osteoclastic resorption of bone, which can lead to significant bone loss. There is also a compensatory increase in osteoblastic bone formation; however, resorption ultimately exceeds formation, leading to bone loss. Osteoblasts release alkaline phosphatase, and serum levels may be elevated in patients with significant bone involvement. In severe cases of hyperparathyroidism, there may be large cystic areas eroded from bone and fibrous tissue in adjacent areas of bone marrow (osteitis fibrosa cystica).

Nephrolithiasis (urinary calculi) occurs in approximately 10% of patients with hyperparathyroidism (Plate 3). Nephrocalcinosis is rarer and is typically seen in severe hyperparathyroidism with bone involvement. PTH increases calcium reabsorption in the renal tubule, but this does not compensate for the increased renal filtration produced by elevated serum calcium levels; the result is hypercalciuria. In some patients, calcium precipitates in the renal tubule and forms urinary calculi, common manifestations of hyperparathyroidism. In severe hypercalcemia, precipitates can occur in the renal interstitium and incite an inflammatory reaction. By the time interstitial calcium deposits are visible on radiography, renal function is already considerably reduced. Hypercalcemia also decreases the capacity to concentrate urine, which frequently leads to polyuria. Pancreatitis and peptic ulcers, although rare, may occur with hyperparathyroidism. Gastrin-secreting tumors associated with hyperparathyroidism may cause ulcers in patients with multiple endocrine neoplasia (MEN) type I syndrome. Prolonged or severe hypercalcemia often results in calcium deposits in the medial and lateral edges of the cornea (band keratopathy).

Hyperparathyroidism was considered a rare disorder until 2 decades ago, when routine screening of the serum calcium level by automated techniques greatly increased its recognition. Most patients have mild hypercalcemia (serum calcium level <12 mg/100 ml) and exhibit either no overt signs or mild, nonspecific signs and symptoms such as fatigue, constipation, and nocturia. More rarely, mental confusion, anorexia, nausea, and vomiting develop in patients with higher calcium levels; unexplained anemia and weight loss may also occur. In a small number of patients, there is clinical or radiographic evidence of hyperparathyroid bone disease. Typical findings include serum calcium levels >12 mg/100 ml, serum PTH levels several times higher than normal, high serum alkaline phosphatase, and diffuse bone pain.

In patients with severe hyperparathyroidism with bone disease, radiographs may show subperiosteal bone resorption (highly specific to hyperparathyroidism) around the phalanges and distal ends of the clavicles, and diffuse decalcification of the skull (salt-and-pepper skull) that resembles multiple myeloma. Bone cysts, if present, are often the sites of pathologic fractures. With bone loss in the spine, the intervertebral discs herniate into the vertebral bodies, creating a "codfish" appearance on radiographs. Even if there is no radiographic evidence of bone disease, mildly excessive, PTH-mediated bone resorption may increase the risk of osteoporosis, a situation of

Clinical Manifestations of Primary Hyperparathyroidism

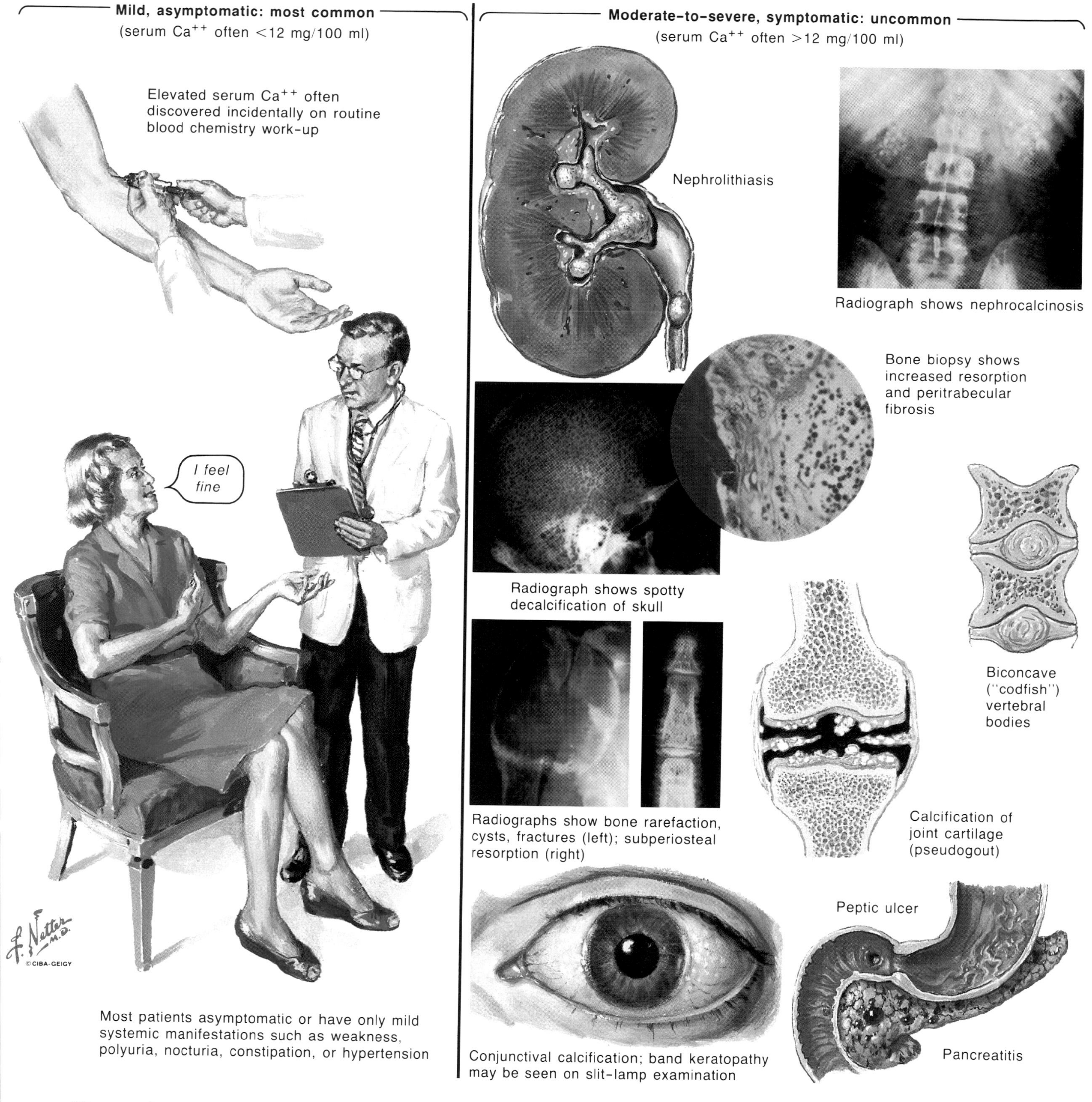

Mild, asymptomatic: most common
(serum Ca^{++} often <12 mg/100 ml)
Elevated serum Ca^{++} often discovered incidentally on routine blood chemistry work-up
I feel fine
F. Netter M.D.
©CIBA-GEIGY
Most patients asymptomatic or have only mild systemic manifestations such as weakness, polyuria, nocturia, constipation, or hypertension
Moderate-to-severe, symptomatic: uncommon
(serum Ca^{++} often >12 mg/100 ml)
Nephrolithiasis
Radiograph shows nephrocalcinosis
Bone biopsy shows increased resorption and peritrabecular fibrosis
Radiograph shows spotty decalcification of skull
Biconcave ("codfish") vertebral bodies
Radiographs show bone rarefaction, cysts, fractures (left); subperiosteal resorption (right)
Calcification of joint cartilage (pseudogout)
Peptic ulcer
Pancreatitis
Conjunctival calcification; band keratopathy may be seen on slit-lamp examination

SECTION IV PLATE 3 Slide 3738

Primary Hyperparathyroidism

(Continued)

particular concern in postmenopausal women because hyperparathyroidism is most prevalent in this group, which is already at risk for osteoporosis.

Treatment

Surgical excision of the single, enlarged parathyroid gland is the treatment of choice for patients with an adenoma. Treatment for hyperplasia is subtotal parathyroidectomy, or removal of all but one gland or one-half. However, excision of too little tissue leads to persistent hypercalcemia, and removal of too much tissue causes hypoparathyroidism.

Three familial autosomal dominant syndromes of hyperparathyroidism have been identified among patients with parathyroid hyperplasia. In two of these syndromes, MEN I and II, neoplasms occur in other endocrine glands. MEN I is characterized by pituitary adenomas and islet cell tumors in the pancreas, while MEN II is characterized by a high incidence of pheochromocytoma and medullary carcinoma of the thyroid. Patients with familial hypocalciuric hypercalcemia, the third and rarest syndrome, exhibit lifelong asymptomatic hypercalcemia that persists even after subtotal parathyroidectomy. This syndrome is distinguished from typical hyperparathyroidism by an absence of hypercalciuria and a high familial incidence of asymptomatic hypercalcemia without other endocrinopathies.

Cancer of the parathyroid glands can rarely be cured by surgical excision, because it has metastasized by the time it is detected. Patients with parathyroid carcinoma usually die of uncontrollable hypercalcemia. □

Differential Diagnosis of Hypercalcemic States

Condition	Serum Ca^{++}	Serum P_i	Serum PTH	Serum 25(OH)D	Serum $1,25(OH)_2D$	Associated findings
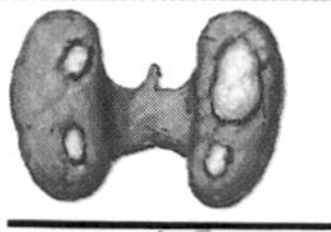 Primary hyperparathyroidism	↑	N or ↓	↑	N	N or ↑	See Plates 2–3
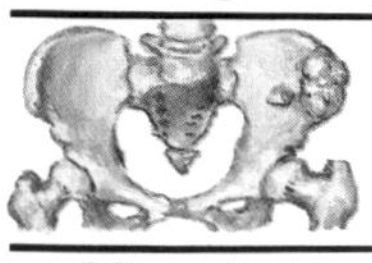 Cancer with extensive bone metastases	↑	N or ↑	N or ↓	N	↓ or N	History of primary tumor, destructive lesions on radiograph, bone scan
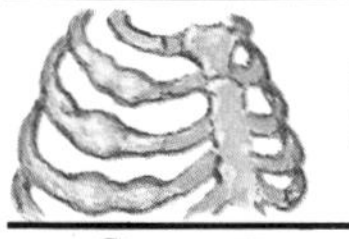 Multiple myeloma and lymphoma	↑	N or ↑	N or ↓	N	↓ or N	Abnormal serum or urine protein electrophoresis, abnormal bone radiographs
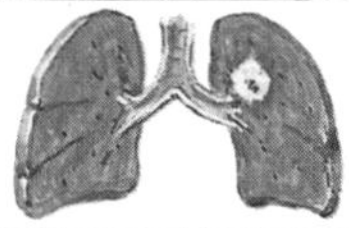 Primary carcinoma (not involving bone)	↑	N or ↓	N or ↓	N	↓ or N	History of primary tumor, chest radiograph, CT, bronchoscopy, IV pyelogram
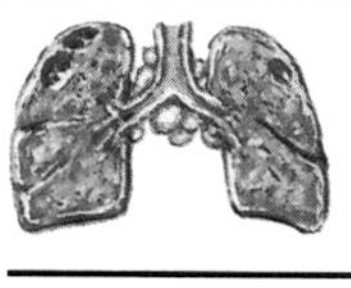 Sarcoidosis and other granulomatous diseases	↑	N or ↑	N or ↓	N	↑	Hilar adenopathy ± interstitial lung disease, elevated angiotensin-converting enzyme
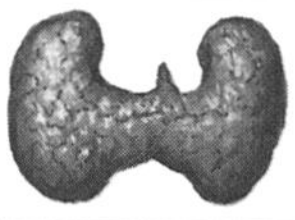 Hyperthyroidism	↑	N	N or ↓	N	N	Symptoms of hyperthyroidism, elevated serum thyroxine
 Vitamin D intoxication	↑	N or ↑	N or ↓	Very ↑	N	History of excessive vitamin D intake
 Milk-alkali syndrome	↑	N or ↑	N or ↓	N	N or ↓	History of excessive calcium and alkali ingestion, heavy use of over-the-counter calcium-containing antacids
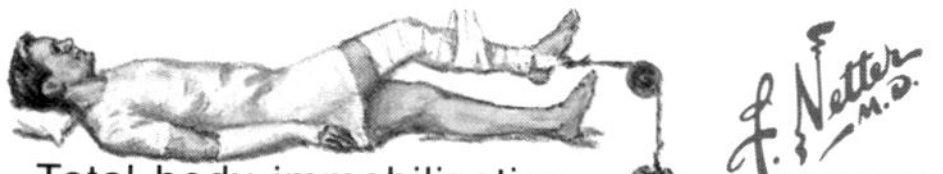 Total body immobilization	↑	N or ↑	N or ↓	N	↓ or N	Multiple fractures, paralysis (children, adolescents, patients with Paget's disease of bone)

Differential Diagnosis of Hypercalcemic States

The cause of hypercalcemia can often be determined on the basis of the patient's history and a careful physical examination, because most of the disorders that give rise to hypercalcemia (aside from primary hyperparathyroidism) are clinically apparent by the time the condition occurs. Tests that measure serum levels of parathyroid hormone (PTH), inorganic phosphate, and vitamin D metabolites aid in the diagnosis.

The serum level of PTH distinguishes primary hyperparathyroidism from other disorders causing hypercalcemia, since it is the only hypercalcemic disorder resulting from excessive production of PTH. Serum phosphate levels are inversely proportional to PTH levels because PTH promotes the urinary excretion of phosphate. Although hypophosphatemia suggests hyperparathyroidism, it may also be present in patients with cancer, since certain tumors secrete phosphaturic factors. High levels of 25-hydroxyvitamin D, or 25(OH)D (also known as calcidiol), suggest an excessive intake of vitamin D. A very high concentration of this usually inactive compound can result in hypercalcemia. A high concentration of 1,25-dihydroxyvitamin D, or $1,25(OH)_2D$ (also known as calcitriol), occurs only in hyperparathyroidism or in granulomatous diseases.

Primary hyperparathyroidism is the most common cause of hypercalcemia; often, the hypercalcemia is the only clinical manifestation (Plates 2–3). In patients with mild-to-moderate hyperparathyroidism, radiographs usually do not reveal bone disease, and the serum phosphate level is often normal. Also, serum $1,25(OH)_2D$ levels are elevated in less than 50% of patients. Therefore, the most specific and reliable tool in the diagnosis of hyperparathyroidism is the measurement of serum PTH level.

Malignancies cause hypercalcemia by increasing bone resorption either locally through skeletal metastases (breast cancer and multiple myeloma) or by secreting hormonal factors that stimulate resorption (lung and renal cell cancer). Hypercalcemia is usually a late manifestation of hypercalcemia due to a nonparathyroid mechanism and usually occurs only after the tumor is apparent. The presence of such a tumor is therefore an important clue to the etiology of hypercalcemia. Hypercalcemia is rarely the first manifestation of occult cancer, but is more typically seen in patients with widespread end-stage disease.

Chronic disseminated sarcoidosis and, rarely, other granulomatous diseases such as tuberculosis and histoplasmosis also lead to hypercalcemia, because of an increase in the synthesis of $1,25(OH)_2D$. Hyperparathyroidism is the only other disease in which the level of $1,25(OH)_2D$ is elevated, but the PTH level is also elevated. In sarcoidosis, by contrast, the PTH level is normal or reduced and the level of the angiotensin-converting enzyme (ACE) activity is usually high. Although this finding is not specific to sarcoidosis, the absence of high ACE activity levels makes the diagnosis very unlikely.

Hyperthyroidism, or thyrotoxicosis, rarely causes mild hypercalcemia. The diagnosis is based on clinical and laboratory evidence of hyperthyroidism and restoration of a normal serum calcium level with antithyroid therapy.

The *milk-alkali syndrome* is typically caused by excessive intake of both milk and absorbable alkalis such as sodium bicarbonate and calcium carbonate. Because absorbable alkalis are not often used as antacids, this syndrome is rare; however, ingestion of large amounts of calcium (usually more than 4 g/day) can produce the syndrome. The diagnosis is made by a careful history and by observing the patient's response to withdrawal of calcium.

Immobilization due to fractures or paralysis increases calcium release from bone. However, hypercalcemia is uncommon except when the rate of bone turnover is high, as in childhood or adolescence, and in extensive Paget's disease. □

Pathologic Physiology of Hypoparathyroidism

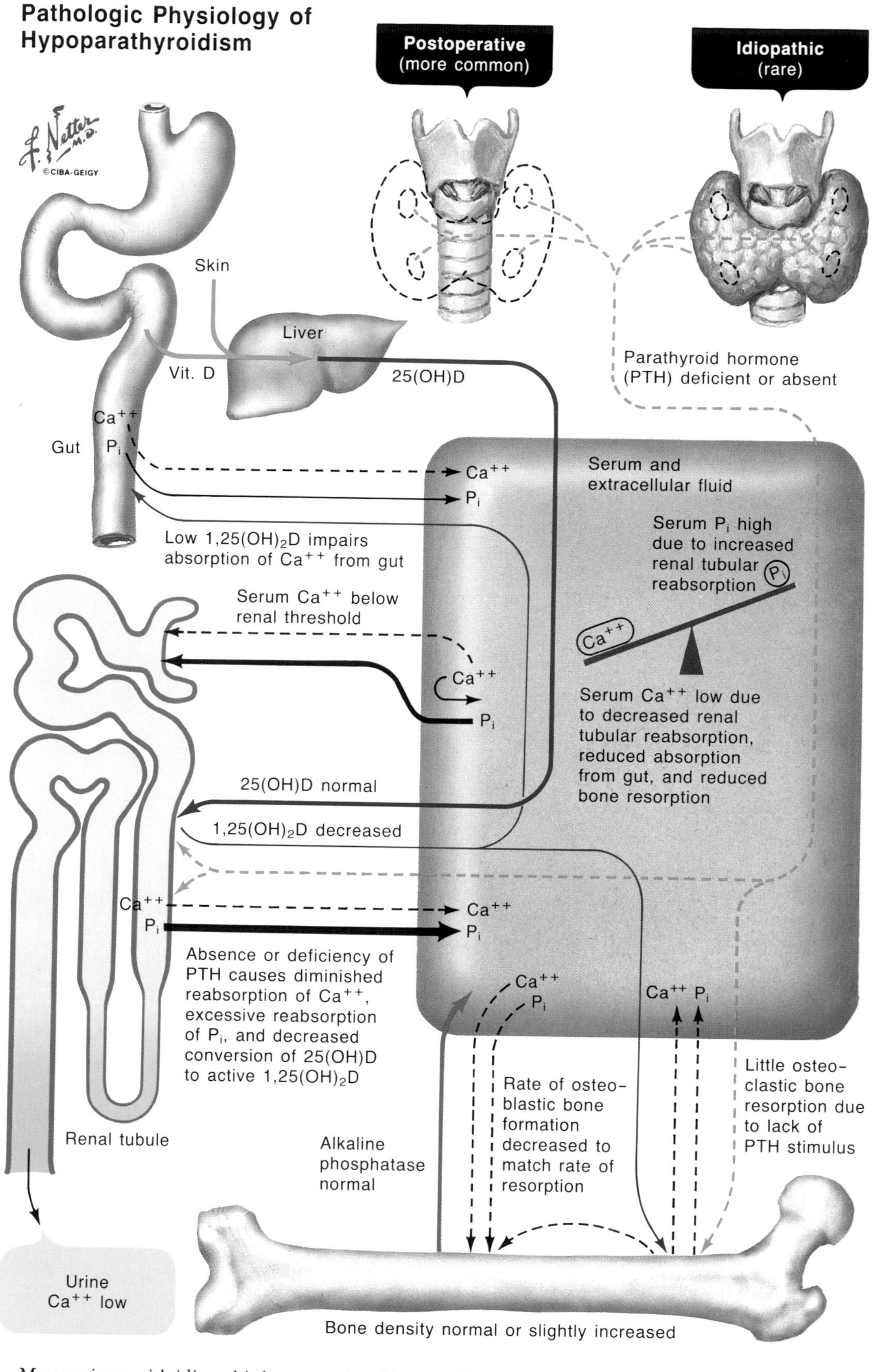
Pathologic Physiology of Hypoparathyroidism
Postoperative (more common)
Idiopathic (rare)
©CIBA-GEIGY
Skin
Liver
Vit. D
25(OH)D
Parathyroid hormone (PTH) deficient or absent
Gut
Ca++
Pi
Serum and extracellular fluid
Low 1,25(OH)2D impairs absorption of Ca++ from gut
Serum Pi high due to increased renal tubular reabsorption
Serum Ca++ below renal threshold
Serum Ca++ low due to decreased renal tubular reabsorption, reduced absorption from gut, and reduced bone resorption
25(OH)D normal
1,25(OH)2D decreased
Absence or deficiency of PTH causes diminished reabsorption of Ca++, excessive reabsorption of Pi, and decreased conversion of 25(OH)D to active 1,25(OH)2D
Rate of osteoblastic bone formation decreased to match rate of resorption
Little osteoclastic bone resorption due to lack of PTH stimulus
Renal tubule
Alkaline phosphatase normal
Urine Ca++ low
Bone density normal or slightly increased

The most common cause of hypoparathyroidism is the inadvertent removal or destruction of the parathyroid glands during thyroid or parathyroid surgery. Idiopathic hypoparathyroidism is rarer and usually begins in childhood; its occurrence may be sporadic or familial. Current evidence suggests that, in many cases, the parathyroid glands are destroyed by an autoimmune mechanism. One form of idiopathic hypoparathyroidism is accompanied by deficiencies of other endocrine glands; thus, patients may also have Addison's disease, diabetes mellitus, primary hypothyroidism, or primary hypogonadism. In some patients, a defect in cellular immunity also leads to chronic mucocutaneous candidiasis (Plate 6).

Transient hypoparathyroidism and mild hypocalcemia are common in the neonatal period, presumably because of the underactivity and immaturity of the parathyroid glands. Maternal hypercalcemia (as seen in hyperparathyroidism) may further suppress the fetal parathyroid gland and produce tetany in the newborn. In patients with alcoholism or malabsorption syndromes, hypomagnesemia leads to functional impairment of the parathyroid glands and hypocalcemia. In these patients, magnesium replacement increases serum levels of both parathyroid hormone (PTH) and calcium.

The biochemical hallmarks of hypoparathyroidism are a low serum calcium level and a high serum phosphate level, which result from a lack of PTH. Hypocalcemia occurs because less calcium is absorbed from the gut and more is cleared by the kidney. Absorption of calcium from the gut is reduced because synthesis of 1,25-dihydroxyvitamin D, or $1,25(OH)_2D$, which enhances absorption, is decreased in the absence of PTH. Because the renal tubular reabsorption of calcium and the threshold for calcium excretion are reduced, when the serum calcium level is therapeutically raised to normal, urinary excretion of calcium is elevated in patients with hypoparathyroidism. However, if the condition is not treated, the serum calcium concentration is usually below the renal threshold, and urinary excretion of calcium is therefore low. Hyperphosphatemia in hypoparathyroidism occurs because phosphate reabsorption by the renal tubule increases when PTH levels are low.

Although acute reduction of PTH diminishes bone resorption, when PTH levels are chronically low, the rate of bone formation falls to match the rate of bone resorption. In chronic hypoparathyroidism, the net results are a reduced rate of bone turnover and a normal or slightly increased bone mass.

Most patients with idiopathic hypoparathyroidism usually exhibit severe hypocalcemia (serum calcium level <7 mg/100 ml), while in patients with postsurgical hypoparathyroidism, there is a wide spectrum of severity. Patients with the mildest form may have latent hypoparathyroidism; they can maintain normal serum calcium and phosphate levels, although physiologic or pathologic stresses, such as pregnancy or diarrhea, may cause the onset of hypocalcemia. Patients with postsurgical hypoparathyroidism may also manifest moderate-to-severe hypocalcemia.

Hypoparathyroidism is treated by a combination of oral calcium supplements and vitamin D; in mild cases, calcium supplementation alone is sufficient. In most patients, calcium absorption from the intestine is too low, necessitating some form of supplemental vitamin D (in addition to calcium salts) to enhance absorption. Because the activation of vitamin D is impaired in these patients, large amounts are required (50,000–100,000 IU/day). In contrast, $1,25(OH)_2D$ (calcitriol) and dihydrotachysterol can be used at much lower doses because they bypass the vitamin D–converting enzyme block in the kidney. Because the renal threshold for calcium is lower in patients with hypoparathyroidism, the total serum calcium concentration should be maintained in the low-to-normal range (8–9 mg/100 ml) to avoid hypercalciuria and nephrolithiasis. □

Clinical Manifestations of Chronic Hypoparathyroidism

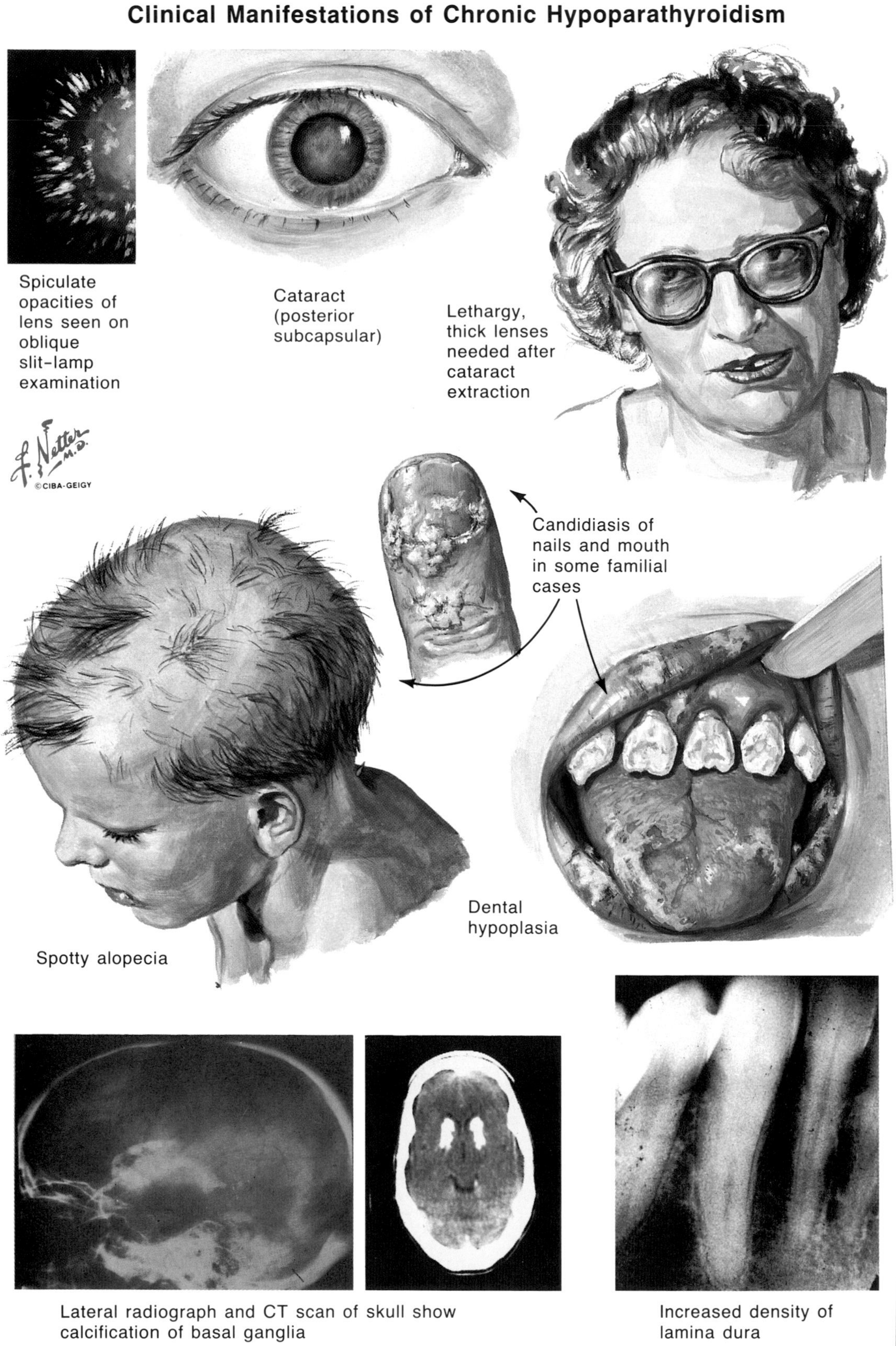
Clinical Manifestations of Chronic Hypoparathyroidism
Spiculate opacities of lens seen on oblique slit-lamp examination
Cataract (posterior subcapsular)
Lethargy, thick lenses needed after cataract extraction
F. Netter M.D.
©CIBA-GEIGY
Candidiasis of nails and mouth in some familial cases
Dental hypoplasia
Spotty alopecia
Lateral radiograph and CT scan of skull show calcification of basal ganglia
Increased density of lamina dura

Although the manifestations of acute hypocalcemia as seen in hypoparathyroidism are not usually overlooked, the clinical signs and symptoms of chronic hypocalcemia are subtle and may be overlooked. For this reason, chronic hypocalcemia is often diagnosed incidentally during investigation of nonspecific symptoms, usually mental, which include lassitude, irritability, depression, or even psychosis. These symptoms are due primarily to the hypocalcemia. There may also be evidence of increased neuromuscular excitability, with signs and symptoms ranging from "pins and needles" sensations around the mouth and in the hands and feet to tetany with muscle cramps and spasms, laryngeal stridor, and seizures. Chvostek's or Trousseau's sign may be elicited even in patients with asymptomatic hypocalcemia.

Despite the hypocalcemia, soft-tissue calcifications may develop in patients with hypoparathyroidism. A patient with poorly controlled, long-standing hypoparathyroidism may develop calcifications in the lens (cataracts) that opacify the lens and can impair vision. Calcifications may also develop in the basal ganglia and, if they are extensive, cause a movement disorder with features of Parkinson's disease. These calcifications can be seen on standard radiographs of the skull or on computed tomography (CT), a more sensitive technique.

The condition of the teeth provides a clue to the patient's age at onset of the disease. Dental hypoplasia with poor dental root formation indicates that the disease occurred before age 6. If onset was during childhood, there is crumbling of the teeth because of poor enamel structure.

In patients with hypoparathyroidism, the skeleton is usually not demineralized; in most cases, bone density is normal or slightly increased.

Hypoparathyroidism can occur as part of a familial tendency to the development of autoimmune destruction of several endocrine glands. Family members have an increased incidence of autoimmune primary hypothyroidism, adrenal insufficiency, diabetes mellitus, and ovarian failure. Alopecia, vitiligo, and pernicious anemia also occur with increased frequency. Another familial autoimmune polyglandular syndrome that produces hypoparathyroidism is associated with a defect in cell-mediated immunity and an absence of delayed cutaneous hypersensitivity reactions to *Candida*. These patients may have chronic *Candida* infections of the skin, especially the hands, toes, and nails, as well as infections of the oral mucosa and vagina, but systemic candidiasis is not a feature of this syndrome. Occasionally, these lesions respond to long-term antifungal therapy. □

Clinical Manifestations of Hypocalcemia

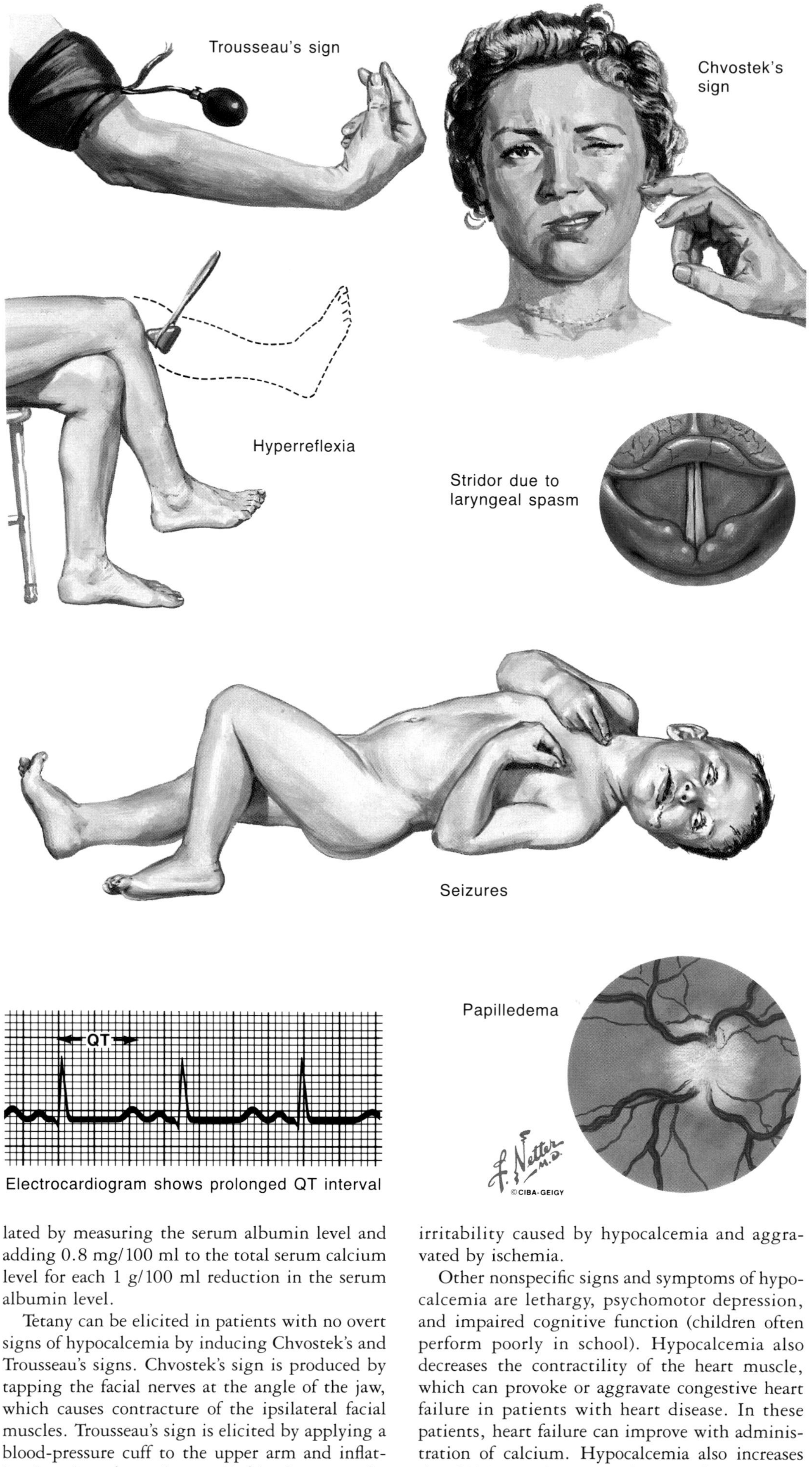
Trousseau's sign
Chvostek's sign
Hyperreflexia
Stridor due to laryngeal spasm
Seizures
QT
Papilledema
Electrocardiogram shows prolonged QT interval
©CIBA-GEIGY

Hypocalcemia increases neuromuscular excitability, which can lead to tetany. The most severe form of tetany is characterized by tonic contractions of the muscles of the forearm and hand and, less commonly, by laryngospasm and seizures ranging from classic types (generalized, or grand mal, and focal) to brief "gray-out" spells. Often it is not clear if hypocalcemia is the direct cause of seizures or if it lowers the seizure threshold in a patient with a predisposition for epilepsy.

More typically, patients with hypocalcemia experience milder symptoms such as muscle cramps and paresthesias. The paresthesias, described as "pins and needles" sensations in the hands, feet, and around the mouth, are episodic and often occur at times of stress, vomiting, or hyperventilation. This can be explained by the fact that metabolic or respiratory alkalosis increases the binding of serum calcium to albumin and decreases the concentration of free ionized calcium that interacts with cells.

Symptoms of hypocalcemia are also more likely to occur when the serum calcium level has fallen abruptly; chronic hypocalcemia, in contrast, can be asymptomatic with very low levels of serum calcium. Asymptomatic hypocalcemia must be differentiated from the low total serum calcium concentration (with a normal ionized calcium level) that occurs with hypoalbuminemia. The corrected total serum calcium concentration can be calculated by measuring the serum albumin level and adding 0.8 mg/100 ml to the total serum calcium level for each 1 g/100 ml reduction in the serum albumin level.

Tetany can be elicited in patients with no overt signs of hypocalcemia by inducing Chvostek's and Trousseau's signs. Chvostek's sign is produced by tapping the facial nerves at the angle of the jaw, which causes contracture of the ipsilateral facial muscles. Trousseau's sign is elicited by applying a blood-pressure cuff to the upper arm and inflating it to just above the systolic blood pressure for 3 minutes. The resulting carpopedal spasm, with contractions of the fingers and inability to open the hand, is a result of increased neuromuscular irritability caused by hypocalcemia and aggravated by ischemia.

Other nonspecific signs and symptoms of hypocalcemia are lethargy, psychomotor depression, and impaired cognitive function (children often perform poorly in school). Hypocalcemia also decreases the contractility of the heart muscle, which can provoke or aggravate congestive heart failure in patients with heart disease. In these patients, heart failure can improve with administration of calcium. Hypocalcemia also increases the QT interval on the electrocardiogram. An unusual ocular manifestation of chronic hypocalcemia is papilledema, caused by increased pressure of the cerebrospinal fluid. □

Pathologic Physiology and Characteristic Signs of Pseudohypoparathyroidism

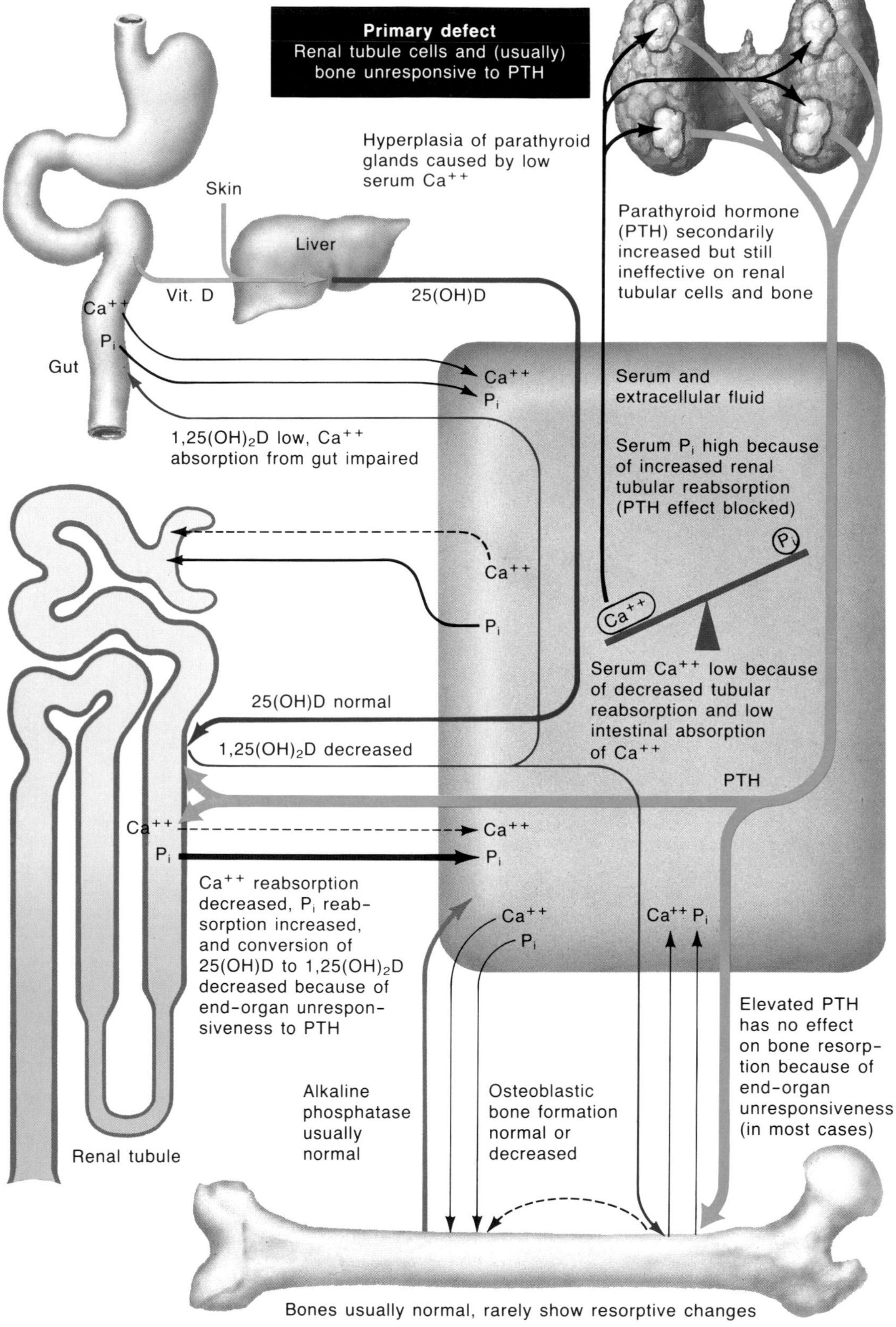

Primary defect
Renal tubule cells and (usually) bone unresponsive to PTH
Hyperplasia of parathyroid glands caused by low serum Ca^{++}
Skin
Liver
Vit. D
25(OH)D
Ca^{++}
P_i
Gut
$1,25(OH)_2D$ low, Ca^{++} absorption from gut impaired
Parathyroid hormone (PTH) secondarily increased but still ineffective on renal tubular cells and bone
Serum and extracellular fluid
Serum P_i high because of increased renal tubular reabsorption (PTH effect blocked)
Serum Ca^{++} low because of decreased tubular reabsorption and low intestinal absorption of Ca^{++}
PTH
25(OH)D normal
$1,25(OH)_2D$ decreased
Ca^{++} reabsorption decreased, P_i reabsorption increased, and conversion of 25(OH)D to $1,25(OH)_2D$ decreased because of end-organ unresponsiveness to PTH
Renal tubule
Alkaline phosphatase usually normal
Osteoblastic bone formation normal or decreased
Elevated PTH has no effect on bone resorption because of end-organ unresponsiveness (in most cases)
Bones usually normal, rarely show resorptive changes

Albright's Hereditary Osteodystrophy

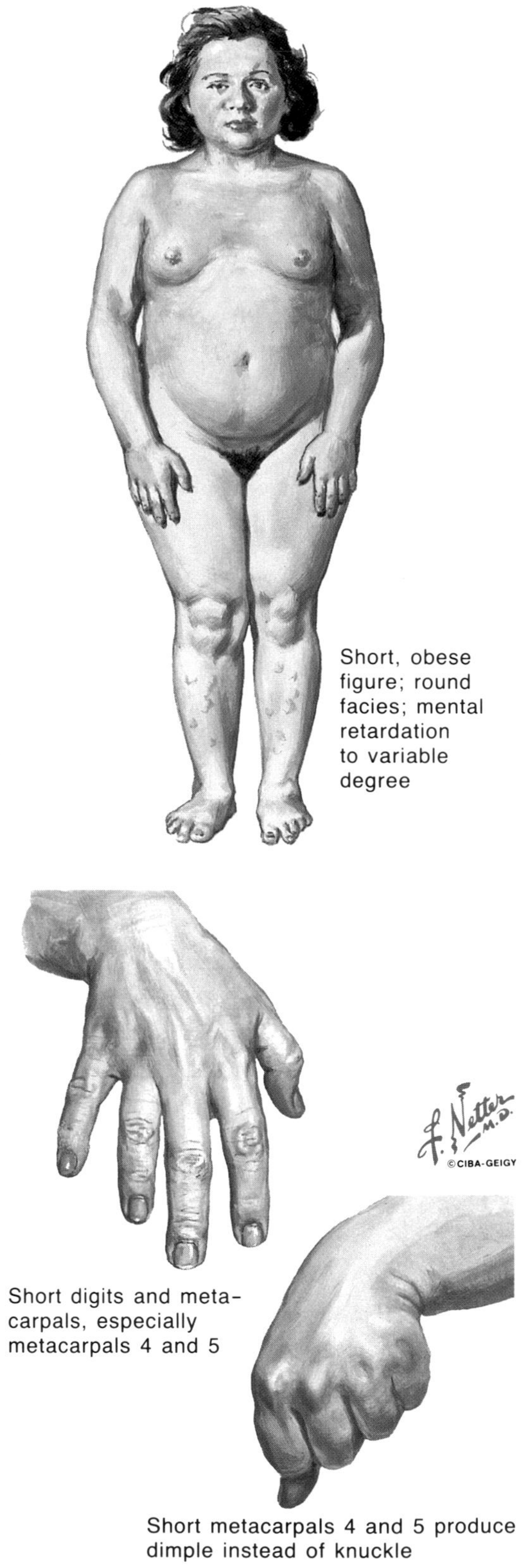

Short, obese figure; round facies; mental retardation to variable degree
Short digits and metacarpals, especially metacarpals 4 and 5
Short metacarpals 4 and 5 produce dimple instead of knuckle
F. Netter M.D.
©CIBA-GEIGY

SECTION IV PLATE 8 Slide 3743

Pseudohypoparathyroidism

In pseudohypoparathyroidism, tissues such as the kidney and bone fail to respond to the action of parathyroid hormone (PTH). Signs, symptoms, and laboratory findings are those of hypoparathyroidism (Plates 5–7). Both hypocalcemia and hyperphosphatemia are present, but the parathyroid glands are enlarged. Many patients also have a characteristic physical appearance known as Albright's hereditary osteodystrophy (AHO). Associated findings are mild primary hypothyroidism and, occasionally, primary hypogonadism.

PTH activates its target cells by increasing cellular levels of cyclic adenosine monophosphate (cyclic AMP), which activates a cascade of proteins that produces the physiologic effect. In most patients, especially in those with AHO, a component of the PTH receptor–adenylate cyclase system in the enzyme complex that synthesizes cyclic AMP is deficient. Although PTH binds to the cell, it fails to elicit an effect because there is no production of its second messenger, cyclic AMP. Therefore, the biochemical abnormalities of hypoparathyroidism develop. Hypocalcemia occurs because of increased calcium clearance by the kidney and decreased calcium absorption from the intestine due to low PTH-mediated synthesis of $1,25(OH)_2D$. The serum phosphate level is high as a result of the decreased renal clearance. In contrast to hypoparathyroidism, the serum PTH level is elevated in response to hypocalcemia.

Bone is also resistant to PTH, although in some cases, osteitis fibrosa cystica occurs as a result of high PTH levels. As in hypoparathyroidism, the hypocalcemia ranges from latent to severe. Treatment for pseudohypoparathyroidism is the same as that for hypoparathyroidism. □

Mechanism of Parathyroid Hormone Activity on End Organ

Mechanism of Parathyroid Hormone (PTH) Activity on End Organ

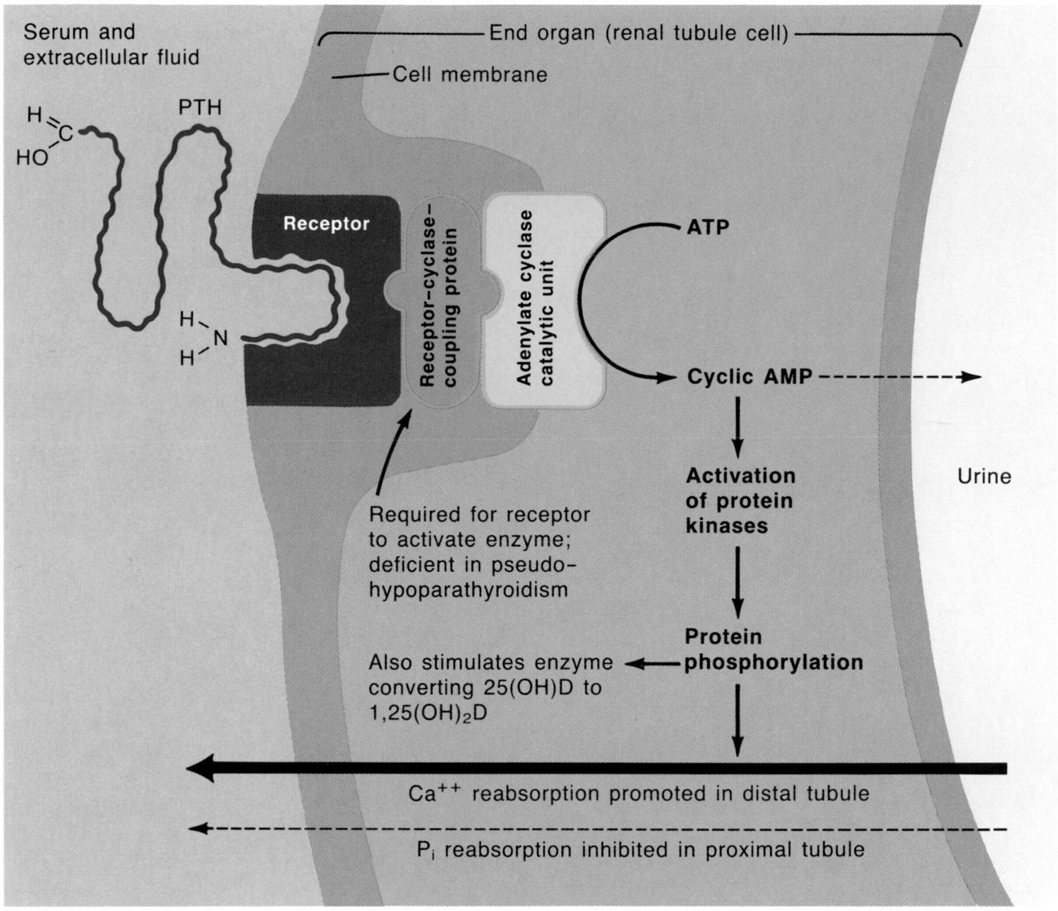

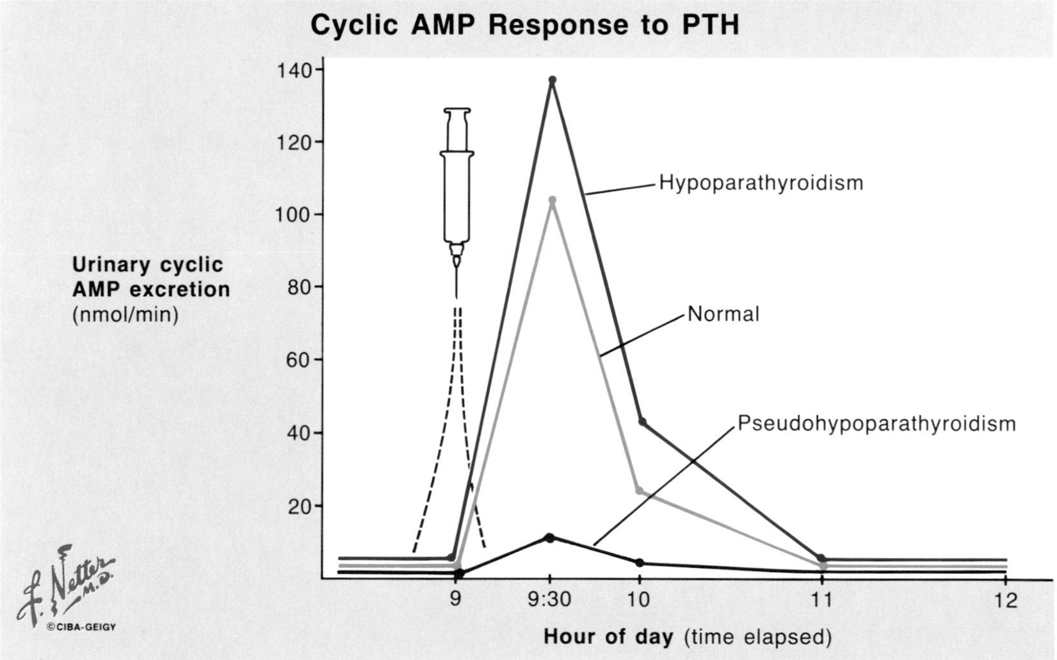

Parathyroid hormone (PTH) has two target organs, kidney and bone. In the kidney, PTH enhances the reabsorption of calcium in the distal tubule and decreases the reabsorption of phosphate in the proximal tubule. It also increases the synthesis of 1,25-dihydroxyvitamin D, or $1,25(OH)_2D$ (calcitriol), the active form of vitamin D, from its precursor 25-hydroxyvitamin D, or 25(OH)D (calcidiol). In bone, PTH stimulates the release of mineral. Initially, there is a rapid activation of existing osteoclasts, the large multinucleated bone cells that resorb bone. These cells resorb mineralized bone and release calcium, phosphate, and fragments of bone protein matrix into the circulation. After this initial phase, new osteoclasts are also recruited. There is also a compensatory increase in bone formation by osteoblasts (the process of bone remodeling is highly coordinated); however, the net effect is bone resorption.

PTH produces its effects on target cells by stimulating the synthesis of cyclic adenosine monophosphate (cyclic AMP) by the enzyme adenylate cyclase. This intracellular second messenger activates the protein kinase, which catalyzes the phosphorylation of several cellular proteins and thereby modifies their activity.

Although the targets of phosphorylation have not been identified, they are likely to include proteins involved in the transport of calcium and phosphate. To activate adenylate cyclase, PTH binds to receptor molecules on the surface of the cell. The first segment contains 33 to 34 of the 84 amino acids in the hormone and is the only component required for binding and activating receptor molecules; the function, if any, of the remainder of the peptide is unknown (Plate 1).

The catalytic unit of adenylate cyclase is a separate molecule on the inner surface of the cell membrane. However, for the catalytic unit to convert adenosine triphosphate (ATP) to cylic AMP, it must interact with the PTH-receptor complex as well as with a third membrane protein, the guanine nucleotide, or receptor-cyclase–coupling protein (this protein is deficient in pseudohypoparathyroidism). Cyclic AMP is rapidly degraded by the enzyme phosphodiesterase, although some of it also leaks out of the cell. In the kidney, cyclic AMP produced under the influence of PTH leaks into the renal tubule and is excreted in the urine.

Normally, about half of the cyclic AMP in urine is derived from the renal action of PTH; the other half comes from circulating cyclic AMP, which is filtered through the glomerulus. Measurement of cyclic AMP excreted in the urine can be used as an index of the level of circulating PTH; excretion of cyclic AMP is increased in hyperparathyroidism and decreased in hypoparathyroidism.

In pseudohypoparathyroidism, cyclic AMP is not synthesized in response to PTH, because of a deficiency of the coupling protein. As a result, little cyclic AMP is produced in target tissues in response to PTH, and functional hypoparathyroidism develops (Plate 7). This defect can be demonstrated in patients by measuring the level of cyclic AMP in urine after an injection of PTH. In normal persons and in patients with hypoparathyroidism, the rise in the excretion of cyclic AMP in urine is rapid and marked; in patients with pseudohypoparathyroidism, it is blunted or absent. □

Clinical Guide to Parathyroid Hormone Assay

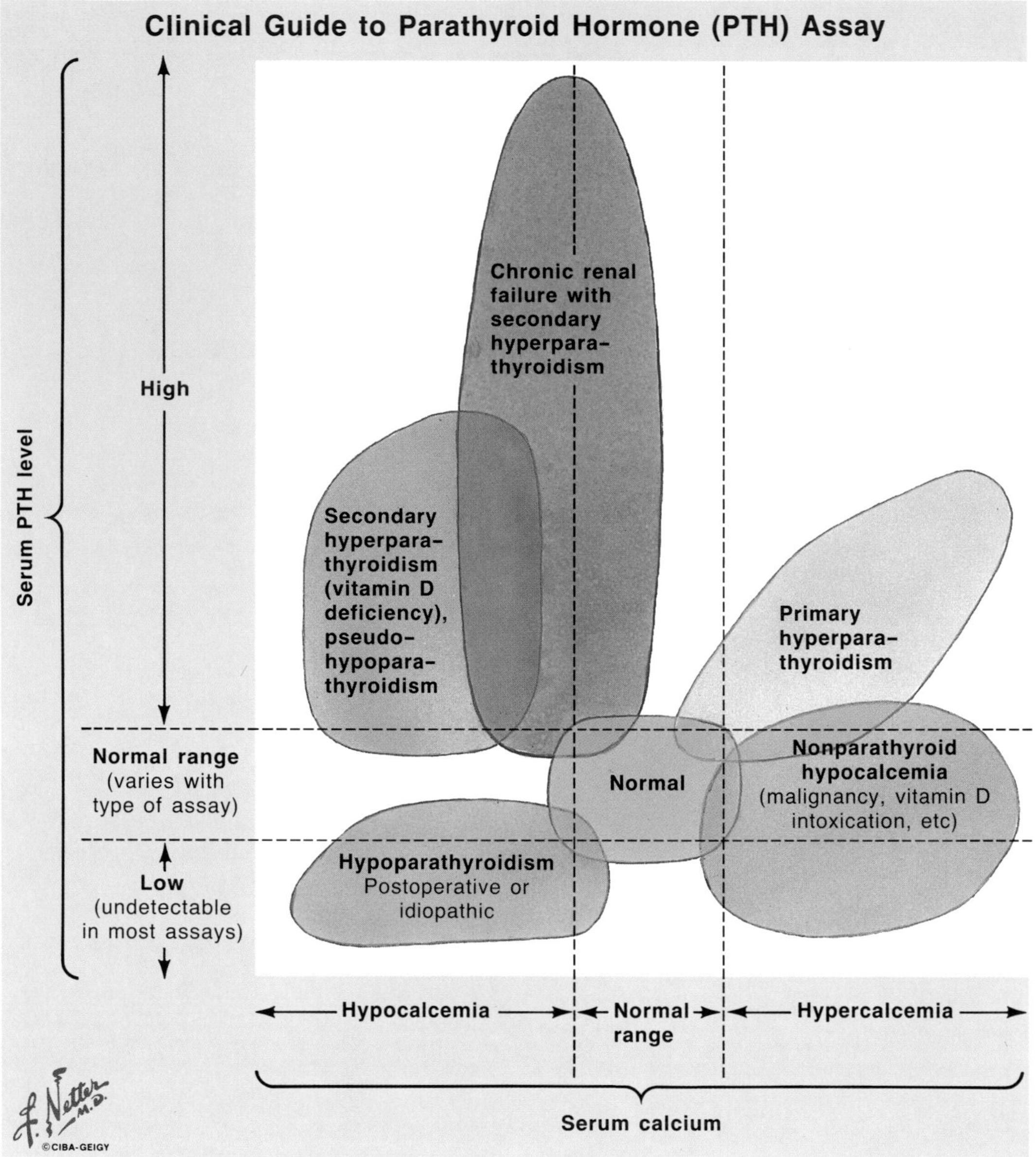

The concentration of parathyroid hormone (PTH) in serum can be measured with a variety of sensitive, commercially available radioimmunoassays. These techniques measure the binding of PTH to specific antibodies that recognize only a portion of the PTH molecule. Since most of the circulating PTH is in the form of fragments, assays usually detect fragments of the peptide rather than the entire hormone.

Within minutes of secretion, PTH is cleaved into two principal fragments. The first, the amino-terminal, or N-terminal, fragment contains the first 33 to 34 amino acids and is the biologically active region of the peptide; the second, the carboxyl-terminal, or C-terminal, fragment contains the remaining 34 to 84 amino acids (Plate 1). The N-terminal fragment, like the whole molecule, has a short half-life and disappears rapidly from the circulation. In contrast, the half-life of the C-terminal is several times longer than that of either the N-terminal or the intact hormone; this fragment is therefore the most abundant form of PTH in serum. Because the C-terminal fragment is cleared by the kidney, renal insufficiency causes this fragment to accumulate in the circulation.

Assays are available to measure the N-terminal fragment, the intact molecule, the C-terminal fragment, or the midregion (a part of the C-terminal). Assays that measure the more abundant C-terminal fragment are usually more sensitive and provide better discrimination than assays that measure the N-terminal or the whole molecule; however, the sensitivity of newer assays to measure the N-terminal is improving. A disadvantage of C-terminal assays is that patients with renal insufficiency may have high serum levels of this fragment without having either increased secretion from the parathyroid gland or high levels of circulating, biologically active PTH.

Primary hyperparathyroidism can be differentiated from nonparathyroid causes of hypercalcemia by means of PTH radioimmunoassays. If hyperparathyroidism is the cause of the hypercalcemia, the serum PTH level is usually increased; in general, the elevation is proportional to the degree of hypercalcemia. In moderate or severe hypercalcemia due to hyperparathyroidism, PTH levels are significantly increased. Hyperparathyroidism is unlikely to be the cause if the serum calcium level is substantially increased and the PTH level is normal.

In hypercalcemia due to a nonparathyroid mechanism, PTH levels are usually in the normal or high-to-normal range. Although the secretion of PTH ought to be suppressed in the presence of hypercalcemia, degraded, predominantly inactive C-terminal fragments of PTH are secreted by the parathyroid gland. These fragments may then accumulate, particularly in patients who have some degree of renal impairment as a result of nephrocalcinosis.

In mild hypercalcemia, normal PTH levels are consistent with either hyperparathyroidism or a nonparathyroid cause. The distinction is usually obvious because hypercalcemia is a late manifestation in patients with nonparathyroid causes of hypercalcemia, such as malignancy. Thus, these disorders are almost always diagnosed on the basis of clinical evidence alone.

The PTH assay can also determine whether hypocalcemia is due to hypoparathyroidism or to a nonparathyroid mechanism such as vitamin D deficiency. In patients with hypoparathyroidism, the serum PTH level is inappropriately low or even normal, despite the presence of hypocalcemia. In hypocalcemia due to nonparathyroid mechanisms, PTH secretion is stimulated (secondary hyperparathyroidism), and thus serum levels are high. In pseudohypoparathyroidism, PTH levels are high but hypocalcemia develops because of resistance to the effects of PTH.

In patients with chronic renal failure, secondary hyperparathyroidism is not uncommon and can produce severe bone disease. Because the clearance of C-terminal fragments and, to some degree, of N-terminal fragments is reduced, PTH levels are extremely high in patients with secondary hyperparathyroidism and may also be high in those without secondary hyperparathyroidism. The distinction can often be made by assessing the degree of PTH elevation. In patients who do not have secondary hyperparathyroidism, PTH levels are only modestly high (the range varies in different assays), whereas in those with secondary hyperparathyroidism, PTH levels are usually more than 10 times normal. □

Childhood Rickets

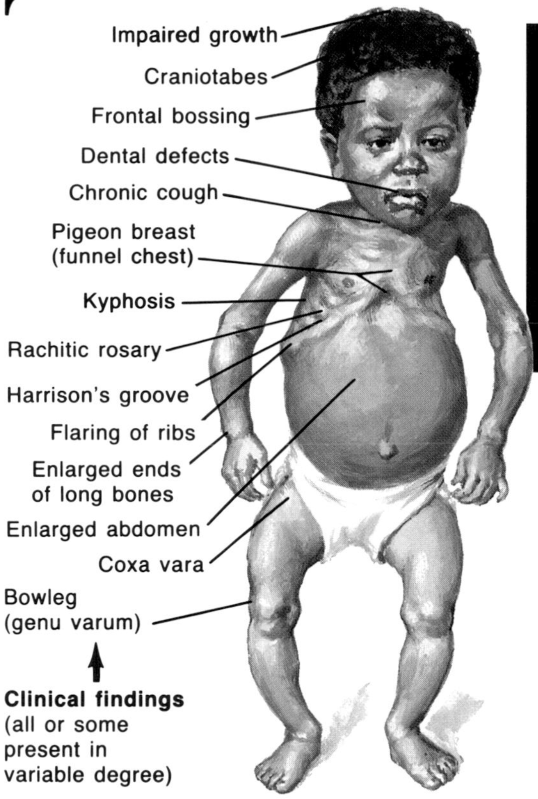
Impaired growth
Craniotabes
Frontal bossing
Dental defects
Chronic cough
Pigeon breast (funnel chest)
Kyphosis
Rachitic rosary
Harrison's groove
Flaring of ribs
Enlarged ends of long bones
Enlarged abdomen
Coxa vara
Bowleg (genu varum)

Clinical findings (all or some present in variable degree)

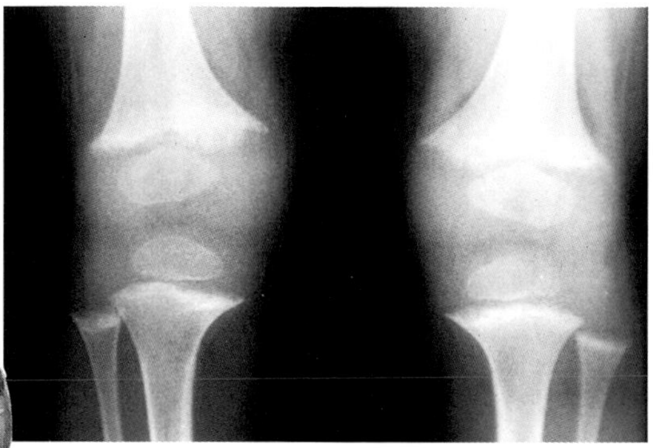

Flaring of metaphyseal ends of tibia and femur. Growth plates thickened, irregular, cupped, and axially widened. Zones of provisional calcification fuzzy and indistinct. Bone cortices thinned and medullae rarefied

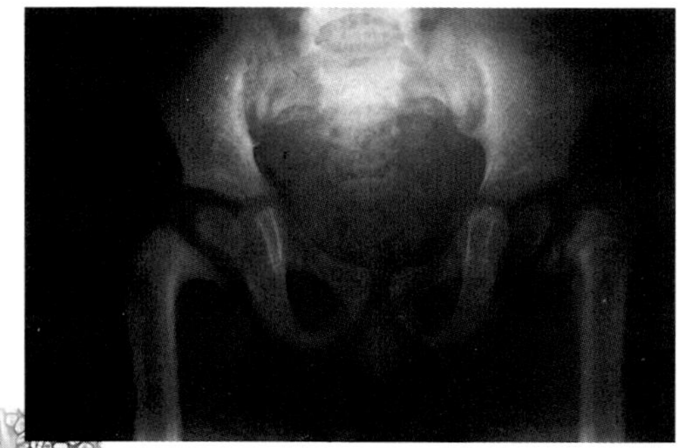

Coxa vara and slipped capital femoral epiphysis. Mottled areas of lucency and density in pelvic bones

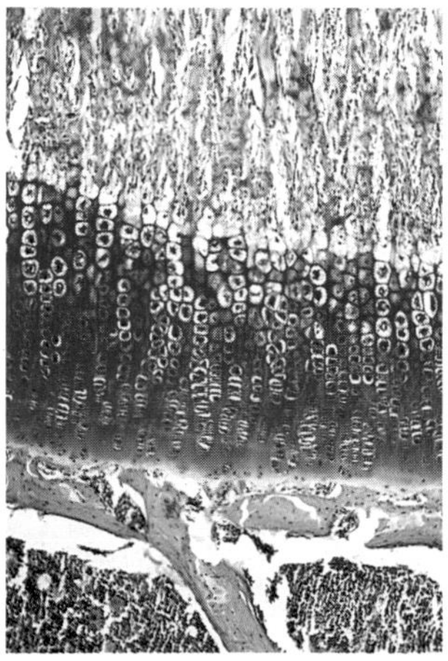

Cartilage of epiphyseal plate in immature normal rat. Cells of middle (maturation) zone in orderly columns, with calcified cartilage between columns

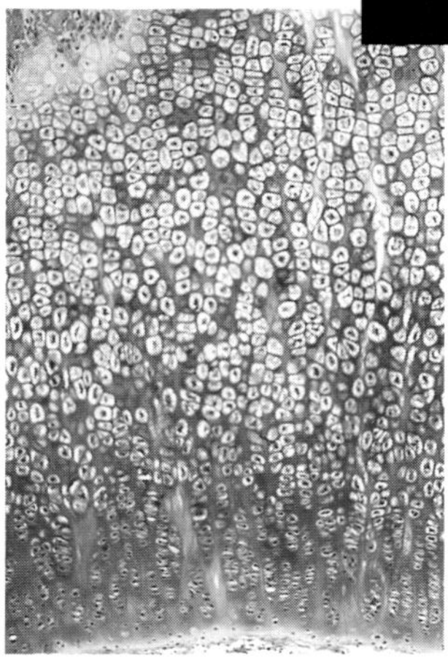

After 6 weeks of vitamin D- and phosphate-deficient diet. Large increase in axial height of maturation zone, with cells closely packed and irregularly arranged

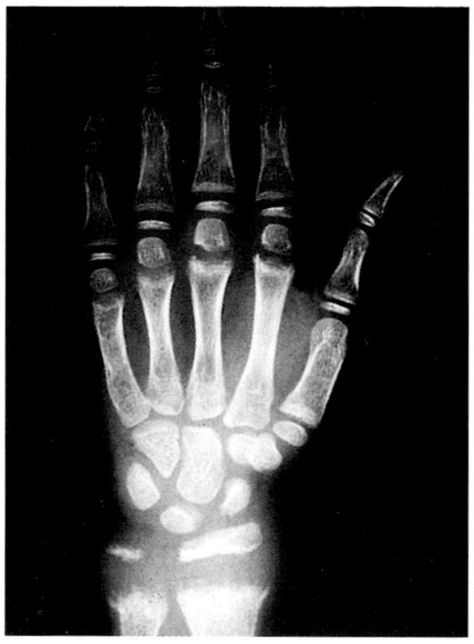

Radiograph of rachitic hand shows decreased bone density, irregular trabeculation, and thin cortices of metacarpals and proximal phalanges. Note increased axial width of epiphyseal line, especially in radius and ulna

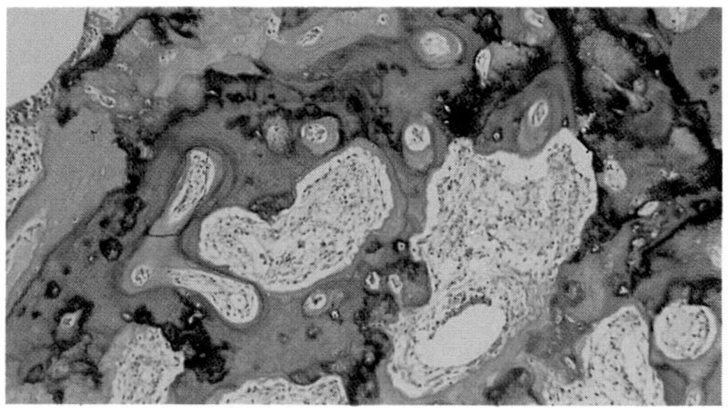

Section of rachitic bone shows sparse, thin trabeculae surrounded by much uncalcified osteoid (osteoid seams) and cavities caused by increased resorption

Adult Osteomalacia

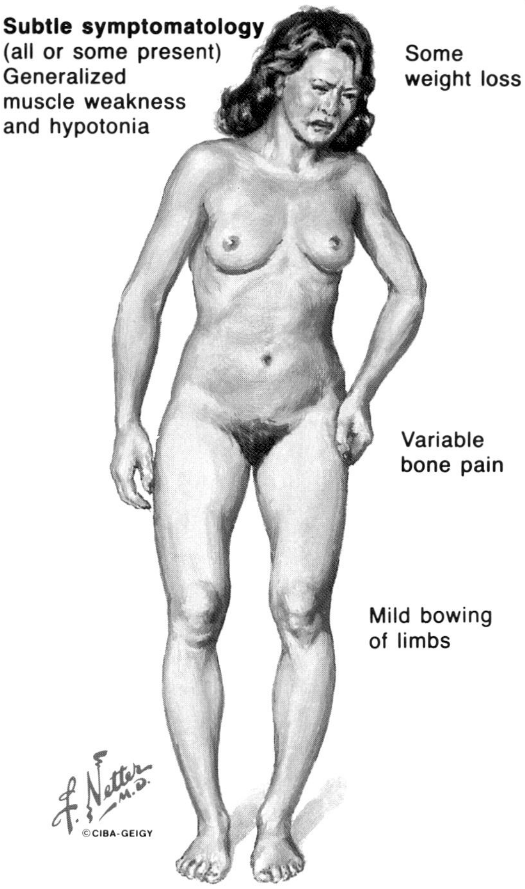
Subtle symptomatology (all or some present)
Generalized muscle weakness and hypotonia
Some weight loss
Variable bone pain
Mild bowing of limbs
F. Netter M.D.
©CIBA-GEIGY

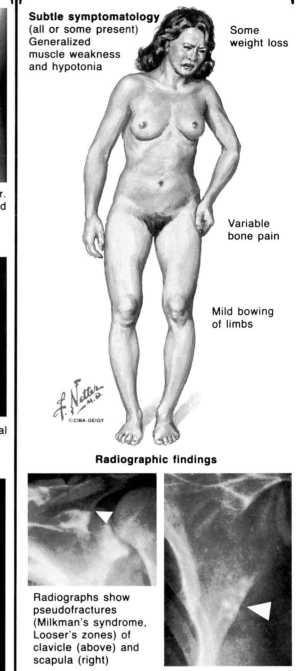

Radiographic findings

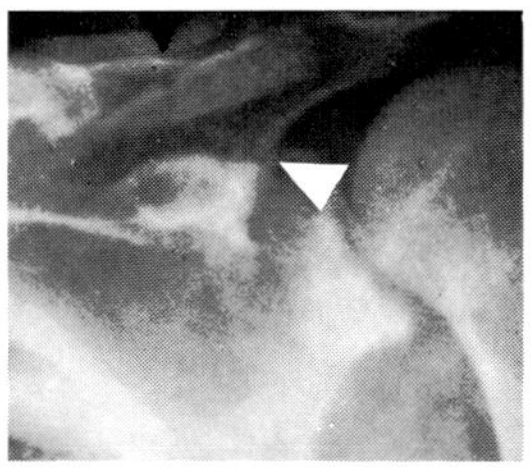

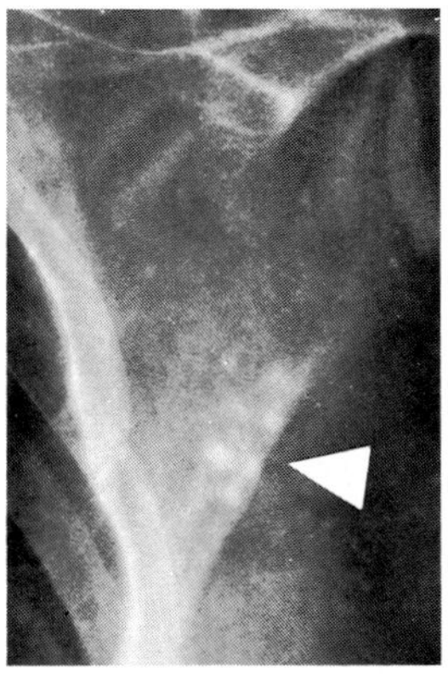

Radiographs show pseudofractures (Milkman's syndrome, Looser's zones) of clavicle (above) and scapula (right)

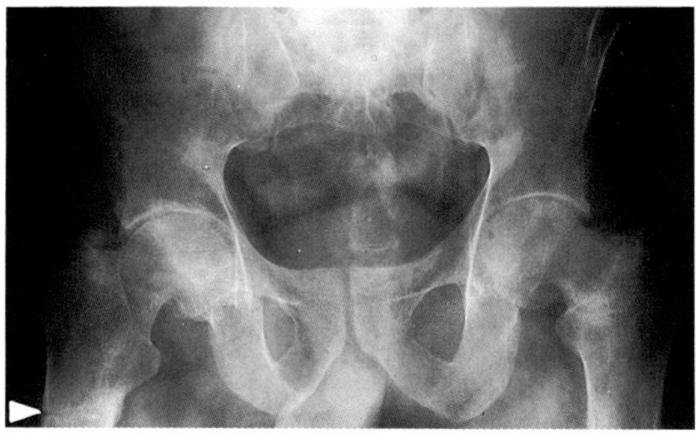

Radiograph shows variegated rarefaction of pelvic bones, coxa vara, deepened acetabula, and subtrochanteric pseudofracture of right femur

SECTION IV PLATE 11 Slide 3746

Rickets, Osteomalacia, and Renal Osteodystrophy

Rickets, osteomalacia, and renal osteodystrophy, far less common now than in the past, are still of great interest to clinicians and scientists. Although a large number of etiologic factors may contribute to these disorders, the basic defect is a deficiency of calcium or phosphate, or both, which impairs the normal mineralization and growth of the skeleton in the child (rickets) or leads to a defective skeletal structure in the adult (osteomalacia).

The causes of rachitic and osteomalacic syndromes are numerous and include a variety of genetic errors, nutritional abnormalities, metabolic disorders, and chronic renal diseases. Quite independent of cause, the clinical manifestations of the disorders are remarkably similar, making it difficult for the physician to solve the often tangled puzzle of causation and introduce the appropriate treatment (Plate 11). However, recent discoveries relating to the disease mechanisms, and the introduction of newer hormonal and drug

Nutritional-Deficiency Rickets and Osteomalacia

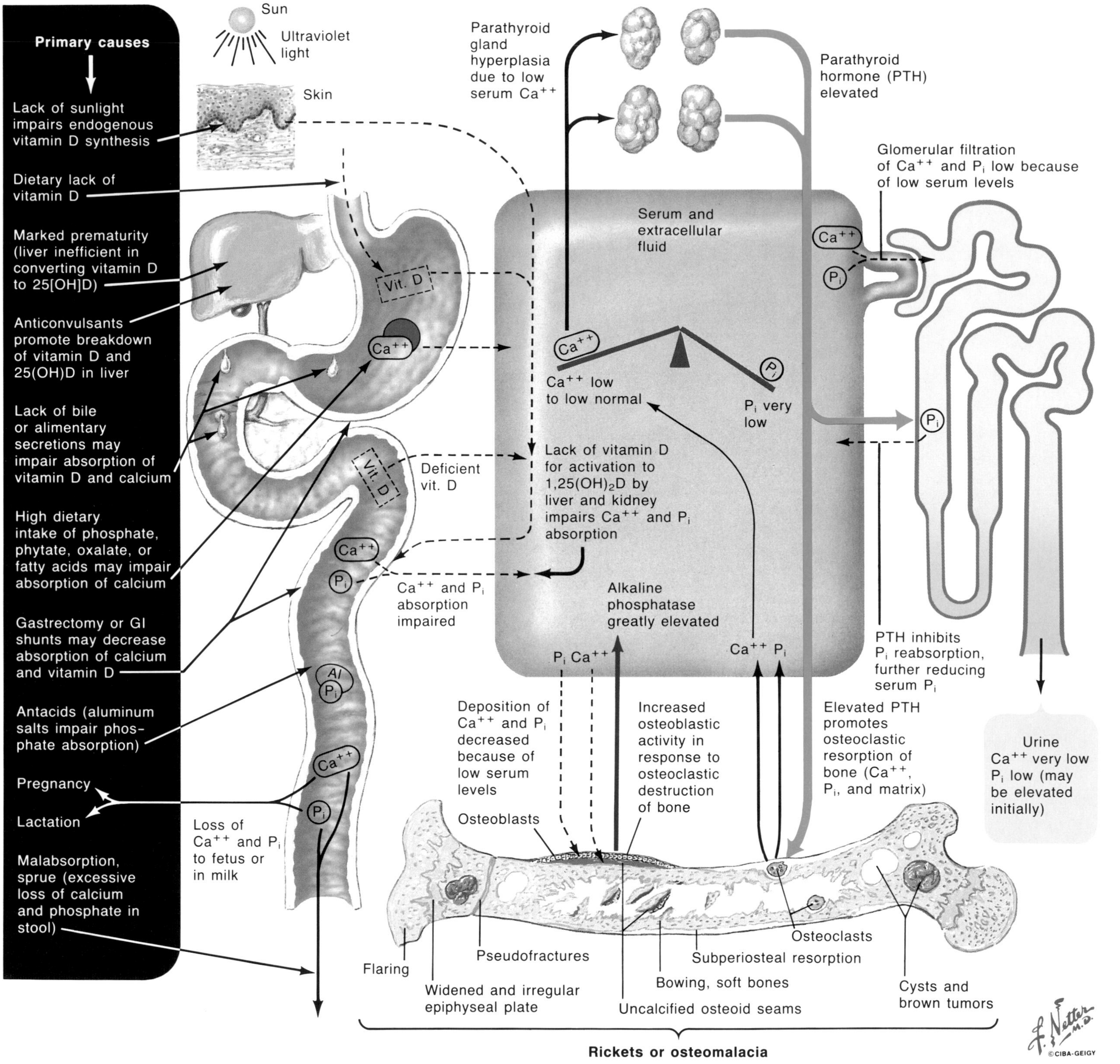

Rickets, Osteomalacia, and Renal Osteodystrophy

(Continued)

treatments are contributing to better management and may lead to a cure.

Besides being interesting, the history of rickets and osteomalacia is important for the classification of the disorders. Both diseases were known in antiquity, but one of the clearest descriptions appeared in a seventeenth-century Latin text by Glisson. Investigations by Schmorl in the late nineteenth century established the role of sunlight in the prevention of the disease, and dietary factors were identified in the first part of the twentieth century. Despite this knowledge, nutritional rickets remained a common occurrence, and many children in working-class families in the temperate zones exhibited the characteristic symptoms of short stature, rib cage deformities, and bowed extremities.

Vitamin D was discovered in the 1920s, and the use of the sterol as a food supplement made nutritional rickets rare in all but the most economically disadvantaged communities. However, within a short time, additional cases were reported, which appeared to be resistant to even massive doses of vitamin D. Scientists such as Albright, Butler, Fanconi, and others, studying the defects that led to this metabolic disorder, identified the mechanisms of calcium transport in the gut, kidneys, and bone cells, as well as the roles played by exogenous factors (the polar metabolites of vitamin D) and endogenous factors (parathyroid hormone [PTH] and phosphate). The identification and synthesis of 1,25-dihydroxyvitamin D, or $1,25(OH)_2D$, and a better understanding of the handling of calcium and phosphate by the kidneys and bones in patients with vitamin D–resistant rickets or chronic renal failure have enabled more

Rickets, Osteomalacia, and Renal Osteodystrophy

(Continued)

Vitamin D–Resistant Rickets and Osteomalacia Due to Proximal Renal Tubular Defects (Hypophosphatemic Rachitic Syndromes)

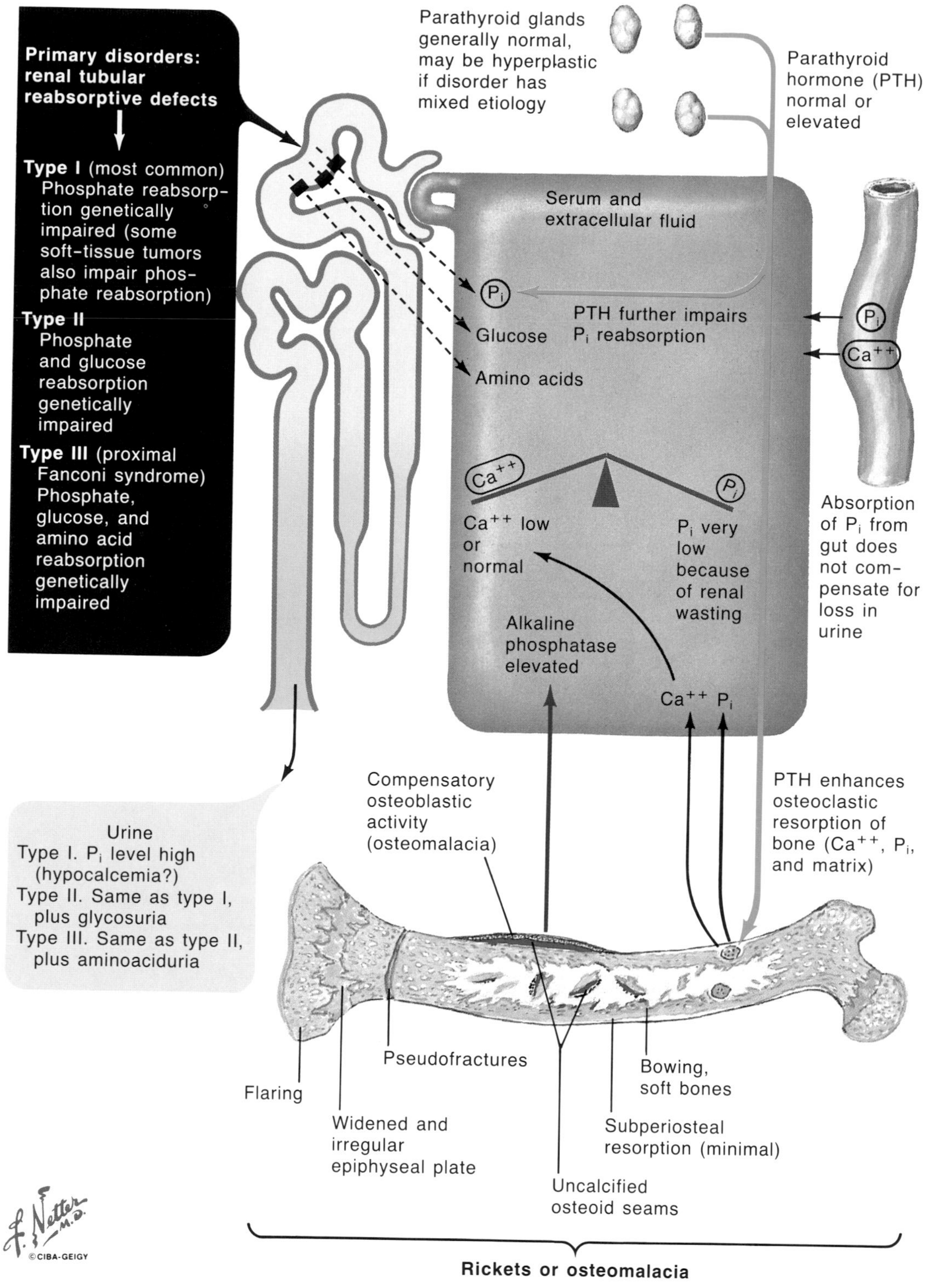
Primary disorders: renal tubular reabsorptive defects
Type I (most common) Phosphate reabsorption genetically impaired (some soft-tissue tumors also impair phosphate reabsorption)
Type II Phosphate and glucose reabsorption genetically impaired
Type III (proximal Fanconi syndrome) Phosphate, glucose, and amino acid reabsorption genetically impaired
Parathyroid glands generally normal, may be hyperplastic if disorder has mixed etiology
Parathyroid hormone (PTH) normal or elevated
Serum and extracellular fluid
P_i
Glucose
Amino acids
PTH further impairs P_i reabsorption
P_i
Ca^{++}
Ca^{++}
P_i
Ca^{++} low or normal
P_i very low because of renal wasting
Absorption of P_i from gut does not compensate for loss in urine
Alkaline phosphatase elevated
Ca^{++} P_i
Urine
Type I. P_i level high (hypocalcemia?)
Type II. Same as type I, plus glycosuria
Type III. Same as type II, plus aminoaciduria
Compensatory osteoblastic activity (osteomalacia)
PTH enhances osteoclastic resorption of bone (Ca^{++}, P_i, and matrix)
Flaring
Widened and irregular epiphyseal plate
Pseudofractures
Uncalcified osteoid seams
Subperiosteal resorption (minimal)
Bowing, soft bones
Rickets or osteomalacia
©CIBA-GEIGY

and more patients to avoid the short stature, skeletal deformities, and multiple systemic problems that once were inevitable.

Childhood Rickets

Clinical Manifestations. In childhood rickets (Plate 11), growth is impaired and height is generally below the third quartile. However, unless there is concurrent severe nutritional disturbance, weight is usually normal. Affected children are apathetic and irritable and frequently remain immobile, sitting in a Buddha-like position. The head displays a number of abnormalities, including softening and deformity of the skull (craniotabes), prominence of the frontal bones (frontal bossing), and caries and enamel defects. Examination of the thorax may reveal flaring and deformity of the ribs, funnel chest (pectus excavatum) or pigeon breast, an indentation at the insertion of the diaphragm into the lower ribs (Harrison's groove), and nodules at the costochondral junctions (rachitic rosary). Frequent manifestations are respiratory infections and a chronic cough.

Children with rickets may also have a gentle thoracic kyphosis (rachitic cat back) and a rachitic potbelly, which, together with the bowed extremities and apathetic facies, emphasize their Buddha-like appearance. Examination of the extremities also uncovers abnormalities such as symmetric enlargement of the ends of the long bones (most prominent at the elbows and wrists), bowleg (genu varum), and, less frequently, knock-knee (genu valgum). Fractures occur frequently.

Histologic Features. In patients with rickets, the histologic appearance of the epiphyseal plate is pathognomonic. Comparison of normal and rachitic epiphyseal plates in rats shows a greatly increased axial height of the epiphyseal plate (sometimes as much as 20 times), principally because of the increased number of cells in the maturation zone; the cells have lost their columnar organization and occur in profligate profusion. Both the zone of provisional calcification of the cartilage and the primary spongiosa of the metaphysis have irregular contours and lack calcific mineral deposition.

Although changes in bone structure are no less pronounced in osteomalacia, they are not specific to that disorder because similar changes may occur in several other metabolic bone disorders (most notably hyperparathyroidism and fibrous dysplasia). The cortices are thin, and the trabeculae are small and irregularly shaped, with evidence of osteoclastic resorption of bone (a mild-to-moderate secondary hyperparathyroidism is characteristic of most rachitic syndromes). The most characteristic histologic feature, however, is the presence of a wide zone of unmineralized bone, or osteoid seam, which surrounds the mineralized trabeculae. In the section shown in Plate 11, the mineralized bone appears dark, and the osteoid seams are pink.

Radiographic Abnormalities. Radiographic findings reflect the histologic changes: thinned cortices and rarefied medullary bone, with indistinct and fuzzy trabecular markings. However, the radiographic hallmarks are the enormously increased axial height of the epiphyseal plate and the poor definition or absence of the zone of provisional calcification, which is normally seen as a dense, white line separating the growth plate, or physis, from the metaphysis. Often noted are cupping and flaring of the ends of the long bones, usually because of a softening of the epiphyseal-metaphyseal region. Slipped capital femoral epiphysis at the widened and severely weakened plate is an occasional finding, particularly in patients with renal osteodystrophy (Plate 20).

Rickets, Osteomalacia, and Renal Osteodystrophy

(Continued)

Vitamin D–Resistant Rickets and Osteomalacia Due to Proximal and Distal Renal Tubular Defects

Type I. Proximal and distal Fanconi syndrome

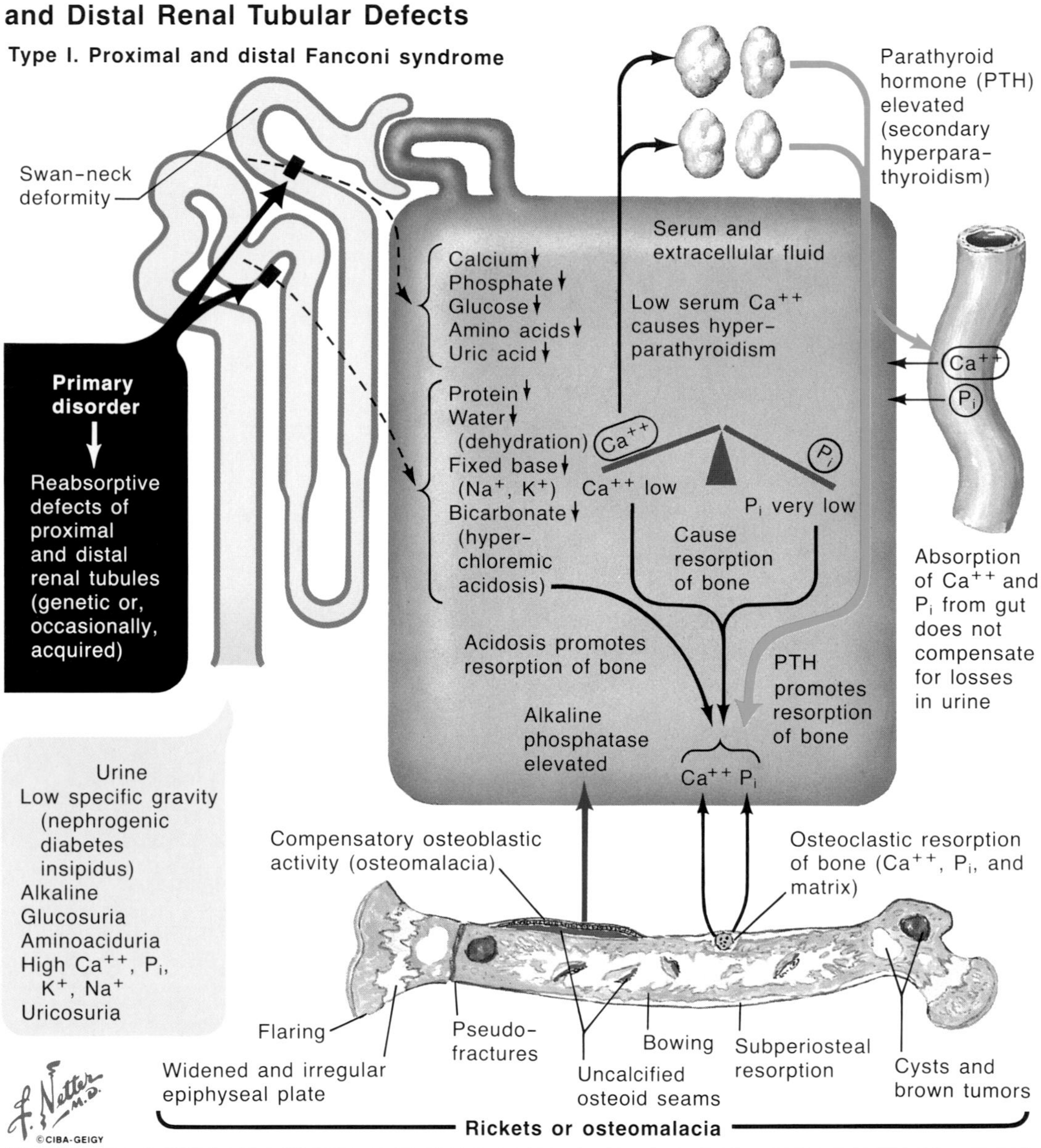
Parathyroid hormone (PTH) elevated (secondary hyperpara-thyroidism)
Swan-neck deformity
Serum and extracellular fluid
Calcium↓
Phosphate↓
Glucose↓
Amino acids↓
Uric acid↓
Low serum Ca++ causes hyper-parathyroidism
Ca++
Pi
Primary disorder
Reabsorptive defects of proximal and distal renal tubules (genetic or, occasionally, acquired)
Protein↓
Water↓ (dehydration)
Fixed base↓ (Na+, K+)
Bicarbonate↓ (hyper-chloremic acidosis)
Ca++ low
Pi very low
Cause resorption of bone
Absorption of Ca++ and Pi from gut does not compensate for losses in urine
Acidosis promotes resorption of bone
PTH promotes resorption of bone
Alkaline phosphatase elevated
Ca++ Pi
Urine
Low specific gravity (nephrogenic diabetes insipidus)
Alkaline
Glucosuria
Aminoaciduria
High Ca++, Pi, K+, Na+
Uricosuria
Compensatory osteoblastic activity (osteomalacia)
Osteoclastic resorption of bone (Ca++, Pi, and matrix)
Flaring
Widened and irregular epiphyseal plate
Pseudo-fractures
Bowing
Uncalcified osteoid seams
Subperiosteal resorption
Cysts and brown tumors
Rickets or osteomalacia
F. Netter M.D.
©CIBA-GEIGY

Type II. Lignac-Fanconi syndrome (cystinosis)
All manifestations of type I, plus generalized disorder of cystine metabolism: deposition of cystine crystals in tissues (cornea, conjunctiva, spleen, bone marrow, liver, lymph nodes)

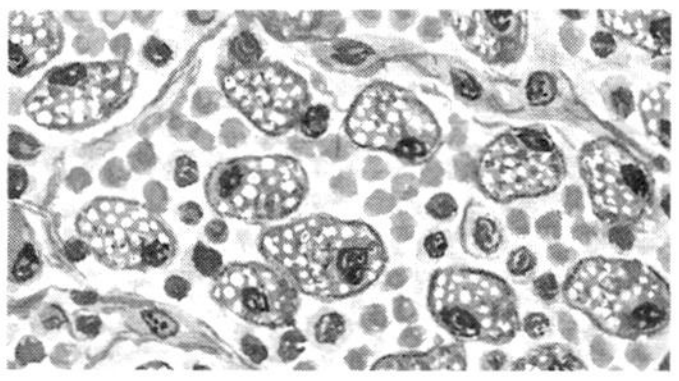
In macrophages of spleen

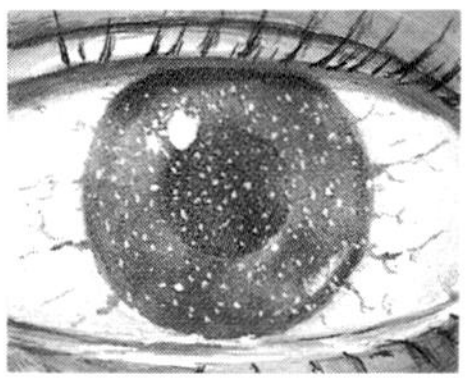
In cornea

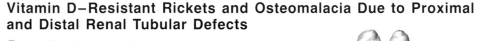

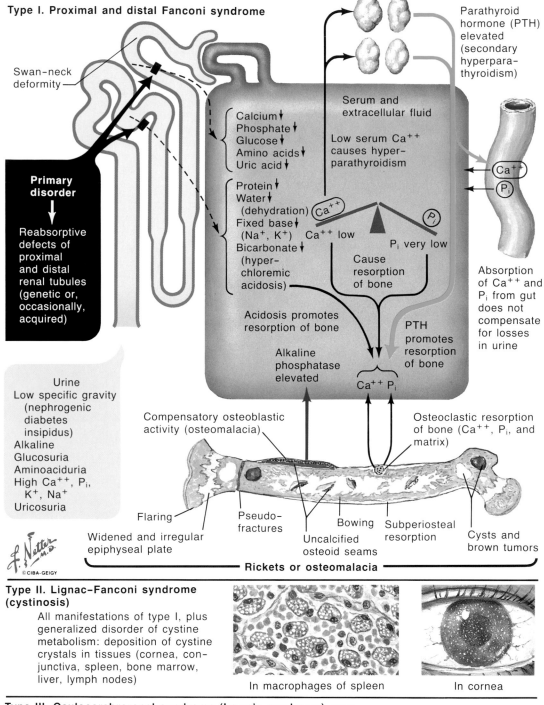

Type III. Oculocerebrorenal syndrome (Lowe's syndrome), rare
Most manifestations of type I, plus congenital glaucoma, cataracts, nystagmus, mental retardation, hyperexcitability, inattentiveness, muscular hypotonia, diminished or absent tendon reflexes, decreased motor power

Type IV. Superglycine syndrome, very rare
Teenage onset; many manifestations of type I, plus severe muscle weakness or atrophy, very high urinary glycine and glycylproline

Unlike osteomalacia, rickets is a disease of growth. If growth slows, either for natural reasons or because the patient becomes ill with other manifestations of the disease, a phenomenon known as the paradox of rickets may occur; that is, the characteristic epiphyseal changes seen on radiography appear to improve. The radiograph of the hand in Plate 11 illustrates this paradox. The hand shows evidence of advanced rachitic changes in the rapidly growing distal radius and ulna, less severe manifestations in the metacarpals, even milder signs in the slowly growing proximal phalanges, and virtually no signs at all in the least active physeal regions of the middle phalanges.

Adult Osteomalacia

Clinical Findings. The diagnosis of adult osteomalacia (Plate 11) may be difficult to establish because the changes are considerably more subtle than those seen in childhood rickets. Recent studies suggest that as many as 25% of elderly patients with hip fractures that have been attributed to postmenopausal or senile osteoporosis have histomorphometric and biochemical manifestations consistent with the diagnosis of mild-to-moderate yet clinically unsuspected osteomalacia, presumably long-standing and nutritional in origin.

Patients with adult osteomalacia complain of generalized weakness, bone pain, easy fatigability, and malaise. The physical findings are minimal: tenderness of bony prominences or, in more serious cases, muscle weakness that is severe enough to cause an abductor-lurch type of gait (the gluteal, or Trendelenburg, gait). In long-standing cases, a bone deformity such as bowleg, coxa vara, or kyphosis may be common.

Radiographic signs are equally subtle, showing for the most part only a diffuse osteopenia, similar to that seen in other metabolic bone diseases such as postmenopausal or senile osteoporosis, hyperparathyroidism, hyperthyroidism, and diffuse skeletal metastatic tumors such as those seen in multiple myeloma. One distinctive feature, present in about 25% of cases, is virtually pathognomonic of osteomalacia. Focal collections of osteoid produce localized, narrow, ribbonlike zones of decreased density in the cortices. These zones are almost always symmetric and are located at right angles to the long axes of the bones. On radiography, they resemble partial fractures. These usually painless pseudofractures—called Looser's zones, Umbauzonen, or Milkman's syndrome—are usually seen on the concave sides of a long bone, the medial side of the femoral neck, the ischial and pubic rami, the clavicle, the ribs, and the axillary border of the scapula. They may serve as stress risers, thus leading to a true fracture (particularly in the femoral neck or in the pubis).

Causes and Mechanisms. The causes of rachitic osteomalacic syndromes, although multiple and varied, can be classified into three groups: (1) nutritional-deficiency rickets and osteomalacia, (2) vitamin D–resistant rickets and osteomalacia, and (3) renal osteodystrophy.

Nutritional-Deficiency Rickets and Osteomalacia

The classic and most clearly understood cause of nutritional-deficiency rickets and osteomalacia (Plate 12) is a dietary deficiency of vitamin D. Regardless of whether the fat-soluble sterol vitamin is deficient in the diet, not activated by appropriate exposure to sunlight, or swept out of the gastrointestinal tract in a variety of spruelike syndromes associated with rapid transit and fatty stools, the diminished concentration of vitamin D in the blood leads to a decrease in the synthesis of 25-hydroxyvitamin D, or 25(OH)D, and a consequent diminution in the synthesis of 1,25$(OH)_2$D. The failure to synthesize sufficient amounts of the active metabolite 1,25$(OH)_2$D results in decreased absorption of calcium from the gastrointestinal tract and in diminished reabsorption of filtered

Rickets, Osteomalacia, and Renal Osteodystrophy

(Continued)

calcium from the renal tubule, both of which lead to hypocalcemia. Often, the hypocalcemia is severe enough to interfere with the mineralization of the epiphyseal plate structures and bones (resulting in rickets and osteomalacia). However, it also causes a mild-to-moderate secondary hyperparathyroidism, which restores the calcium concentration to near normal but at the same time further depletes calcium stores in the skeleton (by osteoclastic resorption of bone). The action of PTH on the tubular reabsorptive mechanism for inorganic phosphate causes a marked phosphate diuresis and a resultant hypophosphatemia.

The biochemical abnormalities shown in Plate 12 lead to a syndrome that manifests all the histologic and radiographic findings of rickets and/or osteomalacia, as well as those of secondary hyperparathyroidism. Biochemical analysis of serum reveals a low or low-to-normal calcium level, low phosphate concentration, elevated serum alkaline phosphatase level (which indicates new bone formation and repair), diminished 25(OH)D and $1,25(OH)_2D$ levels, and increased PTH level. A 24-hour urinalysis usually shows a diminished calcium concentration and a variable but most likely low phosphate level (the low plasma concentration more than offsets the phosphate diuresis induced by PTH). The percentage tubular reabsorption for phosphate (%TRP) is generally low and, depending on the degree of disease, may be less than 60%.

As shown in the left half of Plate 12, other defects or conditions may result in a rachitic or an osteomalacic syndrome. For example, in a premature infant, the immature liver cannot adequately convert vitamin D to 25(OH)D. Chronic use of anticonvulsant medications may lead to a deficiency of 25(OH)D by interfering with the microsomal enzyme systems in the liver. Some nutritional disorders interfering with calcium absorption that may also lead to a similar syndrome are excessive dietary ingestion of phytate (in certain coarse cereals), oxalate (in spinach), citrate or phosphate, and an increased intake of aluminum salts (usually in the form of antacids) that can cause a phosphate deficiency. Any condition in which the gut wall is damaged (tuberculosis, celiac syndromes, sarcoidosis, presence of surgical shunts) or in which rapid transit of gastrointestinal contents occurs (biliary disease, postgastrectomy syndromes) may also cause a rachitic or an osteomalacic syndrome due to a deficiency of either calcium or vitamin D, or both.

Vitamin D–Dependent (Pseudodeficiency) Rickets and Osteomalacia

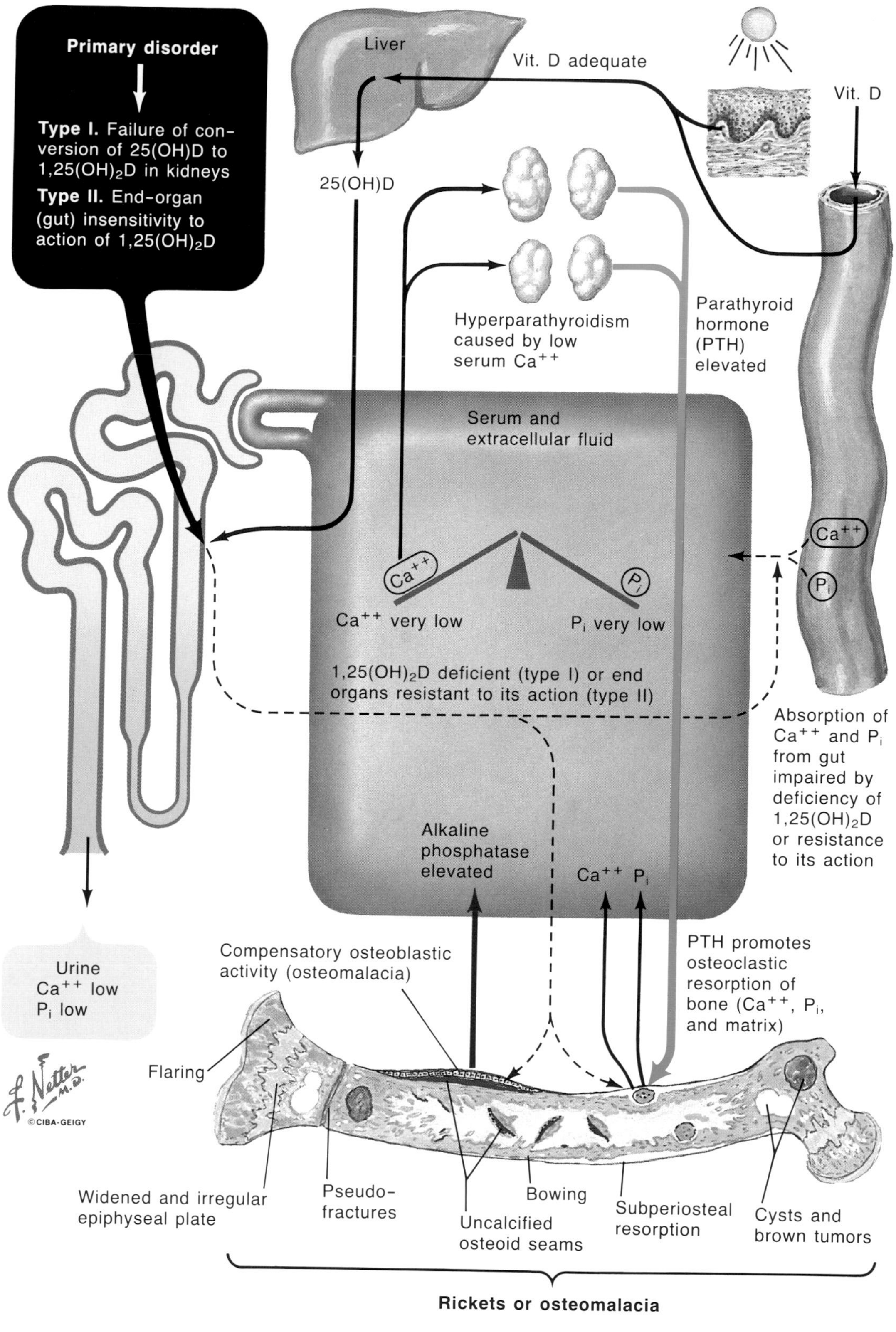
Primary disorder
Type I. Failure of conversion of 25(OH)D to 1,25(OH)₂D in kidneys
Type II. End-organ (gut) insensitivity to action of 1,25(OH)₂D
Liver
Vit. D adequate
Vit. D
25(OH)D
Hyperparathyroidism caused by low serum Ca⁺⁺
Parathyroid hormone (PTH) elevated
Serum and extracellular fluid
Ca⁺⁺
Pᵢ
Ca⁺⁺ very low
Pᵢ very low
1,25(OH)₂D deficient (type I) or end organs resistant to its action (type II)
Absorption of Ca⁺⁺ and Pᵢ from gut impaired by deficiency of 1,25(OH)₂D or resistance to its action
Alkaline phosphatase elevated
Ca⁺⁺ Pᵢ
Urine
Ca⁺⁺ low
Pᵢ low
Compensatory osteoblastic activity (osteomalacia)
PTH promotes osteoclastic resorption of bone (Ca⁺⁺, Pᵢ, and matrix)
F. Netter M.D.
©CIBA-GEIGY
Flaring
Widened and irregular epiphyseal plate
Pseudo-fractures
Uncalcified osteoid seams
Bowing
Subperiosteal resorption
Cysts and brown tumors
Rickets or osteomalacia

Vitamin D–Resistant Rickets and Osteomalacia Due to Proximal Tubular Defects

In affluent, developed communities, genetic or acquired rachitic and osteomalacic syndromes (Plate 13) that are resistant to high therapeutic doses of vitamin D are now more common than those associated with vitamin D deficiencies. Almost all of these syndromes are renal in origin and are associated with a narrow or broad reabsorptive defect in the renal tubule that leads to hypophosphatemia (thus, they are also known as hypophosphatemic vitamin D–resistant rickets, or phosphate diabetes).

The most common of these disorders is type I, a sex-linked dominant genetic disorder in which the renal tubule does not reabsorb phosphate, and as a result, the disease produces all the features of rickets or osteomalacia, as seen on histologic and radiographic examinations. In the purest form of the disorder, there is no abnormality of calcium or vitamin D metabolism, and results of serum analysis and urinalysis demonstrate profound hypophosphatemia, marked lowering of the %TRP, and an increased serum alkaline phosphatase level. Many of the syndromes are impure, however, and either show evidence of a defect in vitamin D metabolism (Plate 15) or some degree of loss of calcium as

Vitamin D–Resistant Rickets and Osteomalacia Due to Renal Tubular Acidosis

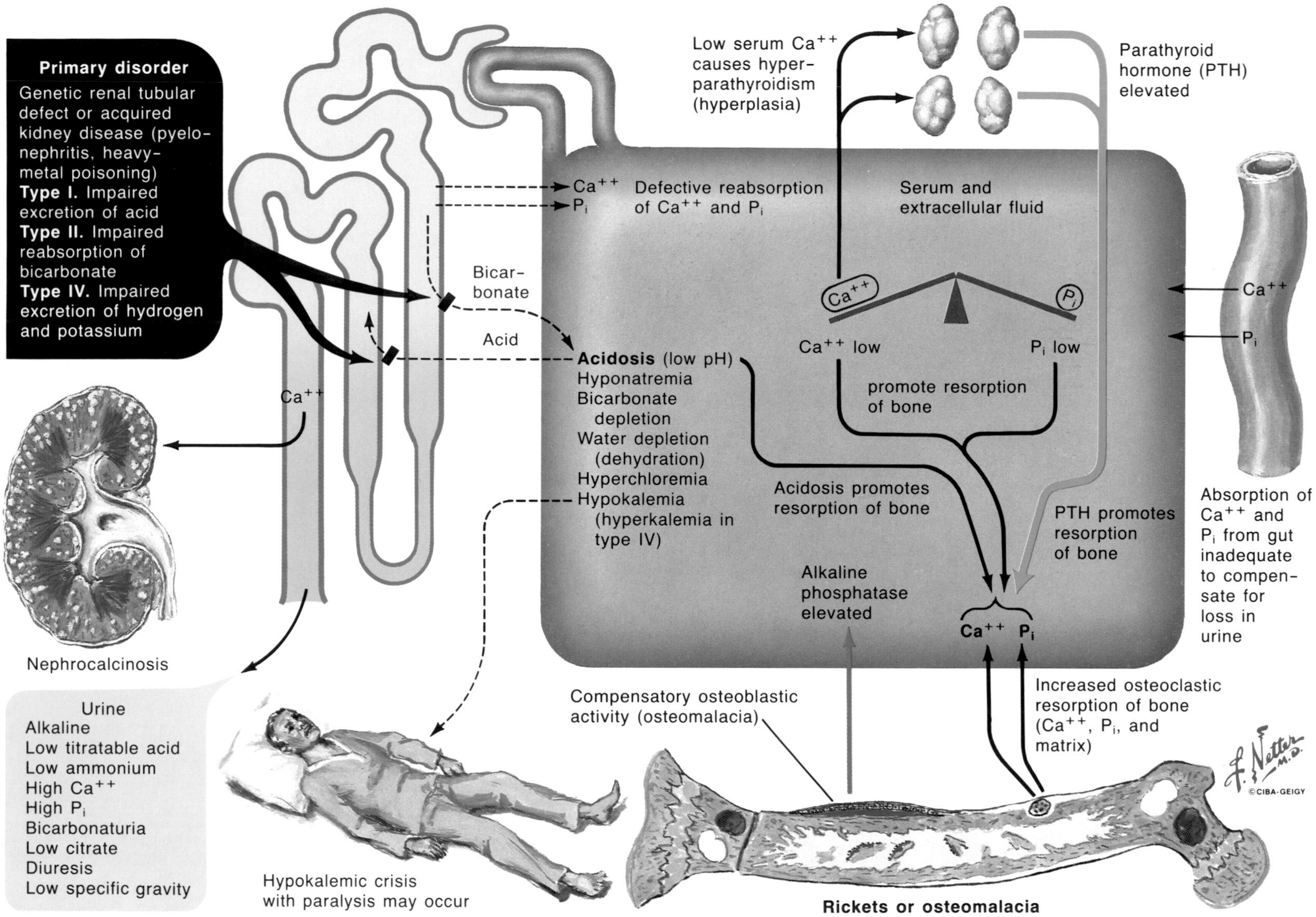

Primary disorder
Genetic renal tubular defect or acquired kidney disease (pyelonephritis, heavy-metal poisoning)
Type I. Impaired excretion of acid
Type II. Impaired reabsorption of bicarbonate
Type IV. Impaired excretion of hydrogen and potassium
Low serum Ca++ causes hyperparathyroidism (hyperplasia)
Parathyroid hormone (PTH) elevated
Ca++
Pi
Defective reabsorption of Ca++ and Pi
Serum and extracellular fluid
Bicarbonate
Acid
Acidosis (low pH)
Hyponatremia
Bicarbonate depletion
Water depletion (dehydration)
Hyperchloremia
Hypokalemia (hyperkalemia in type IV)
Ca++ low
Pi low
promote resorption of bone
Acidosis promotes resorption of bone
PTH promotes resorption of bone
Alkaline phosphatase elevated
Absorption of Ca++ and Pi from gut inadequate to compensate for loss in urine
Nephrocalcinosis
Urine
Alkaline
Low titratable acid
Low ammonium
High Ca++
High Pi
Bicarbonaturia
Low citrate
Diuresis
Low specific gravity
Hypokalemic crisis with paralysis may occur
Compensatory osteoblastic activity (osteomalacia)
Increased osteoclastic resorption of bone (Ca++, Pi, and matrix)
Rickets or osteomalacia
F. Netter M.D.
©CIBA-GEIGY

SECTION IV PLATE 16 Slide 3751

Rickets, Osteomalacia, and Renal Osteodystrophy

(Continued)

fixed base (Plates 14 and 16). Under these circumstances, the serum calcium concentration may be low or low-to-normal and the PTH level may be increased, which aggravates the already severely phosphate-depleted state.

Three phosphate-wasting lesions in the proximal tubule have been identified. *Type I*, first described by Albright, is seen most frequently. It is transmitted as a sex-linked dominant trait, and the reabsorptive defect is confined to phosphate only. In *type II*, the defect is broader and involves both phosphate and glucose. In *type III*, the proximal Fanconi syndrome, the reabsorptive defect is for phosphate, glucose, and various amino acids.

Vitamin D–Resistant Rickets and Osteomalacia Due to Proximal and Distal Renal Tubular Defects

In another group of vitamin D–resistant rachitic and osteomalacic syndromes, the range of renal tubular defects is considerably broader and may interfere more severely with normal metabolism (Plate 14). Reabsorption of phosphate, glucose, and amino acids in the proximal tubule is impaired, and, in addition, the functions of the distal tubule are significantly altered. Consequently, in patients with type I syndrome, known as the proximal and distal Fanconi syndrome, or the Debré–de Toni–Fanconi syndrome, the kidney's ability to reabsorb water, bicarbonate, proteins, and fixed base is to some degree impaired. This represents a severe challenge to the patient, particularly the newborn, and requires major replacement therapy.

Typically, the patient is a very ill child, often dehydrated and hypoproteinemic, with rachitic changes in the bones. The rachitic and osteomalacic patterns in these patients result from the failure of tubular reabsorption of phosphate coupled with the loss of fixed base, including calcium. This unfortunate combination results in both rachitic changes and a mild-to-moderate secondary hyperparathyroidism (which worsens the bone lesions and intensifies the hypophosphatemia). These conditions have little relationship to vitamin D; in fact, treatment with even high doses of the sterol vitamin has little effect.

Biochemical changes in type I disease include hypocalcemia, hypophosphatemia, increased serum alkaline phosphatase level, and normal serum levels of 25(OH)D and $1{,}25(OH)_2D$. The patient is likely to have signs of renal tubular acidosis (hyperchloremia, hyponatremia, and hypokalemia in association with an alkaline urine). Urinalysis reveals a low fixed specific gravity and the presence of excessive concentrations of metabolites, including calcium, sodium, and potassium ions; phosphate (as a result of a greatly lowered %TRP); uric acid; amino acids; and proteins.

Three other less common types of the severe form of vitamin D–resistant rickets and osteomalacia are often included under the proximal and distal Fanconi syndrome. (1) *Type II*, Lignac-Fanconi syndrome, is almost identical to type I but has an additional defect in the metabolism of cystine. This defect leads to the deposition of crystals of the amino acid in the viscera, bone marrow, and eyes (the diagnosis may be made by slit-lamp examination). As a result of the deposits, cirrhosis of the liver and renal failure frequently supervene by puberty. (2) *Type III* disease is known as the oculocerebrorenal, or Lowe's, syndrome. Patients with this condition have many of the manifestations of type I disease but may also have a broad range of ocular and neurologic abnormalities, which include congenital glaucoma, nystagmus, mental retardation, muscular hypotonia, and weakness. (3) *Type IV*, the superglycine syndrome, is rare. It is less severe than the other types and usually has a later onset. Presenting symptoms are profound motor weakness and very high urinary concentrations of glycine and glycylproline.

Rickets, Osteomalacia, and Renal Osteodystrophy

(Continued)

Metabolic Aberrations of Renal Osteodystrophy

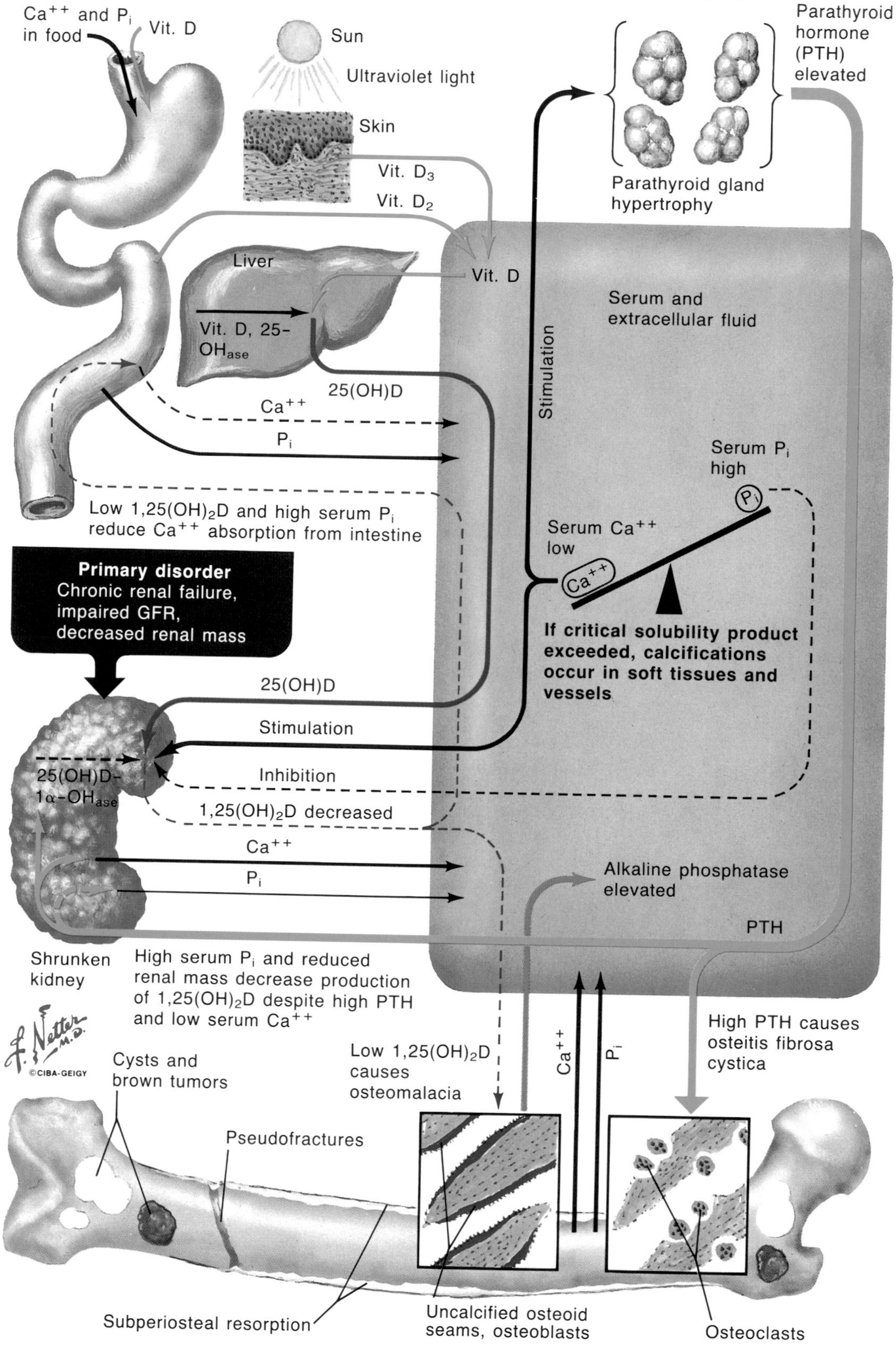

Ca^{++} and P_i in food
Vit. D
Sun
Ultraviolet light
Skin
Vit. D_3
Vit. D_2
Parathyroid hormone (PTH) elevated
Parathyroid gland hypertrophy
Liver
Vit. D, 25-OH_{ase}
Vit. D
Serum and extracellular fluid
Stimulation
25(OH)D
Ca^{++}
P_i
Serum P_i high
Low $1,25(OH)_2D$ and high serum P_i reduce Ca^{++} absorption from intestine
Serum Ca^{++} low
Primary disorder
Chronic renal failure, impaired GFR, decreased renal mass
If critical solubility product exceeded, calcifications occur in soft tissues and vessels
25(OH)D
Stimulation
Inhibition
25(OH)D-1α-OH_{ase}
$1,25(OH)_2D$ decreased
Ca^{++}
P_i
Alkaline phosphatase elevated
PTH
Shrunken kidney
High serum P_i and reduced renal mass decrease production of $1,25(OH)_2D$ despite high PTH and low serum Ca^{++}
High PTH causes osteitis fibrosa cystica
Low $1,25(OH)_2D$ causes osteomalacia
Ca^{++}
P_i
Cysts and brown tumors
Pseudofractures
Subperiosteal resorption
Uncalcified osteoid seams, osteoblasts
Osteoclasts
©CIBA-GEIGY

Vitamin D–Dependent (Pseudodeficiency) Rickets and Osteomalacia

This major category of vitamin D–dependent rickets is characterized by abnormalities of vitamin D metabolism (Plate 15). In this group of syndromes, the error is also almost always inherited but involves either a failure of conversion of 25(OH)D to the potent $1,25(OH)_2D$ or a relative end-organ insensitivity of the gut (and, presumably under certain circumstances, the renal cell as well) to the patient's autogenous $1,25(OH)_2D$. In both of these circumstances, orally administered vitamin D in standard or therapeutic doses does not help to increase reabsorption of calcium by the renal tubule, and hypocalcemia and rachitic manifestations develop. In response to the lowered serum calcium level, the parathyroid glands elaborate PTH, which further depletes the skeleton's calcium reserves and produces hyperphosphaturia and hypophosphatemia.

All of the findings and chemical abnormalities are similar to those seen in the classic nutritional deficiency syndrome, with the exception of the concentrations of the polar metabolites of vitamin D. In the first form of the disorder—failure of conversion of 25(OH)D to $1,25(OH)_2D$ in the kidney—serum levels of 25(OH)D may be very high, while levels of $1,25(OH)_2D$ may be low. In the latter form of the disorder—end-organ insensitivity to the action of $1,25(OH)_2D$—the serum levels of both 25(OH)D and $1,25(OH)_2D$ are usually normal or high. Both forms are often successfully treated with administration of $1,25(OH)_2D$.

Vitamin D–Resistant Rickets Due to Renal Tubular Acidosis

The group of diseases classified under renal tubular acidosis includes metabolic disorders of diverse and multiple etiologies (Plate 16). The basic mechanism common to them all is the kidney's inability to substitute hydrogen ions for fixed base. The diseases are now subclassified into three types. In *type I* (classic or distal), proton exchange is defective; in *type II* (proximal), the cause appears to be a failure of reabsorption of bicarbonate in the proximal tubule; and in *type IV* (generalized distal), through one of several mechanisms, the excretion of both hydrogen and potassium ions in the distal tubule is defective. (Type III is a poorly defined mixture of types I and II and is usually excluded from current classifications.)

All three syndromes are characterized by a hyperchloremic, hyponatremic acidosis and an alkaline urine. Type II is also characterized by a lowered serum bicarbonate level and types I and II, by hypokalemia, which at times may become so severe as to be life threatening. (Hypokalemia does not occur in type IV, which is associated with hyperkalemia.) In about 70% of patients, the urinary citrate concentration is inadequate, which, coupled with the alkaline pH and hypercalciuria, leads to some degree of nephrocalcinosis.

Rickets or osteomalacia of variable degrees is a common manifestation of all the diverse conditions associated with renal tubular acidosis. Some of the rachitic disorders are life threatening, such as vitamin D–resistant rickets and osteomalacia due to proximal and distal tubular defects (Plate 14). Other disorders are mild and, in some cases, do not require prolonged treatment. These include the genetically determined, sometimes self-limiting, Butler-Albright syndrome and disorders associated with altered globulin states or hyperthyroidism.

Rickets, Osteomalacia, and Renal Osteodystrophy

(Continued)

Mechanism of Development of Chemical and Bony Changes in Renal Osteodystrophy

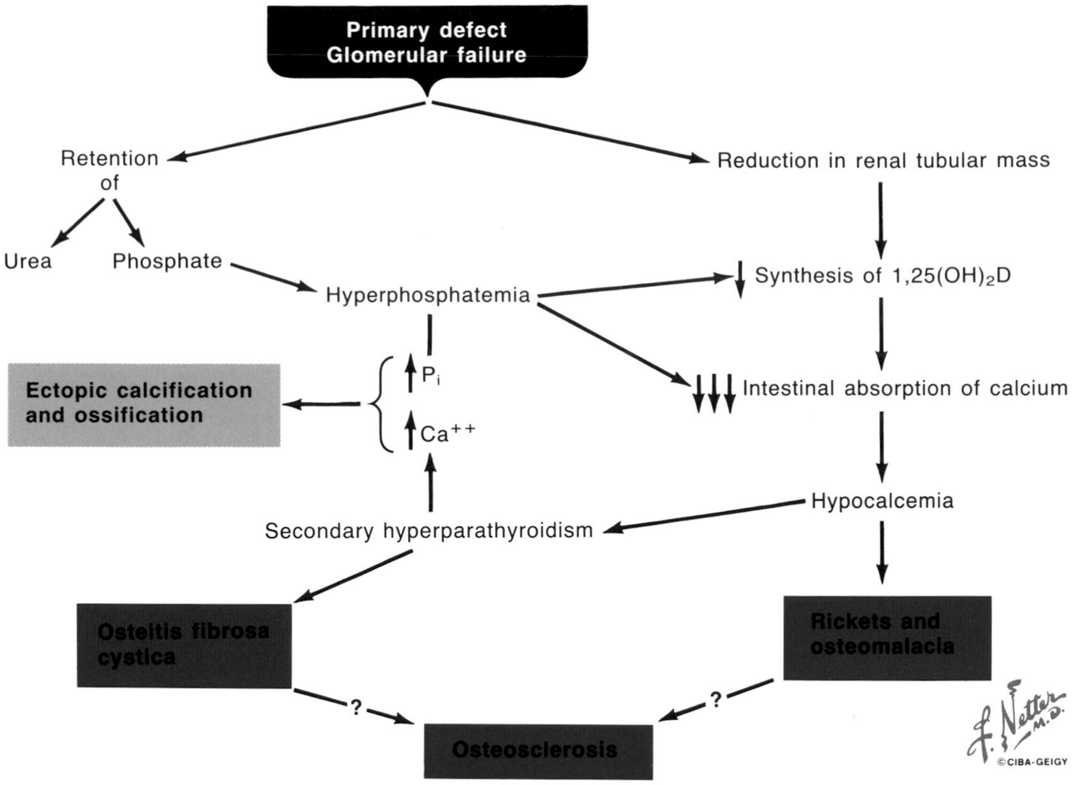

Reduced renal function and glomerular failure cause retention of urea and phosphate, leading to hyperphosphatemia, which, along with reduction in tubular mass, causes profound reduction in synthesis of 1,25(OH)$_2$D. This, plus direct effect of increased concentration of serum phosphate, reduces intestinal absorption of calcium, causing profound hypocalcemia and severe secondary hyperparathyroidism. These changes produce clinical syndromes of rickets and osteitis fibrosa cystica; 20% of patients with this combination of chemical abnormalities also have osteosclerosis. Because phosphate concentration is chronically increased, an occasional increase in serum calcium level can lead to rapid ectopic calcification and ossification in conjunctiva, skin, blood vessels, and periarticular regions

The mechanisms by which renal tubular acidosis contributes to the development of rickets and osteomalacia are not completely understood. Chronic acidosis alone can deplete the bones of calcium and phosphate but is considered to be, at most, a minor mechanism. Many investigators believe that calcium, as well as sodium and potassium, is lost in the urine as fixed base due to a failure to substitute hydrogen or to reabsorb bicarbonate (in types I and II). In some of the syndromes, the %TRP appears to be diminished (either because of a primary renal defect or as a result of secondary hyperparathyroidism and the action of increased concentrations of PTH on the reabsorptive mechanism for phosphate in the renal tubule). The combination of diminished serum calcium and phosphate levels results in rachitic and/or osteomalacic findings that are identical to those seen with other syndromes yet are clearly refractory to even very large doses of vitamin D.

In patients with classic rachitic or osteomalacic changes in the epiphyseal growth plates and bones, renal tubular acidosis should be suspected as the underlying cause of the disease. Characteristic findings on biochemical analysis are hyperchloremia, hyponatremia, hypokalemia (except in type IV), hypophosphatemia, and hypocalcemia, often with a lowered serum bicarbonate level. Serum alkaline phosphatase and PTH concentrations are usually also increased. Urinalysis reveals striking findings: alkaline urine with very low concentrations of acid, ammonia, and citrate; increased levels of fixed base (including calcium); and, occasionally, a fixed low specific gravity. Patients usually fail to respond to an acid-loading test (with ammonium chloride) by acidification of the urine.

Treatment of rachitic and osteomalacic syndromes due to renal tubular acidosis should focus on the primary process rather than on the bone disease. Administration of alkali in the form of sodium bicarbonate or similar materials may be all that is required to correct the metabolic disorder, including the rachitic or osteomalacic syndrome. At times (particularly in type IV), sterol treatment may be necessary.

Renal Osteodystrophy

Metabolic Aberrations. Chronic renal failure causes an extraordinary number of metabolic abnormalities that affect almost all of the body's homeostatic mechanisms (Plate 17). In the patient with azotemia and chronic renal failure, aberrations in water distribution, electrolyte and acid-base balances, protein synthesis, nutrition, and hormonal activities produce extensive changes in bodily structure and functions. The manifestations of chronic renal failure on the body's connective tissues predominate in bone as a multifaceted syndrome. Known in the past as renal rickets, renal hyperparathyroidism, and renal nanism, this syndrome is now generally called renal osteodystrophy (Plates 17, 19–20).

Manifestations of renal osteodystrophy in both children and adults include a number of chronic disorders of epiphyseal cartilage and bone. Among them are rickets and osteomalacia, osteitis fibrosa cystica (secondary hyperparathyroidism), osteosclerosis, and metastatic calcification. In a child with chronic disease, slipped capital femoral epiphysis may be an additional complication. Osteoporosis, osteomyelitis, and (if steroids are administered) osteonecrosis are also seen frequently in both children and adults.

The pathogenetic mechanisms for the bone changes in renal osteodystrophy are complex (Plates 17 and 20). The underlying defect is kidney damage, which includes not only a failure of glomerular filtration that results in azotemia and hyperphosphatemia, but almost always a reduced renal, and thus a reduced tubular, mass. Even when vitamin D intake is normal or increased, the high phosphate concentration and the reduction in tubular function greatly reduce the synthesis of 1,25(OH)$_2$D. Increased serum phosphate levels and the severely lowered concentration of 1,25(OH)$_2$D lead to a markedly reduced absorption of calcium from the gastrointestinal tract and profound hypocalcemia. Despite acidosis, which promotes the solubilization of calcium salts, the hypocalcemia is so severe that it not only causes all the bony and soft-tissue manifestations of rickets or osteomalacia but also induces a secondary hyperparathyroidism. The excessive secretion of PTH leads to osteitis fibrosa cystica (marked osteoclastic resorption of bone and brown tumors). The resorption of bone partially restores serum calcium levels to normal.

In patients with renal osteodystrophy, the critical solubility product for calcium phosphate is in danger of being exceeded. This is a consequence of both the hyperphosphatemia and the reduction in tubular mass, which results in a failure of the increased PTH concentrations to induce a phosphate diuresis. Since calcium salts are more soluble in an acid medium, chronic acidosis helps to prevent deposition of calcific salts. However, the level of calcium or phosphate (or both) may sometimes rise sufficiently or the pH may increase, resulting in ectopic calcification or ossification.

The biochemical alterations in renal osteodystrophy can be summarized as follows:

1. Azotemia, hyperphosphatemia, and changes in acid-base balance and electrolytes that reflect the chronic acidotic state.

Bony Manifestations of Renal Osteodystrophy

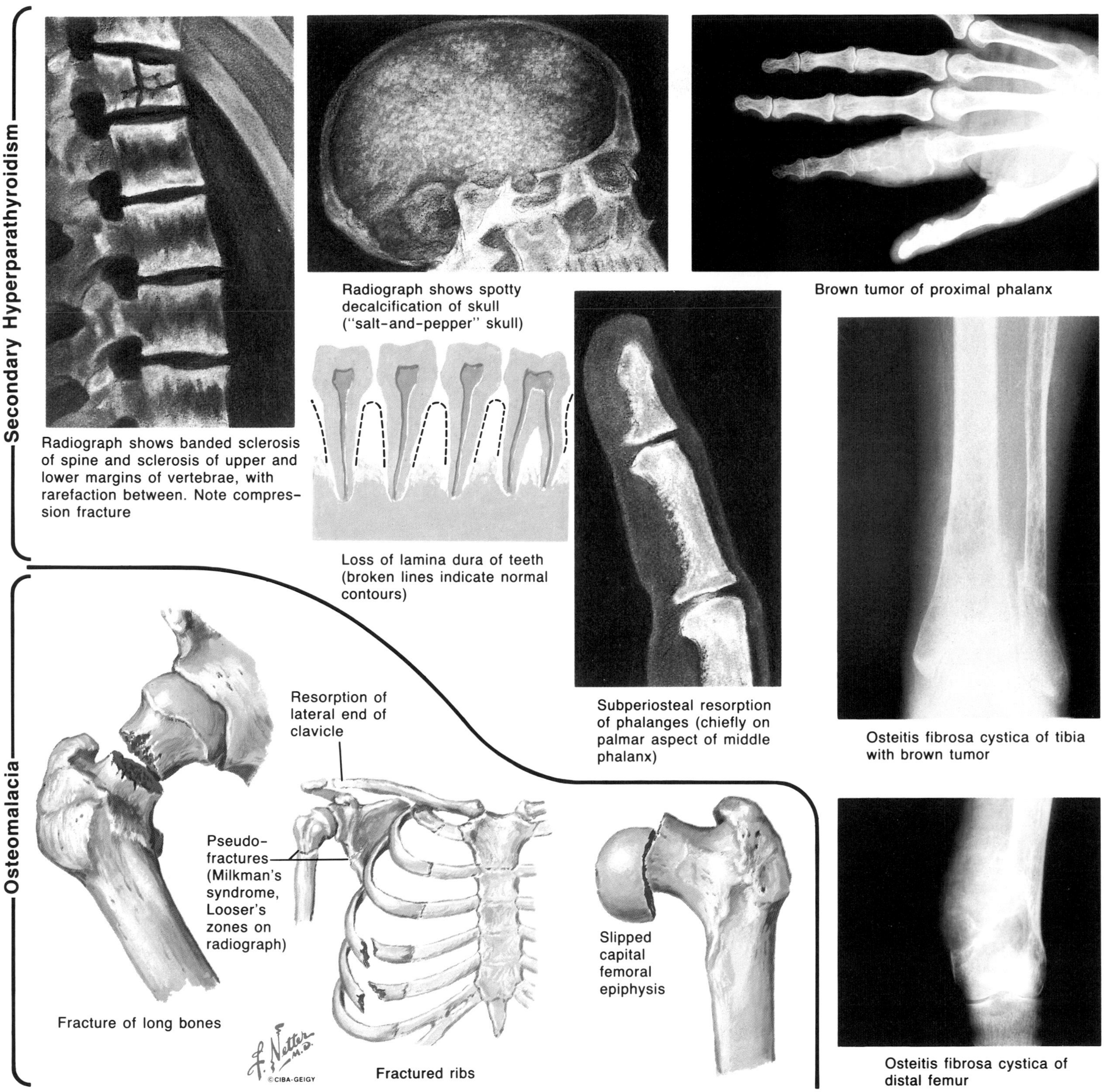
Secondary Hyperparathyroidism
Radiograph shows banded sclerosis of spine and sclerosis of upper and lower margins of vertebrae, with rarefaction between. Note compression fracture
Radiograph shows spotty decalcification of skull ("salt-and-pepper" skull)
Brown tumor of proximal phalanx
Loss of lamina dura of teeth (broken lines indicate normal contours)
Subperiosteal resorption of phalanges (chiefly on palmar aspect of middle phalanx)
Osteitis fibrosa cystica of tibia with brown tumor
Osteomalacia
Resorption of lateral end of clavicle
Pseudo-fractures (Milkman's syndrome, Looser's zones on radiograph)
Slipped capital femoral epiphysis
Fracture of long bones
Fractured ribs
F. Netter M.D.
©CIBA-GEIGY
Osteitis fibrosa cystica of distal femur

SECTION IV PLATE 19 Slide 3754

Rickets, Osteomalacia, and Renal Osteodystrophy

(Continued)

2. Low serum calcium level, in which case a larger percentage of the calcium is ionized because of the acidotic state, but the total amount (including the nonionized calcium) is reduced not only as a result of the factors described above but because of a commonly observed decline in serum proteins.

3. Increased alkaline phosphatase activity due to the increased rate of new bone synthesis.

4. Increased PTH level, which indicates the usually marked secondary hyperparathyroidism.

5. Greatly diminished $1,25(OH)_2D$ levels, even with increased intake of vitamin D, and normal or high levels of 25(OH)D.

Clinical Manifestations. In the growing child with renal osteodystrophy, rachitic changes in the epiphyseal plates are virtually identical to those seen in patients with other forms of rickets (Plates 12 and 18). However, the growth rate of children with renal osteodystrophy is often greatly reduced, with the result that radiographic manifestations of the disease may appear somewhat less severe than the chemical aberrations suggest (see the paradox of rickets, page 208). However, increased axial height of the epiphyseal plates, cupping and flaring, and diminished density in the zone of provisional calcification are characteristic radiographic findings and are indistinguishable from the changes seen in the other forms of rachitic disease. For some unknown reason, slipped capital femoral epiphysis occurs much more frequently in patients with azotemic rickets than it does in patients with vitamin D deficiency or vitamin D–resistant disease.

Rickets, Osteomalacia, and Renal Osteodystrophy

(Continued)

Vascular and Soft-Tissue Manifestations of Renal Osteodystrophy

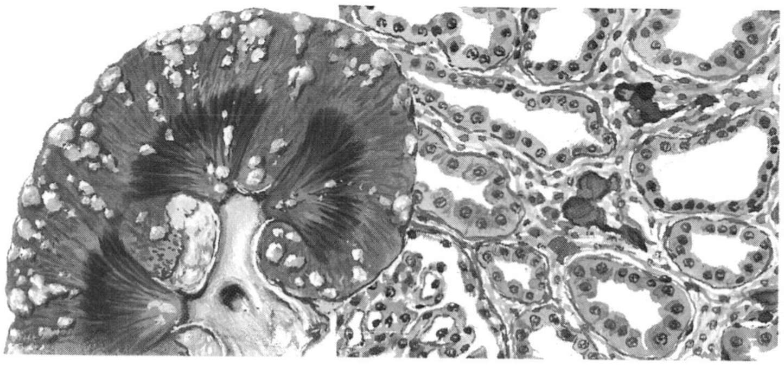

Nephrocalcinosis may be evident on radiograph or only on microscopic examination and can cause interstitial nephritis, further aggravating renal failure

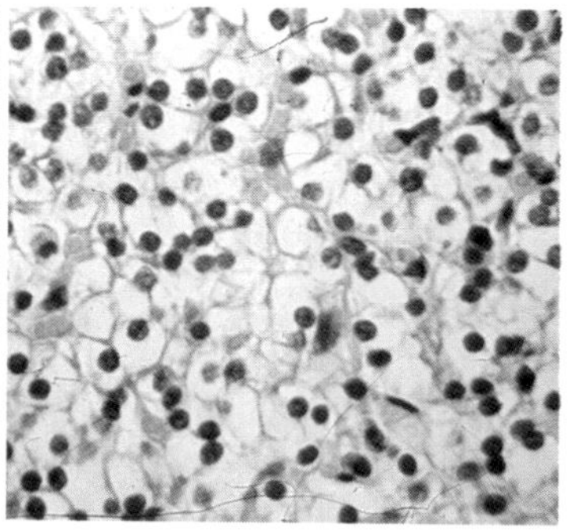

Clear cell hyperplasia of parathyroid gland

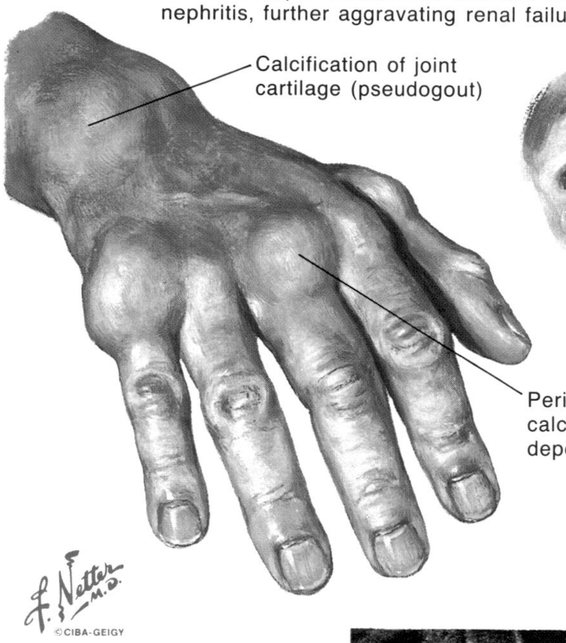

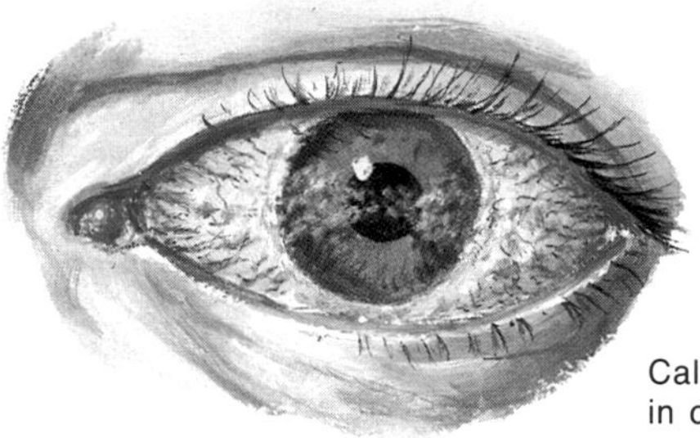

Calcium deposits in conjunctivae. Slit-lamp examination may reveal band keratitis

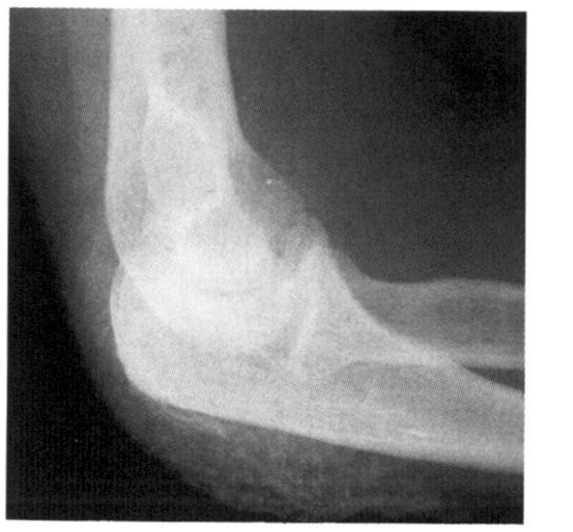

Extraarticular calcification around proximal ulna (elbow essentially fused)

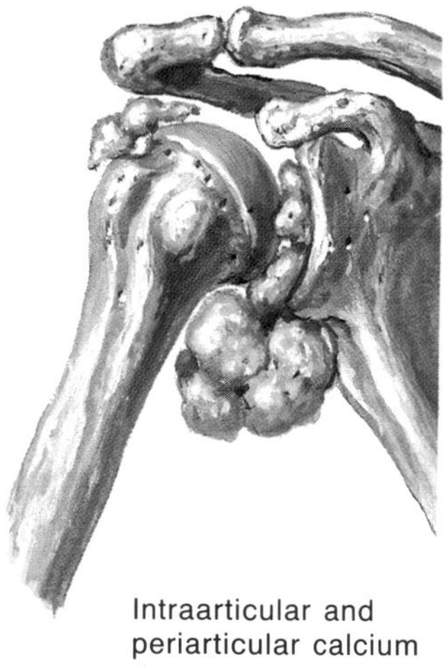

Intraarticular and periarticular calcium deposits in shoulder

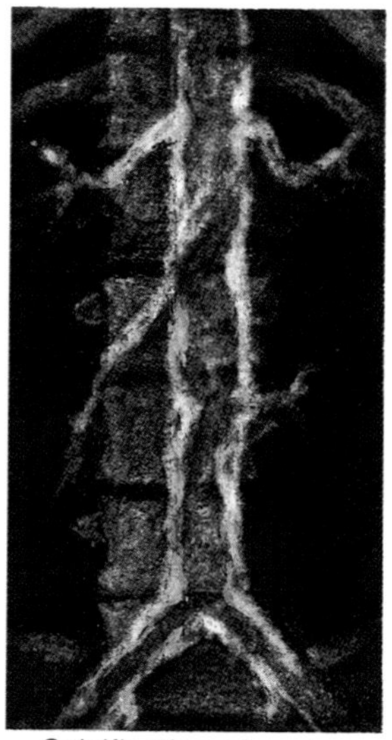

Calcification of aorta and other large vessels

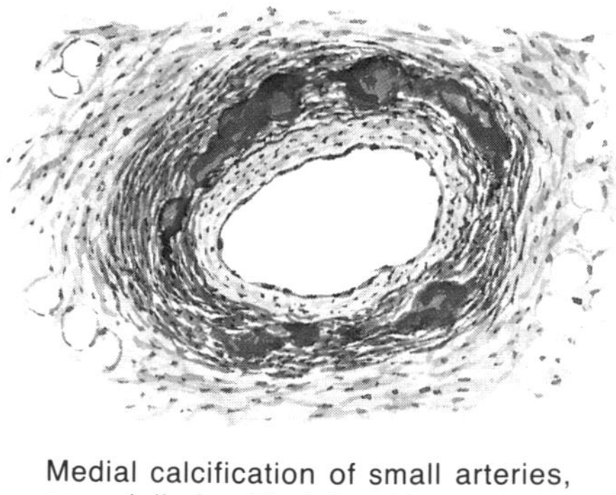

Medial calcification of small arteries, especially in skin (sloughing may occur)

Radiographic examination of the bones reveals all the features of osteomalacia—thin cortices, fuzzy trabecular markings, and Looser's zones—but may also reveal striking features of osteitis fibrosa cystica, which differentiate renal osteodystrophy from the nutritional or vitamin D–resistant forms of the disease. Histologic examination is likely to show more severe degrees of osteoclastic resorption of bone, with fibrosis of the marrow and brown tumors, and large (macroscopic) regions in which resorption is so great that the cortices are enormously thinned and no medullary bone can be found. Islands of osteoblastic activity are frequently observed and account for the increased alkaline phosphatase activity in the serum and the patchy and occasionally significant increment in activity observed on radionuclide bone scans.

Radiographic examination of the skull may show irregular rarefaction of the calvaria ("salt-and-pepper" skull) and loss of the dense white line cast by the lamina dura surrounding the roots of the teeth. Cortical thinning and fuzzy trabeculae characteristic of both osteomalacia and osteitis fibrosa cystica are seen in radiographs of the long bones, with additional findings of small or large, rarefied, rounded lytic lesions characteristic of brown tumors; "disappearance" of the lateral portion of the clavicle; and subperiosteal resorption of the proximal medial tibia. The most noticeable changes are seen in the bones of the hand, with erosion of the terminal phalangeal tufts and subperiosteal resorption of the proximal and distal phalanges most marked on the radial sides.

For reasons not clearly understood, about 20% of the patients with the combination of chronic renal disease, osteomalacia, and osteitis fibrosa cystica also develop a type of osteosclerosis. Histologic findings reveal an increased number of trabeculae per unit volume rather than a healing of the demineralized bone of osteomalacia (osteoid seams) or an alteration in the resorptive changes of osteitis fibrosa cystica. The disease most commonly affects the subchondral cortices of the vertebrae and the shafts of the long bones, producing a radiographic appearance of alternating light and dark shadows (banded sclerosis, or "rugger-jersey spine"). The small bones or digits of the hands and feet are rarely involved.

Metastatic or ectopic calcification and ossification are seen in many sites (Plate 18). The most common articular sites are the articular cartilages and menisci of the knees; the triangular ligaments of the distal radioulnar joints; the soft tissues surrounding the shoulder, elbow, knee, and ankle; and the tunica media or the tunica muscularis of the larger arterial and arteriolar vessels. In many cases, the skin and the conjunctivae (the "red eyes of renal failure") are also involved (Plate 19).

Treatment. Treatment of renal osteodystrophy is complex. The first requirement is appropriate management of the chronic renal disease, which includes procedures such as chronic dialysis and renal transplantation. Additional measures are administration of aluminum to diminish the hyperphosphatemia, judicious administration of $1,25(OH)_2D$ to increase calcium absorption, and parathyroidectomy to control the sometimes autonomous (tertiary) hyperparathyroidism.

Orthopedic problems are sometimes severe and treatment may include internal fixation of slipped capital femoral epiphysis with threaded devices, management of bowleg and knock-knee by bracing or osteotomy, and use of open or closed fixation for the frequent fractures that occur during the course of the disease. □

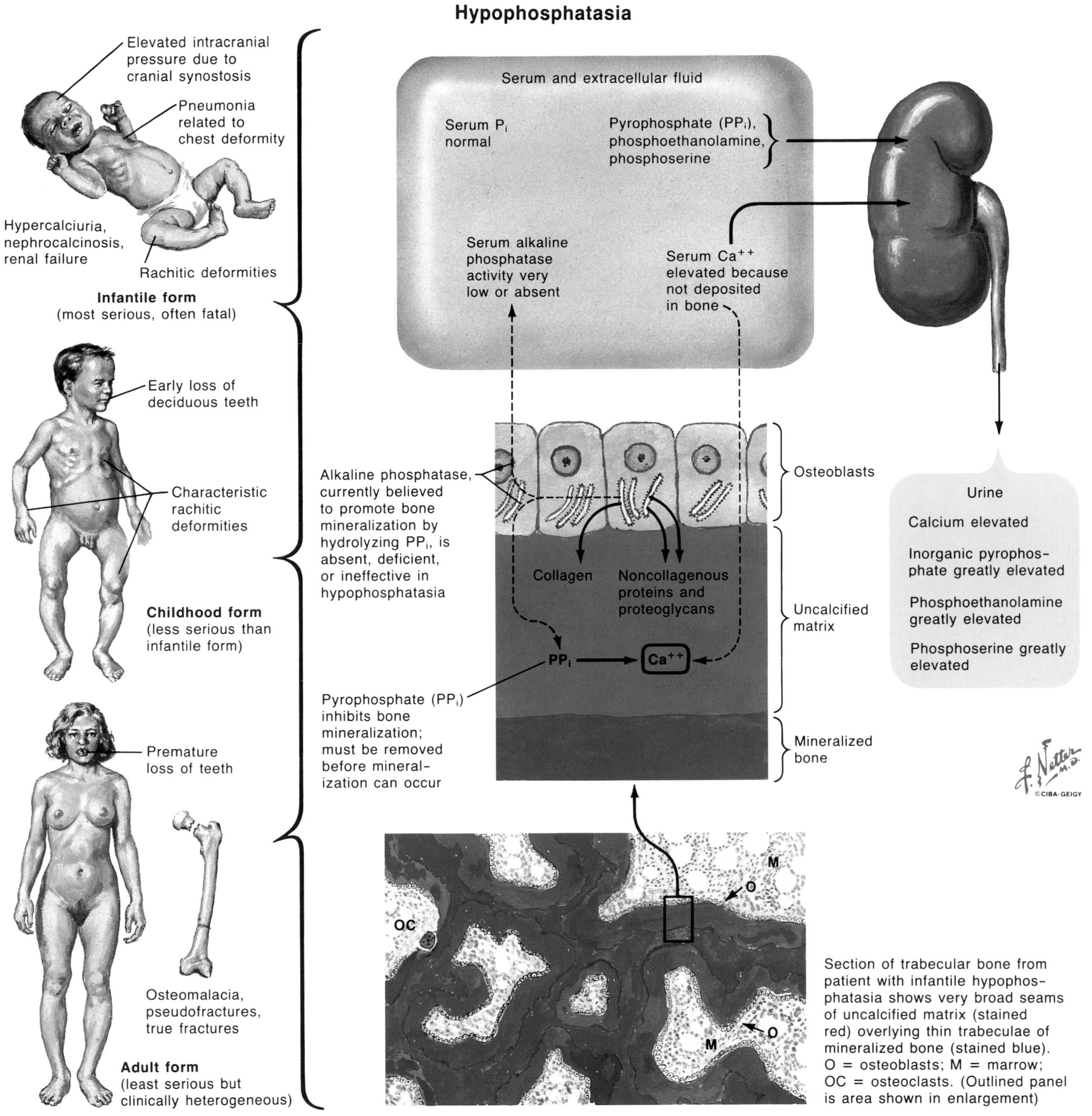
Hypophosphatasia
Elevated intracranial pressure due to cranial synostosis
Pneumonia related to chest deformity
Hypercalciuria, nephrocalcinosis, renal failure
Rachitic deformities
Infantile form (most serious, often fatal)
Early loss of deciduous teeth
Characteristic rachitic deformities
Childhood form (less serious than infantile form)
Premature loss of teeth
Osteomalacia, pseudofractures, true fractures
Adult form (least serious but clinically heterogeneous)
Serum and extracellular fluid
Serum P_i normal
Pyrophosphate (PP_i), phosphoethanolamine, phosphoserine
Serum alkaline phosphatase activity very low or absent
Serum Ca^{++} elevated because not deposited in bone
Alkaline phosphatase, currently believed to promote bone mineralization by hydrolyzing PP_i, is absent, deficient, or ineffective in hypophosphatasia
Osteoblasts
Collagen
Noncollagenous proteins and proteoglycans
Uncalcified matrix
PP_i
Ca^{++}
Pyrophosphate (PP_i) inhibits bone mineralization; must be removed before mineralization can occur
Mineralized bone
Urine
Calcium elevated
Inorganic pyrophosphate greatly elevated
Phosphoethanolamine greatly elevated
Phosphoserine greatly elevated
F. Netter M.D.
©CIBA-GEIGY
OC
M
O

Section of trabecular bone from patient with infantile hypophosphatasia shows very broad seams of uncalcified matrix (stained red) overlying thin trabeculae of mineralized bone (stained blue). O = osteoblasts; M = marrow; OC = osteoclasts. (Outlined panel is area shown in enlargement)

Hypophosphatasia

Hypophosphatasia is a rare heritable metabolic bone disease characterized by low circulating alkaline phosphatase (ALP) activity, increased blood and urine levels of phosphoethanolamine (PEA) and inorganic pyrophosphate (PP_i), edentia, and defective mineralization. It is classified by age of onset into three forms:

The *infantile form* manifests before 6 months of age and is inherited as an autosomal recessive trait. The severe clinical problems result from poor skeletal mineralization and include growth failure, rachitic deformities, and renal compromise, producing a rapidly fatal course.

The *childhood form* occurs after 6 months of age and has a variable but more benign course. Premature loss of deciduous teeth, the most consistent clinical sign, may be accompanied by rachitic defects that sometimes improve spontaneously.

The *adult form* is the least severe but clinically heterogeneous form and is inherited as an autosomal dominant trait with variable penetrance. There may be a childhood history of premature loss of deciduous teeth or of rachitic deformity, but symptoms usually begin with the loss of adult succedaneous teeth, with later recurrent fractures due to osteomalacia.

Skeletal mineralization is defective in the absence of ALP activity. Impairment of cartilage and osteoid calcification is due to the failure of ALP to cleave PP_i molecules, which act as endogenous inhibitors of crystallization. Serum levels of other natural substrates of ALP, such as PEA and phosphoserine (PS), are also elevated. ALP levels are normal in the intestine and placenta and reduced in the liver, kidney, and bone. Because renal ALP may be required for tubular reabsorption of PEA and PS, ALP deficiency may result in increased urinary excretion of these monophosphate esters. The inability of calcium to enter the bone promotes hypercalcemia. Resultant hypercalciuria with nephrocalcinosis leads to renal failure. □

Osteoporosis

Causes of Osteoporosis

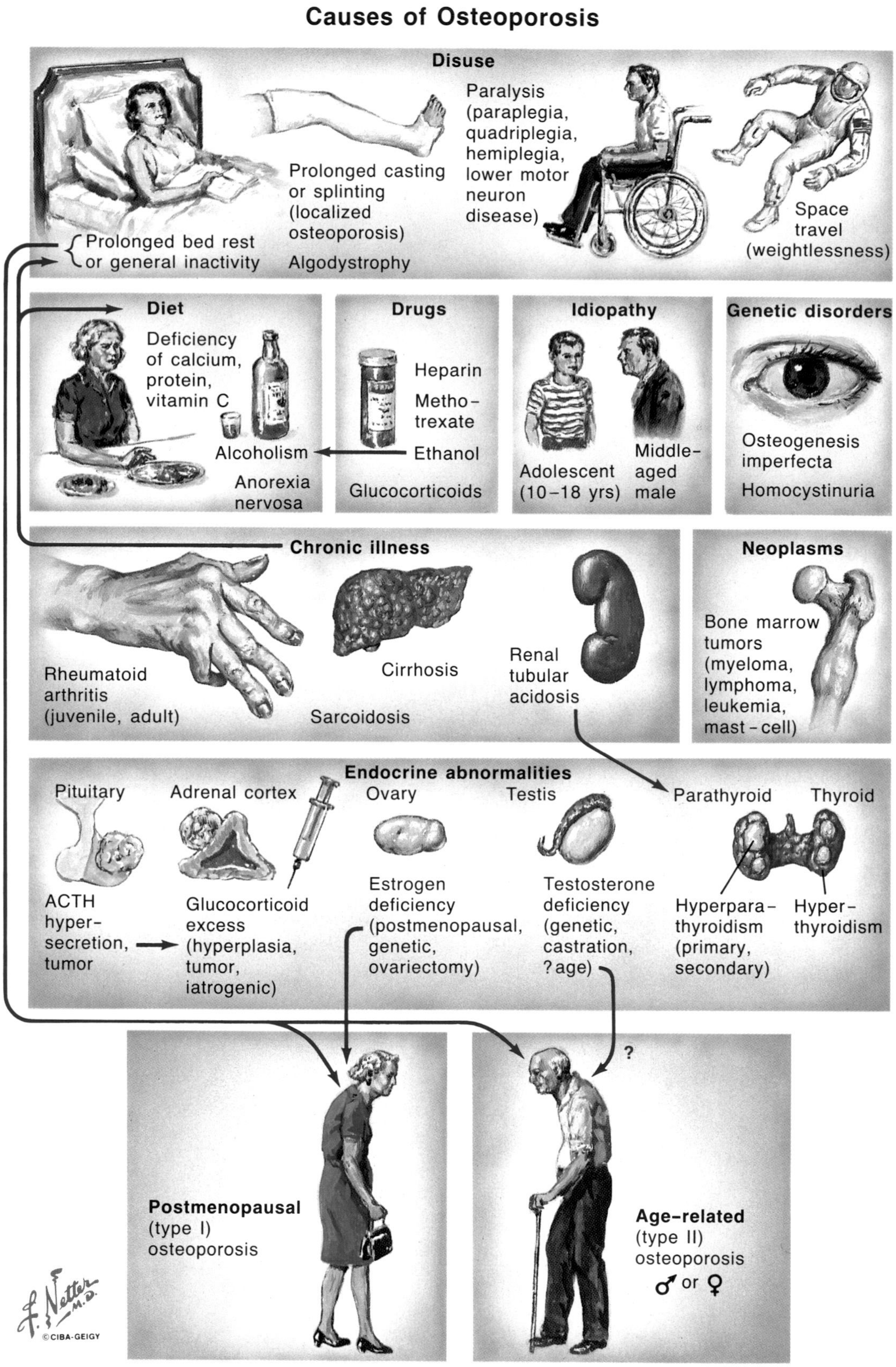

Causes of Osteoporosis

Osteoporosis is a condition characterized by a decreased mass per unit volume (density) of normally mineralized bone matrix (osteoid). It is the most common skeletal disorder in the world and is second only to arthritis as a leading cause of musculoskeletal morbidity in the elderly. Osteoporosis results when bone mass falls below normal for body size, age, sex, and race. It is characterized by structural weakness in the bones, primarily due to enlarged medullary (marrow) and osteonal spaces and reduced cortical thickness (Plate 22).

The most common types of osteoporosis are involutional: type I, postmenopausal osteoporosis, and type II, age-related osteoporosis (Table 1). Other less common but clinically significant disorders also lead to a generalized decrease in bone mass in either the axial or the appendicular skeleton or both. Persons who are affected in childhood or adolescence may never achieve a normal peak adult bone mass.

Age and Sex Factors. More than half of the 40 million American women over 50 years of age have radiographically detectable evidence of osteopenia. Major orthopedic problems related to osteoporosis eventually occur in more than one-third of these women.

The development of osteoporosis in postmenopausal women is linked to a number of factors, the most important being estrogen deficiency. Other factors include lesser bone mass and greater bone loss in women than in men, reduced serum concentration of the active vitamin D metabolite $1,25(OH)_2D$, and a calcium-deficient diet and reduced calcium absorption from the intestine (characteristic of elderly persons). (See Section III, Plates 28–34 for a full discussion of mineral homeostasis.)

Age-related osteoporosis, which occurs in both men and women, is caused by bone loss that normally accompanies aging. After age 40, the rate of bone resorption increases while the rate of bone formation remains constant, and this imbalance ultimately leads to a decrease in bone mass.

Genetic Factors. Although *osteogenesis imperfecta* is an uncommon disorder, it is the most commonly known heritable form of osteoporosis (Plates 34–36).

Homocystinuria, which resembles Marfan's syndrome (Plate 37), is related to a deficiency in the activity of the enzyme cystathionine β-synthetase. The exact cause of the osteoporosis is not certain but is believed to be a defect of collagen metabolism. Clinical signs first appear in childhood and include skeletal fragility, tall stature, kyphosis or scoliosis or both, knock-knee, arachnodactyly, subluxated lenses, capillary fragility, and mental retardation.

Endocrine Abnormalities. Because many hormones affect skeletal remodeling and thus skeletal mass, endocrine-mediated osteoporosis should be suspected in any young or middle-aged person with osteopenia. In the elderly person, endocrine-mediated osteoporosis can occur in conjunction with postmenopausal or age-related osteoporosis. Only the most common endocrine disorders that lead to loss of bone mass are included in this discussion.

Hypogonadism causes bone loss in both men and women. All osteopenia seen in postmenopausal women, whether natural or surgically induced, has a hypogonadal component. Other hypogonadal conditions that lead to bone loss are castration, panhypopituitarism, Klinefelter's syndrome, Turner's syndrome, and idiopathic hypogonadotropic hypogonadism.

Osteoporosis

(Continued)

Table 1. INVOLUTIONAL OSTEOPOROSIS

	Postmenopausal (Type I)	Age-related (Type II)
Epidemiologic Factors		
Age	55 to 75	>70 (F); > 80(M)
Sex Ratio (F/M)	6:1	2:1
Bone Physiology or Metabolism		
Pathogenesis of uncoupling	Increased osteoclast activity; ↑ resorption	Decreased osteoblast activity; ↓ formation
Net bone loss	Mainly trabecular	Cortical and trabecular
Rate of bone loss	Rapid/short duration	Slow/long duration
Bone density	>2 standard deviations below normal	Low normal (adjusted for age and sex)
Clinical Signs		
Fracture sites	Vertebrae (crush), distal forearm, hip (intracapsular)	Vertebrae (multiple wedge), proximal humerus and tibia, hip (extracapsular)
Other signs	Tooth loss	Dorsal kyphosis
Laboratory Values		
Serum Ca^{++}	Normal	Normal
Serum P_i	Normal	Normal
Alkaline phosphatase	Normal (↑ with fracture)	Normal (↑ with fracture)
Urine Ca^{++}	Increased	Normal
PTH function	Decreased	Increased
Renal conversion of 25(OH)D to $1,25(OH)_2D$	Secondary decrease due to ↓ PTH	Primary decrease due to decreased responsiveness of $1\text{-}\alpha\text{-}OH_{ase}$
Gastrointestinal calcium absorption	Decreased	Decreased
Prevention		
High-risk patients	Estrogen or calcitonin supplementation; calcium supplementation; adequate vitamin D; adequate weight-bearing activity; minimization of associated risk factors	Calcium supplementation; adequate vitamin D; adequate weight-bearing activity; minimization of associated risk factors

Hyperthyroidism, whether caused by glandular hyperactivity or by overzealous replacement therapy for hypothyroidism, increases bone turnover and remodeling. Bone resorption exceeds bone formation, resulting in a net decrease in bone mass. Patients with osteopenia, especially those on long-term thyroid hormone replacement therapy, should be examined for symptoms and signs of this disease. Triiodothyronine resin uptake (T_3RU) and thyroxine (T_4) levels should be determined when hyperthyroidism is suspected.

Hyperparathyroidism, either primary or secondary, also increases bone turnover and bone remodeling, causing a net increase in bone resorption (Plates 2–3). Primary hyperparathyroidism should be suspected whenever several of the conditions mentioned are detected in a person with osteopenia and whenever hypercalcemia and hypophosphatemia are discovered incidentally on routine blood chemistry determinations.

Hyperadrenalism, or chronic glucocorticoid excess, whether endogenous (Cushing's syndrome) or iatrogenic, leads to a refractory state of decreased bone mass. Chronic glucocorticoid excess reduces metabolism of bone mineral and connective tissue. Because of their atrophic effect on the lining cells of the small intestine, glucocorticoids decrease intestinal absorption of calcium. This stimulates the PTH–vitamin D endocrine axis to restore the serum concentration of ionized calcium, partly through the mechanism of increased bone resorption. In addition to this indirect catabolic effect on both bone matrix and bone mineral, glucocorticoids exert a direct antianabolic effect on bone metabolism.

The combined direct and indirect effects of glucocorticoids cause a profound decrease in bone mass, which often results in symptomatic osteoporosis. Symptoms are usually more severe in the axial than in the appendicular skeleton. Treatment of glucocorticoid-induced osteopenia continues to be difficult.

Nutritional Deficiencies. Because adequate nutrition plays an essential role in developing and maintaining peak bone mass, various nutritional deficiencies can lead to osteoporosis. Chronic dietary deficiencies in calcium and protein, as well as in vitamin C (an essential cofactor in collagen metabolism), may lead to decreased bone mass.

Alcoholism is the most common cause of bone loss in young men. The development of osteoporosis is most likely to be related to the poor diet of alcoholics. In addition, ethanol may decrease intestinal absorption of calcium and may even be directly toxic to bone-forming cells (osteoblasts).

Anorexia nervosa has recently been recognized as an important cause of osteopenia in young women.

Drug-Induced Bone Loss. Long-term use of the anticoagulant heparin may lead to osteopenia. Although the underlying mechanism is not known, it may be related to the altered metabolism of mucopolysaccharides. Methotrexate, which has both cytotoxic and calciuric effects, has also been reported to cause osteopenia.

Disuse Osteoporosis. Bone mass changes in response to mechanical stress (see Section III, Plate 36). Following immobilization (either focal or general), bone density decreases profoundly and rapidly, with a proportionate loss of both bone matrix and bone mineral. After 6 months of immobilization, total bone mass may be reduced by as much as 30% to 40%. Disuse osteopenia is common in degenerative lower motor neuron disease and in paraplegia and quadriplegia resulting from spinal cord injury.

Movement alone does not protect against osteoporosis, and weight-bearing activity is necessary to maintain skeletal health.

Disease-Related Bone Loss. *Chronic illness* of almost any kind can lead to osteopenia, with malnutrition and disuse the major contributing factors. The glucocorticoid therapy required for treatment of many chronic diseases may also increase bone loss.

Osteopenia is also a common complication of many *bone marrow tumors*. Multiple myeloma, the most common primary malignant bone tumor in adults, may be associated with profound generalized axial and appendicular osteopenia. Myeloma cells produce potent osteoclast-activating factors that stimulate bone resorption. Therefore, myeloma should be suspected in any person over age 50 who has symptomatic osteopenia, anemia, proteinuria, and a sedimentation rate greater than 100 mm/hr. Serum protein electrophoresis helps establish the diagnosis, but if the results are inconclusive, urinary immunoelectrophoresis should be performed. Approximately 1% of myelomas are nonsecretory, and a definitive diagnosis by bone marrow biopsy is therefore required in all patients. Leukemia, lymphoma, and the extremely rare systemic mastocytosis may also be associated with osteoporosis.

Idiopathic Osteoporosis. Both the juvenile and adult forms of idiopathic osteoporosis are rare. The juvenile disorder first becomes evident in late childhood or early adolescence, with the onset of symptomatic osteopenia and skeletal fragility. Symptoms occur predominantly in the axial skeleton. The disease may be severe and may cause vertebral compression fractures, kyphosis, and loss of height. The cause of the disease is unknown. Although remission occurs spontaneously, victims are left with residual disability and seldom achieve a normal peak adult bone mass.

Idiopathic adult osteoporosis, found in middle-aged men, is often accompanied by idiopathic hypercalciuria and active bone remodeling. Symptomatic axial osteopenia predominates.

Osteoporosis

(Continued)

Clinical Manifestations of Osteoporosis

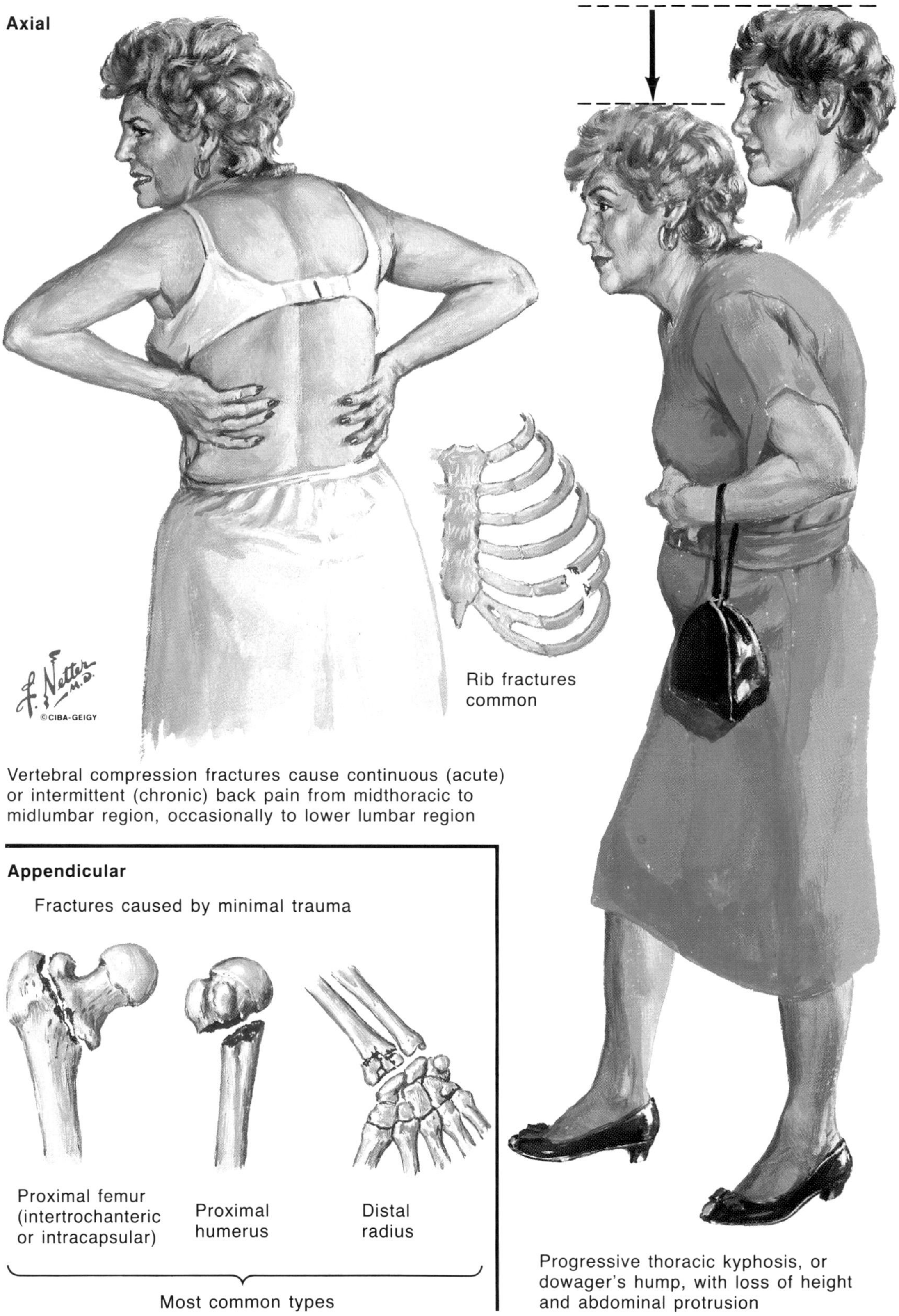
Axial
Rib fractures common
Vertebral compression fractures cause continuous (acute) or intermittent (chronic) back pain from midthoracic to midlumbar region, occasionally to lower lumbar region
Appendicular
Fractures caused by minimal trauma
Proximal femur (intertrochanteric or intracapsular)
Proximal humerus
Distal radius
Most common types
Progressive thoracic kyphosis, or dowager's hump, with loss of height and abdominal protrusion
©CIBA-GEIGY

Clinical Manifestations and Progressive Spinal Deformity

A long latent period often precedes the clinical symptoms or complications of osteoporosis (Plates 23–24). Skeletal resources are depleted, often for decades, before the bone mass is so compromised that the skeletal framework can no longer withstand *everyday* mechanical stresses.

Although the entire skeleton is susceptible to age-related and postmenopausal bone loss, regions of high trabecular bone remodeling such as the thoracic and lumbar vertebral bodies, ribs, proximal femur and humerus, and distal radius sustain the most damage. The most prevalent complications are vertebral compression fractures.

Vertebral Compression Fractures

The earliest symptom of osteoporosis is acute back pain associated with a thoracic or lumbar vertebral compression fracture, which is often precipitated by routine activities—standing, bending, lifting—that under normal circumstances would not be stressful enough to cause a fracture.

The onset of pain is sudden, and most patients can recall the exact moment that it began. The vertebral level where it originates is accurately identified, although pain from a compression fracture in the high lumbar region may be referred to the low lumbar or the lumbosacral region.

Spinal movement is severely restricted. The pain intensifies with sitting or standing and is exacerbated by coughing, sneezing, and straining to move the bowels. Bed rest in the fully recumbent position provides considerable relief. Loss of appetite, abdominal distention, and ileus secondary to retroperitoneal hemorrhage may accompany lower thoracic and upper lumbar vertebral compression fractures.

Although spontaneous vertebral compression fractures are stable injuries, radiculopathies often occur with thoracic or upper lumbar compression fractures and cause either unilateral or bilateral pain that radiates anteriorly along the costal margin of the affected nerve root. Spinal cord or cauda equina involvement is even less common and suggests other conditions, such as infection or tumor (including intradural and extradural lesions), expansile primary bone tumors, Paget's disease, metastases, myeloma, or lymphoma.

During the intervals (often years) between compression fractures, most patients remain pain free. However, approximately 30% continue to be plagued with chronic, dull, aching, postural pain in the midthoracic and upper lumbar regions that is only alleviated by frequent periods of recumbent bed rest.

As each episode of segmental vertebral collapse causes progressive kyphosis, the patient's height may decrease 2 to 4 cm. Both kyphosis and decreased height are reliable clinical signs of the late stage of the disease.

In 95% of postmenopausal patients with symptomatic osteoporosis, more than six radiographically evident vertebral fractures occur over a period of approximately 10 years; 75% of patients lose at least 10 cm in height. Once the spine has collapsed to the point where the lower ribs rest on the iliac crest, further loss in height is unlikely, although loss of bone mass may continue (Plate 23).

Two of the clinically disturbing, long-term side effects of progressive vertebral compression fractures are caused by the decreased size of the thoracic and abdominal cavities. Decreased exercise tolerance results from disease-related postural changes and restrictive hypoventilation. Early satiety, as well as abdominal protrusion secondary to severe lumbar vertebral collapse, is also noted. Circumferential pachydermal skin folds develop at the costal and pelvic margins as the disease progresses. Some persons who sustain an anterior wedge vertebral compression fracture in the midthoracic spine are relatively symptom free except for slight discomfort along the costal margins, incremental loss in height, and mild thoracic kyphosis.

Most persons with an osteoporotic injury seek medical attention within several days of the onset of back pain, but some may wait for months or may not see a physician at all.

Osteoporosis

(Continued)

Progressive Spinal Deformity in Osteoporosis

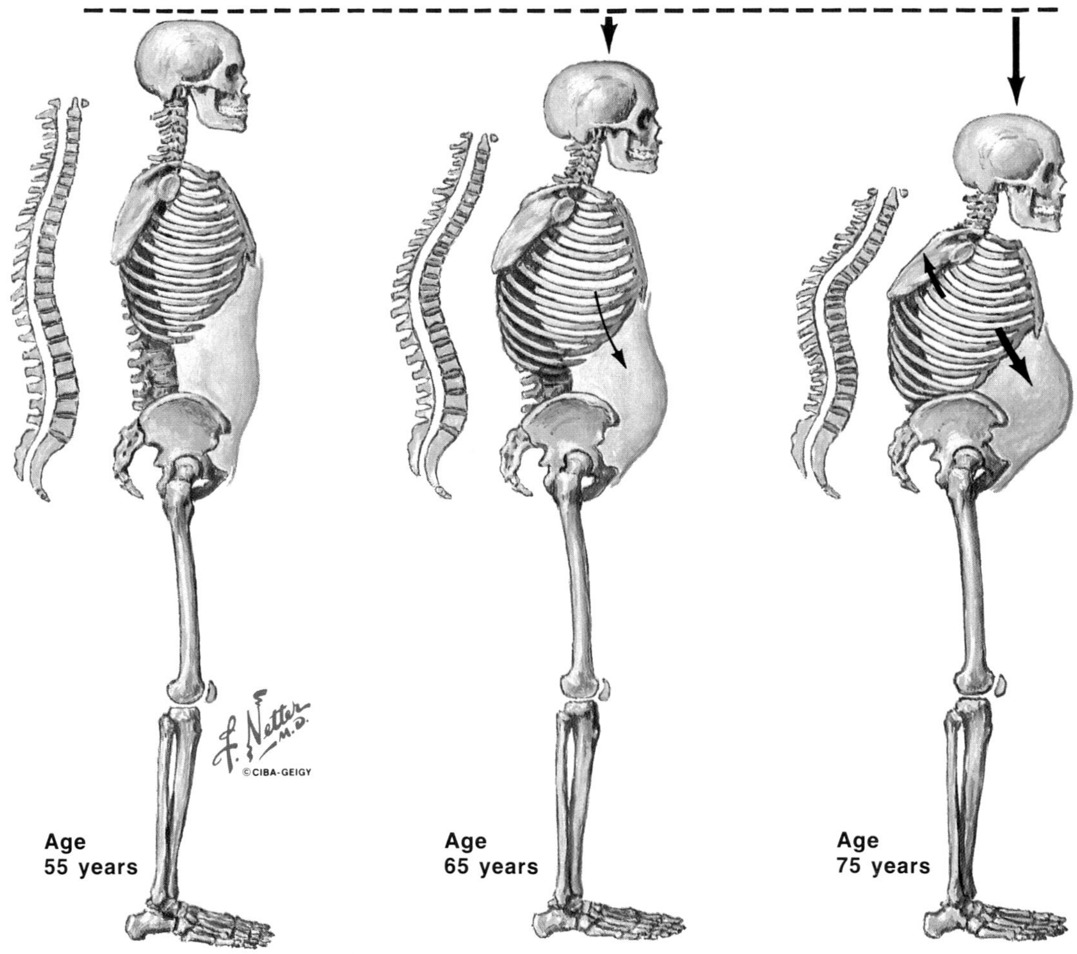

Compression fractures of thoracic vertebrae lead to loss of height and progressive thoracic kyphosis (dowager's hump). Lower ribs eventually rest on iliac crests, and downward pressure on viscera causes abdominal distention

Appendicular Fractures

Although the most common clinical symptom of osteoporosis is back pain due to vertebral compression fracture, sometimes fracture in the appendicular skeleton is the first evidence of the disease. The most frequent appendicular injuries are fractures of the proximal femur (hip) sustained after little or no trauma or a fracture of the distal radius sustained during a fall on the outstretched hand. The incidence of fractures of the proximal femur increases with age and shows a bimodal peak. Intracapsular fractures of the femur occur most often between ages 65 and 75, whereas the incidence of intertrochanteric fractures peaks about 10 years later (see Table 1, page 217).

Work-up in Symptomatic Disease

Other causes of symptomatic bone loss must be systematically excluded. The patient's history may suggest bone loss secondary to primary or iatrogenic hyperthyroidism, primary hyperparathyroidism, chronic glucocorticoid excess (hyperadrenalism or hypercorticism), myeloma, or osteomalacia. Therefore, a thorough history is essential in making the differential diagnosis and should include the following:

*1. **History of Acute Illness***. Onset and duration of symptoms; location and radiation of pain; exacerbating and remitting factors; relationship of pain to posture, activity, time of day.

*2. **Review of Related Symptoms***. Malaise; recent weight loss or weight gain; loss of height; hot flashes; changes in visual fields; purpura or acne; amenorrhea; hypertension; goiter or neck swelling; change in voice, skin texture, or hair consistency; sensitivity to temperature change; palpitations; epigastric pain or burning, change in bowel habits, diarrhea, loose or bulky foul-smelling stools; dysuria, flank pain, fever, renal colic, nephrolithiasis; joint pain or swelling; generalized bone pain or muscle weakness; psychiatric problems.

*3. **Medical and Personal History***. Menstrual history, pregnancy, lactation.

Hospitalizations or operations: oophorectomy; thyroidectomy; pituitary surgery; ulcer or bowel surgery; surgery for cancer, spine or scoliosis, bone or joint fractures.

Medications: all antacids, including over-the-counter preparations; anticonvulsants; tranquilizers; antimetabolites; vitamins and minerals; nutritional supplements; contraceptives; gonadal steroids; thyroid hormone replacement; glucocorticoids.

Dietary history: daily intake of dairy products, protein, alcohol; any discomfort with ingestion of dairy products; abuse of laxatives.

Activity level: daily weight-bearing activity; work history (physical or sedentary); exercise; sports; activity limitations; prolonged generalized or focal immobilization; exposure to sunlight.

*4. **Family History***. Osteoporosis; bone or joint disorders; other growth disturbances; fractures; blue sclerae; deafness; scoliosis; childhood dental problems; joint laxity.

After the history is taken, a complete physical examination is performed, and anteroposterior and standing lateral radiographs of the thoracic and lumbar spine are obtained.

A thorough history also helps in the selection of appropriate baseline tests. Routine laboratory tests include a complete blood count and leukocyte differential; determination of the erythrocyte sedimentation rate; a 24-hour urinalysis to measure excretion of calcium and creatinine; and determination of serum levels of calcium, albumin phosphate, alkaline phosphatase, blood urea nitrogen, and creatinine. The serum concentration of 25(OH)D accurately reflects the total body stores of vitamin D. In uncomplicated postmenopausal osteoporosis, results of routine laboratory tests are normal. Even in severe postmenopausal disease, serum calcium, phosphate, and alkaline phosphatase levels are usually within the normal range.

If bone loss secondary to conditions other than age-related and postmenopausal osteoporosis is suspected, additional tests are performed as necessary. For patients with suspected primary or iatrogenic hyperthyroidism, triiodothyronine resin uptake (T_3RU) and thyroxine (T_4) serum levels are obtained, and the free thyroxine index is calculated. Serum or urine protein electrophoresis is required when multiple myeloma is suspected. In patients with hypercalcemia, the circulating level of carboxyl-terminal (C-terminal) fragments of PTH are measured by radioimmunoassay.

Osteomalacia must be considered in the differential diagnosis of osteopenia (Plate 33). Osteomalacia should be suspected in a patient with generalized myopathy, bone pain and tenderness, and symmetric long-bone fractures. Because abnormalities of vitamin D absorption and metabolism often play a major role in the pathogenesis of the disease, serum levels of 25(OH)D and $1,25(OH)_2D$ should also be ascertained. The diagnosis is confirmed by fluorescent microscopic examination of undecalcified trabecular bone tissue obtained by a transiliac bone biopsy after time-separated double-tetracycline labeling (Plate 30). If available, noninvasive diagnostic techniques to monitor the progression of bone loss and the response to treatment may be more desirable. These techniques include quantitative assessments of bone mineral content (Plates 23 and 39). The technetium 99m methylene diphosphonate bone scan is useful in documenting new compression fractures in persons with preexisting vertebral collapse.

Radiographic Findings in Axial Osteoporosis

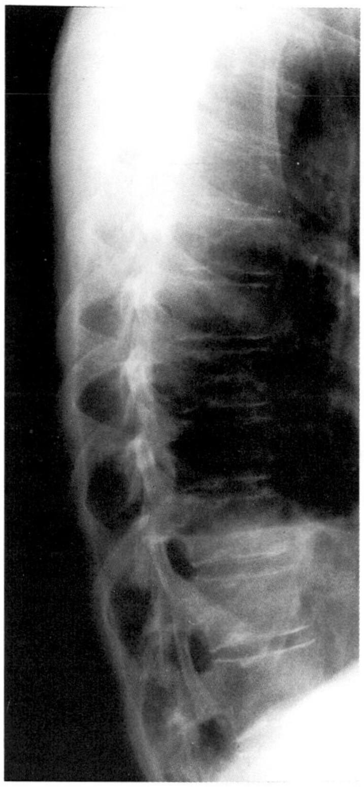

Mild osteopenia in postmenopausal woman. Vertebrae appear "washed-out"; no kyphosis or vertebral collapse

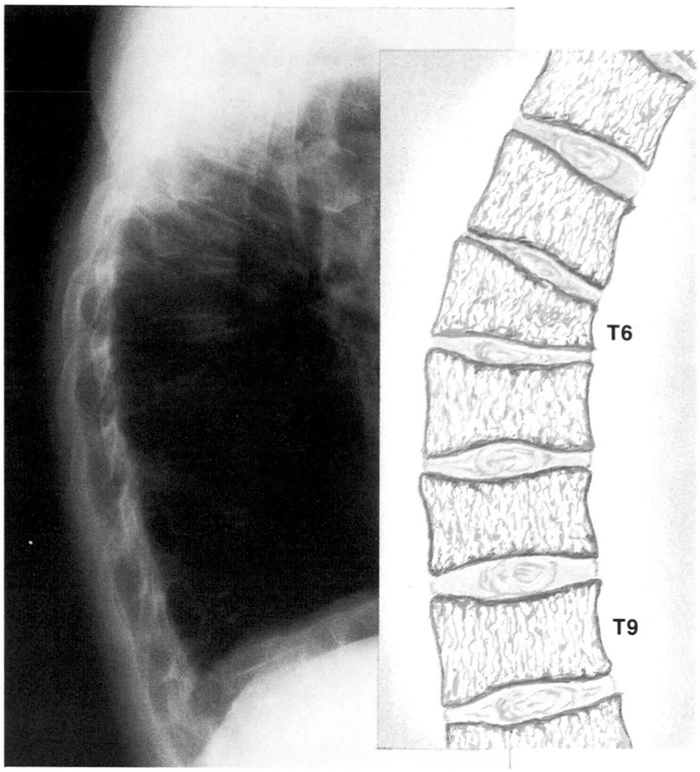

Anterior wedge compression at T6 in same patient 16½ years later. Patient has lymphoma, with multiple biconcave ("codfish") vertebral bodies and kyphosis. Focal lesion at T6 suggests neoplasm

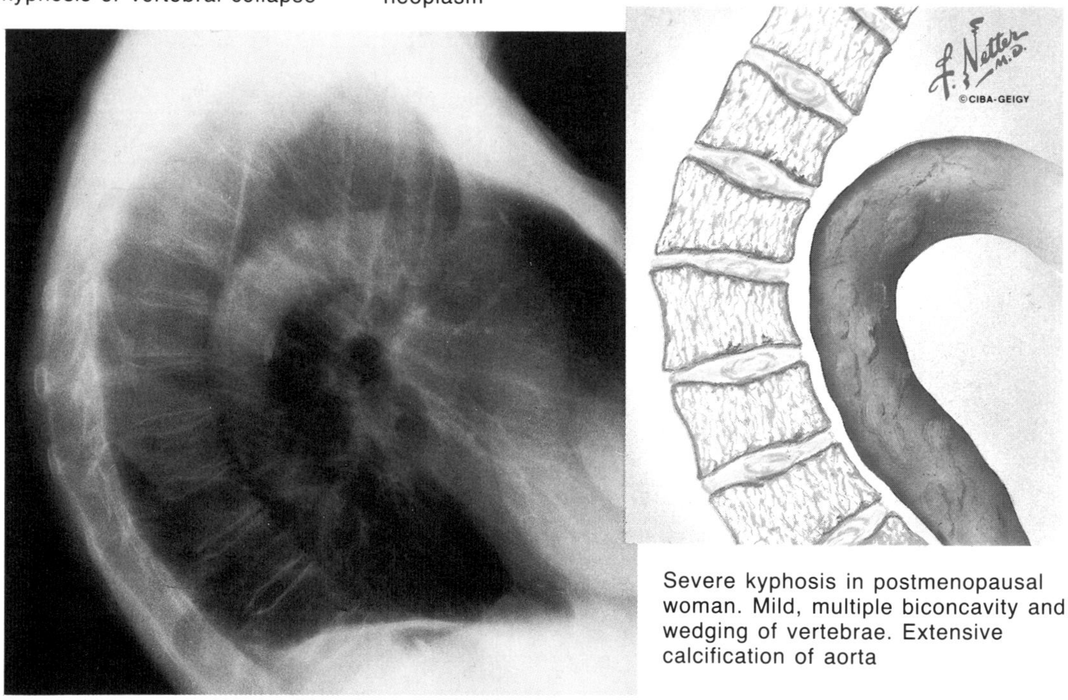

Severe kyphosis in postmenopausal woman. Mild, multiple biconcavity and wedging of vertebrae. Extensive calcification of aorta

Osteoporosis

(Continued)

Radiographic Findings in Axial Osteoporosis

Radiography of the spine is the traditional technique used to assess osteopenic conditions (Plates 25–26). The term "osteopenia" is a generic expression referring to a qualitative paucity of bone; it contributes no information about the etiology or pathogenesis of the condition that leads to this effect or about the status of bone matrix mineralization. The two major categories of osteopenia are osteoporosis and osteomalacia (Plates 22 and 33).

The amount of bone that must be lost before osteopenia is radiographically detectable varies in different bones and in different parts of the same bone, depending on the structural composition of the region involved and on the diagnostic technique employed. In general, as much as 30% to 50% of bone mass must be lost before the decrease is visible on plain radiographs. Therefore, more sensitive, noninvasive densitometric screening techniques are increasingly being used for early detection of osteopenia. However, plain radiographs do help to exclude obvious causes of back pain that are not related to osteopenia, especially in the elderly. They also aid in determining the pattern and extent of osteopenia (either focal or generalized) and in identifying the sites of compression fractures. Frequently, the detection of osteopenia on a plain radiograph directs the physician to investigate an underlying condition, particularly in a young or middle-aged patient. Because the various causes of osteopenia cannot be established by radiography alone, laboratory tests are usually required for a definitive diagnosis. Occasionally, a bone biopsy is necessary.

The earliest radiographic evidence of generalized osteopenia is seen in the spine, especially in the vertebral bodies, which contain a high proportion of metabolically active, high-turnover trabecular bone. Thinning and eventual loss of bone in the vertebral bodies may be striking. The horizontal trabeculae are resorbed first, leading to an early accentuation of the vertical trabeculae or, in more severe cases, a vertebral body that has a hollow appearance. In the worst cases of axial osteopenia, the intervertebral disc may appear denser than the vertebral body, producing a confusing optical illusion.

Lateral radiographs of the thoracic and lumbar spine should be obtained in any patient with a suspected compression fracture. A pathologic compression fracture in a young or middle-aged patient or a fracture above the level of the fifth thoracic vertebra (T5) in a patient of any age necessitates a thorough investigation to identify less common causes of osteopenia.

Radiographic Findings in Axial Osteoporosis (continued)

Osteoporosis
(Continued)

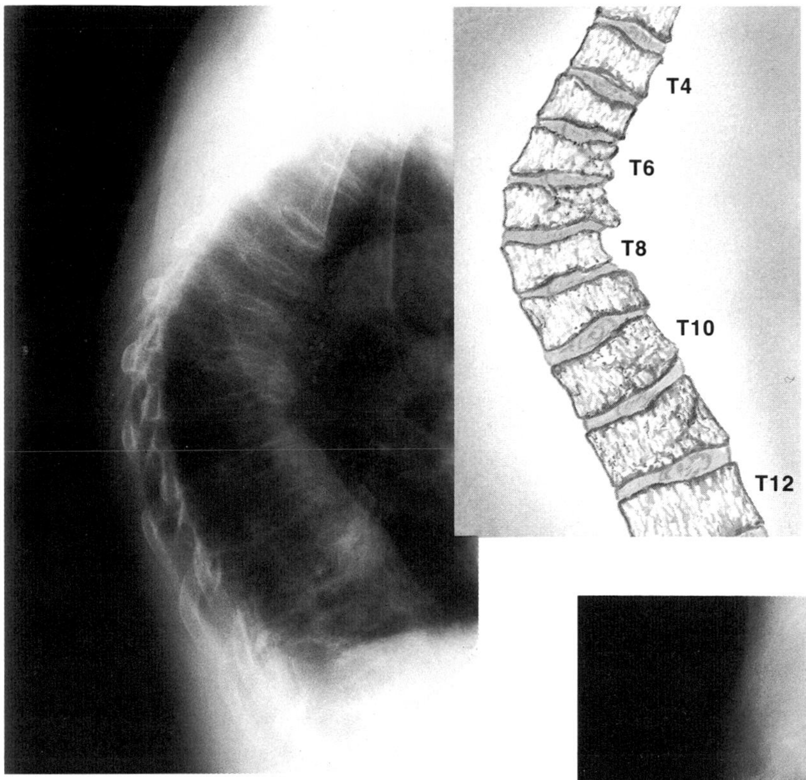

Multiple healing compression fractures of thoracic vertebrae and marked kyphosis in 71-year-old woman with postmenopausal and steroid-induced osteopenia

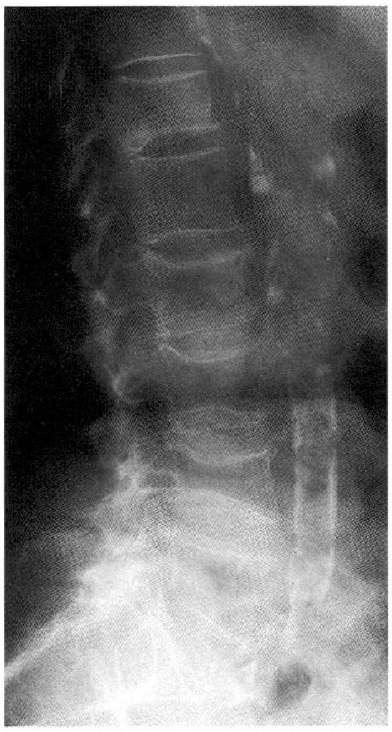

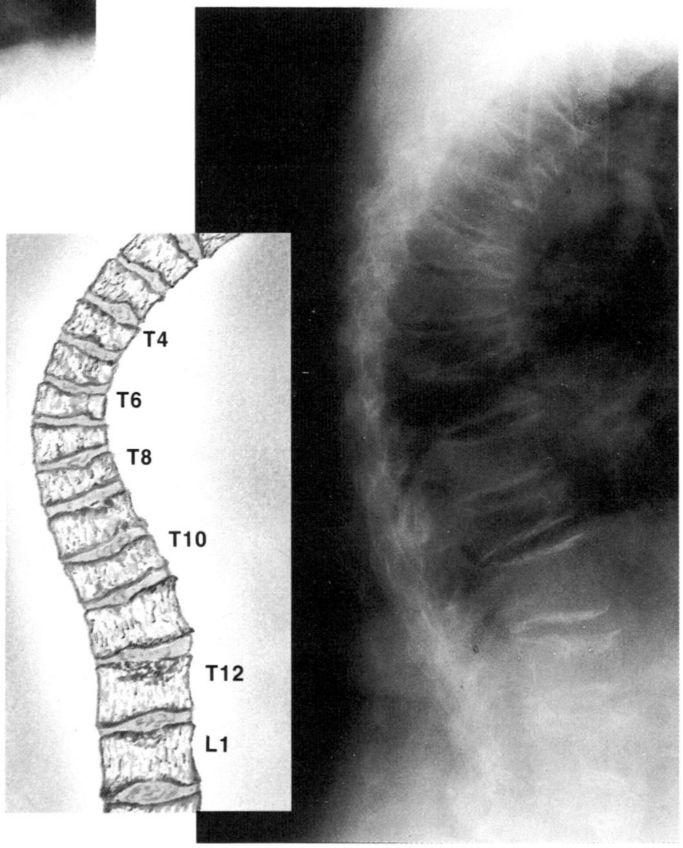

Biconcave compression fractures of L2, 3, and 4 in 72-year-old woman with postmenopausal osteoporosis

Multiple compression fractures of thoracic and lumbar vertebrae in middle-aged man with active osteopenia and hypercalciuria

Prior to gross collapse of the vertebrae, radiographs of the thoracic and lumbar spine affected by osteoporosis reveal a visible loss of bone density, which gives the vertebral column a washed-out appearance (Plate 25). Before clinically evident vertebral compression fractures occur, the vertebral bodies may become increasingly biconcave as the subchondral cortical plates weaken and the intervertebral discs expand. The result is a "codfish" appearance of the vertebrae, best seen on the lateral radiograph (Plates 25–26). The superior and inferior end plates of a vertebral body may or may not be equally affected. However, the inferior end plate is rarely more deformed than the superior end plate. Some regard this picture of biconcave vertebral segments as evidence of a central collapse compression fracture, even though the patient may be asymptomatic.

Types of Fractures

Three types of radiographic findings are common with vertebral compression fractures. The *biconcave central compression fracture* is most commonly seen in the lumbar spine. The *anterior wedge compression fracture* is common in postmenopausal osteoporosis, occurring most often in the thoracic and upper lumbar segments. This type of fracture usually causes a sudden, severe episode of incapacitating back pain and, if numerous fractures occur, may lead to progressive kyphosis.

In the *symmetric transverse compression fracture*, the collapse of the vertebral body leaves a waferlike vertebral segment, which is so compressed that it may be thinner than the intervertebral disc. The collapse, which may occur suddenly and spontaneously, is associated with sudden and severe pain. A single, waferlike vertebral segment with no other radiographic signs of osteopenia suggests a condition other than osteoporosis, such as metastatic disease, eosinophilic granuloma, or hemangioma.

In the most severe cases of osteopenia, a fracture may recur in a previously compressed and healed vertebral body. The deformity and previous collapse make it difficult to ascertain on plain radiographs whether a new fracture has actually occurred. In this situation, a radionuclide bone scan may help to rule out a recent fracture.

Osteoporosis

(Continued)

Radiographic Findings in Appendicular Osteoporosis

Radius and ulna of young woman show normal cortical thickness

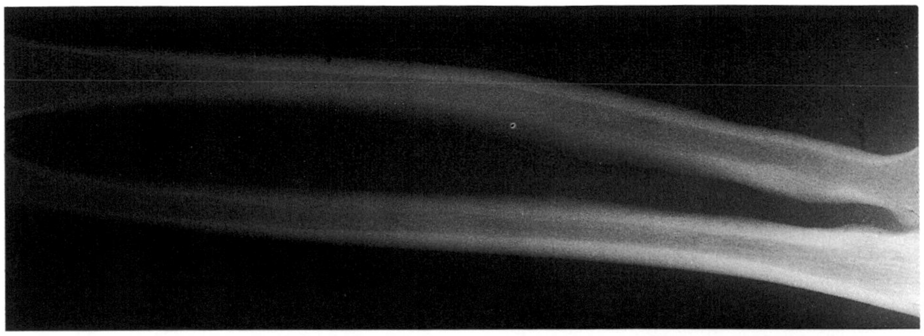

Radius and ulna of 73-year-old woman show cortical thinning, widened medullary canal, and distal fractures of both bones

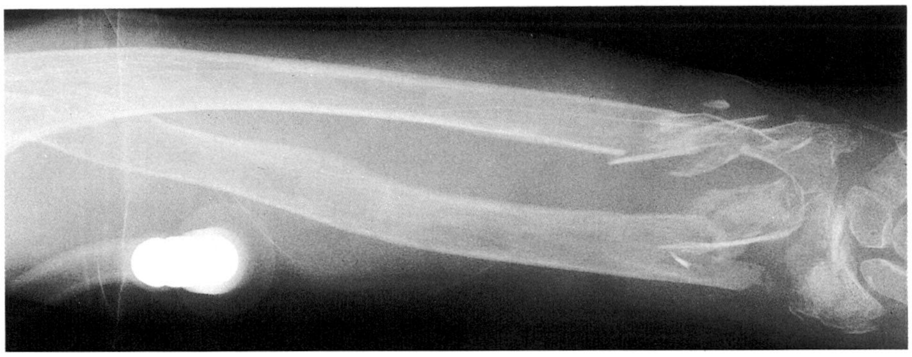

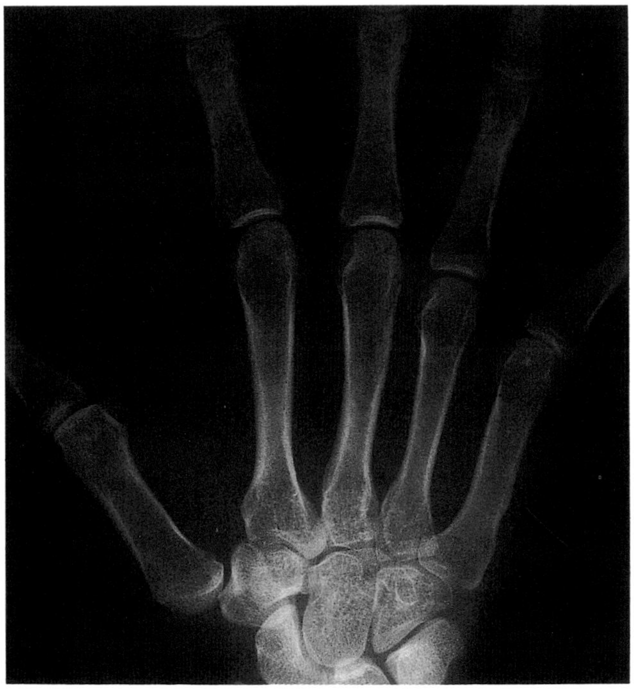

Osteoporotic hand with characteristic cortical osteopenia of metacarpals

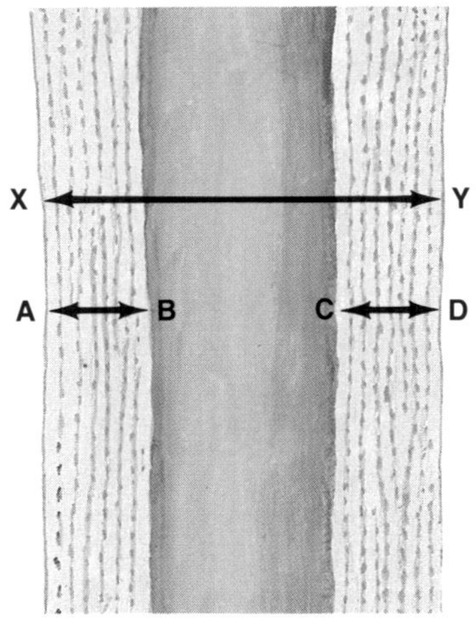

XY = Total width of diaphysis

AB+CD = Combined cortical thickness

$\frac{AB+CD}{XY}$ = Cortical thickness index

Combined cortical thickness of long bone diaphysis (AB+CD) is normally ≥ width of medullary canal, so that cortical thickness index is ≥ ½. In appendicular osteopenia, index is < ½. (Midshaft of index finger metacarpal most commonly used for this determination)

F. Netter M.D.

Radiographic Findings in Appendicular Osteoporosis

The adult axial skeleton consists predominantly of high-turnover, high-remodeling trabecular bone, which is in contact with hematopoietic marrow. Generalized disorders of skeletal homeostasis, such as osteoporosis, affect the axial skeleton, which is more metabolically active than the appendicular skeleton. However, disease and age-related (type II osteoporosis) changes do occur in the appendicular skeleton, which contains a greater proportion of low-remodeling cortical bone than does the axial skeleton (Plate 27).

Gross morphologic changes seen in osteoporosis include increased porosity of both trabecular and cortical bone. Characteristic changes in the long bones include increased intracortical porosity (intracortical resorption), as well as enlargement of the medullary cavities caused by a net increase in endosteal bone resorption over periosteal new bone formation. (See Section III, Plate 39 for a discussion of age-related changes in the geometry of cortical bone.)

Plain radiographs, the traditional diagnostic tools, are the least accurate, least precise method of determining osteopenia, and plain radiographs of the appendicular skeleton are even less sensitive than those of the axial skeleton. For this reason, radiogrammetry, a noninvasive quantitative method of assessing bone mass in the appendicular skeleton is valuable as a primary assessment of the degree of osteopenia. It is relatively simple, highly available, and inexpensive.

Radiogrammetry, popularized by Garn, involves the use of a caliper to measure the thickness of the cortices of a tubular bone on a fine-detail, high-resolution radiograph. Measurements are made at the middiaphysis of the second or third metacarpal; standard tables provide age-specific and sex-specific control values. Commonly used radiogrammetric indices include measurement of combined cortical thickness and the cortical thickness index (combined cortical thickness divided by the total width of the bone at the middiaphysis). A cortical thickness index of below 0.5 at the middiaphysis of any long tubular bone should alert the clinician to the possibility of appendicular osteopenia.

Radiogrammetry has also been used to measure the middiaphysis of the clavicle, distal humerus, radius, femur, and tibia. The major disadvantages of this technique are that only cortical bone and only the thickness of the bone cortex are measured, and that intracortical (haversian) resorption of bone and intracortical porosity are not taken into account.

Radiographic photodensitometry, a similar technique, is used to measure the optical density of bone by means of an optical densitometer, and findings are compared with a standard aluminum step-wedge taken on the same radiograph. The bone mass is determined as a function of optical density. Although this technique is relatively simple and inexpensive, technical problems limit its accuracy and precision.

Osteoporosis

(Continued)

Single-Photon Absorptiometry

Absorption of photons passed through bone from collimated radioactive source varies inversely with mineral density of bone. Collimated scintillation counter detects photons that have passed through bone

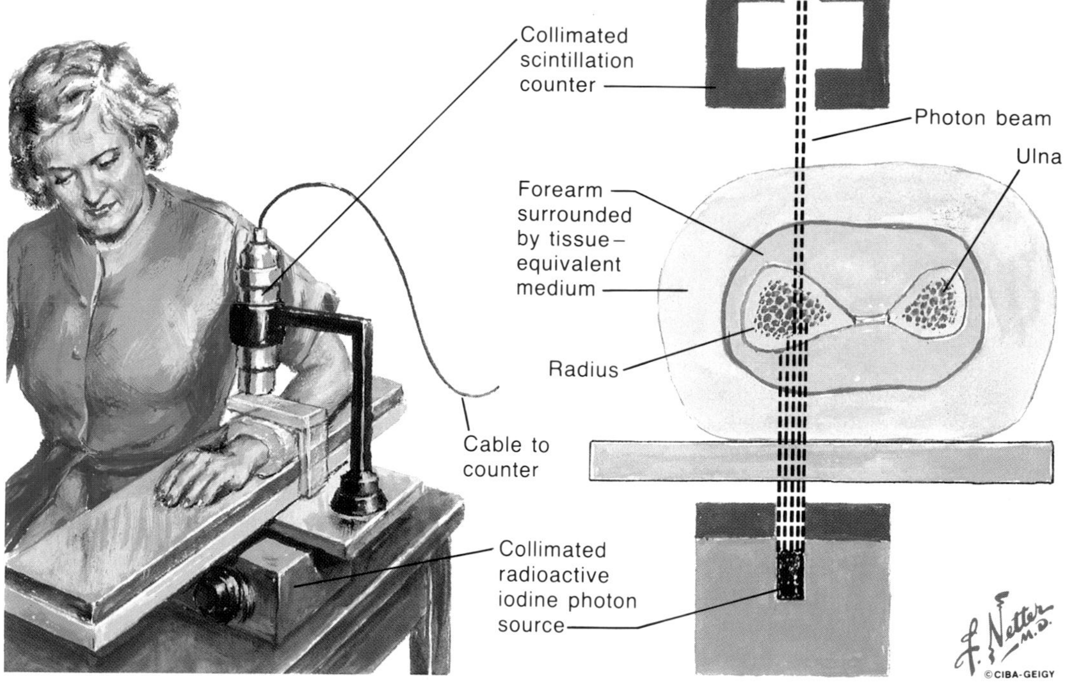
Collimated
scintillation
counter
Photon beam
Ulna
Forearm
surrounded
by tissue-
equivalent
medium
Radius
Cable to
counter
Collimated
radioactive
iodine photon
source
©CIBA-GEIGY

Dual-Photon Absorptiometry

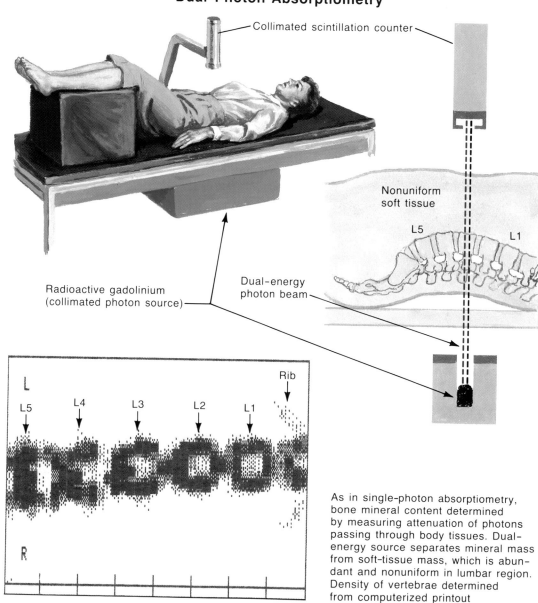

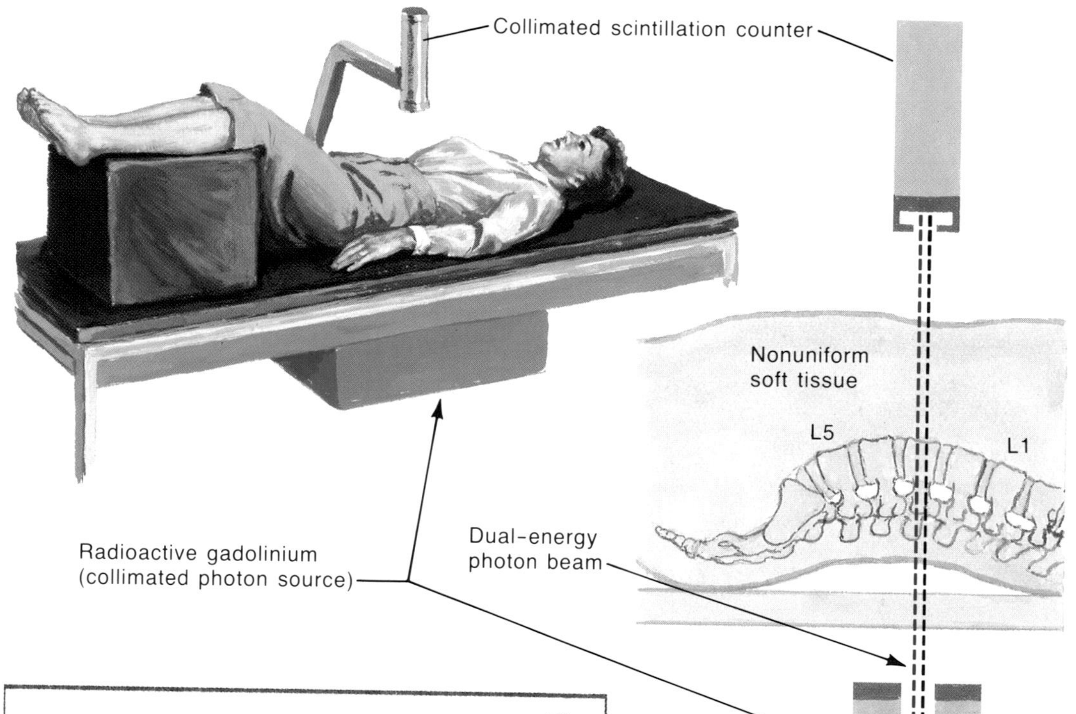
Collimated scintillation counter
Nonuniform
soft tissue
L5
L1
Dual-energy
photon beam
Radioactive gadolinium
(collimated photon source)

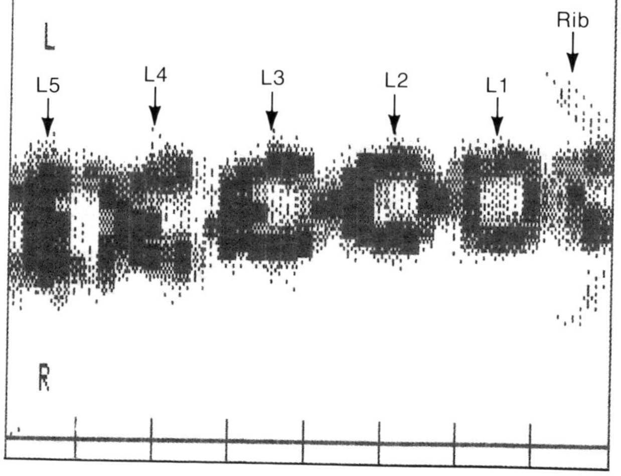
Rib
L
L5
L4
L3
L2
L1
R

As in single-photon absorptiometry, bone mineral content determined by measuring attenuation of photons passing through body tissues. Dual-energy source separates mineral mass from soft-tissue mass, which is abundant and nonuniform in lumbar region. Density of vertebrae determined from computerized printout

Measurements of Bone Mass

Noninvasive radiographic and radioisotope techniques are available to determine bone mass (Plates 28–29). These techniques are precise, sensitive, and safe. Actual quantitation of bone mass in vivo helps to establish the severity of bone loss in an osteopenic patient and serves as a baseline for evaluation of therapy.

Most noninvasive measurements of bone mineral content and density provide site-specific information about the amount of bone in the skeleton at the time of the examination but give no information about the past, current, or future rate of bone remodeling. However, sequential measurement of bone mass or density may provide this information.

The current status of bone remodeling can be determined by various indirect serum and urine laboratory studies performed in conjunction with a nondecalcified transiliac bone biopsy (Plate 30).

At present, three methods are used to assess the mineral content of bone: single-photon absorptiometry, dual-photon absorptiometry, and quantitative computed tomography.

Single-Photon Absorptiometry

Single-photon absorptiometry (SPA) is most commonly used to measure the mineral content of long bones (Plate 28). A monoenergetic photon source of iodine 125 is coupled to a sodium iodide scintillation counter. The radioisotope iodine 125 emits a beam of photons that passes through the forearm surrounded by a soft-tissue equivalent medium. The scintillation counter is moved back and forth on the other side of the forearm to detect transmitted photons. The denser the bone, the more the monoenergetic photon beam is attenuated and the fewer photons pass through to the scintillation counter. The degree of attenuation of the photon beam allows measurement of the mineral content in the limb. SPA is most commonly used to determine the mineral content of the radius, either at the junction of its middle and distal thirds (95% cortical bone and 5% trabecular bone) or in the metaphysis (30% cortical bone and 70% trabecular bone). Although the distal third of the radius contains more trabecular bone, its shape is irregular, which makes reproducibility more difficult and reduces precision.

Single-photon absorptiometry affords precise (within 4%) evaluation of the midradius and provides accurate (within 3%–4%) determination of the density (g/cm^2) of cortical bone at that site because it takes into account the intracortical porosity. The radiation dose is minimal (10 mrem), cost is generally low, and availability and patient acceptance are high. However, this method cannot be used to accurately predict changes in the axial skeleton.

Osteoporosis

(Continued)

Dual-Photon Absorptiometry

This technique is used to measure the mineral content of the axial skeleton, in which trabecular bone predominates (Plate 28).

In dual-photon absorptiometry (DPA), the radioisotope gadolinium 153 is used, which emits photons at two different energy levels, obviates the need for a soft-tissue equivalent medium, and allows measurement of bone density in the hip and spine, areas heavily surrounded by soft tissue. Like SPA, this technique measures areal density. The spine (35% cortical bone and 65% trabecular bone) is usually scanned between L1 and L4. Because all the mineral in the path of the photon beam is measured, possible sources of error include areas of aortic calcification, degenerative osteophytes, and apophyseal joint arthritis.

Dual-photon absorptiometry can also be used for total-body bone density measurement or bone density measurements at other sites. Precision and accuracy are excellent, radiation dose is low, and patient acceptance is high; however, cost is about two to three times that of SPA.

Quantitative Computed Tomography

In quantitative computed tomography (QCT), a cross-sectional view of the vertebral body is obtained, allowing differentiation of cortical and trabecular bone (Plate 29). The rate of turnover in trabecular bone is nearly eight times that of cortical bone, and this unique technique provides a sensitive indicator of early metabolic changes in the axial skeleton. Quantitative computed tomography involves the sequential scanning of the midportion of vertebral bodies T12 through L4 with the simultaneous scanning of a translucent phantom. The phantom comprises tubes that contain standard solutions of mineral, fat, and soft-tissue equivalents.

The CT reading for each tube of the phantom is correlated with the known mineral equivalent of the tube, and a standard curve is constructed by computer. The average CT readings from the midportion of T12 to L4 are also obtained, and the mineral equivalent is determined by the computer from the simultaneously constructed standard curve. The measurement is usually confined to the area of vertebral body posterior to the cortex and anterior to the basivertebral vein, ensuring that only trabecular bone is measured. Osteophytes and aortic calcification are avoided. Precision is within 3% to 5%, although it may be reduced in severely osteopenic and kyphotic patients due to the technical difficulties in locating the exact sites of previous measurements and in keeping the patient still for the examination.

Questions about accuracy arise largely from the variable fat content of the bone marrow in elderly patients with osteoporosis and from the technical problems of patient positioning and beam hardening. Radiation dose is approximately 200 mrem, and the cost is equivalent to that of DPA. Unlike the measurements for SPA and DPA, which are reported as areal densities, this measurement is of true density and is reported in mg/cm^3. The small error in accuracy due to measurement of marrow fat can be reduced with the use of dual-energy CT. However, with the dual-energy technique, the radiation dose increases and precision decreases.

Quantitative Computed Tomography

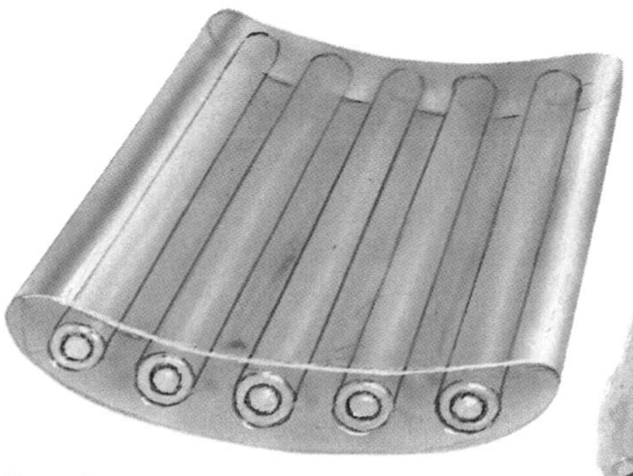

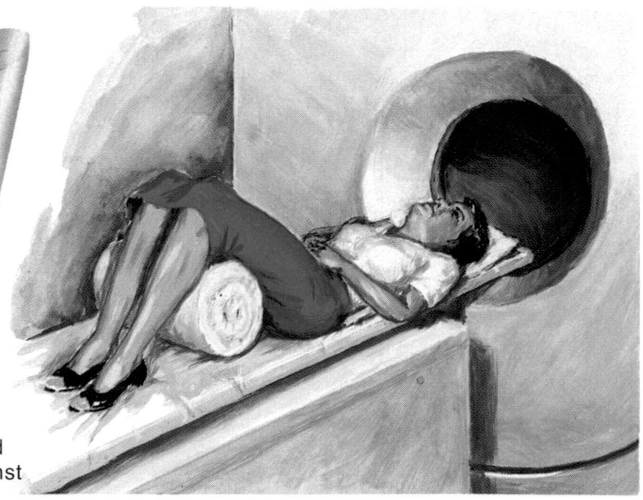

Translucent phantom with tubes containing standard solutions (mg/cm^3) of potassium phosphate (representing minerals), ethanol (fat), and water (soft tissue). Patient lies with knees and hips flexed to flatten lumbar spine against phantom, and is moved into scanner

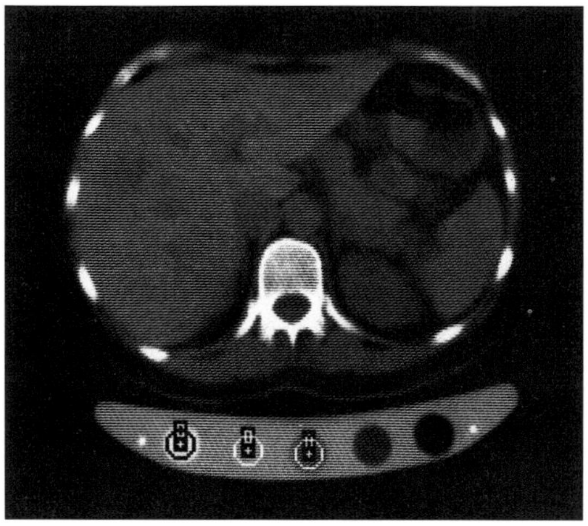

Transverse CT scans made through T12, L1, 2, 3, and 4, with phantom included. Computer gives readings for specific areas of each vertebral body and phantom tubes. Readings are averaged

©CIBA-GEIGY

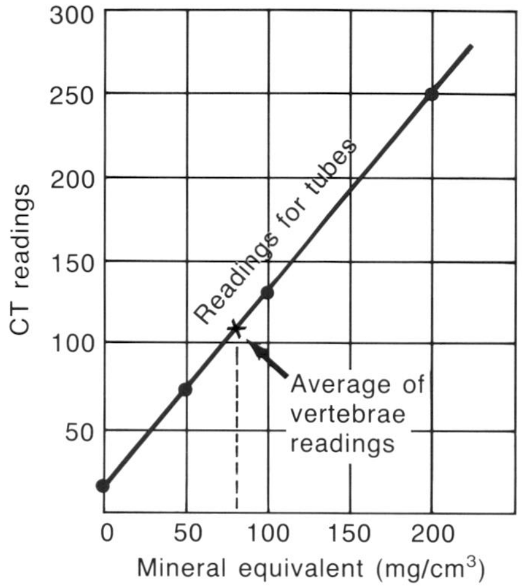

CT readings for known mineral values of tubes (red). Average of readings for vertebrae gives vertebral mineral content in mg/cm^3

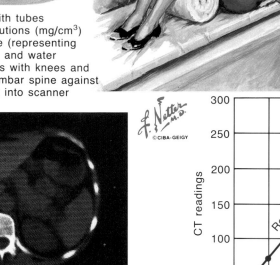

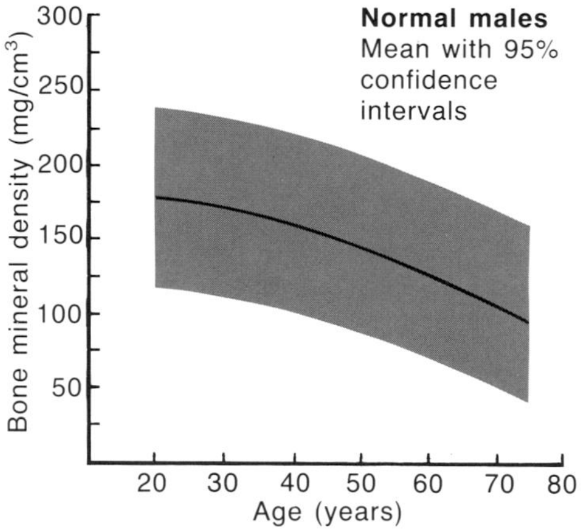

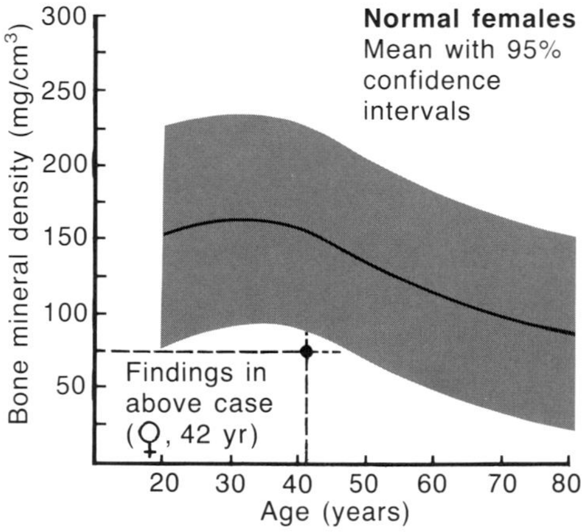

Results compared with normal values by age and sex

Total-Body Neutron Activation Analysis

At present, total-body neutron activation analysis (TBNAA) is used in clinical research to determine the calcium content of the entire body. The body is irradiated with high-energy neutrons that convert the stable isotope calcium 48 into the radioactive isotope calcium 49. The total-body radioactive decay back to the stable calcium 48 is then measured. Since 99% of the body's calcium is sequestered in the skeleton, this technique provides an accurate assessment of total-body bone mass. Regional assessments can also be made. However, the radiation dose is high and the equipment is expensive and generally unavailable.

Osteoporosis

(Continued)

Transiliac Bone Biopsy

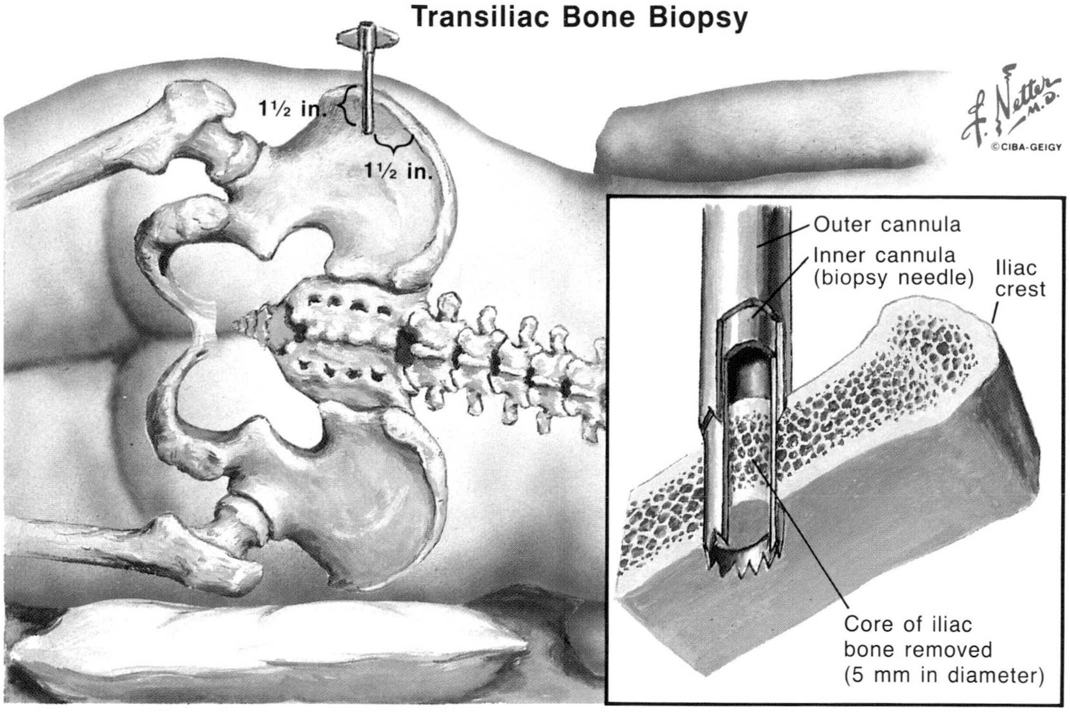
1½ in.
1½ in.
Outer cannula
Inner cannula
(biopsy needle)
Iliac
crest
Core of iliac
bone removed
(5 mm in diameter)
©CIBA-GEIGY

Undecalcified sections of iliac crest bone

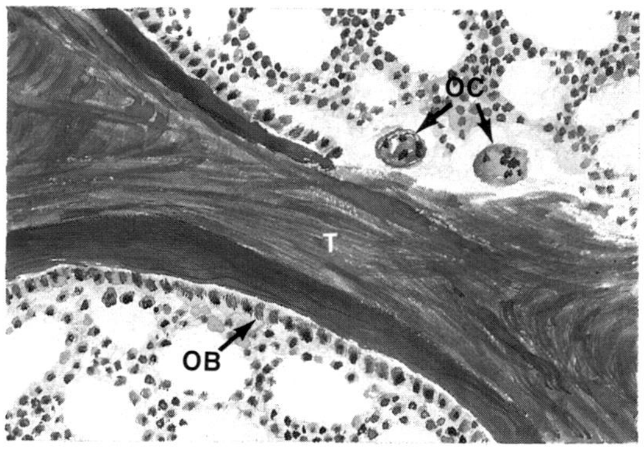

Red-staining osteoid seams (hypomineralized matrix) lined with osteoblasts (OB). Osteoclasts (OC) in resorption bays (Masson's trichrome)

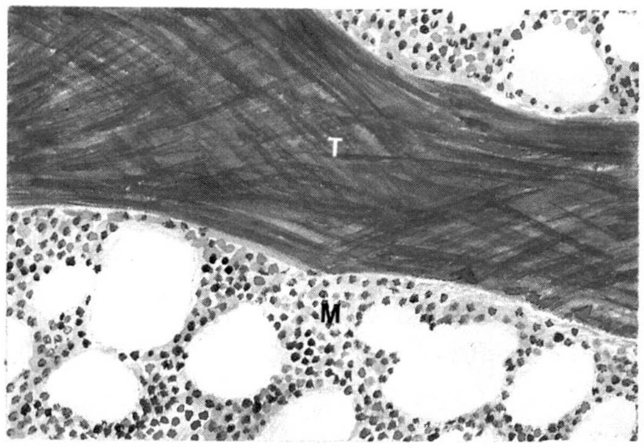

Section shows few osteoid seams, osteoblasts, or osteoclasts, indicating little bone formation or resorption

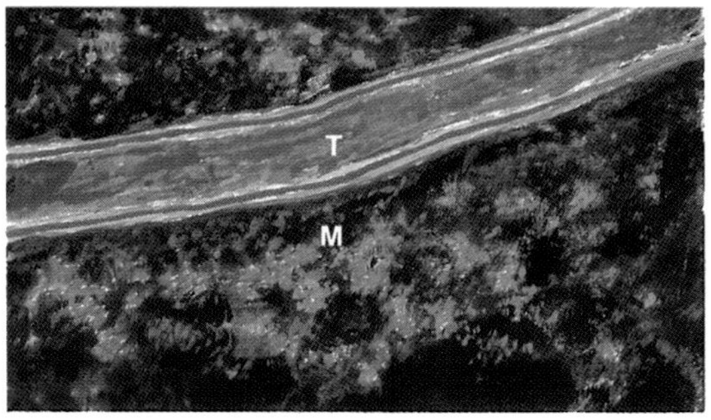

Yellow lines (tetracycline deposition at mineralization front), seen on fluorescent microscopy after 2 courses of tetracycline, indicate normal mineralization

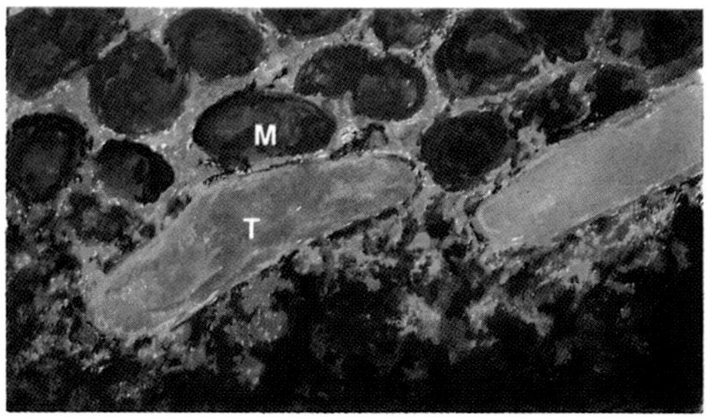

Absence of tetracycline-labeled lines indicates lack of bone formation

T = trabecular bone; **M** = marrow

Transiliac Bone Biopsy

Two technical advances have facilitated the diagnosis of metabolic bone diseases: the simplified bone biopsy and the ability to prepare undecalcified histologic sections of bone (Plate 30). Because metabolic bone diseases are generalized skeletal disorders, a small sample of bone is representative of the entire disease process. The iliac crest region is a readily accessible biopsy site and has been shown to reflect changes that may be occurring at more clinically relevant sites, such as the spine or the long bones. Commercially available trocars ranging from 5 to 8 mm in diameter are used to remove a core, or cylinder, of bone from the anterior portion of the iliac crest. The sample is taken transcutaneously, under local anesthesia.

Histologic Analysis of Undecalcified Bone. Since the differentiation between the two major metabolic bone diseases, osteoporosis and osteomalacia, is based in part on the quantity and quality of bone mineral, the ability to distinguish between calcified and uncalcified bone matrix (osteoid) is critical. The traditional procedures for processing bone require the removal of inorganic matrix to facilitate histologic sectioning. Bone is embedded in plastic, without prior decalcification, sectioned on a heavy-duty microtome, and stained with a trichrome connective tissue stain to produce thin histologic sections suitable for microscopic examination.

In nondecalcified bone sections, osteomalacia is usually characterized by the accumulation of osteoid caused by a defect in the mineralization process. It must be recognized, however, that excess quantities of osteoid may also be seen in osteoporotic conditions caused by an accelerated rate of bone matrix synthesis. Differentiation between these states is based on the determination of mineralization rates, using tetracycline as an in vivo bone marker.

Dynamic Tetracycline Labeling. The binding affinity of autofluorescent tetracycline antibiotics for immature mineral deposits, but not for mature crystal, enables the identification of calcification foci and subsequently permits the determination of the rate of bone mineralization. Tetracycline is administered for 3 days in a dosage of 500 mg twice daily; this is repeated after a 14-day interval. The bone specimen is obtained 3 to 4 days after the last dose of tetracycline.

Tetracycline fluorescence is evaluated on unstained, nondecalcified tissue sections by ultraviolet light. The first course of tetracycline appears as a discrete fluorescent band within the mineralized bone. The second, more recently administered course of tetracycline is located at the current mineralization front.

The mean distance between the midpoints of the double tetracycline labels is measured with a linear reticle. This distance is divided by the number of days between the two courses of tetracycline to determine the mineral appositional rate, which normally ranges from 0.4 to 0.9 μm/day (mean = 0.65 μm/day) and represents the amount of new bone synthesized and mineralized over the tetracycline-free interval.

When the mineral appositional rate increases, the distance between the labels grows wider. In contrast, with a reduced mineralization rate, the parallel bands become narrow and may fuse to produce single labels.

Abnormal patterns of fluorescent label deposition are the hallmark of osteomalacia. The amount of tetracycline fluorescence is proportional to the amount of immature amorphous calcium phosphate deposits in the mineralizing foci of the osteoid seam. If osteoid seams are deficient in mineral, the osteoid is incapable of binding tetracycline, leading to an absence of fluorescence. The mineralization front activity (percentage of osteoid seams bearing *normal* tetracycline labels) is therefore reduced.

Bone Histomorphometry. This technique is the quantitative analysis of undecalcified bone, in which the parameters of skeletal remodeling are expressed in terms of volumes, surfaces, and cell numbers. To obtain these data from a two-dimensional section, the principles of stereology are used to reconstruct the third dimension. This statistical principle states that if measurements are made at random, the ratio of areas is equal to the ratio of volumes.

Osteoporosis

(Continued)

Treatment of Complications of Spinal Osteoporosis

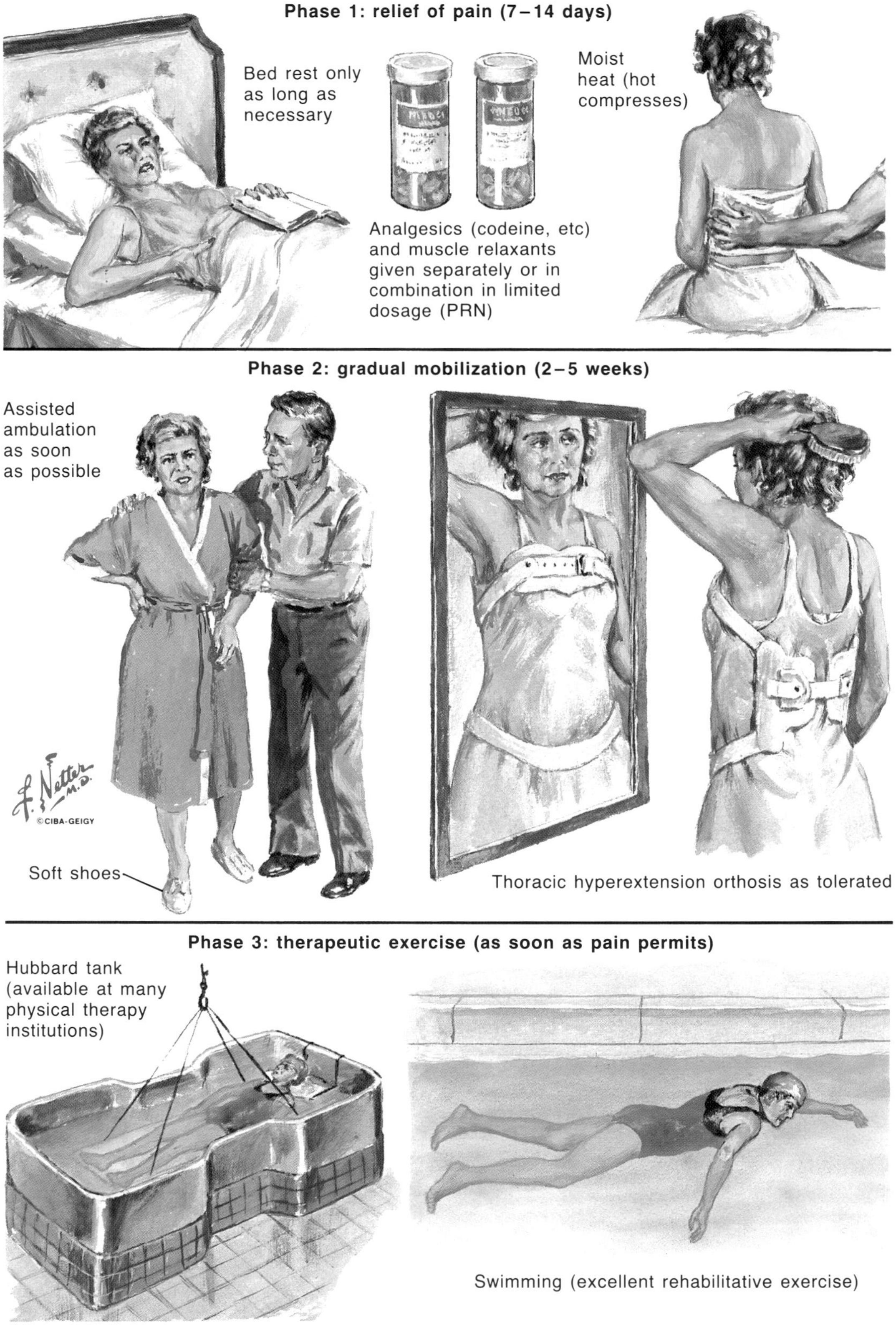
Phase 1: relief of pain (7–14 days)
Bed rest only as long as necessary
Moist heat (hot compresses)
Analgesics (codeine, etc) and muscle relaxants given separately or in combination in limited dosage (PRN)
Phase 2: gradual mobilization (2–5 weeks)
Assisted ambulation as soon as possible
F. Netter M.D.
©CIBA-GEIGY
Soft shoes
Thoracic hyperextension orthosis as tolerated
Phase 3: therapeutic exercise (as soon as pain permits)
Hubbard tank (available at many physical therapy institutions)
Swimming (excellent rehabilitative exercise)

Treatment of Complications of Spinal Osteoporosis

Spinal compression fractures are painful, can cause significant short-term morbidity, but heal quickly, even in severely osteopenic bone. Thus, the goals of therapy are to relieve pain, provide comfortable mechanical support for the spine, arrange assistance in activities of daily living, coordinate a rehabilitation program, and provide encouragement and reassurance to the patient and family (Plate 31).

Following an acute spinal compression fracture, most patients prefer several days of bed rest until the acute pain begins to subside. A firm mattress prevents spinal flexion (which may aggravate the traumatic kyphosis), and a pillow under the knees relieves excessive strain on the lower back. A sheepskin mattress cover improves skin care and reduces the risk of pressure sores over prominent spinous processes.

Short-term oral administration of narcotic analgesics may be helpful. Special care must be taken to avoid constipation, urinary retention, and respiratory depression in elderly patients taking these medications. When necessary, stool softeners and laxatives should be given to prevent fecal impaction.

Deep-breathing exercises are recommended during the period of bed rest, and patients should be encouraged to drink large quantities of fluids. Ileus develops in some patients with high lumbar compression fractures, necessitating parenteral fluid therapy for several days until symptoms and signs resolve. Support services during the period of acute infirmity must be tailored to the individual patient.

Once the acute pain begins to subside and the patient can turn in bed more comfortably, mobilization should begin. The patient should attempt to sit or stand for periods of no more than 10 to 15 minutes several times a day. As the pain subsides, the patient should increase the frequency and duration of mobilization periods but should continue intermittent bed rest.

For severe pain following compression fracture in the middle to lower thoracic regions, a rigid thoracolumbar hyperextension orthosis provides external support, alleviates flexion forces on the affected vertebral segments, and allows easier mobilization. Some patients find the rigid device too restricting, however, and prefer a three-point semirigid thoracolumbar extension orthosis that discourages the kyphotic, or stooped, posture. Still other patients choose to forego orthotic devices altogether.

Low-heeled, soft-soled shoes with foam or plastizote inserts that cushion the concussive forces transmitted to the spine during weight bearing should be worn. A cane may be used for extra support and is especially important for the elderly person with proprioception problems.

A daily program of active spinal extension exercises should be started. The exercises actively rehabilitate the paravertebral musculature, which is important in active spinal extension, and decrease the deforming flexion forces on the spine.

Women with midthoracic compression fracture and progressive kyphosis find pectoral stretching and deep-breathing exercises especially helpful.

Patients should be shown how to avoid unnecessary spinal compressive forces in lifting and bending. They should take great precautions to prevent falls. After 6 to 8 weeks, most patients are relatively pain free and can resume normal activity. If back pain occurs after periods of activity, the patient may benefit from a hyperextension device worn while erect. Short periods of intermittent bed rest, 15 to 20 minutes several times each day, may also help to relieve discomfort.

Osteoporosis

(Continued)

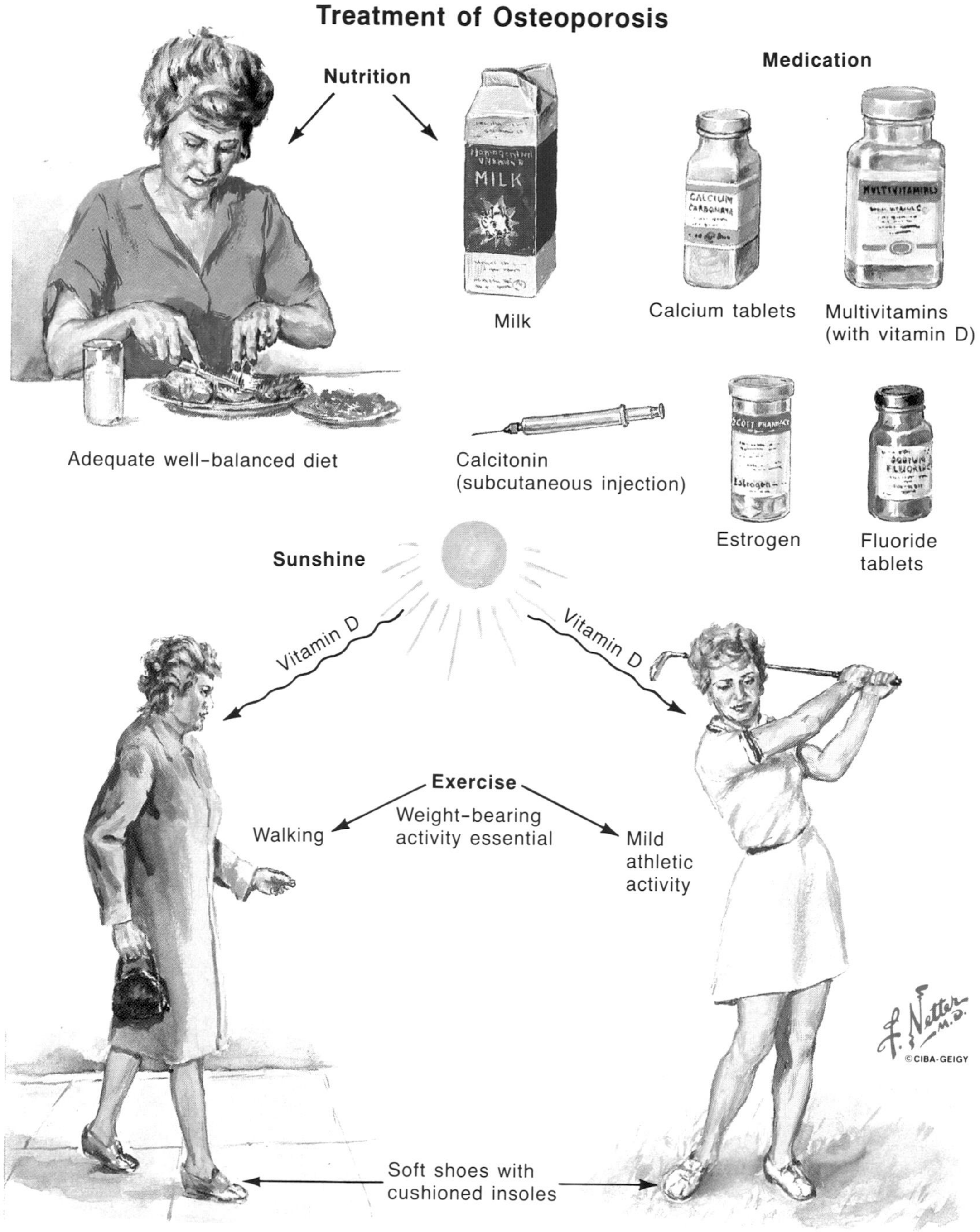
Treatment of Osteoporosis
Nutrition
Medication
Milk
Calcium tablets
Multivitamins (with vitamin D)
Adequate well-balanced diet
Calcitonin (subcutaneous injection)
Estrogen
Fluoride tablets
Sunshine
Vitamin D
Vitamin D
Exercise
Walking
Weight-bearing activity essential
Mild athletic activity
Soft shoes with cushioned insoles
F. Netter M.D.
©CIBA-GEIGY

Treatment of Osteoporosis

Osteoporosis can be more effectively prevented than treated. Use of estrogen supplementation should be considered in postmenopausal women (Plate 32).

The person most at risk for osteoporosis is a sedentary, postmenopausal, white or oriental woman who has a lifelong dietary calcium deficiency. Associated risk factors include small stature, slimness, light-colored hair and freckles, scoliosis, joint hypermobility, family history of osteoporosis, cigarette smoking (estrogen antagonist), and early natural or surgically induced menopause.

Estrogen Replacement Therapy (ERT). The most effective prevention of postmenopausal (type I) osteoporosis is the administration of estrogen, which preserves a positive calcium balance by suppressing the bone remodeling rate. It is most effective in the perimenopausal period, when the rate of trabecular bone loss is highest. However, its exact mechanism of action is unclear. The increased risk of endometrial cancer with ERT (in women with intact uteri) is about 1% per year and may be prevented with the cyclic use of a progestational agent. This regimen may also lower the incidence of breast cancer. Nevertheless, patients receiving ERT require close gynecologic follow-up care.

Oral estrogen preparations expose the liver to high concentrations of the hormone in the portal vein, thereby increasing hepatic production of coagulation factors, renin substrate, and bile acids. These factors may help to account for the increased incidence of venous thrombosis, pulmonary embolism, hypertension, and gallstones in patients using estrogen supplementation. The effect of ERT on coronary artery disease is unclear, but the weight of the evidence indicates a potentially beneficial effect.

Oral Calcium Supplementation. Adequate lifelong calcium intake may be important in the maintenance of cortical bone integrity. The average postmenopausal woman requires 1,500 mg of elemental calcium daily, the equivalent of six glasses of milk. With nonprescription oral calcium supplements, attention must be paid to the elemental calcium content.

Vitamin D Therapy. Vitamin D supplementation within recommended guidelines is essential in persons deficient in this vitamin. In addition, vitamin D and calcium may help to normalize the decreased calcium absorption common in the elderly, a factor that exacerbates age-related (type II) osteoporosis.

The active hormone in the vitamin D pathway, $1{,}25(OH)_2D$, is currently being studied for its role in increasing intestinal calcium absorption in the elderly. However, it has potent bone-resorbing effects as well. Currently, its use for the prevention of age-related (type II) osteoporosis remains investigational. In persons with adequate yearlong exposure to sunlight, vitamin D supplements are not necessary. Routine vitamin D supplementation should not exceed 800 IU daily. (Treatment of osteoporosis with massive doses of vitamin D is no longer recommended because it can lead to dangerous hypercalcemia and hypercalciuria.)

Exercise. Regular weight-bearing activity is essential for the maintenance of bone mass.

Hormone Therapy. The hormone calcitonin has recently been approved by the FDA for use in the treatment (prevention) of postmenopausal osteoporosis. Its major effect is to prevent bone loss by inhibiting osteoclastic bone resorption.

Fluoride Therapy. Fluoride may be beneficial in the treatment of some forms of established osteoporosis, since it both stabilizes the mineral crystal and stimulates the osteoblasts to form new bone matrix. Although the newly synthesized and radiographically denser bone is not structurally or materially normal, early evidence shows a decreased incidence of vertebral body compression fractures in patients taking fluorides, as compared with untreated controls. This suggests that bones may be stronger following fluoride therapy with appropriate calcium and vitamin D intake. Fluoride therapy without the appropriate intake of calcium and vitamin D, however, can lead to a severe osteomalacialike condition.

Other Treatments. Coherence therapy with oral phosphate and calcitonin has recently gained some attention. Phosphate stimulates PTH-mediated bone resorption, thus "turning on" new remodeling units. Before the osteoclasts remove much bone, bone resorption is "turned off" with calcitonin, thus allowing the normally coupled phase of osteoblastic bone formation to act unopposed and to increase the bone mass. After a period of time, the cycle is repeated. This novel approach was proposed by Frost and is currently being studied in several large academic centers. Small amounts of the active amino acid fragment, or N-terminal, of PTH may be used in place of oral phosphate to activate the remodeling system.

The use of capacitively coupled electric fields to prevent osteoporosis and to treat it once it has developed is being investigated. This approach is a direct extension of work with electric fields to stimulate osteogenesis in the healing of fracture nonunion. It has been used successfully to prevent and reverse disuse osteoporosis in laboratory animals. □

Comparison of Osteoporosis and Osteomalacia

Osteoporosis and osteomalacia are two commonly confused osteopenic conditions seen in adults. Osteoporosis is characterized by a decreased density of normally mineralized bone matrix. Osteomalacia refers to an increased, normal, or (most commonly) decreased mass of insufficiently mineralized bone matrix.

A much more common condition than osteomalacia, osteoporosis is age related and occurs most often in the elderly. Occasionally, onset may be due to rare genetic and less common determinants such as hyperglucocorticoidism, hyperthyroidism, hyperparathyroidism, alcohol abuse, tumors, immobilization, and chronic disease.

Unlike its more easily diagnosed childhood counterpart, rickets, adult osteomalacia may be difficult to diagnose clinically. Incidence is evenly distributed throughout all age groups. The most common causes are chronic renal failure, vitamin D deficiency, abnormalities of the vitamin D pathway, and hypophosphatemic syndromes. Rarer causes are renal tubular acidosis, aluminum intoxication, hypophosphatasia, and tumors.

Symptoms of osteoporosis are not usually evident until spontaneous fractures occur, when pain is felt at the fracture site. The most common sites for symptomatic osteoporosis are the spine, ribs, hips, and wrists. Osteomalacia, in contrast, may cause generalized bone pain and bone tenderness, predominantly in the appendicular skeleton.

The radiographic features of the two diseases are often similar, but axial changes predominate in osteoporosis and appendicular changes, in osteomalacia. Osteomalacia should be suspected in anyone with symmetric pathologic fractures, atraumatic fractures, or pseudofractures (Looser's zones), which are small, incomplete cortical fractures perpendicular to the long axis of a bone and often bilaterally symmetric. Common areas of involvement include the medial borders of the scapulae, ribs, ischiopubic rami, femoral necks, lateral borders of the femurs, and distal radii.

Results of routine laboratory studies are normal in osteoporosis but may be abnormal in osteomalacia. Therefore, osteomalacia should be suspected when the product of the serum calcium level multiplied by the serum phosphate level is chronically below 30, especially if accompanied by an elevated bone-specific alkaline phosphatase level and a 24-hour urinary calcium excretion of less than 50 mg/24 hr. Osteomalacia due to vitamin D deficiency should be suspected in a person with bone pain or pathologic fracture who is taking anticonvulsants or has a history of malabsorption syndrome, and also in an elderly person with a fracture of the femoral neck. The serum level of 25(OH)D is an excellent indicator of total body reserves of vitamin D.

Because these two diseases are very similar, a diagnosis of osteomalacia should be excluded by means of a transiliac bone biopsy following a two-course administration of tetracycline (Plate 30). However, the diagnosis of osteomalacia will be expedited if the physician is familiar with the causes and has a high index of suspicion. □

Comparison of Osteoporosis and Osteomalacia

		Osteoporosis	Osteomalacia
Definition	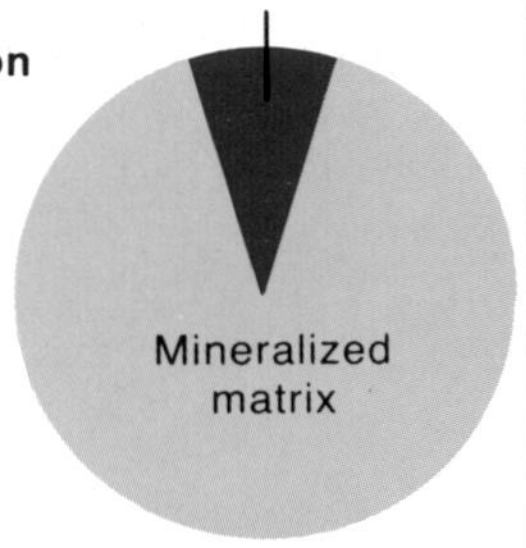 Normal	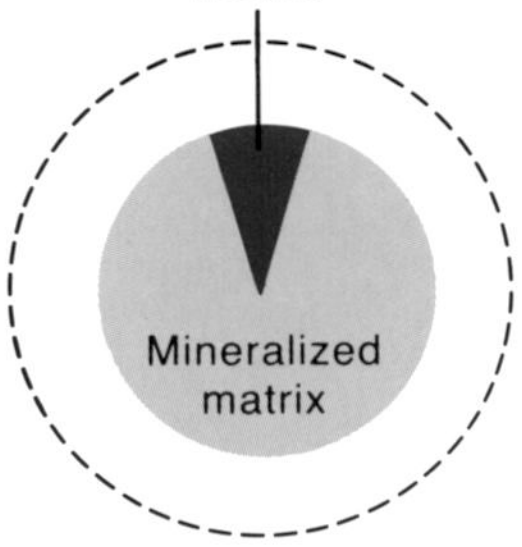 Bone mass decreased, mineralization normal	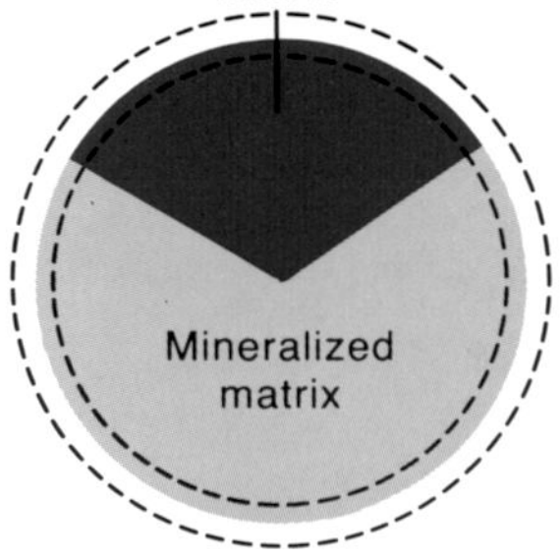 Bone mass variable, mineralization decreased
Age at onset		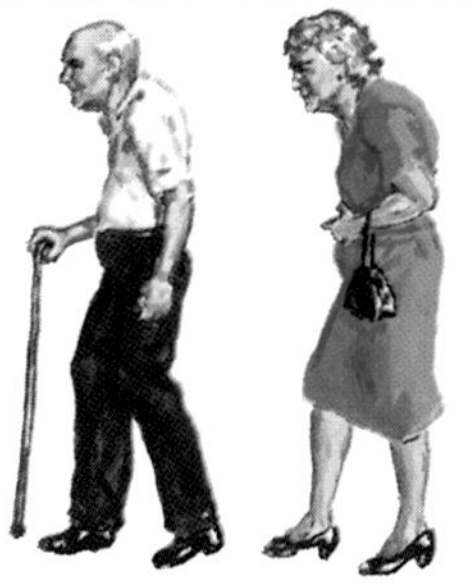 Generally elderly, postmenopause	 Any age
Etiology		Endocrine abnormality, age, idiopathic, inactivity, disuse, alcoholism, calcium deficiency	Vitamin D deficiency, abnormality of vitamin D pathway, hypophosphatemic syndromes, renal tubular acidosis, hypophosphatasia
Symptomatology		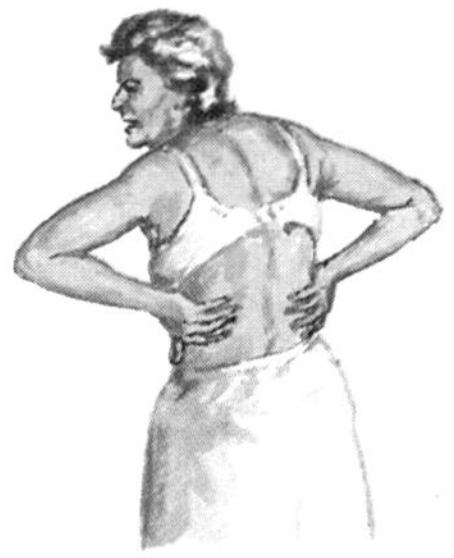 Pain referable to fracture site	 Generalized bone pain
Signs		Tenderness at fracture site	Tenderness at fracture site and generalized tenderness
Radiographic features		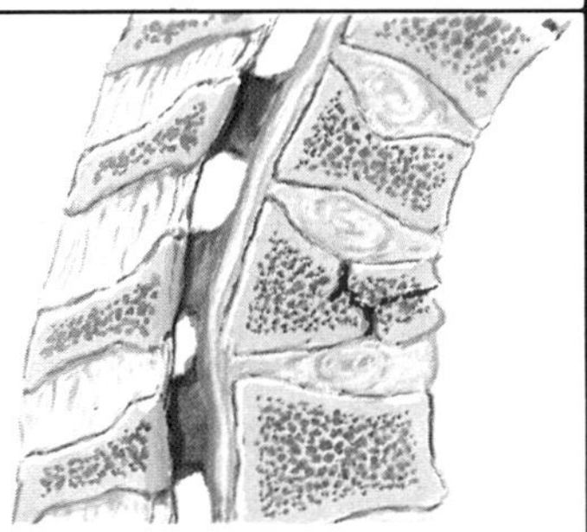 Axial predominance	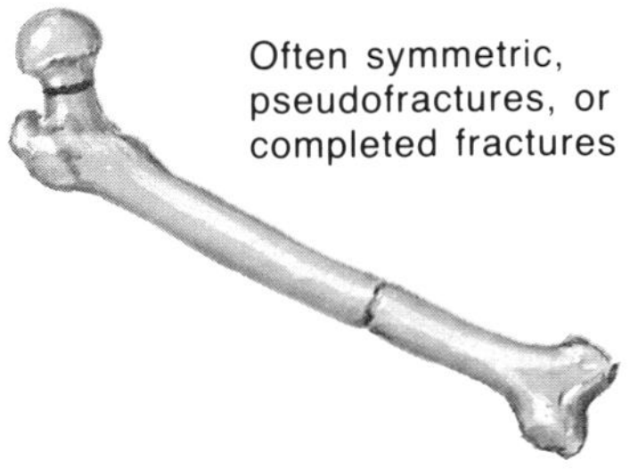 Often symmetric, pseudofractures, or completed fractures Appendicular predominance
Laboratory findings			
	Serum Ca^{++}	Normal	Low or normal (high in hypophosphatasia)
	Serum P_i	Normal $Ca^{++} \times P_i > 30$	Low or normal $Ca^{++} \times P_i < 30$ if albumin normal (high in renal osteodystrophy)
	Alkaline phosphatase	Normal	Elevated, except in hypophosphatasia
	Urinary Ca^{++}	High or normal	Normal or low (high in hypophosphatasia)
	Bone biopsy	Tetracycline labels normal	Tetracycline labels abnormal

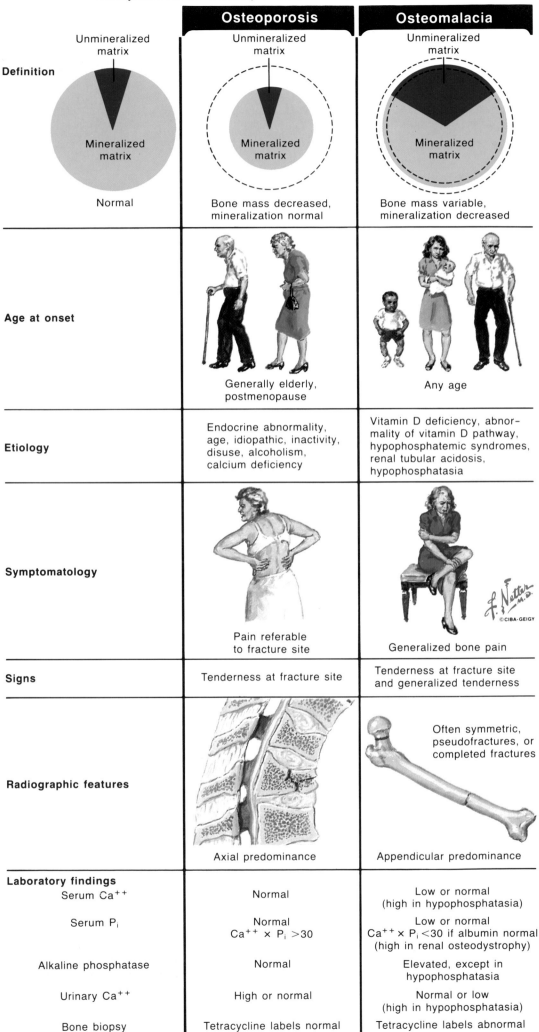

Osteogenesis Imperfecta

Osteogenesis Imperfecta

Familial type (tarda), most common. Mild involvement with blue sclerae

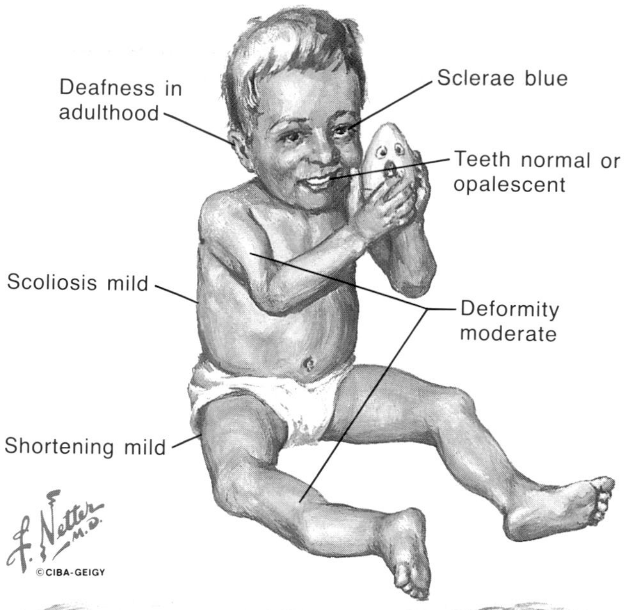
Deafness in adulthood
Sclerae blue
Teeth normal or opalescent
Scoliosis mild
Deformity moderate
Shortening mild
F. Netter M.D.
©CIBA-GEIGY

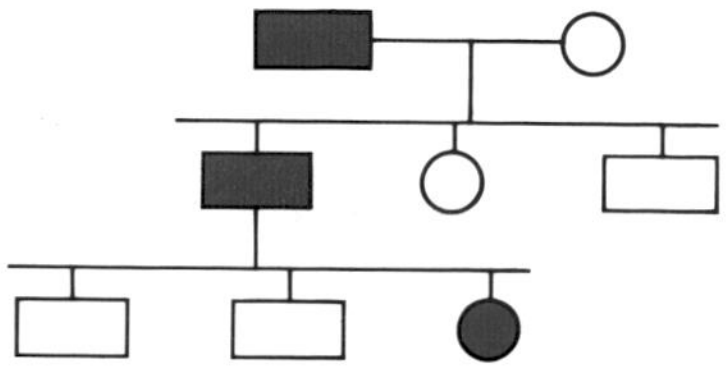

Autosomal dominant

One parent involved

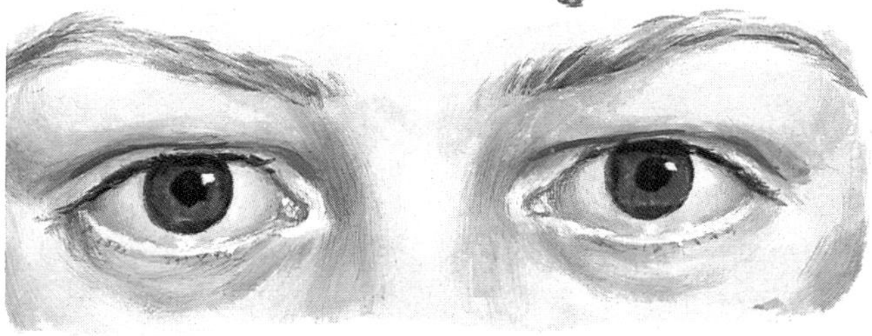

Sclerae usually blue

Teeth normal or opalescent

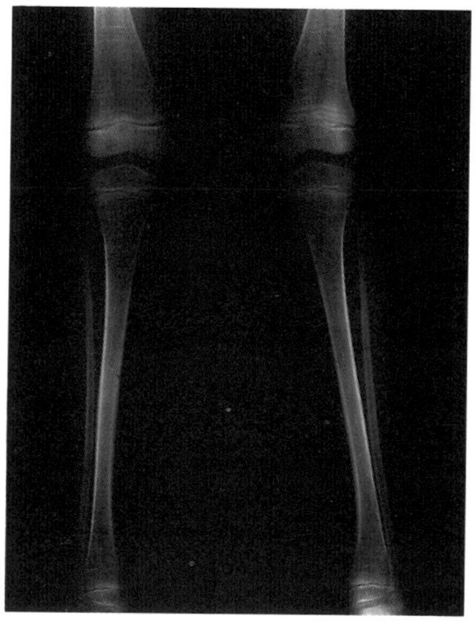

Radiograph shows thin and osteoporotic bones (variable); fracture rate moderate. Deformity mild, often amenable to intramedullary fixation

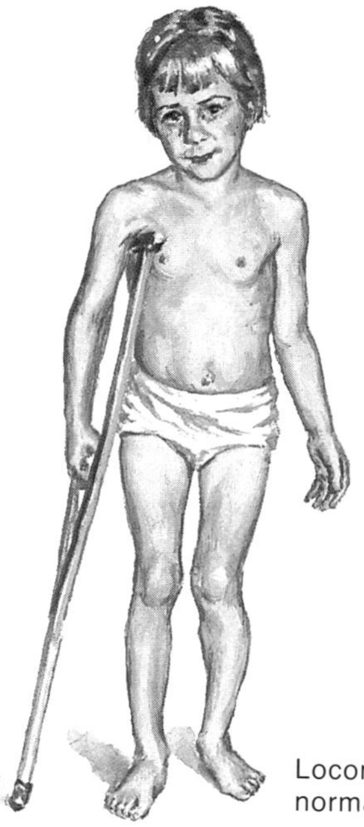

Locomotion normal or with crutch

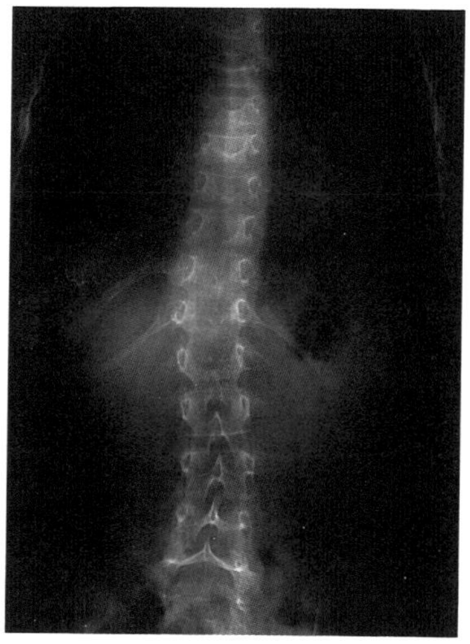

Radiograph shows mild scoliosis

Osteogenesis imperfecta, long thought to be a disease of the bones, is probably a generalized condition involving all the tissues. Although the primary cause remains unknown, many investigations now focus on collagen defects.

Known from antiquity by a variety of terms, osteogenesis imperfecta presents a real diagnostic challenge. It is relatively rare (1 in 20,000 births) and has diverse manifestations, most notably fragile bones and blue sclerae. There are no laboratory tests to establish the diagnosis, which must therefore be based on clinical evaluation alone.

Patients can be readily classified into two groups: those with the familial type, or Sillence type I (previously known as osteogenesis imperfecta tarda), and those with the sporadic type, or Sillence type IV ("congenita" in the old terminology). Sillence types II and III are extremely rare, sometimes fatal conditions characterized by an autosomal recessive inheritance and significant bone involvement.

Familial Type

Sillence type I is inherited as an autosomal dominant trait (Plate 34). The infant is born full term and is usually of normal size. Blue sclerae are a prominent feature. At birth, fractures may or may not be present, and the bones are usually somewhat broad and become slender with growth.

Although the overall bone length is usually diminished, there is no gross alteration in the growth plates. Bone deformities, which range from slight to moderately severe, are attributed to bowing as a result of stress and to recurrent fractures. Osteoporosis is seen in the spine, with a flattening or wedging of the individual vertebral bodies. Mild scoliosis is a common finding. Because of defects in skeletal growth, height is usually less than normal.

Affected children are usually slender, although some may be quite heavy. The muscles tend to be slender, and the ligaments usually exhibit significant laxity. The contours of the upper limbs remain relatively normal, except if repeated fractures occur in a given bone. On the other hand, the lower limbs always show at least a moderate bowing, which is most evident in the tibia. Significant deformities can result from repeated fractures.

Osteogenesis Imperfecta
(Continued)

Osteogenesis Imperfecta (continued)

Sporadic type (congenita) common. Severe involvement with normal or blue sclerae

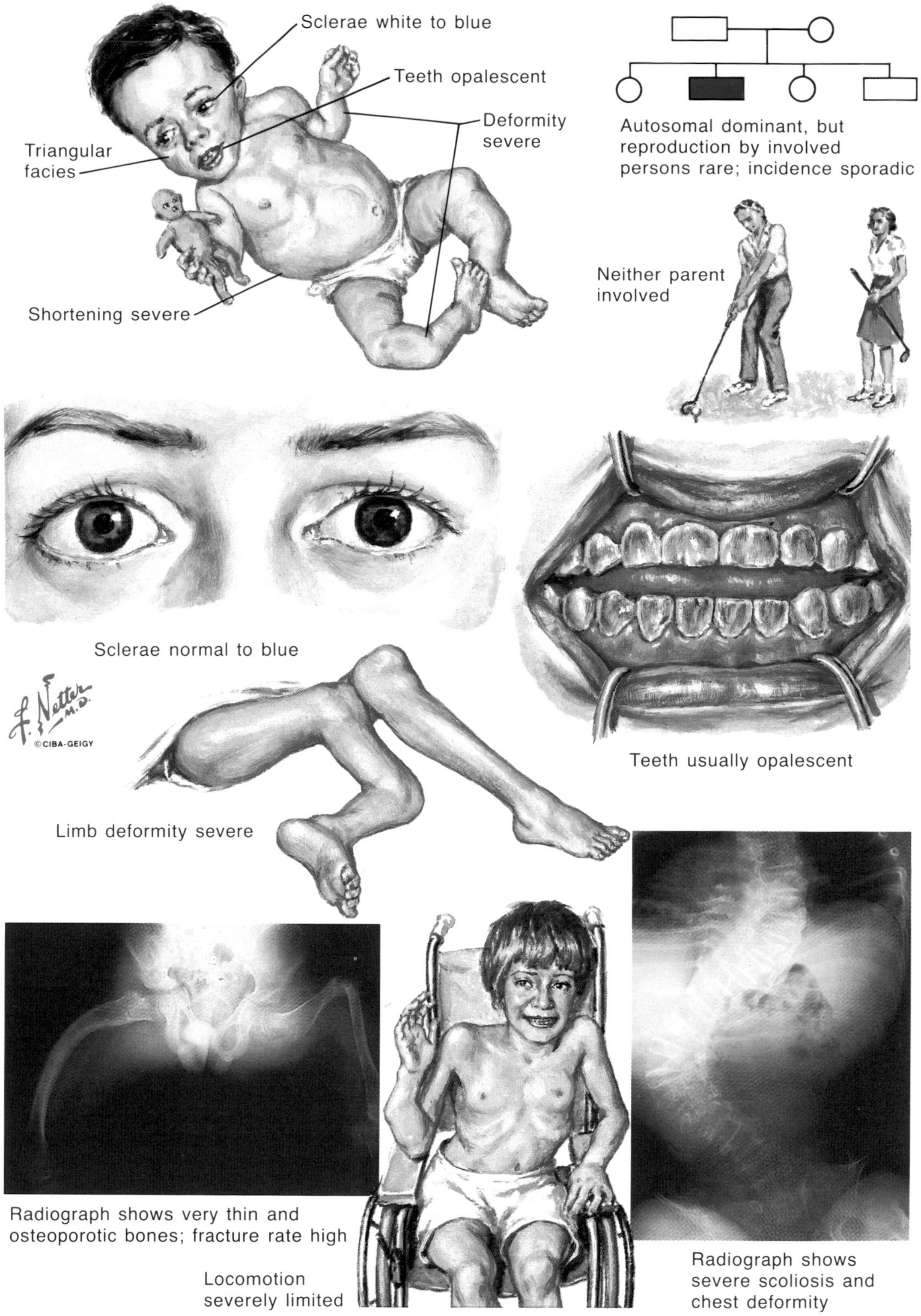
Sclerae white to blue
Teeth opalescent
Deformity severe
Triangular facies
Shortening severe
Autosomal dominant, but reproduction by involved persons rare; incidence sporadic
Neither parent involved
Sclerae normal to blue
Teeth usually opalescent
Limb deformity severe
F. Netter M.D.
©CIBA-GEIGY
Radiograph shows very thin and osteoporotic bones; fracture rate high
Locomotion severely limited
Radiograph shows severe scoliosis and chest deformity

The sclerae usually retain a degree of blueness. The teeth may be normal or opalescent (the appearance of the teeth is the same in all affected members of any given family). Hearing problems usually do not develop until adolescence or adulthood. Occasionally, ambulation is delayed in patients with type I osteogenesis imperfecta, but most patients are able to walk, usually without an aid. Characteristically, they are intelligent, usually very good students, and well able to be fully involved in society. There is a 50% risk that each pregnancy will produce a child with osteogenesis imperfecta. Interestingly, the degree of severity remains about the same from generation to generation.

Sporadic Type

The sporadic type, Sillence type IV, frequently called osteogenesis imperfecta congenita, occurs as a spontaneous mutation. The affected infant may have a low or normal birth weight. The sclerae may be mildly blue at first but tend to fade, or they may be white. At birth, the limbs may be significantly bowed, and numerous fractures may be present. The bones are often broad and poorly contoured, with diminished length, and the growth plates are frequently severely distorted. Although the spine is initially straight, flattening or wedging of the vertebral bodies ultimately leads to a significant spinal curvature. Ligaments are generally lax and muscles are slender.

The hallmark of the sporadic type is failure to grow, which is quickly recognized in severely involved babies. The majority of patients with this type of osteogenesis imperfecta survive; those who do not, usually die of respiratory problems.

The upper limbs may escape the severe deformities that so often occur in the lower limbs. The general habitus, however, includes a short trunk, short limbs, and a relatively large head. The sclerae usually pale with age, and the teeth are usually opalescent. Deafness appears to be less common in the sporadic type than in the familial type.

Some severely affected children are able to walk with the help of crutches or braces, and a few children can walk without external support, although the majority are wheelchair dependent. But, as in the familial type, children with sporadic type osteogenesis are usually intelligent and, despite their small size, function quite well in school. Many maintain a happy outlook on life despite

Osteogenesis Imperfecta

(Continued)

Treatment of Osteogenesis Imperfecta by Intramedullary Fixation and Orthoses

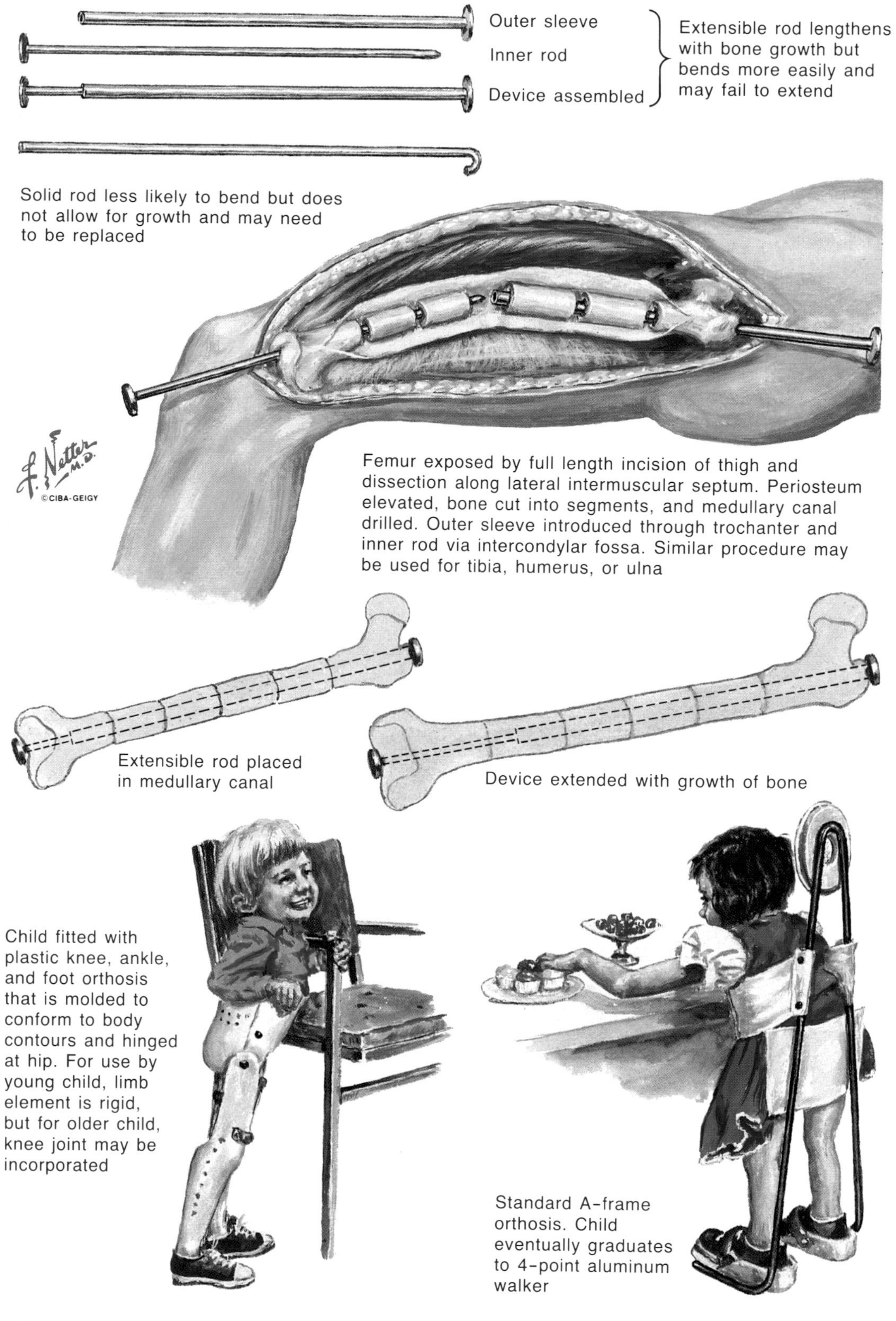
Outer sleeve
Inner rod
Device assembled
Extensible rod lengthens with bone growth but bends more easily and may fail to extend
Solid rod less likely to bend but does not allow for growth and may need to be replaced
F. Netter M.D.
©CIBA-GEIGY
Femur exposed by full length incision of thigh and dissection along lateral intermuscular septum. Periosteum elevated, bone cut into segments, and medullary canal drilled. Outer sleeve introduced through trochanter and inner rod via intercondylar fossa. Similar procedure may be used for tibia, humerus, or ulna
Extensible rod placed in medullary canal
Device extended with growth of bone
Child fitted with plastic knee, ankle, and foot orthosis that is molded to conform to body contours and hinged at hip. For use by young child, limb element is rigid, but for older child, knee joint may be incorporated
Standard A-frame orthosis. Child eventually graduates to 4-point aluminum walker

their infirmities. There is a 50% risk that the affected person will pass on the disorder, although procreation is rare.

Individual cases exhibiting characteristics of both the familial and sporadic types do occur. Sometimes, the familial type first occurs as a spontaneous mutation; the affected person then becomes the first in the line of persons with the familial type. However, the majority of cases can be classified as either familial or sporadic.

Treatment

The goals of orthopedic care are to make the patients comfortable and to enable them to function as well as possible. In infants, the use of soft bulky bandages is generally adequate for the management of fractures; in children and adolescents, more rigid splints or braces are needed. Significant deformities and recurrent fractures require fixation of the long bones with intramedullary rods. This technique is more applicable to the femur, tibia, and humerus than to the smaller bones. Where significant longitudinal bone growth is anticipated, use of extensible rods is desirable. However, intramedullary rods may aid, but do not guarantee, ambulation and are most useful in correcting severe deformities and in managing repeated fractures.

The quality of bones is enhanced by muscle activity, which should be encouraged, and immobilization, with or without surgery, should be kept to a minimum. Scoliosis is a relentless problem in osteogenesis imperfecta, necessitating early, careful, and continued attention to the spine. Because braces are usually ineffective, spinal fusion is essential to avoid severe curvature. Mechanical aids ranging from cradle boards to special automobile controls play a part in helping the patient to cope with activities of daily life. They should be selected with care and with the participation of the patient and/or the parent. Advances in electronic technology have made it possible for many patients to be not only self-sufficient but self-supporting as well.

The genetic patterns in osteogenesis imperfecta are now better understood. Although no reliable antenatal test is yet available, when the potential for osteogenesis imperfecta exists, radiography and sonography may help to determine the presence of the disorder in the fetus. □

Marfan's Syndrome

Marfan's Syndrome

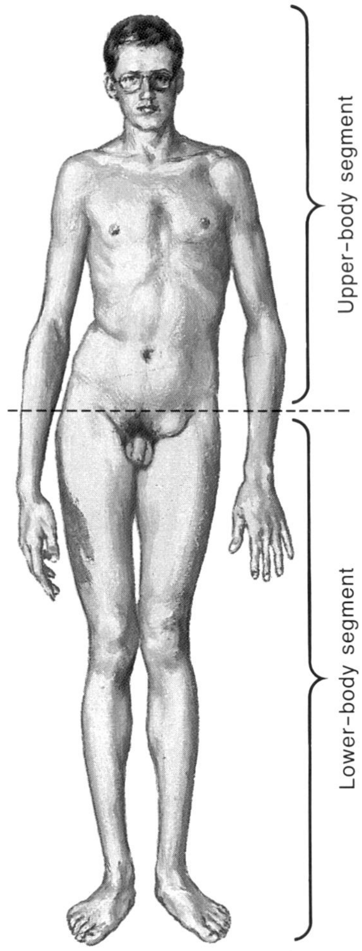
Upper-body segment
Lower-body segment

Tall, thin person with skeletal disproportion. Upper-body segment (top of head to pubis) shorter than lower-body segment (pubis to soles of feet). Fingertips reach almost to knees (span of upper limb greater than total body height). Long, thin fingers (arachnodactyly). Scoliosis, chest deformity, inguinal hernia, flatfoot

Ectopia lentis (upward and temporal displacement of eye lens). Retinal detachment, myopia, and other ocular complications may occur

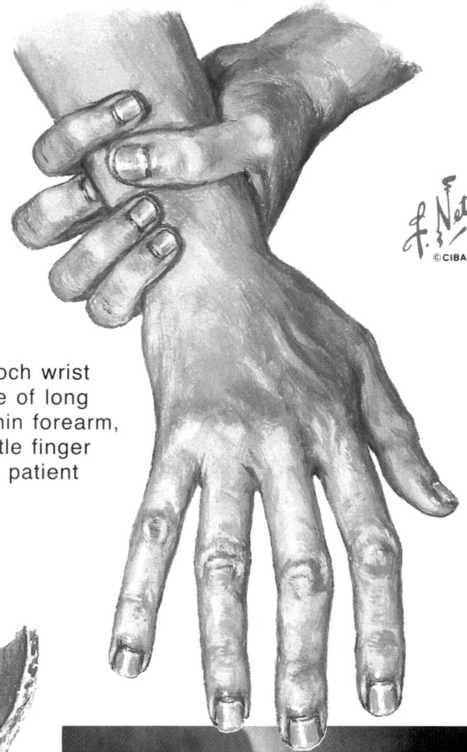

Walker-Murdoch wrist sign. Because of long fingers and thin forearm, thumb and little finger overlap when patient grasps wrist

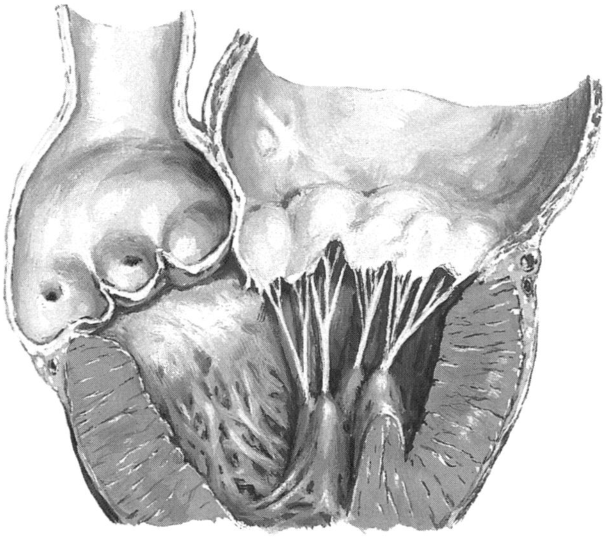

Dilatation of aortic ring and aneurysm of ascending aorta due to cystic medial necrosis causes aortic insufficiency. Mitral valve prolapse causes regurgitation. Heart failure common

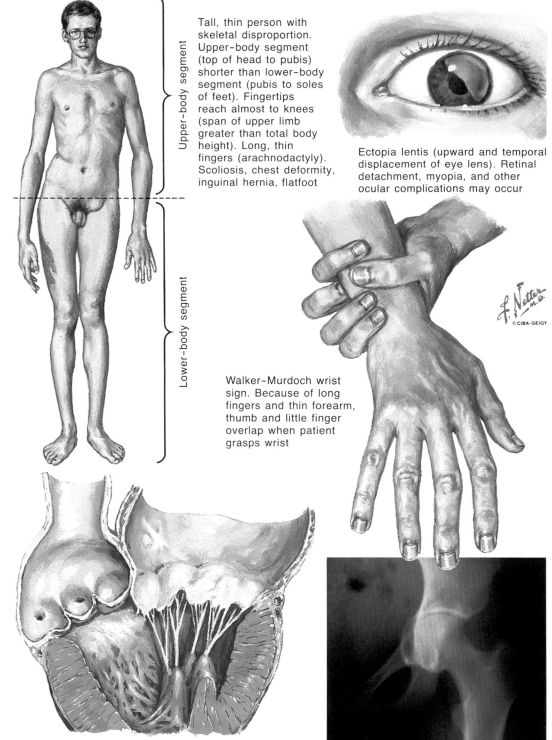

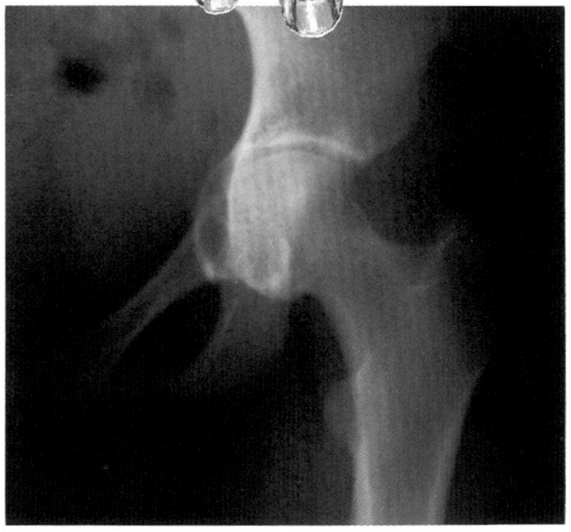

Radiograph shows acetabular protrusion (unilateral or bilateral)

Marfan's syndrome, first described by Antoine Marfan in 1896, is a heritable disorder of connective tissue that is transmitted as an autosomal dominant trait and often arises as a spontaneous mutation. The basic defect probably involves collagen metabolism.

The incidence of classic Marfan's syndrome is between 2 to 5 persons in 100,000, with no sexual, ethnic, or racial predilection. Although nonpenetrance, or skipped generations, does not occur (as in most autosomal dominant disorders), the severity of the phenotypic (clinical) expression of the gene varies greatly.

In its classic presentation, Marfan's syndrome involves three major organ systems: ocular, cardiovascular, and musculoskeletal. The typical ocular problems are superotemporal displacement of the lens (ectopia lentis) and increased axial length of the globe, leading to myopia and retinal detachment. Cardiovascular involvement, which may be life threatening, includes weakness of the aortic media with dilatation of the aortic root, aortic regurgitation, dissecting aortic aneurysm, and pansystolic prolapse of the posterior leaflet of the mitral valve.

Musculoskeletal manifestations are diverse and include a Lincolnesque appearance characterized by excessive height, long spiderlike fingers and toes (arachnodactyly), and disproportionately long limbs (dolichostenomelia); funnel chest (pectus excavatum) or pigeon breast (pectus carinatum); hyperextensible joints; flatfoot (pes planus); kyphoscoliosis; acetabular protrusion; and recurrent inguinal hernias.

Currently, there are no laboratory tests to establish the diagnosis of Marfan's syndrome, which is therefore based solely on clinical evaluation. At least two, and preferably three, of the four characteristic features (familial occurrence and ocular, cardiovascular, and musculoskeletal manifestations) must be present. Although the diagnosis of classic Marfan's syndrome is not difficult, the identification of milder forms of the disease is more complicated and may lead to misdiagnosis.

Prenatal diagnosis is not yet possible, and a patient with Marfan's syndrome has a 50% chance of transmitting the disease to a child. A woman with Marfan's syndrome must be carefully monitored during pregnancy because she is at increased risk of aortic rupture.

Pharmacologic treatment with β-adrenergic blockade may help to decrease the progression of aortic dilatation by reducing the force of ventricular ejection. Early induction of puberty may help to limit excessive height and to decrease scoliosis. Two annual examinations are essential: an ophthalmologic examination that includes slit-lamp illumination with the pupils dilated and a cardiovascular examination with echocardiogram.

Care of a patient with Marfan's syndrome requires a multidisciplinary approach, which includes the ophthalmologist, cardiologist, orthopedist, geneticist, obstetrician and gynecologist, and endocrinologist. However, the treatment program should be directed by a generalist, such as an internist, pediatrician, or family practitioner. Lifelong periodic evaluation is essential. □

Ehlers-Danlos Syndrome

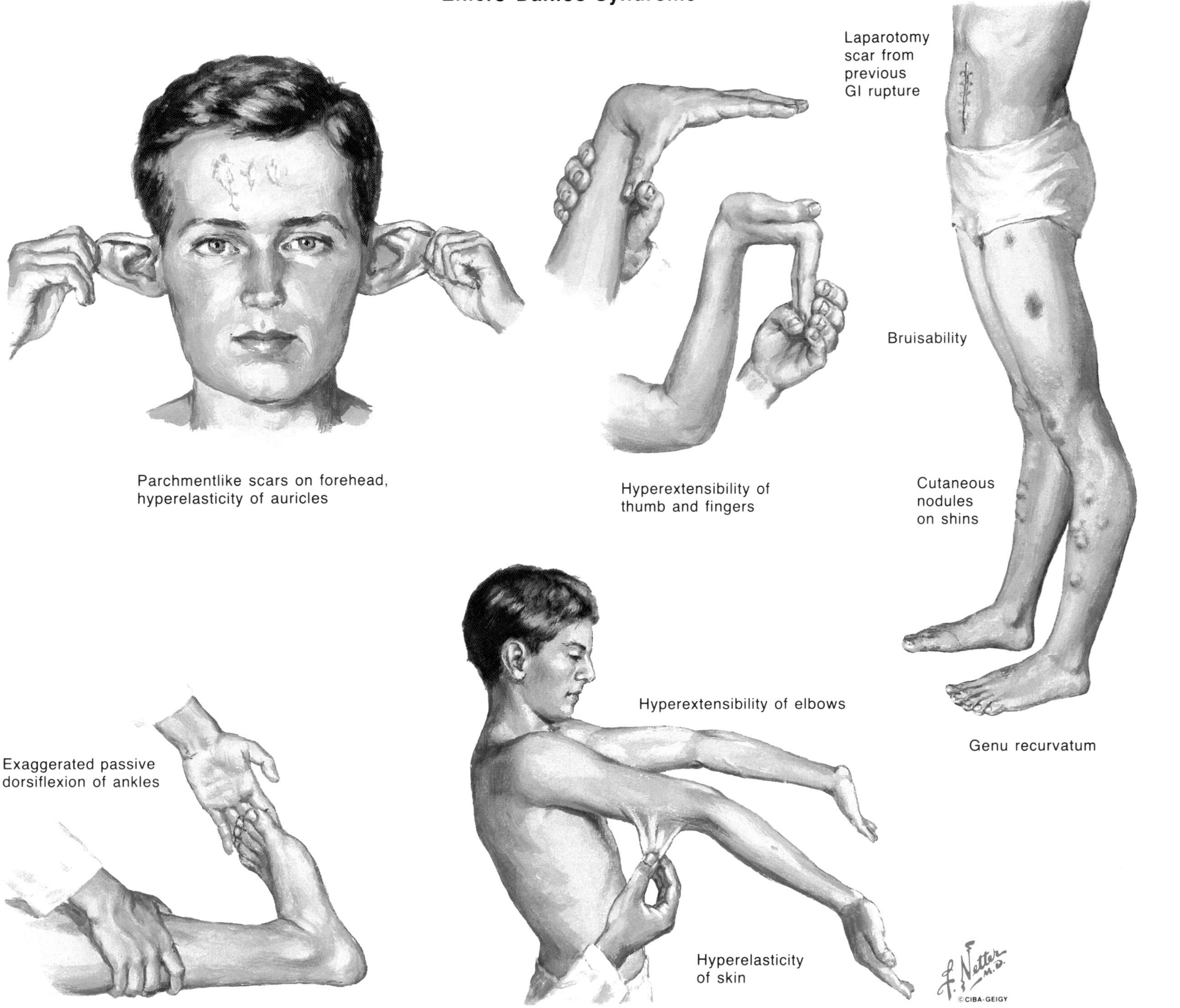

Laparotomy scar from previous GI rupture
Parchmentlike scars on forehead, hyperelasticity of auricles
Hyperextensibility of thumb and fingers
Bruisability
Cutaneous nodules on shins
Hyperextensibility of elbows
Genu recurvatum
Exaggerated passive dorsiflexion of ankles
Hyperelasticity of skin
F. Netter M.D.
©CIBA-GEIGY

SECTION IV PLATE 38 Slide 3773

Ehlers-Danlos Syndrome

Ehlers-Danlos syndrome comprises at least 11 separate clinical entities. Laxity and hypermobility of the joints and increased stretchability, bruisability, and fragility of the skin are common features, but the severity of joint and skin involvement varies dramatically, as does the pattern of inheritance.

Autosomal dominant, autosomal recessive, and X-linked patterns are all seen. In types IV, VI, VII, and IX, basic defects in collagen metabolism have been identified. The complexity of collagen metabolism and the heterogeneity of collagen types help to explain the myriad disorders that can occur. Recognized syndromes are the following:

Type I. Severe classic form. Autosomal dominant inheritance.

Type II. Mild classic form. Autosomal dominant inheritance.

Type III. Benign hypermobility syndrome. Excessive loose-jointedness without skeletal deformity. May be associated with floppy mitral valve. Autosomal dominant inheritance.

Type IV. Ecchymotic, arterial type. Severe, life-threatening syndrome with a predilection to spontaneous rupture of bowel or large vessels. Fragile skin with easy bruisability and large ecchymoses. Paradoxically, hypermobility may be less severe and may be confined to the fingers. Poor tensile properties in all connective tissue; deficiency in production of type III collagen. Autosomal dominant or recessive inheritance.

Type V. Features of type I. Severe mitral regurgitation may occur. X-linked inheritance is probable.

Type VI. Ocular-scoliotic form. Classic features of type I plus severe scoliosis, fragility of the globe with rupture of cornea and sclera, and retinal detachment with minor blunt trauma. Aortic aneurysms may occur. Specific deficiency of lysyl hydroxylase leads to collagen deficient in hydroxylysine and thus to defective collagen cross-linking. Autosomal recessive inheritance.

Type VII. Classic phenotype with either a dominant (defective procollagen) or a recessive (deficiency of the enzyme that cleaves the registration peptide off the amino-terminal, or N-terminal, end of collagen) form.

Type VIII. Classic type I with periodontal disease leading to early tooth loss. Autosomal dominant inheritance.

Type IX. Classic features plus severe skin laxity and occipital exostoses. Deficiency of lysyl oxidase, an important enzyme for collagen cross-linking. X-linked inheritance.

Type X. Classic features with platelet dysfunction from fibronectin abnormality.

Type XI. Recurrent dislocation of major joints. Autosomal dominant inheritance. □

Osteopetrosis (Albers-Schönberg's Disease)

Osteopetrosis (Albers-Schönberg's Disease)

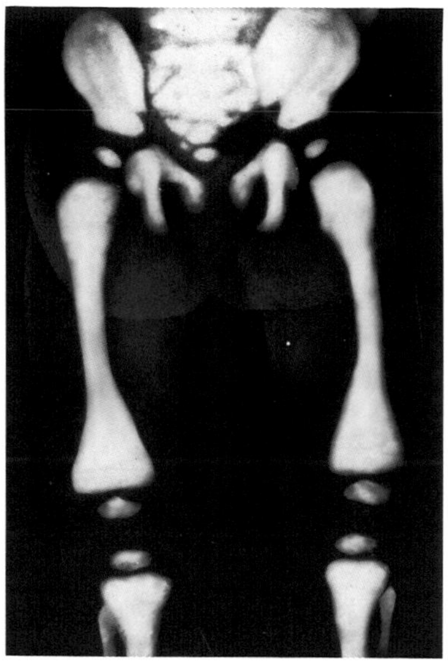

Radiograph shows marked increase in bone density, with medullary cavities obliterated

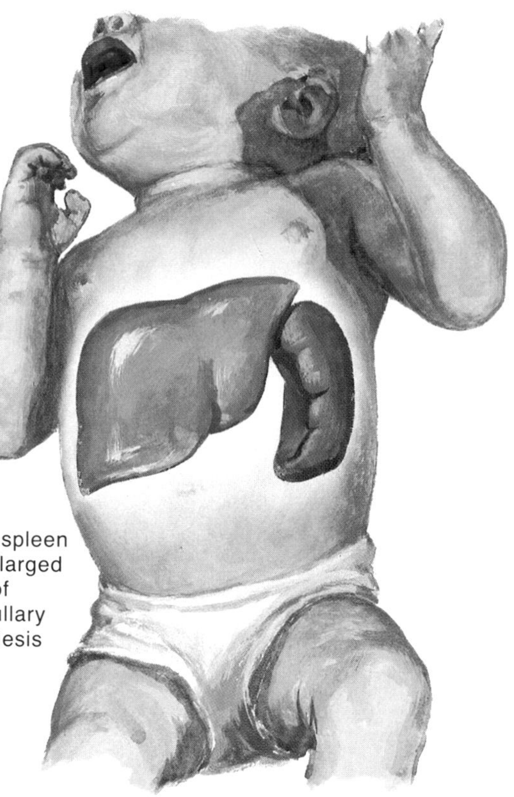

Liver and spleen greatly enlarged because of extramedullary hematopoiesis

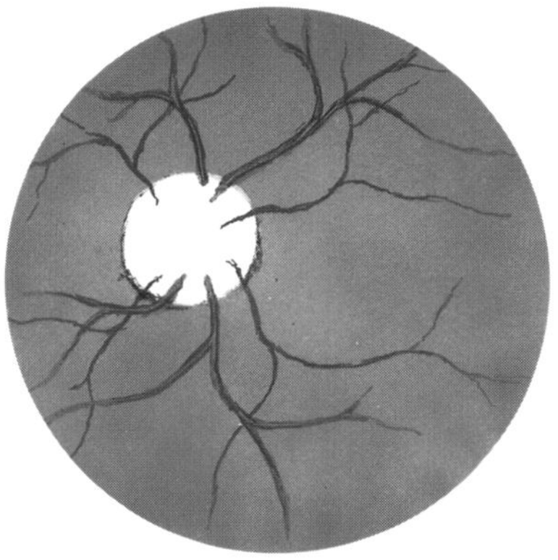

Optic atrophy with blindness due to compression of nerve by bony encroachment on foramen. Deafness may also be caused by similar mechanism

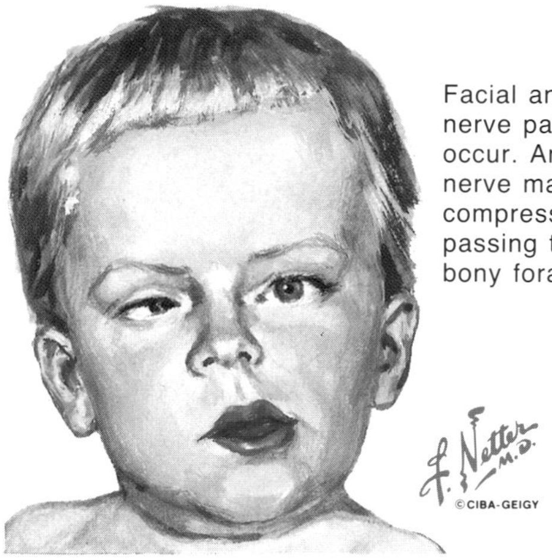

Facial and abducens nerve palsy may occur. Any cranial nerve may be compressed while passing through bony foramina

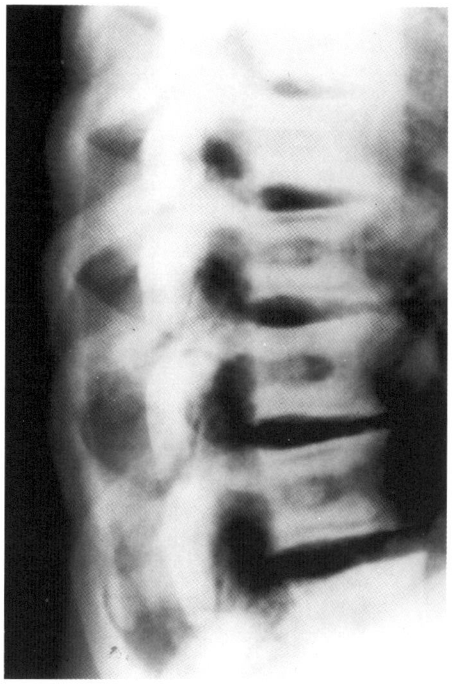

Radiograph shows involvement of thoracic spine. Density of vertebral bodies, spinous processes, and ribs greatly increased

The study of rare genetic diseases often provides an enlightening insight into normal physiology. For example, investigation of the rare heritable disorder osteopetrosis has augmented the understanding of the structure and function of the osteoclast, the large multinucleated cell that resorbs bone and calcified cartilage.

First described by Albers-Schönberg, osteopetrosis is a group of closely related disorders characterized by defective osteoclastic resorption of bone. Two distinct types of osteopetrosis are seen in humans. A benign, or adult, form may be inherited as an autosomal dominant trait or may arise as a spontaneous mutation. In this form of the disorder, disability is minimal and life expectancy is normal. The diagnosis is usually made when radiographs taken for another purpose reveal the dense, radiopaque bones that are characteristic of osteopetrosis. Although most patients are asymptomatic, the risk of fractures is increased because there is a predominance of calcified cartilage rather than of calcified bone.

The most severe form of osteopetrosis is the congenital malignant form, which is inherited as an autosomal recessive disorder. Death usually occurs in infancy or early childhood, and most patients die of the complications of anemia, bleeding, or infection.

The hallmark of congenital malignant osteopetrosis is the complete failure of normal osteoclast activity, which is demonstrated by a variety of hematologic, neurologic, radiographic, histologic, and metabolic abnormalities. Defective osteoclastic bone resorption in the presence of normal osteoblastic bone formation results in an extreme excess of mineralized bone matrix (osteoid) and chondroid, which encroaches on the intramedullary spaces. This leads to excessive extramedullary hematopoiesis with severe hepatosplenomegaly and a resultant myelophthisic anemia, thrombocytopenia, and leukoerythroblastosis. Osteoclastic dysfunction, together with loss of the medullary cavities and thickened bones, leads to encroachment on the cranial nerve foramina and a litany of potential neurologic complications. These include optic atrophy, blindness, deafness, and any of the cranial nerve palsies. Cerebral atrophy and hydrocephalus have also been reported.

Despite the increased density of the skeleton, the bones are brittle because of the predominance of calcified cartilage, and pathologic fractures are common. Paradoxically, rachitic changes have been noted in the bones of children with congenital malignant osteopetrosis. The structural sequestration of mineral and the failure of bone resorption

Osteopetrosis (Albers-Schönberg's Disease)

(Continued)

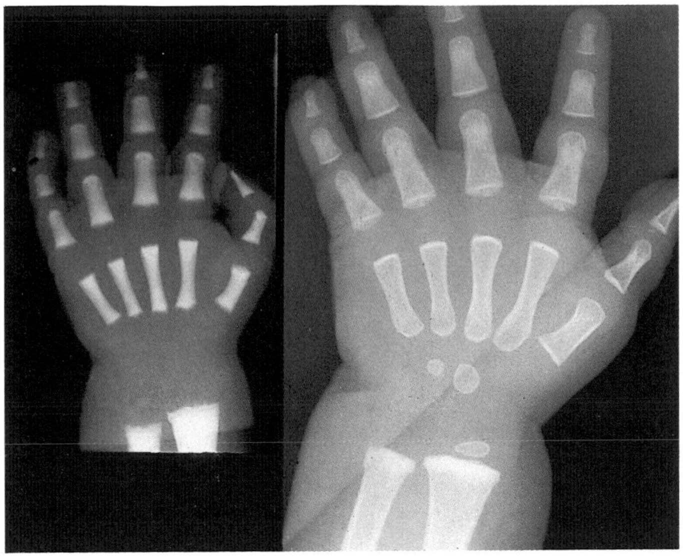

Radiographs taken before bone marrow transplant (left) and 1 year after transplant (right) show distinct decrease in density of medullary canals in metacarpals and phalanges

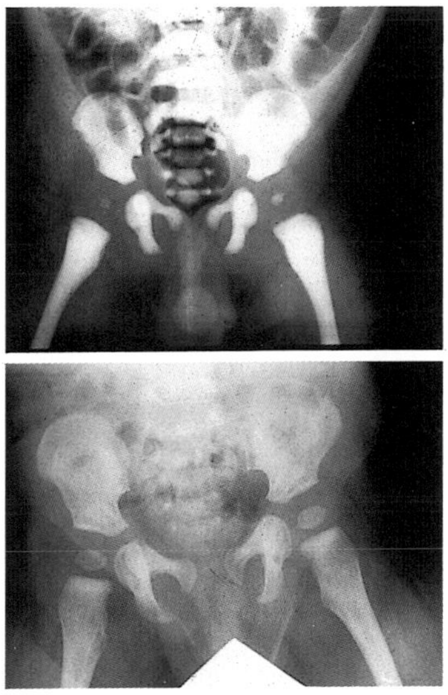

Radiographs taken before transplant (above) and 1 year after transplant (below) show decrease in density of pelvis, femurs, and spine after transplant

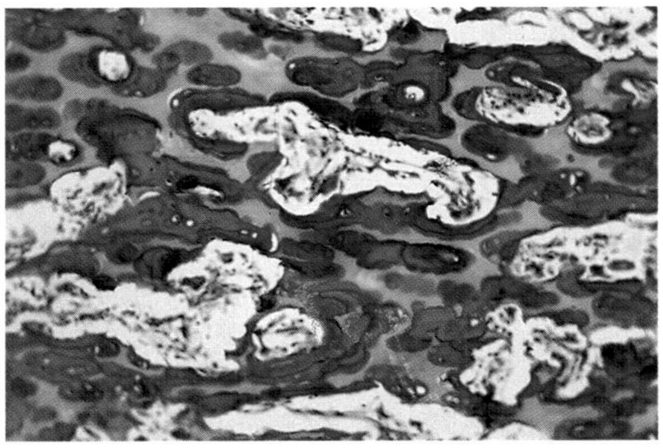

Before transplant, bone biopsy shows marrow spaces almost obliterated by many unresorbed calcified cartilage bars (H and E)

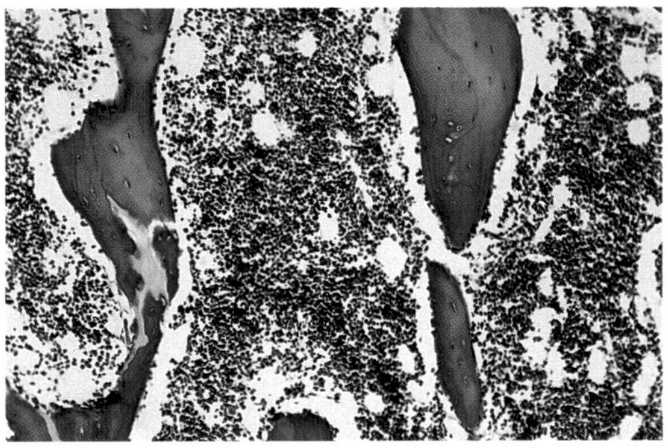

1 year after transplant, bone biopsy shows much narrower trabeculae and greatly enlarged marrow spaces

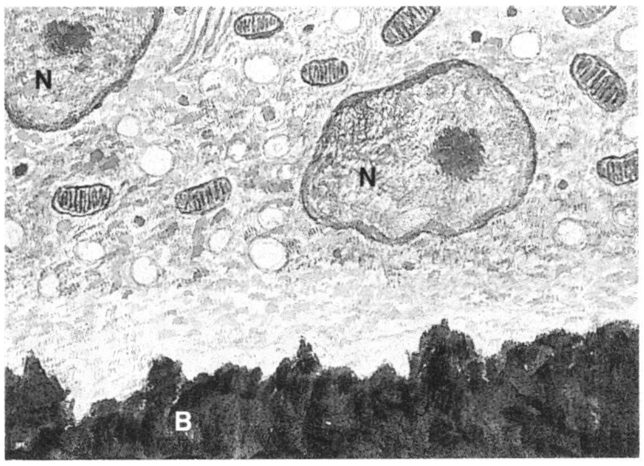

Before transplant, electron microscopic view of segment of osteoclast shows absence of ruffled border (B = bone; N = nuclei)

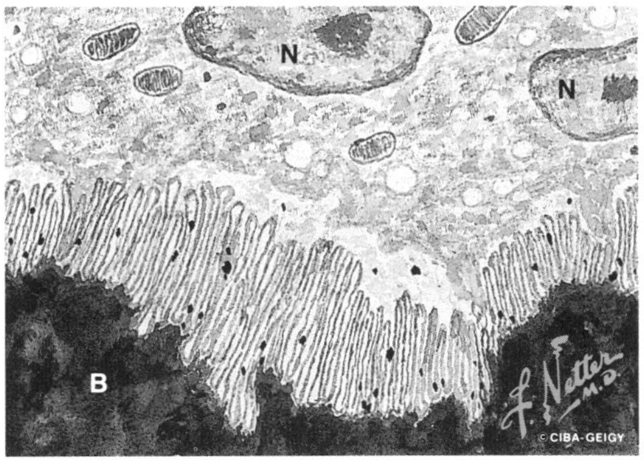

1 year after bone marrow transplant. Ruffled border now apparent (as seen in normal osteoclast actively resorbing bone)

lead to chronic hypocalcemia at times of dietary deficiency, with a consequent mineralization defect in the cartilaginous growth plate and in bone. Although hyperparathyroidism and hypervitaminosis D may develop, the osteoclasts do not respond to the endocrine signals. Rickets in association with osteopetrosis is truly an example of mineral famine in the midst of plenty.

In a series of elegant experiments, Walker showed that osteopetrosis in affected mice can be reversed by intravenous administration (transplantation) of cell suspensions prepared from the spleen and bone marrow of normal littermates. Conversely, osteopetrosis can be induced in normal mice by cell infusions prepared from the spleens of osteopetrotic littermates. These data provided convincing evidence for the hematopoietic origin of the osteoclast. Both osteoclasts and macrophages share similar structural and functional features such as motility, ruffled borders, phagocytic activity, hydrolytic enzymes, and sensitivity to both parathyroid hormone (PTH) and calcitonin. In patients with congenital malignant osteopetrosis, peripheral blood monocytes have been reported to exhibit a decreased capacity to kill bacteria intracellularly, although chemotaxis is normal. The evidence suggests that osteoclasts and phagocytic blood cells share a common hematopoietic precursor.

Walker's classic experiments paved the way for the use of HLA-matched sibling-donor bone marrow transplantation in the treatment of congenital malignant osteopetrosis. At present, bone marrow transplantation is the most promising form of treatment for children afflicted with the severe congenital form of the disease. Results of transiliac bone biopsies carried out before treatment reveal marrow spaces that are nearly obliterated by numerous unresorbed calcified cartilage bars. Osteoclasts are abundant but inactive, as evidenced by a lack of Howship's lacunae seen on light microscopy and a lack of a resorbing surface ruffled border observed on electron microscopy. Within several months of bone marrow transplantation, radiographic and histologic studies show an amelioration of the disease process, with a resumption of normal bone resorption and skeletal remodeling.

In some patients who are not candidates for bone marrow transplantation, administration of 1,25-dihydroxyvitamin D, a potent bone-resorbing agent, may stimulate osteoclastic activity. □

Paget's Disease of Bone

Manifestations of advanced, diffuse Paget's disease of bone (may occur singly or in combination)

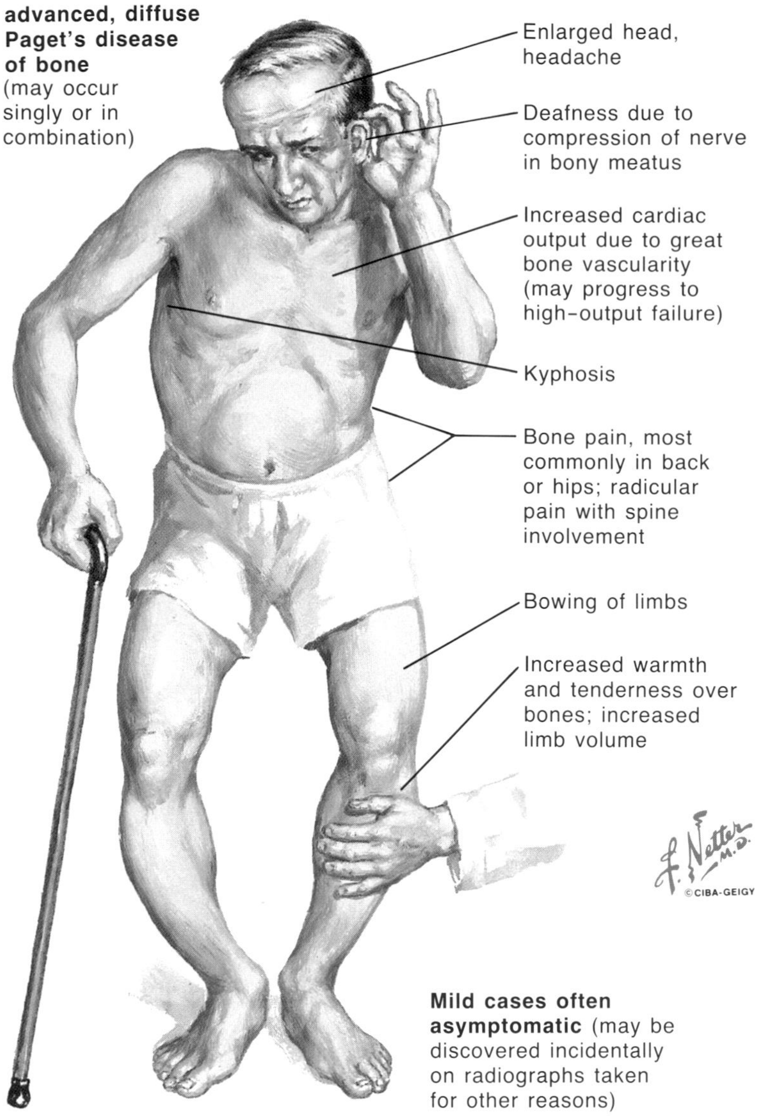

Enlarged head, headache
Deafness due to compression of nerve in bony meatus
Increased cardiac output due to great bone vascularity (may progress to high-output failure)
Kyphosis
Bone pain, most commonly in back or hips; radicular pain with spine involvement
Bowing of limbs
Increased warmth and tenderness over bones; increased limb volume
F. Netter M.D.
©CIBA-GEIGY

Mild cases often asymptomatic (may be discovered incidentally on radiographs taken for other reasons)

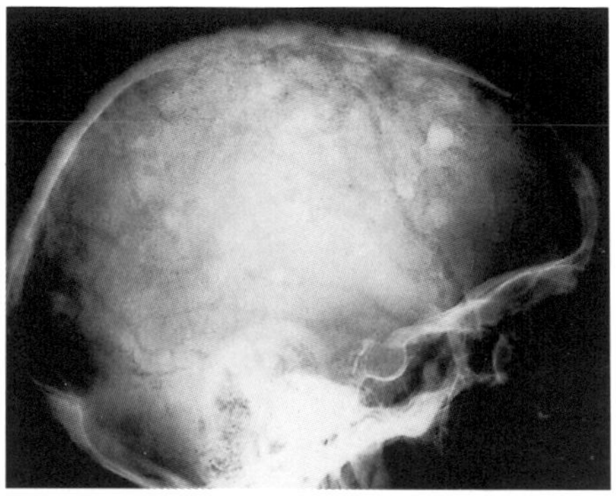

Lateral radiograph shows patchy density of skull, with areas of osteopenia (osteoporosis circumscripta cranii)

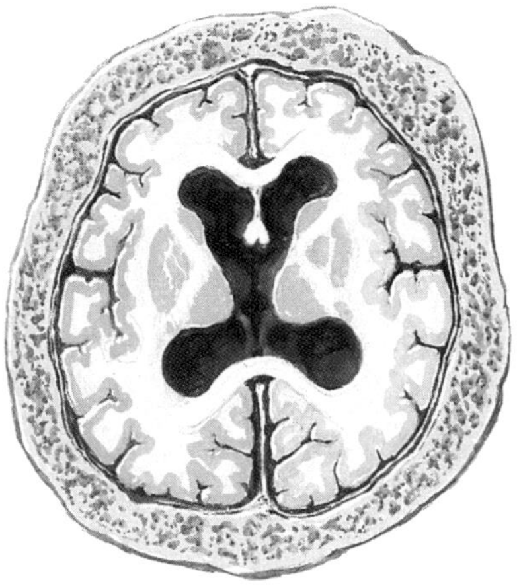

Extremely thickened skull bones, which may encroach on nerve foramina or brainstem and cause hydrocephalus (shown) by compressing cerebral aqueduct

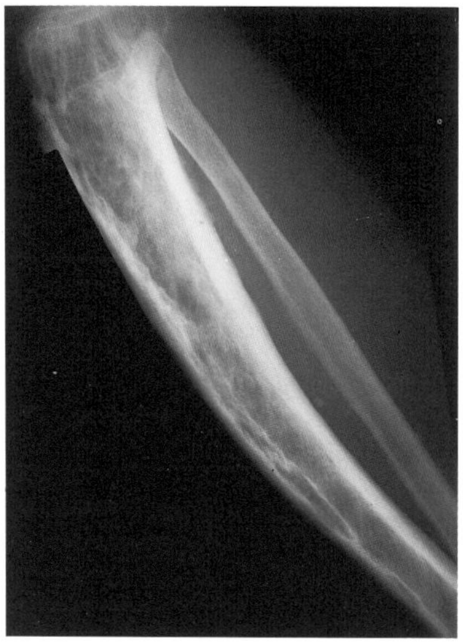

Characteristic radiographic findings in tibia include thickening, bowing, and coarse trabeculation, with advancing radiolucent wedge

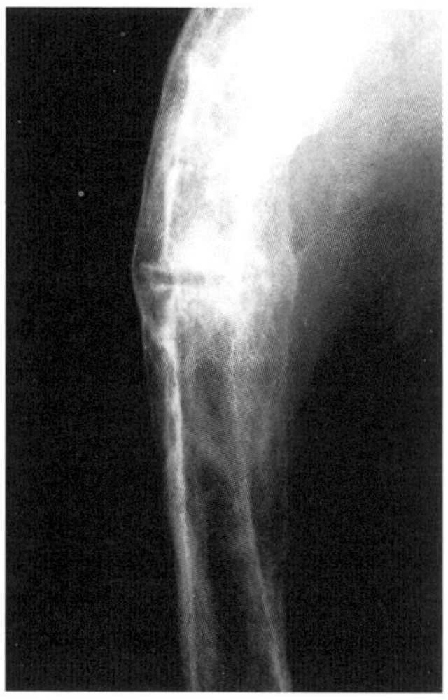

Healing chalk-stick fracture

Paget's Disease of Bone

Paget's disease of bone, or osteitis deformans, was first described in 1876 by Sir James Paget. It is a common disorder occurring in 3% to 4% of middle-aged patients; its incidence increases to 10% by the ninth decade. The disease rarely occurs before age 40 and is slightly more common in men. It may be monostotic or polyostotic, is usually asymmetric, but may often be asymptomatic (Plate 41). There is a striking geographic distribution of the disease, with the greatest prevalence in England, Western Europe, the United States, Australia, and New Zealand.

Paget's disease is an idiopathic focal disorder of skeletal remodeling. The primary abnormality appears to be an increase in osteoclastic resorption of bone. Both the size and the number of active osteoclasts are increased, as is the number of nuclei per cell. Through the process of coupling, increased bone resorption leads to a compensatory increased bone formation. The overall rate of bone remodeling is chaotic, resulting in a predominance of highly vascular, immature woven bone that is often structurally weak and prone to deformities and pathologic fractures.

The etiology of Paget's disease is unknown, but because of its focal nature, it is by definition not a metabolic disease. Recent studies have established the presence of intranuclear inclusions in pagetic osteoclasts that resemble the nucleocapsids of the paramyxovirus family, suggesting that a slow-virus infection may play an important etiologic role. However, virus particles have not been isolated from pagetic bone.

Radiographic Studies

The earliest radiographic evidence of the disease is often a flame-shaped or blade-of-grass–shaped advancing osteolytic wedge seen in the middiaphysis of the long bones (Plate 41). With subsequent compensatory osteoblastic bone formation, the bone becomes grossly enlarged and deformed, with irregular, thickened cortices and coarsened trabeculae. Bowing, incomplete pseudofractures, and complete pathologic fractures are common. In the appendicular skeleton, Paget's disease always involves the metaphysis, often extends into the diaphysis, and may involve the entire bone. It never affects the diaphysis exclusively. Subchondral bone involvement in the metaphysis on one side of a diarthrodial joint may lead to joint incongruity and subsequent arthritis. When the hip is affected, acetabular protrusion together with medial joint space narrowing may result. Most commonly involved are the sacrum, vertebrae, femur, cranium, sternum, pelvis, clavicle, tibia, rib, and humerus. Often, only one bone in a limb is affected—for instance, the radius but not the ulna (Plate 42).

Histologic examination reveals a characteristic mosaic appearance unique to pagetic bone. Irregular areas of hypercellular lamellar bone lined by giant multinucleated osteoclasts (often more than 20 nuclei per cell) with neighboring areas of intense immature osteoblastic fiber bone formation are seen. Cement lines, which indicate a reversal of osteoclastic resorption and subsequent new bone formation, are irregular and erratic.

Clinical Manifestations

Because Paget's disease is usually asymptomatic, it is most often detected incidentally on radiographs taken for other purposes. However, when symptoms do occur, the most common is a deep, aching bone pain. If the lower limb and spine are involved, weight bearing often worsens the pain. The characteristic macroscopic bone enlargement may impinge on the vertebral canal, resulting in cranial neuropathies, spinal radiculopathies, and spinal cord compression and cauda equina syndromes. With increased blood flow to the highly vascular pagetic bone, steal syndromes may also occur, shunting blood away from the neural elements and further exacerbating the neurologic signs.

The olfactory (I), optic (II), trigeminal (V), facial (VII), and vestibulocochlear (VIII) nerves

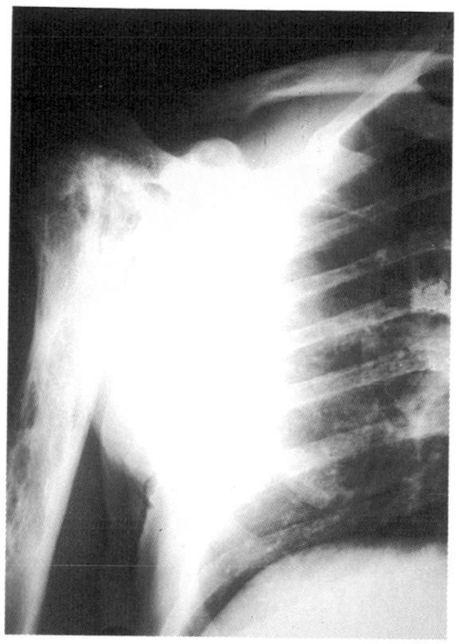

Proximal humerus appears thickened and coarsely trabeculated, with patchy rarefaction

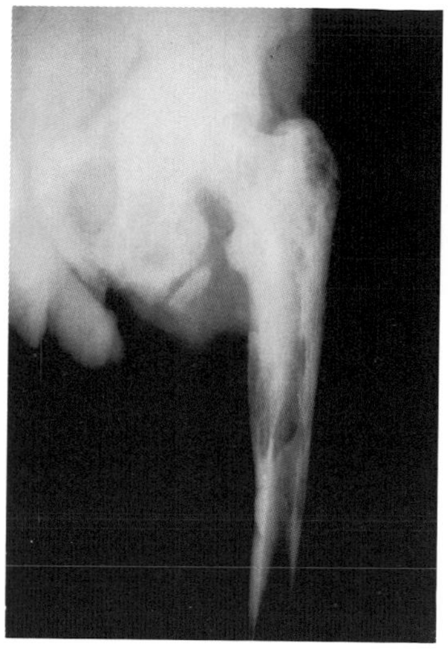

Proximal femur shows typical manifestations of disease

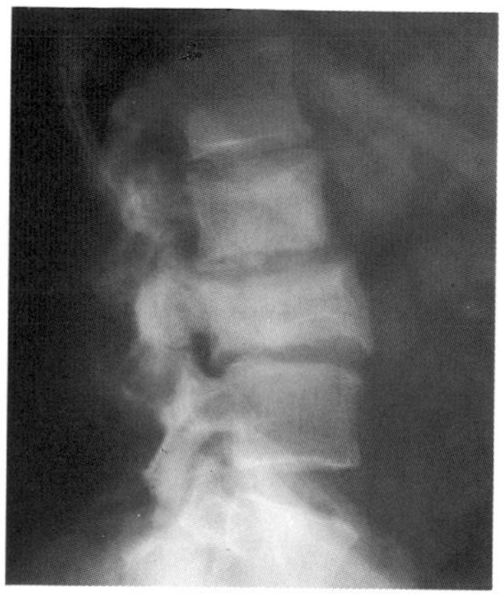

Vertebral body (L3) appears enlarged, with increased density

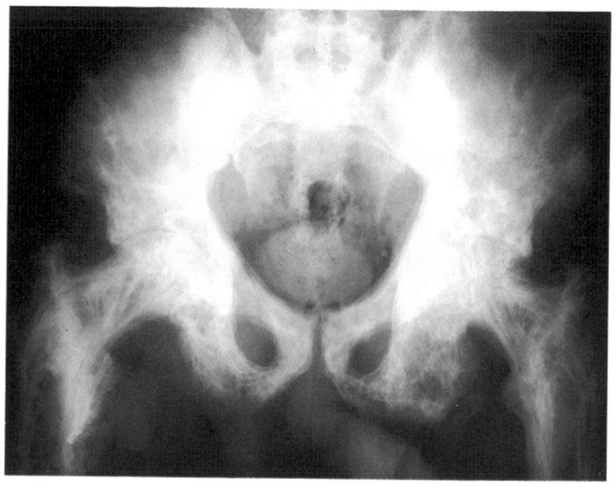

Disease seen throughout pelvis and proximal femurs, with involvement of hip joints

Above radiographs taken from same patient. Findings correspond to areas of increased isotope uptake in bone scan (right)

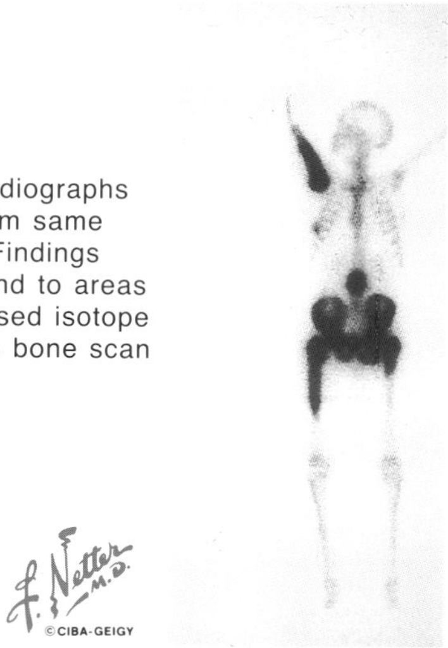

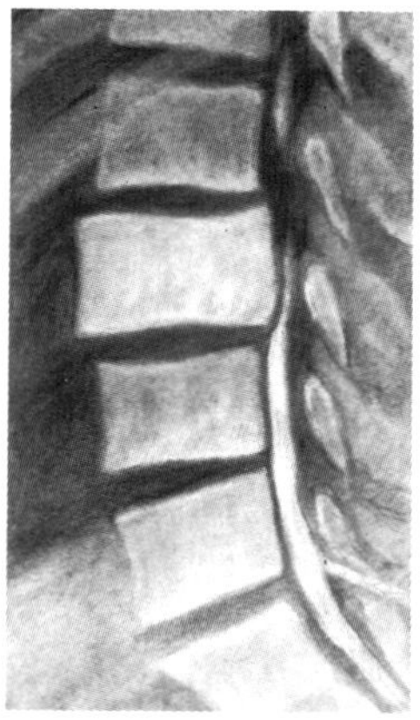

Myelogram shows obstruction of vertebral canal in L3

Disease may be monostotic, affecting only single bone or vertebra. Involvement of right tibia evidenced by increased limb volume and bowing. Fibula not involved

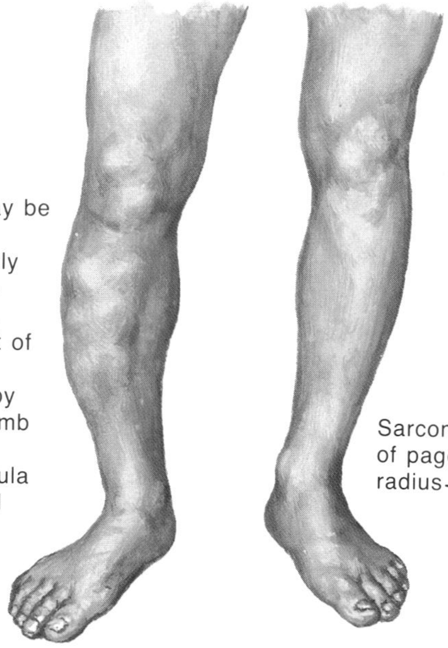

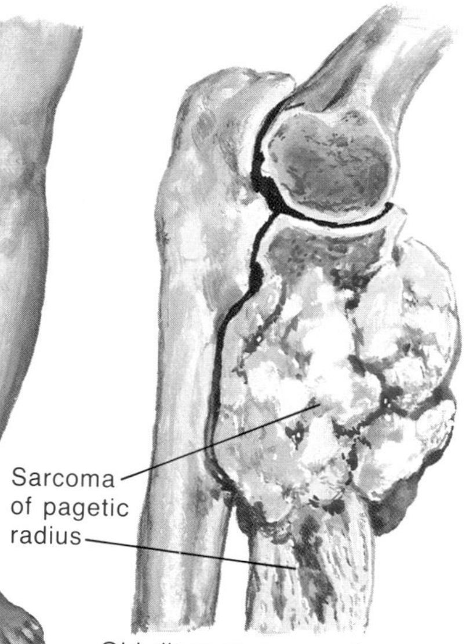

Old disease may undergo sarcomatous degeneration (<1% of symptomatic cases)

SECTION IV PLATE 42 Slide 3777

Paget's Disease of Bone

(Continued)

are most commonly affected. Involvement of the inner ear may result in mixed sensory and conductive deafness. Extreme thickening and involvement of the skull may block cerebrospinal fluid flow, causing hydrocephalus.

Other common signs of Paget's disease include bony deformity, increased skin temperature over affected areas (a useful clinical sign in monitoring the severity of the disease and the response to treatment), and high-output congestive heart failure (in extensive disease).

Painful fissure fractures, or pseudofractures (transverse radiolucent areas on the convex side of a long bone), and completed pathologic fractures (transverse chalk-stick fractures) frequently occur in areas of high stress, particularly in the weight-bearing bones of the lower limbs. Fracture healing is often complicated, and delayed union and nonunion are common.

Paget's sarcoma, a rare complication, occurs in less than 1% of cases and rarely before 70 years of age. A marked and sustained increase in pain in an area of long-standing Paget's disease in an elderly patient suggests this serious sequela. Other manifestations are night pain and radiographic evidence of bone destruction. Serum alkaline phosphatase levels are usually not altered. The prognosis for a patient with sarcomatous degeneration of a pagetic lesion is poor.

Pathophysiology

The pathophysiologic process of Paget's disease may be divided into active and inactive phases. Early in the active phase, intense osteoclastic bone resorption is the predominant activity (lytic phase). Later in the active phase, compensatory bone formation begins (mixed phase), and very late in the active phase, osteoblastic bone formation predominates (sclerotic phase). In the inactive, or "burnt-out" phase, the cellular activity of both osteoclasts and osteoblasts returns to normal, although the structurally ravaged and deformed bone remains for life. Specific radiographic findings vary, depending on the extent of the disease process and the specific areas of skeletal involvement.

Although radiographs are essential to evaluate the involvement of a particular bone, they are much less sensitive than bone-seeking radioisotope scanning in determining the presence, distribution, and extent of the disease (Plate 42). However, because any osseous condition associated with increased bone formation (infection, tumor, fracture) may cause increased radionuclide uptake on the bone scan, if the diagnosis is in question, other possible causes may be established by means of plain radiographs or a bone biopsy.

Laboratory tests to assess the disease activity and to monitor the course of treatment include: (1) a 24-hour urinalysis for the level of hydroxyproline, which indicates the degree of bone collagen breakdown from bone resorption, and (2) determination of the serum alkaline phosphatase level to measure the degree of osteoblastic bone formation. However, because the two laboratory values are essentially equivalent, the latter test is used routinely. Because of the tight coupling between bone resorption and bone formation, calcium levels in serum and urine are usually within normal limits, except when there is concurrent generalized immobilization, fracture, hyperthyroidism, hyperparathyroidism, or an unassociated carcinoma that has metastasized to bone.

Treatment

The most important indication for treatment of Paget's disease is the presence of symptoms, particularly pain, which may be the result of increased metabolic activity, neural compression, steal syndrome, secondary arthritis, incomplete pseudofracture, or completed pathologic fracture. Progressive deformity, anticipated bone surgery, impending fracture, and congestive heart failure also necessitate treatment.

Pagetʼs Disease of Bone

(Continued)

At present, there is no cure for Pagetʼs disease, and the goals of therapy are to relieve the clinical symptoms and to control the metabolic activity of the disease (Plate 43). The two major therapeutic agents currently used are the diphosphonates and calcitonin.

The hormone *calcitonin* directly blocks bone resorption by inhibiting osteoclastic activity. Because bone formation also decreases, the effect of treatment can be monitored by a decrease in the serum alkaline phosphatase level. The highly potent, salmon-derived calcitonin is given daily or every other day by subcutaneous injection. Flushing and nausea are common but tolerable side effects. However, resistance to calcitonin can occur, and antibodies may develop in a large proportion of patients. Human calcitonin, which is not as likely to induce antibody formation, is now available. Approximately 70% of patients respond favorably to calcitonin therapy, usually within 6 weeks.

Diphosphonates are pyrophosphate analogues that interfere with the mineralization of the bone matrix and inhibit osteoclastic activity. The current recommended dose of diphosphonate does not lead to any clinically significant mineralization defects as long as the medication is not used for more than 6 months at a time. However, mild gastrointestinal side effects may occur. The original high doses induced a painful osteomalacia that increased the incidence of pathologic fractures. Because the half-life of diphosphonates is long, disease remission is sustained long after the medication is discontinued.

Although the two medications may be used interchangeably, calcitonin is preferred in patients who are immobilized, have an impending or completed fracture, or have neurologic impairment. Because the diphosphonates impair mineralization, they should not be used in a patient with an impending fracture or a lytic lesion in a weight-bearing bone. If either medication alone fails to provide relief, the two may be used in combination.

The use of the antibiotic mithramycin, which is cytotoxic to osteoclasts and other marrow cells, should be reserved for severe refractory cases because its use requires close medical supervision. Symptoms of Pagetʼs arthritis may be relieved considerably by use of nonsteroidal antiinflammatory medications.

Indications for surgical intervention in Pagetʼs disease include certain fractures and severe arthritis that is refractory to medical treatment. Malalignment of the major weight-bearing bones may be treated with appropriate orthoses and medication or with surgery. Preoperative medical treatment with diphosphonates or calcitonin may help decrease intraoperative bleeding. □

Pathophysiology and Treatment of Pagetʼs Disease of Bone

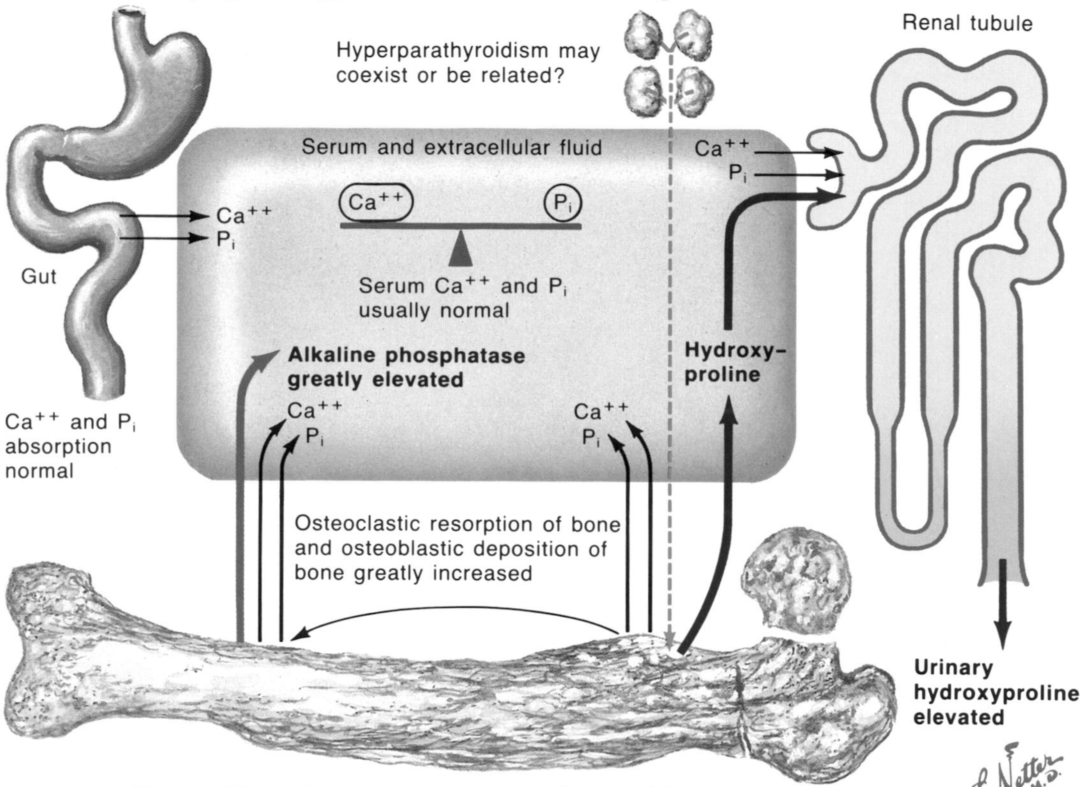

Abnormal bone structure, coarse trabeculation, thickening, bowing, pseudofractures, fractures, hypervascularity

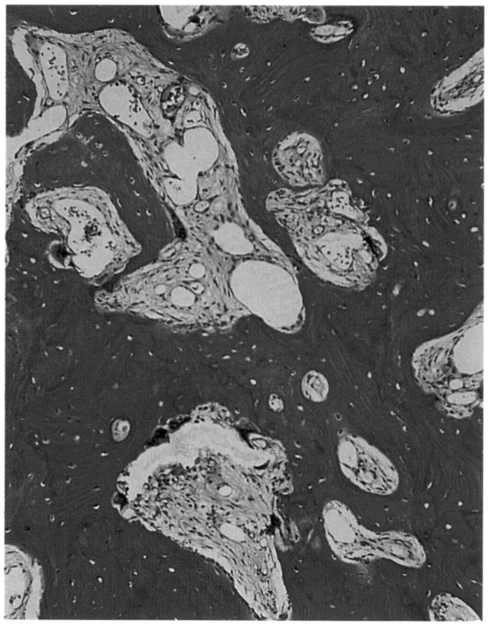

Section of bone shows intense osteoclastic and osteoblastic activity and mosaic of lamellar bone

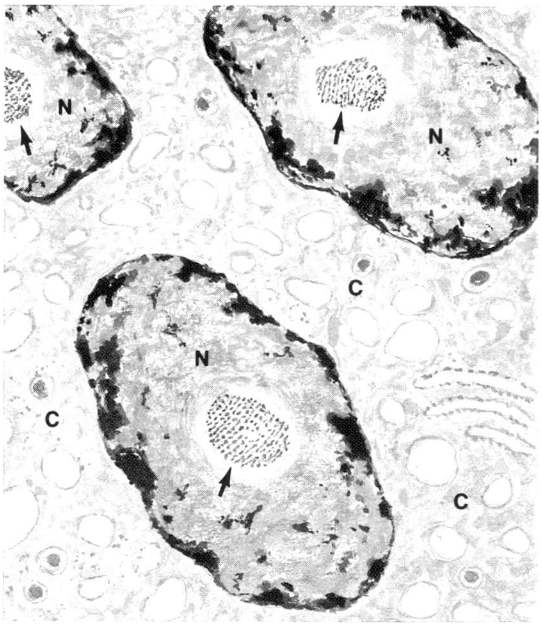

Electron-microscopic view of multinucleated osteoclast with nuclear inclusions that may be viruses (arrows). N = nuclei; C = cytoplasm

Therapeutic agent→	Diphosphonates (synthetic)	Calcitonin (natural hormone)
Dosage	5 mg/kg/day (oral)	0.5 mg/day (subcutaneous)
Mode of action	Indirectly inhibits bone resorption	Directly inhibits bone resorption
Serum alkaline phosphatase reduction ≥30%	80% of patients	66% of patients
Urinary hydroxyproline reduction ≥30%	90% of patients	66% of patients
Disease remission	Up to 1 year after therapy	Not established
Pain reduction	60% of patients	66% of patients
Cost	Less expensive	More expensive
Adverse reactions	GI disturbances, diarrhea	Flushing, nausea
Contraindications	Fractures, pseudofractures	None

Fibrodysplasia Ossificans Progressiva

Fibrodysplasia (Myositis) Ossificans Progressiva

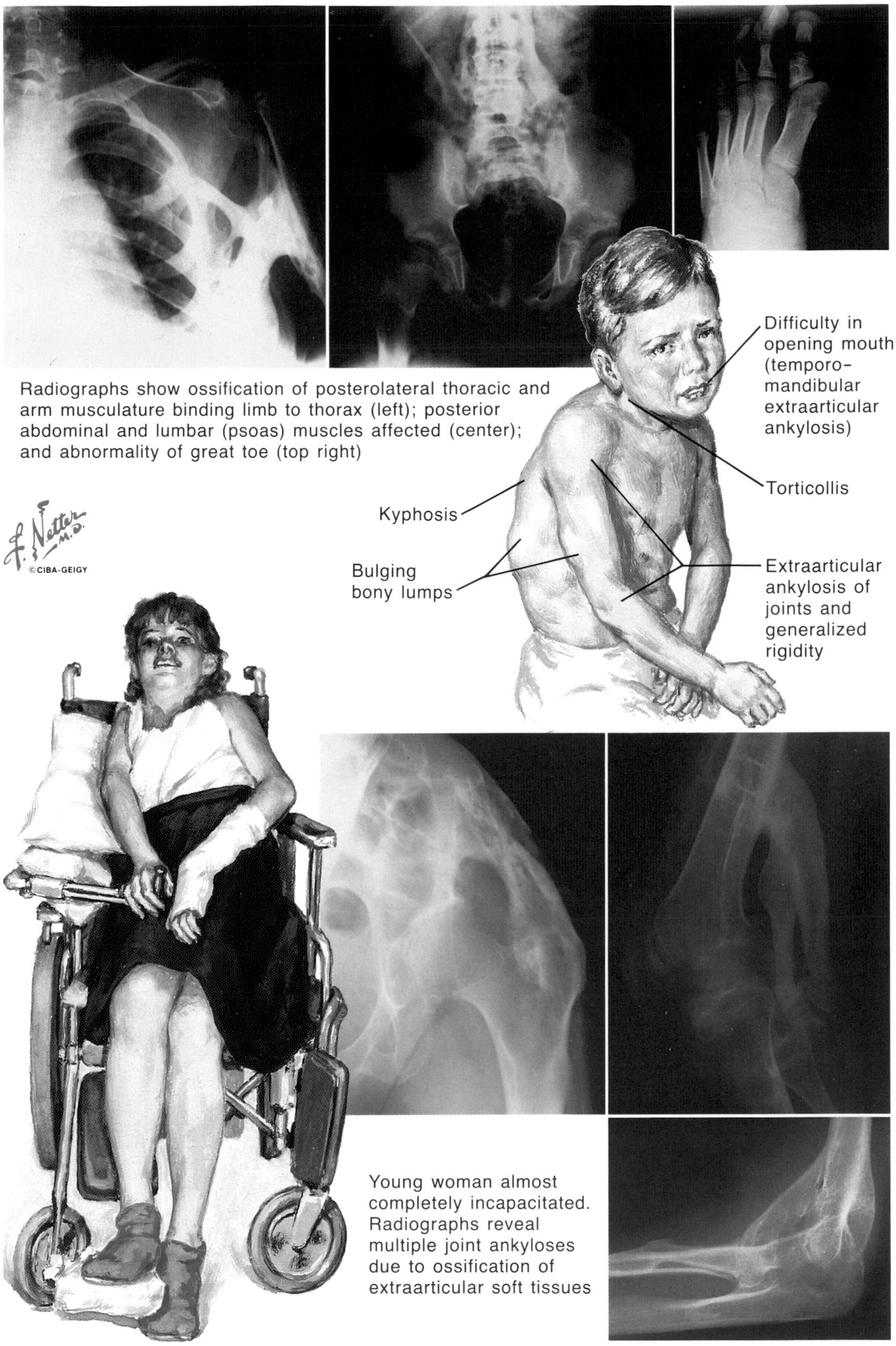
Difficulty in opening mouth (temporomandibular extraarticular ankylosis)
Torticollis
Kyphosis
Bulging bony lumps
Extraarticular ankylosis of joints and generalized rigidity
F. Netter M.D.
©CIBA-GEIGY

Radiographs show ossification of posterolateral thoracic and arm musculature binding limb to thorax (left); posterior abdominal and lumbar (psoas) muscles affected (center); and abnormality of great toe (top right)

Young woman almost completely incapacitated. Radiographs reveal multiple joint ankyloses due to ossification of extraarticular soft tissues

Fibrodysplasia ossificans progressiva, previously known as myositis ossificans progessiva, is a rare heritable disorder of connective tissue. It is characterized by congenital skeletal malformation of the great toes and a severe progressive ossification of the soft tissues that begins in late childhood. The current name reflects the tendency of heterotopic bone to form in soft tissues such as muscles, ligaments, tendons, and joint capsules.

Fibrodysplasia ossificans progressiva may be inherited as an autosomal dominant disorder or may arise as a spontaneous mutation, with full penetrance (no skipped generations) but variable expressivity (variable phenotypic expression of the gene in affected members of the same family) and no sexual or ethnic predilection.

Characteristic skeletal malformations of the feet include reduction defects (absent phalanges) in the toes, most commonly in the great toe. Congenital bunions are also common, and their presence suggests the possibility of the disorder long before heterotopic ossification begins. The limbs may be short, but this deformity is less prevalent than the toe anomalies. The congenital skeletal abnormalities cause few problems, and the affected child remains asymptomatic until heterotopic bone formation begins. This generally occurs by age 10 (the average age at onset is 4 years) with a series of firm, painful, asymmetric soft-tissue lumps in the muscles of the neck and back. These lumps, which vary in size and shape, usually appear spontaneously but may be precipitated by trauma as minor as an intramuscular injection. Axial involvement precedes appendicular involvement; in the limbs, proximal ossification occurs before distal ossification. The muscles of mastication are often affected, but visceral smooth muscles, sphincters, diaphragm, larynx, tongue, extraocular muscles, and the heart are clinically uninvolved. There may be a conductive hearing loss, but ocular problems do not occur. Systemic signs of disease, such as fever and malaise, are usually absent.

As heterotopic ossification develops in the soft tissues throughout the body, extraarticular ankylosis of the joints occurs, beginning proximally and axially, then progressing distally throughout the appendicular skeleton. Although longitudinal growth is normal, it may be masked by deformities caused by bony ankylosis of the spine and limbs. Paradoxically, osteoporosis resulting from immobilization may occur as the disease progresses, most notably about ankylosed joints. Fractures of the osteoporotic bone or the heterotopic new bone occur occasionally.

Although clinical and radiographic findings are dramatic, laboratory studies of serum calcium, phosphate, and alkaline phosphatase; complete blood count; and erythrocyte sedimentation rate are normal.

Prior to clinical involvement, the muscles are histologically normal. Spontaneous edema of the interfascicular muscles occurs first, followed by proliferation of perivascular fibroblastic connective tissue. Involved muscle fibers degenerate rapidly and are replaced by a process of either intramembranous or endochondral ossification. Finally, formation of mature heterotopic lamellar bone that is indistinguishable from normal bone takes place. The disease process is true ossification, not calcification.

The mechanism by which the abnormal gene causes such protean regulatory defects is not known. Although progession of the disease is erratic, spinal and shoulder movement is limited in most patients by 10 years of age, and hip movement is severely limited by age 20. Most patients are confined to a wheelchair by age 30. The average life span is about 35 to 40 years, and pneumonia is the usual cause of death.

The diagnosis of fibrodysplasia ossificans progressiva is based on clinical and radiographic findings. There is no effective treatment, although administration of diphosphonates has been advocated. However, this treatment merely delays the mineralization of bone rather than impairing the production of heterotopic osteoid. Surgery may help a joint to fuse in a more functional position. Even in the late stages of the disease, patients should be considered severely disabled rather than ill. Genetic counseling should be provided to families in which the disease occurs. □

Selected References

Section I

Plates

Abbott LC, Saunders JB deC M, Bost FC, et al: *Injuries to ligaments of the knee joint.* J Bone Joint Surg 1944, 26:503–521 — 1–110

Basmajian JV, Bentzon JW: *An electromyographic study of certain muscles of the leg and foot in the standing position.* Surg Gynec & Obst 1954, 98:662–666 — 1–110

Basmajian JV, DeLuca CJ: *Muscles Alive,* ed 3. Baltimore, Williams & Wilkins, 1974 — 1–110

Basmajian JV, Latif A: *Integrated actions and functions of the chief flexors of the elbow. A detailed electromyographic analysis.* J Bone Joint Surg 1957, 39A:1106–1118 — 1–110

Basmajian JV, MacConaill MA (eds): *Muscles and Movements: A Basis for Human Kinesiology,* ed 2. Huntington NY, RE Krieger Publishing, 1977 — 1–110

Beaton LE, Anson BJ: *The sciatic nerve and the piriformis muscle: their interrelation a possible cause of coccygodynia.* J Bone Joint Surg 1938, 20:686-688 — 1–110

Bradley KC, Sunderland S: *The range of movement at the wrist joint.* Anat Rec 1953, 116:139–145 — 1–110

Brantigan OC, Voshell AF: *The tibial collateral ligament: its function, its bursae and its relation to the medial meniscus.* J Bone Joint Surg 1943, 25:121–131 — 1–110

Brantigan OC, Voshell AF: *The mechanics of the ligaments and menisci of the knee joint.* J Bone Joint Surg 1941, 23:44–66 — 1–110

Brown RW: *Composition of Scientific Words.* Washington, Smithsonian Institution Press, 1956 — 1–110

Chandler SB, Kreuscher PH: *A study of the blood supply of the ligamentum teres and its relation to the circulation of the head of the femur.* J Bone Joint Surg 1932, 14:834–846 — 1–110

Codman EA: *The Shoulder.* Boston, Thomas Todd, 1934 — 1–110

Crock HV: *The arterial supply and venous drainage of the bones of the human knee joint.* Anat Rec 1962, 144:199–217 — 1–110

Dempster WT, Finerty JC: *Relative activity of wrist moving muscles in static support of the wrist joint; an electromyographic study.* Am J Physiol 1947, 150:596–606 — 1–110

DePalma AF: *Surgery of the Shoulder.* Philadelphia, JB Lippincott, 1950 — 1–110

Edwards EA: *Anatomy of the small arteries of the foot and toes.* Acta Anat 1960, 41:81–96 — 1–110

Edwards EA, Roebuck JD Jr: *Applied anatomy of the femoral vein and its tributaries.* Surg Gynec & Obst 1947, 85:547–557 — 1–110

Evans FG, Lissner HR: *"Stresscoat" deformation studies of the femur under static vertical loading.* Anat Rec 1948, 100:159–190 — 1–110

Finerty JC: *Persistent ischiatic artery.* Anat Rec 1947, 98:587–595 — 1–110

Section I *(continued)*

Plates

Friedman L, Munro RR: *Abduction of the arm in the scapular plane—scapular and glenohumeral movements. A roentgenographic study.* J Bone Joint Surg 1966, 48A:1503–1510 — 1–110

Gardner E: *The nerve supply of muscles, joints and other deep structures.* Bull Hosp Jt Dis Orthop Inst 1960, 21:153–161 — 1–110

Gardner E: *The anatomy of joints. Development of joints.* Amer Acad Orthoped Surg Lect #9, 1952 — 1–110

Gardner E: *The innervation of the elbow joint.* Anat Rec 1948, 102:161–174 — 1–110

Gardner E: *The innervation of the hip joint.* Anat Rec 1948, 101:353–371 — 1–110

Gardner E: *The innervation of the knee joint.* Anat Rec 1948, 101:109–130 — 1–110

Gardner E: *The innervation of the shoulder joint.* Anat Rec 1948, 102:1–18 — 1–110

Gardner E, Gray DJ, O'Rahilly R: *Anatomy: A Regional Study of Human Structure,* ed 3. Philadelphia, WB Saunders, 1969 — 1–110

Girgis FG, Marshall JL, Monajem A: *The cruciate ligaments of the knee joint. Anatomical, functional and experimental analysis.* Clin Orthop 1975, 106:216–231 — 1–110

Goodfellow J, Hungerford DS, Zindel M: *Patello-femoral joint mechanics and pathology. 2. Chondromalacia patellae.* J Bone Joint Surg 1976, 58(3):287–290 — 1–110

Gray DJ, Gardner E: *The innervation of the joints of the wrist and hand.* Anat Rec 1965, 151:261–266 — 1–110

Greenwold AS, O'Connor JJ: *The transmission of load through the human hip joint.* J Biomech 1971, 4:507–528 — 1–110

Greulich WW, Pyle SI: *Radiographic Atlas of Skeletal Development of the Hand and Wrist,* ed 2. Stanford, Stanford University Press, 1959 — 1–110

Grodinsky M: *A study of the fascial spaces of the foot and their bearing on infections.* Surg Gynec & Obst 1929, 49:737–751 — 1–110

Grodinsky M, Holyoke EA: *The fasciae and fascial spaces of the palm.* Anat Rec 1941, 79:435–451 — 1–110

Haines RW: *Mechanism of rotation at the first carpometacarpal joint.* J Anat 1944, 78:44–46 — 1–110

Harris W: *The Morphology of the Brachial Plexus.* London, Oxford University Press, 1939 — 1–110

Hicks JH: *The mechanics of the foot. II. The plantar aponeurosis and the arch.* J Anat 1954, 88:25–30 — 1–110

Hicks JH: *The mechanics of the foot. I. The joints.* J Anat 1953, 87:345–357 — 1–110

Hollinshead WH: *Anatomy for Surgeons,* ed 2. Vol 3: *The Back and Limbs.* Hagerstown MD, Harper & Row, 1969 — 1–110

Hollinshead WH, Jenkins DB: *Functional Anatomy of the Limbs and Back,* ed 4. Philadelphia, WB Saunders, 1976 — 1–110

Hollinshead WH, Markee JE: *The multiple innervation of limb muscles in man.* J Bone Joint Surg 1946, 28:721–731 — 1–110

Huber JF: *The arterial network supplying the dorsum of the foot.* Anat Rec 1958, 132:81–92 — 1–110

Huelke DF: *Variation in the origins of the branches of the axillary artery.* Anat Rec 1959, 135:33–41 — 1–110

Huelke DF: *A study of the transverse cervical and dorsal scapular arteries.* Anat Rec 1958, 132: 233–246 — 1–110

Huelke DF: *A study of the formation of the sural nerve in adult man.* Am J Phys Anthropol 1957, 15(1):137–147 — 1–110

Hughston JC, Andrews JR, Cross MJ, et al: *Classification of knee ligament instabilities. Part II. The lateral compartment.* J Bone Joint Surg 1976, 58(2):173–179 — 1–110

Hughston JC, Andrews JR, Cross MJ, et al: *Classification of knee ligament instabilities. Part I. The medial compartment and cruciate ligaments.* J Bone Joint Surg 1976, 58(2): 159–172 — 1–110

Inman VT, Saunders JB deC M: *Observations on the function of the clavicle.* California Med 1946, 65:158–166 — 1–110

Inman VT, Saunders JB deC M, Abbott LC: *Observations on the function of the shoulder joint.* J Bone Joint Surg 1944, 26:1–30 — 1–110

International Anatomical Nomenclature Committee: *Nomina Anatomica,* ed 4. Amsterdam, Excerpta Medica, 1977 — 1–110

Jones RL: *The human foot. An experimental study of its mechanics, and the role of its muscles and ligaments in support of the arch.* Am J Anat 1941, 68:1–39 — 1–110

Jonsson B, Steen B: *Function of the gracilis muscle. An electromyographic study.* Acta Morphol Neerl Scand 1966, 6:325–341 — 1–110

Jonsson B, Steen B: *Function of the hip and thigh muscles in Romberg's test and "standing at ease." An electromyographic study.* Acta Morphol Neerl Scand 1963, 5:269–276 — 1–110

Kaplan EB: *The fabellofibular and short lateral ligaments of the knee joint.* J Bone Joint Surg 1961, 43A:169–179 — 1–110

Kaplan EB: *The iliotibial tract; clinical and morphological significance.* J Bone Joint Surg 1958, 40A:817–832 — 1–110

Kaplan EB: *Functional and Surgical Anatomy of the Hand.* Philadelphia, JB Lippincott, 1953 — 1–110

Kaplan EB: *The relation of the extensor digitorum communis tendon to the metacarpophalangeal joint.* Bull Hosp Jt Dis Orthop Inst 1945, 6:149–154 — 1–110

Keegan JJ, Garrett FD: *The segmental distribution of the cutaneous nerves in the limbs of man.* Anat Rec 1948, 102:409–437 — 1–110

Keen JA: *A study of the arterial variations in the limbs, with special reference to symmetry of vascular patterns.* Am J Anat 1961, 108: 245–261 — 1–110

Kennedy JC, Weinberg HW, Wilson AS: *The anatomy and function of the anterior cruciate ligament. As determined by clinical and morphological studies.* J Bone Joint Surg 1974, 56(2):223–235 — 1–110

Kuczynski K: *Carpometacarpal joint of the human thumb.* J Anat 1974, 118(1):119–126 — 1–110

Lambert KL: *The weight-bearing function of the fibula. A strain gauge study.* J Bone Joint Surg 1971, 53A:507–513 — 1–110

Landsmeer JM: *The coordination of finger-joint motions.* J Bone Joint Surg 1963, 45:1654–1662 — 1–110

Landsmeer JMF: *Anatomical and functional investigations on the articulation of human fingers.* Acta Anat (Suppl 24) 1955, 25:1–69 — 1–110

Last RJ: *Specimens from the Hunterian collection: the genicular arteries (specimen S 142 A).* J Bone Joint Surg 1951, 33B:264–267 — 1–110

Last RJ: *The popliteus muscle and the lateral meniscus, with note on the attachment of the medial meniscus.* J Bone Joint Surg 1950, 32B:93–99 — 1–110

Last RJ: *Some anatomical details of the knee joint.* J Bone Joint Surg 1948, 30B:683–688 — 1–110

Lawrence M, Strachan JCH: *The dynamic stability of the knee.* Proc R Soc Lond 1970, 63:34–35 — 1–110

Section I *(continued)* — Plates

Linell EA: *The distribution of nerves in the upper limb, with reference to variabilities and their clinical significance.* J Anat 1921, 55:79–112 — 1–110

Lockhart RD: *Living Anatomy. A Photographic Atlas of Muscles in Action and Surface Contours*, ed 2. Glasgow, University Press, 1949 — 1–110

Lockhart RD: *A further note on the movements of the shoulder joint.* J Anat 1933, 68:135–137 — 1–110

Lockhart RD: *Movements of the normal shoulder joint and of case with trapezius paralysis studied by radiogram and experiment in living.* J Anat 1930, 64:288–302 — 1–110

Lucas DB: *Biomechanics of the shoulder joint.* Arch Surg 1973, 107:425–432 — 1–110

MacConaill MA: *The movements of bones and joints; function of musculature.* J Bone Joint Surg 1949, 31B:100–104 — 1–110

MacConaill MA: *The mechanical anatomy of the acromio-clavicular joint in man.* Proc Roy Acad Irish 1944, 508:159–166 — 1–110

MacConaill MA: *The function of intra-articular fibrocartilages with special reference to the knee and inferior radio-ulnar joints.* J Anat 1932, 66:210–227 — 1–110

McCormack LJ, Cauldwell EW, Anson BJ: *Brachial and antebrachial arterial patterns; study of 750 extremities.* Surg Gynec & Obst 1953, 96:43–54 — 1–110

Markee JE, Logue JT Jr, Williams M, et al: *Two-joint muscles of the thigh.* J Bone Joint Surg 1955, 37:125–142 — 1–110

Martin BF: *The annular ligament of the superior radio-ulnar joint.* J Anat 1958, 92(3):473–482 — 1–110

Martin BF: *The oblique cord of the forearm.* J Anat 1953, 92:609–615 — 1–110

Martin CP: *A note on the movements of the shoulder-joint.* Br J Surg 1932, 20:61–66 — 1–110

Mayor D: *Anatomical and functional aspects of the knee-joint.* Physiotherapy 1966, 52:224–228 — 1–110

Moosman DA, Hartwell SW Jr: *The surgical significance of the subfascial course of the lesser saphenous vein.* Surg Gynec & Obst 1964, 118:761–766 — 1–110

Napier JR: *The form and function of the carpometacarpal joint of the thumb.* J Anat 1955, 89:362–369 — 1–110

Napier JR: *The attachments and function of the abductor pollicis brevis.* J Anat 1952, 86:335–341 — 1–110

Ogden JA: *The anatomy and function of the proximal tibiofibular joint.* Clin Orthop 1974, 101:186–191 — 1–110

Powell T, Lynn RB: *The valves of the external iliac, femoral, and upper third of the popliteal veins.* Surg Gynec & Obst 1951, 92:453–455 — 1–110

Pyle I, Sontag LW: *Variability in onset of ossification in epiphyses and short bones of the extremities.* Amer J Roentgen 1943, 49:795–798 — 1–110

Pyle SI, Hoerr N: *Radiographic Atlas of Skeletal Development of the Knee. A Standard of Reference.* Springfield IL, Charles C Thomas, 1955 — 1–110

Roberts JM: *The surgical knee.* Surg Clin North Am 1974, 54:1313–1326 — 1–110

Röhlich K: *Ueber die Arteria Transversa Colli des Menschen.* Anat Anz 1934, 79:37–53 — 1–110

Rouviére H: *Anatomie des Lymphatiques de l'Homme.* Paris, Masson, 1934 — 1–110

Scheldrup EW: *Tendon sheath patterns in the hand; anatomical study based on 367 hand dissections.* Surg Gynec & Obst 1951, 93:16–22 — 1–110

Sevitt S, Thompson RG: *The distribution and anastomoses of arteries supplying the head and neck of the femur.* J Bone Joint Surg 1965, 47B:560–573 — 1–110

Section I *(continued)* — Plates

Shephard E: *Tarsal movements.* J Bone Joint Surg 1951, 33B:258–263 — 1–110

Shrine NG, O'Connor JJ: *Load bearing in the knee joint.* Clin Orthop 1978, 131:279–287 — 1–110

Simon SR, Mann RA, Hagy JL, et al: *Role of the posterior calf muscles in normal gait.* J Bone Joint Surg 1978, 60:465–472 — 1–110

Stevenson PH: *Age order of epiphyseal union in man.* Am J Phys Anthropol 1924, 7:53–93 — 1–110

Sunderland S: *The distribution of sympathetic fibres in the brachial plexus in man.* Brain 1948, 71:88–102 — 1–110

Sunderland S: *Metrical and non-metrical features of the muscular branches of the radial nerve.* J Comp Neurol 1946, 85:93–111 — 1–110

Sunderland S: *The actions of the extensor digitorum communis, interosseous and lumbrical muscles.* Am J Anat 1945, 77:189–217 — 1–110

Sunderland S, Hughes ESR: *Metrical and non-metrical features of the muscular branches of the sciatic nerve and its medial and lateral popliteal divisions.* J Comp Neurol 1946, 85:205–222 — 1–110

Sunderland S, Hughes ESR: *Metrical and non-metrical features of the muscular branches of the ulnar nerve.* J Comp Neurol 1946, 85:113–125 — 1–110

Sunderland S, Ray LJ: *Metrical and non-metrical features of the muscular branches of the median nerve.* J Comp Neurol 1946, 85:191–203 — 1–110

Todd JW: *Atlas of Skeletal Maturation.* Part I. *Hand.* St Louis, CV Mosby, 1937 — 1–110

Travill A, Basmajian JV: *Electromyography of the supinators of the forearm.* Anat Rec 1961, 139:557–560 — 1–110

Tubiana R, Valentin P: *The anatomy of the extensor apparatus of the fingers.* Surg Clin North Am 1964, 44:897–906 — 1–110

Vann HM: *A note on the formation of the plantar arterial arch of the human foot.* Anat Rec 1943, 85:269–275 — 1–110

Walker PS, Erkman MJ: *The role of the menisci in force transmission across the knee.* Clin Orthop 1975, 109:184–192 — 1–110

Warwick R, Williams PL (eds): *Gray's Anatomy*, ed 35. Philadelphia, WB Saunders, 1973 — 1–110

Weathersby HT: *The artery of the index finger.* Anat Rec 1955, 122:57–64 — 1–110

Williams GD, Martin CH, McIntire LR: *Origin of the deep and circumflex femoral group of arteries.* Anat Rec 1934, 60:189–196 — 1–110

Wood Jones F: *The Principles of Anatomy as Seen in the Hand*, ed 2. London, Bailliére, Tindall & Cox, 1942 — 1–110

Woodburne RT: *Essentials of Human Anatomy*, ed 7. New York, Oxford University Press, 1983 — 1–110

Woodburne RT: *The accessory obturator nerve and the innervation of the pectineus muscle.* Anat Rec 1960, 136:367–369 — 1–110

Wright RD: *A detailed study of movement of the wrist joint.* J Anat 1935, 70:137–142 — 1–110

Section II

Arey LB: *Developmental Anatomy*, ed 7. Philadelphia, WB Saunders, 1965 — 1–21

Basmajian JV, DeLuca CJ: *Muscles Alive*, ed 3. Baltimore, Williams & Wilkins, 1974 — 1–21

Section II *(continued)* — Plates

Bourne GH (ed): *The Biochemistry and Physiology of Bone*, Vols I–III, ed 2. New York, Academic Press, 1972 — 1–21

Crelin ES: *Development of the upper respiratory system.* Clin Symp 1976, 28(3):3–30 — 1–21

Crelin ES: *Development of the lower respiratory system.* Clin Symp 1975, 27(4):3–28 — 1–21

Crelin ES: *Development of the nervous system. A logical approach to neuroanatomy.* Clin Symp 1974, 26(2):2–32 — 1–21

Crelin ES: *Functional Anatomy of the Newborn.* New Haven, Yale University Press, 1973 — 1–21

Crelin ES: *Anatomy of the Newborn: An Atlas.* Philadelphia, Lea & Febiger, 1969 — 1–21

Davis JA, Dobbing J (eds): *Scientific Foundations of Paediatrics.* Philadelphia, WB Saunders, 1974 — 1–21

Guyton AC: *Textbook of Medical Physiology*, ed 5. Philadelphia, WB Saunders, 1976 — 1–21

Hamilton WJ, Boyd JD, Mossman HW: *Human Embryology*, ed 3. Baltimore, Williams & Wilkins, 1964 — 1–21

Hyman LH: *Comparative Vertebrate Anatomy*, ed 2. Chicago, University of Chicago Press, 1942 — 1–21

Keibel F, Mall FP (eds): *Manual of Human Embryology*, Vol II. Philadelphia, JB Lippincott, 1912 — 1–21

Keibel F, Mall FP (eds): *Manual of Human Embryology*, Vol I. Philadelphia, JB Lippincott, 1910 — 1–21

Patten BM: *Human Embryology*, ed 3. New York, McGraw-Hill, 1968 — 1–21

Warwick R, Williams PL (eds): *Gray's Anatomy*, ed 35. Philadelphia, WB Saunders, 1973 — 1–21

Section III

General References

Avioli LV (ed): *The Osteoporotic Syndrome. Detection, Prevention, and Treatment.* New York, Grune & Stratton, 1983

Avioli LV, Krane SM (eds): *Metabolic Bone Disease*, Vol II. New York, Academic Press, 1978

Avioli LV, Krane SM (eds): *Metabolic Bone Disease*, Vol I. New York, Academic Press, 1977

Lane JM (ed): *Symposium on metabolic bone disease. The Orthopedic Clinics of North America.* Philadelphia, WB Saunders, 1984

Levine MA, Foster GV: *Disorders of bone and mineral metabolism.* In Harvey AM, Johns RJ, McKusick VA, et al (eds): *The Principles and Practice of Medicine*, ed 21. Norwalk CT, Appleton-Century-Crofts, 1984, pp 861–888

Alberts B, Bray D, Lewis J, et al: *The Molecular Biology of the Cell.* New York, Garland Publishing, 1983, pp 549–559 — 1–15

Aloia JF, Cohn SH, Ostuni JA, et al: *Prevention of involutional bone loss by exercise.* Ann Intern Med 1978, 89(3):356–358 — 34, 36

Anderson DR: *The ultrastructure of elastic and hyaline cartilage of the rat.* Am J Anat 1964, 114:403–434 — 21

Anderson HC: *Vesicles associated with calcification in the matrix of epiphyseal cartilage.* J Cell Biol 1969, 41:59–72 — 16–18

Asboe-Hansen G (ed): *Connective Tissue in Health and Disease*. Copenhagen, A Munksgaard, 1954 — 27
Bloom W, Fawcett DW: *A Textbook of Histology*, ed 10. Philadelphia, WB Saunders, 1975, chap 11 — 2
Boskey AL, Posner AS: *The structure of bone*. Orthop Clin North Am 1984, 15(4):597–612 — 20
Boucek RJ, Noble NL, Gunja-Smith Z, et al: *The Marfan syndrome: a deficiency in chemically stable collagen cross-links*. N Engl J Med 1981, 305(17):988–991 — 24
Bourne GH (ed): *The Structure and Function of Muscle*, Vol 2. New York, Academic Press, 1973 — 2
Bourne GH (ed): *The Biochemistry and Physiology of Bone*, Vol 1, ed 2. New York, Academic Press, 1972 — 21
Bourne GH (ed): *The Structure and Function of Muscle*, Vol 1. New York, Academic Press, 1972 — 2
Brighton CT: *The growth plate*. Orthop Clin North Am 1984, 15:571–595 — 16–18
Brighton CT: *Present and future of electrically induced osteogenesis*. In Straub LR, Wilson PD Jr (eds): *Clinical Trends in Orthopaedics*. New York, Thieme-Stratton, 1981, pp 1–15 — 37–38
Brighton CT: *Clinical problems in epiphyseal plate growth and development*. Instr Course Lect 1974, 23:106 — 16–18
Brighton CT, Heppenstall RB: *Oxygen tension in zones of the epiphyseal plate, the metaphysis and diaphysis. An in vitro and in vivo study in rats and rabbits*. J Bone Joint Surg 1971, 53A:719–728 — 16–18
Brighton CT, Hunt RM: *Mitochondrial calcium and its role in calcification. Histochemical localization of calcium in electron micrographs of the epiphyseal growth plate with K-pyroantimonate*. Clin Orthop 1974, 100:406–416 — 16–18
Buckwalter J: *Fine structural studies of human intervertebral disc*. In The American Academy of Orthopaedic Surgeons: *Symposium on Idiopathic Low Back Pain*. St Louis, CV Mosby, 1982 — 19, 25
Buckwalter JA: *Articular cartilage*. Instr Course Lect 1983, 32:349–370 — 19, 25
Buckwalter JA: *Proteoglycan structure in calcifying cartilage*. Clin Orthop 1983, 172:207–232 — 16–18
Burmester GR, Dimitriu-Bona A, Waters SJ, et al: *Identification of three major synovial lining cell populations by monoclonal antibodies directed to Ia antigens and antigens associated with monocytes/macrophages and fibroblasts*. Scand J Immunol 1983, 17:69–82 — 26
Caplan AI: *Cartilage*. Sci Am 1984, 251(4):84–95 — 19, 25
Carter DR, Hayes WC: *The compressive behavior of bone as a two-phase porous structure*. J Bone Joint Surg 1977, 59A:954–962 — 23
Carter DR, Spengler DM: *Mechanical properties and composition of cortical bone*. Clin Orthop 1978, 135:192–217 — 20, 22
Castor CW: *The microscopic structure of normal human synovial tissue*. Arthritis Rheum 1960, 3:140–151 — 26
Chesney RW: *Current clinical applications of vitamin D metabolite research*. Clin Orthop 1981, 161:285–314 — 32
Coccia PF: *Cells that resorb bone* (editorial). N Engl J Med 1984, 310(7):456–458 — 20
Coccia PF, Krivit W, Cervenka J, et al: *Successful bone-marrow transplantation for infantile malignant osteopetrosis*. N Engl J Med 1980, 302:701–708 — 20
Connor JM: *Soft Tissue Ossification*. New York, Springer-Verlag, 1983, chap 8 — 20
DeLuca HF: *The biochemical basis of renal osteodystrophy and post-menopausal osteoporosis: a view from the vitamin D system*. Curr Med Res Opin 1981, 7:279–293 — 11–20
Dowben RM: *Contractility, with special reference to skeletal muscle*. In Mountcastle VB (ed): *Medical Physiology*, ed 14. St Louis, CV Mosby, 1980, pp 82–119 — 1–15
Elias SB, Appel SH: *Biochemistry of the myoneural junction and of its disorders*. In Siegel GJ, Albers RW (eds): *Basic Neurochemistry*, ed 3. Boston, Little, Brown, 1981, pp 513–528 — 1–15
Eyre DR: *Collagen: molecular diversity in the body's protein scaffold*. Science 1980, 207:1315–1322 — 24
Fawcett DW: *The sarcoplasmic reticulum of skeletal and cardiac muscle*. Circulation 1961, 24:336–348 — 3, 7
Freeman MA (ed): *Adult Articular Cartilage*, ed 2. Marshfield MA, Pitman Publishing, 1979 — 19, 24–25
Fujita T: *Calcium and aging* (editorial). Calcif Tissue Int 1985, 37:1–2 — 29, 33
Gallagher JC, Riggs BL: *Current concepts in nutrition. Nutrition and bone disease*. N Engl J Med 1978, 298:193–195 — 29, 31, 33
Garn SM: *Bone loss and aging*. In Goldman R (ed): *Physiology and Pathology of Human Aging*. New York, Academic Press, 1975, pp 39–57 — 39
Garn SM: *The course of bone gain and the phases of bone loss*. Orthop Clin North Am 1972, 3:503–520 — 39
Goodman LS, Gilman A (eds): *The Pharmacological Basis of Therapeutics*, ed 7. New York, Macmillan, 1985, pp 100–129 — 1–15
Gross J: *Collagen biology: structure, degradation, and disease*. Harvey Lect 1974, 68:351–432 — 27
Gross J: *Collagen*. Sci Am 1961, 204:121–130 — 27
Hall BK: *Developmental and Cellular Skeletal Biology*. New York, Academic Press, 1978 — 21
Ham AW: *Histology*, ed 6. Philadelphia, JB Lippincott, 1969, chap 18 — 21
Hamerman D, Rosenberg LC, Schubert M: *Diarthrodial joints revisited*. J Bone Joint Surg 1970, 52A:725–774 — 26
Heaney RP, Recker RR, Saville PD: *Menopausal changes in calcium balance performance*. J Lab Clin Med 1978, 92(6):953–963 — 28–31, 33
Heaney RP, Recker RR, Saville PD: *Calcium balance and calcium requirements in middle-aged women*. Am J Clin Nutr 1977, 30(10):1603–1611 — 28–31, 33
Jaworski ZF, Liskova-Kiar M, Uhthoff HK: *Effect of long-term immobilisation on the pattern of bone loss in older dogs*. J Bone Joint Surg 1980, 62B:104–110 — 34, 36
Kandel ER, Schwartz JH (eds): *Principles of Neural Science*, ed 2. New York, Elsevier Biomedical, 1981, pp 25–154 — 1–15
Kember NF: *Cell division in endochondral ossification. A study of cell proliferation in rat bones by the method of tritiated thymidine autoradiography*. J Bone Joint Surg 1960, 42B:824–839 — 16–18
Kleerekoper MB, Tolia K, Parfitt AM: *Nutritional endocrine and demographic aspect of osteoporosis*. Orthop Clin North Am 1981, 12:547–558 — 28–31, 33
Kuffler SW, Nicholls JG, Martin AR: *From Neuron to Brain*, ed 2. Sunderland MA, Sinauer Associates, 1984, pp 97–291 — 1–15
Lehninger AL: *Principles of Biochemistry*. New York, Worth Publishers, 1982 — 1–15
MacIntyre I, Evans IM, Hobitz HH, et al: *Chemistry, physiology, and therapeutic applications of calcitonin*. Arthritis Rheum 1980, 23(10):1139–1147 — 32
McLean FC, Urist MR: *Bone. An Introduction to the Physiology of Skeletal Tissue*, ed 2. Chicago, University of Chicago Press, 1961, p 24 — 16–18
Mauro A (ed): *Muscle Regeneration*. New York, Raven Press, 1979 — 2
Minor RR: *Collagen metabolism. A comparison of diseases of collagen and diseases affecting collagen*. Am J Pathol 1980, 98:227–280 — 24
Murray JM, Weber A: *The cooperative action of muscle proteins*. Sci Am 1974, 230:58–71 — 1–15
Murray PD: *Bones. A Study of the Development and Structure of the Vertebrate Skeleton*. London, Cambridge University Press, 1936 — 21
Nastuk WL: *Neuromuscular transmission*. In Mountcastle VB (ed): *Medical Physiology*, ed 14. St Louis, CV Mosby, 1980, pp 151–183 — 1–15
Nicogossian AE, Parker JF Jr: *Bone and mineral metabolism*. In Nicogossian AE, Parker JF Jr (eds): *Space Physiology and Medicine*. Washington, NASA, 1982, pp 204–209 — 35
Norman AW: *Evidence for a new kidney-produced hormone 1,25-dihydroxycholecalciferol, the proposed biologically active form of vitamin D*. Am J Clin Nutr 1971, 24:1346–1351 — 32
Owen M: *The origin of bone cells in the postnatal organism*. Arthritis Rheum 1980, 23(10):1073–1080 — 20
Pace N: *Weightlessness: a matter of gravity*. N Engl J Med 1977, 297:32–37 — 35
Palfrey AJ, Davies DV: *The fine structure of chondrocytes*. J Anat 1966, 100:213–226 — 21
Parfitt AM: *Dietary risk factors for age-related bone loss and fractures*. Lancet 1983, 2(8360):1181–1185 — 28, 30, 33
Parfitt AM: *The coupling of bone formation to bone resorption: a critical analysis of the concept and of its relevance to the pathogenesis of osteoporosis*. Metab Bone Dis Relat Res 1982, 4(1):1–6 — 30–31
Peachey LD (ed): *Handbook of Physiology: Section 10, Skeletal Muscle*. Bethesda MD, American Physiological Society, 1983 — 1–15
Peachey LD: *The sarcoplasmic reticulum and transverse tubules of the frog's sartorius*. J Cell Biol 1965, 25(suppl):209–231 — 3, 7
Peachey LD, Franzini-Armstrong C: *Structure and function of membrane systems of skeletal muscle cells*. In Peachey LD (ed): *Handbook of Physiology: Section 10, Skeletal Muscle*. Bethesda MD, American Physiological Society, 1983, pp 23–71 — 1–15
Porter KR: *The sarcoplasmic reticulum in muscle cells of Amblystoma larvae*. J Biophys Cytol 1956, 2(4):163–170 — 3, 7
Prockop DJ: *Mutations in collagen genes. Consequences for rare and common diseases*. J Clin Invest 1985, 75:783–787 — 24
Prockop DJ, Kivirikko KI: *Heritable diseases of collagen*. N Engl J Med 1984, 311(6):376–386 — 24
Prockop DJ, Kivirikko KI, Tuderman L, et al: *The biosynthesis of collagen and its disorders* (second of two parts). N Engl J Med 1979, 301(2):77–85 — 24

Section III *(continued)* Plates

PROCKOP DJ, KIVIRIKKO KI, TUDERMAN L, et al: *The biosynthesis of collagen and its disorders* (first of two parts). N Engl J Med 1979, 301(1):13–23 — 24

PYERITZ RE, MCKUSICK VA: *Basic defects in the Marfan syndrome* (editorial). N Engl J Med 1981, 305:1011–1012 — 24

PYERITZ RE, MCKUSICK VA: *The Marfan syndrome: diagnosis and management*. N Engl J Med 1979, 300(14):772–777 — 24

RAISZ LG: *Bone metabolism and calcium regulation*. In AVIOLI LV, KRANE SM (eds): *Metabolic Bone Disease*. New York, Academic Press, 1978 — 32

RAMBAUT P, JOHNSTON J: *Prolonged weightlessness and calcium loss in man*. Acta Astronautica 1979, 6:1113–1122 — 35

RANVIER L: *Quelques faits relatifs au développement du tissu osseux*. C R Seances Acad Sci 1873, 77:1105 — 16–18

RECKER RR, SAVILLE PD, HEANEY RP: *Effect of estrogens and calcium carbonate on bone loss in postmenopausal women*. Ann Intern Med 1977, 87(6):649–655 — 28–31, 33

REEVES J, ARNAUD S, GORDON S, et al: *The pathogenesis of infantile malignant osteopetrosis: bone mineral metabolism and complications in five infants*. Metab Bone Dis Relat Res 1981, 3:135–142 — 20

REILLY DT, BURSTEIN AH: *The mechanical properties of cortical bone*. J Bone Joint Surg 1974, 56:1001–1022 — 22

REVEL JP, HAY ED: *An autoradiographic and electron microscopic study of collagen synthesis in differentiating cartilage*. Z Zellforsch 1963, 61:110–144 — 21

RIGGS BL, WAHNER HW, DUNN WL, et al: *Differential changes in bone mineral density of the appendicular and axial skeleton with aging: relationship to spinal osteoporosis*. J Clin Invest 1981, 67(2):328–335 — 39

ROSS R, BORNSTEIN P: *Elastic fibers in the body*. Sci Am 1971, 224:44–52 — 27

RUFF CB, HAYES WC: *Subperiosteal expansion and cortical remodeling of the human femur and tibia with aging*. Science 1982, 217:945–947 — 39

SALTIN B, GOLLNICK PD: *Skeletal muscle adaptability: significance for metabolism and performance*. In PEACHEY LD (ed): *Handbook of Physiology: Section 10, Skeletal Muscle*. Bethesda MD, American Physiological Society, 1983, pp 555–631 — 1–15

SAMAHA F, GERGELY J: *Biochemistry of muscle and of muscle disorders*. In SEGAL GJ, ALBERS RW (eds): *Basic Neurochemistry*, ed 3. Boston, Little, Brown, 1981, pp 529–561 — 1–15

SCHUMACHER HR JR: *Ultrastructure of the synovial membrane*. Ann Clin Lab Sci 1975, 5:489–498 — 26

SHAPIRO F, HOLTROP ME, GLIMCHER MJ: *Organization and cellular biology of the perichondral ossification groove of Ranvier: a morphological study in rabbits*. J Bone Joint Surg 1977, 59(6):703–723 — 16–18

SILBERBERG R: *Ultrastructure of articular cartilage in health and disease*. Clin Orthop 1968, 57:233–257 — 21

SLOVIK DM, ADAMS JS, NEER RM, et al: *Deficient production of 1,25-dihydroxy-vitamin-D in elderly osteoporotic patients*. N Engl J Med 1982, 305:372–374 — 29, 33

STAMBAUGH JE, BRIGHTON CT: *Diffusion in the various zones of the normal and the rachitic growth plate*. J Bone Joint Surg 1980, 62(5):740–749 — 16–18

Section III *(continued)* Plates

STANESCU V, STANESCU R, MAROTEAUX P: *Pathogenic mechanisms in osteochondrodysplasias*. J Bone Joint Surg 1984, 66A:817–836 — 16–18

TRUETA J, MORGAN JD: *The vascular contribution to osteogenesis. I. Studies by the injection method. A zonal analysis of inorganic and organic constituents of the epiphysis during endochondral calcification*. J Bone Joint Surg 1960, 42B:97–109 — 16–18

UHTHOFF HK, JAWORSKI ZFG: *Bone loss in response to long-term immobilisation*. J Bone Joint Surg 1978, 60B:420–429 — 34, 36

URIST MR, DELANGE RJ, FINERMAN GAM: *Bone cell differentiation and growth factors*. Science 1983, 220:680–686 — 20

WARWICK R, WILLIAMS PL (eds): *Gray's Anatomy*, ed 35. Philadelphia, WB Saunders, 1973, pp 474–486 — 1–15

WHEDON GD: *Disuse osteoporosis: physiological aspects*. Calcif Tissue Int 1984, 36:S146–S150 — 34–36

WHEDON D, LUTWAK L, RAMBAUT P, et al: *Mineral and nitrogen balance studies, experiment M071*. In JOHNSTON RS, DIETLEIN LF (eds): *Biomedical Results From Skylab*. Washington, NASA, 1977, pp 164–174 — 35

Section IV

General References

AVIOLI LV (ed): *The Osteoporotic Syndrome. Detection, Prevention, and Treatment*. New York, Grune & Stratton, 1983

AVIOLI LV, KRANE SM (eds): *Metabolic Bone Disease*, Vol II. New York, Academic Press, 1978

AVIOLI LV, KRANE SM (eds): *Metabolic Bone Disease*, Vol I. New York, Academic Press, 1977

LEVINE MA, FOSTER GV: *Disorders of bone and mineral metabolism*. In HARVEY AM, JOHNS RJ, MCKUSICK VA, et al (eds): *The Principles and Practice of Medicine*, ed 21. Norwalk CT, Appleton-Century-Crofts, 1984, pp 861–888

MURRAY LW, BLUMBERG B, SHAMBAN A: *Genetic diversity of collagen structure*. In UITTO J (moderator): *Biochemistry of Collagen in Diseases*. Ann Intern Med 1986, 105:740–756

RIIS B, THOMSEN K, CHRISTIANSEN C: *Does calcium supplementation prevent postmenopausal bone loss? A double-blind, controlled clinical study*. N Engl J Med 1987, 316(4):173–177

AGUS ZS, ATTIE MF, GOLDFARB S, et al: *Mineral metabolism in health and disease*. In KLAHR S, MASSRY SG (eds): *Contemporary Nephrology*. New York, Plenum Publishing, 1983, pp 241–350 — 1–10

ALBRIGHT F, BUTLER AM, BLOOMBERG E: *Rickets resistant to vitamin D therapy*. Am J Dis Child 1937, 54:529–547 — 11–20

ALBRIGHT JA: *Management overview of osteogenesis imperfecta*. Clin Orthop 1981, 159:80–87 — 34–36

ALOIA JF, COHN SH, OSTUNI JA, et al: *Prevention of involutional bone loss by exercise*. Ann Intern Med 1978, 89(3):356–358 — 32

ANDERSEN J, NEILSEN HJ: *Renal osteodystrophy in nondialysed patients with chronic renal failure*. Acta Radiol 1980, 21:803–806 — 11–20

Section IV *(continued)* Plates

APONTE CJ, PETRELLI MP: *Anticonvulsants and vitamin D metabolism*. JAMA 1973, 225:1248 — 11–20

AURBACH GD, MARX SJ, SPIEGEL AM: *Parathyroid hormone, calcitonin and the calciferols*. In WILLIAMS RH: *Textbook of Endocrinology*. Philadelphia, WB Saunders, 1985, pp 1218–1255 — 1–10

BACHRACH S, FISHER J, PARKS JS: *An outbreak of vitamin D deficiency rickets in a susceptible population*. Pediatrics 1979, 64(6):871–877 — 11–20

BIJOVET OL: *Kidney function in calcium and phosphate metabolism*. In AVIOLI LV, KRANE SM (eds): *Metabolic Bone Disease*, Vol II. New York, Academic Press, 1978 — 32

BOUCEK RJ, NOBLE NL, GUNJA-SMITH Z, et al: *The Marfan syndrome: a deficiency in chemically stable collagen cross-links*. N Engl J Med 1981, 305(17):988–991 — 22, 37

BROADUS AE, RASMUSSEN H: *Clinical evaluation of parathyroid function*. Am J Med 1981, 70(3):475–478 — 4, 9–10

BROOKS MH, BELL NH, LOVE L, et al: *Vitamin-D-dependent rickets type II. Resistance of target organs to 1,25-dihydroxyvitamin D*. N Engl J Med 1978, 298:996–999 — 11–20

BROWN DJ, DAWBORN JK, THOMAS DP, et al: *Assessment of osteodystrophy in patients with chronic renal failure*. Aust NZ J Med 1982, 12(3):250–254 — 11–20

COCCIA PF: *Cells that resorb bone*. N Engl J Med 1984, 310(7):456–458 — 39–40

COCCIA PF, KRIVIT W, CERVENKA J, et al: *Successful bone-marrow transplantation for infantile malignant osteopetrosis*. N Engl J Med 1980, 302:701–708 — 39–40

COHN SH: *Techniques for determining the efficacy of treatment of osteoporosis*. Calcif Tissue Int 1982, 34:433–438 — 28–29

COHN SH (ed): *Non-Invasive Measurements of Bone Mass and Their Clinical Application*. Boca Raton FL, CRC Press, 1981 — 28–29

CONNOR JM: *Soft Tissue Ossification*. New York, Springer-Verlag, 1983, chaps 1, 4 — 44

CONNOR JM, EVANS DAP: *Fibrodysplasia ossificans progressiva. The clinical features and natural history of 34 patients*. J Bone Joint Surg 1982, 64B:76–83 — 44

CONSENSUS CONFERENCE: *Osteoporosis*. JAMA 1984, 252:799–802 — 22–24, 26, 28–30, 32

CUMMINGS SR, BLACK D: *Should perimenopausal women be screened for osteoporosis?* Ann Intern Med 1986, 104(6):817–823 — 22–24, 26, 28–30, 32

CUNNINGHAM J, FRAHER LJ, CLEMENS TL, et al: *Chronic acidosis with metabolic bone disease. Effect of alkali on bone morphology and vitamin D metabolism*. Am J Med 1982, 73(2):199–204 — 32

DALINKA MK, ARONCHICK JM, HADDAD JG JR: *Paget's disease*. Orthop Clin North Am 1983, 14(1):3–19 — 41–43

DAVID DS: *Calcium metabolism in renal failure*. Am J Med 1975, 58(1):48–56 — 11–20

DELUCA HF: *The biochemical basis of renal osteodystrophy and post-menopausal osteoporosis: a view from the vitamin D system*. Curr Med Res Opin 1981, 7:279–293 — 11–20

DOUGLAS DL, DUCKWORTH T, KANIS JA, et al: *Spinal cord dysfunction in Paget's disease of bone. Has medical treatment a vascular basis?* J Bone Joint Surg 1981, 63B:495–503 — 41–43

DUNN WL, WAHNER HW, RIGGS BL: *Measurement of bone mineral content in human vertebrae and hip by dual photon absorptiometry*. Radiology 1980, 136(2):485–487 — 28

Eastwood JB: *Renal osteodystrophy—a radiological review.* CRC Crit Rev Diagn Imaging 1977, 9:77–104 11–20

Elkinton JR, Huth EJ, Webster GD Jr, et al: *The renal excretion of hydrogen ion in renal tubular acidosis. I. Quantitative assessment of the response to ammonium chloride as an acid load.* Am J Med 1960, 29:554–575 11–20

Evans GA, Arulanantham K, Gage JR: *Primary hypophosphatemic rickets. Effect of oral phosphate and vitamin D on growth and surgical treatment.* J Bone Joint Surg 1980, 62:1130–1138 11–20

Fallon MD: *Nontumor pathology of bone.* In Spicer SS (ed): *Histochemistry in Pathologic Diagnosis.* Chemical and Biochemical Analysis Series, Vol 22. New York, Marcel Dekker, 1986, pp 864–882 30

Fallon MD, Teitelbaum SL: *The interpretation of fluorescent tetracycline markers in the diagnosis of metabolic bone diseases.* Hum Path 1982, 13(5):416–417 30

Fallon MD, Teitelbaum SL, Weinstein RS, et al: *Hypophosphatasia: clinicopathologic comparison of the infantile, childhood, and adult forms.* Medicine 1984, 63:12–24 21

Farfel Z, Brickman AS, Kaslow HR, et al: *Defect of receptor-cyclase coupling protein in pseudohypoparathyroidism.* N Engl J Med 1980, 303(5):237–242 8

Fisken RA, Heath DA, Somers S, et al: *Hypercalcemia in hospital patients. Clinical and diagnostic aspects.* Lancet 1981, 1(8213):202–207 4

Frost HM (ed): *Symposium on the osteoporoses. The Orthopedic Clinics of North America.* Philadelphia, WB Saunders, 1981 22–24, 26, 28–30, 32

Fujita T: *Calcium and aging* (editorial). Calcif Tissue Int 1985, 37:1–2 22

Gallagher JC, Melton LJ, Riggs BL, et al: *Epidemiology of fractures of the proximal femur in Rochester, Minnesota.* Clin Orthop 1980, 150:163–171 23

Gallagher JC, Riggs BL: *Current concepts in nutrition. Nutrition and bone disease.* N Engl J Med 1978, 298:193–195 22, 32–33

Garabédian M, Vainsel M, Mallet E, et al: *Circulating vitamin D metabolite concentrations in children with nutritional rickets.* J Pediatr 1983, 103:381–386 11–20

Garn SM: *The course of bone gain and the phases of bone loss.* Orthop Clin North Am 1972, 3:503–520 22, 27

Genant HK, Cann CE: *Quantitative computed tomography for assessing vertebral bone mineral.* In Genant HK, Chafetz N, Helms CA (eds): *Computed Tomography of the Lumbar Spine.* San Francisco, University of California Printing Department, 1982, pp 289–314 29

Genant HK, Cann CE, Ettinger B, et al: *Quantitative computed tomography of vertebral spongiosa: a sensitive method for detecting early bone loss after oophorectomy.* Ann Intern Med 1982, 97(5):699–705 29

Genant HK, Ettinger B, Cann CE, et al: *Osteoporosis: assessment by quantitative computed tomography.* Orthop Clin North Am 1985, 16:557–568 22–24, 26, 28–30, 32

Goldman AB, Lane JM, Salvati E: *Slipped capital femoral epiphyses complicating renal osteodystrophy: a report of three cases.* Radiology 1978, 126:333–339 11–20

Greenfield GB: *Roentgen appearance of bone and soft tissue changes in chronic renal disease.* Amer J Roentgen 1972, 116:749–757 11–20

Habener JF, Potts JT Jr: *Clinical features of primary hyperparathyroidism.* In DeGroot LJ (ed): *Endocrinology.* New York, Grune & Stratton, 1979, 693–701 2–3

Habener JF, Segre GV: *Parathyroid hormone radioimmunoassay.* Ann Intern Med 1979, 91(5):782–785 10

Harrison HE, Harrison HC: *Rickets then and now.* J Pediatr 1975, 87:1144–1151 11–20

Health and Public Policy Committee, American College of Physicians: *Radiologic methods to evaluate bone mineral content.* Ann Intern Med 1984, 100:908–911 27–29

Heaney RP: *Management of osteoporosis: nutritional considerations.* Clin Invest Med 1982, 5(2–3):185–187 32

Heaney RP, Recker RR, Saville PD: *Menopausal changes in calcium balance performance.* J Lab Clin Med 1978, 92(6):953–963 22, 32

Heaney RP, Recker RR, Saville PD: *Calcium balance and calcium requirements in middle-aged women.* Am J Clin Nutr 1977, 30(10):1603–1611 22, 32

Heath H 3d, Hodgson SF, Kennedy MA: *Primary hyperparathyroidism. Incidence, morbidity, and potential economic impact in a community.* N Engl J Med 1980, 302(4):189–193 2–3

Johnston CC Jr, Altman RD, Canfield RE, et al: *Review of fracture experience during treatment of Paget's disease of bone with etidronate disodium (EHDP).* Clin Orthop 1983, 172:186–194 41–43

Kanis JA, Russell RGG (eds): *Diphosphonates and Paget's disease of bone.* Metab Bone Dis Relat Res 1981, 3(4–5):219–343 41–43

Kaplan FS: *Osteoporosis: pathophysiology and prevention.* Clin Symp 1987, 39(1):2–32 22–33

Key L, Carnes D, Cole S, et al: *Treatment of congenital osteopetrosis with high-dose calcitriol.* N Engl J Med 1984, 310(7):409–415 39–40

Kleerekoper MB, Tolia K, Parfitt AM: *Nutritional endocrine and demographic aspect of osteoporosis.* Orthop Clin North Am 1981, 12:547–558 22

Krieger DT, Bardin CW (eds): *Current Therapy in Endocrinology and Metabolism.* Philadelphia, BC Decker, 1985 1–10

Lane JM (ed): *Symposium on metabolic bone disease. The Orthopedic Clinics of North America.* Philadelphia, WB Saunders, 1984 22–33

Lane JM, Vigorita VJ: *Osteoporosis.* J Bone Joint Surg 1983, 65(2):274–278 22–33

Lehman RAW, Reeves JD, Wilson WB, et al: *Neurological complications of infantile osteopetrosis.* Ann Neurol 1977, 2:378–384 39–40

Lemann J Jr, Gray RW: *Calcitriol, calcium, and granulomatous disease* (editorial). N Engl J Med 1984, 311(17):1115–1117 4

Levine MA, Downs RW Jr, Moses AM, et al: *Resistance to multiple hormones in patients with pseudohypoparathyroidism. Association with deficient activity of guanine nucleotide regulatory protein.* Am J Med 1983, 74(4):545–556 8

Lubs HA, Travers H: *Genetic counseling in osteogenesis imperfecta.* Clin Orthop 1981, 159:36–41 34–36

McBroom RJ, Hayes WC, Edwards WT, et al: *Prediction of vertebral body compressive fracture using quantitative computed tomography.* J Bone Joint Surg 1985, 67A:1206–1214 29

McKenna MJ, Freaney R, Casey OM, et al: *Osteomalacia and osteoporosis: evaluation of a diagnostic index.* J Clin Pathol 1983, 36(3):245–252 33

McKusick VA: *Mendelian disorders.* In Harvey AM, Johns RJ, McKusick VA, et al (eds): *The Principles and Practice of Medicine,* ed 21. Norwalk CT, Appleton-Century-Crofts, 1984, pp 427–451 37–38

McKusick VA: *Heritable Disorders of Connective Tissue,* ed 4. St Louis, CV Mosby, 1972 34–44

Mallette LE, Bilezikian JP, Heath DA, et al: *Primary hyperparathyroidism: clinical and biochemical features.* Medicine 1974, 53:127–146 2–3

Mankin HJ: *Rickets, osteomalacia, and renal osteodystrophy. Part II.* J Bone Joint Surg 1974, 56A:101–128, 352–386 18

Marx SJ, Attie MF, Levine MA, et al: *The hypocalciuric or benign variant of familial hypercalcemia: clinical and biochemical features in fifteen kindreds.* Medicine 1981, 60(6):397–412 2–3

Matthews JL, Talmage RV: *Influence of parathyroid hormone on bone cell ultrastructure.* Clin Orthop 1981, 156:27–38 11–20

Merkow RL, Lane JM: *Current concepts of Paget's disease of bone.* Orthop Clin North Am 1984, 15(4):747–763 41–43

Mills BG, Singer FR, Weiner LP, et al: *Evidence for both respiratory syncytial virus and measles virus antigens in the osteoclasts of patients with Paget's disease of bone.* Clin Orthop 1984, 183:303–311 41–43

Minor RR: *Collagen metabolism. A comparison of diseases of collagen and diseases affecting collagen.* Am J Pathol 1980, 98:227–280 37–38

Morgan B: *Osteomalacia, Renal Osteodystrophy and Osteoporosis.* Springfield IL, Charles C Thomas, 1973 11–20

Neufeld M, Maclaren NK, Blizzard RM: *Two types of autoimmune Addison's disease associated with different polyglandular autoimmune (PGA) syndromes.* Medicine 1981, 60(5):355–362 5–6

Nordin BE, Heyburn PJ, Peacock M, et al: *Osteoporosis and osteomalacia.* Clin Endocrinol Metab 1980, 9(1):177–205 33

Nusynowitz ML, Frame B, Kolb FO: *The spectrum of the hypoparathyroid states: a classification based on physiologic principles.* Medicine 1976, 55(2):105–119 5

Ott S: *Should women get screening bone mass measurements?* (Editorial). Ann Intern Med 1986, 104(6):874–876 22–24, 26, 28–30, 32

Pace N: *Weightlessness: a matter of gravity.* N Engl J Med 1977, 297:32–37 22

Paget J: *On a form of chronic inflammation of bones (osteitis deformans).* Trans R Med Chir Soc Lond 1877, 60:37–64 41–43

Parfitt AM: *Dietary risk factors for age-related bone loss and fractures.* Lancet 1983, 2(8360):1181–1185 22

Parfitt AM: *Surgical idiopathic and other varieties of parathyroid hormone-deficient hypoparathyroidism.* In DeGroot LJ (ed): *Endocrinology.* New York, Grune & Stratton, 1979, pp 755–768 5–6

Parfitt AM: *Hypophosphatemic vitamin D refractory rickets and osteomalacia.* Orthop Clin North Am 1972, 3:653–680 11–20

Potts JT Jr, Kronenberg HM, Rosenblatt M: *Parathyroid hormone: chemistry, biosynthesis, and mode of action.* Adv Protein Chem 1982, 35:323–396 1, 9

Prockop DJ: *Mutations in collagen genes. Consequences for rare and common diseases.* J Clin Invest 1985, 75:783–787 34–38

Prockop DJ, Kivirikko KI: *Heritable diseases of collagen.* N Engl J Med 1984, 311(6):376–386 22, 34–38

PROCKOP DJ, KIVIRIKKO KI, TUDERMAN L, et al: *The biosynthesis of collagen and its disorders* (second of two parts). N Engl J Med 1979, 301(2):77–85 — 37–38

PROCKOP DJ, KIVIRIKKO KI, TUDERMAN L, et al: *The biosynthesis of collagen and its disorders* (first of two parts). N Engl J Med 1979, 301(1):13–23 — 37–38

PYERITZ RE, MCKUSICK VA: *Basic defects in the Marfan syndrome.* N Engl J Med 1981, 305:1011–1012 — 37

PYERITZ RE, MCKUSICK VA: *The Marfan syndrome: diagnosis and management.* N Engl J Med 1979, 300(14):772–777 — 22, 37

RAISZ LG: *Osteoporosis.* J Am Geriatr Soc 1982, 30(2):127–138 — 22–33

RECKER RR, SAVILLE PD, HEANEY RP: *Effect of estrogens and calcium carbonate on bone loss in postmenopausal women.* Ann Intern Med 1977, 87(6):649–655 — 32

REEVES J, ARNAUD S, GORDON S, et al: *The pathogenesis of infantile malignant osteopetrosis. Bone mineral metabolism and complications in five infants.* Metab Bone Dis Relat Res 1981, 3:135–142 — 39–40

REEVES JD, AUGUST CS, HUMBERT JR, et al: *Host defense in infantile osteopetrosis.* Pediatrics 1979, 64(2):202–206 — 39–40

RIGGS BL, MELTON LJ 3D: *Evidence for two distinct syndromes of involutional osteoporosis.* Am J Med 1983, 75(6):899–901 — 22

RIGGS BL, WAHNER HW, DUNN WL, et al: *Differential changes in bone mineral density of the appendicular and axial skeleton with aging: relationship to spinal osteoporosis.* J Clin Invest 1981, 67(2):328–335 — 27–28

RUDE RK, OLDHAM SB, SHARP CF JR, et al: *Parathyroid hormone secretion in magnesium deficiency.* J Clin Endocrinol Metab 1978, 47(4): 800–806 — 5

RUFF CB, HAYES WC: *Subperiosteal expansion and cortical remodeling of the human femur and tibia with aging.* Science 1982, 217:945–947 — 22, 27

SCHNEIDER R: *Radiologic methods of evaluating generalized osteopenia.* Orthop Clin North Am 1984, 15(4):631–651 — 25–28, 33, 39

SCHROEDER HW JR, ZASLOFF M: *The hand and foot malformations in fibrodysplasia ossificans progressiva.* Johns Hopkins Med J 1980, 147:73–78 — 44

SCRIVER CR, READE TM, DELUCA HF, et al: *Serum 1,25-dihydroxyvitamin D levels in normal subjects and in patients with hereditary rickets or bone disease.* N Engl J Med 1978, 299:976–979 — 11–20

SEEMAN E, MELTON LJ 3D, O'FALLON WM, et al: *Risk factors for spinal osteoporosis in men.* Am J Med 1983, 75(6):977–983 — 22

SILLENCE D: *Osteogenesis imperfecta: an expanding panorama of variants.* Clin Orthop 1981, 159:11–25 — 34–36

SINGER FR, MILLS BG: *Evidence for a viral etiology of Paget's disease of bone.* Clin Orthop 1983, 178:245–251 — 41–43

SIRIS ES, CANFIELD RE: *Paget's disease of bone. Current concepts as to its nature and management.* Orthopaedic Review 1982, XI(12): 43–49 — 41–43

SLOVIK DM, ADAMS JS, NEER RM, et al: *Deficient production of 1,25-dihydroxy-vitamin-D in elderly osteoporotic patients.* N Engl J Med 1982, 305:372–374 — 22

SORELL M, KAPOOR N, KIRKPATRICK D, et al: *Marrow transplantation for juvenile osteopetrosis.* Am J Med 1981, 70:1280–1287 — 39–40

STANBURY SW: *Bone disease in uremia.* Am J Med 1968, 44:714–724 — 11–20

STEWART AF, ADLER M, BYERS CM, et al: *Calcium homeostasis in immobilization: an example of resorptive hypercalciuria.* N Engl J Med 1982, 306(9):1136–1140 — 4

SUNDARAM M, WOLVERSON MK, HEIBERG E, et al: *Erosive azotemic osteodystrophy.* Amer J Roentgen 1981, 136:363–367 — 11–20

THOMPSON NW, ECKHAUSER FE, HARNESS JK: *The anatomy of primary hyperparathyroidism.* Surgery 1982, 92(5):814–821 — 2–3

UHTHOFF HK, JAWORSKI ZFG: *Bone loss in response to long-term immobilisation.* J Bone Joint Surg 1978, 60B:420–429 — 22

WAHNER HW: *Assessment of metabolic bone disease: review of new nuclear medicine procedures.* Mayo Clin Proc 1985, 60:827–835 — 28

WAHNER HW, DUNN WL, MAZESS RB, et al: *Dual-photon Gd-153 absorptiometry of bone.* Radiology 1985, 156:203–206 — 22–24, 26, 28–30, 32

WALKER DG: *Bone resorption restored in osteopetrotic mice by transplants of normal bone marrow and spleen cells.* Science 1975, 190:784–785 — 39–40

WALKER DG: *Spleen cells transmit osteopetrosis in mice.* Science 1975, 190:785–787 — 39–40

WELLER M, EDEIKEN J, HODES PJ: *Renal osteodystrophy.* Amer J Roentgen 1968, 104:354–363 — 11–20

WENGER DR, DITKOFF TJ, HERRING JA, et al: *Protrusio acetabuli in Marfan's syndrome.* Clin Orthop 1980, 147:134–138 — 37

WHEDON GD: *Disuse osteoporosis: physiological aspects.* Calcif Tissue Int 1984, 36:S146–S150 — 22

WHEDON GD: *The New England journal of medicine* (editorial). N Engl J Med 1981, 305(7): 397–399 — 22–33

WHYTE MP, BERGFELD MA, MURPHY WA, et al: *Postmenopausal osteoporosis. A heterogeneous disorder as assessed by histomorphometric analysis of iliac crest bone from untreated patients.* Am J Med 1982, 72(2):193–202 — 30

WHYTE MP, MAGILL HL, FALLON MD, et al: *Infantile hypophosphatasia: Normalization of circulating bone alkaline phosphatase activity followed by skeletal remineralization.* J Pediatr 1986, 108:82–88 — 21

WYNNE-DAVIES R, GORMLEY J: *Clinical and genetic patterns in osteogenesis imperfecta.* Clin Orthop 1981, 159:26–35 — 34–36

ZERWEKH JE, GLASS K, JOWSEY J, et al: *An unique form of osteomalacia associated with end organ refractoriness to 1,25-hydroxyvitamin D and apparent defective synthesis of 25-hydroxyvitamin D.* J Clin Endocrinol Metab 1979, 49:171–175 — 11–20

Subject Index

Boldface numbers refer to terms that appear on plates

A